STRUCTURAL GEOLOGY OF ROCKS AND REGIONS

STRUCTURAL GEOLOGY OF ROCKS AND REGIONS

GEORGE H. DAVIS
The University of Arizona

JOHN WILEY & SONS
New York
Chichester
Brisbane
Toronto
Singapore

Library of Congress Cataloging in Publication Data:

Davis, G. H. (George Herbert), 1942–
Structural geology of rocks and regions.

Bibliography
Includes indexes.
1. Geology, Structural. I. Title.
QE601.D3 1984 551.8 83-17076
ISBN 0-471-09267-3

Printed in the United States of America

10 9 8 7 6 5 4 3 2 1

PREFACE

My hopes for this text lie in communicating the elegance of the physical and geometrical architecture of the Earth's crust, and the degree to which that architecture reflects the dynamics of structural movements. There is both philosophical and historical value in trying to reconstruct the physical evolution of the Earth's outer shell. Our perceptions of who we are and where we are in time and space are shaped by facts and interpretations regarding the historical development of the crust of the planet on which we live. And knowing fully the extent to which our planet is dynamic, not static, is a reminder of the lively and special environment we inhabit.

We have come to understand that our earthly foundations are not fixed. Instead we live on continental-sized plates that are in a continual state of slow motion. The interference of these plates has played a dominant role in both the formation and deformation of rock bodies in the Earth's crust. Knowledge of present-day plate tectonic processes aids us in interpreting past structural movements. Furthermore, evaluating and interpreting past dynamics may help us to predict what the present actions may hold for the future.

Applications of structural geology are broad ranging, and herein lies the practical incentive for understanding Earth structure. The exploration for hydrocarbons and metals, the evaluation of daring but dangerous proposals for subterranean disposal of radioactive waste, and the prediction of geological events like earthquakes and volcanic eruptions are intellectually demanding activities that require knowledge derived from intimacy with Earth architecture and dynamics.

Field relations are the primary sources for structural geologists. What value the field has for those intent on understanding Earth dynamics! Dr. Howard Lowry, a brilliant scholar in the field of English literature, captured the significance of primary sources. His words have special meaning to geologists:

> By [primary sources] we mean the first-hand things, the authentic ground of facts and ideas, the original wells and springs out of which all the rest either is drawn or flows . . . . Regard for the primary sources makes one forever the enemy of preconceptions, of manipulated data, raw opinion, and guesswork—of all the sleek shortcuts to wisdom in ten easy lessons. . . . Exclusive reliance on second-hand things makes second-hand men and women. It deludes us into thinking we are wiser than we are. . . . Breadth of knowledge, even knowing a little about a lot, has its obvious value. But breadth that perpetually sends down no clean, strong roots in the primary sources—into the deep earth and "the hidden rivers murmuring in the dark" of the rocks—such breadth clarifies very little. It merely puts our bewilderment on a broader basis. It leads us into incredible naivete and gullibility. It makes us too quick to believe all we read. (From *College Talks* by H. F. Lowry, edited by J. R. Blackwood, p. 86–87. Published with permission of Oxford University Press, New York, copyright © 1969).

This text in structural geology is based on and directed to the primary sources. It is written with the conviction that most geologists will benefit from a practical understanding of structural geology. The slant of this book is toward applications in regional tectonics and exploration geology, but at all times it emphasizes the importance first of understanding fundamental concepts and principles.

George H. Davis

ACKNOWLEDGMENTS

Merrily Davis helped enormously in the preparation of this book. She typed all drafts of the manuscript, organized the compilation of the bibliography, and administered the seeking of author and copyright permissions. In all she advanced the publication of the text by many months and kept tireless watch for errors and inconsistencies. I deeply appreciate the help she gave so generously.

All drafting and photography were carried out through The University of Arizona's Graphic Services, managed by Josh Young. Wallace P. Varner generated most of the line drawings and artwork. I am grateful to Wally and value the time spent with him in the planning and preparation of illustration materials. David A. Fischer rendered the artwork for the cover and most of the cartoons. Others in Graphic Services who contributed so much to this project were Marilyn Hodges, Betty Hupp, Jim Chapman, George Kew, and Dottie Larson.

I express special appreciation to students and colleagues who helped in various phases of the work, especially Robert W. Krantz, Scott R. Showalter, Paige W. Bausman, Lee D. DiTullio, Stephen J. Naruk, Gregory L. Cole, Stephen L. Wust, Ji Xiong, Daniel J. Lynch, Peter Kresan, and Edgar J. McCullough, Jr. Furthermore, I thank Susie Lebsack of Hughes-Calihan Corporation for providing technical support and encouragement in the use of the NBI Word Processor.

The manuscript was formally reviewed by Edward C. Beutner, Frederick W. Cropp, Randall D. Forsythe, Donal M. Ragan, and David V. Wiltschko. Their comments and suggestions were very constructive and helpful. I thank Susan Archias, Albert F. Siepert, and Margaret G. Siepert for helping to proof the manuscript.

I appreciate immensely the quality of support rendered by John Wiley & Sons, Inc., in transforming manuscript and figures to printed page. I extend my deepest thanks and heartfelt praise for scrupulous care in step-by-step editing and production to Susan Winick, copy editor; Lilly Kaufman, production manager; and Ruth Greif, production supervisor. I offer my gratitude to Joan Willens for creating a design so well suited to the purpose and personality of the book. I express my thanks to Donald H. Deneck, Editor, for guidance, judgment, and encouragement throughout the project.

Finally, Merrily and I thank our sons, Mike, Matt, and Drew, for their help in duplicating manuscript drafts and for their patience and support during the intrusion of "the book" into our family life.

G.H.D.

CONTENTS

I FUNDAMENTALS

Chapter 1 NATURE OF STRUCTURAL GEOLOGY 3
Chapter 2 CONCEPT OF DETAILED STRUCTURAL ANALYSIS 16
Chapter 3 DESCRIPTIVE ANALYSIS 36
Chapter 4 KINEMATIC ANALYSIS 87
Chapter 5 DYNAMIC ANALYSIS 132
Chapter 6 PLATE TECTONICS 163

II STRUCTURES

Chapter 7 CONTACTS 203
Chapter 8 PRIMARY STRUCTURE 231
Chapter 9 FAULTS 261
Chapter 10 JOINTS 325
Chapter 11 FOLDS 354
Chapter 12 CLEAVAGE, FOLIATION, AND LINEATION 401

CONCLUDING THOUGHTS 446

REFERENCES 448

AUTHOR INDEX 469

SUBJECT INDEX 473

STRUCTURAL GEOLOGY OF ROCKS AND REGIONS

I FUNDAMENTALS

chapter 1 NATURE OF STRUCTURAL GEOLOGY

ARCHITECTURE AND STRUCTURE

The start of any journey into unfamiliar territory is often spurred by dreaming, a kind of dreaming that spawns not lightheadedness but intense curiosity and the setting of goals. Our journey will explore the architecture of the crust of the planet on which we live. We will be primarily concerned with architectural forms that have developed as a response to crustal deformation. **Deformation** is that which produces a change in the original form and/or volume of a body of rock. Deformation commonly produces a change in location and/or orientation of the body of rock as well. Every body of rock, no matter how hard, has a point at which it will fracture or flow. This concept emerges in photographs of a fenceline located at the site of the Hebgen Lake earthquake, a destructive quake that wracked southwesternmost Montana in 1959. As a result of shifts in the ground surface, the fence was forced to shorten. Where shifts were modest, shortening was accommodated by bending (Figure 1.1*A*). But where shortening exceeded the bending limit of the wooden slats, the fence fractured and splintered explosively (Figure 1.1*B*).

More interested in rocks than in fences, we want to learn how structural changes in the initial state of a rock body reflect ways that parts of the Earth's crust have shifted positions and changed shape through time. The folded and

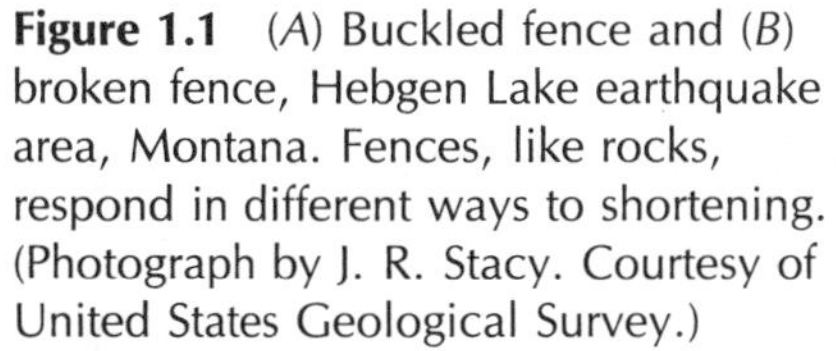

Figure 1.1 (*A*) Buckled fence and (*B*) broken fence, Hebgen Lake earthquake area, Montana. Fences, like rocks, respond in different ways to shortening. (Photograph by J. R. Stacy. Courtesy of United States Geological Survey.)

Figure 1.2 Faulted and folded metamorphic rocks in Upper Convict Canyon, Sierra Nevada of California. (Photograph by E. B. Mayo.)

faulted rocks shown in Figure 1.2 did not start out in the location, shape, and condition that we view them today. Rather they reflect deformation of the Earth's crust.

Jacob Bronowski, in his superb essays entitled *The Ascent of Man*, suggests that our conception of science today is a

> description and exploration of the underlying structures of nature, and words like "structure," "pattern," "plan," "arrangement," and "architecture" constantly occur in every description that we try to make. (From *The Ascent of Man* by J. Bronowski, p. 112. Published with permission of Little, Brown and Company, Boston, copyright © 1973.)

He believes

> The notion of discovering an underlying order in matter is man's basic concept for exploring nature. The architecture of things reveals a structure below the surface, a hidden grain, which when it is laid bare, makes it possible to take natural formations apart. . . .(From *The Ascent of Man* by J. Bronowski, p. 95. Published with permission of Little, Brown and Company, Boston, copyright ©1973.)

Bronowski's remarks apply beautifully to structural geology. **Structural geology** is most succinctly defined as the study of the architecture of the Earth's crust, insofar as it has resulted from deformation (Billings, 1972, p. 2). The expression "architecture of the Earth" is very appropriate because structural geology addresses the form, symmetry, geometry, and certainly the elegance and artistic rendering of the components of the Earth's crust on all scales (Figure 1.3). At the same time, structural geology focuses on the strength and mechanical properties of crustal materials, both now and at the time they were formed and deformed. The architecture of the Earth is chiefly fashioned by large-scale fault and flow movements in the crust. Geologists have concluded that such structural movements may be generated in many ways. Examples include compressional forces that concentrate stresses at the margins of colliding plates, leading to the development of major fault systems and fold belts; gravitational collapse of volcanoes to produce enormous craterlike calderas; rifting and extension of the oceanic or continental crust, forming deep depressions and concentrated volcanic activity; and persistent buoyant ascent of hot magma through rock from great depths toward the Earth's surface. To interpret such movements on the basis of scientific data that can never be complete is the challenge and intrigue of structural geology.

Although architecture and structural geology have much in common, the challenges of the architect and the structural geologist are quite different. The architect designs a structure, perhaps a building or a bridge, giving due attention to function, appearance, geometry, material, size, strength, cost, and other such factors. Then he or she supervises the process of construction daily, or perhaps weekly, making changes where necessary. In the end, the architect may be the only person who is aware of discrepancies between the original plan and the final product.

Figure 1.3 White House Ruin in Canyon de Chelly, Arizona, a sublime blend of the architecture of Nature and that of the Ancient Ones. (Photograph by G. H. Davis.)

Figure 1.4 Geologist confronting the structure of Nature, in this case an "exfoliation" jointing in granite near Shuteye Peak in the Sierra Nevada. (Photograph by N. K. Huber. Courtesy of United States Geological Survey.)

In contrast, the structural geologist is greeted in nature by what looks like a finished product, like the structural product shown in Figure 1.4, and is challenged to ask a number of questions. What is the structure? What starting materials were used? What is the geometry of the structure? How did the materials change shape during deformation? And what was the sequence of steps in construction? Attempts to answer these questions generate even more questions. When was the job done? How long did it take? What were the temperature and pressure conditions? How strong were the materials? And, why "on earth" was it done?

Figure 1.5 Fault offset of a fine-grained dike. (Photograph by J. P. Lockwood. Courtesy of United States Geological Survey.)

FUNDAMENTAL STRUCTURES

Some preliminary descriptions of the most common structures might serve to illustrate why words like form and geometry, style and symmetry, and orientation and order have a special place in the vocabulary of structural geologists. The basic geologic structures in nature are faults, joints, folds, cleavage, foliation, and lineation.

Faults are fractures that have accommodated displacements of rocks in the crust (Figure 1.5). They are the only geologic structures that regularly make

Figure 1.6 Physiographic expression of the San Andreas fault, as viewed northerly along the Elkhorn scarp, San Luis Obispo County, California. (Photograph by R. E. Wallace. Courtesy of United States Geological Survey.)

headlines, for they are among the ordinary surficial expressions of earthquake shocks. During deformation, rocks are commonly forced to move past one another to achieve a better fit. The faulting permits parts of the crust to be shortened or lengthened. Some faults are fundamental breaks in the crust (Figure 1.6). They exert their influence in broad-ranging ways—from creating sites of catastrophic earthquakes to determining locations of boom town precious-metal camps where mineralization is fault controlled.

Joints are fractures or cracks in rocks along which there has been negligible discernible movement parallel to the plane of the fracture. They often reflect in their patterns a systematic geometry and symmetry (Figure 1.7). Where filled with minerals like quartz, calcite, or copper, joints serve as **veins** that have accommodated an extension or lengthening of the rocks they occupy (Figure 1.8). In other cases, micro-offsets along the joints reveal tiny faultlike displacements. Joints are simple to produce experimentally in the laboratory, but they can be produced in so many ways that it is difficult to interpret the exact origin of the specific joint system that is under investigation.

Folds are systematically curved layers and surfaces in rocks (Figure 1.9). They come in all sizes and shapes, and they reveal in their internal form something of the conditions under which they developed. Folds form in

Figure 1.7 Jointing in sandstone in Canyonlands, Utah. The larger joint-bounded blocks are about 50 m on a side. Eroded aisles are zones of closely spaced joints. (Photograph by G. F. McGill; taken through a hole in the floor of a low-flying small plane.)

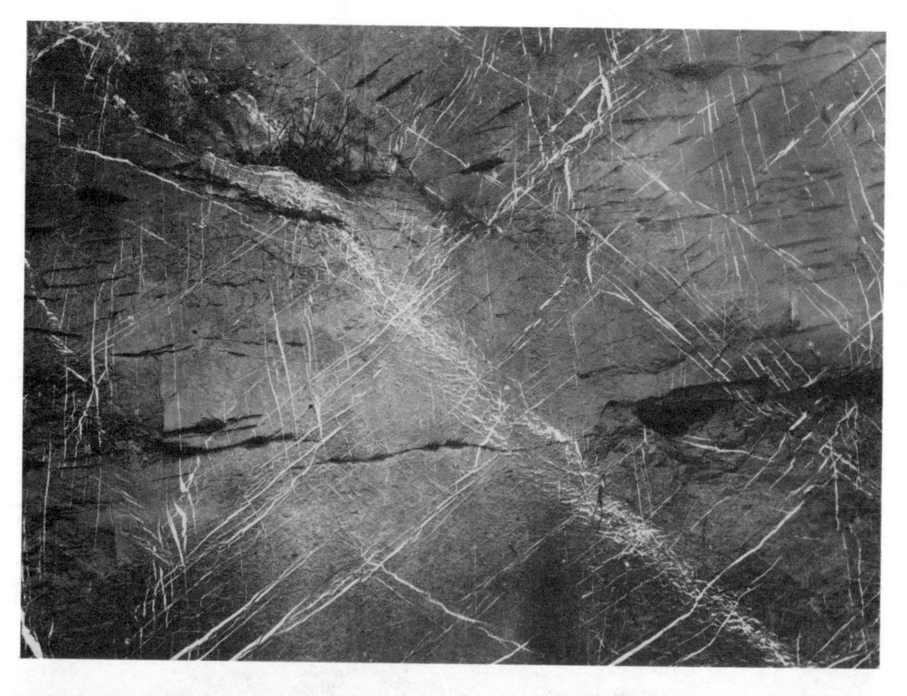

Figure 1.8 Calcite veins in limestone, near Limekiln Point, Highgate Springs, Vermont. (Photograph by C. D. Walcott. Courtesy of United States Geological Survey.)

Figure 1.9 Folded dolomite and limestone near Danby, Vermont. (Photograph by A. Keith. Courtesy of United States Geological Survey.)

many ways. They commonly reflect end-on, vicelike buckling and shortening of originally horizontal layers (Figure 1.10). Some folds result from regional steplike draping of layers along the flanks of differentially uplifted fault blocks. Under high-temperature conditions, gravitational forces acting on plastically deforming layers can produce cascades of folds that pile up like taffy candy (Figure 1.11). Folds, however formed, offer a wonderful geometric challenge in three-dimensional visualization and analysis.

Cleavage, **foliation**, and **lineation** constitute a final major category of common structures. They form under conditions of elevated temperature and/or pressure, commonly through deformation and recrystallization during ductile flow in igneous or metamorphic environments. Foliations are very closely spaced parallel planar alignments of features like micaceous minerals, crystals, microfaults, and flattened pebbles (Figure 1.12). A special category of foliation is cleavage, closely spaced subparallel surfaces that impart a splitting property to highly deformed rocks. Cleavage forms in response to shortening during folding (Figure 1.13). Some cleavage accom-

Figure 1.10 Photomicrograph of chevron folds in schist. (Photograph by A. L. Albee. Courtesy of United States Geological Survey.)

Figure 1.11 Folded interlayers of marble and quartzite. [India-ink rendering by D. O'Day of photograph by G. H. Davis. From Davis (1980). Published with permission of Geological Society of America and the author.]

Figure 1.12 Foliation defined by alignment of deformed cobbles in the Purgatory Conglomerate, Rhode Island. [After Mosher (1981). Copyright ©1981 by The University of Chicago. All rights reserved.]

Figure 1.13 Cleavage in banded slate near Walland, Tennessee. (Photograph by A. Keith. Courtesy of United States Geological Survey.)

modates nearly imperceptible faulting. But most cleavage reflects discontinuities along which rock has been partially removed by stress-induced dissolution to permit even greater shortening of the rock mass. Lineations are preferred linear alignments of elements like hornblende needles, mineral aggregates, bundles of tiny folds, or striations and grooves that pervade rock bodies (Figure 1.14). Under exceedingly rare circumstances lineation is large enough for children and structural geologists to ride (Figure 1.15). The presence of foliation and/or lineation in an igneous or metamorphic rock reflects significant changes in the shape of the rock in which these structures are found.

Rocks that acquire pervasive cleavage, foliation, and/or lineation are known as **tectonites**. These are rocks that have been distorted by flow in the solid state. The flow movements that produce foliation and lineation pervade the entire mass of rock, just like the internal movements that are generated within a penny when it is flattened by a moving train. The dramatic change in the shape of the rock or coin can be measured and defined quantitatively, as long as reference markings of known original size and shape are recognizable, like fossils in the deformed rock, or Abe Lincoln's profile in the flattened penny (Figure 1.16). Even though the change in size and shape of the flattened penny can be described, it is not easy to tell which way the train was going, how much the engine weighed, or how fast it was traveling, unless we were there when it all happened. To interpret the stress conditions and movements that produced foliations and lineations in distorted rocks in nature is equally difficult.

Figure 1.14 Lineation in metamorphic rocks, San Juan Islands, Washington. The lineation reflects the expression of pervasive, tiny folds within the foliated rock. (Photograph by G. H. Davis.)

Figure 1.15 Lineation in the "extraordinary striated outcrop" at Saqsaywaman, Peru (see Feininger, 1978). To kids in Cusco this lineated rock is known as *el rodadera* (the rollercoaster). The ride is fast! To slow down, riders spit on their hands and then use the hands as brakes. (Photograph by L. A. Lepry.)

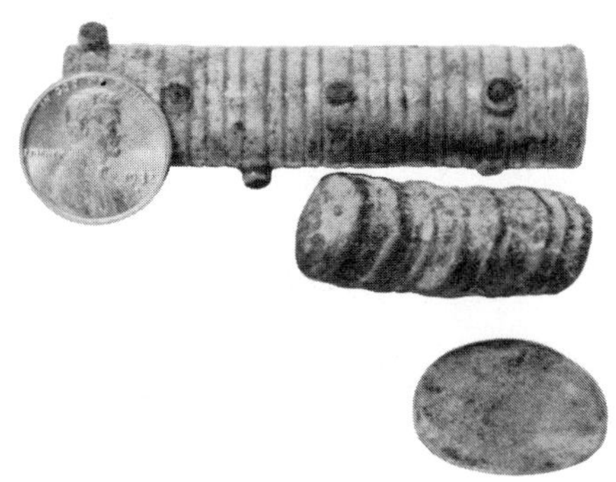

Figure 1.16 Deformed and undeformed crinoid stems. Deformed and undeformed pennies. (Photograph by G. Kew.)

PURPOSE OF STRUCTURAL GEOLOGY

The study of structures is fundamental to any historical understanding of how crustal rocks have responded to physical processes that have been at work through time. Speck-sized to continent-sized masses of crustal rocks have been shifted, rotated, uplifted, and internally distorted as a response to Earth forces. Faults, joints, folds and cleavage, and foliation and lineation are abundant in the Earth's crust because the rocks in which these structures are found were forced to change size and/or shape and/or position.

As students of structural geology we learn to recognize, describe, and measure the orientations of deformational structures, and we learn methods to interpret the displacements, rotations, and distortions that give rise to the structures. Integrating this knowledge with theoretical and experimental concepts, we can analyze the nature of crustal deformation. The fruits of dynamic structural analysis bear importantly on our understanding the formation and control of strategic natural resources, especially energy (oil, gas, geothermal) and metallic mineral deposits.

BARRIERS TO UNDERSTANDING

If our work is to lead to fresh discoveries and understanding, it cannot be carried out in a casual, generalized, unimaginative way. The complexity of structural systems and processes demands much more. To paraphrase Howard Lowry,

> Excellence and learning are not commodities to be bought at the corner store. Rather they dwell among rocks hardly accessible, and we must almost wear our hearts out in search of them. (From *College Talks* by H. F. Lowry, edited by J. R. Blackwood, p. 116. Published with permission of Oxford University Press, New York, copyright © 1969.)

This "wearing our hearts out" begins as soon as we recognize some of the barriers to understanding.

For example, we are free to map and sample a pitifully small percentage of the Earth's volume. The Earth's spectacular topographic relief, from the highest mountains, like the Alps, Andes, or Himalayas, to the deepest parts of the sea, like the Mariana Trench, is inconsequential when viewed at the scale of the Earth as a whole. Three fourths of that skin is covered by water, and much of the rest is masked by forest and dense brush, deep soil, and even man's own concrete testaments. And the samples that we do have an opportunity to map and collect are not really representative of the Earth as a whole. We sample directly only the outermost film of the upper crust. Samples from the depths of the Earth are provided only rarely through belching, gas-charged volcanoes (Figure 1.17), deep drills, and deep-sea dredging.

Then there is the vastness of geologic time. Who can really deal with immeasurable, colossal dimensions of time? Geological events that we consider to be instantaneous, like the outpouring of the Columbia River Plateau basalts, take millions of years. Geologists must develop the ability to speak about time in 10-digit figures with the passive, frozen expression of a banjo picker.

Yet a third barrier is the utter complexity of natural systems. For me, this reality was mirrored in a small pool that I encountered amid dense underbrush—within the bush of eastern Canada. Rock exposure is poor in this region, and thus clues regarding structural history are meager. Yet the surface waters of this pool were marked by foam patterns that conveyed

Figure 1.17 Mount St. Helens in eruption on May 18, 1980. (Photograph courtesy of United States Geological Survey.)

geological insight. Delicately fashioned, these patterns resembled the layering of rocks deformed under hot, deep conditions—exactly the kind of rocks that are exposed in that region from place to place. This pool became my one-day laboratory. Patterns of movement on the surface of the pool were both complex and ever-changing. Seeing the patterns come and go, my mind shifted to what would happen to these structures when winter set in. Some single pattern would be frozen; one of an infinite number of patterns would be preserved; and yet, that pattern might or might not be representative of the kinds of motion I had watched. I began to realize more fully that every geologic record we examine is built out of millions of frozen records, stop-action points, tiny scenarios from a much longer and more complex drama that we never will know in full.

I was reminded of all this not long ago in Utah when I looked down into a pool beneath a bridge and saw flow patterns of the same kind (Figure 1.18).

Figure 1.18 Deformed foam layers in a Rocky Mountain pool. Paper cup (upper right corner) is being blown by the wind from right to left, while the water is being pulled by gravity toward the lower right. (Photograph by G. H. Davis.)

But this time, the pattern carried more symbolism. The ordered patterns, guided ultimately by the flow of water under the compelling tug of gravity, were being modified simultaneously by the competing movement of a paper cup, blown by the wind, superimposing the imprint of its wake. We see in matters of geologic record as well the complex, interfering effects of competition among the agents of heat, pressure, and gravity in fashioning architectural form. And we see in the wadded-up paper cup the symbol of the influence of man on natural systems.

SCOPE OF TEXT

I have arranged this text in two parts. *Fundamentals* (Part I) provides essential background for analyzing *Structures* (Part II).

Chapter 2, "Concept of Detailed Structural Analysis," provides the basic philosophical approach to structural geology as used throughout the text. The forms of analysis are descriptive (both physical and geometric), kinematic (evaluation of translations, rotations, dilations, and distortions), and dynamic (interpretation of forces and stresses responsible for deformation). These three forms of analysis provide the leverage for unraveling structures and structural systems **at any scale**, from **rocks to regions.**

"Descriptive Analysis" is the title of Chapter 3. Learning to recognize structures and to describe their geometry is essential groundwork to the mastery of structural geology. Chapter 4, "Kinematic Analysis," begins with a concept that I hold dear—**that the very presence of structures and structural systems in the Earth's crust reflects changes in the size and/or shape of the rock bodies in which the structures are found.** If no such changes were required, the structures would have no reason to exist. Kinematic analysis is the process of evaluating the structures in terms of the displacements, rotations, and distortions they reflect.

"Dynamic Analysis" (Chapter 5) probes the origin of deformation in terms of forces, stresses, and rock strength. "Plate Tectonics" (Chapter 6) provides the larger view in which dynamics and ultimate origin of structures can be assessed. It expresses our modern concept that much, if not most, crustal deformation is directly or secondarily linked to processes of sea-floor spreading and plate tectonics. It also builds the concept that plates and plate movements produce not only deformation, but also the chief rock-forming environments.

Part II, *Structures*, begins with "Contacts" (Chapter 7), a discussion of the guidelines for recognizing the fundamental kinds of contact relationships: contacts of deposition, faulting, intrusion (both igneous and sedimentary), and ductile shearing. Chapter 8, "Primary Structure," explores primary structures of depositional and deformational origin. Primary structures are original structures that form concurrently with the formation of the rocks in which they are contained. Primary structures are considered at all scales, from cross-beds to volcanoes, from columnar jointing to growth faults.

The chief kinds of secondary structure in the Earth's crust are presented in Chapters 9 to 12. The chapters are entitled "Faults," "Joints," "Folds," and "Cleavage, Foliation, and Lineation." Structures are described and interpreted in the pattern of thought and analysis developed in *Fundamentals*. Examples are selected that blend field, experimental, theoretical, and subsurface studies. A major focus is interpreting the strain significance of structures.

"Concluding Thoughts" emphasizes that structures do not exist in isolation, but are integrated in a network of interconnecting structures to accomplish strain. Restoring rocks to their original locations and configurations requires unraveling structures of all kinds at all levels of observation. This emphasis provides a point of departure from structural geology of rocks and regions to the tectonic analysis of regional geologic structure.

Throughout the book, techniques, methods, experiments, and calculations are described in detail, with the aim of inviting active participation and discovery through laboratory and field exercises and through outside readings. Indeed, the first definition of "knowledge" in Webster's emphasizes familiarity gained through experience (*Webster's New Collegiate Dictionary*, 1973, p. 639).

CONCEPT OF DETAILED STRUCTURAL ANALYSIS

chapter 2

THE CONCEPT

We base our work on a branch of structural geology known as **detailed structural analysis**, with particular emphasis on strain analysis (Sander, 1930; Knopf and Ingerson, 1938; Turner and Weiss, 1963; Ramsay, 1967). The three fundamental strategies of detailed structural analysis are descriptive, kinematic, and dynamic analysis. Each of these looks at geologic structure from a different standpoint. **Descriptive analysis** is concerned with recognizing and describing structures and measuring their orientations. **Kinematic analysis** focuses on interpreting the deformational movements responsible for the development of the structures. The basic movements are those of translation, rotation, dilation, and distortion (Figure 2.1). **Dynamic analysis** interprets deformational movements in terms of forces and stresses responsible for the formation of structures. Dynamic analysis is generally the most interpretive part of detailed structural analysis, but it derives remarkable strength from thoughtful experimental and theoretical studies. Both kinematic and dynamic analysis are anchored in descriptive fact.

Basic to detailed structural analysis is discovering and emphasizing the degree to which deformed rock bodies are marked by profound geometric order, at all levels of observations. Detailed structural analysis leads us to recognize that the geometry and symmetry of the internal structure of rocks reflect the geometry and symmetry of movements and stresses responsible for the deformation (Sander, 1930).

Knopf and Ingerson (1938) helped introduce these perceptions. They illustrated the main points of the **symmetry principle** with homely but provocative illustrations. For example, birds on a telephone line commonly face the same direction (Figure 2.2). The geometric order is striking, but puzzling kinematic questions are raised: What, if anything, does the geometric alignment of these structures tell us about the movement of each bird during landing? What do we conclude about the single bird among fifty that faces in the opposite direction? Different flight paths? Different destinations?

Turner and Weiss (1963) summarized valuable methods that can be used to pick apart the geometric order within deformed rocks, at any scale. John Ramsay (1967), in turn, led modern structural geologists to recognize the degree to which the geometric order in deformed rocks reflects the mathematical dictates of strain theory.

Throughout this book there are recurrent demonstrations that geologic structures and structural systems are marked by a high degree of geometric

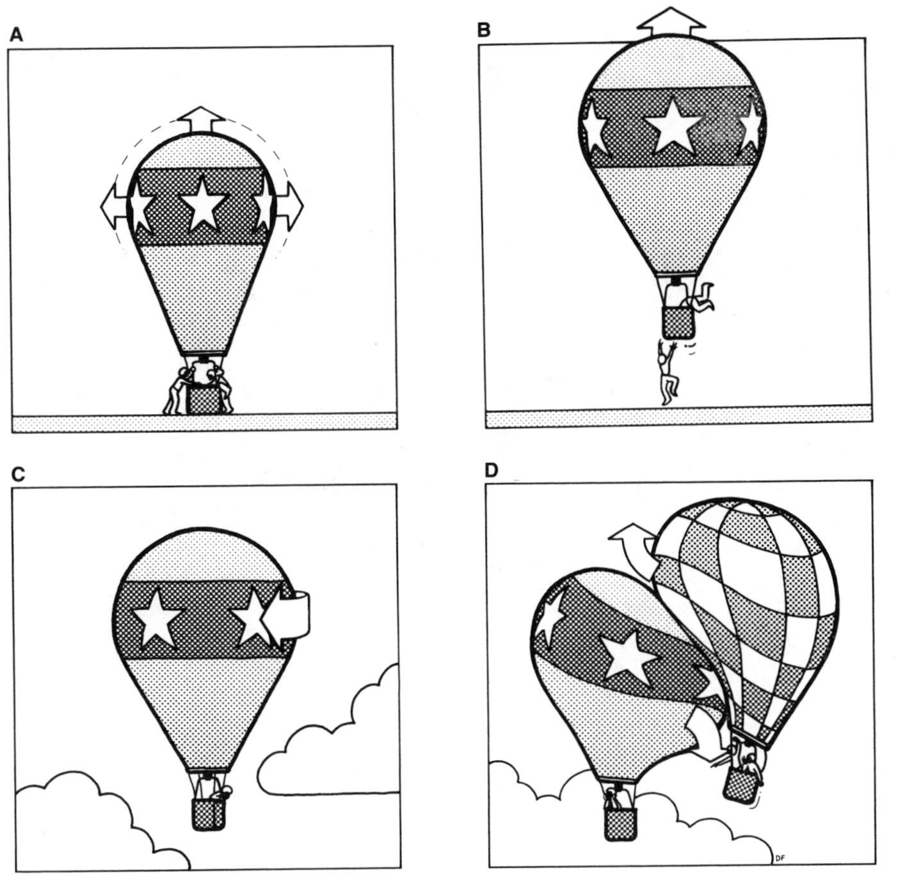

Figure 2.1 The basic movements of (*A*) **dilation** (change in volume), (*B*) **translation** (change in position), (*C*) **rotation** (change in orientation), and (*D*) **distortion** (change in shape). (Artwork by D. A. Fischer.)

Figure 2.2 Preferred orientation of birds on a wire. (Photograph by B. J. Young.)

order; that the geometric order expresses the kinematics of deformation, especially changes in shape and/or size; and that the application of experimental and theoretical principles is helpful in explaining the relation of the geometric order to ultimate origin.

DESCRIPTIVE ANALYSIS

GENERAL APPROACH

In descriptive analysis we recognize structures, measure their orientations, and describe, literally inside and out, their physical and geometric components. Descriptive analysis results in facts regarding the physical properties, orientations, and internal configuration of the structures. The basis for descriptive analysis is broad ranging: direct observation of field relationships, as featured in the cover illustration; examination of rocks deformed experimentally in the laboratory; drilling into the subsurface; geophysical monitoring and probing of the subsurface; and studying the stratigraphy and/or petrography of the rocks in which the structures occur.

In a practical sense, descriptive analysis may require geologic mapping of rocks and structures; measuring the physical dimensions of structures, using tape, rule, surveying instrument, micrometer, or topographic map; measuring the orientations of structures with a compass; plotting graphically the

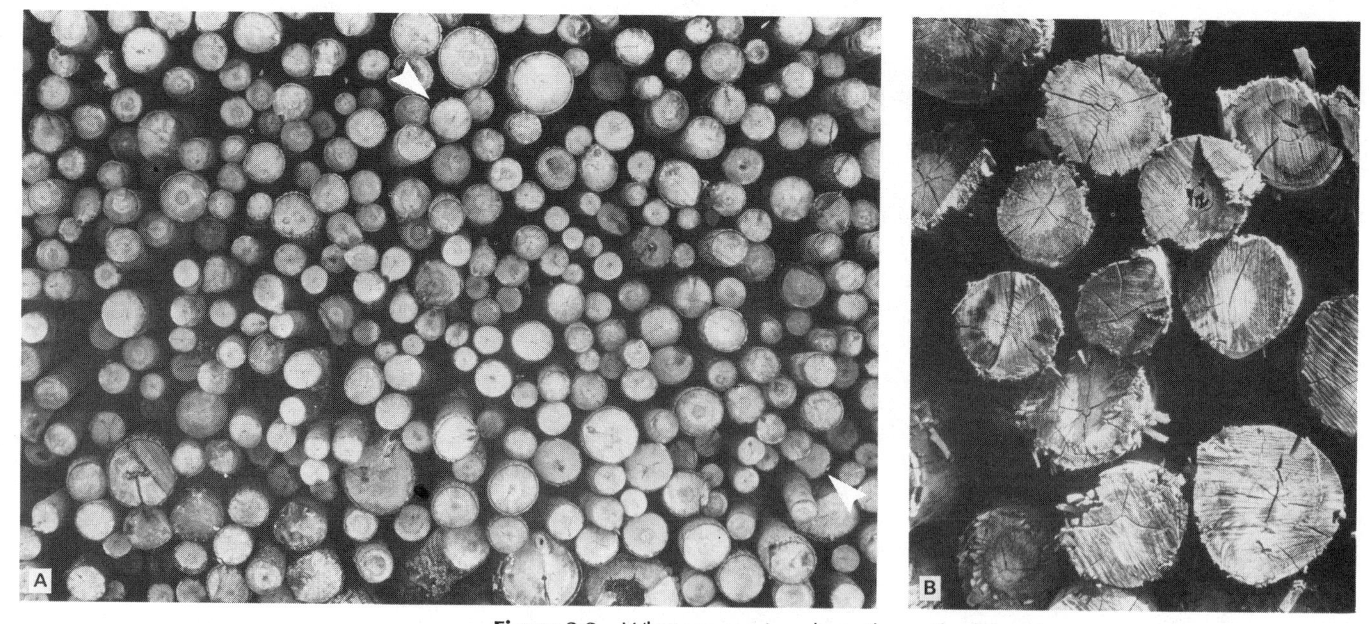

Figure 2.3 What we see in a log pile partly depends on scale of observation. The same is true of rocks. (*A*) Middle-distant view. Arrows point out linear alignment of apparently circular faces of logs. (*B*) Close-up view reveals that the circular faces are not so circular. The close-up view also discloses some interesting fracture patterns. (Photographs by B. J. Young.)

orientations of structures; determining preferred orientations of minerals in thin sections of deformed rocks using a microscope; and examining photogeologic characteristics of fracture patterns as expressed in satellite imagery. Ideally, this phase of work is free of interpretation, but realistically the mapping and measuring operations are nudged by interpretation where structures are excessively complex or poorly exposed.

The specific structural features that can be recognized and described depend importantly on the **scale of observation**. Figure 2.3, which provides a useful nongeologic example of the influence of scale on geologic observation, features two different views of a log pile. As I look at the logs from a distance (Figure 2.3*A*), I am struck by the circular cross-sectional forms of the logs and by differences in diameter. In addition, I see linear alignments of the circular faces of the logs, such as the obvious alignment marked with arrows. In close-up view (Figure 2.3*B*), I am struck immediately by the irregularity of the cross-sectional shapes of the logs. They are by no means uniformly circular. Furthermore, what pops into view is a fascinating internal structure marked by patterns of radial fractures displayed by each of the logs. And with further study I begin to see that the cross-sectional face of each log is marked by striations produced by sawing.

Application of these woodpile principles to the world of structural geology is easy to imagine. What must be remembered above all is that shifts in scale of observation are absolutely necessary to make a complete inventory of the rock and structural components of any deformed body of rock.

There are a number of geologic studies that serve as superb models for descriptive analysis. For example, in reference to their structural investigations of the Canadian Rockies, Price and Mountjoy (1970) concisely summarized the results of an enormous amount of data. Their descriptions, plus the geologic cross section they constructed (Figure 2.4), leave little to

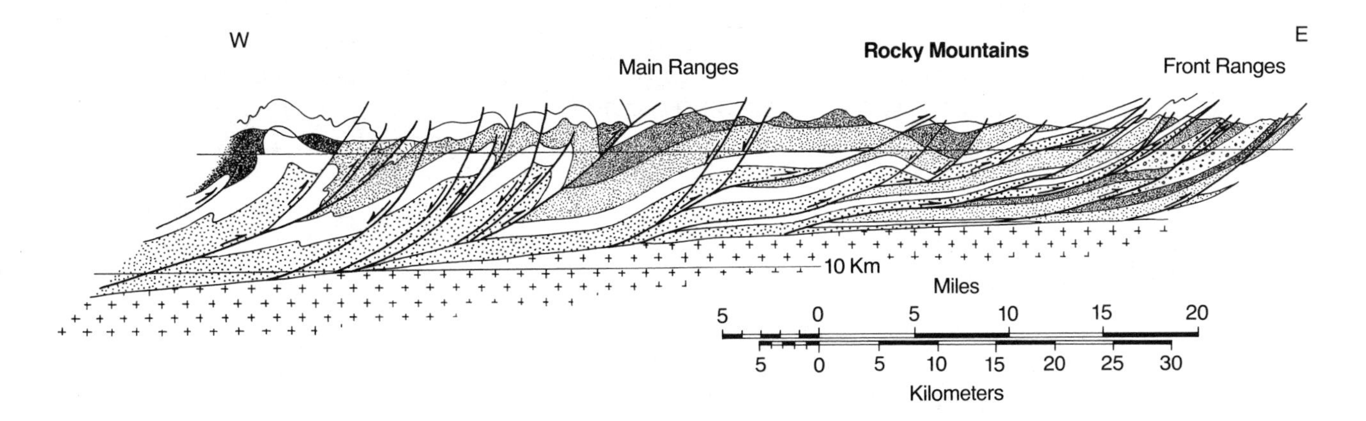

Figure 2.4 Geologic cross section of folded and faulted strata in a part of the Canadian Rockies. [After Price and Mountjoy (1970). Published with permission of Geological Association of Canada.]

the imagination. They specifically call attention to types of structure (thrust faults and folds), structural orientations (southwest-dipping faults, upright folds), shapes of structures (concave-upward faults), relation of faults to bedding (faults cut up-section), and relation of the folds to the faults (thrusts die out as folds). The important descriptive statements and phrases are here shown in boldface type to emphasize the extent of descriptive information and the economy of word choice.

> The structure of this part of the Canadian Cordillera is dominated by **thrust faults** which are generally **southwest-dipping** and **concave upward in profile**. The faults **flatten with depth** and have the **upper side displaced relatively northeastward and upward**. They gradually **cut up through the stratigraphic layering northeastward, but commonly following the layering** over large areas. . . . Many of the faults **bifurcate [split] upwards into numerous splays**, and the total **displacement along them becomes distributed among these splays**. **Folds are widespread** and have developed in conjunction with the thrusting. . . . **Many of the thrust faults themselves are folded** along with the sedimentary layering. . . . The **folds generally are inclined to the northeast or are upright**. **Many of the thrusts die out as folds**. . . . [From Price and Mountjoy (1970), p. 10. Published with permission of Geological Association of Canada.]

STRUCTURAL ELEMENTS

Learning to recognize and describe the main classes of structures is not quite enough. Many different types of each major structure exist in nature. Furthermore, each structure is composed of **structural elements** that in turn are identified and described, to permit us to carry out a complete descriptive analysis.

Structural elements are the physical and geometric components of structures. The **physical elements** are real and tangible, and they have measurable geometry and orientation. The **geometric elements** are imaginary lines and surfaces, invisible but identifiable in the field; they too have measurable geometries and orientations. A troublesome gate I built along a stretch of fence in my side yard not only may illustrate the approach to analyzing structural elements, but also may clarify how descriptive analysis leads naturally to kinematic analysis.

I first isolated the part of the fence to become the gate by sawing through the two-by-fours forming the horizontal support for the fenceline. Then I fastened strap hinges to one side of the gate and to the supporting, adjacent fenceline. To the other side of the gate I attached a "catch" (Figure 2.5*A*). So that the bottom of the gate would clear the top of the adjacent cement driveway, I trimmed the base of the gate with my saw, leaving about 1 in. (25 mm) of clearance. In just a few days, a structural problem had developed. The bottom of what was now a sagging gate no longer cleared the driveway (Figure 2.5*B*). Impulsively, I trimmed 2 in. (50 mm) from the base of the gate. In three more days, the gate was again dragging on the cement. Clearly it was time for descriptive and kinematic analysis.

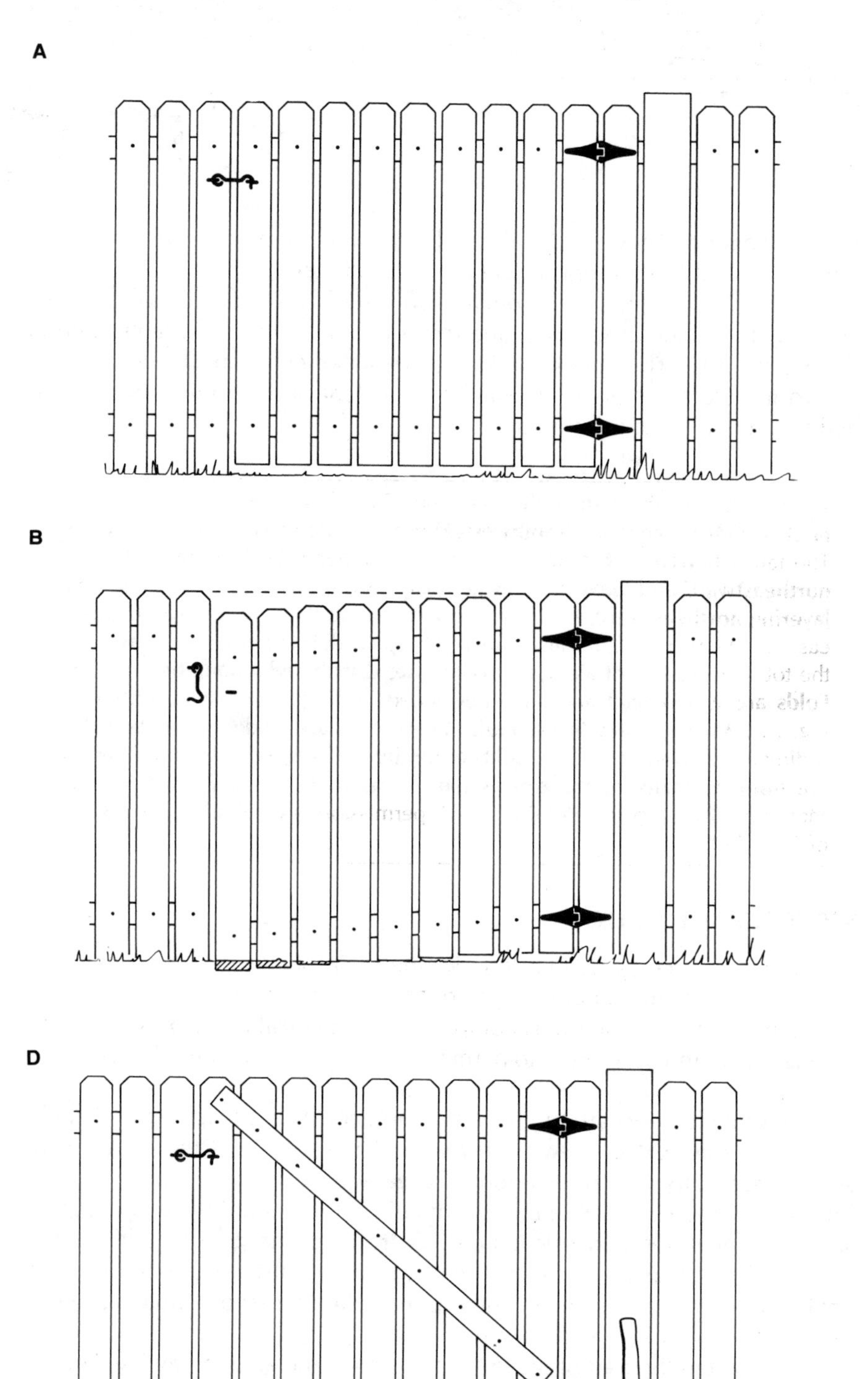

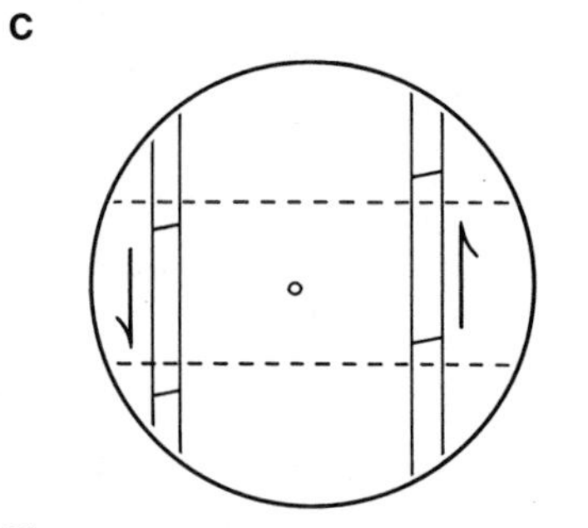

Figure 2.5 Saga of the troublesome gate. (*A*) The perfect gate. (*B*) The structural problem. (*C*) The kinematic solution. (*D*) The remedy.

The gate consists of both physical and geometric elements. The physical elements are two two-by-fours, nine slats, two hinges, and the metal catch. All the elements lie in the same plane. The strap hinges and the two-by-fours are oriented horizontally, the slats vertically. Between the slats are thin, rectangular cracks, or in the language of structural geology, **discontinuities**. These constitute the only important geometric elements within the gate. Like the slats, they are vertically inclined.

The kinematic solution to the structural problem was discovered by lifting the catch and thus raising the gate off the pavement. The gate did not rotate at all where hinged. Rather, the "uplift" was absorbed within the gate by small translational movements along each of the discontinuities between the slats. In effect, the small translations allowed the gate to be restored from its deformed shape, a parallelogram, to its original undeformed shape, a rectangle. The small translational movements permitted distortion of the gate to take place, creating a change in shape (Figure 2.5*C*). The practical solution was to eliminate future possibility of movements along the discontinuities by nailing a brace across the face of the gate (Figure 2.5*D*).

In matters of geology too, descriptive analysis takes into consideration geometric elements. Angular, chevron-shaped folds, like the ones shown in Figure 2.6, may be subdivided into folded layers, folded bedding surfaces, hinge points, and axial surfaces. The folded layers are physical and real, composed of the rocks that have been folded. The bedding surfaces, hinge points, and axial surfaces are geometric and imaginary. The bedding surfaces are discontinuities that separate the folded layers. Hinge points and axial surfaces are conveniences for helping to define the orientation and form of the fold. Focusing attention on the bedding-plane discontinuities between the rock layers rather than on the rock layers themselves is quite a shift in emphasis. It is like suggesting, perhaps quite correctly, that the key elements of venetian blinds are the openings between adjacent slats, not the slats. Whether the physical or geometric elements are emphasized in structural analysis depends on the problem being addressed. In my west-facing office in Tucson, I appreciate the geometric elements (the openings) of venetian blinds in the morning, but I turn to the physical elements (the slats) in the late afternoon.

The degree of geometric order in a deformed rock body is evaluated by measuring the orientations of large numbers of structural elements. The orientations are plotted graphically and evaluated statistically to discern the quality of preferred orientation. Through this process, **sets** of structures or

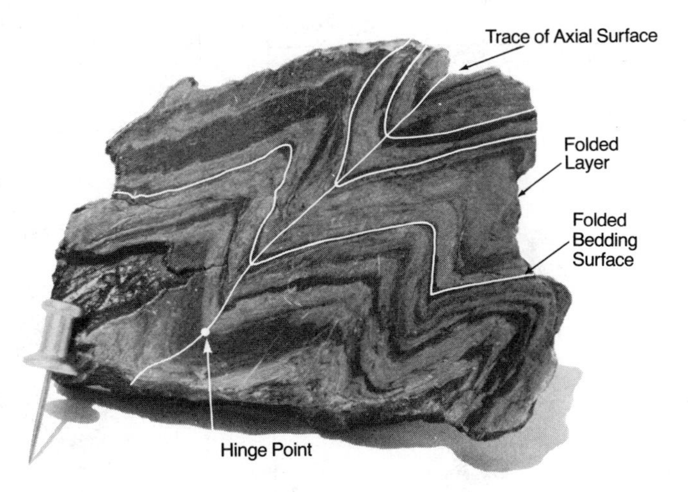

Figure 2.6 Physical and geometric elements of a chevron fold. (Photograph by G. Kew.)

Figure 2.7 Three **sets** of barrels containing Spanish olives. (Photograph by B. J. Young.)

structural elements can be defined. Sets are composed of elements sharing common geometric and/or physical appearance and parallel orientation. For example, the barrels featured on the loading dock shown in Figure 2.7 may be subdivided into three mutually perpendicular sets on the basis of size and orientation. Two or more sets of like structures or structural elements constitute a **system**. All such systems, taken together, plus all structures that do not conveniently arrange themselves into sets, comprise the total **structural system**.

KINEMATIC ANALYSIS

GENERAL APPROACH

Kinematic analysis depends on data provided by careful descriptive studies. The goal of kinematic analysis is to interpret the combination of translations, rotations, dilations, and distortions that altered the location, orientation, size, and shape of a body of rock. Translations and rotations change the location and/or orientation of a rock body without changing the form or size of the body (Figure 2.8*A*, *B*). However, a rock body may be dilated and/or distorted during translation and rotation, in which case the rock undergoes changes in size and/or shape (Figure 2.8*C*, *D*). Evaluating the changes in a body that result from dilation and distortion is the focus of **strain analysis**.

Strain analysis is basic to modern structural analysis. It requires a quantitative appraisal of the degree to which the original sizes and shapes of

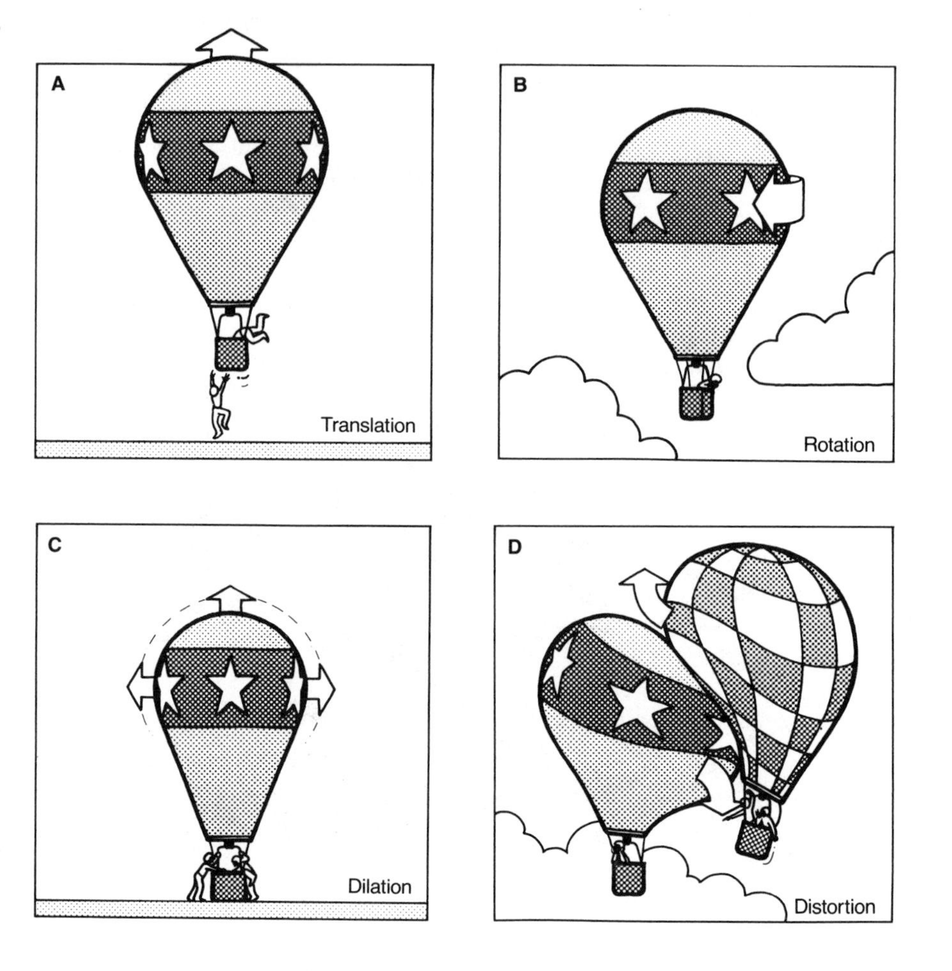

Figure 2.8 (*A*) Translations and (*B*) rotations change location and orientation without necessarily changing size or shape. (*C*) Dilations and (*D*) distortions change size and shape without necessarily changing location and orientation. (Artwork by D. A. Fischer.)

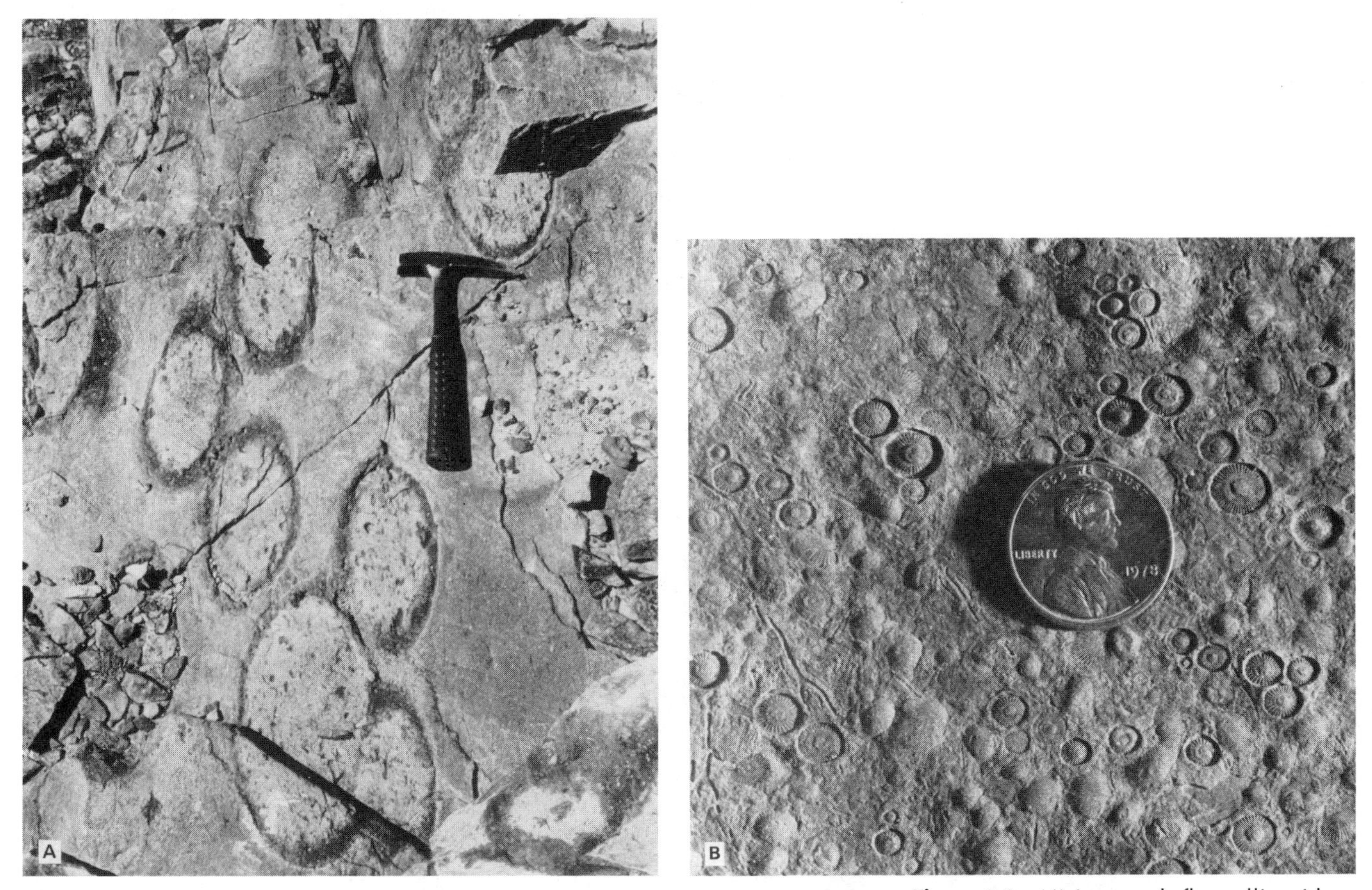

Figure 2.9 (*A*) Large ash-flow ellipsoids in highly deformed volcanic rocks near Bishop, California. The ellipsoids were derived from distortion of primary volcanic structures, which were probably originally spherical. [From Tobisch and others (1977). Published with permission of Geological Society of America and the authors.] (*B*) Slightly distorted crinoid stems in limestone, Appalachian Plateau sector of New York State. [From Engelder and Engelder (1977). Published with permission of Geological Society of America and the authors.]

geologic objects or geologic bodies have changed during deformation, like the geologic objects shown in Figure 2.9. The evaluation of strain is prerequisite to stress analysis, but it is carried out irrespective of the forces or stresses that caused the strain-producing movements.

Distortion generally produces a high degree of order within a deformed body. The internal order reflected in deformed rocks is achieved by systematic movements that affect the entire mass of rock. The movements may take some advantage of preexisting surfaces like bedding, joints, faults, cleavage, or foliation. Ordinarily the deformational process creates some new surfaces as well, to accommodate fully the movements required for distortion. Newly created structural surfaces can include joints, faults, cleavage, and foliation.

PENETRATIVE NATURE OF STRUCTURES

Structural analysis is influenced by the degree to which structural elements penetrate the rock body under study (Turner and Weiss, 1963). For structures to be **penetrative**, they must be spaced so closely with respect to the size of the rock body under consideration that they appear to be everywhere. Figure 2.10*A* provides a useful image: individual logs within the wooden islands are considered to be penetrative because they are very small and very closely spaced compared to the size of the floating bodies they comprise. Figure 2.10*B*, a geological example, shows tiny fractures that are penetrative at the outcrop scale.

What is learned about structural systems depends partly on the scale at which the structures are penetrative. In Figure 2.11*A*, siltstone and shale are pervasively cut by a closely spaced cleavage. The cleavage is penetrative at

Figure 2.10 (*A*) **Penetrative**, parallel logs in a river. (Photograph by B. J. Young.) (*B*) Penetrative, parallel fractures in an outcrop. [Tracing of photograph by G. H. Davis, rendered by D. O'Day. From Davis (1980). Published with permission of Geological Society of America and the author.]

the scale of single handspecimens. The cleavage largely masks the original textures of the sedimentary rock. Only thin relics of rock containing the original primary structure are preserved. Given this situation, numerous data bearing on the kinematics of cleavage formation can be drawn from single outcrops of this rock. But clues regarding the exact nature of the original rock would have to be sought at the microscopic scale. Conversely, primary stratigraphic relationships, if they are to be recognized at all, would have to be explored in a more regional view (Figure 2.11*B*).

In yet another example of this concept (Figure 2.11*C*, *D*), two faults, spaced kilometers apart, cut and displace a sequence of shale, limestone, and siltstone. Outcrops and small area investigations of this geologic system would yield stratigraphic and petrologic information regarding the primary

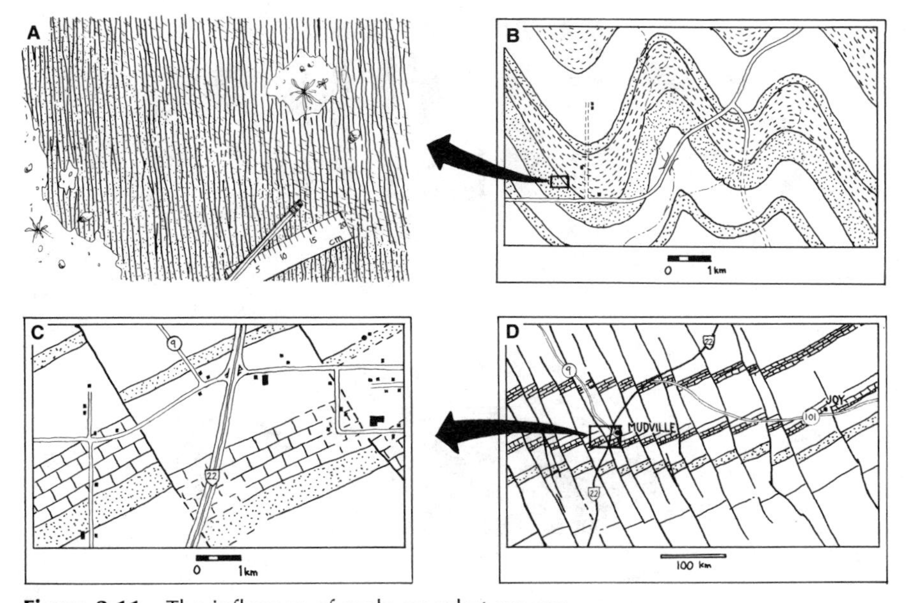

Figure 2.11 The influence of scale on what we see. (*A*) Penetrative cleavage masks original bedding in the outcrop view, (*B*) but not in the regional view. (*C*) At a relatively large scale of observation, the faults appear to be widely spaced. (*D*) At a smaller scale of view, the faults can be considered to be penetrative. (Artwork by R. W. Krantz.)

rock assemblage. But only limited structural kinematic insight would be derived from the study of single faults. To really understand this structural system, large regional domains would have to be studied. Seen in a regional view, faults would be considered to be penetrative (Figure 2.11*D*). The structural kinematic evaluation of the system of faults, viewed as a whole, would aid enormously in interpreting the deformational history of the region.

SLIP, FLOW, AND DISTORTION

Kinematic movements are often described in terms of **slip** or **flow**. The distinction between flow and slip is scale dependent. In some cases a layer that appears to have flowed can be seen upon close examination to have achieved its folded form through a myriad of small slip displacements on "microfaults." No experiments demonstrate this better than those carried out by O'Driscoll (1962, 1964a, b). By drawing parallel lines representing layers of undeformed rock on the sides of card decks, and systematically translating the cards along their close-spaced slip surfaces, he was able to produce dramatic flow patterns (Figure 2.12).

Price and Mountjoy understood the scale dependence of slip and flow in their evaluation of the Canadian Rockies.

> All of the individual thrust faults are small when the area within which they occur is compared to the total area of the Rocky Mountains. . . . on the appropriate scale of observation, the overall structural pattern within the Rocky Mountains reflects a type of "plastic deformation" or flow involving large-scale strain and translation within a coherent mass of layered sedimentary rock. Slip within the mass was concentrated in a myriad array of discrete, but discontinuous, interleaved, and interlocked shear surfaces, all of which were contained within the mass and are now represented by the various thrust faults. [From Price and Mountjoy (1970), p. 10. Published with permission of Geological Association of Canada.]

Figure 2.12 O'Driscoll-like experiment carried out by Paige Bausmann. (A) Computer card deck with stripes simulating layering. (*B*) Distortion of the deck by penetrative slip achieves "flow" of the layers. (Photographs by G. Kew.)

The major kinematic lesson that emerges from both experimental studies and field work is this: systematic movements on relatively close-spaced surfaces of slip can produce significant distortion (i.e., strain) of a body of rock.

DYNAMIC ANALYSIS

GENERAL APPROACH

Dynamic analysis interprets forces, stresses, and mechanical processes that give rise to structures. For dynamic analysis to be meaningful, it must explain the kinematic movement plan and the physical and geometric characteristics of the structures. The major aim of the analysis is to describe the relative magnitudes and absolute orientations of the stresses that were responsible for the deformation. This is a difficult step in detailed structural analysis, for significant inferences must be made regarding the strength and physical state of the materials during deformation, the rate at which deformation proceeded, and the boundary conditions for the structural system under investigation.

Geological and engineering literature is full of dynamic models that are helpful in interpreting the origin of structures. The models are descriptions of the conditions under which deformational structures form. The basis for dynamic analysis is theoretical and experimental research. Most dynamic models are valid in principle, but most structural systems can be explained satisfactorily by more than one dynamic model. Choosing among alternative models is challenging but speculative science.

Think of dynamic models as "recipes" for making structures and structural systems. Recipes describe the amount, the kind, and the condition of the ingredients as well as the conditions to which the ingredients are subjected. If the recipe is valid, the final product is guaranteed, provided the conditions are met. Such recipes can be prepared on the basis of experimental or theoretical studies in which the conditions for deformation are predetermined, measured, and/or arbitrarily established. Whether a particular model applies to a particular geological system depends largely on interpretation of the conditions under which deformation was achieved. For structural systems formed approximately 2 billion years ago, interpretation of the conditions of deformation is very difficult.

One of the interesting arenas of dynamic modeling is the experimental deformation of small cylindrical cores of rock under conditions of regulated temperature, confining pressure, rate of deformation, and fluid pressure. Empirical observations derived from experimental testing of materials have produced broad-ranging guidelines regarding the conditions under which rocks will fail by faulting, and even the orientations and characteristics of the faults thus formed. Important results of such work are "recipes" for producing faults. One such recipe might read this way (Figure 2.13).

Figure 2.13 Some of the ingredients and products of rock-deformation experiments. *Left to right*: a copper jacket; a 1-in. (2.54 cm) core of marble; a 1-in. core of siltstone; a faulted, jacketed core of siltstone; a faulted, slightly barrel-shaped core of marble. (Photograph by R. W. Krantz.)

Use a core-drill device to cut a 1/2-in. diameter (13 mm), 1-in. (25 mm) cylinder from a specimen of fine-grained sandstone. Bevel ends of cylinder so that the top and bottom surfaces are planar and parallel. Insert specimen in copper jacket and place in deformation apparatus. Pressurize its environment with 20,000 lb/in.^{2} (1406 kg/cm^2) confining pressure, thus simulating deep burial. Load the ends of the specimen with a steadily increasing force by mechanically moving a piston into the deformation chamber. Add the force at a rate of 1000 lb/min. The specimen should fracture by faulting when the force reaches approximately 8000 lb (3600 kg). Expect one or two faults to form, each inclined at about 28° to the direction of loading.

Each of the major geologic structures has been the subject of decades of dynamic analysis. Faults and fault patterns have been modeled experimentally, not only using real rock materials (Figure 2.14), but also using soft materials like clay (Figure 2.15). The fault patterns can be interpreted in the light of known stress and/or movement conditions. Fold structures and fold patterns are experimentally replicated routinely in soft layered materials (Figure 2.16). Moreover, theoretical, mathematical analysis of structures has been pursued effectively from the perspective of engineering and fluid mechanics. The equations that "picture" the deformation are hardly Kodachromes of outcrop features. They are more like abstract art.

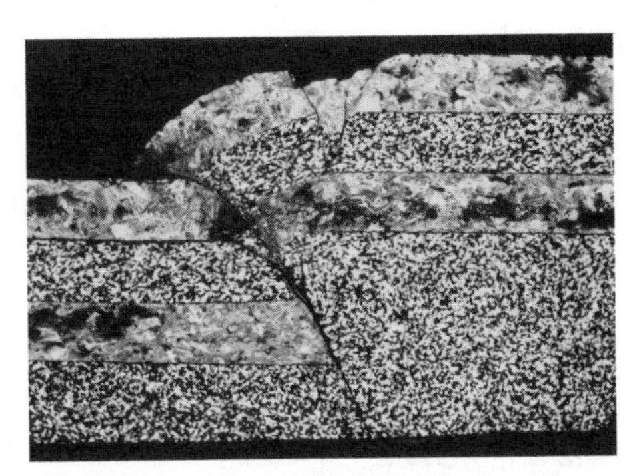

Figure 2.14 Photomicrograph of rock specimen experimentally deformed in the Laboratory of Tectonophysics at Texas A & M University. The specimen was built from alternating layers of sandstone (fine-grained layers with salt-and-pepper appearance) and limestone (coarse-grained layers, including the uppermost layer). Each of the upper three layers is 0.3 cm thick. Faulting was achieved by compressing the lowermost, thick basal layer. [From Friedman and others (1976). Published with permission of Geological Society of America and the authors.]

$$L = 2\pi t \sqrt[3]{\frac{\eta}{6\eta_1}} \quad (2.1)$$

But equations describe dynamic relationships in ways that words and photographs never could. Decoding the equations simply requires knowledge of what the variables represent. When Equation 2.1 is decoded, it reads:

The wavelength of a fold in a layer of viscosity η_1 and thickness t is equal to the product of (1) 6.2832 times thickness and (2) the cube root of the ratio of the viscosity of the layer to 6 times the viscosity of the rock in which the layer is contained.

A

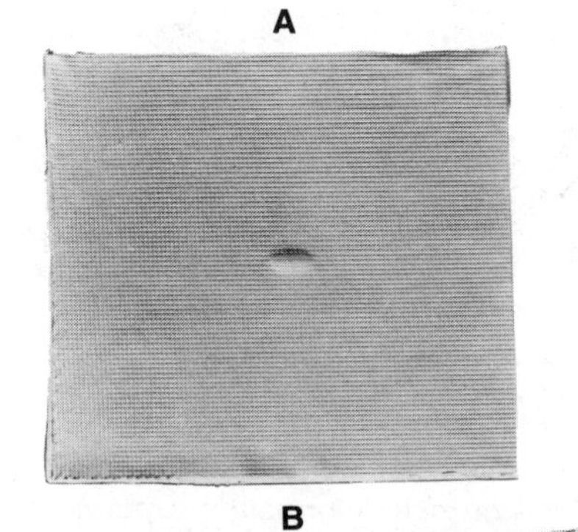

B

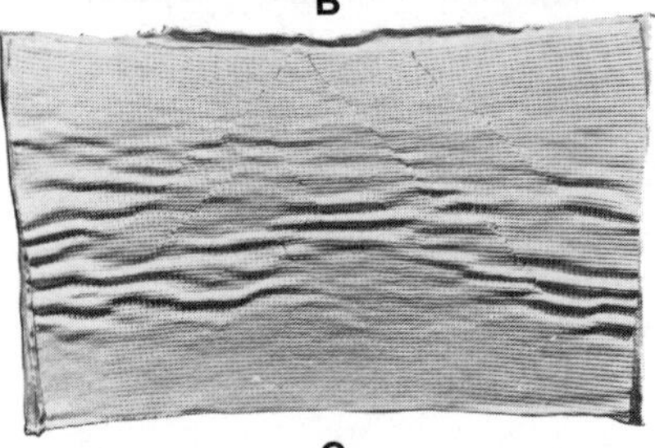

C

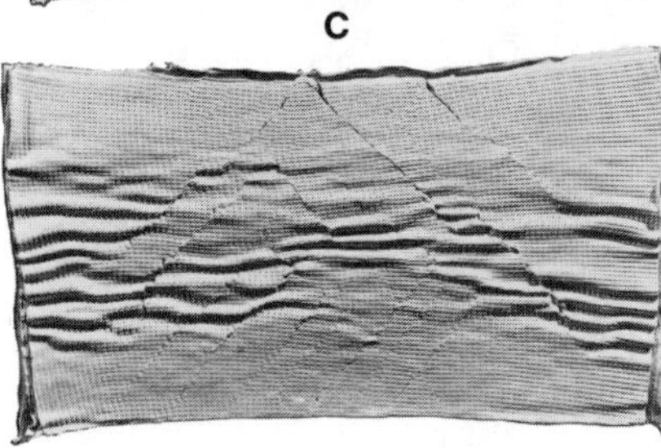

D

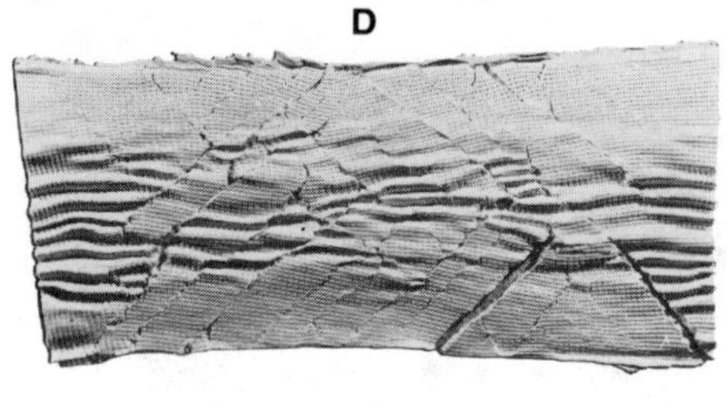

5mm

Figure 2.15 Experimental deformation of a plasticine model. Folding and faulting were achieved by a one-sided compression from the top. Original block is about 10 cm square. The three stages of deformation record 25, 30, and 42% shortening. [From Dubey (1980), Tectonophysics, v. 65. Published with permission of Elsevier Scientific Publishing Company, Amsterdam.]

Figure 2.16 An example of one of Bailey Willis's classic experimental deformation models. This one, made of wax, accommodated layer-parallel shortening by folding. Much of the work of Willis, carried out in the nineteenth century, was directed toward understanding mountain building in the Appalachians (see Willis, 1894). (Photographs by J. K. Hillers. Courtesy of United States Geological Survey.)

DETAILED STRUCTURAL ANALYSIS OF A PIZZA

We study many examples of dynamic analysis in the ensuing chapters, to understand better the origin of structures. For now let me summarize with a nongeologic example. The example, although a little bizarre, may serve to clarify kinematic analysis and show the bridge that kinematic analysis builds between descriptive analysis and dynamic analysis.

Going to the freezer for a midnight snack, I pulled out a pizza whose form is portrayed in the geologic map shown in Figure 2.17*A*. The pizza, as shown, is arbitrarily oriented with respect to north. In most respects this was a normal pizza. A thin stratum of cheese rested atop tomato sauce and crust.

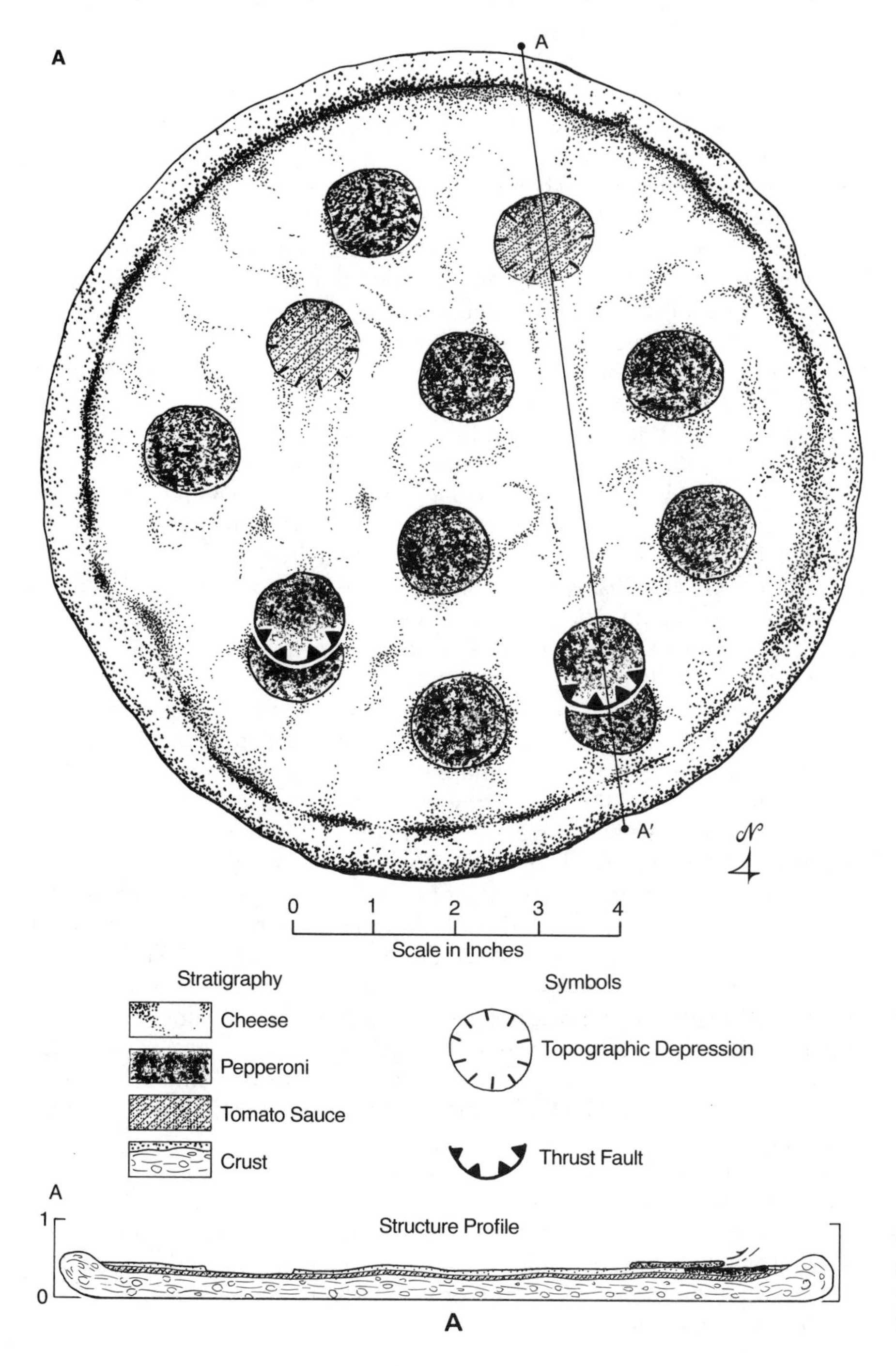

Figure 2.17 (*A*) Geologic map and structure profile of a medium-sized pepperoni pizza. (*B*) Kinematic model of the translation and rotation of pepperonis.

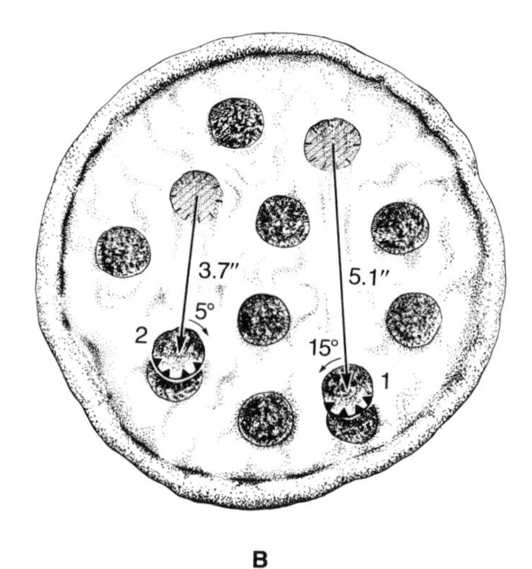

Figure 2.17 *(cont.)*

And pepperonis, lightly dusted with cheese, were distributed across the face of the pizza. The diameter of the pizza was 9 in. (23 cm). Topographic relief of the pizza, as revealed in cross-sectional profile (Figure 2.17*A*), was merely 1/2 in. (13 mm). My delight as a structural geologist came upon observing that two pepperoni-sized circular depressions existed near the edge of the pizza in the northeast and northwest quadrants. There was not even a trace of cheese in these depressions, let alone pepperoni. Furthermore, opposite these depressions, in the southeast and southwest quadrants, two pepperonis overlapped each other in low-angle (overthrust) fault contact. The cross section of Figure 2.17*A* clarifies this relationship.

The physical and geometric properties of the pizza demanded the following kinematic interpretation (Figure 2.17*B*): somehow the pepperonis that had once occupied the circular depressions now devoid of cheese had translated 5.1 and 3.7 in. (12.9 and 9.4 cm), respectively, to their present locations. The path of movement for each pepperoni was determined by measuring the orientation of a line connecting the exact center of the circular depression and the exact center of the closest faulted pepperoni. Orientations for pepperonis 1 and 2 were S5°E and S4°W, respectively. Careful matching of the outlines of each of the faulted pepperonis with each of the outlines of the circular depressions revealed that pepperoni 1 underwent about 15° of counterclockwise rotation during translation, and that pepperoni 2 underwent about 5° of clockwise rotation during its translation. No microscopic study of the frozen sauce or cheese was undertaken to evaluate internal distortion due to the translation.

What can be concluded about the dynamics of origin of this structural system? By far, this is the most speculative part of the analysis, but the most enjoyable. (You will note that the language used in dynamic analysis is always cautious.) First, it would seem that the force(s) that triggered movement of the pepperonis did not violate the general integrity of the pizza.The box was not crushed in any way, nor was the crust distorted beyond primary kneading and shaping. The probable force causing deformation was gravity. Gravitational forces would not have triggered movement of the pepperonis, however, unless the pizza had been tilted at some stage in its history. My working model is that the manufacturer, after preparing the pizza, chose not to stack the boxed pizzas horizontally in freezer compartments. The manufacturer may have concluded, perhaps on the basis of experimentation, that tall stacking of pizza-filled boxes might have the adverse affect of flattening cardboard to cheese and tomato sauce before freezing set in. Instead the pizzas may have been filed vertically. If stacked vertically while cheese and tomato sauce were yet warm and/or moist, the pepperonis, under the influence of gravitational forces, might have been vulnerable to translation along the low-viscosity tomato sauce discontinuity. Each of the pepperonis would have ceased moving when it encountered the frictional resistance of another pepperoni. What was not clear to me then, nor is it now, is the rate at which the pepperonis may have moved—rapid or sluggish? The magnitude of the stresses that were required to initiate movement is also a puzzle. In fact, interpreting the strength of the various materials as a function of temperature would constitute a major study in itself.

My working model is, of course, only one interpretation. Maybe the structural event that dislodged the pepperonis was of an entirely different nature. Maybe my interpretation is correct except for timing: after all, the pizza, stacked vertically, may have been undeformed and solidly frozen . . . until the power failure. Interpreting the timing of structural events is often very difficult.

THE TIME FACTOR

TIME AND SPACE

Kinematics and dynamics feed naturally into broader geologic synthesis, but not without a grasp of the dimension of time. Analysis of crustal movements in relation to the overall physical evolution of the Earth simply cannot be pursued outside a time reference framework. Interpreting the timing of deformational movements provides a way to recognize discrete periods of deformation. It allows **diachronous** events to be recognized: like the movement of a cold front, these are events that steadily migrate across a region through time. Interpreting timing permits the fruits of focused structural investigations to be integrated within a framework of broader geological processes and events. It promotes an understanding that structural deformations are just a small part of a much larger orchestration, the knowledge of which serves to clarify why certain structural relationships developed in the first place.

Trying to comprehend the scope of time is as awesome as effectively considering the magnitude of space. The nature of each varies with position and perspective. Carl Becker (1949), upon viewing human progress from the Olympian heights, shifted his position to a point beyond the world. What he described from there is a chilling reminder of the relative size of our field areas, no matter how large they might seem, within the whole of the spatial dimension.

> Thus conveniently placed, . . . we look out upon a universe that comprises perhaps a billion galaxies, each galaxy comprising perhaps ten thousand stars. If we look long and attentively we may detect, within one of the lesser galaxies, one of the lesser stars which is called the sun; and circling around the sun, one of the lesser planets which is called the earth. (From *Progress and Power* by C. L. Becker, p. 113. Published with permission of Alfred A. Knopf, Inc., New York, copyright ©1949.)

James Michener, in *Hawaii*, provides a complementary account of the dimension of time. In narrating the birth of the first volcanic island in the Hawaiian chain, he writes:

> For . . . an extent of time so vast it is meaningless, only the ocean knew that an island was building in its bosom, for no land had ever appeared above the surface of the sea. . . .from that extensive rupture in the ocean floor, small amounts of liquid rock seeped out, each forcing its way up through what had escaped before, each contributing some small portion to the accumulation that was building on the floor of the sea. Sometimes a thousand years, or ten thousand, would silently pass before any new eruption of material would take place. At other times gigantic pressures would accumulate beneath the rupture and with unimaginable violence rush through the existing apertures, throwing clouds of steam miles above the surface of the ocean. Waves would circle the globe and crash upon themselves as they collided twelve thousand miles away. Such an explosion, indescribable in its fury, might in the end raise the height of the subocean island a foot. (From *Hawaii* by J. A. Michener, p. 4. Published with permission of Random House, Inc., New York, copyright ©1959.)

In spite of our inherent awe for the duration of Earth time, we soon learn as students of geology to deal comfortably from a deck of 4,500,000,000

Figure 2.18 Cliff Palace in Mesa Verde National Park. White circles on beams in lower left are the tops of plugs that fill holes from which cores of wood were extracted for tree-ring dating. (Photograph by G. H. Davis.)

years. And wherever possible, we attempt to define carefully in real numbers the age limits for specific rocks and events.

TIMEPIECES

The timepieces, the instruments with which to tell the geologic time, are varied. The one timepiece that provides the opportunity of absolutely dating an object or event, in terms of a single year, is dendrochronology, more specifically, tree-ring dating. The tree-ring calendar now available to us is fully 8600 years in duration. Although the precision of this timepiece is best known in matters of archaeology (Figure 2.18), geological applications are no less significant. Major ancient flood events have been dated. Shifts in the position of timberline in mountainous regions have given clues to fluctuations in paleoclimate. The step-by-step chronology of sedimentation processes in the burial of forests in ancient alluvial-filled valleys in the Southwest has been monitored. Additionally, the absolute timepiece that tree rings provide is being used to calibrate the ^{14}C radiometric timepiece more accurately.

Radiometric timepieces tell time on the basis of calibrated radioactive decay of certain atoms. The radiometric clocks yield more than a relative placement such as that furnished by paleontology and stratigraphy. However, they do not provide a single-year chronology in the fashion of tree-ring dating. Age determinations provided by the use of radiometric clocks incorporate statistical error, and so results are provided in what Smiley (1964) has referred to as globs of time, not points in time.

Radiometric clocks make use of the property of certain isotopes to decay spontaneously to some other element. In the language of geochronology, a parent isotope, like ^{14}C, progressively decays to daughter-isotope products, like ^{14}N. Different parent isotopes have different rates or constants of decay. Thus the time (i.e., the half-life) required for a parent isotope to decay to half its original volume varies from system to system. The half-life for the decay ^{14}C to ^{14}N is approximately 5500 years. The half-life for the decay of ^{40}K to ^{40}Ar is about 1 billion years. The isotope ^{238}U decays to ^{206}Pb in a half-life of approximately 4.5 billion years; ^{87}Rb decays to ^{87}Sr, and the half-life of this system is about 50 billion years.

The different radiometric timepieces require different kinds and amounts of sample material. Use of the ^{14}C method requires samples of materials like charcoal, wood, shell, peat, and bone in amounts ranging from about 10 to 100 g. Given the short half-life of this isotopic system, the limit to dating by this method is about 70,000 years. Radiometric analysis by $^{87}Rb/^{87}Sr$ requires careful selection of individual mineral separates (e.g., biotite, muscovite, chlorite, potassium feldspar), or crushed whole-rock samples. For the latter, fresh samples on the order of 30 kg are required. Similarly, the $^{40}K/^{40}Ar$ method uses mineral separates or whole-rock samples. For $^{238}U/^{206}Pb$ isotopic analysis, 50-kg samples of fresh unaltered rock must be collected. From these, mineral separates of zircon, or less commonly, monazite and sphene, are hand picked for the isotopic analysis.

There are significant problems in making age determinations on the basis of radiometric methods. Smiley (1964) has emphasized that all dating is a matter of interpretation, whether it be a date on a structure in an archaeological site, the date of a particular rock formation, or the date of a specific geological event. He stresses that geochronological interpretations can be strengthened greatly by knowing the exact physical makeup of the sample to be dated, by determining the complete geological history of the rock body from which the sample was removed, and by learning the precise association between the material studied and the event whose age is being analyzed. In other words, meaningful age determinations are derived from paying close attention to the primary sources.

Modern radiometric isotopic determinations are providing globs of ages for rocks and events that are nicely consistent with relative age relationships determined independently on the basis of geology. In such cases, it is inferred that the original rock contained only parent isotopes (no daughter products), and that the decay systems remained completely "closed" and isolated through time, thus preventing entrance or exit of daughter or parent isotopic materials (Jaeger and Hunziker, 1979). However, age determinations sometimes are completely inconsistent with known or inferred geological relationships. In these cases it is suspected that the thermal effects of metamorphism and igneous intrusion(s), or the abrupt cooling effects of regional uplift, altered the ratios of parent to daughter products in the isotopic systems. In structurally complex areas, it is not uncommon to face problems of interpreting whether a given radiometric age determination reflects an age of crystallization, an age of metamorphism, a time of uplift and cooling, or perhaps nothing geological at all.

In addition to radiometric timepieces, two modern timepieces have emerged recently from paleomagnetic research. The first paleomagnetic timepiece yields approximate ages on the basis of paleomagnetic pole positions frozen in sedimentary and volcanic rocks. Sometimes the use of paleomagnetic analysis to determine age is the only way to unlock the approximate time of formation of thick piles of sedimentary assemblages that completely lack fossils. A second paleomagnetic timepiece has evolved from recognition of magnetic reversals of the Earth's dipole field through time. A time scale of magnetic reversals has been constructed by defining stratigraphic intervals marked by **normal** versus **reverse polarity**, and dating the ages of these reversals both through radiometric and paleontological methods.

The wide variety of timepieces that is available to use in geological research affords great leverage in picking apart the dynamic evolution of any region. Figure 2.19 captures the ways in which different geochronological methods can be used together to maximize our understanding of geological history. Ages of rocks and events in sedimentary, igneous, and metamorphic environments can all be probed (see Table 2.1).

THE GEOLOGICAL COLUMN

In frontier regions where the rock/stratigraphic succession is poorly understood, structural analysis cannot proceed without concurrent geochronological investigations using all timepieces that are available, in the fashion displayed in Figure 2.19. But the timepiece most of us use in structural studies is the established **geological column** for the area or region of study. Knowledge of the time-reference framework depends largely on familiarity with established stratigraphy, combined with an awareness of radiometric age determinations for igneous and metamorphic rocks (Figure 2.20).

The most direct approach to gaining familiarity for the timing of events and the ages of rocks within a specific region of interest is to search out and study geologic maps. The explanation of any good geologic map provides the time frame for the local or regional geological column. The column is, in effect, the timepiece to be used in investigating the geology of the area or region of interest.

The geological column is integrated within the context of the **geological time scale** (Figure 2.21). Learning this time scale places us in the proper

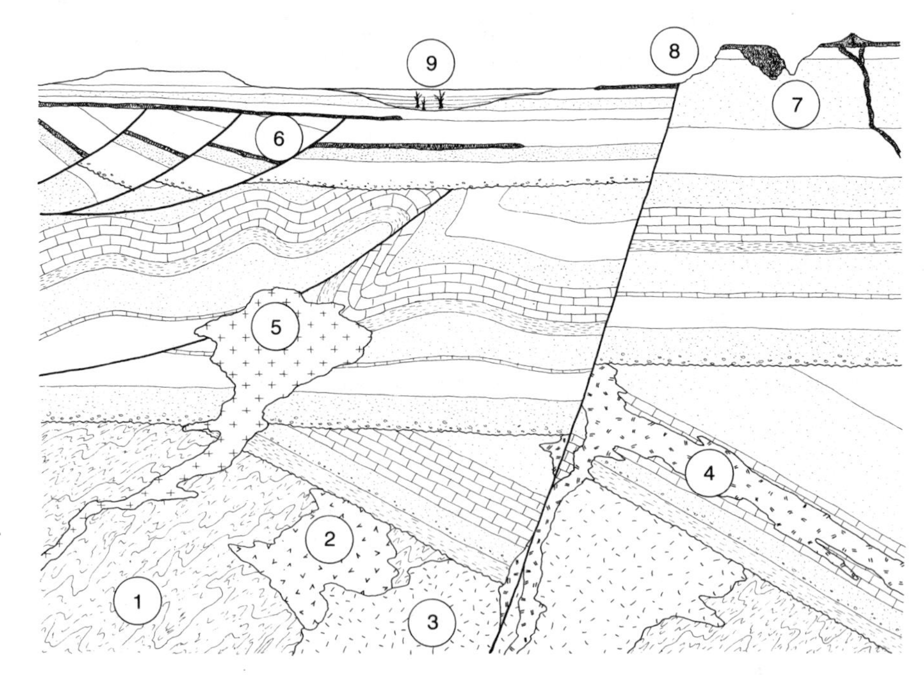

Figure 2.19 Schematic geologic cross section representative of Arizona's geological history. Dates of the rock bodies (keyed by numbers) provide the basis for establishing the timing of the many events that have affected the region. Table 2.1 recounts the history of events and presents the specific age-date information that is the basis for control. (Artwork by R. W. Krantz.)

Table 2.1
Interpreting History with the Help of Age Dates

1. *1680–1700 m.y.* (U-Pb). Deposition of volcanics and sediments. (Silver, 1963).
2. *1625 ± 10 m.y.* (U-Pb). Intrusion of granodiorite. (Silver, 1963).
3. *1420 ± 10 m.y.* (U-Pb). Intrusion of quartz monzonite. (Silver, 1978).
4. *1120 ± 10 m.y.* (U-Pb). Intrusion of diabase sills. (Silver, 1960).
5. *51 ± 3 m.y.* (K-Ar). Intrusion of Laramide quartz monzonite, after thrusting. (Marvin and others, 1973).
6. *~20 to 15 m.y.* (K-Ar). Normal faulting. Timed bracketed by youngest tilted volcanics and oldest flat volcanics. (Shafiqullah and others, 1978).
7. *6.85 ± .20 m.y.* (K-Ar). Extrusion of basalt. (Hamblin, Damon, and Bull, 1981).
8. *<0.29 m.y.* (K-Ar). Recent faulting. (Hamblin, Damon, and Bull, 1981).
9. *1300–1550 A.D.* (Tree rings). Burial of forest by sediments. (Euler and others, 1979).

framework for learning, for contributing, and for communicating (Figure 2.21). It is first-order business. The day after graduation from high school, I entered the office of J. VonFeld, Exploration Geologist for Consolidation Coal Company. His first statement to me was, "I understand you want to be a geologist." His second statement was in the form of a question: "Have you learned the geologic time scale?"

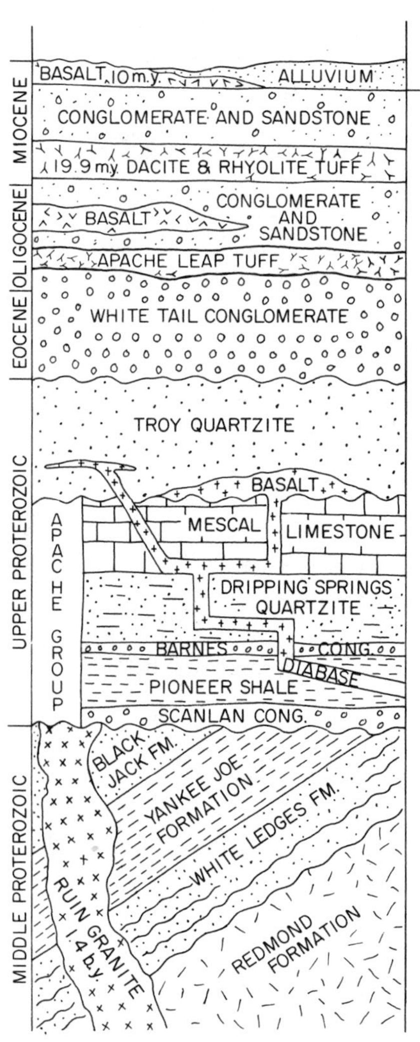

Figure 2.20 Composite geological column for the Salt River Canyon region of central Arizona (m.y. = million years). [From Davis and others (1981), fig. 8, p. 56. Published with permission of Arizona Geological Society.]

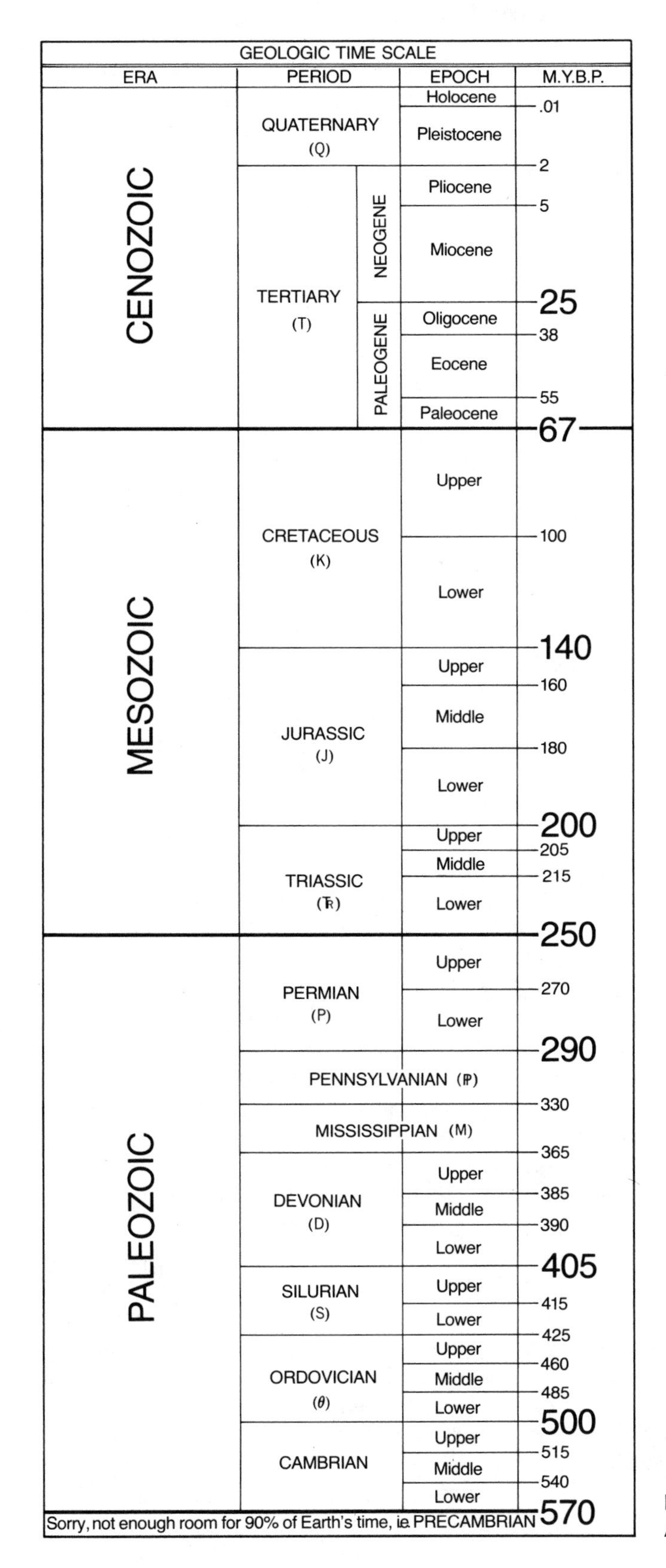

Figure 2.21 Geologic time scale. Although not digital, it keeps great time.

DESCRIPTIVE ANALYSIS *chapter* 3

Field Procedures

INTRODUCTION

Descriptive analysis, both physical and geometric, is the basis for kinematic and dynamic interpretation. The seeking, collecting, and recording of data provide the facts and stimulation that promote new ideas. Much excitement is felt during the data-collection stage when traverses are made into dark canyon recesses, leading to fresh scientific discoveries.

FIELD NOTEBOOK AND ROCK DESCRIPTIONS

Fundamental descriptive data collected in the field or laboratory are recorded in notebooks, which become the depositories for descriptive analysis. As a matter of course, rock and/or sample descriptions are generally recorded first. Consider a field project. Preparation for a field excursion should include the "homework" of studying what has been written about rocks in the pertinent geologic column or the part of the geological column that is likely to be encountered. This information can be summarized ahead of time in the field notebook. Notes regarding the ages and general lithologic characteristics of the rocks serve to anchor the study, especially in its early stages. On-the-outcrop work brings us in touch with the full range of rock types, and for most structural studies it is essential to describe the hand-specimen and formational characteristics of each mappable unit. Where needed, samples are collected for microscopic studies. Care in rock description is essential, for rock type strongly influences the types of structure that evolve during deformation.

Turner and Weiss (1963) recommended a notebook layout I have found extremely practical for field work (Figure 3.1). In structural studies, many measurements of structural elements are recorded, both orientations and dimensions. It is convenient to reserve one side of the notebook for such information, keeping tabulations simple and orderly for efficient retrieval at a later stage of the research. Extracting such data for entry into computers, for example, is much easier when measurements are arranged systematically, not scattered within paragraphs. Rock descriptions should be placed on facing pages, along with sketches or cross sections of important relationships, code numbers for specimens collected, photograph numbers, and most important, *ideas* that come to mind during field work.

Hardback bound notebooks, approximately 5 x 8 in., are traditionally used in geologic studies. But an attractive alternative approach is to carry a small two-ring loose-leaf notebook containing punched 3 x 5-in. cards. The cards are durable. They may be coded, catalogued, and filed. When it is

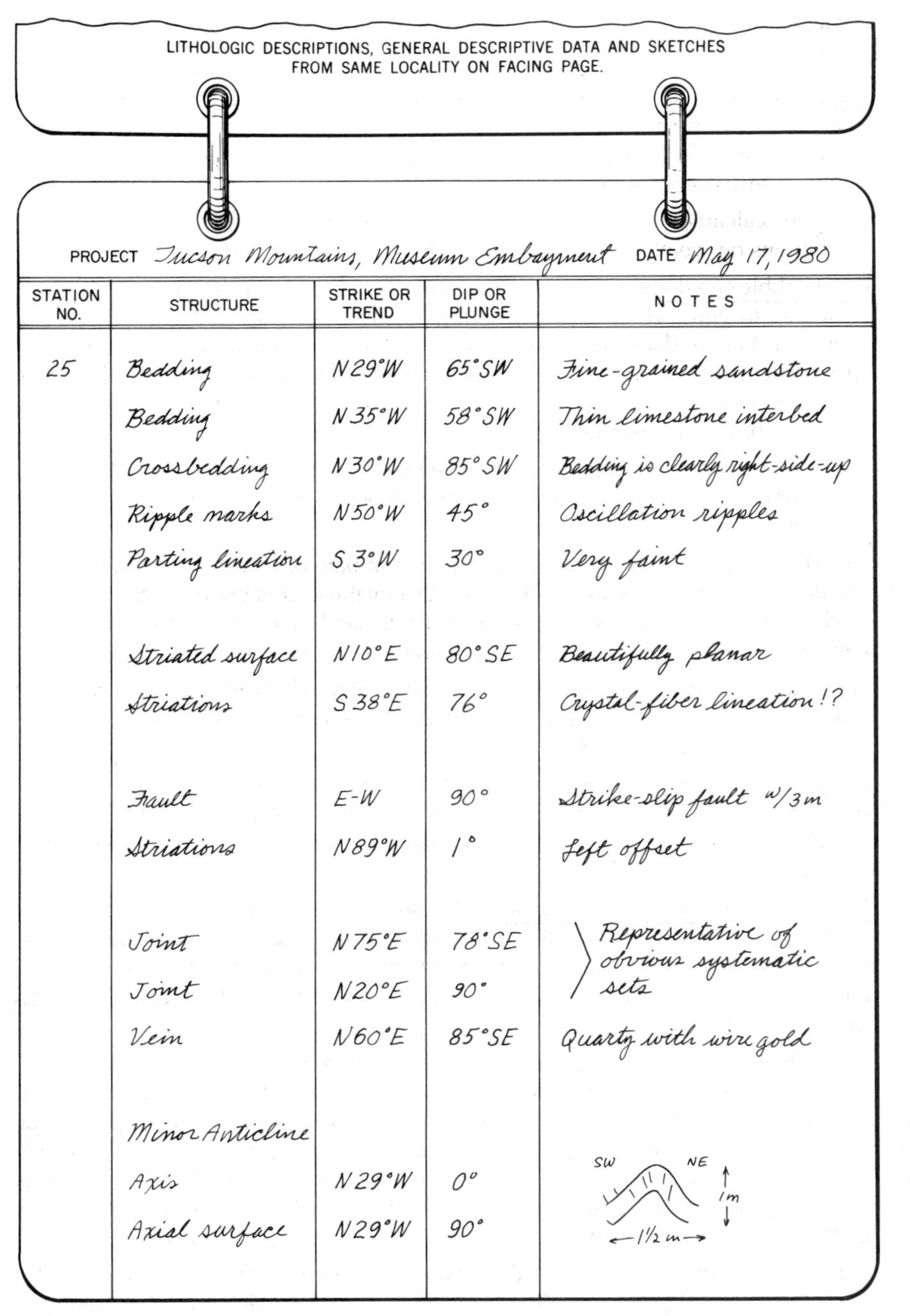
LITHOLOGIC DESCRIPTIONS, GENERAL DESCRIPTIVE DATA AND SKETCHES FROM SAME LOCALITY ON FACING PAGE.

PROJECT Tucson Mountains, Museum Embayment DATE May 17, 1980

STATION NO.	STRUCTURE	STRIKE OR TREND	DIP OR PLUNGE	NOTES
25	Bedding	N29°W	65°SW	Fine-grained sandstone
	Bedding	N35°W	58°SW	Thin limestone interbed
	Crossbedding	N30°W	85°SW	Bedding is clearly right-side-up
	Ripple marks	N50°W	45°	Oscillation ripples
	Parting lineation	S3°W	30°	Very faint
	Striated surface	N10°E	80°SE	Beautifully planar
	Striations	S38°E	76°	Crystal-fiber lineation!?
	Fault	E-W	90°	Strike-slip fault w/3m
	Striations	N89°W	1°	Left offset
	Joint	N75°E	78°SE	Representative of obvious systematic sets
	Joint	N20°E	90°	
	Vein	N60°E	85°SE	Quartz with wire gold
	Minor Anticline			
	Axis	N29°W	0°	
	Axial surface	N29°W	90°	

Figure 3.1 Notebook arrangement recommended by Turner and Weiss. (From *Structural Analysis of Metamorphic Tectonites* by F. J. Turner and L. E. Weiss. Published with permission of McGraw-Hill Book Company, New York, copyright ©1963.)

time to write a geologic report, the cards can be combined with 3 x 5-in. cards bearing notes from library or experimental research, and arranged according to an appropriate organizational scheme. If the cards are filed regularly, all is not lost on the tragic day the notebook is left on a remote mountain peak or is swallowed by torrential canyon whitewater.

Descriptions of rocks examined at a given outcrop can be presented in the field notebook in a short paragraph. The rock description ought to be organized in some standard format for writer efficiency and reader understanding. A useful sequence is **rock name**, **color**, **mineralogy**, **texture**, **primary features**, **gross characteristics**, **formation name**, and **contacts** (Schreiber, 1974). Examples of some well-written rock descriptions are presented in Table 3.1.

Table 3.1
Some Rock Descriptions

Sandstone (quartzarenite); white and very pale orange, weathers light brown and moderate reddish brown; very fine grained, subangular, well sorted; laminated; locally crossbedded; bedding thickness as much as 1 ft (30 cm), mostly covered with rubble; forms steep, rounded slope. Bolsa Quartzite.

Siltstone, calcareous; grayish red, weathers pale red; coarse silt; faint horizontal laminations on weathered surface; thin bedded. Earp Formation.

Chert pebble conglomerate; grayish red purple, weathers grayish red; 70% red chert, 30% gray to white chert, ranging from 0.4 cm in diameter to 6 cm in diameter, average 1.3 cm in diameter; angular to subround, very poor sorting; cement mostly sparry calcite, minor micritic cement; rare microscopic limonite; tabular-planar and wedge-planar cross bedding; medium bedded; forms resistant bed on slope; disconformity at base. Earp Formation.

Clay, calcareous; moderate brown, numerous limy nodules and concretions near top, crumbly and blocky; forms moderate slopes; top surface irregular. Pantano Formation.

Limestone, light gray to light yellowish gray and pale olive gray, commonly sandy and dolomitic, beds generally 1–4 ft (30–120 cm) thick, contains pelecypods, brachipods, and echinoid spines, grades laterally into sandstone. Forms steep slopes and cliffs with distinctive bench at the top. Kaibab Limestone.

Limestone (micrite); pale red, weathers pale yellowish brown; red chert flecks; slightly silty; medium bedded; fossils replaced by red chert, crinoid stems, echinoid spines, small brachiopods. Earp Formation.

Dolomite; light olive gray, weathering same; medium crystalline; unit characterized by abundant (6 in; 15 cm) chert nodules (oolites with fossil hash), white to very light gray, becoming non-cherty at top; pock-marked, gash-like weathered surface; thin-bedded.

Porphyritic and amygdaloidal basalt flows and sills, dark olive green to dark gray; weather olive brown to light olive gray. Phenocrysts of olivine and augite partly to wholly altered to iddingsite and serpentine-chlorite. Amygdules consist of calcite, natrolite, analcite, or quartz. Flows are interbedded with younger gravel. Sills are in the Naco Formation.

Schist, light gray to dark gray, weathers brown or greenish-brown. Comprised of a variety of types from coarse-grained quartz-sericite schist to fine-grained quartz-sericite-chlorite schist. Low grade metamorphism greenschist facies; higher-grade metamorphism occurs locally. Relict bedding of sedimentary protolith is generally recognizable in outcrop; plunging overturned tight to isoclinal folding is pervasive. Poorly exposed, forming subdued outcrops covered with flaky chips. Overlain unconformably by younger Precambrian rocks of the Apache Group. Pinal Schist.

Quartz monzonite, light gray or light pink, usually deeply weathered to light brown. Typically coarse-grained, containing large phenocrysts of pale-pink orthoclase up to 3 in (7.6 cm) long. Coarse-grained groundmass consists of pale-pink orthoclase, chalky plagioclase (albite or andesine), quartz, and books of black biotite. Probably underlies diabase and sedimentary formations in most of the region. Ruin Granite.

The nomenclature and style of describing petrography and texture, as well as the actual assignment of a specific rock name, may vary from worker to worker, depending on past training and preferred classification schemes. Colors describing fresh and weathered surfaces are normally selected qualitatively, but in certain studies it is valuable to use a rock color chart designed for quantitative, reproducible results. Primary features are structures, textures, and objects that form in the rock at the time of its inception. Examples are fossils, ripple marks, gas vesicles, volcanic bombs, inclusions, and bedding. Commonly the primary features observed in outcrops may

Table 3.2
Bedding and Splitting Properties of Layered Rocks

Bedding	*Splitting Property*	*Thickness*
Very Thick	Massive	>120 cm (4 ft)
Thick	Blocky	60–120 cm (2–4 ft)
Thin	Slabby	6–60 cm (2 in–2 ft)
Very Thin	Flaggy	1–5 cm (½ in–2 in)
Laminated	Shaly (mudstone)	2 mm–1 cm
	Platy (limestone and sandstone)	(1/10 in–½ in)
Thinly Laminated	Fissile	<2 mm (1/10 in)

shed light on important aspects of the environment of rock formation and the degree of internal distortion of the rock body. "Gross characteristics" is a catchall category featuring descriptions of outcrop characteristics. Under this heading may be included quality of the rock exposures, topographic/physiographic expression of rocks and structures, weathering characteristics, and the nature of bedding (or layering or foliation) (Table 3.2). Finally, the name of the formation is entered, if it is known. If it is not known, interpretations of formation assignment may be suggested. Such interpretations are to be boldly labeled as such, pending discovery of facts that help pin down the formation assignment.

If the outcrop under study discloses a contact with some other rock formation, the contact relationship should be described as an important ingredient of the notebook record. As I emphasize again later (Chapter 7), it is essential to try to distinguish between contacts of deposition, faulting, intrusion, and unconformity.

Laboratory work demands the same degree of concern for description of rock type and materials as is required in field research. Experimental deformation of natural rocks should proceed with full awareness of the petrography of the sample to be deformed. Descriptions of thin sections of samples, based on microscopic examination of the rocks before and after experimental deformation, are essential. Experimental deformation of soft materials like clay, putty, wax, and paste should include a reporting of the properties of the materials, especially **viscosity** (i.e., resistance to flow). Complete descriptions of starting materials, mixing recipes, and temperature conditions under which the experiments were conducted are valuable to other workers who might wish to pursue such work.

GEOLOGIC MAPPING

PHILOSOPHY

Geologic mapping is the heart of descriptive analysis in structural geology. The geologic map provides an image of the distribution of rock formations within an area. At the same time, it discloses the form and structural geometry of rock bodies. Even the internal structures of the rocks are portrayed on geologic maps. In effect, the geologic map is a three-dimensional description of rocks, structures, and contacts in a given area. The "flattening" of the description into a two-dimensional sheet of paper is achieved through the use of standard symbols.

Geologic mapping is more than an activity. It is a powerful method for systematic structural analysis and scientific discovery. A proper geologic map cannot have loose ends. If a geologist decides that one of the essential

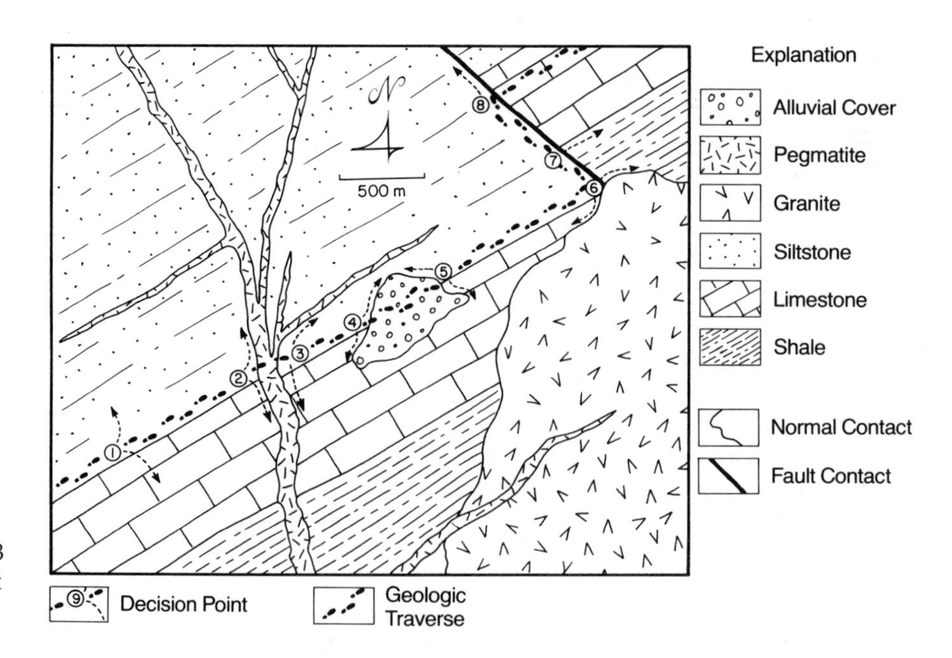

Figure 3.2 Decisions, decisions, decisions in geologic mapping. Table 3.3 presents the specific nagging worries that might surface at the numbered locations.

formations to be mapped is the Nugget Sandstone, then the Nugget must be tracked across the entire map area. Its upper and lower contacts with adjacent formations must be followed (figuratively speaking) step by step. Where the contacts are covered with alluvium, this must be shown. Where the contacts are offset by faults, or intruded by dikes, the offsets or intrusions also must be shown. If the formation "suddenly disappears, "the geologist must cope until the formation is "found" or until a satisfactory explanation is discovered that can be portrayed through the use of standard geological symbols.

Hundreds of decisions have to be made in the course of a single day of geologic mapping. Consider the map area shown in Figure 3.2: the geology consists of northeast-trending sedimentary rocks that first were faulted, then intruded by granite and pegmatite. But what if the geology of this area had never been mapped, and we chose to map it? As we begin traversing the area, we might as well be walking through a corridor with walls so high that we are cut off from all geologic relationships except those exposed directly under foot. Assume that we start our mapping at the crack of dawn in the southwestern corner of the area. By lunchtime, our stomachs would be empty and our minds would be overflowing with nagging worries (Table 3.3).

Problems and nagging worries can accumulate exponentially in even the simplest geologic circumstances. Imagine how they would build during mapping of an area whose structure was comparable to that shown in Figure 1.18! We need to remember to keep a sense of humor and a sense of purpose during the initial stages of mapping and not be overwhelmed by loose ends.

If all rock bodies were absolutely predictable in form, continuity, and distribution, the geologic mapping process might not be essential to detailed structural analysis. However, given the unpredictable twists in geometry of rock formations, geologic mapping is fundamental to understanding. Through mapping, the investigator who is concerned about tying up all loose ends is led into outcrop areas of real discovery. The map is the tangible descriptive product of the process. The fruit of the process, however, lies in what was learned through the force of the method along the way. The record of the intellectual growth is revealed in the ideas and insights posted along with data in the field notebook.

Table 3.3
Nagging Worries While Mapping the Area Shown in Figure 3.2

Point	*Decision*	*Nagging Worries*
1	To map the contact between the limestone and siltstone units, following it to the NE.	When should I map the contact to the SW? When should I map northwestward across the siltstone? When should I traverse southeastward across the limestone?
2	To search for the contact between limestone and siltstone on the east side of the pegmatite dike.	When should I follow the west edge of the dike to the north? When should I follow the western contact of the dike to the south?
3	To continue to follow the contact between the limestone and siltstone to the northeast.	When should I map the east edge of the dike to the north? When should I map the east edge of the dike to the south? Where does the dike branch?
4	To cross the alluvial cover and rediscover the contact between the limestone and the siltstone.	When should I map-out the contact between alluvium and the siltstone unit? When should I map the contact between alluvium and the limestone unit?
5	To continue to follow the contact between limestone and siltstone to the northeast.	When should I traverse northwestward across the siltstone? When should I traverse southeastward across the limestone?
6	To map the fault northwestward and to search for the offset limestone unit.	When should I map the contact between limestone and granite to the southwest? When should I map the contact between shale and granite to the southeast? Where is the offset limestone bed? What if I don't find it??
7	To continue to map the fault contact and to search for the siltstone/limestone contact.	When should I map the contact between limestone and shale to the east?
8	To continue to follow the contact between limestone and siltstone.	When should I map the fault northward? When is lunch??

BASE MAPS

The initial step in preparing for geologic mapping is selecting a suitable base on which mapping can be carried out. **Control** for establishing the exact locations of contacts and structures can be achieved in many ways. The measurement of compass bearings and the pacing (or taping) of distances can afford reasonable location control in some projects. Paced distances over irregular topography are converted to true horizontal map distances (Figure 3.3).

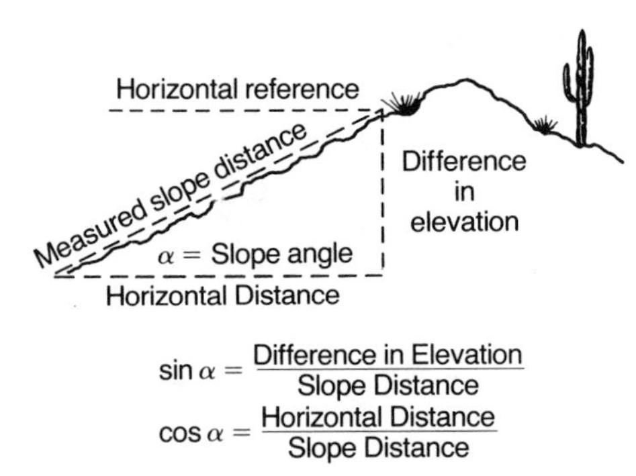

$$\sin \alpha = \frac{\text{Difference in Elevation}}{\text{Slope Distance}}$$

$$\cos \alpha = \frac{\text{Horizontal Distance}}{\text{Slope Distance}}$$

Figure 3.3 Trigonometric relationships for converting slope distance to true map distance (i.e., horizontal distance).

Mapping by the **pace-and-compass** or **tape-and-compass** system is reserved primarily for mapping very small areas at **large scale** (e.g., 1:100 or 1:1000). However, these methods may also be used as a complement to other mapping approaches. One of my first remote mapping projects was in northern Ontario, Canada, where the only base available to me was aerial photography. Usually, aerial photographs provide a sufficient basis for control. But in this case the terrain was so monotonously uniform in photogeologic expression (i.e., unending swampy expanses) and so closed in by vegetation (i.e., bush) that I found it impossible to locate myself on the basis of photos alone.

The solution to the dilemma was traversing into the bush from some known location on a lake shore, with my partner pacing along a fixed bearing for control (Figure 3.4). Maintaining a straight and accurate pace-and-compass traverse for distances up to 3 km or so is practically impossible in thick bush. Consequently, traverses were planned so that large, distinctive control checks (like a clearing or an isolated hill) would be crossed along the way. Outcrops were positioned on the map according to bearing and paced distance, and modified slightly (or substantially) according to the accuracy of the traverse as determined through meeting or missing the checkpoints.

Mapping within a **grid-line** or **picket-line system** is common practice in the mining industry. It is the chief method used around mining camps when companies are in the early stage of developing an ore body or in the advanced stages of exploring a prospect. It is an essential mapping approach for detailed work in areas of heavy vegetation. Picket-line paths are cut along north–south and east–west orientations, guided by engineering surveying (Figure 3.5*A*). Flagged picket posts are positioned and marked according to location every 50 ft (15 m) or so. Upon completion of the gridwork, the geologist traverses the complete course, mapping the locations and geologic characteristics of bedrock outcrops and **float** (rocks out of bedrock position but inferred to be short traveled). The map begins as an outcrop map (Figure 3.5*A*) but may be expanded into a geologic map by projecting and following contacts through the interior parts of the gridwork (Figure 3.5*B*). Geo-

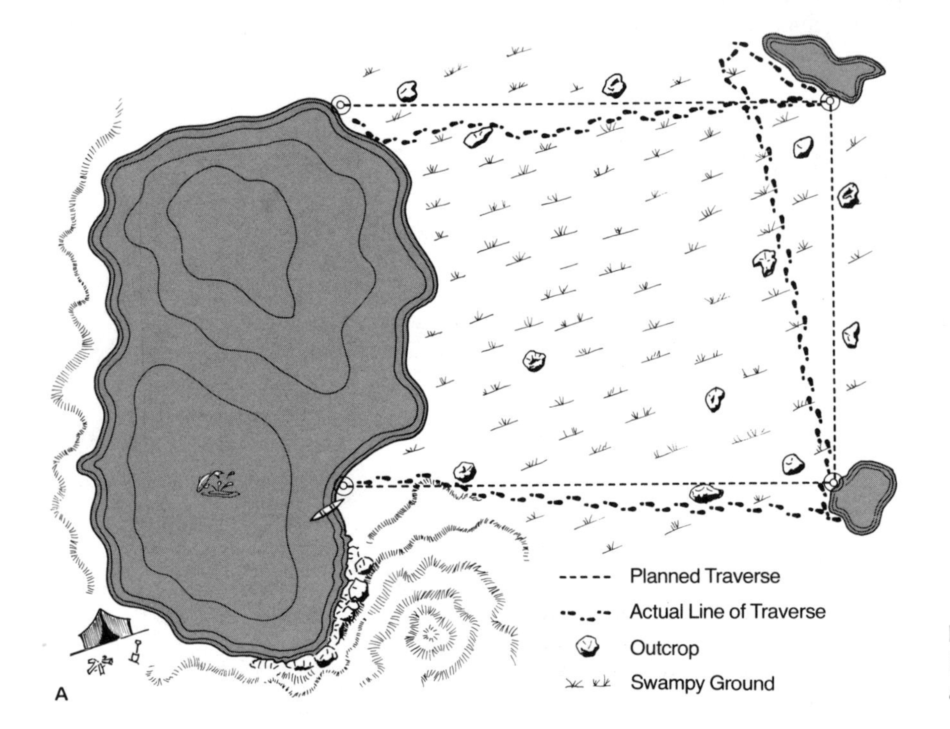

Figure 3.4 Pace-and-compass mapping in northern Ontario, Canada. (Cartoon artwork by D. A. Fischer.)

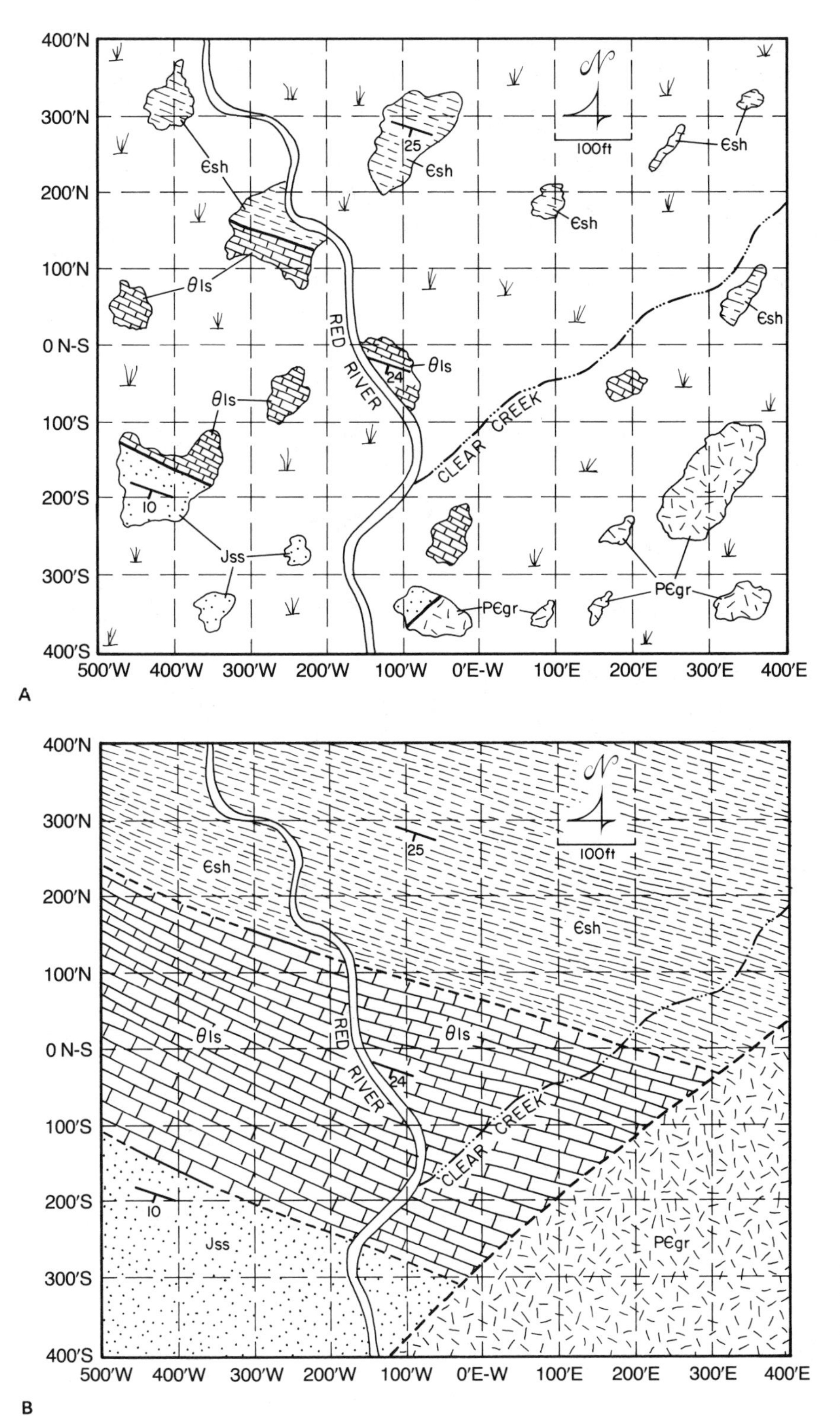

Figure 3.5 Gridline mapping. (*A*) Outcrop map showing locations of rock exposures relative to distances within surveyed grid. (*B*) Geologic map interpreted on the basis of the outcrop relationships shown in *A*.

physicists walk the same lines with black-box instruments to monitor magnetic, electrical, and gravitational properties of the rocks in the subsurface. Additionally, soil samples for geochemical analysis are collected along the grid. The combined geological, geophysical, and geochemical data, all assembled and plotted on the same grid, yield useful integrated descriptive data regarding the ore potential of an area.

Underground mapping proceeds in a similar fashion, except that traverses are made through solid earth. Mining engineers direct the tunneling of drifts, cross-cuts, raises, and shafts and furnish vertical and horizontal map control for the whole mine complex (Figure 3.6). Survey markers are positioned at close intervals throughout the mine. Mapping is carried out commonly at scales of 1 in. = 100 ft (1 : 1200) and 1 in. = 50 ft (1 : 600). Exposures available to the geologist are on the walls and ceiling (**back**) of each tunnel, the floor being covered with water, mud, or dust. Underground mapping is made difficult by the dirty and/or oxidized nature of the rocks, the dim lighting, and the unconventional mode of projection that is used in the mapping. Since rock relationships on the floor are not exposed, the mapped

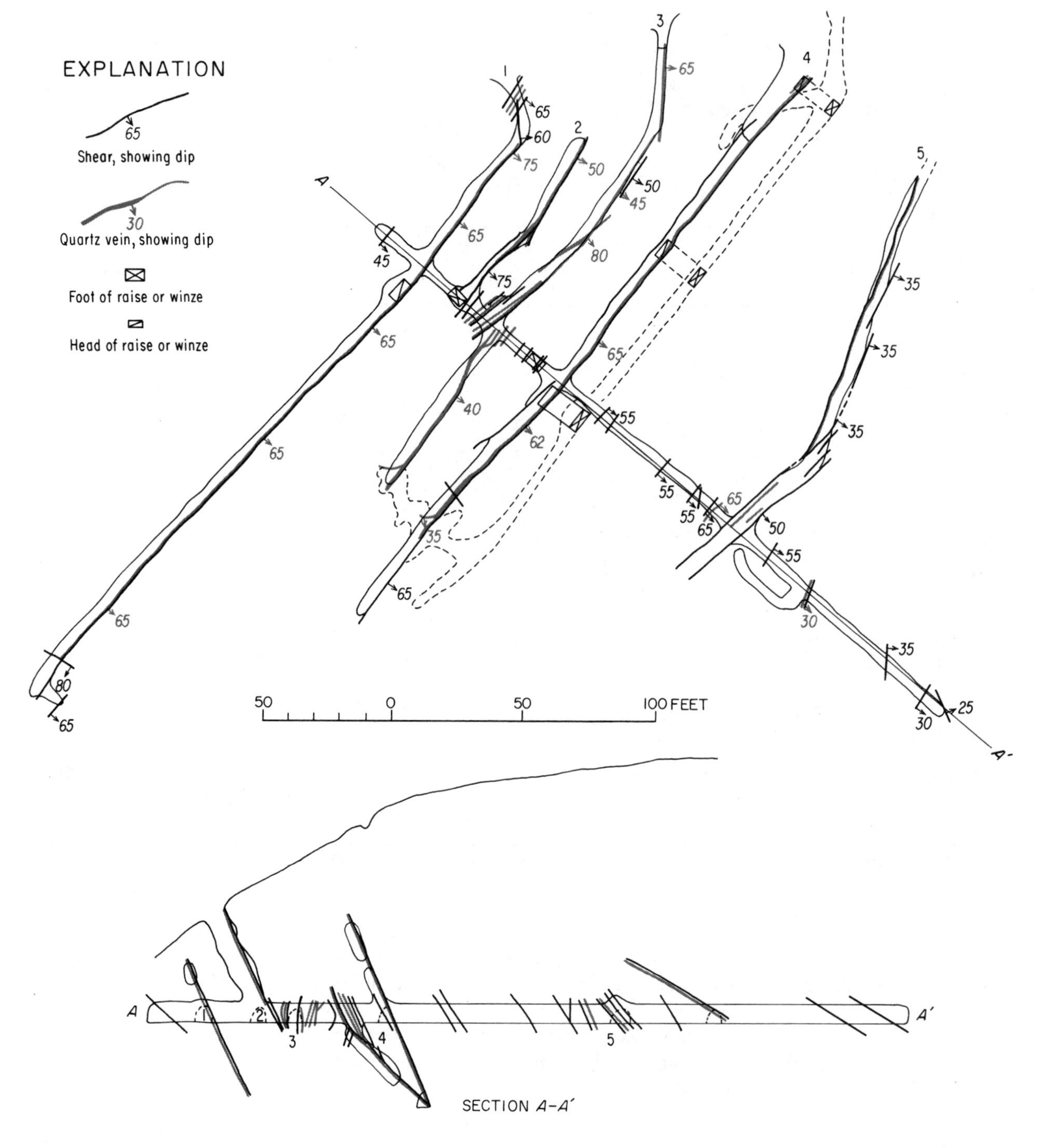

Figure 3.6 Underground geologic map of the Bluebird Mine, Cochise County, Arizona. [Based on maps by K. Krauskopf and R. Stopper, 1943. From J. R. Cooper and L. T. Silver (1964). Courtesy of United States Geological Survey.]

A

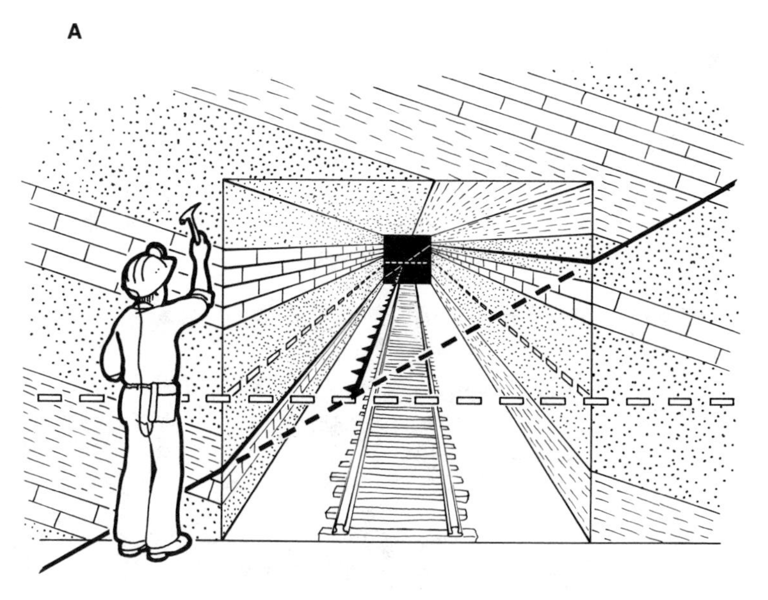

B

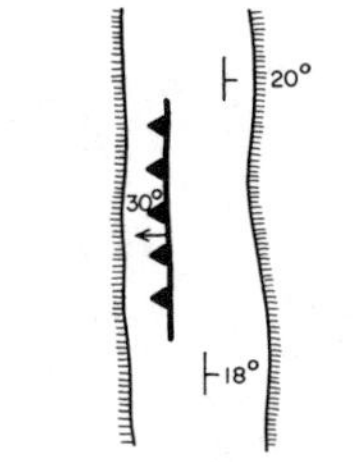

Figure 3.7 (*A*) The convention in underground mapping is to project contacts to a waist-high projection plane. (*B*) Map showing the location and orientation of the fault (heavy line with teeth) relative to the walls of the tunnel. (Art design by R. W. Krantz.)

locations of contacts and structures are projected from walls and the back to an imaginary waist-high horizontal plane. Thus an inclined fault whose trace is exposed near the top of a wall might project to the center of the drift (Figure 3.7). Results of underground mapping, at all levels inside the mine and in all vertical shafts and raises, provide a three-dimensional skeleton of descriptive and geometric facts. When supplemented by selective drilling, it furnishes the basis for a geological model of the deposit.

The traditional base for geologic mapping is the **topographic map.** Topographic maps show details of physiography and culture, which permit accurate positioning of contacts and structures. The United States Geological Survey publishes both 15- and 7.5-min maps, scaled at 1:62,500 and 1:24,000, respectively. These maps cover areas corresponding to 15 or 7.5 min of latitude and longitude. For regional mapping and compilation, 2° topographic sheets scaled at 1:250,000 are extremely valuable. Topographic maps at all these standard scales are sold by the United States Geological Survey, Denver Federal Center, Denver, Colorado 80225, and by their office in Reston, Virginia 22092. Folders describing topographic maps and symbols are available on request.

In some cases, topographic maps must be prepared as part of the geologic mapping process. This is a time-consuming process, warranted only by the need to carry out large-scale mapping of some small area containing structures for which elevational/topographic control is necessary to unravel the three-dimensional structural forms. Large-scale topographic maps can be constructed through the use of the **plane table and alidade** (see Compton, 1962). The basis for this approach is surveying the elevations and locations of an array of points that afford optimum geological and topographic control. The map is prepared directly in the field on a tripod-supported table (Figure 3.8).

In many instances the most valuable base for geologic mapping is a set of **aerial photographs,** preferably with enough overlap to afford stereoscopic coverage. For some areas, aerial photographs disclose the bedrock distribution and structure so explicitly that the mapping of contacts can be done effortlessly (Figure 3.9). More commonly, the chief advantage of aerial photographs is their total display of rock, vegetation, and physiographic and cultural features. This control allows the locations of control points and contacts to be posted easily and accurately.

Figure 3.8 Plane table mapping, in progress. Target of mapping is strata in Aikens Corner in the Mecca Hills of southeastern California. Geologists are Ken Yeats (left) and Steve Wust. (Photograph by G. H. Davis.)

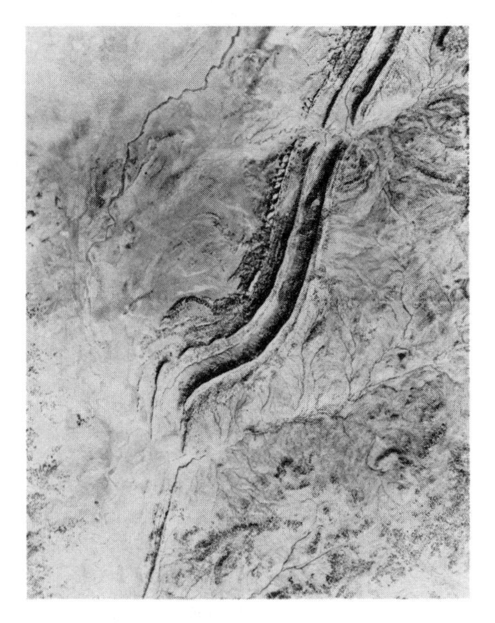

Figure 3.9 Aerial photograph showing conspicuous expression of steeply inclined resistant beds on the east side of the Defiance uplift, northeastern Arizona–northwestern New Mexico.

Aerial photographs are available through a number of sources. Large-scale aerial photographs in color and in black and white can be obtained through the National Forest Service and other agencies. The National Aeronautics and Space Administration (NASA) is a major source of high-quality space satellite imagery of large regional tracts (Figure 3.10*A*). Regional mosaics are especially revealing in structural studies (Figure 3.10*B*). Transparencies and paper prints, in color or in black and white, can be purchased through the NASA office in Sioux Falls, South Dakota.

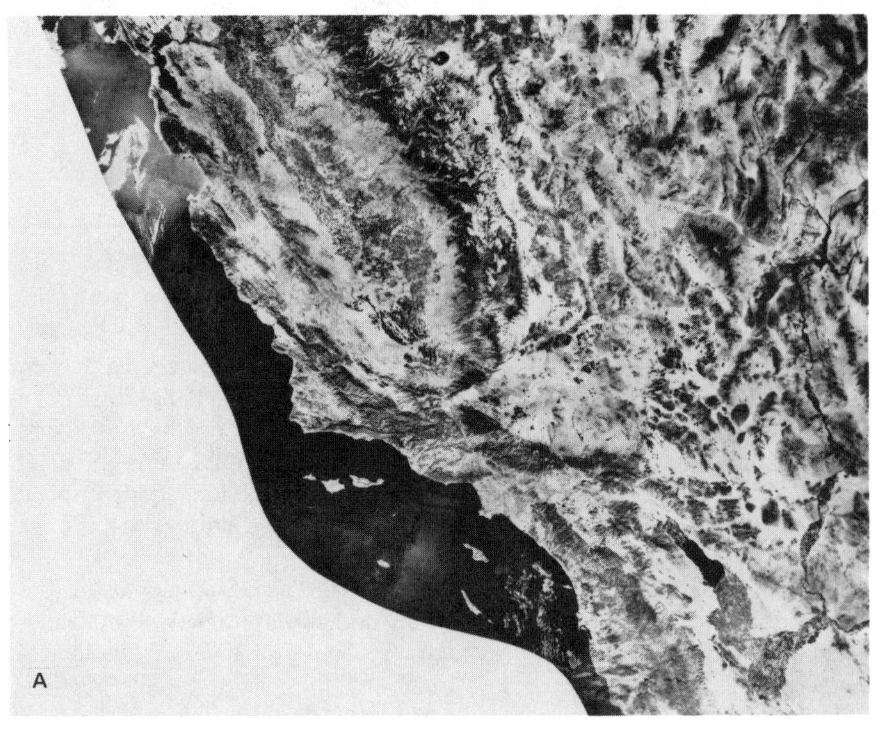

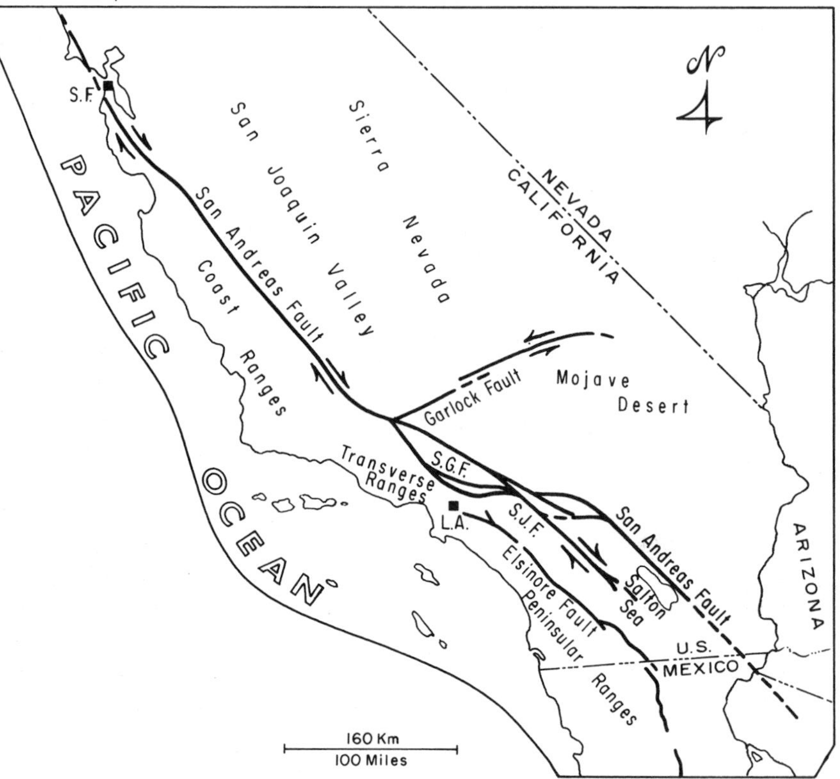

Figure 3.10 (*A*) LANDSAT-1 mosaic of the southwestern United States. (*B*) Geographic index map for some of the structures revealed in the LANDSAT mosaic. S.G.F. = San Gabriel fault; S.J.F. = San Jacinto fault. [From Lowman (1981). Courtesy of Goddard Space Flight Center, National Aeronautics and Space Administration.]

For special studies of small areas, I have found it useful to contract low-altitude flyovers. Through this approach it is possible to obtain high-quality imagery of the study areas at scales as large as 1:1200. Enlargement is achieved through preparing 1 x 1-m Mylar positives. Blue-line or black-line Ozalid copies can be produced inexpensively from a Mylar positive.

Ideal control for geologic mapping is the combination of topographic maps and aerial photographs. The photographs reveal rock relationships in ways that aid in planning traverses and in interpreting structures photogeologically. The topographic map, on the other hand, provides a stable base of uniform scale and permits the critical third dimension, the vertical dimension, to be evaluated.

COMPONENTS OF A GEOLOGIC MAP

Learning how to make a geologic map is aided by learning the fundamental components of such a map. Ridgeway (1920) long ago described the components (Figure 3.11). The map shows the distribution of rock formations by means of color and/or patterns, and letter symbols. Each color, pattern, and symbol on the map should coincide exactly with the distribution of the corresponding rock formation. The meaning of the colors, patterns, and symbols is provided in the **Explanation**: the series of boxes, suitably colored, patterned, and labeled, which provide the identification of for-

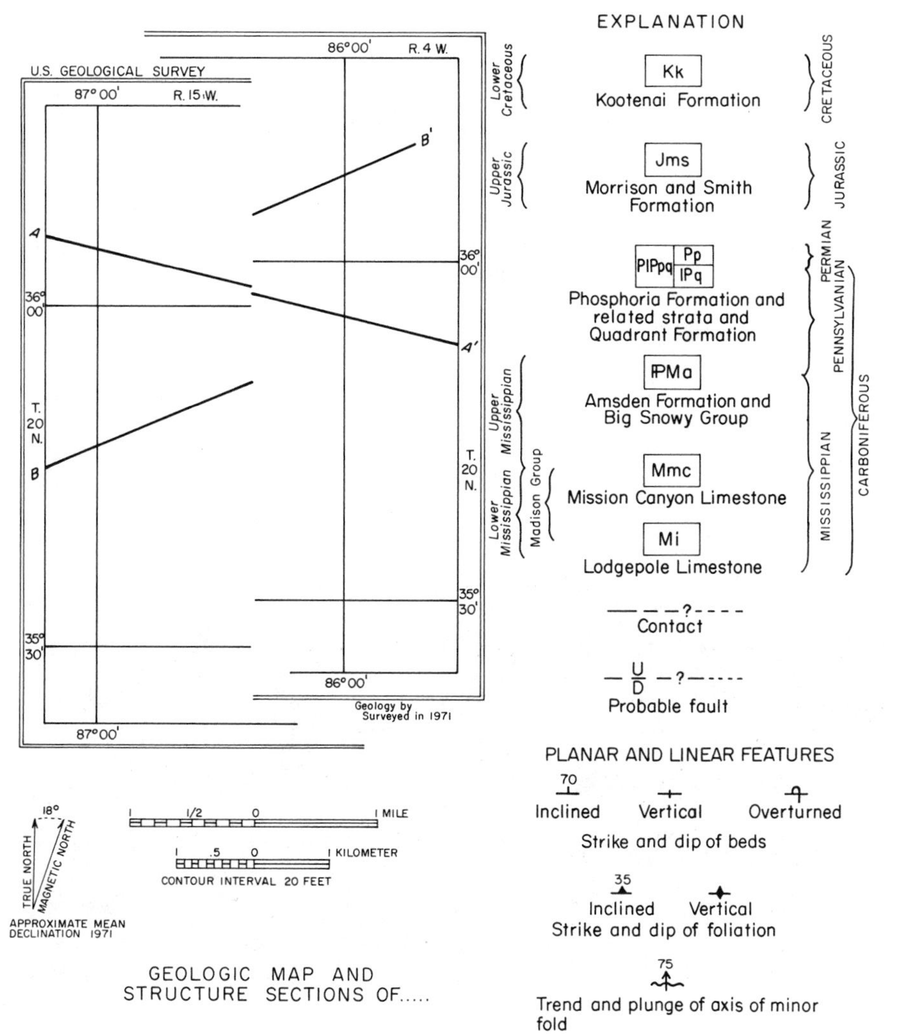

Figure 3.11 Elements of the layout of a geologic map. [From Ridgeway (1920). Courtesy of United States Geological Survey.]

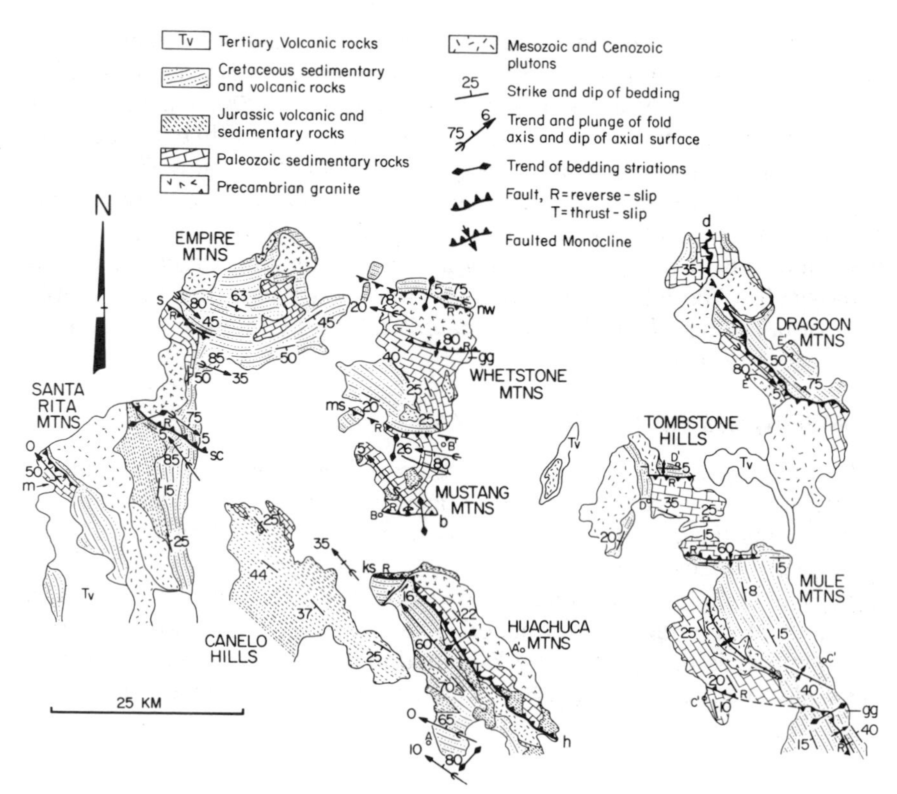

Figure 3.12 Structural geologic map of southeastern Arizona showing the use of lithological symbols to distinguish among the mapped units. [From Davis (1979), *American Journal of Science*, v. 279.]

Table 3.4
Official United States Geological Survey Map Colors

Age of Rock Unit	*Map Color*
Quaternary	orange
Tertiary	yellow ocher
Cretaceous	olive-green
Jurassic	blue-green
Triassic	bluish gray-green
Permian	blue
Pennsylvanian	grey
Mississippian	blue-violet
Devonian	heliotrope (helio*what??*)
Silurian	purple
Ordovician	red-violet
Cambrian	brick red
Precambrian	pink or gray-brown

mation name and usually a brief lithologic description. The official list of map colors used by the United States Geological Survey is presented in Table 3.4.

If geologic maps are printed in black and white, line and symbol patterns must be used in lieu of color to show the distribution of rock formations. Line symbols are difficult to choose in structural studies because most patterns interfere with and/or mask the fundamental **line work** of structural information and symbology. Therefore, it is useful to represent the various mappable units with **lithologic symbols** that convey an impression of the nature of the dominant bedrock units. An example of the use of lithologic symbols is shown in Figure 3.12, a map of part of southeastern Arizona that I compiled in the course of regional structural analysis (Davis, 1979).

Figure 3.13 In among the trees is a continuous resistant sandstone marker unit that reveals the structural deformation by folding. Fluted rocks near Great Cacapon, West Virginia. (Photograph by W. G. Stose. Courtesy of United States Geological Survey.)

The formation symbols and ages are arranged vertically in the Explanation, in the order of relative age (see Figure 3.11). Astride the vertical array of boxes are entered the ages of the rock formations, as known or as inferred.

For purposes of detailed structural analysis, it is often helpful to identify and map **marker units** as an aid in clarifying the structural style of deformation. Markers are chosen on the basis of distinctiveness, resistance to erosion, and continuity (Figure 3.13). Marker beds are like a godsend in complexly folded sequences of rock (Figure 3.14*A*). The marker units are identified in the Explanation of the map, in boxes corresponding to the formation(s) within which they lie (Figure 3.14*B*).

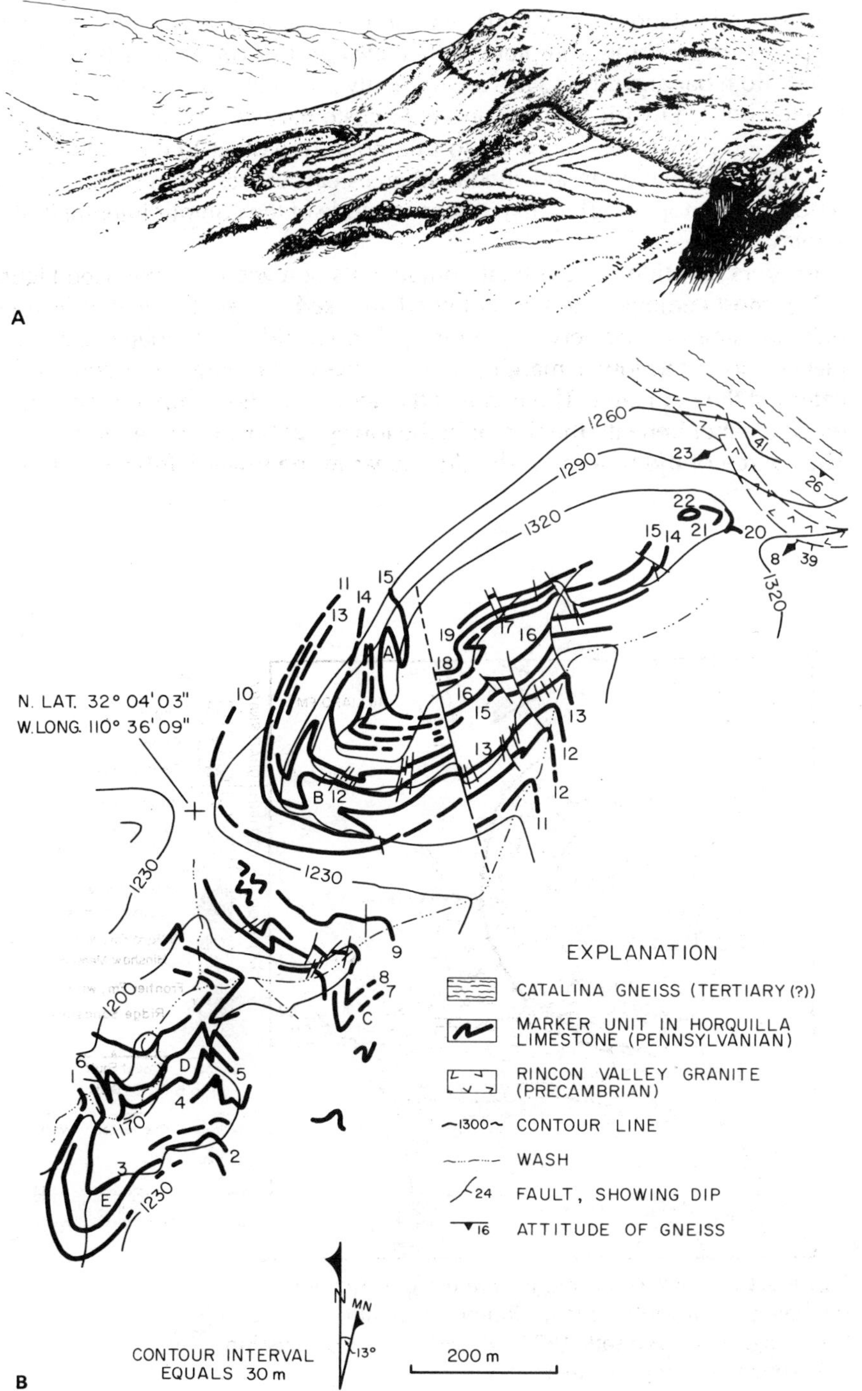

Figure 3.14 Complexly folded sedimentary strata in the Rincon Mountains near Tucson, Arizona. (*A*) India-ink rendering of photograph showing marker units that allow the structure to be unraveled. (*B*) Geologic map of the area and its markers. [From Davis and others (1974). Published with permission of National Association of Geology Teachers.]

Table 3.5
Map symbols for contacts

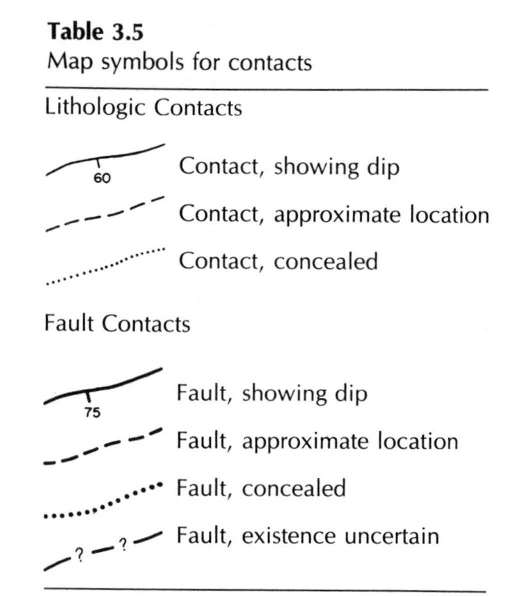

Lithologic Contacts
Contact, showing dip (60)
Contact, approximate location
Contact, concealed
Fault Contacts
Fault, showing dip (75)
Fault, approximate location
Fault, concealed
Fault, existence uncertain (? — ?)

Each of the formations shown on a geologic map must be enclosed by a contact, except of course where the formations extend outside the map area. There they are intercepted (cut off) by the border of the base map. The **contact lines** are usually of two types—depositional or intrusive contacts, and fault contacts. Fault contacts are drawn with heavy lines, whereas depositional or intrusive contacts are represented by thin lines (Table 3.5).

Showing the distribution and contacts of rock formations is not enough. To these must be added symbols that disclose the major structures and the internal geometry of the rock system. The most informative structural geologic maps display abundant **structural symbology** (Figure 3.15), which conveys the geometric and physical nature of the structures present, like bedding and/or foliation, faults, joints, folds, cleavage, and lineation (Table 3.6).To understand the size and orientations of features shown on a geologic map, all maps come complete with **scales** and **north arrows** (see Figure 3.11). Both bar and ratio scales are normally presented. Bar scales should be labeled both in metric (km, m) and U.S. customary (mi, ft) units. Two north arrows are shown, one pointing to true north, the other to magnetic north. Magnetic declination is specified in degrees east or west of true north. If the base map is a topographic map, the magnitude of the contour interval is also shown.

Borders and **title** are the final components of a geologic map (see Figure 3.11). Most geologic maps do not need an inked border; this feature is used only for maps that are very large or irregular. The title of the map is generally placed along the lower margin, and it conveys the name and general location of the map area. The name of the geologist who compiled the map is entered either beneath the title or in the lower right-hand corner of the map. The source of the base map should appear in the lower left-hand corner.

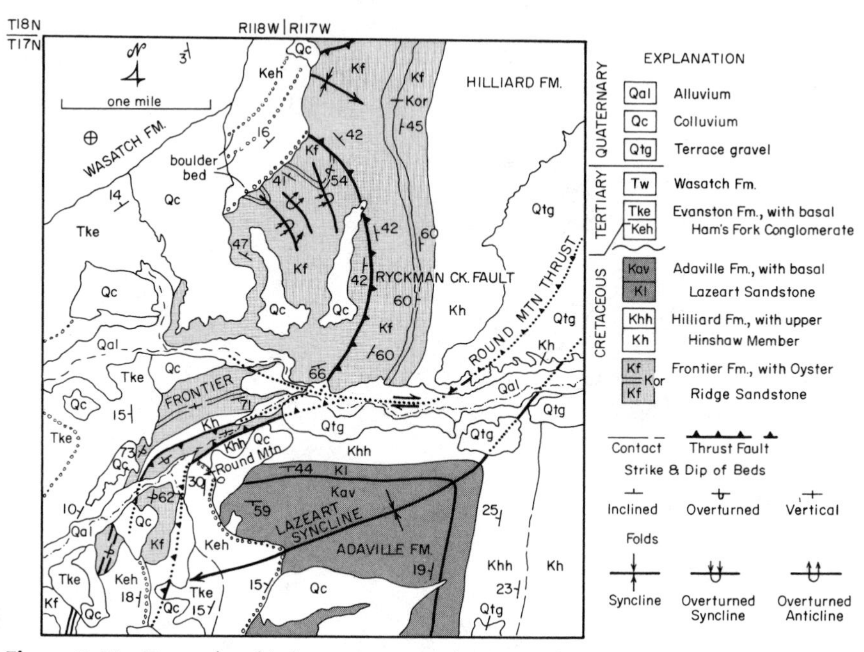

Figure 3.15 Example of informative use of structural symbology on a geologic map, Round Mountain area, Wyoming. [From Worrall (1977). Published with permission of Wyoming Geological Association.]

Much can be learned about geologic maps and geologic mapping by studying those that have been published. Time spent in the library examining geologic maps, map explanations, and map patterns is time well spent.

MAPPING PROCEDURES

The geologic mapping process is aided by keen observational skills, breadth of geological background, attention to geometric details, accuracy, neatness, and patience. Some maps are better than others because of the attention paid to detail and the quality of the line work. The geologic map is a description. The more complete and accurate the description, the greater the impact of the map.

The basic tools for geologic mapping include covered clipboard, hard-back field notebook, protractor scale, pencil(s) (hardness No. 4- 5), colored pencils, drafting pen, hand lens, rock hammer, 2-m tape, and compass. All these items can be carried handily on a belt and in a day pack, along with other gear that might be useful (canteens, camera, chisel, binoculars, first-aid kit, pocket altimeter, 50-ft tape, sunscreen, raincoat).

The actual process of geologic mapping begins with establishing **map units** appropriate to the project at hand. For large-scale mapping of small areas, the map units might be distinctive markers within a single formation. In mapping a 7.5- or 15-min quadrangle, it is customary to map the formally established formations that characterize the region. For small-scale mapping (1:125,000 or smaller) and/or compilation of large regional tracts, each map unit may consist of a number of rock formations, combined for reasons of stratigraphic or tectonic significance. Selecting map units involves both reconnaissance field work and the reading of pertinent literature. Through these activities, the lithologic characteristics and contact relationships in the geologic system of interest become reasonably well known. A specific knowledge of lithologic characteristics of each map unit is achieved through visiting **type localities** and through measuring sections of the rock in the area of study.

Mapping may proceed in a number of ways, but significant time and effort will always be invested in tracing out contacts of each of the map units. It is useful early in the mapping to traverse across the grain of the rocks and structures, to become familiar with the total spectrum of rock formations, their contacts, and their internal structures. Traverses are also directed *along* specific contacts, to carry the units to the limits of the area.

Rock and structural data are collected along each traverse. Localities of data collection are assigned **station numbers** for convenience of reference. These numbers are posted on the base map at their exact locations. Data are entered in the field notebook under the appropriate station number. So that station numbers do not interfere with the geologic data, it is useful to poke a tiny pinhole through the base map at each station location, turn the map over, circle the pinhole, and write the station number next to the circled pinhole.

Contact lines are drawn on the base as mapping proceeds, and map units are colored progressively as their distribution becomes established. Orientations of bedding, foliation, folds, and other structures are entered in the field notebook, and representative readings are posted directly on the map, as well. The map is made in the field (Figure 3.16), not constructed in the office at the end of the day on the basis of notebook data and memory. Geologic mapping is a scientific method, and the careful study of the map as it develops will provide direction in determining where to go next for geologic insight.

Table 3.6
General map symbols for orientations of bedding, foliation, cleavage, and lineation

Symbol	Meaning
Bedding	
20	Strike and dip of bedding
60	Strike and dip of overturned bedding
	Strike and dip of vertical bedding
⊕	Horizontal bedding
Foliation and Cleavage	
30	Strike and dip of foliation
	Strike of vertical foliation
80	Strike and dip of cleavage
	Strike of vertical cleavage
Lineations	
30	Trend and plunge of lineation
25 40	Strike and dip of foliation, and trend and plunge of lineation in the plane of foliation.
65	Strike and dip of bedding, and trend and plunge of lineation in the plane of bedding.

Figure 3.16 Debby Currier mapping thrusted sedimentary rocks in the Swisshelm Mountains, Cochise County, Arizona. Students in background are tracking a black marker bed. (Photograph by G. H. Davis.)

USE OF THE COMPASS

A compass is used to measure the orientations of structures. The Brunton compass is the standard compass used by geologists (Figure 3.17*A*), but the Silva compass (Ranger) is becoming increasingly popular (Figure 3.17*B*). Each instrument is equipped with the means to set **magnetic declination**, the angle between true north and magnetic north for a specific locality.

Compasses are used in structural analysis to measure trend and inclination. **Trend** refers to the azimuth or bearing of a line. **Azimuth** is measured in degrees clockwise from north (e.g., 120°; 267°). **Bearing** is measured in degrees east or west from north or south (e.g., N60°E; S21°W). **Inclination** is the angle, measured in degrees, between an inclined line and horizontal. Its value may range from 0° to 90°.

Figure 3.18 depicts the trend and inclination of a rather unusual line, the tallest human ladder ever constructed on a steeply inclined fault surface. The trend is determined by projecting both the "foot" of the ladder and the "head" of the ladder **vertically upward** into a common horizontal plane, and connecting these points of projection. The azimuth of this line is measured with a compass. The inclination of the human ladder is the angle between the line of bodies and horizontal, *as measured in a vertical plane*.

In practice, the orientation of a line in space is expressed in terms of **trend and plunge**, where plunge is a measure of inclination. Geologic lines, like

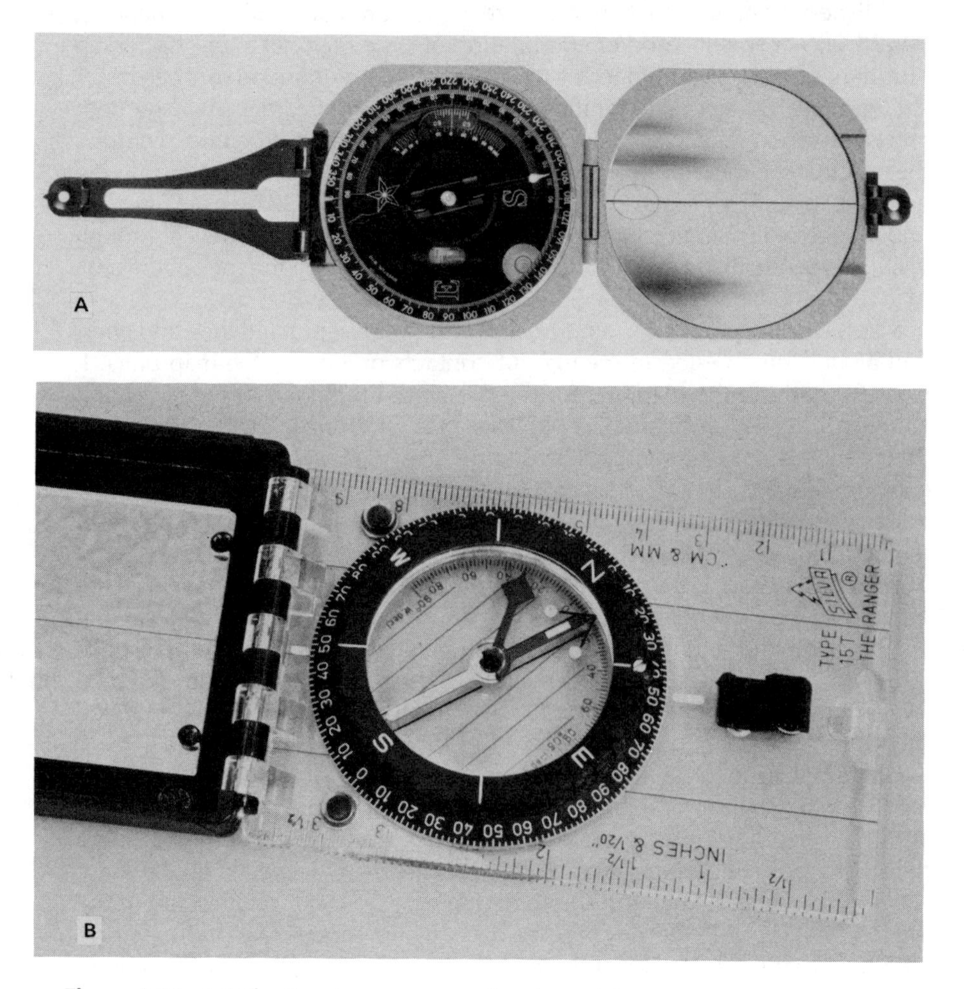

Figure 3.17 (*A*) The Brunton compass. Courtesy of the Brunton Company. (*B*) The Silva compass (Ranger). (Photograph by G. Kew.)

Figure 3.18 (*A*) Human ladder constructed on face of a steeply inclined fault, near Patagonia, Arizona. (Photograph by G. H. Davis.) (*B*) Azimuth and inclination of human ladder.

grooves on a fault surface (Figure 3.19), are called **linear elements**. The trend of a linear element is measured by holding the compass level while aligning its edge parallel to the direction of the line (Figure 3.20*A*). The compass is pointed parallel to the **vertical projection** of the line onto an imaginary horizontal plane. Trend is read directly from the Brunton compass after the compass needle has come to rest (Figure 3.20*B*). To measure the azimuth of trend using the Silva compass, the calibrated outer ring on the face of the compass must be rotated until the rotatable outline of the compass arrow coincides with the actual free-spinning magnetic needle (Figure 3.20*C*).

Plunge is measured by turning the compass on its side and aligning its edge along the linear element, or parallel to it (Figure 3.20*D*). If the Silva compass is used, a plumb-boblike inclination needle points automatically to the value of the plunge when the compass is thus oriented (Figure 3.20*E*). The Brunton compass does not have a free-swinging inclination needle. Instead, a calibrated scale known as a **clinometer** and located inside the compass is used to measure plunge. The clinometer, attached to a small carpenter's level, can be moved back and forth by means of a lever on the outside base of the compass (Figure 3.20*D*). Holding the Brunton such that its edge is positioned parallel to the line being measured, the clinometer lever is moved until the

Figure 3.19 Chris Menges (moustache and glasses) proudly shows off the exceptionally well-exposed, grooved fault surface that he mapped near Patagonia in southern Arizona (see Menges, 1981). Chris' hands rest on crest of a convex groove. Richard Gillette's foot (*far left*) is squarely placed in trough of concave groove. Nancy Riggs measures the strike azimuth of a part of the fault surface. (This is the same fault surface on which the human ladder of Figure 3.18 was constructed.) (Photograph by G. H. Davis.)

Figure 3.20 Steps in measuring trend and plunge.

bubble in the carpenter's level becomes centered (Figure 3.20*F*). The value of plunge is read directly using the inner scale embossed on the inside base of the compass.

The full description of the trend and plunge of a line in space can be recorded in two different ways. For example, 20° N60°E refers to the orientation of a line that plunges 20° along an azimuth 60° east of north; N60°E is the **sense** of direction of the down-plunge end of the line. Similarly, 45° S20°E describes the orientation of a line plunging 45° in a direction that is 20° east of south. For a compass calibrated in azimuth from 0° to 360°, these measurements would be recorded as 20°/60° and 45°/160°, respectively.

Measuring the orientation of a planar feature is handled differently. If the orientations of two lines are known, the orientation of the plane that contains these lines is also known. Two lines determine a plane. The measuring of orientations of bedding planes, fault planes, dikes (Figure 3.21*A*), and other geological **planar elements** is based on this relationship. If the orientations of two lines that lie in a plane can be established, the orientation of the plane itself is established as well. Any two lines will do, as long as they are not parallel or close to being parallel. For convenience, the two lines in a plane that are chosen are a horizontal line and the line of steepest inclination (Figure 3.21*B*). These two lines are at right angles to each other. The first is called the **line of strike**; the second is the **line of dip**. For the special case of a strictly horizontal plane, all lines are strike lines.

Figure 3.20 (*cont.*)

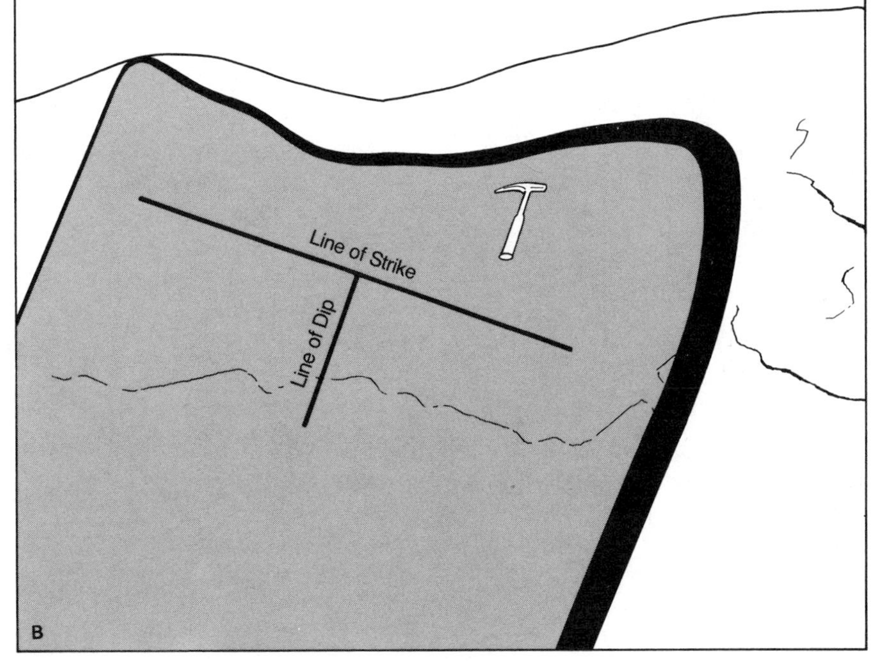

Figure 3.21 (*A*) Aplite dike (white) in granite. The planar dike is exposed both in plan and cross-sectional views. (Photograph by G. H. Davis.) (*B*) Schematic view into the granite within which the dike occurs. The orientation of the dike is expressed in terms of orientations of the lines of strike and dip.

Strike and dip are the measurements required to define the orientation of a plane. **Strike** is the trend of the line of strike. Because its inclination is by definition 0°, the value of strike is recorded simply in terms of degrees of azimuth or bearing. For compasses with trend calibrated by quadrant, strike is always expressed in terms of north: N72°E, N68°W, N1°W. For compasses calibrated in azimuth from 0° to 360°, azimuth is presented as degrees between 0° and 90°, or between 270° and 360°. The **dip** of a plane is the inclination of the line of dip. It is recorded in terms of inclination angle and the dip of the plane (SW, NW, NE, SE, N, S, E, W). The specific azimuth of dip direction is not directly measured because its value can be determined from the strike value of the plane. For example, a N35°W-striking plane dips either N55°E or S55°W. Only the general value of dip direction is recorded in the notebook: SW or NE. This distinction permits the two possible dip directions to be distinguished.

The procedure for measuring the strike and dip of a plane using a Brunton compass is as follows (Figure 3.22). To find the line of strike, first set the clinometer to 0° so that the compass may be used as a carpenter's level (Figure 3.22*A*). Place a side or edge of the compass flush against the plane (Figure 3.22*B*), or against a field notebook or nonmagnetic clipboard held parallel to the plane; rotate the compass until the carpenter's level bubble is centered. When centered, the edge of the compass held against the plane is horizontal and oriented parallel to the line of strike. To determine azimuth of the strike line, rotate the compass downward (Figure 3.22*C*), still keeping the lower side edge of the compass fixed against the surface until the bulls-eye bubble is centered. The compass needle now swings freely. Dampen the compass needle to stop it from swinging and read the azimuth (Figure 3.22*D*).

To measure the strike of a plane using a Silva compass, it is useful to carry an auxiliary level, or to attach a small level to the compass itself. The level can be used to quickly identify the line of strike. Once found, its orientation can be measured by aligning the edge of the Silva parallel to the line of strike and rotating the calibrated outer ring until the reference compass needle is aligned with the actual magnetic needle (Figure 3.22*E*).

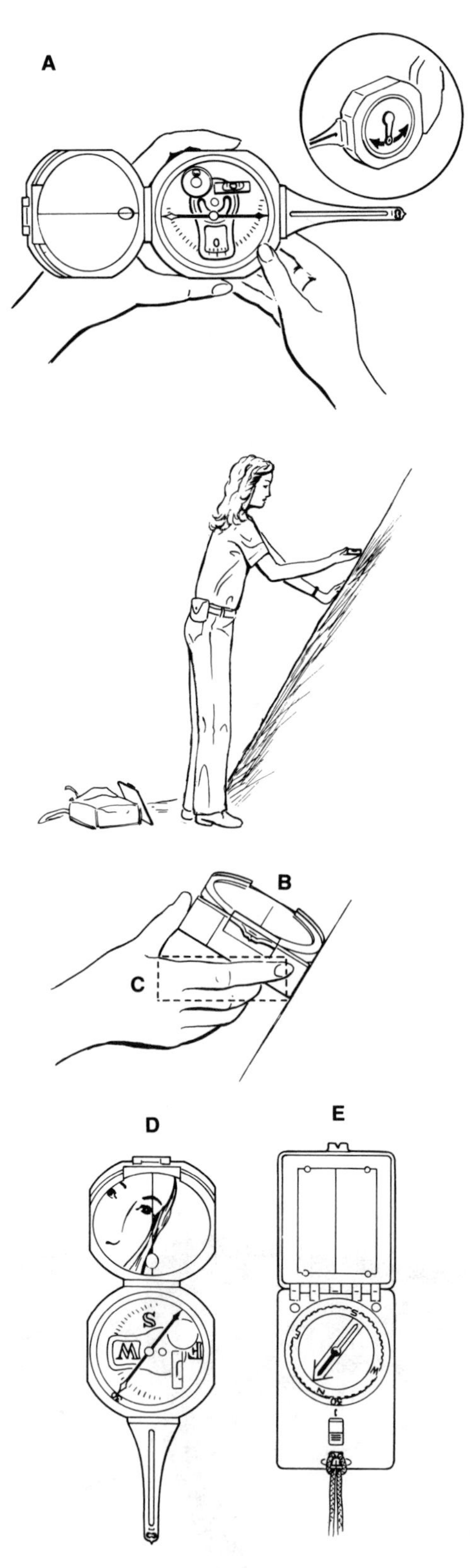

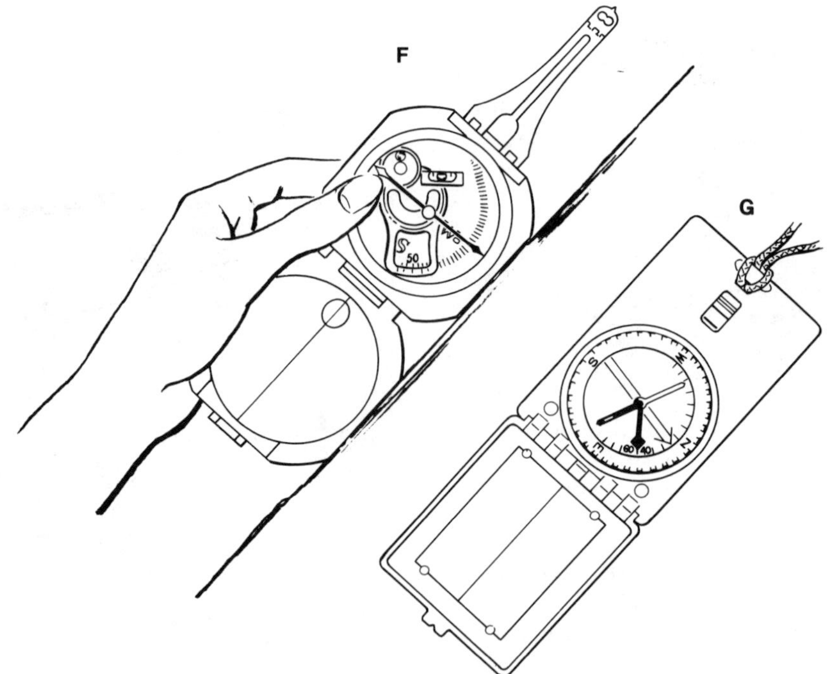

Figure 3.22 Steps in measuring strike and dip.

Figure 3.23 The do's and don't's of the sighting method for measuring strike and dip.

To measure dip, place the compass on a side face on the inclined plane such that the compass points in the direction of the line of dip (Figure 3.22*F*). Then measure the inclination of the line by rotating the clinometer until the carpenter's level bubble is centered. The Silva compass has an inclination needle for measuring dip (Figure 3.22*G*).

Strike-and-dip readings can be taken by the **sighting method** as well. It is an especially useful method when beds or layers do not crop out as convenient resistant planes for direct measurement, and/or when attempting to measure the average strike and dip of an area of rock that is larger than outcrop size. The method is illustrated in Figure 3.23. The trick is to position yourself in the proper location so that your line of sight is a strike line in the plane of the dipping layering whose orientation is being determined. When viewed in this way, the dipping layer appears in strict cross-sectional view, with no expression of the surface of the layer. The azimuth of the line of sight constitutes the strike of the dipping layer. The inclination of the layer as seen from this unique line of sight is true dip. Watch where you stand.

Strike-and-dip and trend-and-plunge orientations measured in this way are recorded in the field notebook. Representative readings are placed on the geologic map as well, using the kind of symbology shown in Table 3.6. A protractor or protractor scale is used for accurate plotting. The base map is scribed lightly with penciled N–S guidelines so that strike or trend can be measured and plotted with relative efficiency. The Silva compass has the added advantage of being functional as a protractor for plotting strike or trend on the base map. Without disturbing the compass setting for trend or strike, the compass is placed on the map in such a way that the guidelines on the base of the compass coincide with the N–S guidelines scribed on the map. When this is accomplished, the straight edge of the compass matches the azimuth of strike or trend.

Measurements collected and plotted in this way give geometric life to the geologic map. There emerges from the map a physical and geometric expression of the form and internal structure of the rock formations. Furthermore, the orientation measurements stored in the field notebook become the basis for subsequent analysis and interpretation.

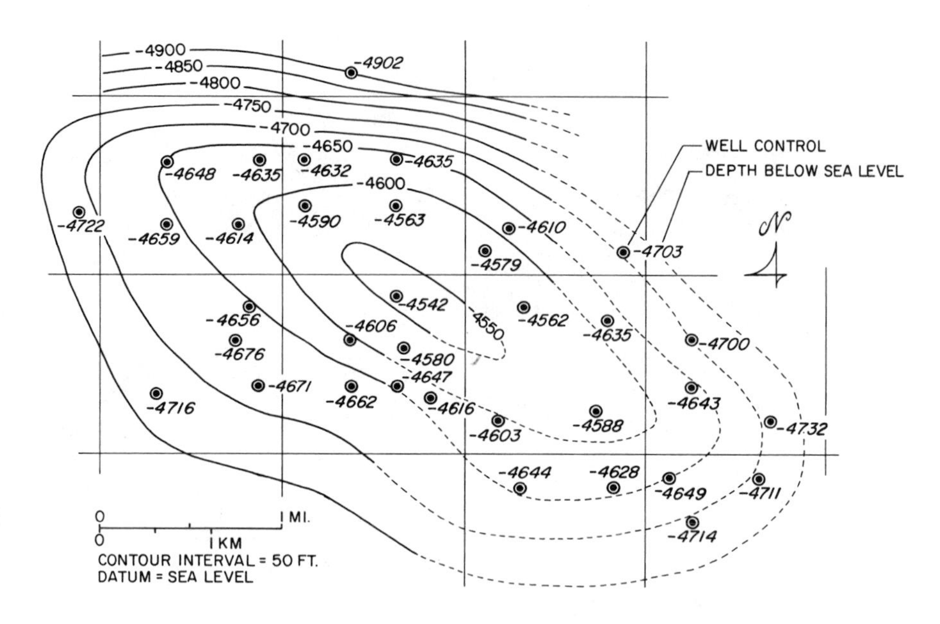

Figure 3.24 Contouring of a structure contour map on the basis of raw data giving elevations of the top of a marker bed in the subsurface. [Modified from Dutton (1982). Published with permission of American Association of Petroleum Geologists.]

SUBSURFACE MAPPING

Subsurface exploration in the search for petroleum, metals, and other natural resources provides yet another source of data for descriptive analysis. Drilling into the subsurface yields information regarding structure and lithology of the rock column in third dimension. Depths at which specific rock

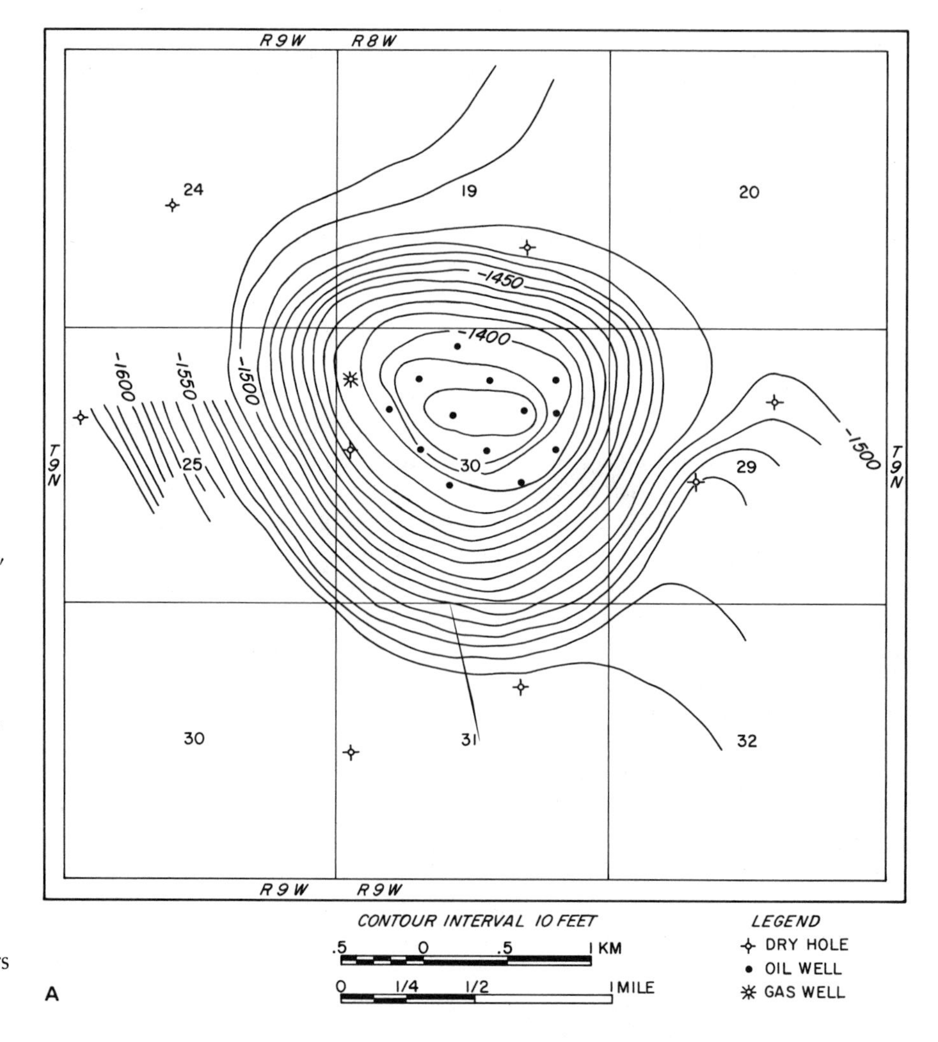

Figure 3.25 Representative structure contour patterns. (*A*) Dome, Wilfred pool, Indiana. [From Dana (1980). Published with permission of American Association of Petroleum Geologists.] (*B*) Basin. Contours represent the top of Middle Silurian strata of the Michigan Basin. (From *The Evolution of North America* by P. B. King, fig. 17C, p. 30. Published with permission of Princeton University Press, Princeton, New Jersey, copyright ©1959.) (*C*) Anticline (1), syncline (2), homocline (3), fault (4), and steeply dipping strata (5), Canyonlands, Utah. [From Huntoon and Richter (1979). Published with permission of Four Corners Geological Society.]

formations are encountered become the basis for constructing **structure contour maps**. Structure contour maps describe the structural form of rock bodies at depth. Usually elevation values corresponding to the top of a particular formation are used to define a reference datum. The elevations of **the datum** are posted on a base map and are contoured in the same way that topographic maps are contoured on the basis of surface elevation data (Figure 3.24). Most often such maps are used to describe the structure of sedimentary formations. Domal and basinal patterns are marked by concentrically arranged, closed contours (Figure 3.25*A*, *B*). Anticlines or synclines might display combinations of crescent-shaped, straight-lined, and closed contour patterns (Figure 3.25*C*, 1 and 2). **Homoclines**, which are simple tilted structures where bedding dips uniformly in a single direction, are distinguished by subparallel contour lines that steadily decrease (or increase) in elevation value across the face of the map (Figure 3.25*C*, 3). Faults are marked by offset contour lines (Figure 3.25*C*, 4). Where a rock formation gradually steepens to vertical along strike, the contour lines become closer and closer spaced, producing patterns of convergence and divergence (Figure 3.25*C*, 5).

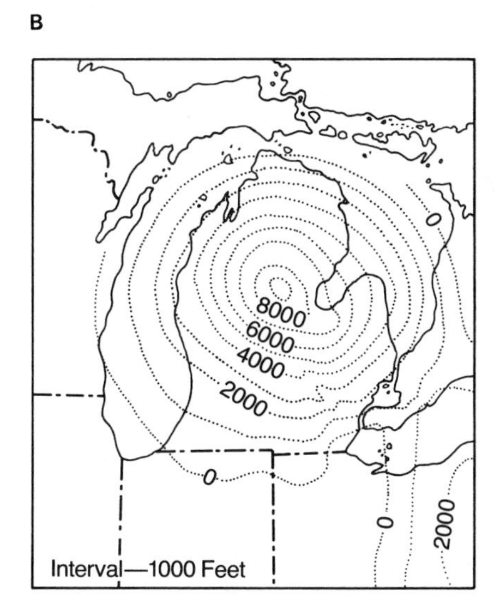

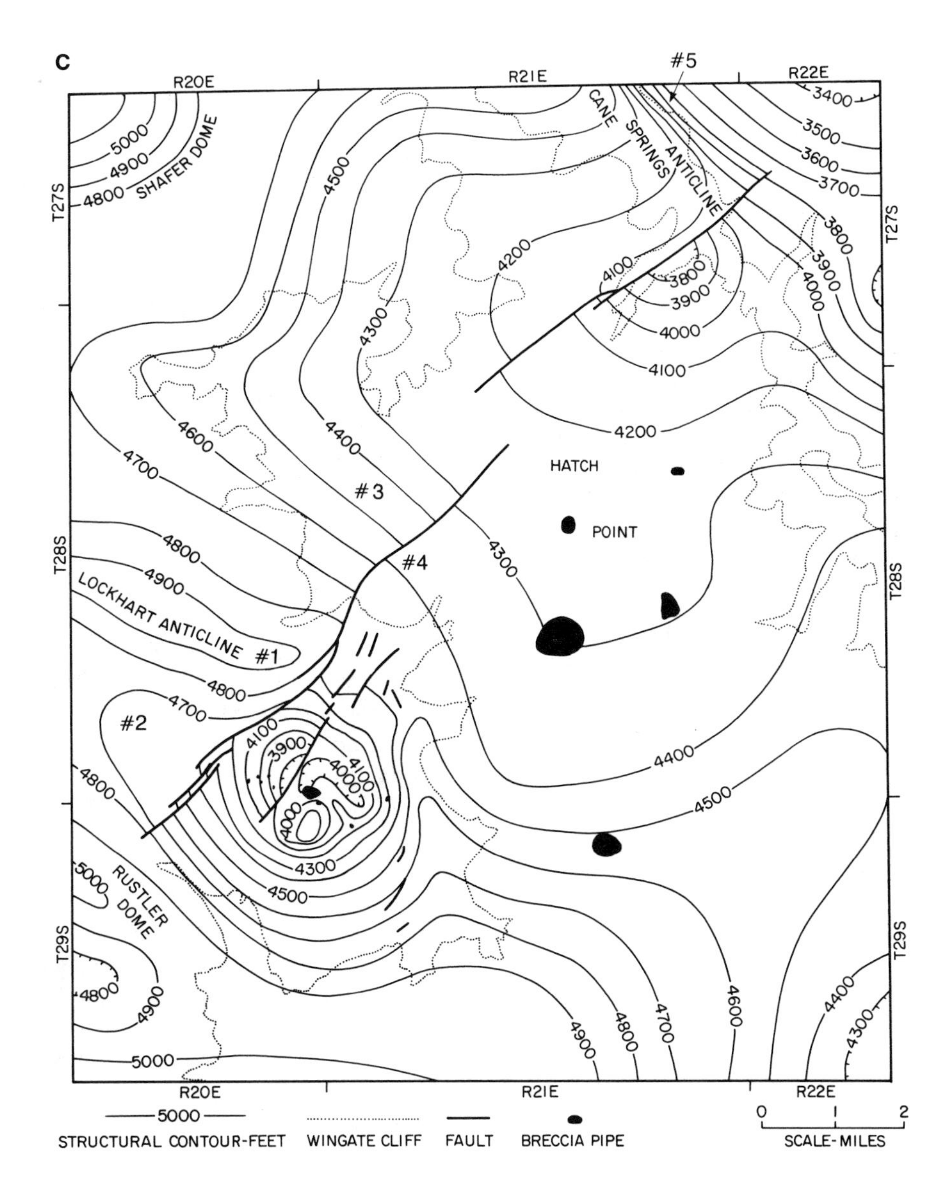

Figure 3.25 (*cont.*)

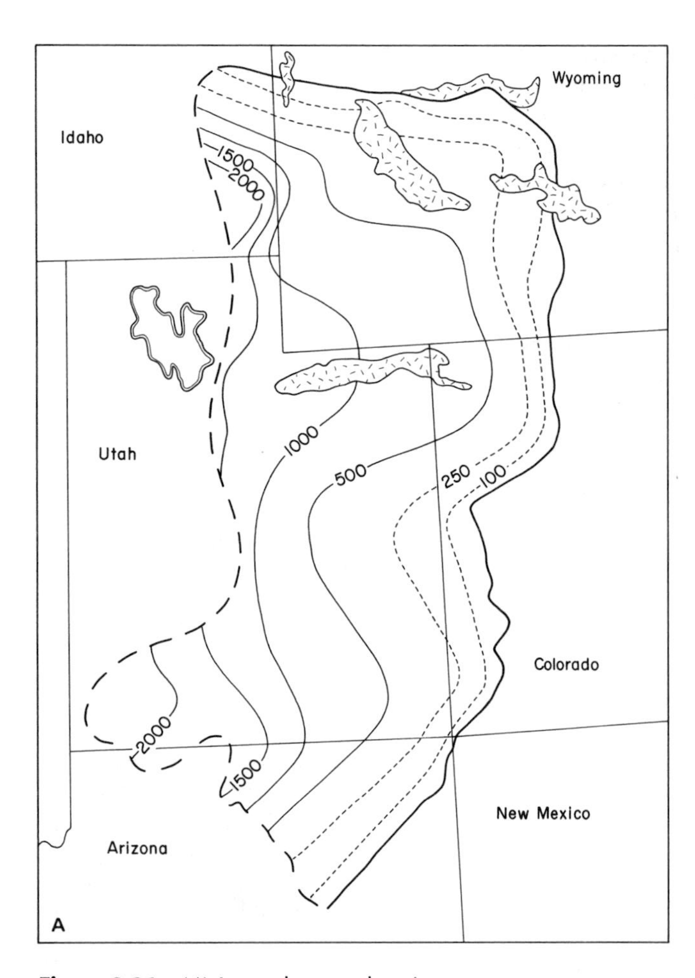

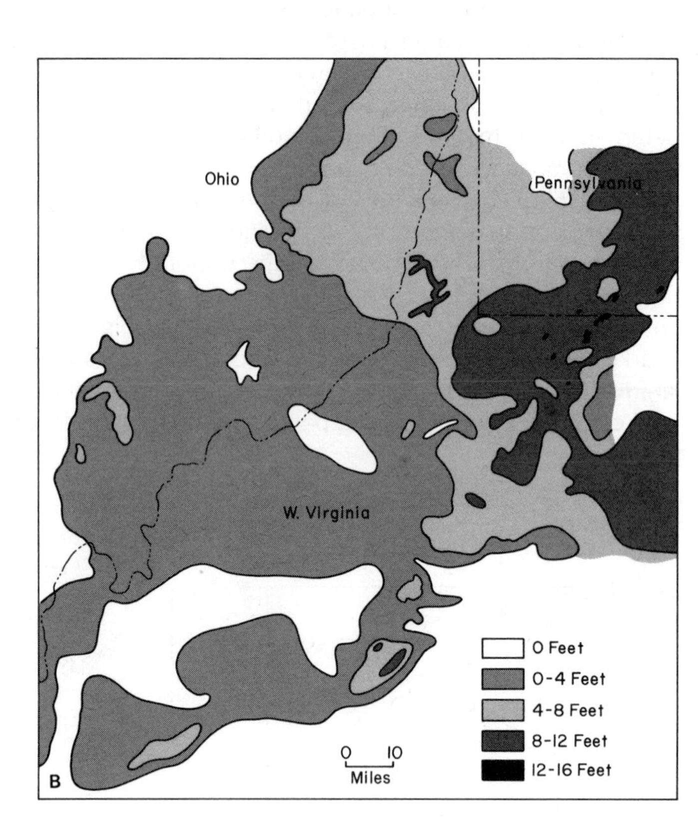

Figure 3.26 (*A*) Isopach map showing variations in thickness of the Navajo and Nugget Sandstones (Jurassic). [From Jordan (1965).] (*B*) Isopach map showing variations in the Pittburgh Coal. [From Hoover and others (1969). Published with permission of Geological Society of America and the authors.]

Isopach maps are contour maps that describe formation thicknesses. Thicknesses are compiled on the basis of geologic mapping, underground mapping and mining, and subsurface drilling. Values of thicknesses are posted on a base map and contoured. The resulting patterns describe the variations in thickness of a particular formation and/or series of formations (Figure 3.26*A*, *B*). The forms revealed in such maps significantly influence interpretations of paleogeography.

A good example of the regional structural insight afforded by isopach maps is found in maps of sedimentary assemblages of youngest Precambrian and Paleozoic ages in the hingeline region of the western United States (Figure 3.27*A*). The hingeline is generally regarded as the belt in western Utah and southern Nevada that coincides with a remarkable change in thickness of late Precambrian and Paleozoic sedimentary rocks (Figure 3.27*B*). The location of the hingeline coincides closely with the edge of the North American continent as it was fashioned by faulting in late Precambrian time.

Because of its sedimentary record and its structure, the hingeline is a region of active oil and gas exploration (Hill, 1976). In west-central Utah, less than 1000 ft (300 m) of late Cambrian to middle Cambrian sandstones crops out on the east side of the hingeline. Yet a thickness of 15,000 ft (4500 m) marks the same section just west of the hingeline (Armstrong, 1968a, b). Paleozoic formations east of the hingeline are thin and of shallow-marine, intertidal, and nonmarine origin. The presence of numerous unconformities, and the lack of a sedimentary record for certain periods of geologic time, reveal that transgressions onto the platform were relatively short-lived.

By way of contrast, Paleozoic formations west of the hingeline are thick sequences of deeper water marine sediments. All periods are represented, and long time periods were marked by apparent continuous sedimentation. The influence of the hingeline on sedimentation patterns is clearly and sensitively revealed in isopach maps of parts or all of the total younger Precambrian and Paleozoic sedimentary record in that region (Figure 3.27*C*).

SUMMARY

Geologic maps and subsurface maps are integral to descriptive structural analysis. They are the images that summarize carefully collected and recorded geological data.

Orthographic and Stereographic Projection

GEOMETRIC ANALYSIS

Map relationships and measurements of structure data are the the raw materials for geometric analysis. The major purpose of geometric analysis is to describe comprehensively the geometric attributes of structures and structural systems. Geometric analysis is used to answer many questions. What is the average orientation of bedding in an area? What is the overall orientation and shape of a sedimentary formation? What is the trend and plunge of the intersection of a fault and a bed?

Geometric questions of this type and countless others are solved routinely through a number of standard graphical and stereographic operations. The fundamental methods are introduced here as a basis for specific applied operations that will be encountered later in this text.

ORTHOGRAPHIC PROJECTION

A traditionally useful method for solving geometric problems is a kind of descriptive geometry known as **orthographic projection**. In essence, line-drawing constructions are prepared as a means of determining angular and spatial relationships in three dimensions. The constructions are difficult to visualize at first because the drawings convert map relationships into mixtures of maps and cross sections. Fundamental to the procedure is constructing structure profiles and structure contour lines.

CONSTRUCTING SIMPLE STRUCTURE PROFILES

Cross-sectional profiles, or **structure profiles**, are generally drawn at right angles to the trend or strike of structural features. In these profiles are shown the **traces** of plunging or dipping structures, as they would appear in a vertical "cut" through the uppermost part of the earth. Consider a limestone bed that crops out in a perfectly flat area at the location shown in Figure 3.28*A*. The bed strikes N40°E and dips 60°SE. A structure profile view of the bed is constructed in a vertical cut along *A*- *A′* at right angles to the line of strike (Figure 3.28*B*). The true dip of the bed is exposed to full view, as is the true thickness of the bed.

Using orthographic projection, let us construct step by step a structure profile for the limestone layer shown in Figure 3.28*A*. First choose the orientation and location of the **profile line** along which the structure section

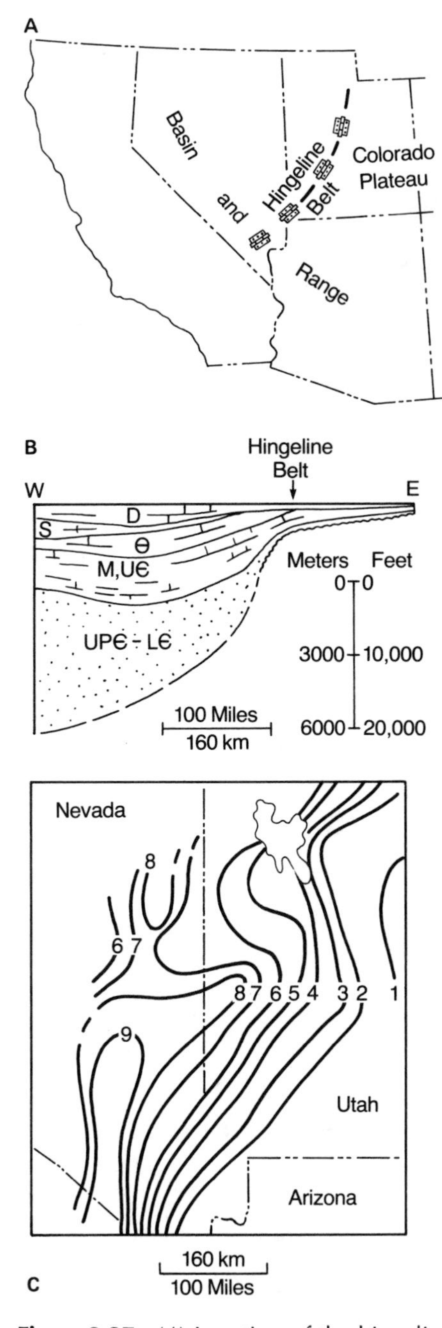

Figure 3.27 (*A*) Location of the hingeline belt of the western United States. (*B*) Cross section showing the radical thickening of strata west of the hingeline. [From Armstrong (1964).] (C) Isopach maps of younger Precambrian and Paleozoic strata in the hingeline region. Contour values expressed in kilometers. [From Armstrong (1964).]

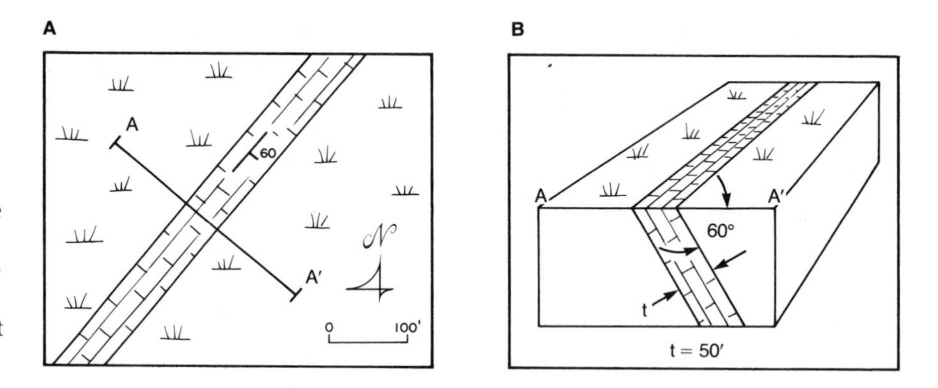

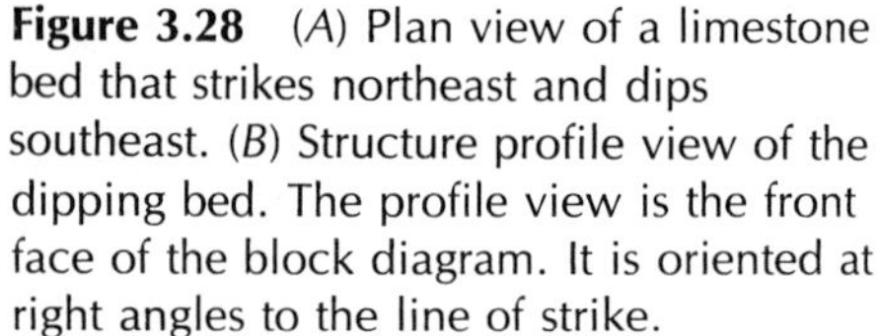

Figure 3.28 (*A*) Plan view of a limestone bed that strikes northeast and dips southeast. (*B*) Structure profile view of the dipping bed. The profile view is the front face of the block diagram. It is oriented at right angles to the line of strike.

is to be constructed (Figure 3.29*A*). Points *A* and *B* are identified along the profile line such that *A* is on the lower contact of the limestone and *B* is on the upper contact at a location directly along the dip direction from *A*. Project points *A* and *B* from the interior of the map toward the edge of the map (or off the map onto another sheet of paper), where there is more available working space (Figure 3.29*B*). Project each reference point by the same distance and in a direction strictly parallel to the strike of the limestone bed.

Draw a line through the **projected points** *A′* and *B′* (Figure 3.29*B*). This line represents the topographic **surface profile** for the location of the profile line where the limestone crops out. In this special case the surface profile is perfectly horizontal. In the general case, the surface profile would be marked by some **topographic relief**. Showing such relief in profile would be part of the construction process. Topographic control would be afforded by topographic contour lines on the base map.

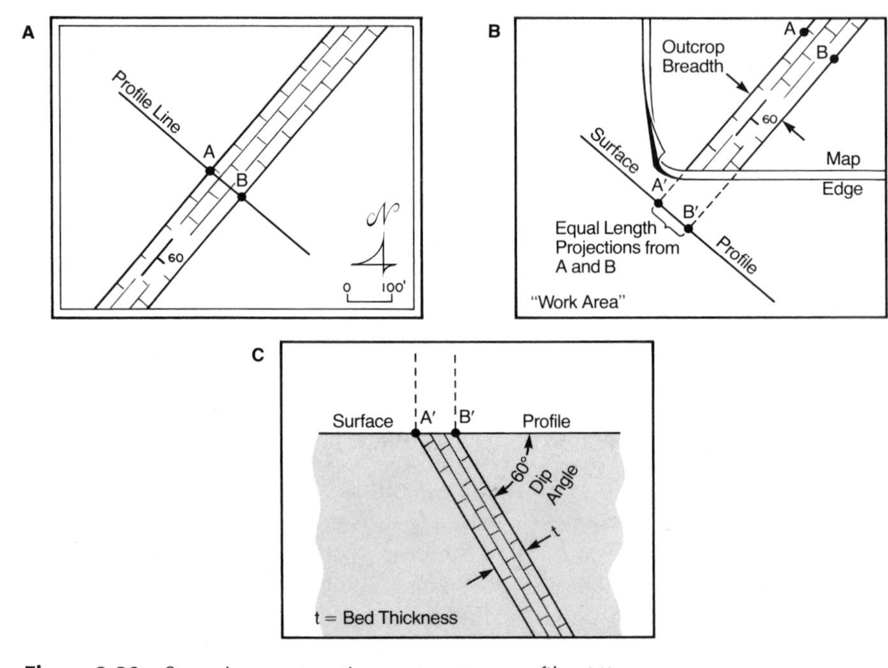

Figure 3.29 Steps in constructing a structure profile. (*A*) Draw the profile line (*AB*) for which the structure profile is to be constructed. (*B*) Step the profile line to the edge of the map, or off the map sheet onto another piece of paper, where there is working space. (C) Plot the true dip of the bed and draw the subsurface expression of the top and bottom of the bed.

Using a protractor, construct the angle of dip of the limestone bed in the subsurface, beneath the surface profile (Figure 3.29*C*). For this example the dip is 60°SE. From *A′* and *B′* draw the 60°-dipping lines that correspond to the lower and upper contacts of the limestone bed. This completes the structure profile (Figure 3.29*C*). It represents an approximation of the structural form and the **attitude** (i.e., orientation) of the limestone bed at depth. It is a "projection" of the form and the dip of the limestone unit based on surface exposures.

DETERMINING THE THICKNESS OF A BED

The thickness of a bed can be measured directly in structure profile view, provided the profile is constructed at right angles to strike. The map scale is used as the guide in determining thickness. Thickness is measured perpendicular to the upper and lower contacts for the bed in question. Figure 3.29*C* shows how thickness is measured in structure profile view. Note that the measured thickness of the bed is not the same as **outcrop breadth** (see Figure 3.29*B*). Only where a bed dips vertically are outcrop breadth and the thickness the same.

REPRESENTING DIPPING PLANES BY STRUCTURAL CONTOUR LINES

In problems of applied geology, it is commonly necessary to predict the location of rock layers, contacts, and structures at depth. For example, if the surface outcrop of the limestone bed in the preceding example were found to be mineralized, the location of the limestone in the subsurface could be of economic interest. It would then be beneficial to prepare a structural contour map of the limestone bed, using the upper contact of the limestone bed as a datum. Structure contour lines would connect points of equal elevation on the upper surface of the limestone. Since the contour lines, by definition, connect points of equal elevation, they are lines of strike. Each contour line represents a strike line on the limestone bed at some specified elevation.

If the designated contour interval for the structure contour map is, for example, 100 ft (30 m), the pattern of the corresponding contour lines can be found by a series of orthographic construction steps. First, construct a structure profile for the limestone bed, and add horizontal lines to the structure profile below the surface profile such that the lines are spaced at 100-ft intervals (Figure 3.30*A*). The map scale is used as the reference for positioning the lines. Each of the lines represents the intersection of the plane of the structure profile with a horizontal plane of some given elevation. The horizontal planes are called topographic reference planes in the subsurface, or simply **reference planes**.

The next step is to identify the points of intersection of the upper surface of the limestone bed and the trace of each of the reference planes (Figure 3.30*B*). These **structural intercepts** become the basis for positioning the structural contour lines.

Project each of the structural intercepts vertically to the surface profile (Figure 3.30*C*). These projected points are the **vertical projections** of the structural intercepts of the upper contact of the limestone bed with each of the horizontal reference planes. Vertical projections are fundamental to orthographic projection. Consider vertical projection points *M* and *O* as examples. Point *M* lies directly above the point where the top of the limestone bed is exactly 200 ft (61 m) below the surface. Point *O* lies directly above the point where the upper contact of the limestone lies exactly 400 ft (122 m) below the surface.

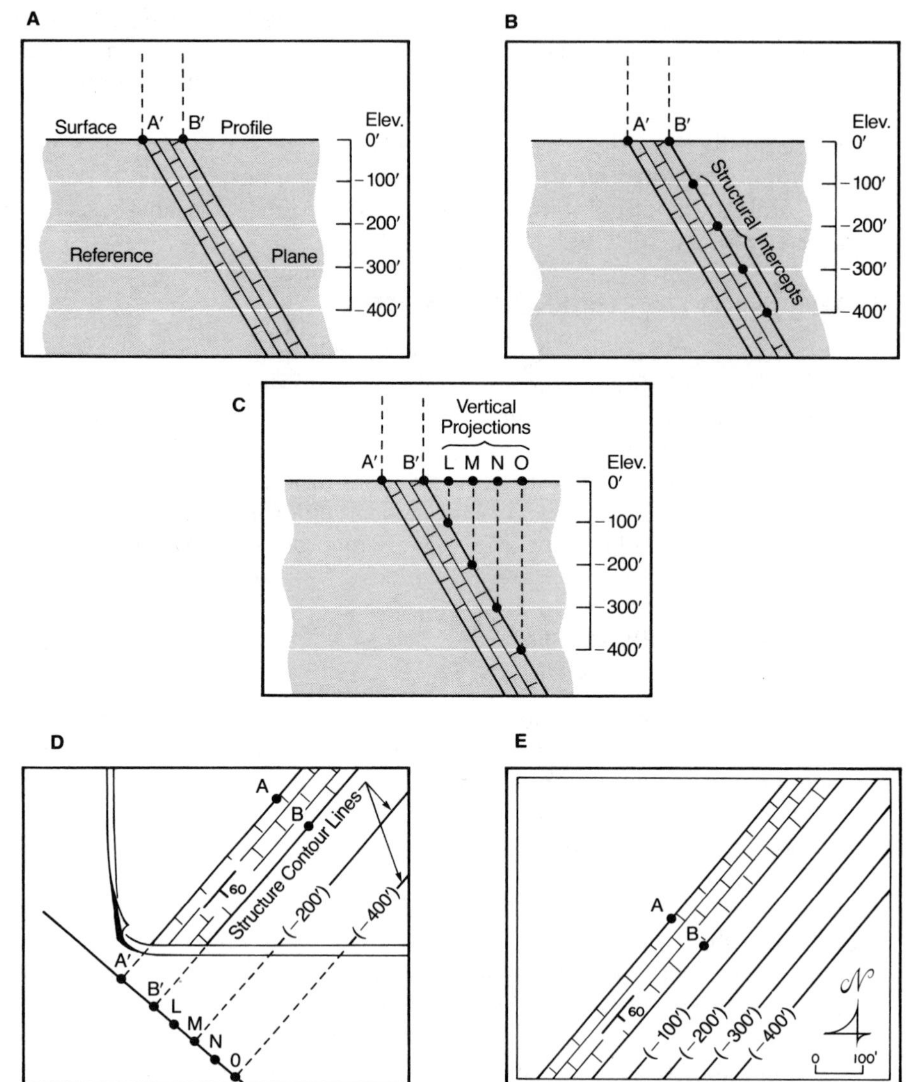

Figure 3.30 Construction of structure contour lines representing the upper surface of a dipping bed. (*A*) Starting information: the structure profile view of a dipping bed. (*B*) Using the structure profile view, identify the structural intercepts of the top of the bed with the elevation reference planes shown in the subsurface. (*C*) Plot the vertical projections of each of the structural intercepts. (*D*) Project the vertical projections parallel to the line of strike. (*E*) The finished structure contour map.

Finally we shift our construction from structure profile to map view (Figure 3.30*D*), and we project **strike lines** from *M* and *O* into the interior of the map. These lines are **structure contour lines**, one of value -200 ft (i.e., 200 ft below the surface), the other of value -400 ft (-122 m). Since point *M* is the vertical projection of a point on the top of the limestone bed at elevation -200 ft, every point on the strike line through *M* must also lie 200 ft below the surface. The structure contour map is completed by drawing strike lines through all the vertical projections (Figure 3.30*E*).

The ability to construct structure profiles and structure contour maps on the basis of surface or mine map patterns provides the means to solve a number of practical structural geologic problems.

MEASURING APPARENT DIP

Apparent dip is the inclination of the trace of a plane in a direction other than the true dip direction. Using the structure contour map displayed in Figure 3.30*E*, we can solve for apparent dip of the limestone bed in any direction, for example, along a north–south line. First draw a north-trending profile line from one structure-contour line to another (Figure 3.31*A*). The profile line in Figure 3.31*A* connects a point on the -400-ft contour line with

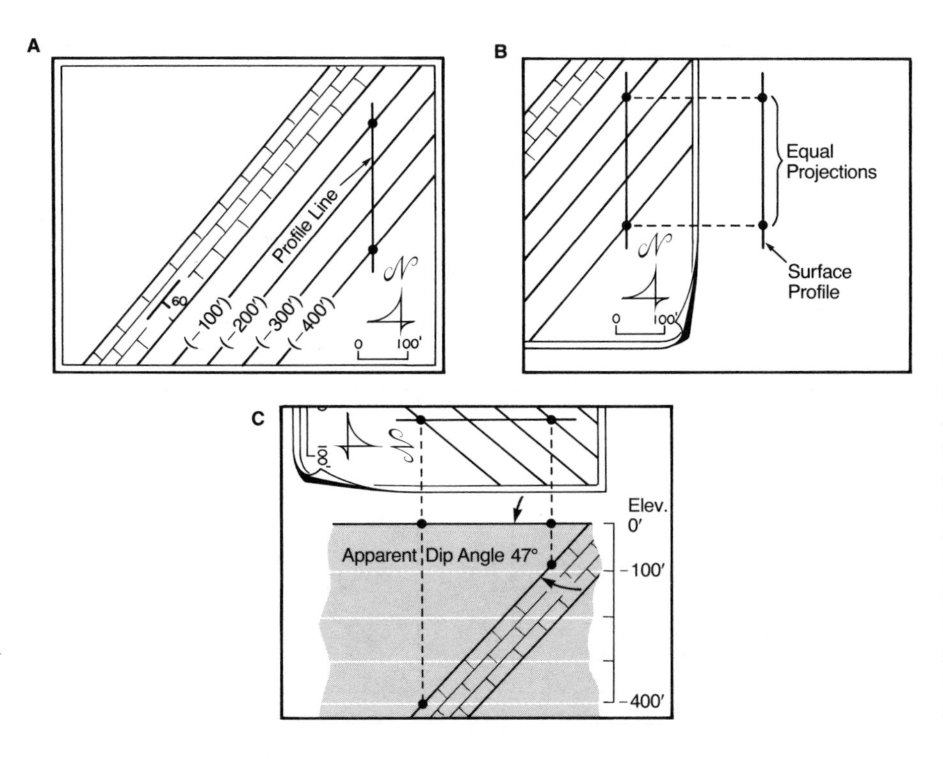

Figure 3.31 Orthographic construction method for determining the apparent dip of a bed. (*A*) Designate the location and trend of the profile line along which apparent dip is to be determined. (*B*) Step the profile line to a location where working space is available. (*C*) Construct horizontal reference planes for the -100-ft (-30 m) and -400-ft (-122 m) levels. Project reference points on profile line to corresponding reference planes. Connect these structural intercepts to display apparent dip.

a point on the -100-ft (-30 m) contour line. Project the end points of this profile line to the edge of the map, or off it, and draw the surface profile (Figure 3.31*B*). The end points of the surface profile are vertical projections from the upper surface of the limestone bed at elevations corresponding to the values of the structure contour lines.

Begin to fashion the structure profile by constructing the horizontal reference planes that correspond to the contour lines on which the reference points of the profile line rest (Figure 3.31*C*). Then project lines vertically down from the reference points on the surface profile to the corresponding structural intercepts of the limestone bed and the horizontal reference planes. A line connecting the structural intercepts at the -100- and -400-ft levels represents the upper contact of the limestone bed. Its angle of inclination, as measured from the horizontal, is the apparent dip. Its value, as measured with a protractor, is 47° (Figure 3.31*C*).

The standard orthographic solution to the apparent dip problem is a shortcut to the orthographic method just described. Here is how to do it. On a sheet of paper designate a **control point** that lies on the upper (or lower) contact of the limestone bed (Figure 3.32*A*). Through it draw a strike line (N40°E) representing the strike attitude of the bed. At right angles to the strike line, construct a surface profile and draw the structural profile view of the dipping bed (Figure 3.32*B*). Add to this profile view a horizontal reference plane that is positioned some arbitrary but known distance beneath the surface, for example, -100 ft (Figure 3.32*C*). Plot the vertical projection of the structural intercept of the reference plane and the dipping bed (Figure 3.32*D*), and project a structure contour line across the map from the vertical projection point.

To find apparent dip as viewed in a north-trending vertical exposure, draw a north–south-trending profile line northward from the original control point (Figure 3.32*E*). Where this profile line crosses the strike line of value -100 ft, there lies the vertical projection of the structural intercept of the upper contact of the limestone bed and the -100-ft structure contour (Figure 3.32*F*). The apparent dip of the limestone is found by constructing a structure profile

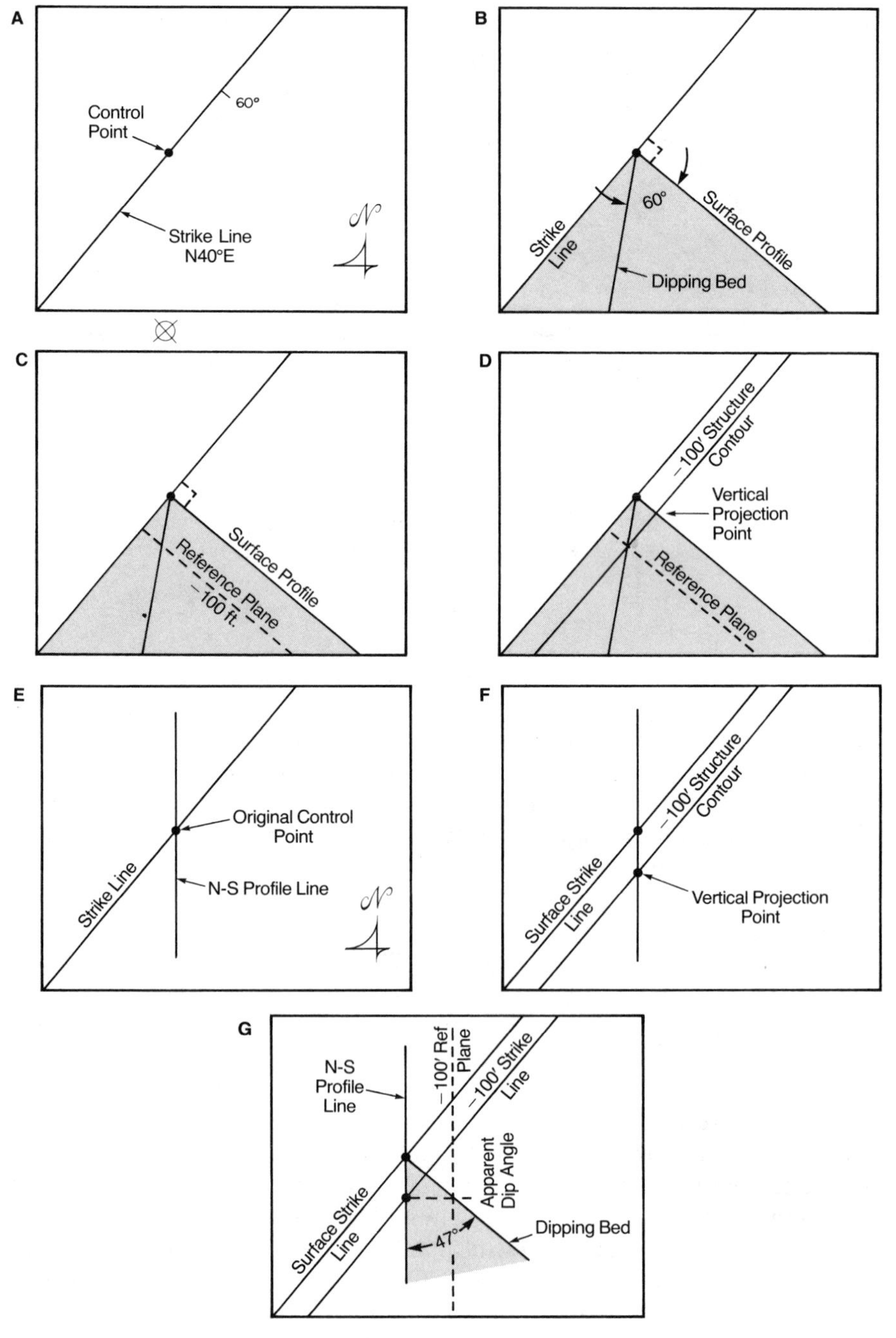

Figure 3.32 "Shortcut" method for determining apparent dip. (*A*) Designate control point at convenient location on lower contact of limestone. (*B*) Construct structural profile view of dipping bed. (*C*) Plot the structural intercept representing the vertical projection of the top of the bed at the -100-ft (-30 m) level. (*D*) Identify the vertical projection of the structural intercept. (*E*) Through the control point established in *A*, draw the north–south profile line along which apparent dip is to be determined. (*F*) Identify the location where the north–south profile line intersects the vertical projection of the dipping plane at the -100-ft level. (*G*) Construct a structure profile on the basis of the known elevation of the surface control point (*A*) and the -100-ft structural intercept. Measure the apparent dip with a protractor.

along the north-trending surface profile (Figure 3.32*G*). The value of the apparent dip is 47°S. Apparent-dip constructions of this type make it clear that true dip of bedding or any other planar structure can be viewed only in profiles oriented perpendicular to strike. Exposures of structure profiles oriented parallel to strike reveal 0° apparent dip. Sections oriented obliquely to strike disclose intermediate values of dip between 0° and the true dip.

CONSTRUCTING THE LINE OF INTERSECTION OF TWO PLANES

Determining the trend and plunge of the line of intersection of two planes is fundamental to a number of geometric and geologic problems. Let us solve for the trend and plunge of the line of intersection of a dike and a limestone bed (Figure 3.33*A*). Map relationships show that the dike strikes N68°E and

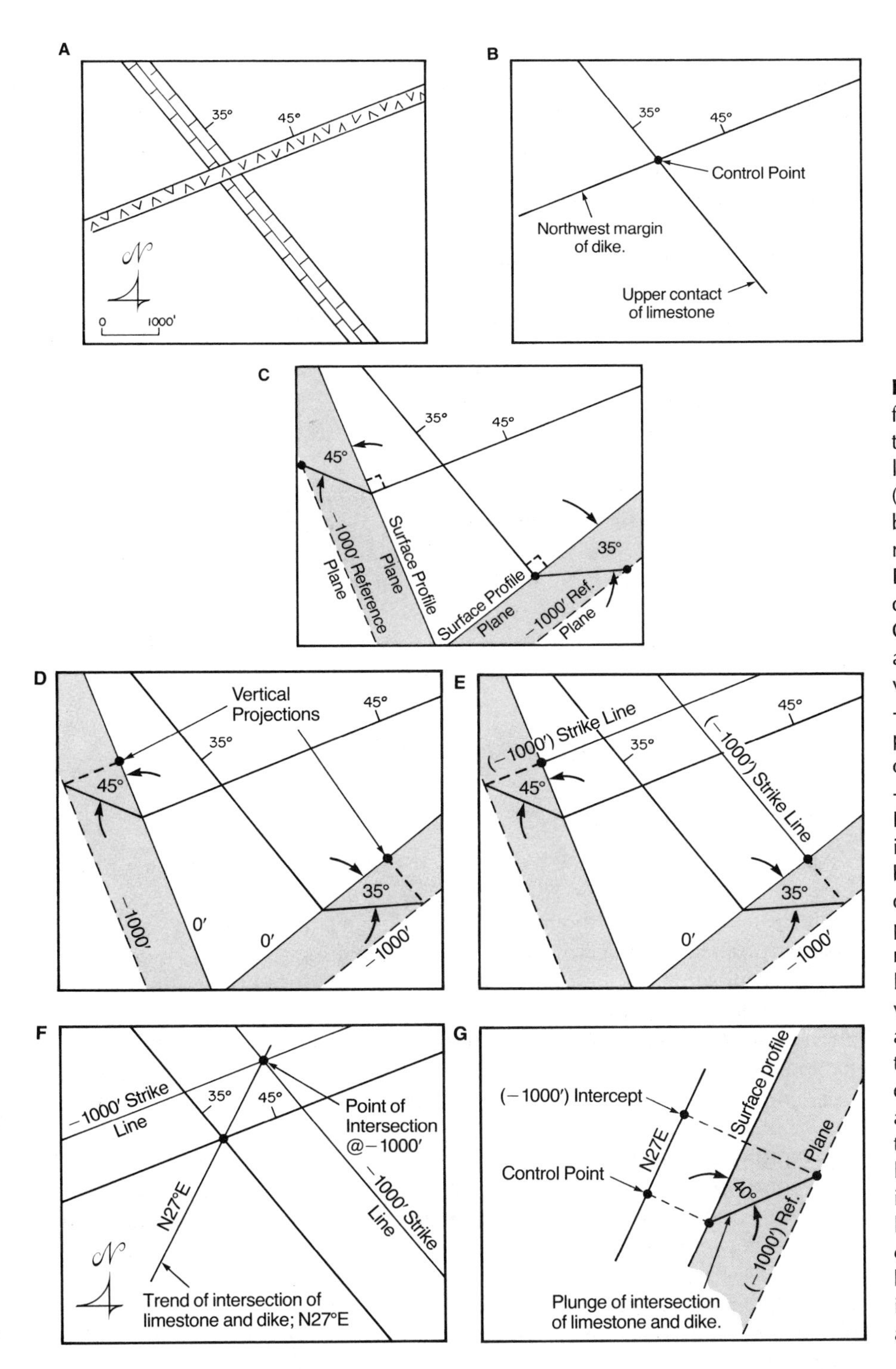

Figure 3.33 Orthographic construction for determining the trend and plunge of the intersection of two planes. (*A*) Map of limestone bed that is intruded by a dike. (*B*) Simplified map of the same limestone bed, showing only the elements that are required to solve the problem at hand. (*C*) Draw two structure profiles: one for the dike, another for the limestone bed. Construct each profile, as always, at right angles to the line of strike. (*D*) Identify the vertical projection of the dike with the -1000-ft (-300 m) elevation reference plane. Also identify the vertical projection of the top of the limestone bed with the -1000-ft elevation reference plane. (*E*) Identify the vertical projection of the intersection of the top of the limestone bed with the northwestern margin of the dike at the -1000-ft level. (*F*) Connect the point of intersection of the northwest margin of the dike and the top of the limestone bed at the surface with the vertical projection of the same intersection at the -1000-ft elevation level. This line is the trend of the line of intersection of the dike and the limestone bed. (*G*) Construct a structure profile parallel to the trend of the line of intersection of the dike and the limestone bed. The plunge of the line of intersection is determined by drawing a line in profile view that connects the point of intersection of the dike and limestone bed at the surface level with the point of intersection of the dike and limestone bed at the -1000-ft level.

dips 45°NW; and the limestone bed strikes N39°W and dips 35°NE. We assume that the dike and the sandstone bed are perfectly planar and that the land surface is perfectly flat. For simplicity, no shifting of the limestone bed due to dike emplacement is shown in Figure 3.33*A*.

Let the intersection of the northwest margin of the dike with the upper contact of the limestone bed be a control point for the constructions that follow (Figure 3.33*B*). The control point is one of the two intersection points that we need to define the trend and plunge of the line of intersection of the two planes. The second control point will be found in the subsurface, where the top of the limestone bed and the northwest margin of the dike intersect at some known depth.

To find the second control point, first construct structure profile views of both the limestone bed and the dike (Figure 3.33*C*). For each structure profile, construct a horizontal reference plane at some specified elevation [e.g., -1000 ft (-300 m)] below each surface profile (Figure 3.33*C*).

Identify the structural intercept of the upper surface of the limestone bed and the horizontal reference plane, and then define its vertical projection to the surface profile (Figure 3.33*D*). In the same manner, plot the vertical projection of the structural intercept of the dike and the horizontal reference plane.

Construct a strike line through the vertical projection of the limestone/reference plane structural intercept. This line is a map view of the vertical projection of the intersection of the top of the limestone bed with the -1000-ft reference plane (Figure 3.33*E*). It is a structure contour on the limestone bed at elevation -1000 ft. In the same fashion, establish a -1000-ft contour line for the dike.

The intersection of the -1000-ft structure contour lines for the dike and the top of the limestone bed, respectively, is the vertical projection of the intersection of the dike and the limestone bed at elevation -1000 ft (Figure 3.33*F*). Connect this intersection point with the original control point to define the trend of the intersection of the two planes. Its value is N27°E. To determine the plunge, construct a structure profile of the line of intersection of the dike and the limestone bed (Figure 3.33*G*). The plunge measures 40°NE.

STEREOGRAPHIC PROJECTION

INTRODUCTORY COMMENTS

Stereographic projection is a powerful method for solving geometric problems in structural geology (Bucher, 1944; Phillips, 1971). Stereographic projection differs from orthographic projection in a fundamental way: orthographic projection preserves spatial relations among structures, but stereographic projection displays geometries and orientations of lines and planes without regard to spatial relations.

The use of stereographic projection is preferable to orthographic projection in solving many geometric problems, simply because of ease of operations. Solving for apparent dip, the trend and plunge of the intersection of two planes, and angles between lines and planes in space can be carried out rapidly and accurately using stereographic projection. Orthographic projection, in contrast, requires the slow, careful construction of line drawings. But, orthographic projection remains the only effective way to solve geometric problems when topographic relief, map relationships, and depth(s) to structures in the subsurface are integral to the solution of structural problems. In practice, we combine orthographic and stereographic projection techniques in ways that are practical, efficient, and complementary.

GEOMETRY OF PROJECTION

Think of stereographic projection as a procedure comparable to using a three-dimensional protractor. A two-dimensional protractor is simple to use. Using a protractor we can plot trends of lines, measure angles between lines, construct **normals** (i.e., perpendiculars) to lines, and rotate lines by specified angles. Stereographic projection permits the same kinds of operation, but in three-dimensional space. Moreover, both lines and planes can be plotted

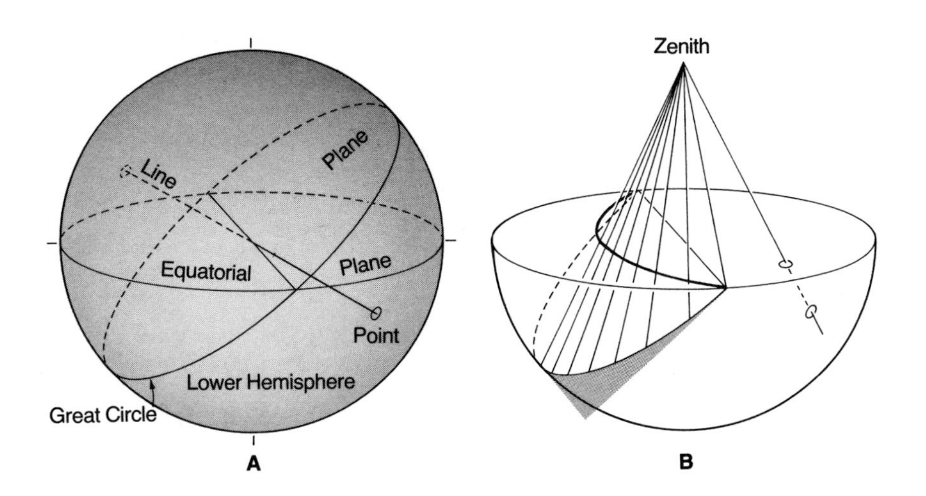

Figure 3.34 The inherent three-dimensional geometry of stereographic projection. (*A*) Projection of a plane and a line through the center of a reference sphere. The plane intersects the lower hemisphere of the reference sphere as a great circle. The line intersects the lower hemisphere as a single point. (*B*) Projection of intersection points from the lower hemisphere of the reference sphere to the zenith of the projection.

and analyzed. Equipped with a three-dimensional protractor, we can do the following: plot orientations of lines; plot orientations of planes; determine the orientation of the intersection of two planes; determine the angle between two lines; determine the angle between two planes; measure the angle between a line and a plane; and rotate lines and planes in space about vertical, horizontal, or inclined axes.

All the operations above would be simple if it were possible to assemble real lines and planes in space, like tinker toys, and measure their geometric properties directly. Using stereographic projection techniques, we figuratively assemble lines and planes within a reference sphere.

The line or plane to be stereographically represented can be thought of as passing through the center of a reference sphere and intersecting its lower hemisphere (Figure 3.34*A*). Planes intersect the lower hemisphere in the form of **great circles**; lines intersect the lower hemisphere in **points**. Stereographic projection of lines and planes to points and great circles constitutes a systematic reduction of three-dimensional geometry to two dimensions. The "flattening" to two dimensions is achieved by projecting the lower hemisphere intersections to an **equatorial plane** of reference that passes through the center of the sphere (Figure 3.34*B*). This is the plane of stereographic projection. The lower hemisphere intersections are projected as rays upward through the horizontal reference plane to the **zenith** of the sphere. Where the rays of projection pass through the horizontal reference plane, point or great-circle intersections are produced, and these are **stereograms** or **stereographic projections** of lines and planes. Details of the projection geometry are presented in Phillips (1971).

Steep-plunging lines stereographically project to locations close to the center of the horizontal plane of projection; shallow-plunging lines project to locations near the perimeter of the plane of projection (Figure 3.35*A*). Steeply dipping planes stereographically project as great circles that pass near the center of the plane of projection; gently dipping planes project as great circles passing close to the perimeter of the horizontal plane of projection (Figure 3.35*B*). The distance that a great circle or point departs from the center of the plane of projection is a measure of the degree of inclination of the plane or line that has been stereographically plotted.

The trend of a great circle across the plane of stereographic projection corresponds to the strike of the plane that the great circle stereographically portrays (Figure 3.35*C*). And the trend of a line connecting the center of the stereographic projection to a point representing a stereographically plotted line is the same as the trend of the line in space (Figure 3.35*D*).

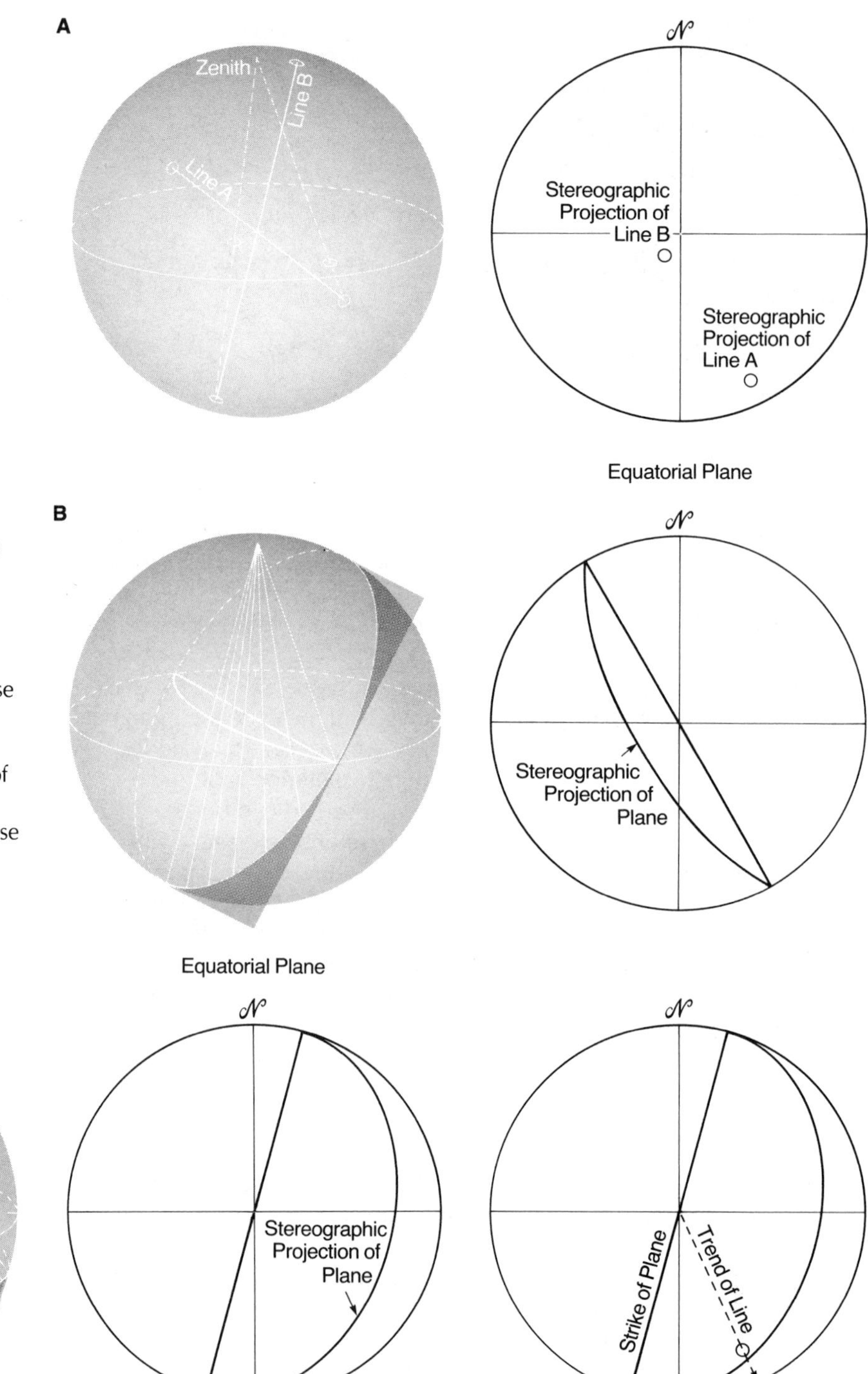

Figure 3.35 The distance that a great circle or point lies from the center of the equatorial plane of projection is a measure of the inclination of a plane or line. (*A*) Shallow plunging lines project close to the perimeter of the equatorial plane; steeply plunging lines project close to the center. (*B*) Great circles that represent the orientation of steeply dipping planes pass close to the center of projection; shallow dipping planes are represented by great circles that pass close to the perimeter of the equatorial plane. (*C*) Stereographic representation of the strike of a plane and the trend of a line.

STEREOGRAPHIC NET OR STEREONET

In actual practice, stereographic projection of lines and planes is carried out through the use of a **stereographic net** (Figure 3.36). A **stereonet**, as it is called, displays a network of great-circle and small-circle projections that occupy the equatorial plane of projection of the reference sphere. Both the great circles and small circles are spaced at 2° intervals; every fifth one is darkened so that 10° intervals can be readily counted. The great circles represent a family of planes of common strike whose dips range from 0° to 90°. The planes intersect in a horizontal line represented by the north–south

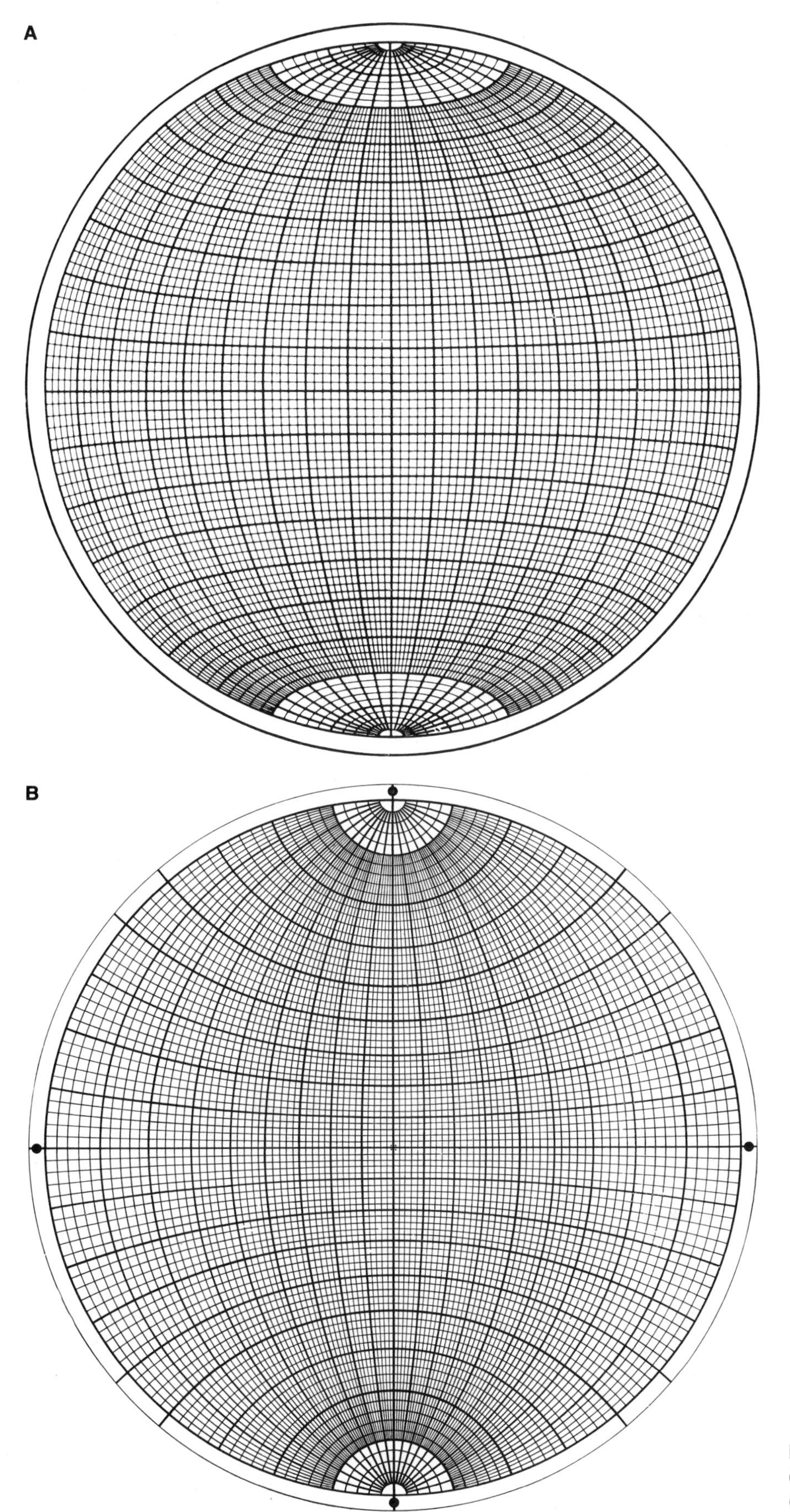

Figure 3.36 Types of stereographic net. (*A*) The Schmidt net, an equal-area net. (*B*) The Wulff net, an equal-angle net.

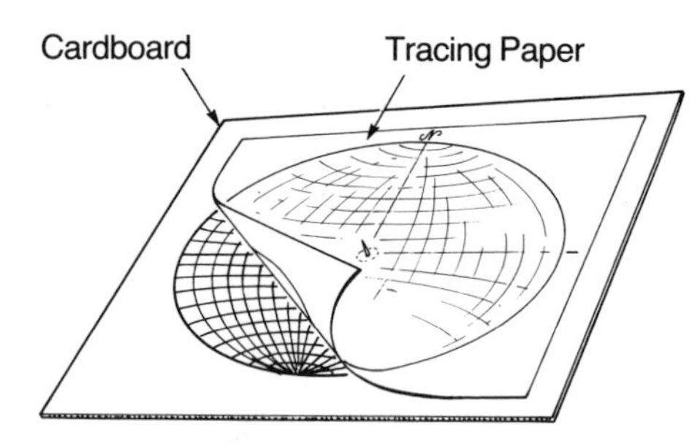

Figure 3.37 Stereographic net ready for action. (From *Structural Geology of Folded Rocks* by E. T. H. Whitten. Originally published by Rand-McNally and Company, Skokie, Illinois, copyright ©1966. Published with permission of John Wiley & Sons, Inc., New York.)

line of the net. The small circles may be thought of as the paths along which lines would move when rotated about a horizontal axis oriented parallel to the ordinate of the net. The combination of small and great circles constitutes an orientation framework for stereographically plotting lines and planes.

There are two different kinds of stereonet: Wulff nets and Schmidt nets (Figure 3.36). Constructions are carried out the same way on each (Phillips, 1971, p. 61). Structural geologists find the Schmidt net (Figure 3.36A) to be the most versatile, for reasons to be explained later.

To prepare the Schmidt net for use, tape or glue it to a heavy backing, like cardboard or masonite. Insert a thumbtack through the backing and through the exact center of the net, taping the tack to the underside of the net so it cannot fall free (Figure 3.37). A sheet of tracing paper is placed on the net so that the paper, punctured by the thumbtack, can rotate about the tack. All construction work is carried out on the tracing paper, which is oriented with respect to north by marking a **north index** at a point corresponding to the top of the north–south line. This geographically orients the overlay for the constructions to be carried out.

PLOTTING THE TREND AND PLUNGE OF A LINE

Before any problems can be solved stereographically, it is necessary to learn how to represent the orientations of lines and planes on a stereonet, without any immediate thought of setting records or solving problems. It is just like learning to pole vault: before vaulting 17 ft (5 m), it is necessary to learn to run with the pole. The first attempts will feel very awkward.

Lines are easiest to plot. Consider a line that plunges 26° N40°E. To represent the trend and plunge of this line stereographically, first plot the trend, in degrees, on the outer perimeter of the stereographic net. To do this, measure east from north by 40° (Figure 3.38*A*). This can be accomplished simply by using the stereographic net as you would a protractor, counting clockwise from the north index on the tracing paper along the periphery of the net to 40°. The 40° azimuth corresponds to the 40° small-circle intercept on the perimeter of the net. Mark the point at 40° with a **trend-index mark** (*t*) (Figure 3.38*A*).

To plot the 26° plunge, first rotate the overlay clockwise until *t* comes to rest on the right end of the east–west line of the net (Figure 3.38*B*). This is one of two lines (the *y* axis being the other) where inclination can be directly measured and plotted. The plunge is measured by counting 26° from the perimeter of the net along the east–west line toward the center of the net. Point *L* represents the 26° plunging line (Figure 3.38*B*). By rotating the tracing paper counterclockwise such that the north index again becomes aligned with the top of the north–south line ("home position"), point *L* can be viewed in its proper orientation framework (Figure 3.38*C*). As a general check, it can be seen that point *L* lies in the northeast quadrant, corresponding to a northeast trend. Furthermore, it falls relatively close to the perimeter, reflecting a rather shallow plunge.

One shortcut is available. Plotting the N40°E trend can be achieved simply by rotating the tracing paper counterclockwise such that the north index comes to rest on 40°W (Figure 3.38*D*). This automatically orients the trend line along the north–south line of the net, along which the 26° plunge can be directly measured.

As a second example, let us plot stereographically the orientation of a line plunging 75° S65°W. First define the trend by measuring 65° west of south (Figure 3.39*A*). Point *t* represents the trend of this line. Next rotate *t* until it coincides with the left end of the east–west line (Figure 3.39*B*). Measure the

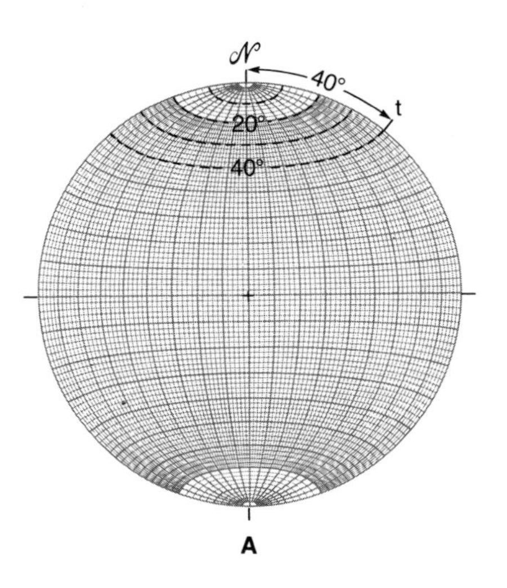

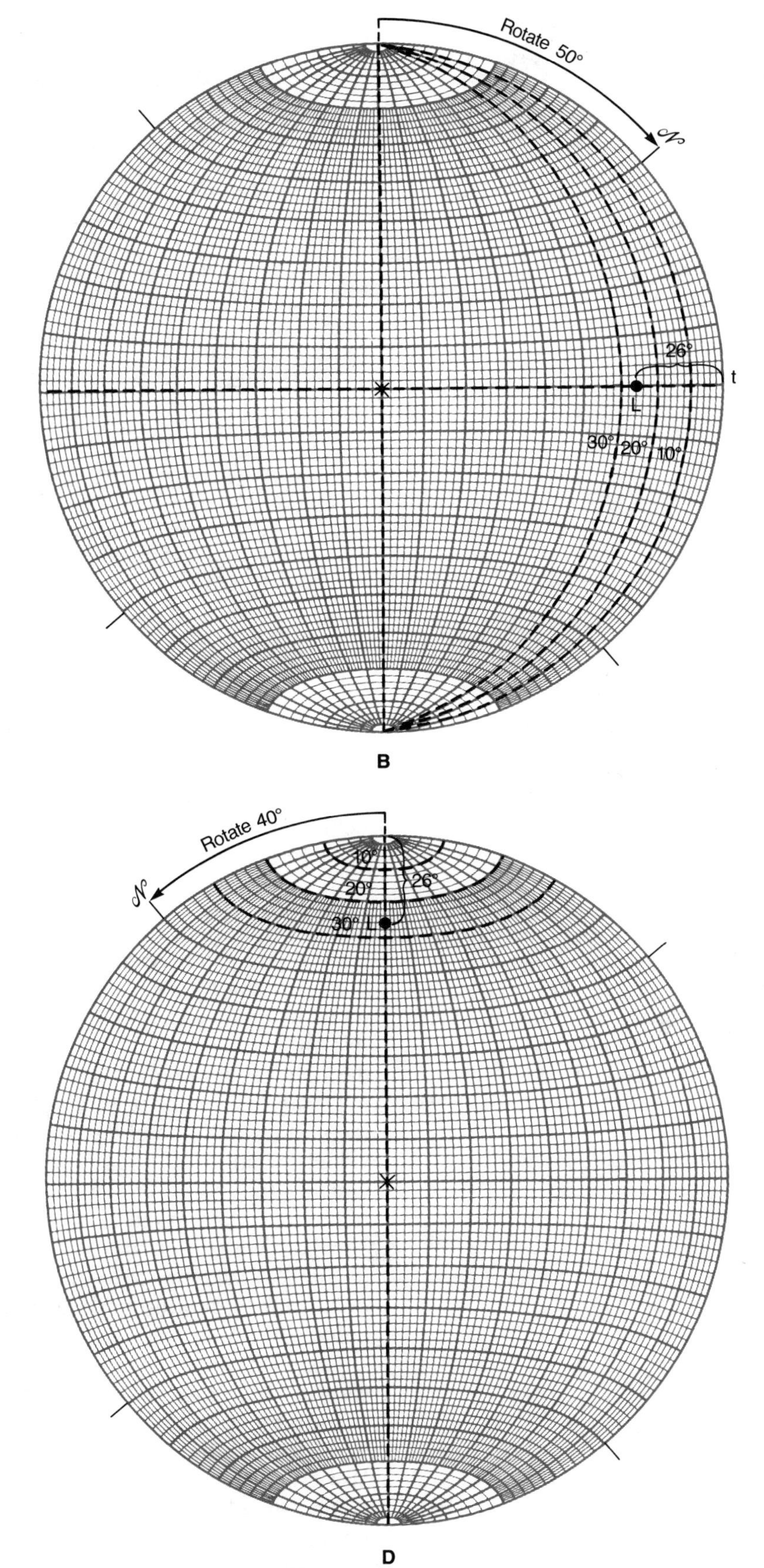

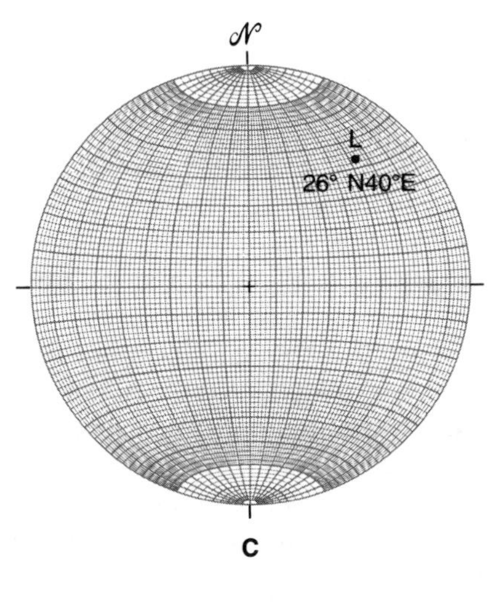

Figure 3.38 Stereographic projection of a line that plunges 26° N40°E. (*A*) Measure the azimuth of the trend along a line that lies 40° east of north. Mark its orientation, *t* = trend. (*B*) Measure the 26° inclination of plunge along the east–west line of the net. (*C*) Stereographic representation of the orientation of a line as a single point. (*D*) Shortcut method for plotting the trend and plunge of a line.

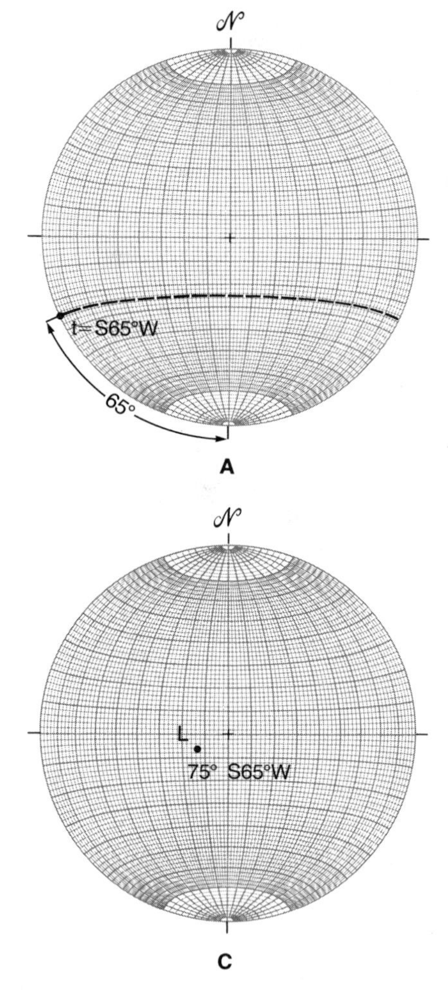

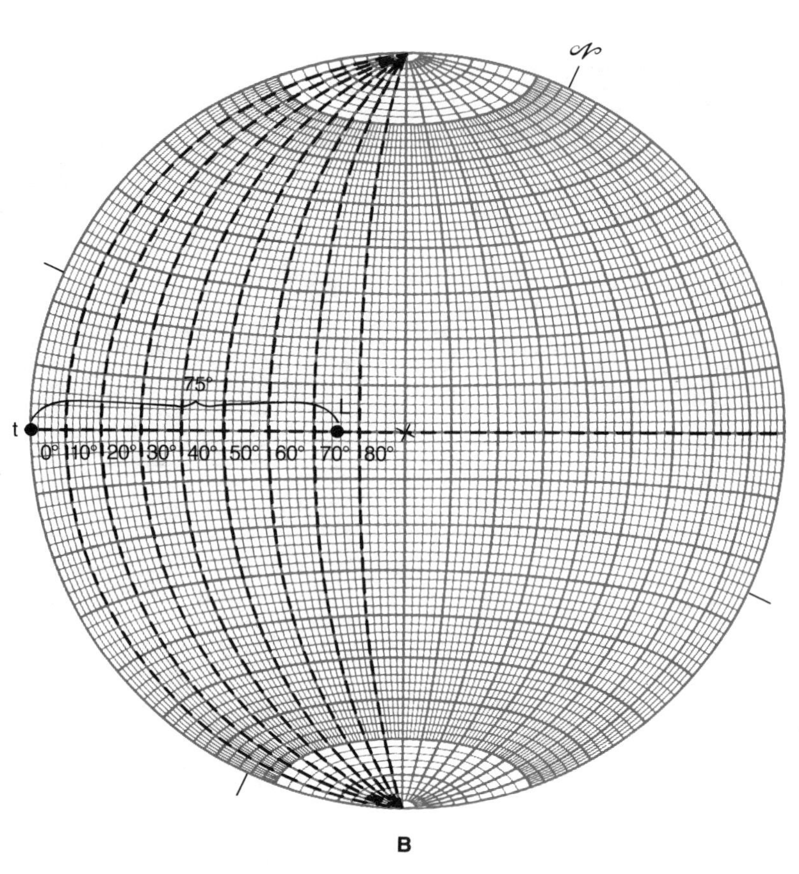

Figure 3.39 Stereographic projection of a line that plunges 75° S65°W. (*A*) Measure the azimuth of the trend of the line. (*B*) Rotate the trend line to the east–west line of the net. Measure the plunge inclination along the east–west line, moving from the edge of the net toward the center. (*C*) Stereographic result as seen when overlay is in home position.

value of the plunging line by counting inward 75° from the perimeter of the net (*L* marks the 75°-plunging line). Rotate the tracing paper back to home position and view *L* in its proper orientation (Figure 3.39*C*).

PLOTTING THE STRIKE AND DIP OF A PLANE

Let us now stereographically plot the orientation of a plane. Consider a plane striking N40°W and dipping 30°SW. The strike line of this plane is found by rotating the tracing paper clockwise until the north index comes to rest on 40°E (Figure 3.40*A*, *B*). The strike can be visually shown by tracing the north–south line of the net (Figure 3.40*B*), although this is not strictly necessary. When the strike line of the plane is rotated into coincidence with the north–south line of the stereonet, the dip of the plane can be plotted. Count inward 30° from the perimeter of the net along the east–west line, which is the line of dip when the strike line of the plane coincides with the north–south line of the net (Figure 3.40*C*). Then trace the great circle that coincides with the 30° dip. By rotating the north index back to home position (Figure 3.40*D*), the strike line and the great circle become aligned in an orientation that corresponds stereographically to a plane striking N40°W and dipping 30°SW.

There is another way to represent the orientation of a plane stereographically. The orientation of any plane in space can be described uniquely by the orientation of a line perpendicular to the plane. If the trend and plunge of a normal (**pole**) to a plane is known, the orientation of the plane itself is also established. The pole to a vertical plane is horizontal, and it stereographically plots as a point on the perimeter of the stereonet. The pole to a horizontal plane is vertical, and it plots stereographically as a point at the very center of the stereonet. The pole to an inclined plane plots as a point somewhere in the interior of the net, but not at its center. When large

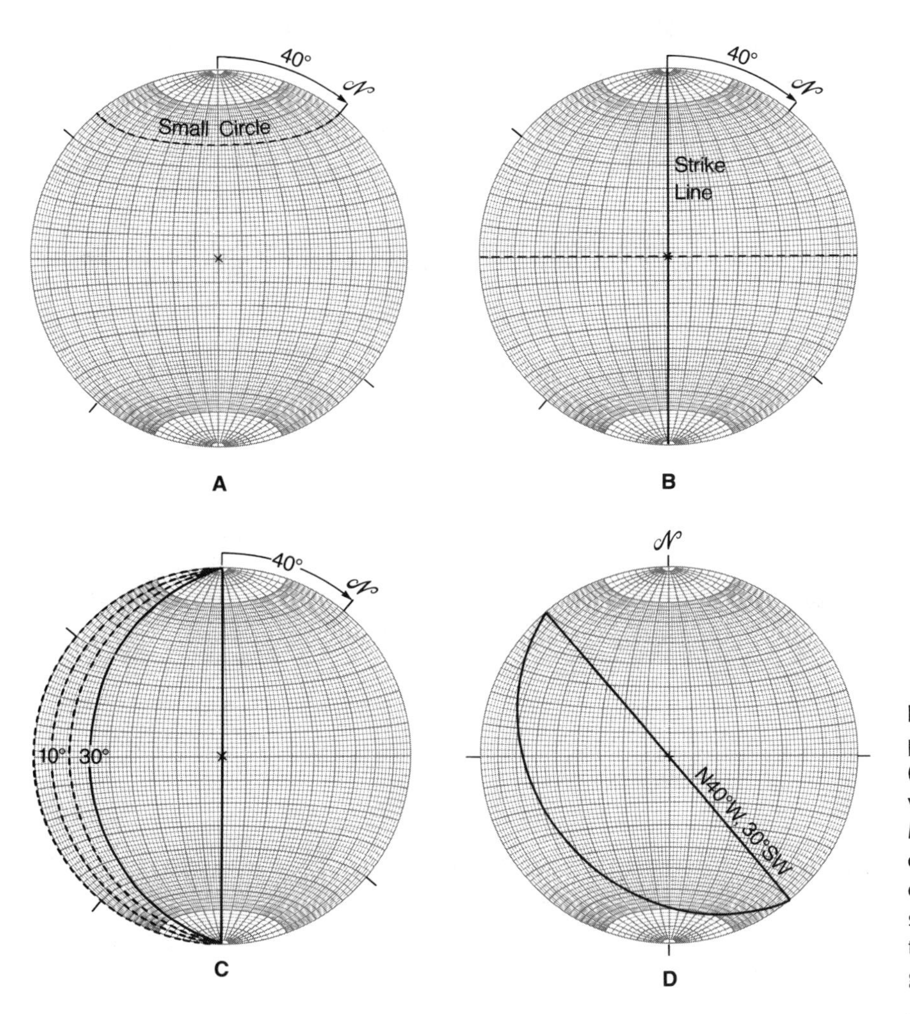

Figure 3.40 Stereographic projection of a plane that strikes N40°W and dips 30°SW. (*A*) Measure the azimuth of strike, 40° west of north. (*B*) Draw the strike line. (*C*) Measure the 30° dip inclination along the east–west line of the net, moving from the edge of the net toward its center. (*D*) The stereographic representation of a plane is the combination of a strike line and a great circle.

numbers of planes must be plotted stereographically on a single projection, it is far more practical to plot each plane as a pole, not as the combination of a strike line and a great circle. The resulting diagram is cleaner and more conducive to interpretation of preferred orientations.

The procedure for plotting poles to planes stereographically is reasonably straightforward. Consider the pole to a plane that strikes N80°E and dips 20°SE. Figure 3.41*A* shows the orientation of the plane plotted stereographically as the combination of a strike line and a great circle. By definition, the pole to this plane is oriented 90° to the plane measured in a vertical plane that contains the line of true dip of the plane in question. To plot this pole stereographically, rotate the strike line of the plane so that it coincides with the north–south line of the stereographic net (Figure 3.41*B*). In this orientation, the line of true dip in the plane lies on the east–west line of the net. From the point representing the line of true dip of the plane, measure 90° along the east–west line (Figure 3.41*C*). The position of the pole is 20° beyond the center point of the net and is plotted as point *P*. Rotating the north index back to home position results in placement of the pole in its proper orientation (Figure 3.41*D*).

In practice, plotting a pole to a plane need not include plotting the plane as a great circle. Rather, the strike line of the plane is rotated so that it coincides with the north–south line of the stereographic net (as in Figure 3.41C); then the dip of the plane is measured along the east–west line outward from the center of the net, into the quadrant opposite the dip direction of the plane.

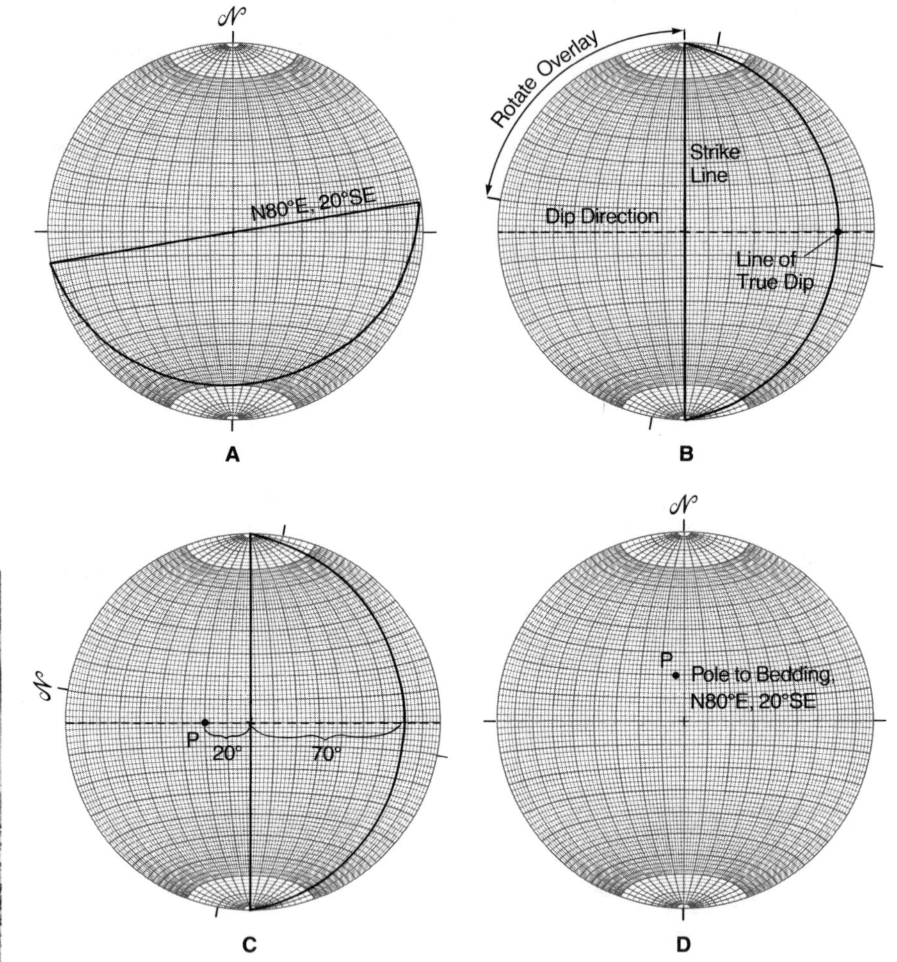

Figure 3.41 Stereographic projection of the pole to a plane. (*A*) Stereographic representation of a plane that strikes N80°E and dips 20°SE. (*B*) Rotate strike line into parallelism with the north–south line of the net. (*C*) Identify the pole to the plane by measuring 90° along the east–west line of the net from the point that represents the inclination of true dip. (*D*) Final portrayal of the stereographic representation of the plane as a pole.

Figure 3.42 Striations on a fault surface, a structural geologic example of lines in a plane. (Photograph by G. H. Davis.)

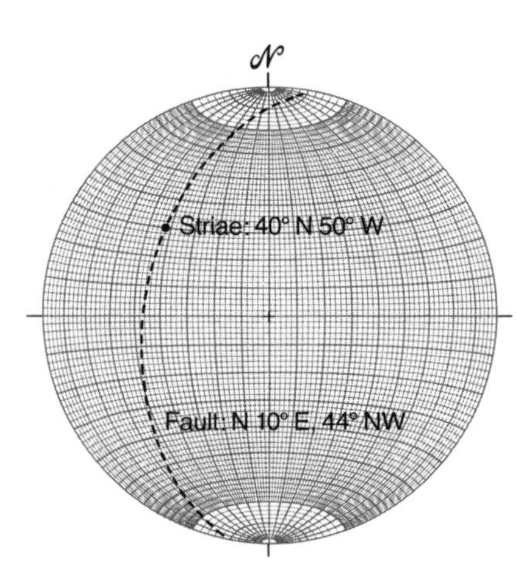

Figure 3.43 If a line lies in a plane, the stereographic projection of the line as a point must fall on the great circle that stereographically represents the orientation of the plane.

PLOTTING THE ORIENTATION OF A LINE IN A PLANE

Many structural geologic relationships involve the presence of a line in a plane, like striations on a fault surface (Figure 3.42). Stereographically, the point representing the trend and plunge of a line in a plane must lie on the great circle representing the strike and dip of the plane. Consider the geometry of a fault that strikes N10°E and dips 44°NW, containing striations (striae) that plunge 40° N50°W. If the stereographic orientations of these two elements are plotted independently, it is found that the trend and plunge of the striations are represented by a point that falls on the great circle corresponding to the strike and dip of the fault (Figure 3.43).

Another way to describe the orientation of a line in a plane is to measure the **rake** (or **pitch**) of the line. Rake is the angle between a line and the strike line of the plane in which it is found (Figure 3.44). If the orientation of a line, as measured in the field, is expressed in the field notebook as a rake angle within a plane of known orientation, its stereographic portrayal can be plotted readily. Consider, for example, a line that rakes 62°SE in a plane whose orientation is N40°W, 45°SW. The orientation of the plane is plotted stereographically as a great circle (Figure 3.45*A*). Since the line lies in this plane, the point representing the trend and plunge of the line must lie on the great circle representing the plane. The rake angle of 65°SE is measured from the S40°E end of the strike line. To show the line stereographically, simply measure 65° from the SE quadrant of the tracing paper along the great

Figure 3.44 Barnyard conversation about tools and recreation. (Artwork by D. A. Fischer.)

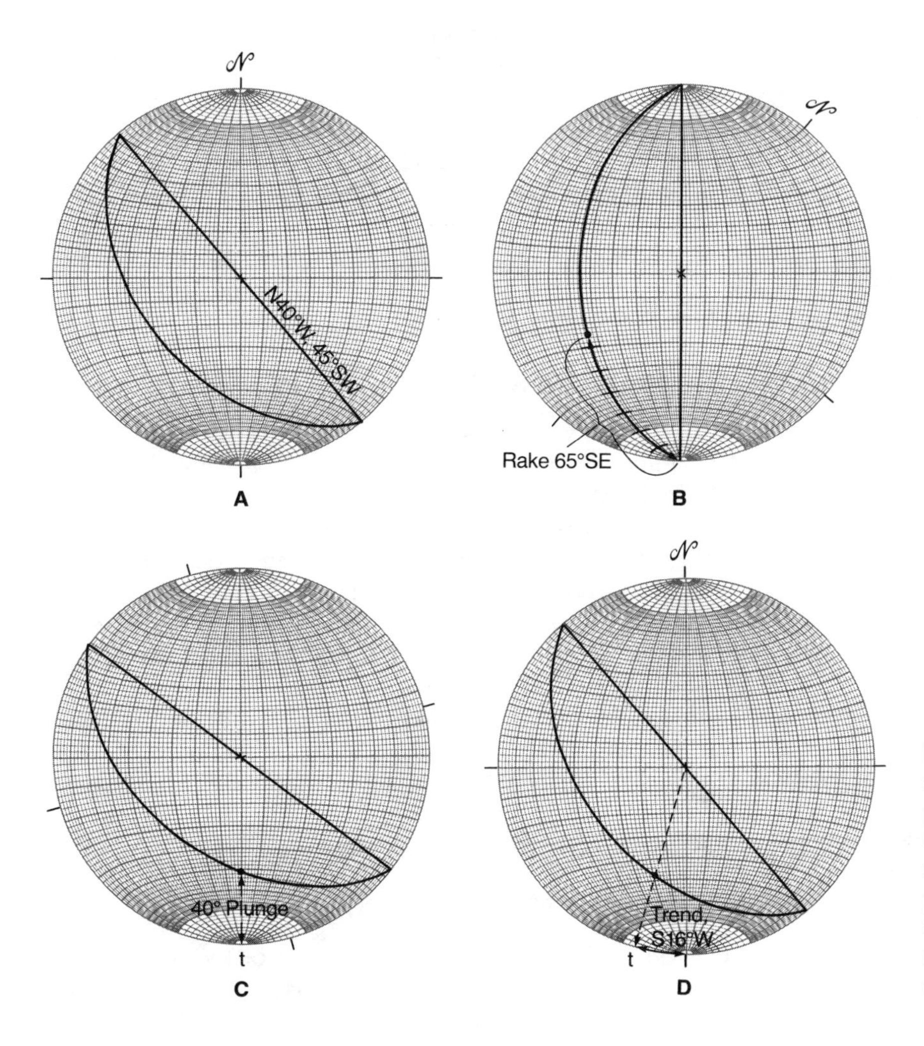

Figure 3.45 Stereographic meaning of rake. (*A*) Stereographic representation of a plane that strikes N40°W and dips 45°SW. (*B*) Counting the 65° rake angle that describes the orientation of a line in the plane. (*C*) Measurement of the plunge of the line contained in the dipping plane. (*D*) Measurement of the trend of the line contained in the dipping plane.

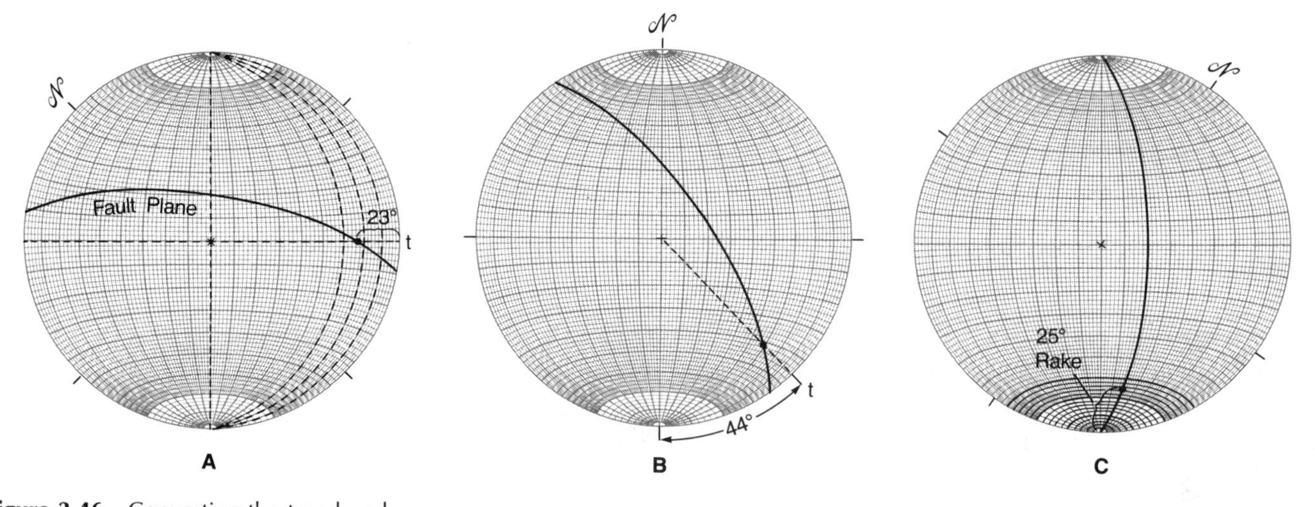

Figure 3.46 Converting the trend and plunge of a line to rake in a plane. (*A*) Stereographic representation of a fault surface and the striations on the fault. The great circle portraying the orientation of the fault plane is oriented such that the plunge of the striations can be measured directly. (*B*) With the north-index mark in home position, the trend of striations can be measured as S44°E. (*C*) The rake of striations (25°S) can be measured directly when the great circle that represents the fault orientation is rotated to coincide with an appropriately oriented great circle on the stereographic net.

circle representing the plane in which the line is found (Figure 3.45*B*). Small circle/great circle intercepts, spaced at 10° and 2° intervals, are the basis for measuring. The trend and plunge values for this line are interpreted by rotating the stereographically plotted point to the east–west or north–south line of the projection, and measuring the inclination of the point from the horizontal, in this case 40° (Figure 3.45*C*). While the tracing paper is in the same position, the trend index *t* of the point can be marked on the perimeter of the net. Rotating the overlay to home position, the trend can be interpreted, in this case S16°W (Figure 3.45*D*).

Converting trend-and-plunge to rake is a reasonably smooth operation as well. The fault surface shown in Figure 3.46*A* is positioned stereographically in such a way to emphasize the 23° **plunge** value of striations. The trend of the striations is S44°E (Figure 3.46*B*). **Rake** of the striations is found by first aligning the great circle representing the fault with the corresponding great circle on the stereographic net (Figure 3.46*C*). The rake of 25°SE is measured along the great circle, inward from the perimeter to the point representing the trend and plunge of striations.

MEASURING THE ANGLE BETWEEN TWO LINES IN SPACE

If someone walked up to you on the street and asked you to compute the angle between two lines in space, one plunging 16° N42°E and the other plunging 80° S16°E, how would you do it? The stereographic solution is based on knowledge that two lines define a plane, and that the angle between the two lines is measured in the plane common to both. The orientations of the lines are given as 16°/42°E (line 1) and 80°/164° (line 2). We can plot these two lines stereographically, as shown in Figure 3.47*A*. The plane defined by these two lines is found by rotating the tracing paper overlay until the stereographic points representing the lines lie on a common great circle (Figure 3.47*B*). (The plane strikes 40° and dips 81°SE.) The angle between the lines is measured by counting 2° small-circle intercepts along the great circle representing the common plane (Figure 3.47*C*). The acute angle separating these points is 80°; the obtuse angle is 100°. Always carry a stereonet in the street. You never know who you might meet.

MEASURING THE ANGLE BETWEEN TWO PLANES IN SPACE

The angle between any two planes, like two faults or two joints, is the same as the angle between the poles to planes. Consequently, if the orien-

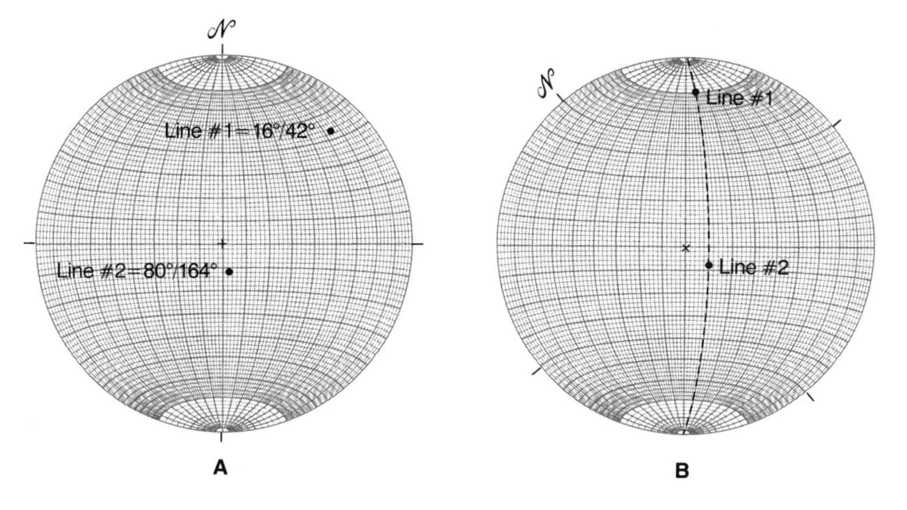

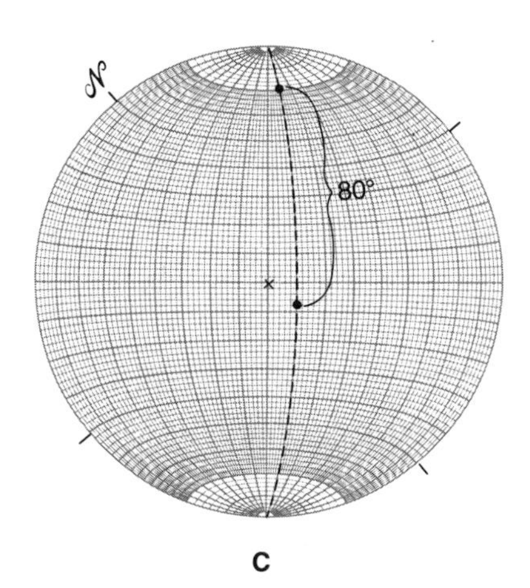

Figure 3.47 Stereographic determination of the angle between two lines. (*A*) Plot the orientation of the lines stereographically as points. (*B*) Fit the lines to a common great circle (i.e., to a common plane). (*C*) Measure the acute angle between the two lines, by counting along the common great circle that connects the points.

tations of two planes are plotted as poles, measuring the angle between the planes reduces the problem to measuring the angle between two lines. And we have seen how this is done.

Figure 3.48*A* shows two planes, plane 1 striking 305° and dipping 26°SW, plane 2 striking 10° and dipping 41°NW. The poles to these planes are shown stereographically. To measure the angle between the planes, simply align the two poles on the same great circle (Figure 3.48*B*) and measure the acute angle between the poles by counting 2° small-circle intercepts along the great circle. For this example, the angle is 36°.

DETERMINING THE ORIENTATION OF THE INTERSECTION OF TWO PLANES

The payoff for learning the principles of stereographic projection is derived from the ease with which certain geometric problems can be solved. One of the best examples of the effectiveness of stereographic projection is determining the trend and plunge of the intersections of two planes. Consider two

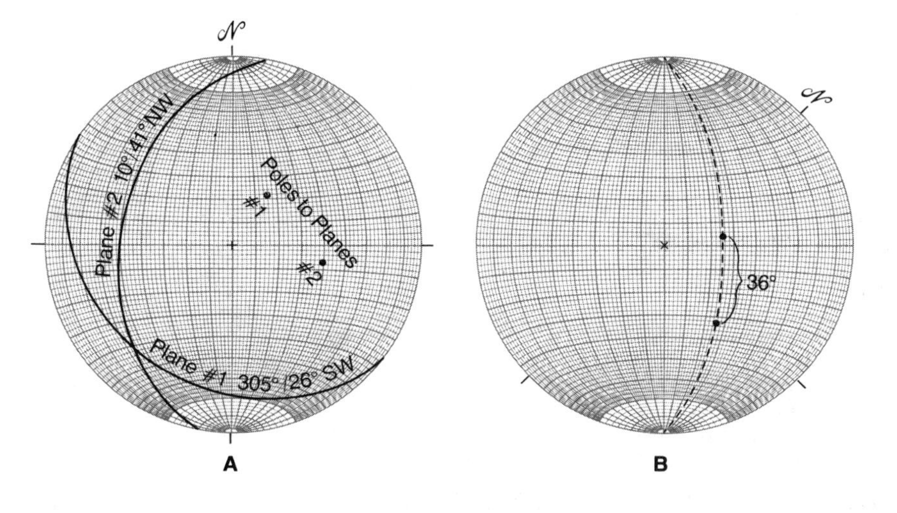

Figure 3.48 Stereographic determination of the angle between two planes. (*A*) Plot the orientations of the planes both as great circles and as poles. (*B*) Fit the poles of the planes to a common great circle, then measure the acute angle between the poles.

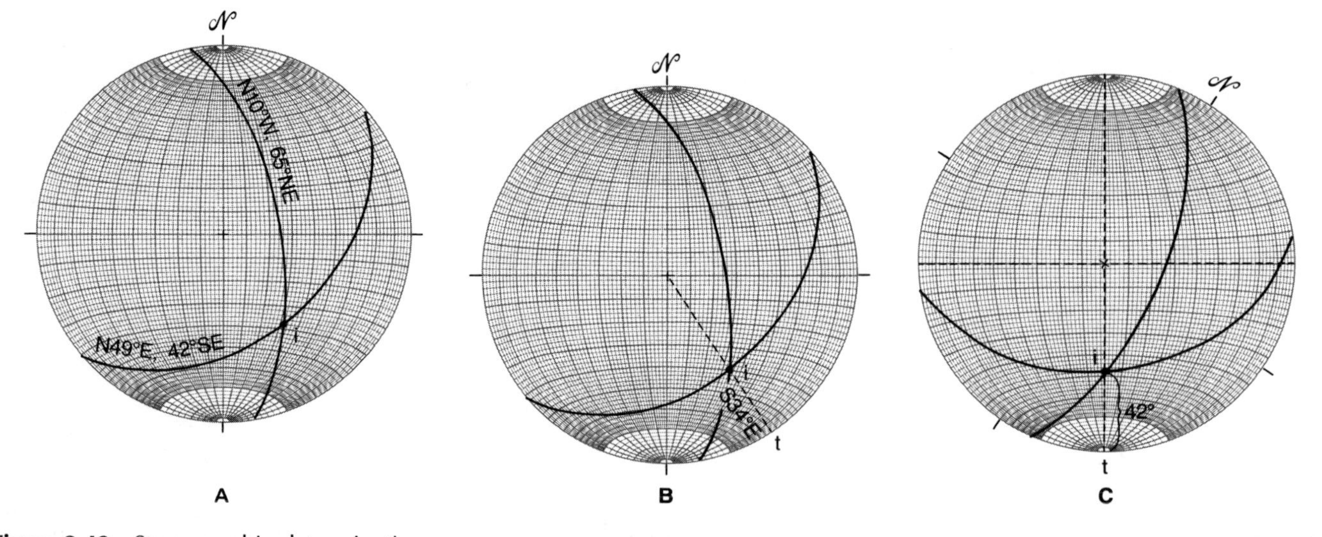

Figure 3.49 Stereographic determination of the trend and plunge of the line of intersection of two planes. (*A*) Plot the two planes stereographically as great circles. Identify the intersection (*i*) of the two great circles, recognizing that its orientation reflects the trend and plunge of the intersection of the two planes. (*B*) Interpret the trend of the line of intersection. (*C*) Measure the plunge of the line of intersection along either the east–west line or the north–south line of the net.

planes, one striking N49°E and dipping 42°SE, the other striking N10°W and dipping 65°NE (Figure 3.49*A*). The intersection of the great circles is a point *i* whose orientation is that of the line of the intersection of the two planes. The trend of the line *i* can be determined by drawing (projecting) a straight line from the center of the projection through line *i* to the perimeter, and measuring the orientation of this line with respect to north or south. For this example, the trend is S34°E (Figure 3.49*B*). The plunge of the line is found by rotating line *i* to the east–west or north–south lines and measuring its inclination from the perimeter (Figure 3.49*C*). The plunge as measured in this example is 42°.

Apparent dip problems are a special case of determining the line of intersection of two planes. Figure 3.50*A* shows bedrock in a seacoast exposure. Bedding strikes N24°E and dips 79°SE. What would be the apparent dip for these beds, as observed in a vertical cliff face that trends N40°E?

To solve, plot the orientation of the bed and the orientation of the vertical

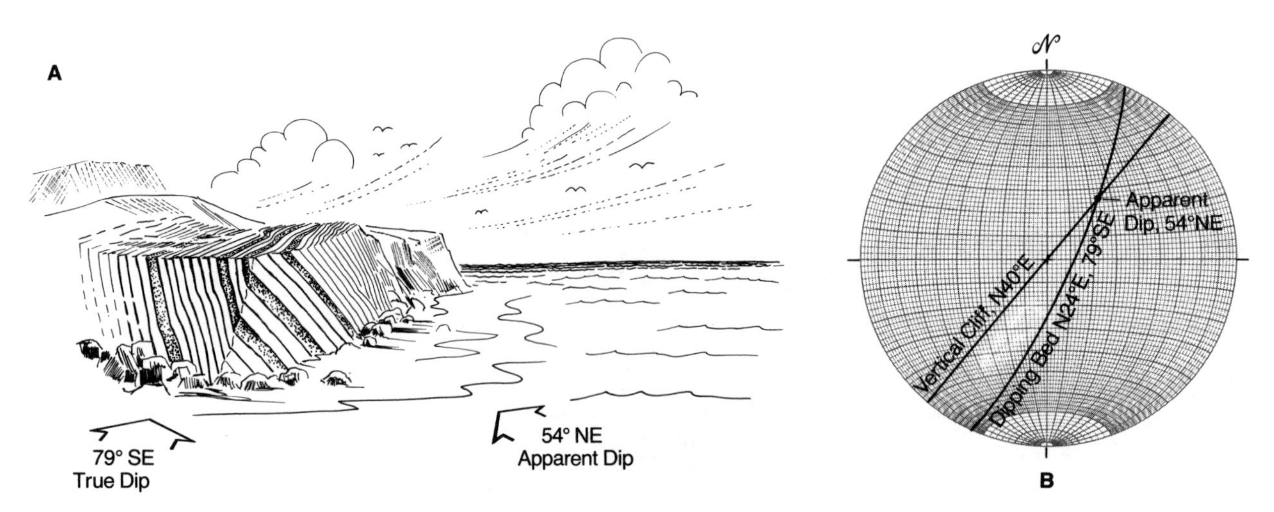

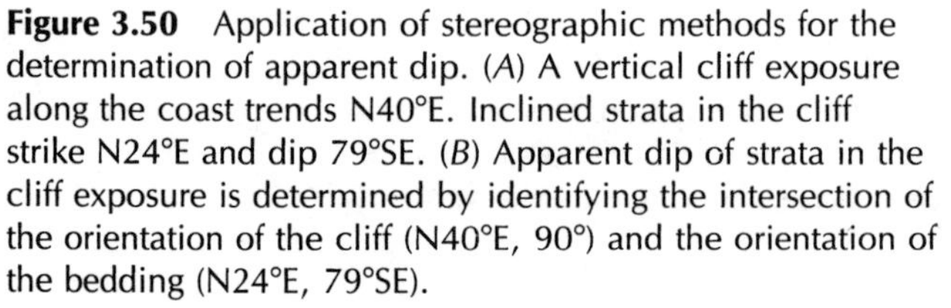

Figure 3.50 Application of stereographic methods for the determination of apparent dip. (*A*) A vertical cliff exposure along the coast trends N40°E. Inclined strata in the cliff strike N24°E and dip 79°SE. (*B*) Apparent dip of strata in the cliff exposure is determined by identifying the intersection of the orientation of the cliff (N40°E, 90°) and the orientation of the bedding (N24°E, 79°SE).

face stereographically (Figure 3.50*B*). The intersection of the two planes is a point that represents the trend and plunge of the line of intersection of the bed and the vertical cliff face. The plunge value is in fact the apparent dip of the plane, namely 54°NE. Examples like this begin to scratch the surface of the power of the three-dimensional protractor known as the stereographic net.

STEREOGRAPHIC PROJECTION AS A STATISTICAL TOOL

PROPERTIES OF WULFF AND SCHMIDT NETS

We are often required in structural analysis to determine the average orientation of a certain structural element, or to determine whether the range of orientations of a particular structure is in some way systematic. One way to identify **preferred orientations** of structures might be to plot lines or poles to planes stereographically on a Wulff net (see Figure 3.36*B*) and to evaluate the extent to which the plotted points tend to cluster or to spread in systematic ways. But were we to do this, we would find that the Wulff net, otherwise known as the equal-angle net, has an undesirable built-in bias. Two-degree areas bounded by great and small circles on a Wulff net are not of equal size. Those toward the periphery of the net are larger than those toward the center (see Figure 3.36*B*). The central part of the Wulff net takes in a greater range of orientations than the peripheral parts. Consequently, even orientations gathered from a table of random numbers and plotted stereographically would show an uneven concentration of points across the face of the circular net. The distribution of points might convey the incorrect impression that most of the lines, plotted as points, reflect relatively steeply plunging orientations. This geometric peculiarity invalidates the equal-angle net as a useful statistical device for evaluating preferred orientations. Instead, the Schmidt net, the equal-area net, is used.

The geometry of projection of the Schmidt net (see Figure 3.36*A*) is such that 2° areas bounded by great and small circles are the same size across the net (Phillips, 1971). Since 2° great circle/small circle areas are the same size across the entire face of a Schmidt net, nonrandom concentrations of stereographically plotted points reflect preferred orientations. Contouring the values of the **density distribution** of plotted points provides a measure of the degree of preferred orientation.

PLOTTING POINTS

In using the equal-area net as a statistical tool, the first step is to plot the orientations of the structural elements that were measured in the field, or in the laboratory. The plotting of orientations is a slow, tedious operation, and measurements number in the hundreds or thousands. Consequently, it is common practice to harness a computer in the plotting and processing of orientation data. In lieu of ready access to computer plotting, a plotting device of the type shown in Figure 3.51 is recommended. This device, known as a Biemesderfer counter, eliminates the need to rotate the tracing overlay time and time again during plotting. Instead, the east–west line of the equal-area net is rotated. This is achieved by calibrating the edges of the counter in the same way as that of the east–west line of the net. In plotting the orientation of a line, the counter is rotated so that one of its two edges is aligned parallel to trend. Then the plunge of the structure can be measured and plotted directly.

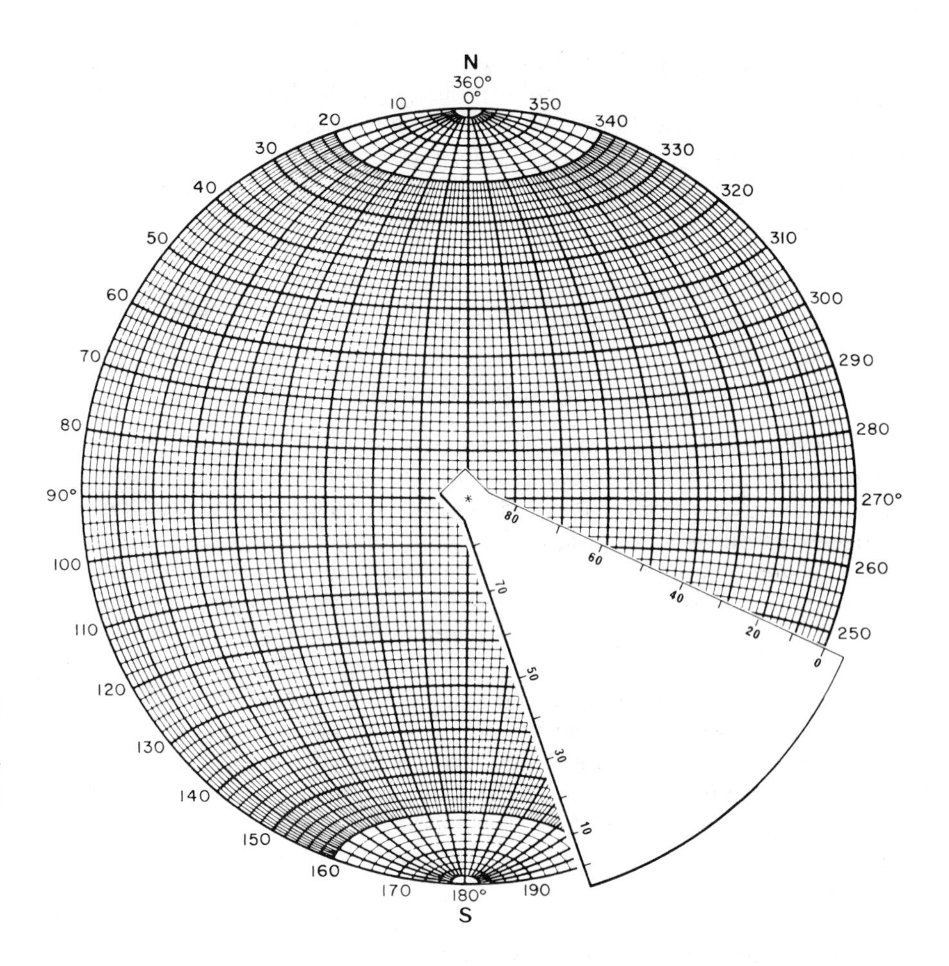

Figure 3.51 Biesmesderfer counter, used in faster plotting of points and poles. (From *Structural Geology of Folded Rocks* by E. T. H. Whitten. Originally published by Rand-McNally and Company, Skokie, Illinois, copyright ©1966. Published with permission of John Wiley & Sons, Inc., New York.)

As always, plunge is measured from the perimeter of the net inward. In plotting the orientation of a pole to a plane, the edge of the counter is rotated to an azimuth perpendicular to the strike of the plane under consideration. Moreover, the edge of the counter is brought to rest in the quadrant of the stereographic projection *opposite* the dip direction of the plane. When thus oriented, the inclination of the pole to the plane may be measured and plotted directly, counting outward from the center of the stereographic net.

DENSITY DISTRIBUTION OF ORIENTATIONS

In the stereographic projection of 65 poles to bedding in Cretaceous strata in the Mule Mountains near Bisbee, Arizona shown in Figure 3.52*A*, the general concentration of points is near the center of the projection, signifying that the beds whose orientations are plotted are rather gently dipping. But what is the specific orientation of the bedding, as expressed in strike and dip? And what is the strength of the preferred orientation? Answering these questions requires contouring the density distribution of the plotted points on the face of the stereonet.

To evaluate density distribution, the equal-area net is subdivided into a gridwork of many overlapping circular areas, each of which corresponds to 1% of the area of the stereographic projection. Density is described in terms of percentage of total data points falling within a given 1% area of the stereographic projection.

$$\text{Density (\%)} = \frac{\text{number of data points in 1\% area of net}}{\text{total data points}} (100)$$

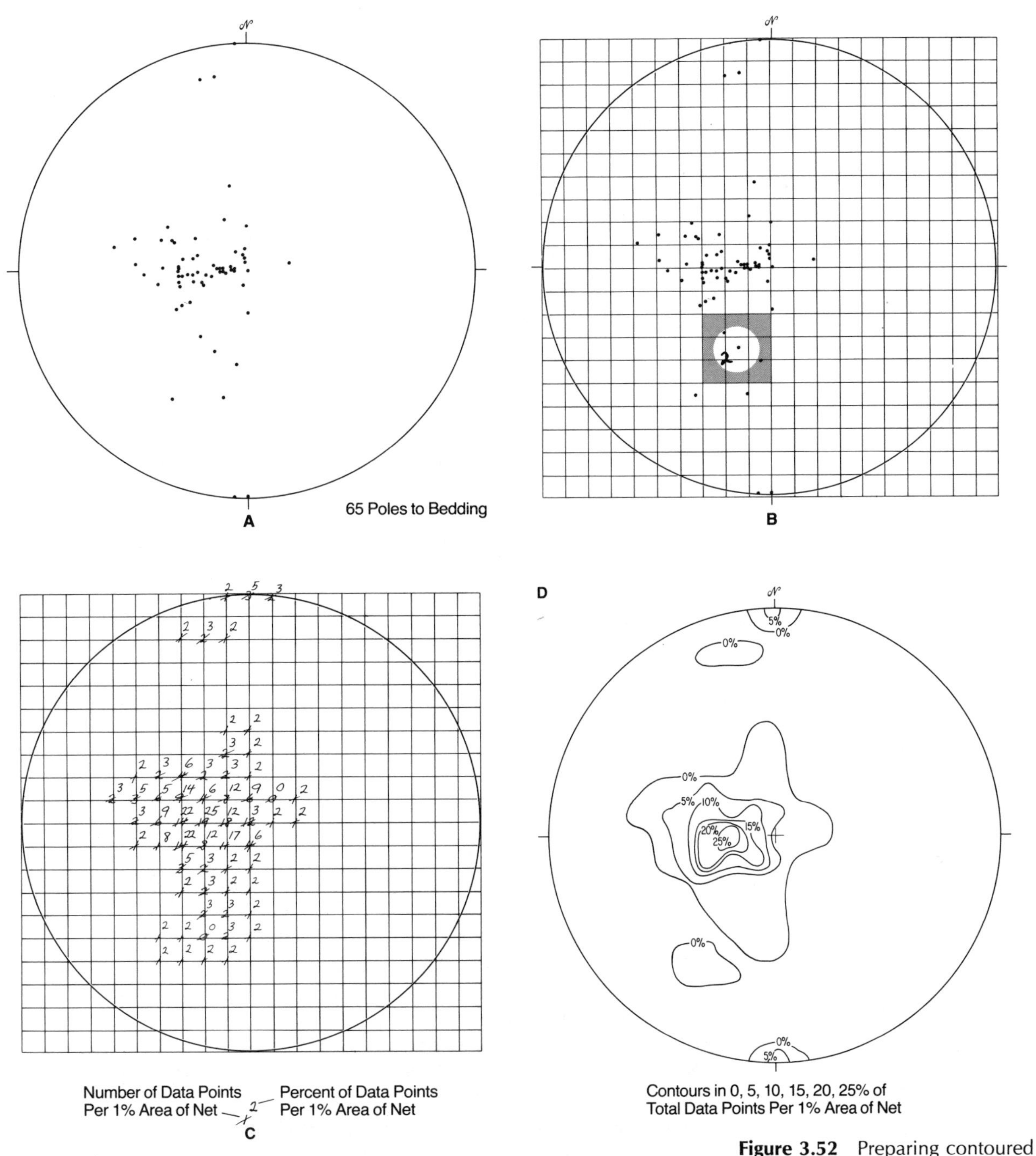

Figure 3.52 Preparing contoured pole-density diagrams. (*A*) Plot the orientation of bedding as poles. (*B*) Use a center counter and determine the number of data points that lie within each of the overlapping 1% counting areas of the net. (*C*) Convert the number of data points within each counting area to the percentage of total data points within each counting area. (*D*) Contour the percentage values. (*E*) A finished version of the contoured pole-density diagram.

POLE-DENSITY DISTRIBUTION

To subdivide a 20-cm-diameter stereogram into overlapping 1% circular areas, a square grid is constructed such that the spacing of grid intersections is 1 cm. The square gridwork is overlain by the tracing paper on which the data points were originally stereographically plotted (Figure 3.52*B*). The grid intersection points are used as control points for systematically moving a **center counter** whose area is 1% that of the stereogram (Whitten, 1966, pp. 20-26). In this manner the **number** of data points that lie within each of the overlapping 1% areas is counted. The counts are posted on yet another overlay (Figure 3.52C).

A calculator is used to convert the numbers of data points in each of the 1% areas to percentage of total data points (Figure 3.52C). It is the **percentage values** of density distribution that are contoured (Figure 3.52*D*). Since the 1% areas are overlapping, and because most data points are counted more than once, the density distribution values represent a smoothed portrayal of the raw orientations.

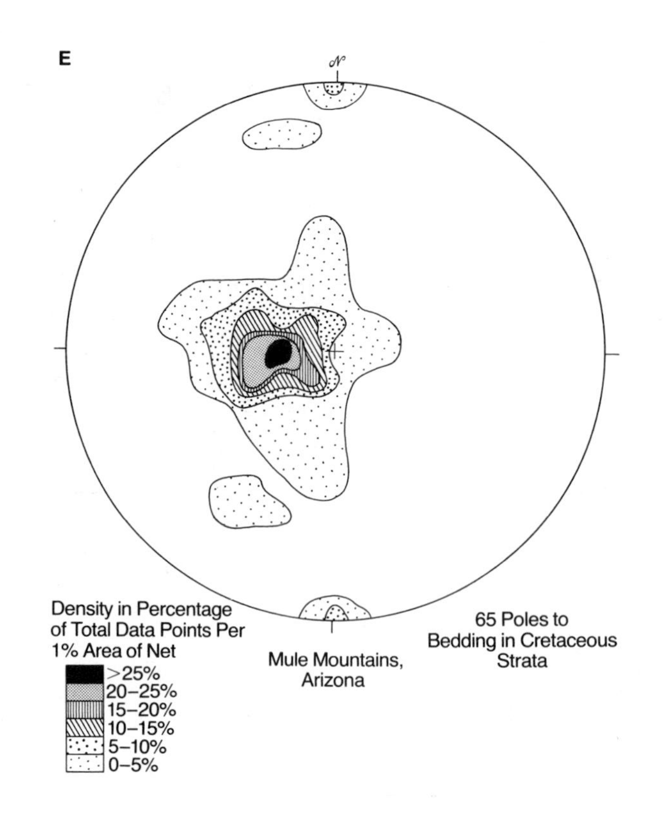

Figure 3.52 (*cont.*)

Ideally, the finished contour diagram (Figure 3.52*E*) should be marked by some constant contour interval, with the number of contour line values not exceeding five or six. To emphasize the density distribution, it is common practice to shade the diagrams in such a way that the zones of highest density are darkest and most pronounced. For the example of bedding orientations in the Mule Mountains, the completed diagram discloses a bulls-eye, **unimodal distribution** of points. The eye or center of the contoured pattern corresponds to the preferred orientation of the bedding.

The counting and contouring process is relatively straightforward, except when considering density distribution values close to or on the perimeter of the projection. Consider the point plot of 85 measurements of mineral lineation in gneiss in the Rincon Mountains of southern Arizona (Figure 3.53*A*). The lineations are low plunging, as can be judged by their closeness to the perimeter of the projection. The number of data points falling within each 1% area in the interior of the stereographic projection can be counted using a center counter. But in this example individual 1% area counting circles sprawl beyond the perimeter of the projection, and thus the points near or at the periphery occupy areas less than 1% of the area of the projection. To count these points, a **peripheral counter** is used (Figure 3.53*B*). The orientations of structural elements represented by points lying in peripheral counting areas correspond closely to those that fall within the peripheral counting area diametrically (180°) opposite. In fact, each point on the perimeter of the net corresponds exactly in orientation to the point that lies on the perimeter 180° away. Recognizing this equivalence in orientation, and given that each peripheral counting area is less than 1% of the area of the net, an extra step is required to convert from "number of data points" to "percentage of data points." It is necessary to add the number of data points that fall within each pair of supplementary partial circles on the perimeter of the net and to assign this sum to *each* of the partial circles. These values are converted to per-

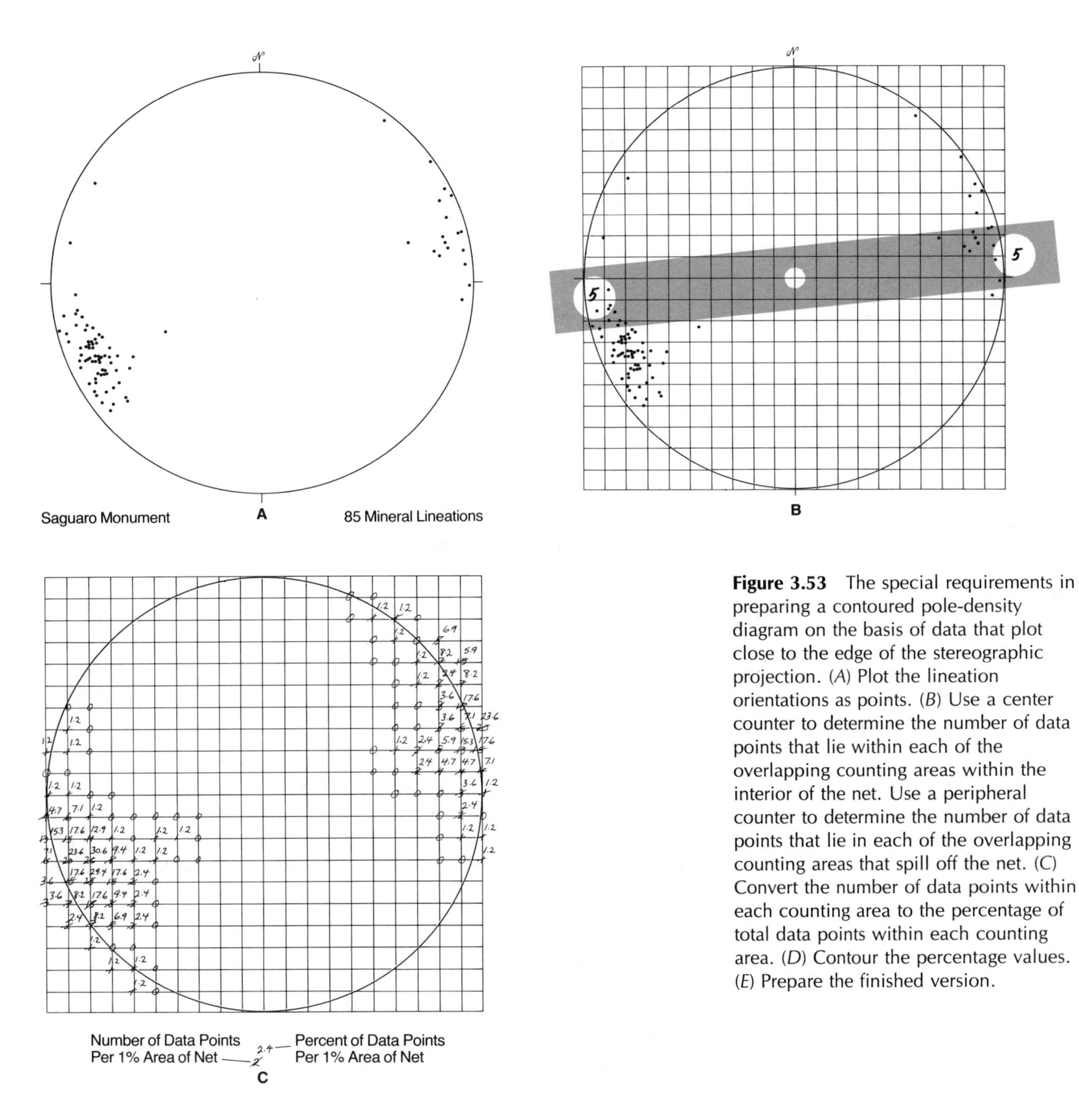

Figure 3.53 The special requirements in preparing a contoured pole-density diagram on the basis of data that plot close to the edge of the stereographic projection. (*A*) Plot the lineation orientations as points. (*B*) Use a center counter to determine the number of data points that lie within each of the overlapping counting areas within the interior of the net. Use a peripheral counter to determine the number of data points that lie in each of the overlapping counting areas that spill off the net. (*C*) Convert the number of data points within each counting area to the percentage of total data points within each counting area. (*D*) Contour the percentage values. (*E*) Prepare the finished version.

centage, along with the values assigned to center counting areas in the interior of the net (Figure 3.53*C*). When the values are contoured (Figure 3.53*D*), special care must be made to assure that each point of intersection of a contour line with the perimeter of the net is matched by a corresponding intersection point on the perimeter 180° away. The finished product is shown in Figure 3.53*E*.

Contour diagrams prepared in this way are called **pole-density diagrams**, where "pole" refers loosely to a stereographically plotted point, regardless of whether it represents the trend and plunge of a linear element or the trend and plunge of a pole to a plane. Pole-density diagrams describe the range in distribution and the preferred orientation(s), if any, of the structures whose orientations have been measured. They are useful in summarizing large

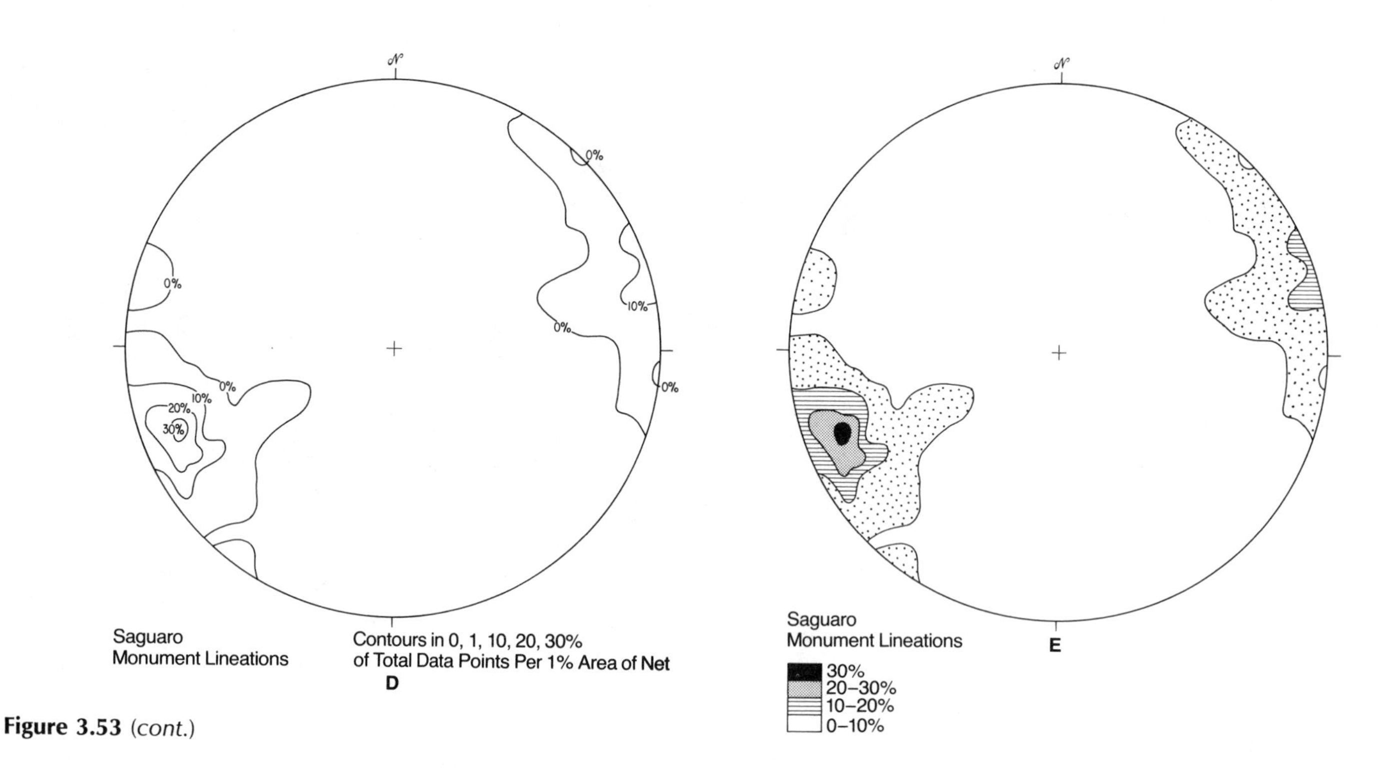

Figure 3.53 *(cont.)*

quantities of geometric data. Each diagram should be clearly labeled according to area of study (within which the data were collected), structural element, number of measurements, and contour line values, (see Figures 3.52*E* and 3.53*E*).

Commonly an array of such diagrams is necessary to describe the full orientation range for a system of structures. Where many diagrams are presented, individual diagrams should be prepared so that contour lines are of a common interval, in percentage. When there are fewer than 50 orientation measurements, pole-density diagrams are not particularly meaningful statistically and thus point diagrams suffice.

Pole-density diagrams should be viewed critically in regard to number of data points, values of pole density, and patterns. Statistical tests are available to evaluate the significance of pole-density values in light of the number of data points (Kamb, 1959). Many types of pattern are possible. Chapters that follow demonstrate that the nature and the symmetry of pole-density diagrams have important implications for the geometry and kinematics of structural systems.

SUMMARY

The breadth of descriptive analysis goes far beyond the bounds of what has just been presented. Methods and concepts introduced in this chapter are applied throughout the book, and discussions of specific structures and structural systems will bring us in contact with additional forms of descriptive analysis. There is no question that success in structural analysis depends on thoroughness and on the development of creativity and versatility in carrying out descriptive analysis. By the end of our journey into rocks and regions, I hope that this message is clear.

chapter 4 KINEMATIC ANALYSIS

STRATEGY

Kinematic analysis is the reconstruction of movements that take place during the formation and deformation of rocks. Of main interest is the analysis of **secondary structures** formed in solid rock as a result of crustal deformation. But we also apply kinematic analysis to the interpretation of movements that accompany the primary formation of rocks. Clues to formational movements are preserved in **primary structures** that arise in sediments before the sediments are consolidated into sedimentary rocks, or in extrusions or intrusions of magma before the magma becomes frozen into volcanic or intrusive igneous rocks. Kinematic movements, whether formational or deformational, are evaluated at all scales of observations.

There are two different strategies for kinematic analysis of deformational structures, and the one that is used depends on whether the rocks under consideration have behaved as rigid or as nonrigid bodies during the deformation. The distinction between rigid and nonrigid rests on whether the original configuration of points in the body of rock has been preserved or modified.

During **rigid body deformation**, rock is translated and/or rotated in such a way that original size and shape are preserved. A schematic example of rigid body translation without rotation is portrayed in Figure 4.1*A*. Block *abde* has shifted its position in such a way that its original size and shape are maintained. There is no change in the configuration of points within the block.

A geologic example of rigid body deformation is pictured in Figure 4.2, a photomicrograph of a shattered phenocryst of quartz in a fine-grained igneous rock from the Mariana Islands in the western Pacific. The brittle, rigid mass of quartz has responded to deformation like ice breaking up on a frozen sea. There is no alteration of the configuration of points within each of the offset pieces. Space between the quartz fragments (white) is occupied by fine-grained groundmass material (black), which flowed into place.

During **nonrigid body deformation**, rock undergoes a change in shape (**distortion**) and/or in size (**dilation**). Both distortion and dilation result from a change in configuration of points within a body. A single body, during a single deformational event, may experience both distortion and dilation.

Pure distortion is a change in shape without a change in size. As shown in Figure 4.1*B*, the change in shape of a body from a square to a rhomb is made possible by systematic changes in spacing between points in the body. The distance between points *a* and *e* increases from 10 units to 23 units. Angular relations between alignments of points change as well. For example, angle *aed* is reduced from 90° to 26°.

When nonrigid body deformation results in the systematic distortion of what we normally regard as solid rock, the results can be spectacular. An

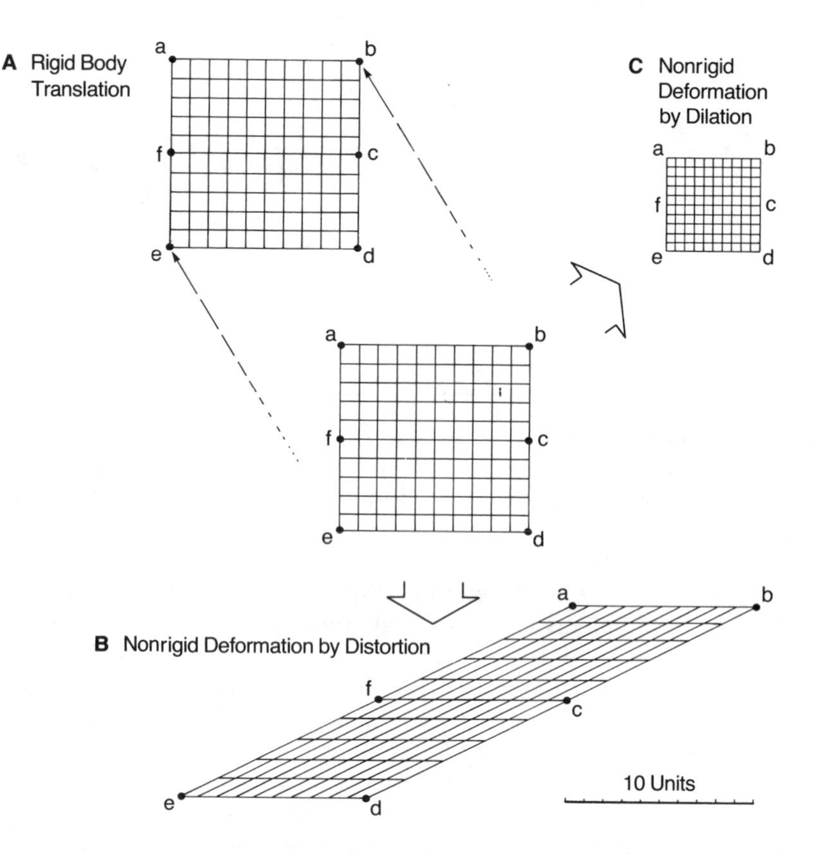

Figure 4.1 Originally undeformed body in center of diagram (i.e., square *abde*) is deformed by (*A*) rigid body deformation, (*B*) nonrigid body distortion, and (*C*) nonrigid body dilation.

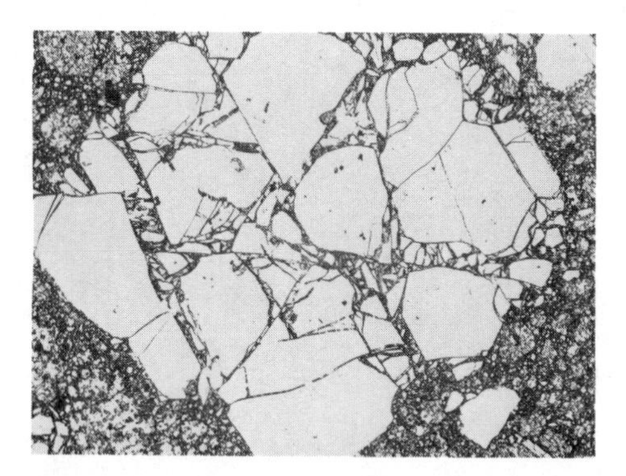

Figure 4.2 Photomicrograph of shattered quartz crystal. The fragmented crystal displays the results of rigid body deformation. (Photograph by R. G. Schmidt. Courtesy of United States Geological Survey.)

Figure 4.3 Distorted trilobite in Cambrian shale, Maentwrog, Wales. The width of the fossil is 3 cm. (From *The Minor Structures of Deformed Rocks: A Photographic Atlas* by L. E. Weiss. Published with permission of Springer-Verlag, New York, copyright ©1972.)

unusually fine example is pictured in Figure 4.3: this distorted trilobite bears little resemblance to its former self. Better than words, the shape of the fossil communicates the full extent of the nonrigid body distortion that the trilobite and the rocks in which it was found were forced to endure.

Pure dilation is a change in size but without a change in shape. Figure 4.1C pictures a negative dilation (decrease in size) wherein the spacing of points within the original body is cut in half. Significant dilation accompanies nonrigid structural processes, like the compaction of sediments, the cooling of lava, the emplacement of igneous intrusions, and the folding of rocks. Small dilation changes in nature are probably common as well, but they are hard to detect because of the difficulty of precisely reconstructing the original size of deformed geologic bodies.

In summary, kinematic analysis describes translation and rotation of bodies of rocks that have moved in rigid fashion during deformation, and describes changes in size and shape of bodies of rock that have behaved nonrigidly. In practice, these two strategies of analysis are integrated, for nonrigid and rigid body deformation commonly operate together. Moreover, rigid body translation and rotation observed in a small area of view may contribute to dilation and/or distortion of the large rock body of which the small area is a part (remember the gate; see Figure 2.5).

RIGID BODY MOVEMENTS

TRANSLATION

GENERAL CONCEPT. During pure translation, a body of rock is displaced in such a way that all points within the body move along parallel paths. The sliding of a hockey puck across the ice is a good example of pure translation, provided that the puck does not spin. If the puck is perfectly rigid, all points within it move exactly in the same direction and cover the same distance. The points within the puck retain their original internal configuration.

Rigid bodies translate past one another along **discontinuities**, surfaces where the integrity of the rock is interrupted and along which movement is possible. Examples of geologic discontinuities include faults, joints, and bedding planes. The geologic record is full of examples of the translation of rigid rock bodies along fault discontinuities, even small fault discontinuities like those shown in Figure 4.4*A*. Veins occupy fractures, the fractures having served as discontinuities that permitted dilational movement to take place between adjacent rigid blocks (Figure 4.4*B*). Bedding planes typically serve as discontinuities that accommodate translations that contribute to the process of folding (Figure 4.4*C*).

Translation can be considered at the world scale as well. The concept of plate tectonics (Chapter 6) is a view of the entire outer shell of the Earth composed of an array of broad, thin, rigid plates of crust and uppermost mantle (**lithosphere**). The rigid plates translate (and rotate) with respect to one another, moving on hot, ductile, nonrigid mantle material (**asthenosphere**). The boundary between each rigid plate and the nonrigid mantle beneath constitutes a major tectonic discontinuity. The plate motions are described mathematically as rigid body translations and rotations.

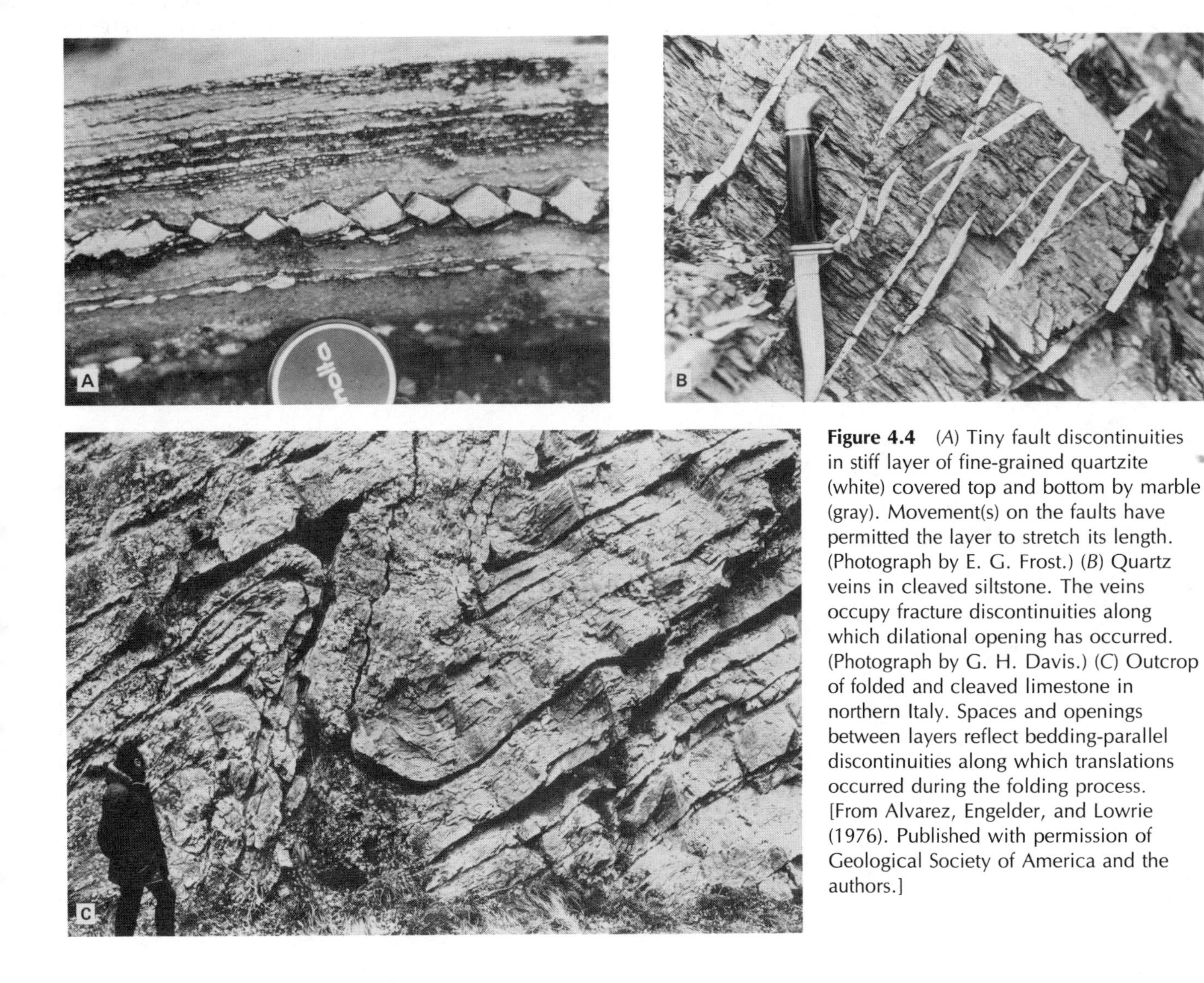

Figure 4.4 (*A*) Tiny fault discontinuities in stiff layer of fine-grained quartzite (white) covered top and bottom by marble (gray). Movement(s) on the faults have permitted the layer to stretch its length. (Photograph by E. G. Frost.) (*B*) Quartz veins in cleaved siltstone. The veins occupy fracture discontinuities along which dilational opening has occurred. (Photograph by G. H. Davis.) (*C*) Outcrop of folded and cleaved limestone in northern Italy. Spaces and openings between layers reflect bedding-parallel discontinuities along which translations occurred during the folding process. [From Alvarez, Engelder, and Lowrie (1976). Published with permission of Geological Society of America and the authors.]

DISPLACEMENT VECTOR. Rigid body translations can be expressed conveniently in terms of **displacement vectors**. Displacement vectors describe translation in terms of three parameters (Ramsay, 1969): distance of transport, direction of transport, and sense of transport. **Distance of transport** is the magnitude of transport (e.g., 5 m). **Direction of transport** is the trend and plunge of the line of movement (e.g., horizontal, northeast/southwest). **Sense of transport** is the sense of movement (e.g., *from* southwest *to* northeast). These three parameters are distinguished in a photograph and in an interpretive diagram of one of the sliding stones on Racetrack Playa in California (Sharp and Carey, 1976). The stones translate across slippery wet mud surfaces by the force of the wind (Figure 4.5*A*). Trails left by the stones allow displacement vectors to be constructed. The displacement vector constructed for the stone (Figure 4.5*B*) describes the net translation of a stone that slid smoothly along most of its course, but then encountered a small-stone obstacle that caused it to flip over several times and come to rest.

I find it useful to illustrate the concept of displacement vector with a nongeological example as well. The bus route I take from my home to the east entrance of the university is shown in Figure 4.6. Although the direction of transport of the bus changes from place to place, no transfer to another bus is required. The displacement vector describing my translation to work yields a direction of transport of N72°E/S72°W, a S72°W sense of transport, and a distance of transport of 7.6 mi (12.2 km). The displacement vector expresses the location of my destination with respect to my original starting point. (Note: If collision with another vehicle produces nonrigid body deformation

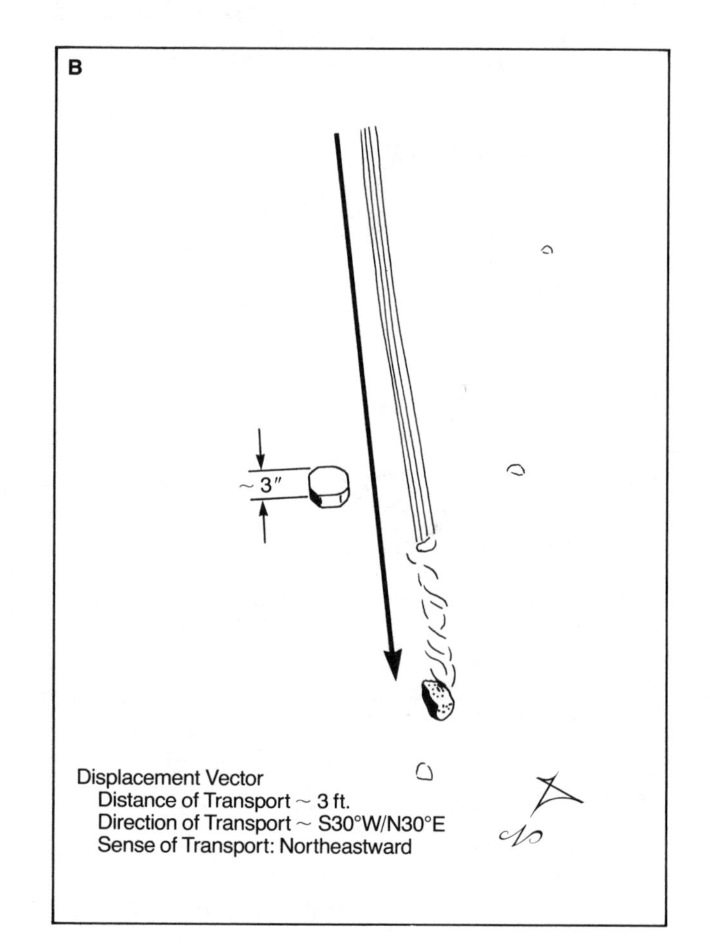

Figure 4.5 (*A*) Sliding stone on Racetrack Playa, California. [From Sharp and Carey (1976). Published with permission of Geological Society of America and the authors.] (*B*) Kinematic description of the sliding stone.

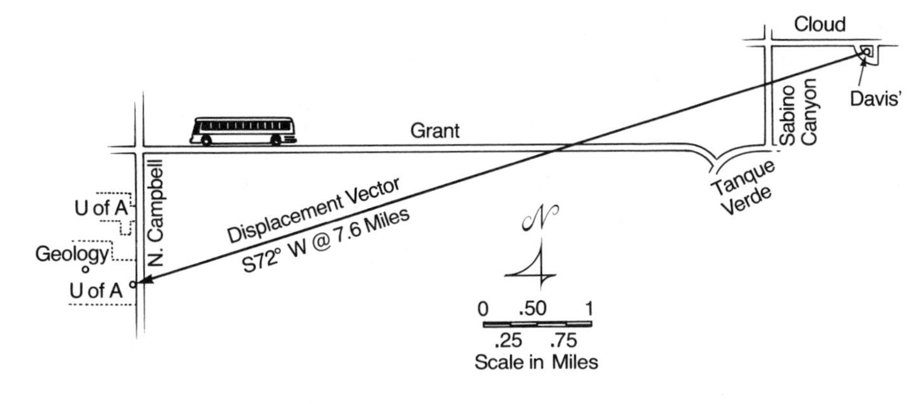

Figure 4.6 Displacement vector describing Davis's ride (translation) to the university by bus.

of the bus, resulting in distortion and a change in spacing of the seats on the bus due to overall shortening, a minor correction would have to be made).

Actually, my bus ride displacement vector of 7.6 mi/S72°W is a two-dimensional approximation of the true, three-dimensional solution. The true displacement vector must take into account the difference in elevation between where I board and where I get off, a difference in elevation of about 400 ft (122 m). The true displacement vector is oriented in the same direction as that of the two-dimensional solution, but its plunge is about 0.5° southwest, not horizontal. The sense of transport remains S72°W, but the distance of transport is greater by 80 ft (24 m). In this example, the difference between the two- and three-dimensional solutions is negligible, but in most geological problems the third dimension cannot be ignored.

Although I know where I get on the bus each morning, I seldom know the exact departure points of geological bodies that have been translated. For example, India is inferred to have been positioned in deep southerly latitudes during the early Cenozoic era, and later the plate in which India is embedded was translated far to the north (Figure 4.7). The values describing the displacement vector for India's ride northward depend crucially on the accuracy of early Cenozoic reconstructions of plates in the southern hemisphere. Assuming that the reconstruction shown in Figure 4.7 is reasonably accurate, the direction of transport for India is N12°E/S12°W, the sense of transport is NNE, and the distance of transport is 7000 km. This is an approximation of the true displacement vector, the exact values of which may never be known precisely.

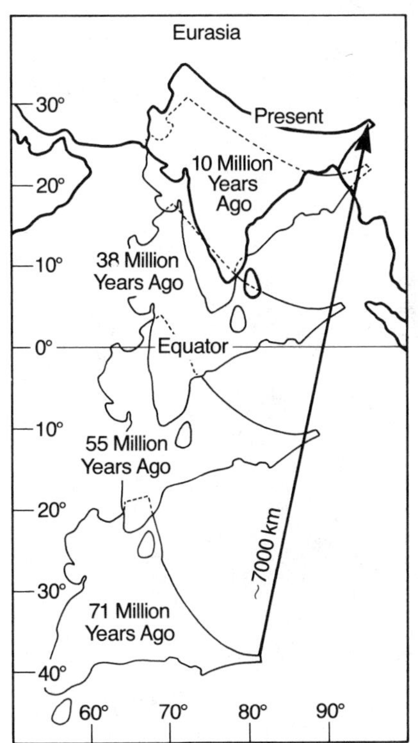

Figure 4.7 Reconstruction of the northward translation of India during the Cenozoic era. Calculation of the displacement vector depends upon the interpretation of the starting position of India. (From Molnar and Tapponnier, *Science*, v. 189, p. 419–425, copyright ©1977 by American Association for the Advancement of Science.)

SLIP ON FAULTS. The concept of the displacement vector as a description of translation can be applied to fault analysis. To construct displacement vectors for faulting, it is necessary to identify two reference points, one on each side of a given fault, that shared a common location in the body before faulting. Translation is described in terms of **slip**, the actual relative displacement between two points that before faulting occupied the same location.

Figure 4.8 shows a rigid block cut by a vertical fault that strikes N20°E. Translation on the fault is strictly horizontal and left-handed. **Left-handed** means that a faulted marker, when followed to the trace of the fault, is offset to the left. Since points *A* and *B* initially occupied the same location, they provide a means to evaluate slip along the fault. The direction of transport (direction of slip) of *A* relative to *B* is S20°W; the distance of transport (net slip) is 3 cm. A wonderful real-life example of small left-slip movements is

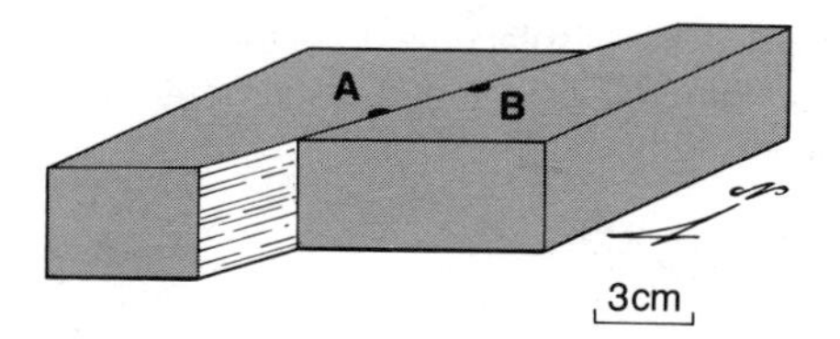

Figure 4.8 Schematic portrayal of horizontal, left-handed slip on a vertical fault. The magnitude of translation is 3 cm.

Figure 4.9 Real-life examples of horizontal, left-handed slip on vertical faults. The sidewalk panels were shifted by faulting during the earthquake in Managua, Nicaragua, in 1972. Faulting of the sidewalk took advantage of preexisting discontinuities. Maximum translation is about 3 cm. Aggregate displacement across the 12-m wide exposed zone is 28.6 cm. (Photograph by R. D. Brown, Jr. Courtesy of United States Geological Survey.)

pictured in Figure 4.9, sidewalk panels shifted in left-slip fashion during the Managua earthquake of 1972.

A more general illustration of translation along a fault surface is shown in Figure 4.10*A*. The fault surface strikes N60°E and dips 50°SE. Translation is such that the southeast block of the fault moves down relative to the northwest block. Additionally, the fault accommodates left-handed movement. The actual path of slip during fault translation rakes 35°NE on the fault plane. The displacement vector describing this faulting connects *B* to *A*, two points that occupied the same location before faulting. The orientation of the displacement vector can be determined stereographically by plotting both the fault plane and the rake of the slip direction, and then interpreting trend and plunge (Figure 4.10*B*). The sense of slip of *B* relative to *A* is 26° N85°E. The magnitude of the total displacement (i.e., the **net slip**), is 5.0 cm. The horizontal component of the slip (**strike–slip** component) is 4.1 cm; and the down-the-dip component of translation (**dip–slip** component) is 2.9 cm (Figure 4.10*C*).

In almost all geological examples, it is impossible to be sure whether a given fault block was stationary or in motion during faulting. All that can be reported is **slip**, the **actual relative displacement** that takes place as a result of translation. The translation portrayed in Figure 4.10*A* could have been achieved in a number of ways: movement of the southeast block downward along a northeastward trend; movement of the northwest block upward along a southwestward trend; movement of the southeast block down and to the northeast accompanied by simultaneous movement of the northwest block up and to the southwest; movement of both blocks downward and northeastward, but such that the southeast block moves a greater distance.

In regions of active faulting, such as the San Andreas fault system in California, there are many natural and man-made reference points that permit fault translation(s) to be measured. Displacement vectors can be calculated on the basis of faulted streams (Figure 4.11), faulted fencelines and roads, and even faulted city streets (Figure 4.12).

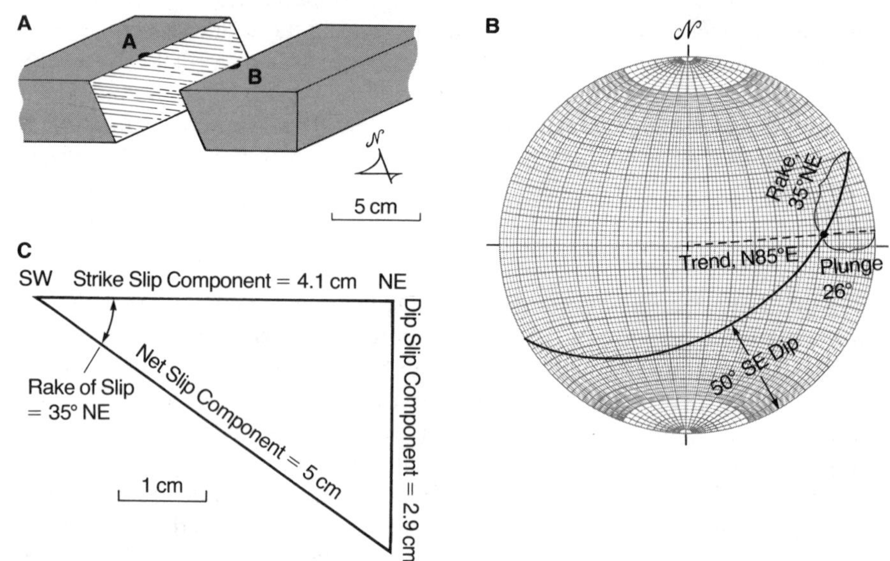

Figure 4.10 (*A*) Left-handed oblique slip (5 cm) on steeply dipping fault. (*B*) Stereographic determination of the trend and plunge of the displacement vector. (*C*) Net-slip, strike-slip, and dip-slip components of the displacement vector.

Figure 4.11 Right-handed offset of stream due to movement(s) on the San Andreas fault as exposed in the Carrizo Plains of California. (Photograph by R. E. Wallace. Courtesy of United States Geological Survey.)

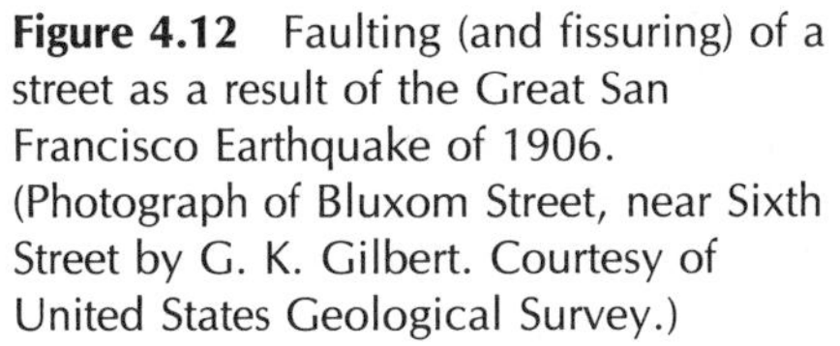

Figure 4.12 Faulting (and fissuring) of a street as a result of the Great San Francisco Earthquake of 1906. (Photograph of Bluxom Street, near Sixth Street by G. K. Gilbert. Courtesy of United States Geological Survey.)

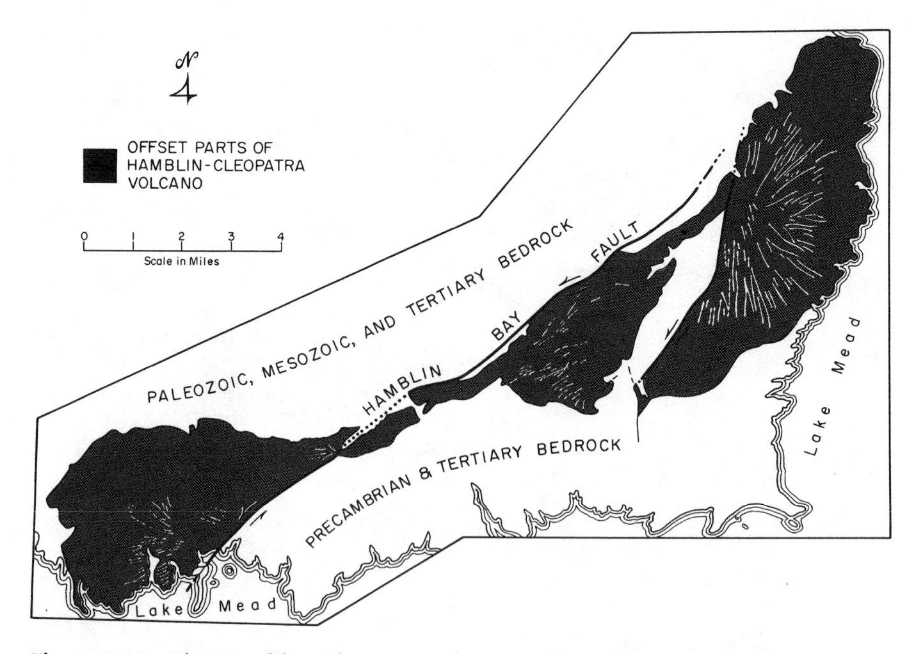

Figure 4.13 The Hamblin–Cleopatra volcano, Miocene in age, was cut in half and rearranged by faulting during the time interval 15 m.y. to 10 m.y.b.p. The offset parts of the volcano, including its once-radial dike swarm, permit the magnitude of the displacement vector for the faulting to be calculated. [From Anderson (1973). Courtesy of United States Geological Survey.]

Where faults have long been inactive, it is very difficult to locate reference points that can be used to analyze net translation. The "*A*'s" and "*B*'s" of Figures 4.8 and 4.10 are difficult to find in nature. However, chance relationships have permitted some elegant kinematic appraisals. Anderson (1973), working in the Lake Mead region of southern Nevada, recognized a Miocene andesitic stratovolcano, the Hamblin–Cleopatra volcano, which was cut in half between 15 m.y. and 10 m.y. ago by the Hamblin Bay fault (Figure 4.13). The northwest part of the volcano was translated approximately 12 mi (19.3 km) relative to the southeast half, in a left-handed sense.

DETERMINING SLIP THROUGH ORTHOGRAPHIC PROJECTION. Discovering offset geological features like igneous plugs and volcanoes is the rare exception and not the rule. Evaluating fault translation requires a creative assessment of all geologic facts on hand. A classic method in evaluating slip on faults is one made famous in *Structural Geology* by Billings (1942, 1954, 1972). The heart of the problem is to identify and visualize a real or geometric line that has been cut and displaced by the fault whose kinematics are being investigated.

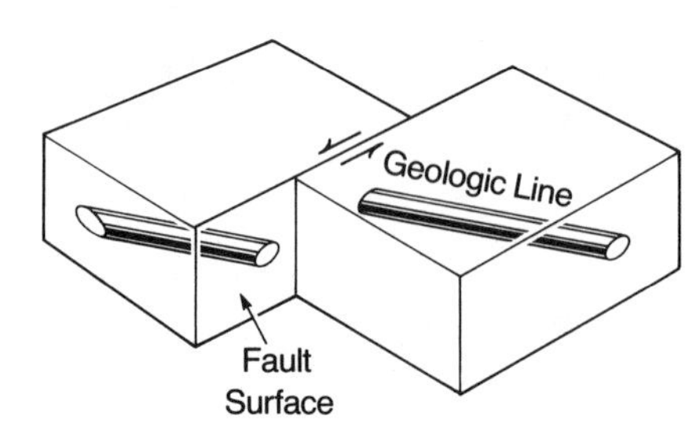

Figure 4.14 Schematic portrayal of the offset of a once-continuous line by faulting. Reconstruction of the line permits the direction, sense, and magnitude of the translation vector due to faulting to be calculated.

Not many real, physical geological lines exist in nature. On the other hand, **geometric lines** exist in deformed rock systems. Billings drew attention to the kind of geometric line that is formed by the intersection of two geologic planes, like a dike and a bed. If such a line is cut and displaced by a fault, a basis for evaluating translation is readily available. The concept is relatively simple (Figure 4.14). The actual solution is geometrically complicated.

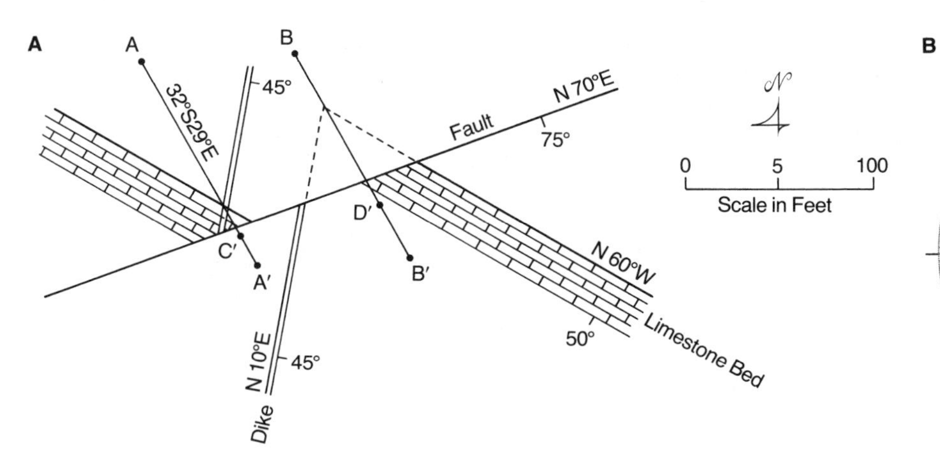

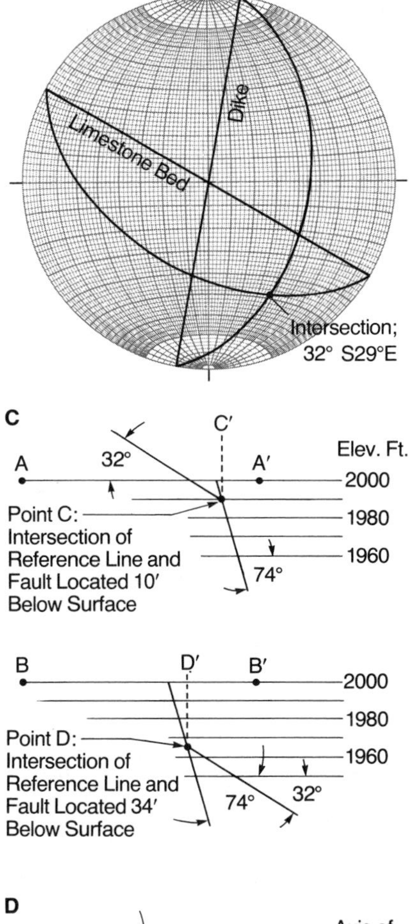

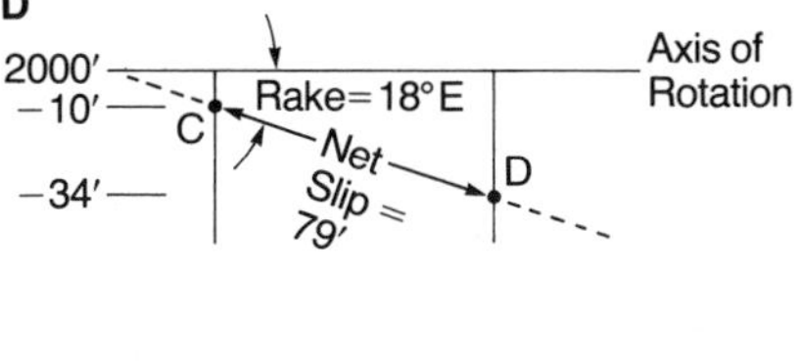

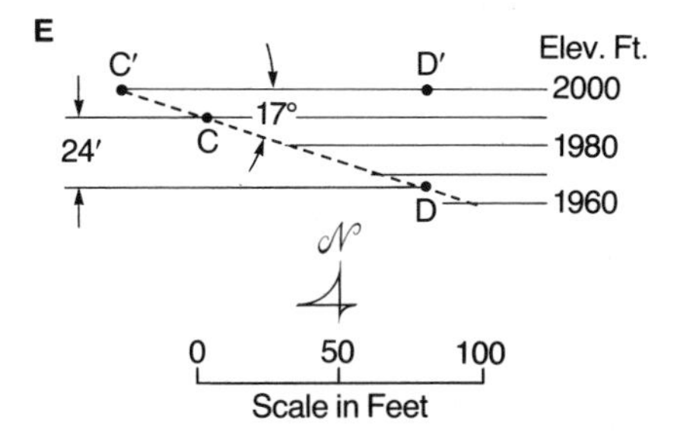

Figure 4.15 Determination of the displacement vector for a fault by reconstructing the faulted line of intersection of a dike and a limestone bed. (*A*) Faulted limestone bed and faulted dike. Lines *AA′* and *BB′* are the vertical projections of the limestone/dike intersections on the north and south sides of the fault, respectively. (*B*) Stereographic determination of the trend and plunge of the line of intersection of the limestone bed and the dike. (*C*) Cross sections showing the relation of the fault trace to the line of intersection of the limestone bed and dike. (*D*) View in the plane of the fault showing net slip and rake of net slip. (*E*) Cross section showing plunge of the displacement vector.

Consider a fault that strikes N70°E and dips 75°SE (Figure 4.15*A*). Assume that it is exposed in a perfectly flat area whose elevation is 2000 ft (609 m). A dike and a distinctive marker bed (a limestone bed) crop out on both the north and south sides of the fault. The dike strikes N10°E and dips 45°SE. The limestone bed strikes N60°W and dips 50°SW. What is the displacement vector describing the movement of the south block of the fault relative to the north block?

The intersection of the dike and the limestone bed provides the geometric line of reference for assessing translation on this fault. The trend and plunge of the reference line is determined stereographically by plotting the intersection of the great circles representing the dike and the limestone bed, respectively (Figure 4.15*B*). Its orientation is 32° S29°E. The vertical projection of the line of intersection of the dike and the limestone bed on the north block of the fault is plotted as line *A–A′* (see Figure 4.15*A*). This line passes through the surface intersection of the trace of the dike and the trace of the limestone bed as seen on the surface in outcrop. The counterpart of this line on the south block of the fault is *B–B′*.

To find the displacement vector for fault translation, it is necessary to determine where the two geometric lines of intersection, one on each side of the fault, pierce the fault surface. This step is ordinarily the most difficult to visualize and to construct graphically. To see the relationship, construct vertical cross sections that pass through the vertical projections (*A–A′* and *B–B′*) of the intersection of the dike and the limestone bed (Figure 4.15C). The 32° plunge of the line of intersection of the dike with the limestone bed is plotted on each of the cross sections. The **apparent dip** of the fault is also plotted, and its value is determined stereographically. In this example, the apparent dip of 74° happens to be merely 1° less than true dip. Where the line of intersection of the dike and the limestone bed meets the inclined trace of the fault, there lies the reference point. For the north block of the fault, *C* marks the intersection point and point *C′* its vertical projection (Figure 4.15*C*). For the south block of the fault, the intersection point is *D*, and the vertical projection is *D′* (Figure 4.15*C*). Points *C* and *D* originally occupied the same position before faulting; they are equivalent points that can be connected by a displacement vector. Elevations of these reference points are evident in the cross sections: the elevation of *C* is 1990 ft (606 m); the elevation of *D* is 1966 ft (599 m).

To determine the value of the net slip, points *C* and *D* must be viewed in the plane of the fault, for slip is measured in the plane of movement. To achieve this view, the fault surface is figuratively rotated to the surface about its strike line (Figure 4.15*D*). The fault plane is thus laid out in the plane of the paper so that the net slip can be directly measured. The axis of rotation, the strike line, is the intersection of the fault with the horizontal surface of the terrain. Its elevation is 2000 ft. The trace of this axis of rotation is shown in Figure 4.15*D*. Reference points *C* and *D* are plotted along lines perpendicular to the axis of rotation, and at distances from the axis corresponding to the elevations of *C* and *D*. Once *C* and *D* have been plotted, the net slip can be measured. The value of the net slip is 79 ft (24 m). The rake of the net slip on the fault plane is 18°E. The trend of the displacement vector is the azimuth of the line *C′–D′* (see Figure 4.15*A*) connecting the vertical projections of reference points *C* and *D*. The azimuth measures N78°E. The plunge of the displacement vector is found by constructing a structure profile along *C′–D′* (Figure 4.15*E*), showing the relative positions of *C* and *D* at depth. The plunge measures 17°. To sum up, the southeast block moved down and to the east with respect to the northwest block, along a displacement vector of 79 ft, 17° N78°E.

In solving problems of this type, perhaps in the search for a displaced ore body, it is important to avoid memorizing recipes of attack. The geometric conditions from one problem to the next are highly variable. Successful problem solving results from a clear image of the offset reference line or point that is sought and a solid grasp of basic graphic and stereographic methods.

COMPUTING DILATIONAL SPREADING OF DIKE WALLS. Another useful operation in kinematic analysis is evaluating translation accompanying the spreading apart of wall rock during emplacement of a dike or vein (Figure 4.16*A*). Two-dimensional patterns are simplest. Figure 4.16*B*, a cross section, shows a sandstone layer and the trace of a vertical fracture. The fracture is about to become occupied by a dike. Figure 4.16*C* depicts the sandstone bed after emplacement of the dike. Points *A* and *A′* occupied the same location before intrusion; points *B* and *B′* also occupied a common location. Translation of the northeastern block relative to the southwestern block was perpendicular to the wall of the dike, thus northeast–southwest and horizontal. The magnitude of translation equals the thickness of the dike, which in this example is 6 m. An example of translation of rigid rock bodies due to the filling of a fracture with vein material is shown in Figure 4.17.

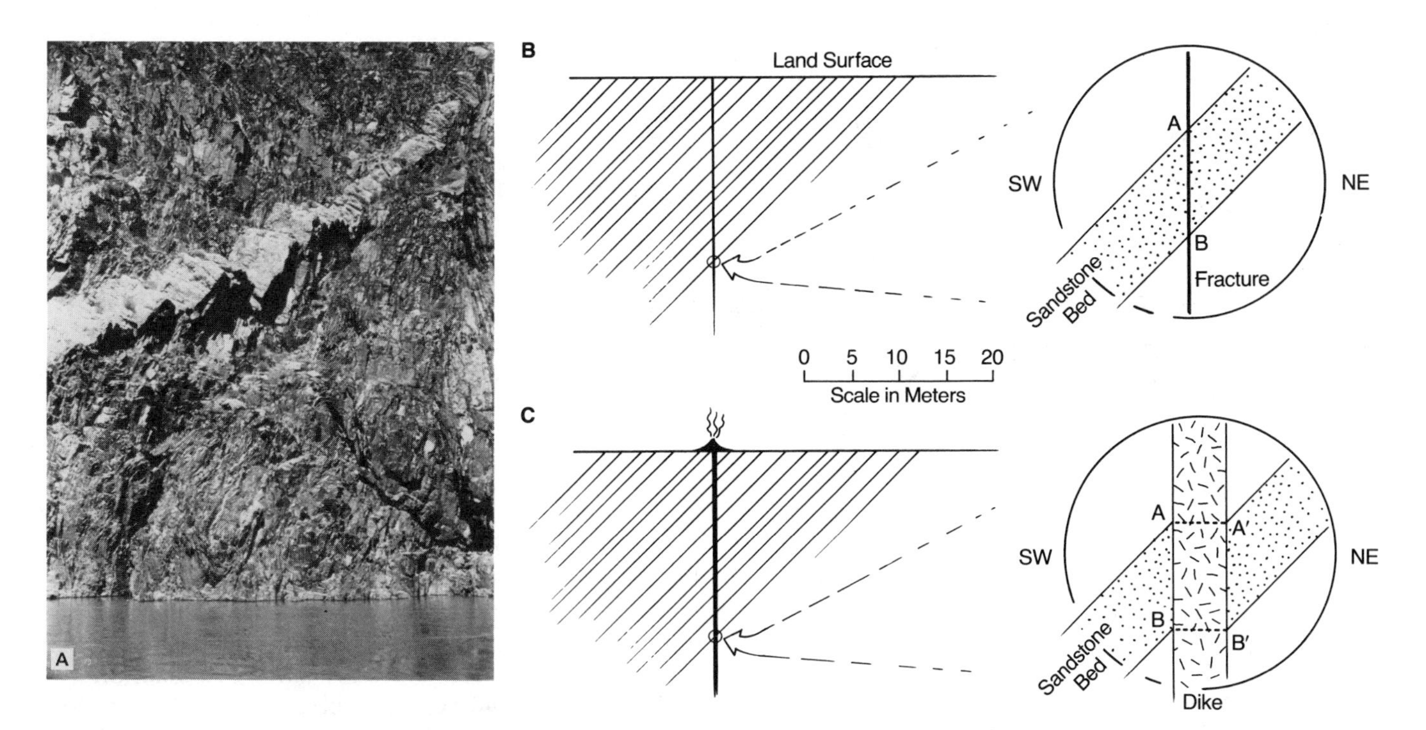

Figure 4.16 (*A*) The emplacement of a dike accommodates a spreading apart of wall rock along a fracture discontinuity. This dike is exposed in metamorphic rocks in the Inner Gorge of the Grand Canyon. (Photograph by A. F. Siepert.) (*B*) Vertical fracture soon to be occupied by a dike. (*C*) Offset of a sandstone bed as a result of slip accompanying the spreading apart of wall rock.

Figure 4.17 Vein of stilpnomelane and quartz. The vein invaded thin-bedded wall rock in the North Hillcrest mine area, Minnesota. The very center of the vein bears a faint line, perhaps a vestige of the former fracture trace that guided the hydrothermal solutions that gave rise to the vein. Spreading apart of the walls was directed at right angles to the centerline of the vein and to the contact of the vein with wall rock. Note that the conspicuous parting (*p*) in wall rock on the left wall of the vein is offset in a way that perfectly matches the kinematic pattern predicted in Figure 4.16C. (Photograph by R. G. Schmidt. Courtesy of United States Geological Survey.)

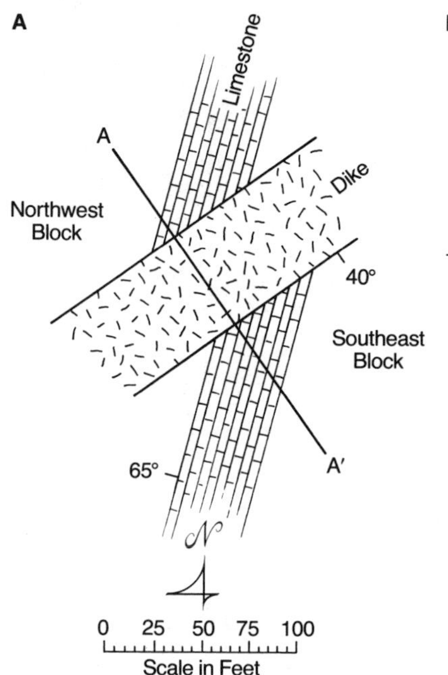

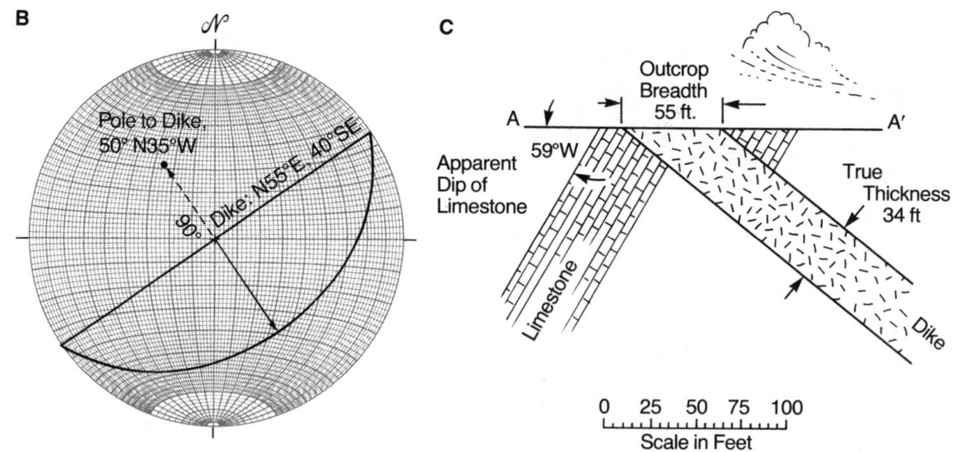

Figure 4.18 Determination of translation (slip) due to dike emplacement. (*A*) Map of limestone bed intruded by diabase dike. (*B*) Stereographic construction of pole to the plane of the dike. Trend and plunge of pole coincides with direction of spreading apart of dike walls. (*C*) Structure profile drawn to show true thickness of dike.

The three-dimensional evaluation of translation due to spreading along a dike, sill, or vein is more interesting and uses a number of the stereographic and geometric operations introduced earlier. Figure 4.18*A*, a map view this time, shows a limestone sequence into which has been intruded a diabase dike. The limestone layer strikes N15°E and dips 65°W. The dike, on the other hand, strikes N55°E and dips 40°SE. The problem is to determine the displacement vector for translation brought about by dike intrusion.

Let us evaluate the translation of the northwest block relative to an arbitrarily fixed southeast block. Assuming that the direction of translation is perpendicular to the walls of the dike, the pole to the plane of the dike must coincide with the direction of transport. The orientation of the pole is plotted stereographically in Figure 4.18*B*. It plunges 50° N35°W. The sense of transport, again assuming that the southeastern block is fixed, is the same: 50° N35°W. The distance of transport equals the thickness of the dike. The geologic map (see Figure 4.18*A*) shows the outcrop breadth of the dike, and not its true thickness. True thickness is determined by constructing a structure profile (*A–A′*) at right angles to the strike of the dike (Figure 4.18*C*). True thickness, and thus the magnitude of the displacement vector, is 34 ft (10 m).

ROTATION

GENERAL CONCEPT. Rotation is a rigid body operation that changes the configuration of points in a way best described by rotation about some common axis. Amusement parks thrive on rotational operations (Figure 4.19*A*). Axes of rotation may be horizontal (Ferris wheel), vertical (merry-go-round), and inclined (rotor), or all of the above (tilt-a-whirl, roll-o-plane, the hammer, and other ghastly rides). The changes in locations of points are described by the orientation of the **axis of rotation** (trend and plunge), the **sense of rotation** (clockwise versus counterclockwise), and the **magnitude of rotation** (in degrees). Based on this set of facts, the locations of points before and after rotation can be calculated.

Designating a clockwise versus a counterclockwise sense of rotation depends partly on the direction of view. The operator in front of a Ferris wheel sees the cars rotating clockwise about the horizontal center axis (Figure 4.19*B*). The bystander behind the ferris wheel, on the other hand, observes counterclockwise motion. Even more contradictory, the sense of motion of the hands of a clock would be counterclockwise if a clock with a transparent housing were viewed from behind. To avoid such ambiguity in geological studies, a special convention is adopted to specify sense of rotation: we

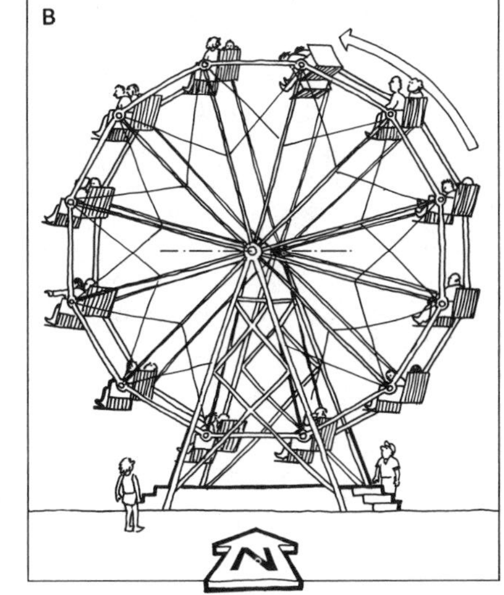

Figure 4.19 (*A*) Amusement parks thrive on rotational operations. Kinematic analysis of rotation includes describing the orientation of the axis of rotation, the sense of rotation, and the magnitude of rotation. (*B*) Sense of rotation depends on the direction of view.

describe the sense of rotation while looking *down* the axis of rotation. If the axis is strictly horizontal, we describe not only the sense of motion but also the direction in which the structure is observed.

GEOLOGIC EXAMPLES. There are a number of geological examples that serve to illustrate why rotation is an important kinematic operation. Originally horizontal layered rocks can become inclined as a result of folding and faulting processes. Consider strata in the middle limb of the Hunter's Point monocline shown in Figure 4.20. The overall form of the fold is revealed in the distant view, where horizontal bedding on the west (left) abruptly bends to very steep dips (upper right corner). The bold white rock exposures in the foreground show that bedding has been rotated to dips as steep as 90°. The inclined strata can be thought of as having rotated as a nearly rigid body about a horizontal axis. The orientation of the axis of rotation of bedding in the Hunter's Point monocline is 0° N5°W; the sense of rotation of middle limb strata is clockwise when viewed from south to north; and the magnitude of rotation is as much as 90°.

Figure 4.20 Rotation of bedding in the Hunter's Point monocline, northeastern Arizona. Bedding in middle background is horizontal, but it is rotated clockwise to a steep dip in the right background. White outcrops in foreground display near-vertical bedding. Strata in the foreground represent the strike projection of the steeply dipping strata in the fold in the distance. (Photograph by G. H. Davis.)

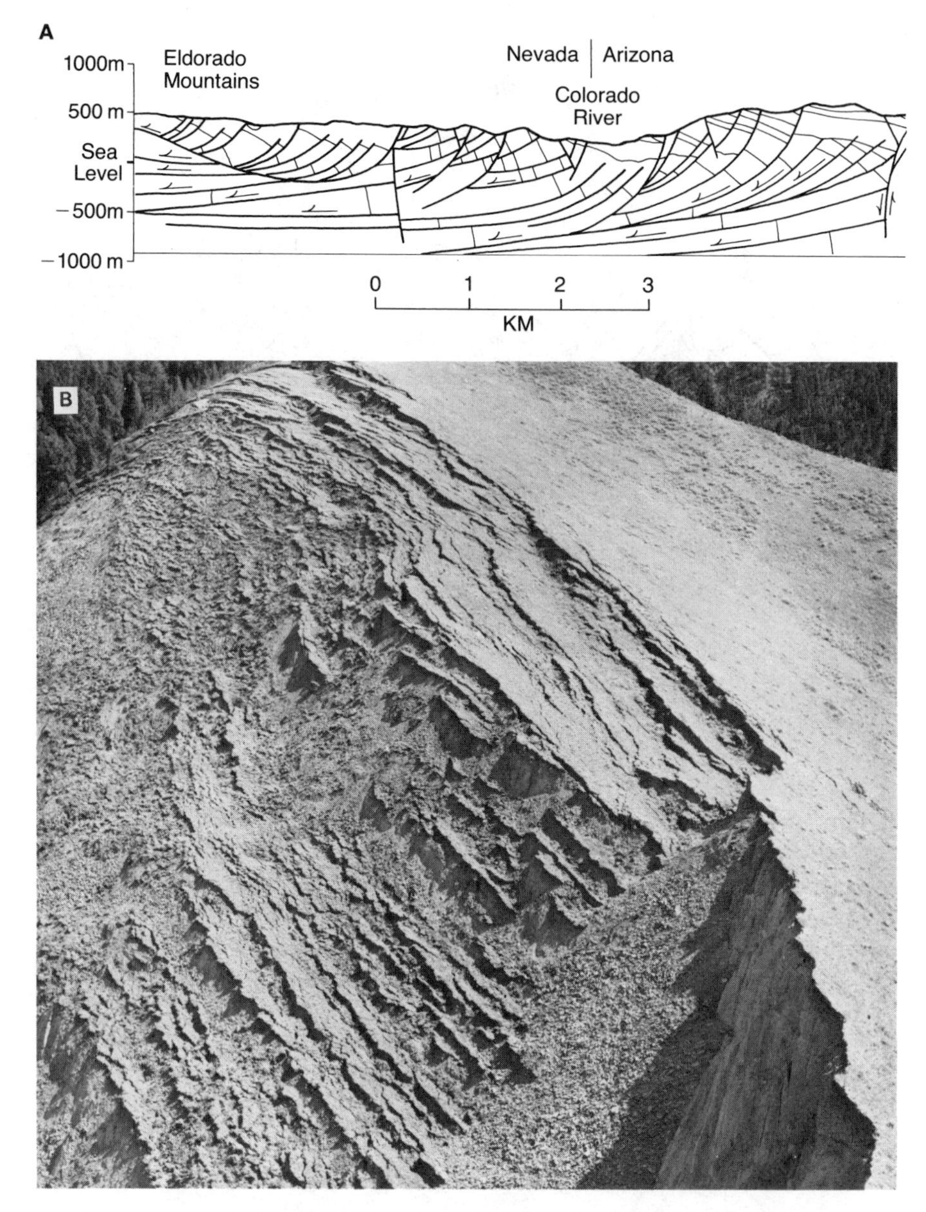

Figure 4.21 (*A*) Listric normal faulting in the Lake Mead region. Tertiary and Precambrian rocks are rotated to steep dips along curved faults. [From Anderson (1973). Courtesy of United States Geological Survey.] (*B*) Example of the back rotation that accompanies slumping along curved fault surfaces, Hebgen Lake earthquake area, Montana. (Photograph by J. R. Stacy. Courtesy of United States Geological Survey.)

Faulting too can result in the rotation of strata. Dramatic examples of fault-induced rotation are found in the southern part of the Basin and Range province of the western United States, notably in southeastern California, southern Nevada, and southern Arizona (Anderson, 1971; Proffett, 1977; Davis and others, 1980; Davis and Hardy, 1981). Some of the faulting there is a **listric normal faulting** (Figure 4.21*A*), wherein blocks of rock move downward along curved concave-upward faults. The geometric consequence of translation of originally horizontal layered strata along curved fault surfaces is the rotation of strata to moderate or steep dips, not unlike rotation due to slumping on oversteepened hillslopes (Figure 4.21*B*). For the example of listric faulting shown in Figure 4.21*A*, the axis of rotation is the strike orientation of the fault, the sense of rotation is clockwise when viewed from the south, and the magnitude of rotation ranges from 10° to 90°.

An example of rotational kinematics on the regional scale is evident in comparing the orientation of the North American continent in Pennsylvanian time to its orientation today. In addition to the translational movement of the continent by some 1000 km from equatorial to northern latitudes, the continent underwent a counterclockwise rotation of at least 20° (Van der Voo, Mauk, and French, 1976).

STEREOGRAPHIC EVALUATION. Certain stereographic techniques are indispensable in the kinematic analysis of rotational operations. Stereographic techniques can be used to picture the rotation of lines and planes in geometric space. A stereonet is used to plot the path or **locus** of points representing lines and planes at various stages of rotation. The plotting procedure is based on facts regarding the orientation of the axis of rotation, sense of rotation, and the magnitude of rotation in degrees. Let us consider some examples.

Figure 4.22*A* shows strata in various stages of rotation during listric normal faulting. Orientation of strata at each stage is portrayed stereographically in Figure 4.22*B*, both through the use of great circles representing strike and dip of bedding and through poles to bedding. Poles provide the most convenient representation. The cluster of poles, taken together, is the locus of points representing progressive rotation of the strata by faulting.

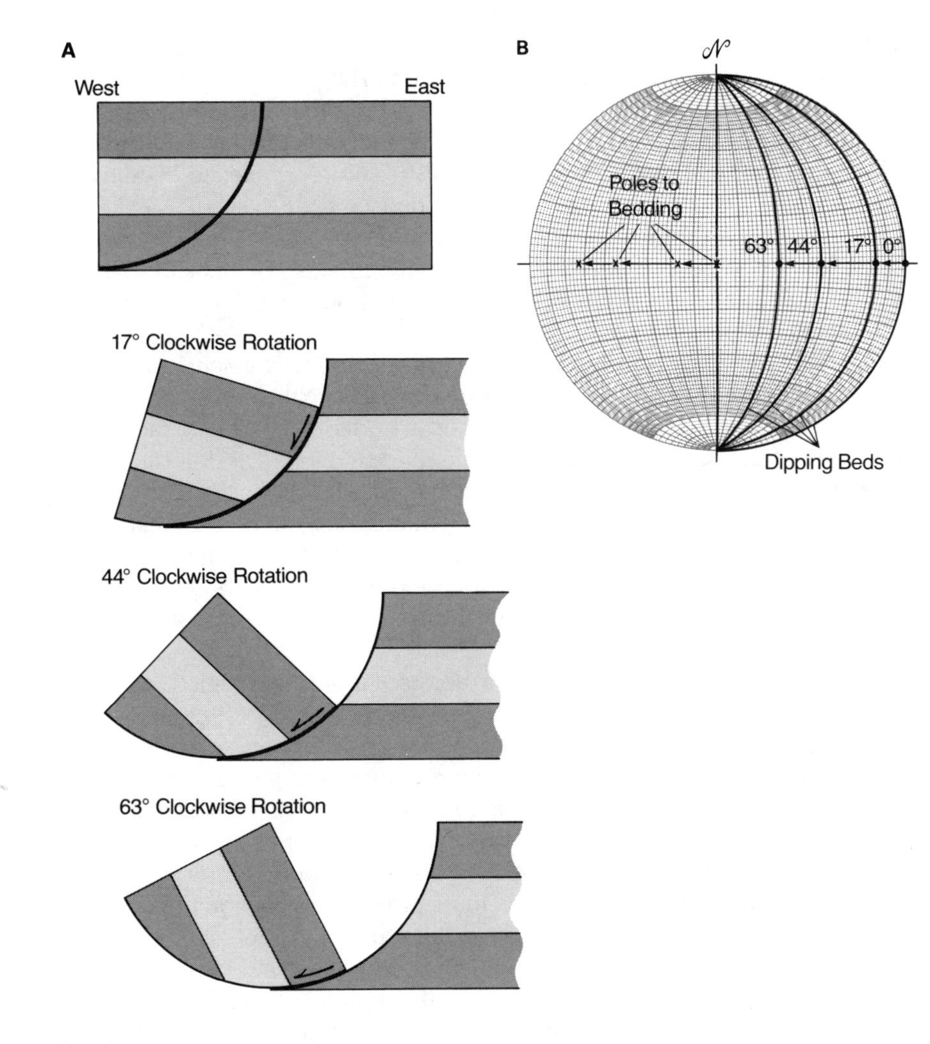

Figure 4.22 (*A*) Rotation of strata by listric faulting, portrayed at different stages. (*B*) Stereographic representation of the rotation of strata.

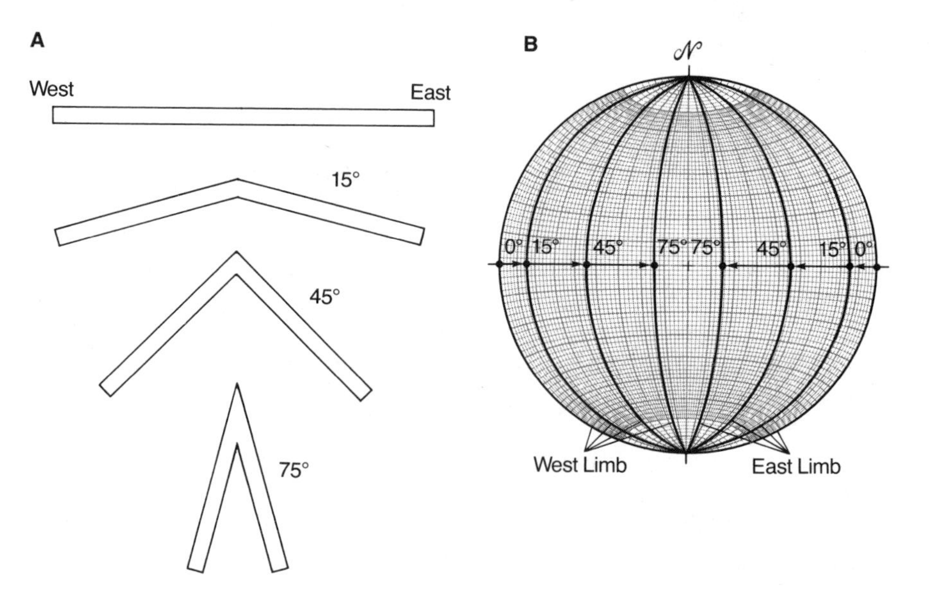

Figure 4.23 (*A*) Rotation of strata during progressive folding. (*B*) Stereographic representation of the rotation of bedding during the folding.

Progressive folding of a layer of rock is portrayed in Figure 4.23*A*. In Figure 4.23*B*, the strike and dip of bedding on the limbs of the fold is shown stereographically at each stage of deformation. The poles to bedding, taken together, represent the locus of points describing the rotation of bedding during folding. The locus is in the form of a great circle that lies 90° from the axis of rotation, just like the great-circle distribution of poles to bedding in the example of rotation during listric faulting (see Figure 4.24*B*).

The general case of rotation of strata by folding is shown in Figure 4.24, where layered strata are flexed about an inclined axis. The axis of rotation plunges 30° S45°E. The locus of poles to folded bedding describes a great circle lying 90° from the axis.

The ability to stereographically rotate lines in space is essential in structural studies. The geological lines we focus on now are the crests and troughs of ripple marks (Figure 4.25*A*). Consider a horizontal layer of sandstone containing parallel-aligned current ripple marks (Figure 4.25*B*). The ripple marks lie in the plane of bedding, and thus a point that stereographically portrays the trend and plunge of the ripple marks must fall on the great circle representing the orientation of bedding (Figure 4.25*C*). If the horizontal sandstone layer is subjected to 48° of southeastward tilting about a horizontal, N70°E-trending axis (Figure 4.25*D*), how will the orientation of the ripple marks change?

When the bed is tilted 48° about a horizontal axis of rotation, poles to bedding at the various stages of tilting trace out part of a great circle, which lies at right angles to the axis of rotation (Figure 4.25*E*). In contrast, points describing the orientation of ripple marks during tilting trace out a small circle on the stereonet. At each stage of tilting, the point describing the orientation of the ripple marks remains on the great circle representing the strike and dip of bedding. These stereographic relations can be seen more clearly by rotating the fold axis to the north–south line of the net and observing that poles to bedding all lie on the *x* axis of the net, and noting the correspondence between the movement path of ripple mark orientations and one of the small-circle traces embossed on the underlying net (Figure 4.25*F*). The angle between the ripple mark orientation and the axis of rotation, at each stage of tilting, remains constant, matching the angle between the trend of the ripple marks and the trend of the axis of rotation before the tilting commenced (see Figure 4.25*B*).

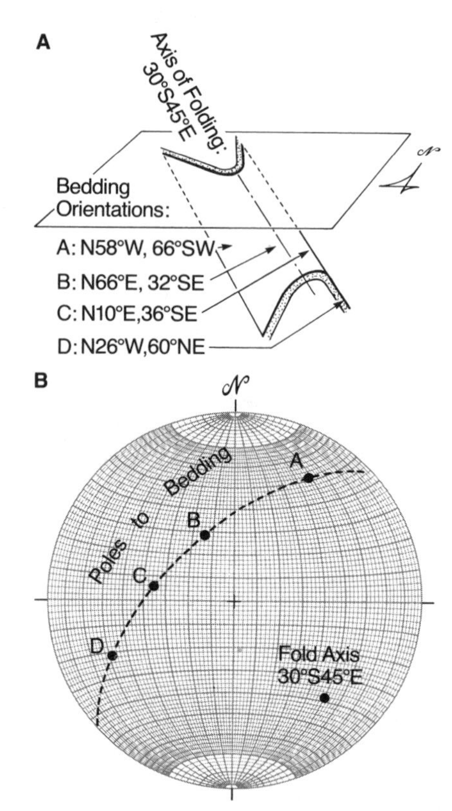

Figure 4.24 (*A*) Rotation of bedding by folding about an inclined axis. (*B*) Stereographic portrayal of poles to the rotated, folded bedding, measured at locations *A–D*. Note that the poles lie on a great circle whose pole is the fold axis.

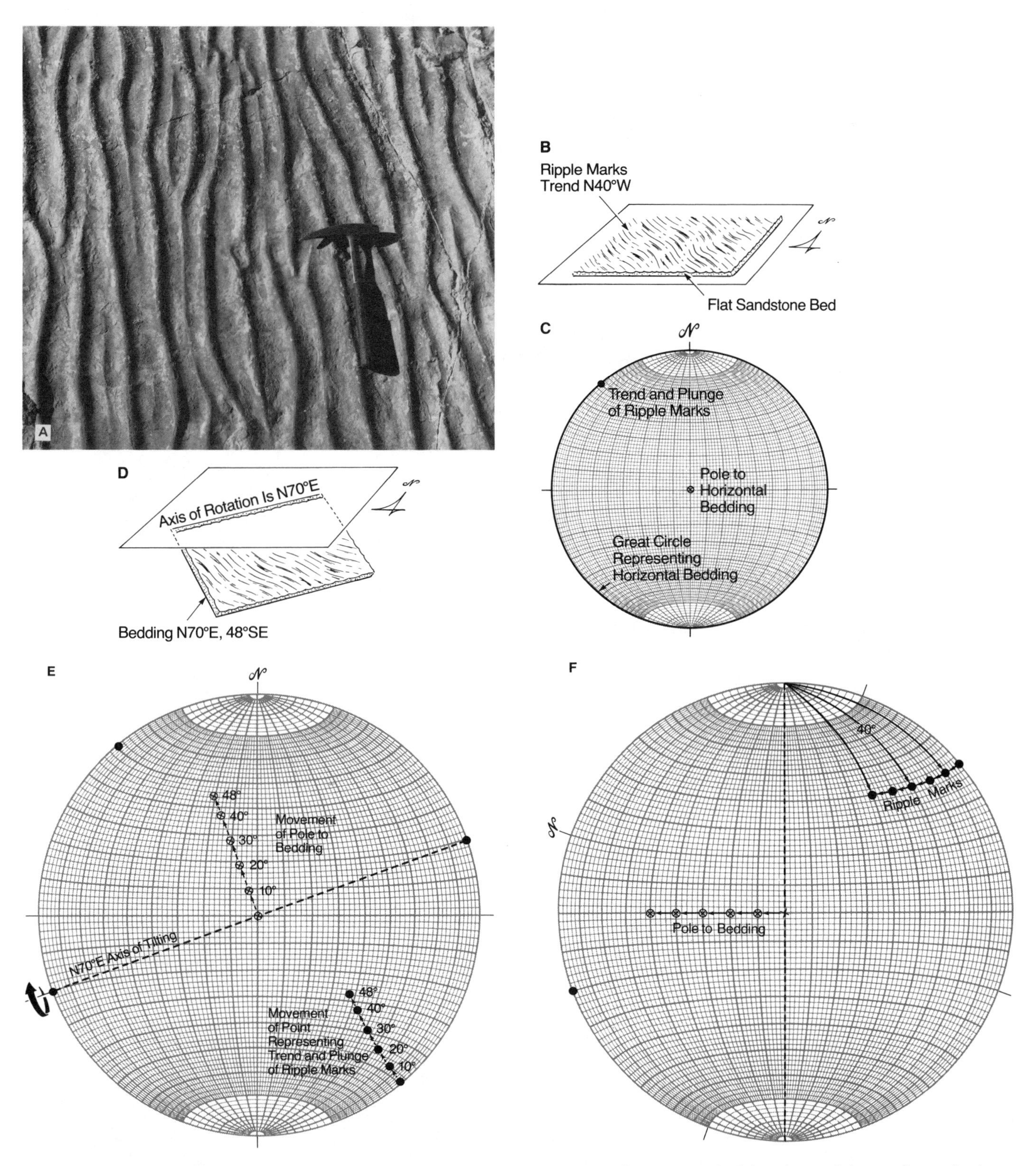

Figure 4.25 (*A*) Geological lines in a plane: namely, crests and troughs of current ripple marks in the Dakota Sandstone. (Photograph by J. R. Stacy. Courtesy of United States Geological Survey.) (*B*) Ripple marks trending N40°W in horizontal sandstone bed. (*C*) Stereographic portrayal of the tilting of the sandstone bed as well as the ripple marks it contains. (*D*) Bedding and ripple marks after tilting. (*E*) Stereographic portrayal of the tilting of the sandstone bed and the consequent change in orientation of the ripple marks. (*F*) Stereographic view showing how the movement path of the pole to bedding follows the great-circle trace along the east–west line of the net, whereas the movement path of the point representing the orientation of the ripple marks follows a small-circle path.

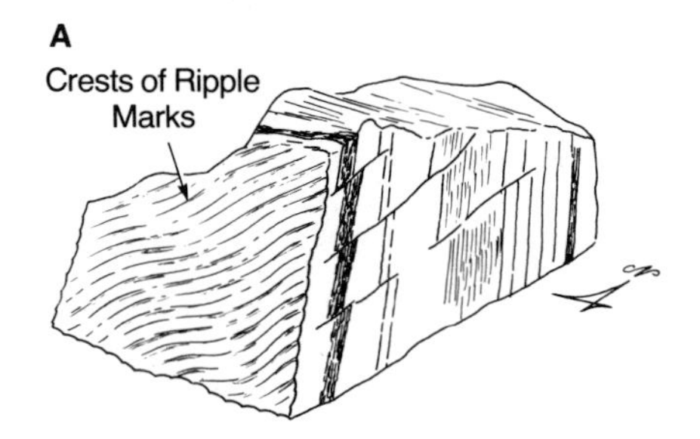

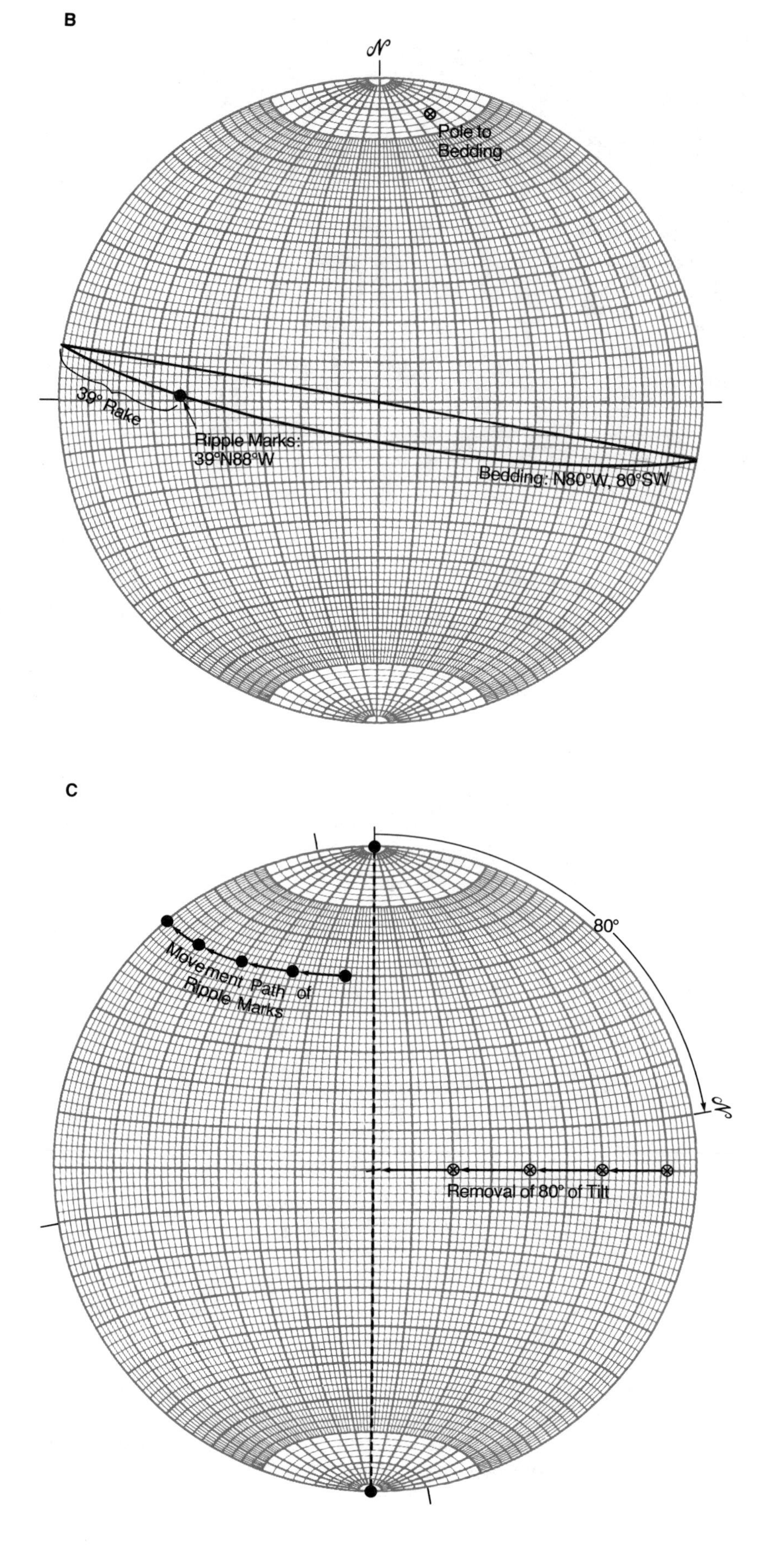

Figure 4.26 (*A*) Outcrop of steeply dipping faulted sedimentary strata. Front face of outcrop is the plane of bedding, revealing ripple mark lineation. (*B*) Stereographic representation of the orientations of bedding and ripple marks, as measured in outcrop. (*C*) Stereographic restoration of bedding and ripple marks to their orientations before tilting.

Suppose as field geologists we encounter an outcrop of tilted strata revealing both bedding and ripple marks (Figure 4.26*A*). How can we determine the original orientation of the ripple marks? We begin by measuring the strike and dip of bedding (N80°W, 80°SW) and the trend and plunge of ripple marks (39° N88°W). Alternatively, we measure the strike and dip of bedding (N80°W, 80°SW) and the rake of the ripple marks in the plane of bedding (39°W). When these data are plotted stereographically (Figure 4.26*B*), the point representing the trend and plunge of the ripple marks is seen to lie on the great circle that describes the strike and dip of bedding. The original orientation of the ripple marks is found by rotating bedding to its inferred original horizontal orientation. To do this an axis of rotation must be chosen.

For this problem, the line of strike of bedding is the right choice. Its orientation is N80°W. To rotate stereographically about any horizontal axis, the point(s) representing the selected axis of rotation must be brought into alignment with north (or south) on the perimeter of the stereonet. This is achieved simply by rotating the tracing overlay clockwise by 80° (Figure 4.26*C*). Thus positioned, the axis of rotation lies as the central axis to the small-circle paths of rotation. By rotating the pole to bedding and the point representing trend and plunge of ripple marks about this axis, and *in the proper sense,* the bed containing the ripple marks is, in effect, lifted from its steeply inclined orientation to horizontal (Figure 4.26*C*). The points traverse small circles from the interior of the net to the perimeter. Once accomplished, the original trend of the ripple marks can be measured: namely, S61°W.

MULTIPLE ROTATIONS. A final example of rotational kinematics is required to present the general operation(s) that can be adapted to all rotational problems, regardless of complexity. The goal is to be able to carry out successive rotational operations about axes of different orientations. The order in which rotational operations are performed vitally influences the final orientation of the rotated body. Multiple rotations are simply a series of single rotations applied in sequence. If we can do one, we should be able to do more than one.

Suppose we are interested in determining the original orientation of ripple marks that lie in a steeply inclined limb of a plunging fold (Figure 4.27*A*). The fold axis plunges 30°N36°E. Bedding on the west limb of this fold strikes N50°E and dips 65°NW. The ripple marks rake 80°SW. These geometric data are plotted stereographically in Figure 4.27*B*. The first step in the restoration is to rotate the fold axis to horizontal, and at the same time to rotate the pole to bedding and the ripple marks by the same amount. The rotation axis chosen to achieve this plunges 0° N60°W, at right angles to the trend of the fold axis. The rotation axis is brought to the north position on the perimeter of the net (Figure 4.27*C*) and is lifted to horizontal, traversing 30° along the east–west line of the net to the perimeter. The pole to bedding and the point defining the trend and plunge of ripple marks move 30° in the same sense but along small-circle paths.

Once this initial step has been accomplished, the great circle perpendicular to the pole to bedding is plotted (Figure 4.27*D*). It passes through the rotated position of the ripple mark axes. The strike line of the bedding is then aligned along the north–south line of the net (Figure 4.27*E*), serving as the axis of rotation about which the final operation is completed. Bedding and the ripple marks are lifted to horizontal, traversing small-circle paths to the perimeter of the net (Figure 4.27*E*). Returning the tracing paper to home position (Figure 4.27*F*), the original trend of the ripple marks can be directly measured. Its value is N30°W, substantially different from N63°W, the trend of the ripple marks as they appear in outcrop.

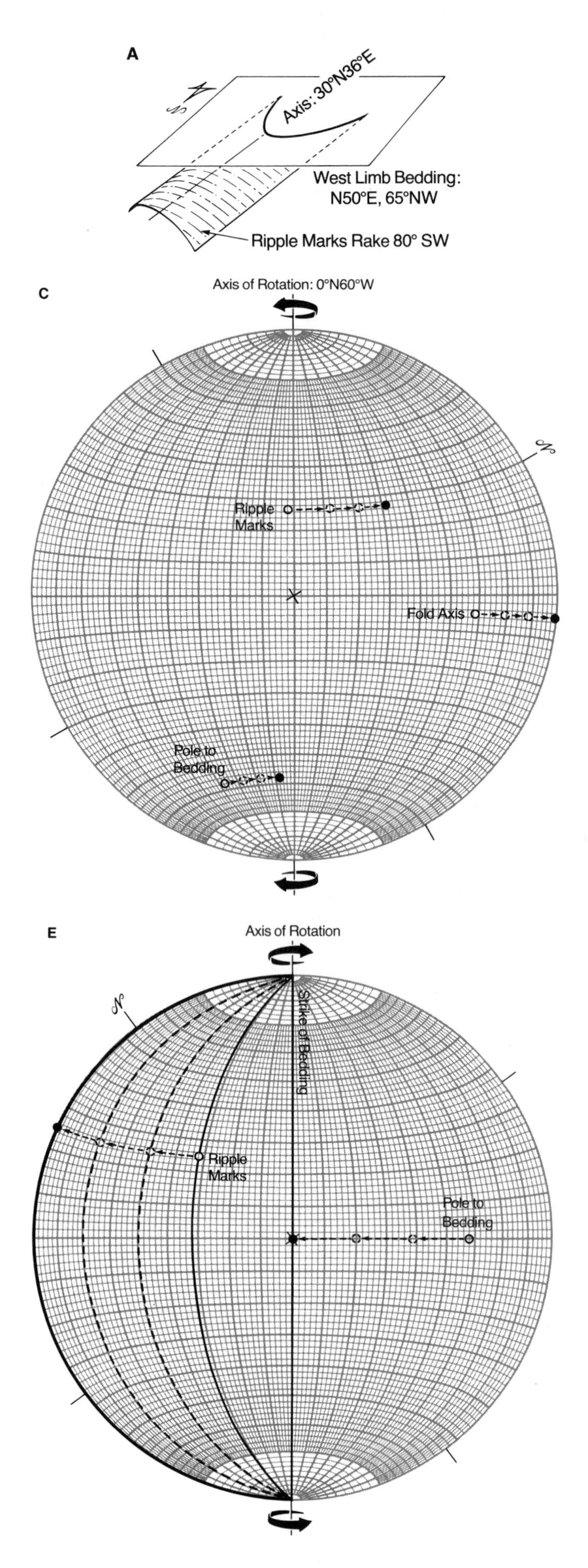

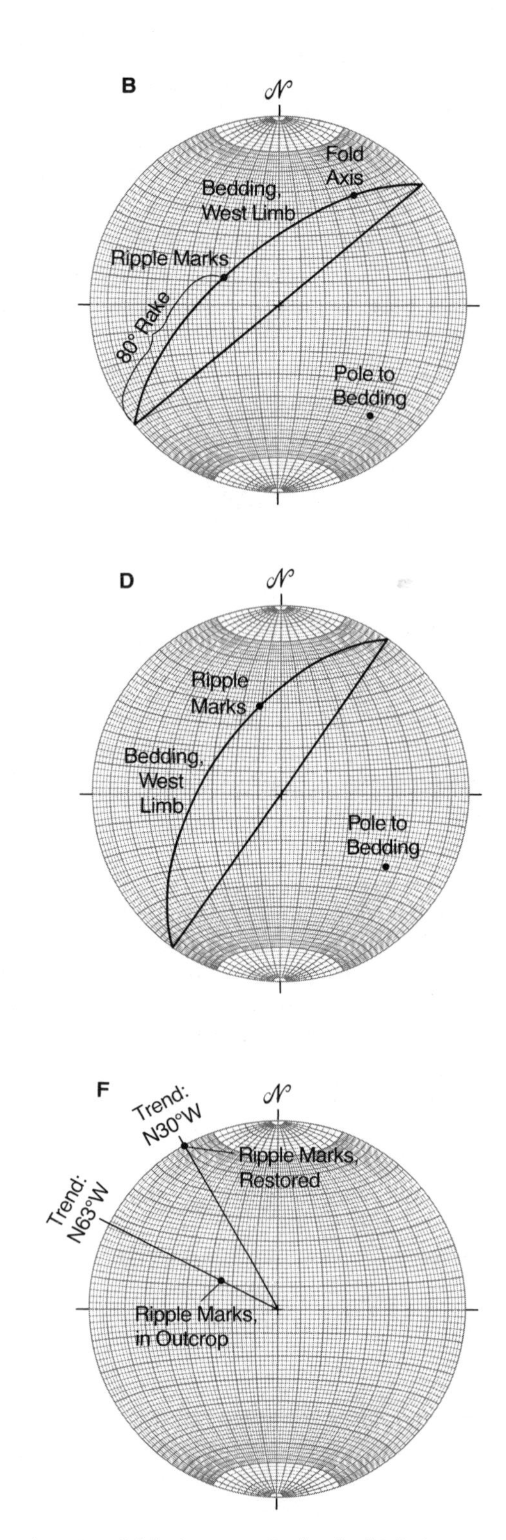

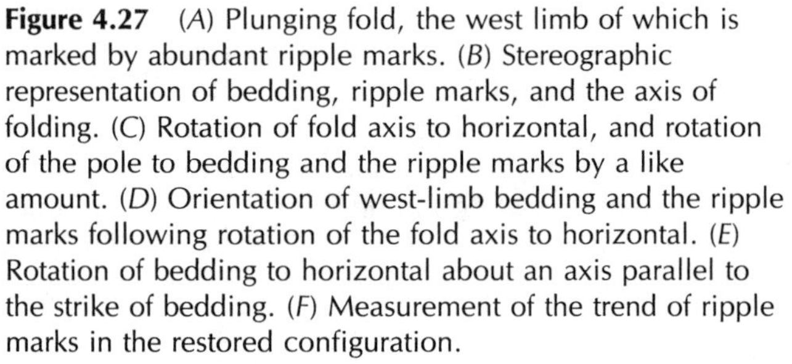

Figure 4.27 (*A*) Plunging fold, the west limb of which is marked by abundant ripple marks. (*B*) Stereographic representation of bedding, ripple marks, and the axis of folding. (*C*) Rotation of fold axis to horizontal, and rotation of the pole to bedding and the ripple marks by a like amount. (*D*) Orientation of west-limb bedding and the ripple marks following rotation of the fold axis to horizontal. (*E*) Rotation of bedding to horizontal about an axis parallel to the strike of bedding. (*F*) Measurement of the trend of ripple marks in the restored configuration.

STRAIN

GENERAL CONCEPT. Nonrigid body movements involving dilation and/or distortion are referred to as **strain**. Points within strained bodies do not retain their original configuration during the structural disturbance. Points are translated and/or rotated by amounts that change the original spacing of points within the body. This leads to the **dilation** (change in size) and/or **distortion** (change in shape) that is observed.

During dilation, internal points of reference in a body spread apart or pack closer together in such a way that line lengths between points become uniformly longer or shorter while shape remains the same. During distortion, the changes in spacing of points in a body are such that the overall shape of the body is altered, with or without a change in size (see Figure 4.1).

As structural geologists we are indebted to John G. Ramsay, whose text *Folding and Fracturing of Rocks* (1967) underscores the importance of strain analysis and presents the fundamentals and applications of strain analysis in a thorough and enlightening way. Although mathematical in nature, strain analysis is fundamentally a geometrical challenge (Ramsay, 1967, p. 50). The emphasis on geometry in this and previous chapters is preparation for understanding the geometric problem called strain.

The object of strain analysis is a formidable one. We are asked to describe the changes in size and/or shape that have taken place in a nonrigid body during deformation, to describe how each and every point within a body has changed position, and to describe how every line in a body has changed length and relative orientation.

THE GROUND RULES

We simplify our work in strain analysis by studying the theory in two, not three, dimensions. For advanced structural geological applications, the three-dimensional approach to strain analysis is available in other texts, notably Ramsay (1967), Jaeger and Cook (1976), and Means (1976).

A customary simplification is to restrict strain analysis to the description of **homogeneous deformation**. It is almost impossible to apply mathematical theory to unwieldy, irregularly deformed, **heterogeneous** structural systems (Ramsay, 1967). Instead, strain analysis focuses on the properties of bodies, or parts of bodies, that have deformed in a regular, uniform manner, like the folded rocks shown in Figure 4.28. The chief constraints that the "homogeneous deformation" clause imposes are (1) straight lines that exist in the nonrigid body before deformation remain straight after deformation, and (2) lines that are parallel in the body before deformation remain parallel after deformation. For these conditions to hold, the strain must be systematic and uniform across the body that has been deformed.

Figure 4.28 Systematic folding of metasedimentary rocks in the Salt River Canyon region, Arizona. Note handlens for scale in left-center part of photo. [Photograph by F. W. Cropp. From Davis and others (1981), fig. 32, p. 83. Published with permission of Arizona Geological Society.]

COMPUTING CHANGES IN LINE LENGTH

Strain can be described fully if changes in lengths and orientations of all lines in the nonrigid body can be measured. Three conventional parameters permit changes in lengths of lines to be described. One is **extension**, symbolized by **e**; the second is **stretch** (***S***); and the third is **quadratic elongation**, symbolized by lambda (**λ**) (Ramsay, 1967; Means, 1976). Consider the line *L* whose **original length** (l_o) is 5 cm (Figure 4.29*A*). During deformation, the nonrigid body in which *L* is contained changes shape and/or size such that the line stretches to a **final length** (l_f) of 8 cm (Figure 4.29*B*). The **change in length** (Δl) is 3 cm. The value of extension (e) along the line of this deformation is the change in unit length of the line,

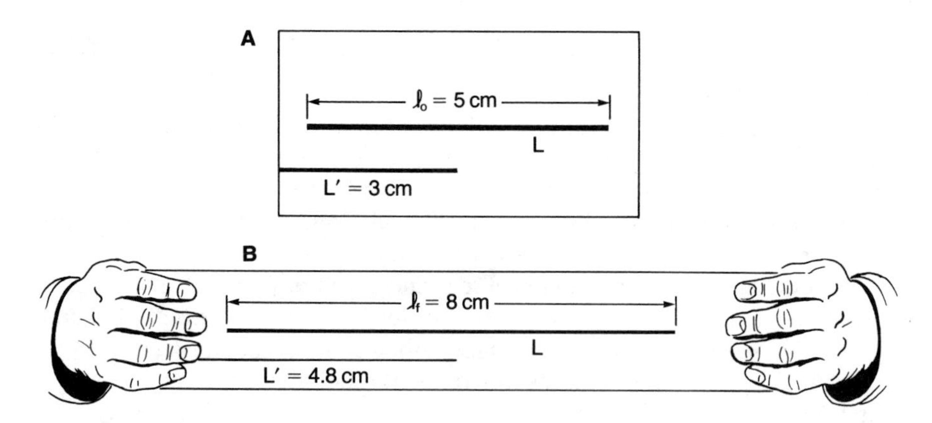

Figure 4.29 (*A*) Lines *L* and *L′* before stretching. (*B*) Lines *L* and *L′* after stretching.

$$e = \frac{l_f - l_o}{l_o} \tag{4.1}$$

In this example,

$$e = \frac{8 \text{ cm} - 5 \text{ cm}}{5 \text{ cm}} = 0.6$$

The fact that the calculated value of extension in this example is positive (+) records lengthening, not shortening, of the line. The **percent lengthening** is 60%, a value found by multiplying extension (e) times 100%.

The possible values for e range from -1 (severe shortening) through zero (no change in length) to $+\infty$ (severe stretching). This range in variation is schematically pictured in Figure 4.30.

If line *L* in Figure 4.29*B* lies within a body that has undergone homogeneous deformation, the value of $e = 0.6$ must hold for all the lines in the body that are parallel to *L*. Line *L′* is such a line (see Figure 4.29*A*). If the length of *L′* before deformation is 3 cm, its length after deformation can be computed by using Equation 4.1,

$$e = \frac{l_f - l_o}{l_o}$$

$$0.6 = \frac{l_f - 3 \text{ cm}}{3 \text{ cm}}$$

$$l_f = 4.8 \text{ cm (see Figure 4.29}B\text{)}$$

An even simpler way to compute l_f for this line is to multiply l_o by the stretch (*S*). Stretch is equal to the value of extension plus 1,

$$S = (1 + e) = \frac{l_f}{l_o} \tag{4.2}$$

In this example,

$$S = (1 + e) = 1.6$$

If we multiply this 1.6 stretch value times the original length of *L′*, we obtain a final length for *L′* of 4.8 cm (see Figure 4.29*B*),

$$l_f = l_o S = 3 \text{ cm}(1.6) = 4.8 \text{ cm}$$

Figure 4.30 A ribbonlike banner that is neither stretched nor shortened has an e value of zero. If the banner is stretched toward infinite length, its e value approaches infinity. If the banner is shortened toward zero, its e value approaches -1,. . . the airplane to which it is attached undergoes nonrigid deformation. (Artwork by R. W. Krantz.)

Quadratic elongation (λ), a parameter that sounds complicated, is equal to the square of the stretch. It is the squared value of $(1 + e)$,

$$\lambda = S^2 \tag{4.3}$$

$$\lambda = (1 + e)^2 \tag{4.4}$$

$$\lambda = S^2 = (1 + e)^2$$

Quadratic elongation can be thought of as the square of the length of a line (l_f) whose original length (l_o) was 1 (Ramsay, 1967). Figure 4.31 shows a line (L) whose original length (l_o) is indeed 1, namely one cm. Assume that the body in which L is contained is deformed in such a way that L becomes shorter. The amount of shortening can be described in terms of the value of extension (e), which for this example is -0.4. Given the starting facts of $l_o = 1.0$ cm and $e = -0.4$, we can calculate the final length (l_f) of the line,

$$l_f = l_o S$$

$$l_f = l_o(1 + e)$$

$$l_f = 1 \text{ cm } (1 - 0.4) = 0.6 \text{ cm}$$

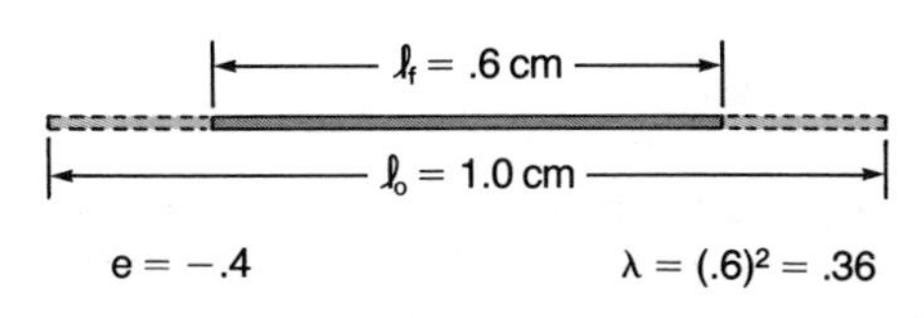

Figure 4.31 Quadratic elongation (λ) is the square of the length of a line whose original length was one. It can readily be determined by dividing final length (l_f) by original length (l_o) and squaring the result.

Quadratic elongation is the square of this final length,

$$\lambda = (0.6)^2 = 0.36 = (1 + e)^2$$

The value of quadratic elongation, 0.36, is indeed the same as the "square of the length of a line whose original length was 1.0." **The simplest way to determine quadratic elongation is to divide final length (l_f) by original length (l_o) and square the result**. Try it.

The value of quadratic elongation of a line provides an immediate image of whether a line in a deformed body has been shortened or lengthened, and by how much. If the quadratic elongation (λ) of a line is 1.0, there has been no change in the length of the line. If λ is greater than 1.0, the line has been lengthened; if less than 1.0, the line has been shortened. The range of quadratic elongation is from zero to infinity.

GEOLOGIC EXAMPLE OF COMPUTING CHANGE IN LINE LENGTH

The value in computing extension, stretch, and quadratic elongation stems from analyzing stretched and shortened geological lines. The stretched belemnite fossil featured in Figure 4.32 provides a good example. This fossil, discovered in folded rocks in the western Alps by Albert Heim in the nineteenth century (Milnes, 1979), was stretched into an array of rigid shell fragments of approximately equal size. Spaces that developed during deformation were simultaneously filled by calcite. The original length (l_o) of the belemnite fossil can be determined by measuring and summing the thickness of the individual shell fragments. The final length (l_f) of the belemnite is simply the total length of the fossil in its present state, including the calcite filling. The actual size of the deformed belemnite is not revealed in Figure 4.32. However, knowledge of the relative values of l_o and l_f is all that is necessary to calculate extension, stretch, and quadratic elongation. Using a Xerox copy of the belemnite illustration as presented in Milnes (1979), I measured an l_o value of approximately 82 mm and an l_f value of approximately 185 mm. On the basis of these measurements, I calculated the values of extension, stretch, and quadratic elongation for the rock in which the fossil was found, as measured parallel to the trend of the deformed fossil,

$$e = \frac{l_f - l_o}{l_o}$$

Thus

$$e = \frac{185 \text{ mm} - 82 \text{ mm}}{82 \text{ mm}} = 1.25$$

$$\text{Percent lengthening} = e(100\%) = 125\%$$

$$S = 1 + e = 2.25$$

$$\lambda = (1 + e)^2 = 5.06$$

Not only was the belemnite stretched by 125%, but the rock in which the belemnite is encased was stretched 125% as well.

Figure 4.32 Stretched belemnite, broken into an array of separated fragments (dark) between which calcite (white) has precipitated. Lengthening is approximately equal to 125%. [From Milnes (1979). Published with permission of Geological Society of America and the author.]

CHANGES IN THE ANGLE BETWEEN LINES

Although strain parameters of extension (e), stretch (S), and quadratic elongation (λ) effectively describe changes in lengths of lines in deformed bodies, they provide no insight regarding changes that take place in the angles between lines. In most geologic problems, the original absolute orientations of lines in space cannot be determined. Consequently, it is seldom possible to interpret changes in the **absolute orientations** of lines during strain. Given this practical limitation, the standard approach is to describe the changes in **relative orientations** of lines in a body during deformation, particularly the relative orientations of lines that were originally perpendicular.

A parameter called **angular shear**, symbolized by psi (ψ), describes the degree to which two originally perpendicular lines are deflected from 90° (Ramsay, 1967). The shack shown in Figure 4.33*A* has an angular shear of −20°, assuming that the angle between the walls and the foundation was originally exactly 90°. The angular shear is given a negative sign, not because of the quality of the housing but because the sense of movement (rotation) is counterclockwise as we view the front of the shack. The collapsing line of dominos shown in Figure 4.33*B* displays angular shear values ranging from 70° to 0°. The sense of rotation is clockwise.

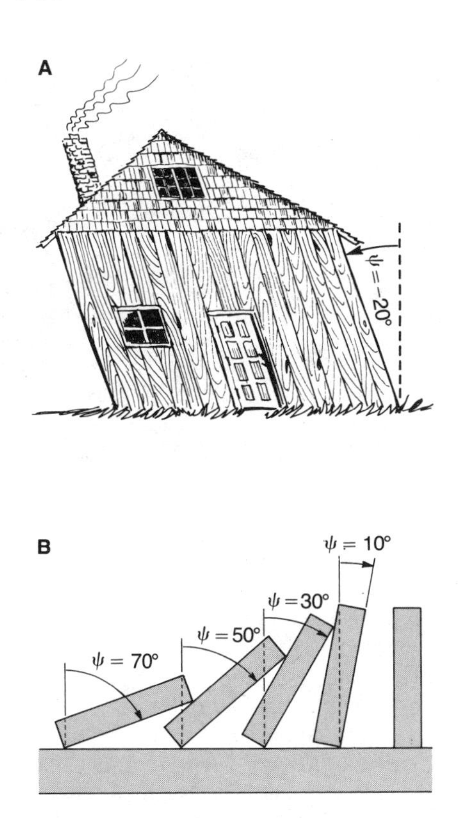

Figure 4.33 (*A*) Angular shear of shack. (*B*) Angular shear of collapsing line of dominos.

Our frame of reference in each of the examples of Figure 4.33 is a horizontal line—the ground surface in the creeping shack example, the table top in the collapsing dominos example. To be precise, we might say that the angular shear of the shack walls *with respect to the ground surface* is −20°, and that the angular shear of the dominos *with respect to the table top* ranges from 0° to +70°. Conversely, we could change our frame of reference and describe an angular shear of +20° for the ground surface with respect to the shack wall, and an angular shear ranging from 0° to −70° for the table top with respect to the dominos. We mentally resist this alternative description because it is difficult for anyone except Californians and compulsive domino players to imagine the ground surface or the table top catastrophically shifting underneath "fixed" houses and domino lines. However, it remains true that the angular relationships shown in Figure 4.33 can be described in two ways.

GEOLOGIC EXAMPLE OF COMPUTING CHANGE IN ANGLE BETWEEN LINES

Where primary objects have been homogeneously distorted, they sometimes provide a means for assessing the angular shear of the rock body in which the objects are contained. For example, fossils sometimes contain perpendicular lines in the functional and/or decorative makeup of their shells. The modification of originally perpendicular lines by distortion can be used as a measure of angular shear.

The distorted trilobite featured in Figure 4.34*A* readily lends itself to appraisal of angular shear. Lines parallel to the original length (line *L–L′*) and to the original width (line *W–W′*) of the trilobite are assumed to have been perpendicular before distortion. Now they intersect not at 90°, but at an angle of 60°. The angular shear of line *L–L′* with respect to line *W–W′* is +30° (Figure 4.34*B*). Angular shear of *W–W′* with respect to *L–L′* is –30° (Figure 4.34*C*).

ANGULAR SHEAR

The parameter ψ is referred to as angular **shear** because changes in orientations of lines can commonly be produced by shearing movements on closely spaced parallel discontinuities like bedding planes, foliation planes,

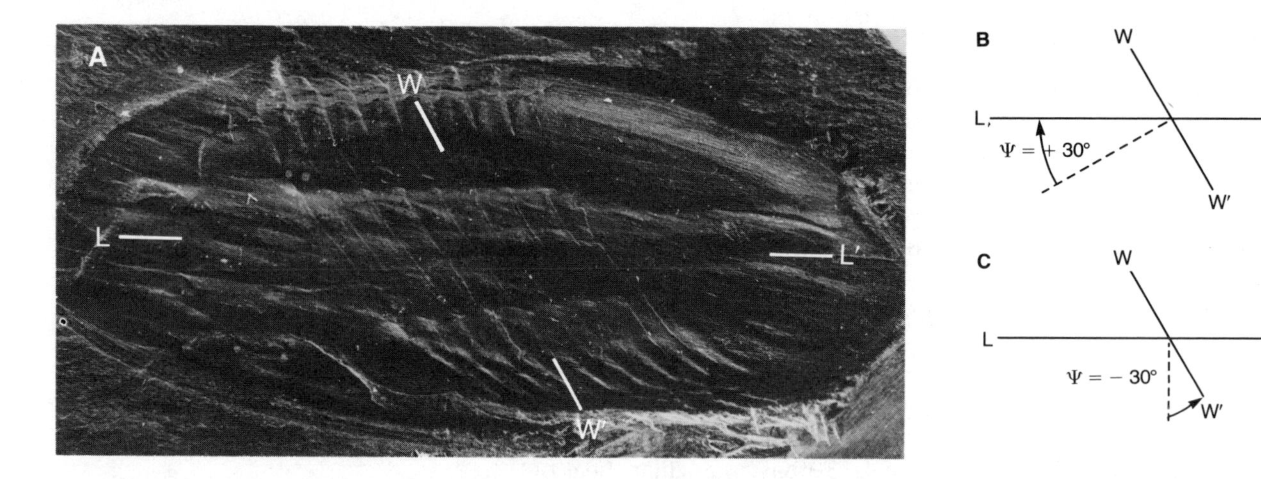

Figure 4.34 (*A*) Distorted trilobite in Cambrian shale, Caernarvonshire, Wales. Angular shear of rock within which this fossil is found can be determined by measuring the angular relationship between lines *L–L'* and *W–W'*, lines that were perpendicular before the deformation. (From *The Minor Structures of Deformed Rocks: A Photographic Atlas* by L. E. Weiss. Published with permission of Springer-Verlag, New York, copyright ©1972.) (*B*) Angular shear of *L–L'* with respect to *W–W'* is +30°. (*C*) Angular shear of *W–W'* with respect to *L–L'* is −30°.

and microfaults. Shearing causes two parts of a body to slide past each other parallel to their surface of contact. A wonderful illustration of this action is seen in the shearing of decks of computer cards (Ragan, 1969). As illustrated in Figure 4.35, the relative orientation of two perpendicular lines drawn on the side of a deck of computer cards can be changed by shearing. In the example shown, the angular shear of line A_f relative to line B_f is +30°. It is −30° for line B_f relative to line A_f. Line A_f has an extension (e) of 0.16. The length of line *A* before shearing was 1.6 in. (4.06 cm). To determine its length after shearing, original length (l_o) is multiplied by stretch (*S*),

$$l_o S = 1.86 \text{ in. } (4.72 \text{ cm})$$

Line *B* did not change in length, for it is aligned parallel to the direction of shearing.

SHEAR STRAIN

Let us consider how points on a line move as a response to shearing. Points 1 to 4 on line A_o in Figure 4.35*A* are translated by various distances as a result

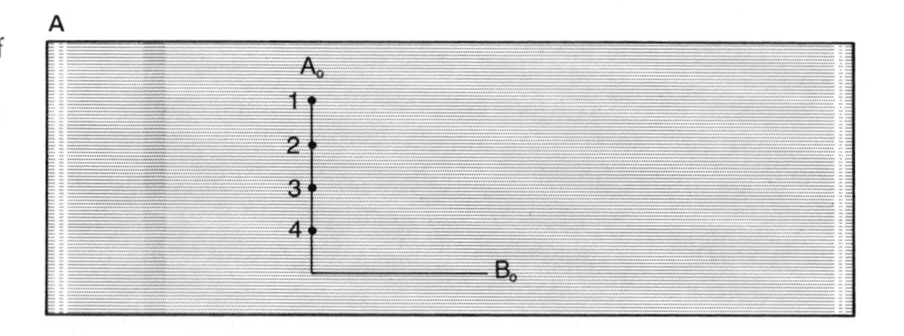

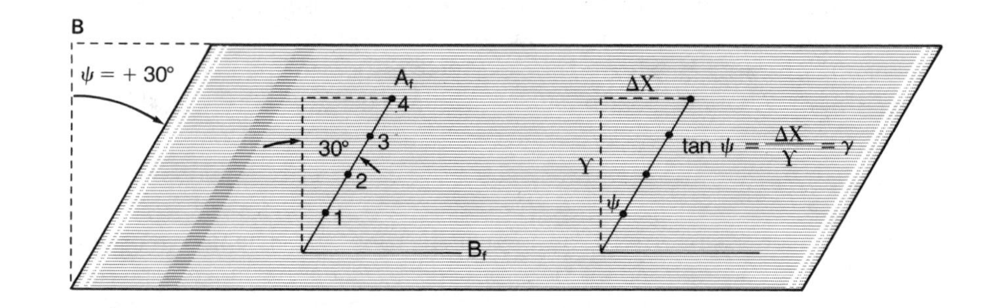

Figure 4.35 Simulation of the shearing of a computer card deck. (*A*) Deck embossed with lines A_o and B_o and points 1 to 4 before deformation. (*B*) Configuration of the deck, including the reference lines and points, after shearing.

of the shearing. Line A_o is the locus of points 1 to 4. Line A_f is the locus of the same points in their deformed locations (Figure 4.35*B*). Since shearing was systematic and deformation was homogeneous, lines A_o and A_f are straight. Points 1 to 4 move a distance that is directly related to the angular shear and to the distance of each point above the base of the deck. If the vertical distance of each point above the base of the deck is denoted as y (Figure 4.35*B*), the horizontal distance of translation can be found as follows (Ramsay, 1967).

$$\tan \psi = \frac{\Delta x}{y} \tag{4.5}$$

$$\Delta x = y \tan \psi$$

Thus **tan ψ** is another way of describing shifts in orientations of lines. It is called **shear strain**, symbolized by the Greek letter gamma (γ),

$$\gamma = \tan \psi \tag{4.6}$$

Shear strain may be positive or negative, depending on the sense of rotation (deflection) of the two mutually perpendicular reference lines. The range of shear strain (γ) is zero to infinity. For the example shown in Figure 4.35*B*, the shear strain (γ) of line A_f is $+\tan 30°$, or 0.58.

A major goal of strain analysis is to identify directions in a deformed body where extension is greatest and least, and where angular shear is maximum and zero.

THE MAGIC OF STRAIN

The elegance and systematic rendering of Earth structure is a gift of the "rules" of homogeneous strain. The study and application of strain theory results in surprising disclosures regarding the regularity and predictability of changes in lengths and orientations of lines.

A useful way to visualize the two-dimensional properties of strain is through homogeneous distortion of a circle. If a body containing a perfectly circular reference marker is subjected to homogeneous deformation, the reference circle will be transformed into a perfect ellipse. Furthermore, if the body is subjected to a second homogeneous deformation, the elliptical form of the deformed reference marker will be transformed into yet another perfect ellipse, regardless of the magnitude and orientation of the second deformation. There is only one exception, the special case of the second deformation being the exact **reciprocal** of the first, returning the ellipse to its original circular form. In general, however, no matter how many times an ellipse is distorted homogeneously, it will be repeatedly transformed into a new ellipse.

To explore the geometric elegance of strain, it is useful to keep on hand a deck of blank computer cards, the flank of which is embossed with a perfect circle and a perfect ellipse (Figure 4.36*A*). Through simple shear of the deck of cards, the outline of the starting circle can be transformed into a perfect ellipse. This is achieved by holding tight to the right end of the card deck with the right hand (Figure 4.36*A*), flexing the deck into the form of a fold (Figure 4.36*B*), and grasping tight the left end of the deck with the left hand and then releasing the right hand (Figure 4.36*C*) (Ragan, 1969). It even works left handed, though the sense of shear is opposite.

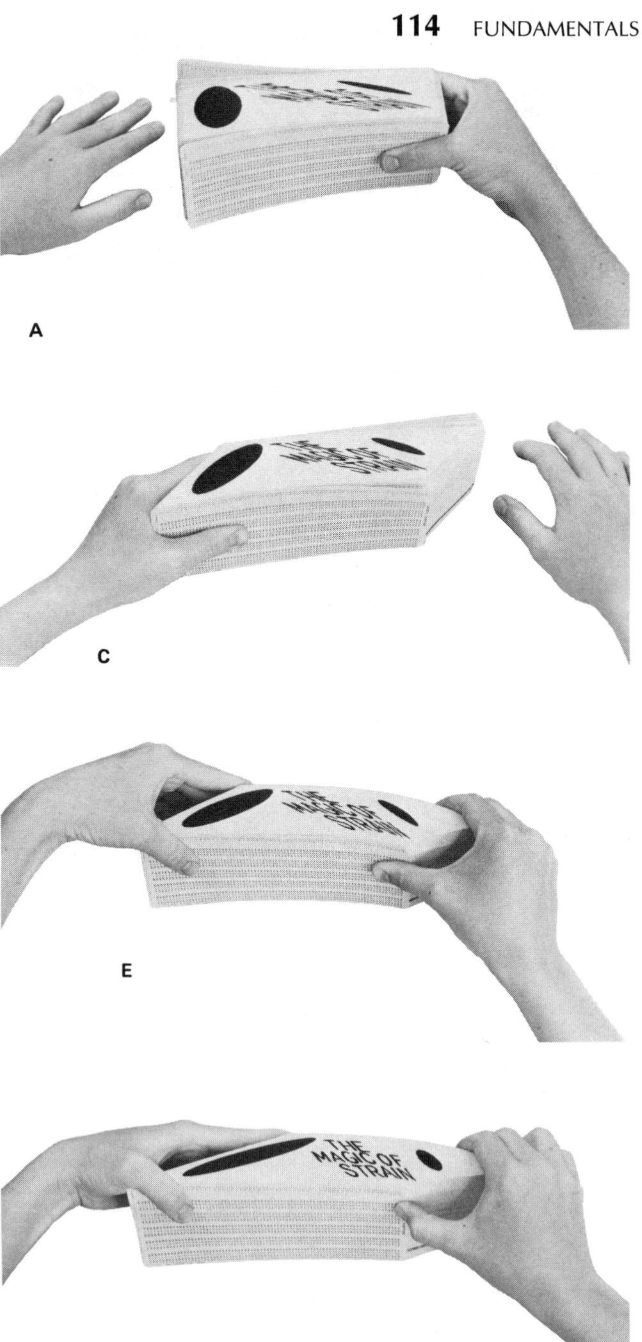

A

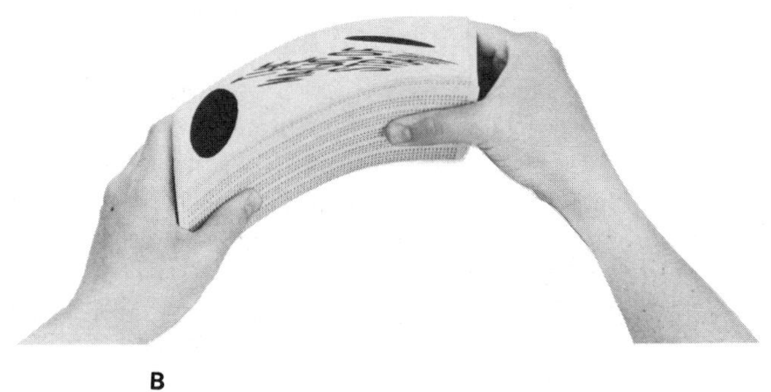

B

C

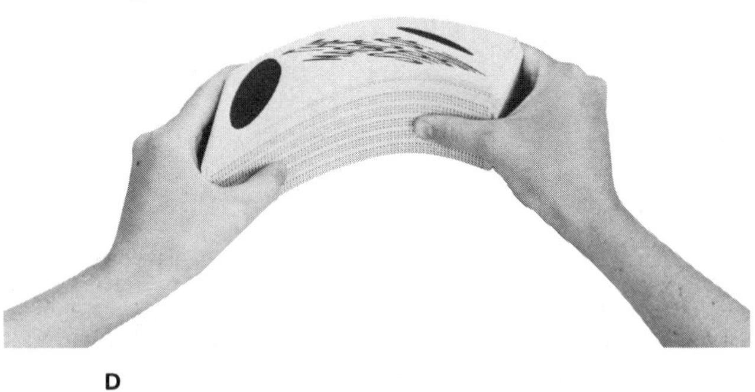

D

E

F

G

H

Figure 4.36 Computer deck demonstration of the magic of strain, carried out by Paige Bausman and friends. (*A*) Undeformed deck on which has been drawn a circle and an ellipse, as well as some indecipherable script. Deck is grasp firmly in right hand to begin. (*B*) Flex of the deck. (*C*) Presto! Two ellipses. (*D*) One more flex. (*E*) Original circle is now strongly elliptical. Original ellipse is now much less elliptical. Indecipherable script begins to become decipherable. (*F, G*) Can't stop. (*H*) Original ellipse is now a circle. Original circle is as elliptical as original ellipse. (*I*) Deck becomes so thinned and stretched that it is hard to support without help. Indecipherable script is indecipherable once again, but the slant of the writing has changed. The original ellipse is an ellipse once again, but its direction of slant has also reversed. The original circle is now profoundly distorted. (Photographs by G. Kew.)

By repeating the flex-and-shear drill a number of times, any ellipse can be made more and more elongate (Figure 4.36*D–H*). The only limit to flattening and extending an ellipse through shearing is the difficulty of trying to hang on to the deck as it progressively thins and lengthens (Figure 4.36*I*). Nature has no such limit.

The outline of the ellipse traced onto the flank of the deck (see Figure 4.36*A*) is transformed by shearing into a variety of ellipses, and in one special case into a perfect circle (see Figure 4.36*H*). Try it.

THE STRAIN ELLIPSE

Given the perfection of ellipses derived from homogeneous deformation (distortion) of circles and ellipses, it is no wonder that the strain within geologic bodies is conventionally described through the image of a **strain ellipse**. A strain ellipse pictures the magnitude and configuration of the degree of distortion accommodated by a geologic body. It pictures how the shape of an imaginary circular reference object, or perhaps a not-so-imaginary geologic object (Figure 4.37), would be changed as a result of distortion.

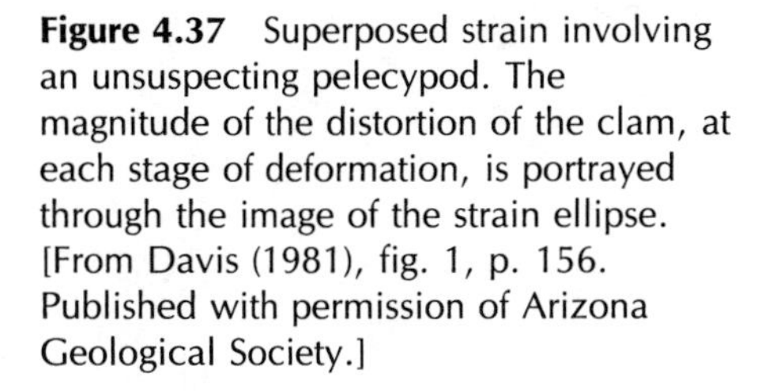

Figure 4.37 Superposed strain involving an unsuspecting pelecypod. The magnitude of the distortion of the clam, at each stage of deformation, is portrayed through the image of the strain ellipse. [From Davis (1981), fig. 1, p. 156. Published with permission of Arizona Geological Society.]

The **principal axes (X** and **Z)** of a strain ellipse are mutually perpendicular, parallel to the directions of **maximum and minimum extension** within the deformed body (Figure 4.38). Of all the lines drawn within an originally circular (or elliptical) body before deformation, the lines that end up parallel to the long axis of the strain ellipse are marked by the greatest extension (e) value; the lines that end up parallel to the short axis of the strain ellipse are marked by the smallest possible value of extension (e). This is illustrated in Figure 4.39. Lines *X′* and *Z′*, drawn through the center of an undeformed reference circle (Figure 4.39*A*), are the two lines that become parallel to the long (*X*) and short (*Z*) axes of the distorted circle, the strain ellipse (Figure 4.39*B*). Line *X′* changes in length from l_o = 19 units to l_f = 31.5 units. Line *Z′* changes in length from l_o = 19 units to l_f = 11.5 units. The extension value

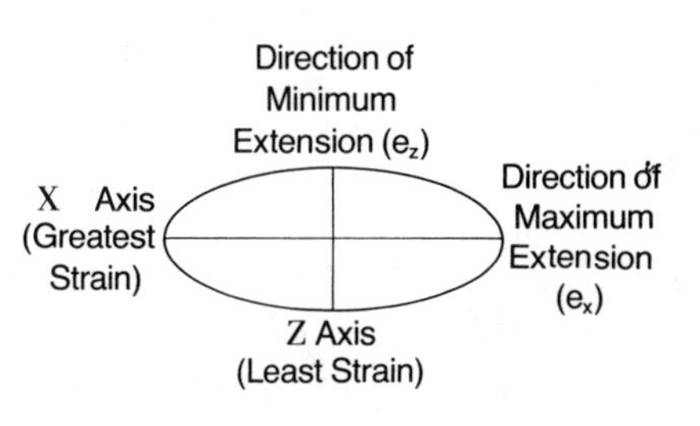

Figure 4.38 Elements of a strain ellipse.

A

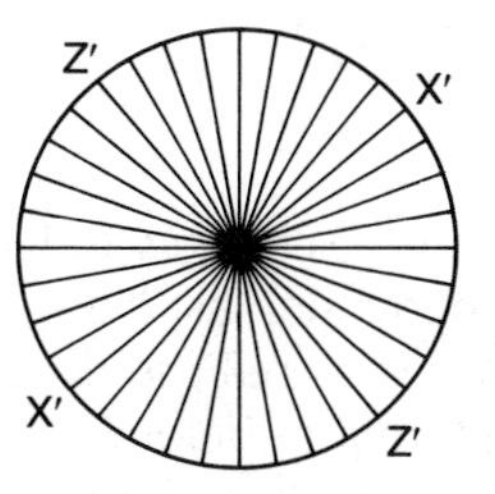

B

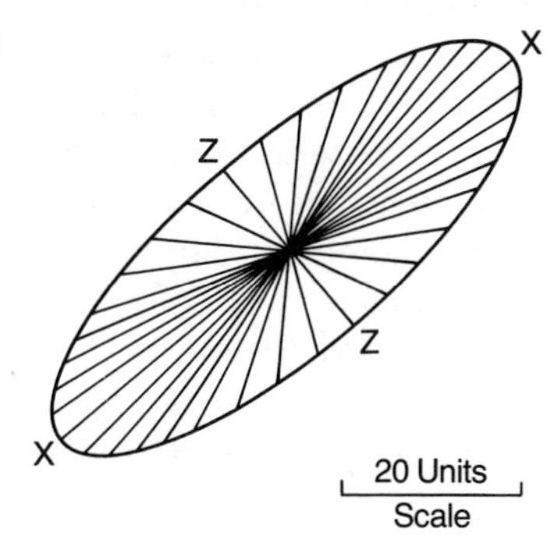

Figure 4.39 (*A*) Undeformed circular body inscribed with lines of common length but different orientations. (*B*) After deformation, almost all the lines have changed in length and orientation. Line *X* was stretched the most; line *Z* was shortened the most. (The original locations and orientations of *X* and *Z* were *X′* and *Z′*, respectively.) Lines *X* and *Z* constitute the axes of the strain ellipse that describe the deformation.

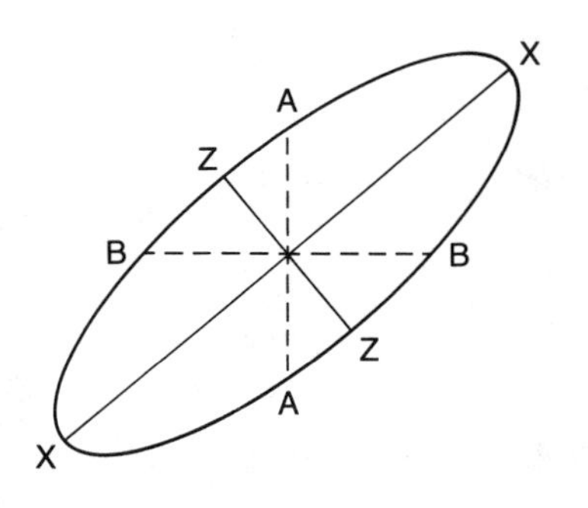

Figure 4.40 Ellipse, derived from homogeneous deformation of a circle. Lines *X* and *Z*, the mutually perpendicular axes of this strain ellipse, are the only two mutually perpendicular lines in the deformed body that were originally perpendicular, before deformation. All other perpendicular lines in the ellipse, like lines *A* and *B*, were not perpendicular in the original reference circle, before deformation.

for line *X* is thus $e_X = 0.66$. $e_Z = -0.39$. Quadratic elongation values for λ_X and λ_Z are 2.75 and 0.37, respectively. If, patiently, we proceeded to measure the original and deformed lengths of each line shown in Figure 4.39, computing the values of extension for each, we would find that no extension value exceeded that calculated for line *X*. And no extension value was less than that computed for line *Z*. Thus, of all lines in the original starting circle, the one that ends up parallel to the long axis (*X*) of the strain ellipse lengthens the most. And the one that ends up parallel to the short axis (*Z*) of the strain ellipse shortens the most.

The axes of a strain ellipse are special in yet another way. They are **directions of zero angular shear**. This means that the lines that end up parallel to the principal axes of the ellipse must have been perpendicular before deformation as well. In fact, they are the only perpendicular lines in the distorted body that were perpendicular before distortion. Lines *X* and *Z* in Figure 4.40 are the mutually perpendicular axes of the strain ellipse. They started off at right angles before deformation. All other sets of mutually perpendicular lines in the distorted body, like the set composed of lines *A* and *B*, were not perpendicular before deformation.

CALIBRATING THE STRAIN ELLIPSE

The strain ellipse can be calibrated for purposes of evaluating changes in lengths and relative orientations of lines during nonrigid body distortion. If the radius of an original unstrained reference circle is considered to be 1 unit (Figure 4.41*A*), the lengths of the principal axes of the strain ellipse can be described in a very convenient manner (Ramsay, 1967). The final length (l_f) of a line, we have learned, is equal to its original length (l_o) multiplied by stretch (*S*). Thus the semiaxis length of *X* (Figure 4.41*B*) is equal to the radius of the undeformed reference circle ($l_o = 1.0$) multiplied by the factor $(1 + e_X)$, which is the value of stretch in the *X* direction. Since the radius of the reference circle is taken to be 1.0, the length of the semiaxis of the strain ellipse in the *X* direction is simply S_X, stretch in the *X* direction, otherwise known as the square root of quadratic elongation in the *X* direction ($\sqrt{\lambda_x}$). Similarly, the length of the semiaxis of the ellipse measured in the *Z* direction is equal to $(1 + e_Z) = S_Z = \sqrt{\lambda_Z}$ (Figure 4.41*B*).

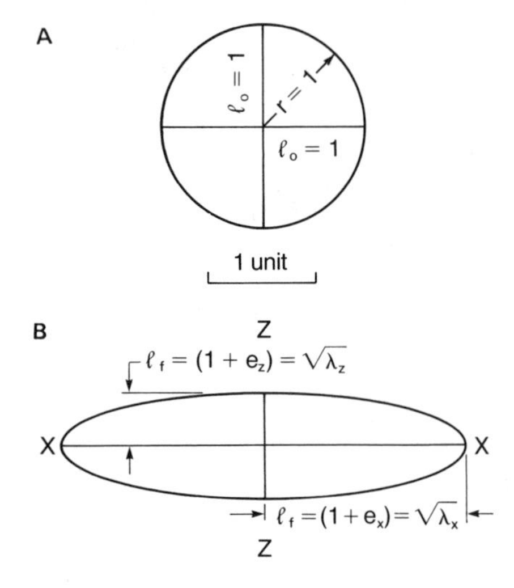

Figure 4.41 (*A*) Original unstrained reference circle. Radius of the circle is 1 unit. (*B*) Strain ellipse resulting from homogeneous deformation of the circle. The lengths of the axes of the ellipse are calibrated with respect to extension (e) and quadratic elongation (λ).

EVALUATING THE STRAIN OF LINES IN A BODY

In carrying out strain analysis, we must evaluate changes in lengths and relative orientations of *all* lines in a geologic body, not just the special lines that end up parallel to the principal axes of the strain ellipse describing the deformation. At Mount St. Helens, the monitoring of deformation has become a way of life. As I write, displacements of the Mount St. Helens volcanic edifice are being monitored by laser beam surveying of relatively widely spaced marker targets in an outer net of survey stations (Figure 4.42*A*). Moreover, the finer details of displacements are being monitored, using a wide variety of methods, within an inner survey net in the crater itself adjacent to an emerging central volcanic dome (Figure 4.42*B*). The floor of the crater is shortening by thrust faulting as a response to the emplacement of the dome (Figure 4.42*C*). The thrusts are by no means insignificant. Up to several hundred meters long, they are accommodating translations that over a period of several months amount to tens of meters. Just before times of eruption, translation on individual faults has exceeded 50 cm/day. It is quite a show!

Monitoring and mapping the changes in lengths and relative orientations of lines in the crater floor of Mount St. Helens is dangerous business. For our

Figure 4.42 (*A*) Keeping vigil at Mount St. Helens, and measuring all suspicious movements. (Photograph by K. Cashman. Courtesy of United States Geological Survey.) (*B*) An emerging volcanic dome in the center of the crater. (Photograph by K. Cashman. Courtesy of United States Geological Survey.) (*C*) The curved traces of thrust faults formed by shortening of the floor of the crater in response to the growth of the dome. Scarp in lower right is about 5 m high. (Photograph by D. A. Swanson. Courtesy of United States Geological Survey.)

purposes, it is more practical, though admittedly less exciting, to monitor the strain of lines and learn the geometry of distortion mathematically. An advantage above and beyond the safety of environment is the opportunity to model perfectly homogeneous deformation, thus avoiding the heat given off by the heterogeneous display of strain at Mount St. Helens.

Figure 4.43 pictures the hypothetical experimental deformation of a clay cake. It shows how a perfectly deformable clay cake ought to distort, given the rules of ideal strain. A reference circle with internal reference lines (Figure 4.43*A*) shows the undistorted nature of the clay cake before deformation. The deformed state of the original reference circle is shown in Figure 4.43*B*. Given this special opportunity to view both the starting materials and the finished product, we can describe the strain of lines of any orientation in the model.

Describing the **strain of lines** means examining the three ways that any reference line can respond to distortion: by undergoing a **change in length**; by experiencing a **shear strain** (i.e., changing its orientation relative to an originally perpendicular reference line); and by experiencing **internal rotation** (i.e., changing its orientation relative to the direction of greatest principal strain, X). Let us start out by examining the strain of line L (Figure 4.43*B*).

Before deformation, the length of line L was 1 unit; in the strained state, L is 1.11 units. Using Equation 4.4, the value of quadratic elongation (λ) for L can be found as follows.

$$\lambda = (1 + e)^2 = (1.11)^2 = 1.23$$

Using Equation 4.2, the stretch (S) and extension (e) values for L in the deformed state can be determined.

$$S = (1 + e) = \sqrt{\lambda} = 1.11$$

$$e = S - 1 = 0.11$$

In its undeformed state (see Figure 4.43*A*), line *L* made an angle of $\theta = 50°$ with the *X* axis. In its strained state (see Figure 4.43*B*), it makes an angle of 26.5° with the *X* axis. This change in absolute orientation is the **internal angle of rotation (α)**. The internal angle of rotation (α) for *L* is +23.5° (clockwise rotation). *This value is not angular shear,* for angular shear (ψ) is the deflection of *L* from a 90° intersection with some other reference line. We can directly measure the angular shear of line *L* by studying its relationship with line *M*, which was perpendicular to *L* before deformation (see Figure 4.43*A*). After deformation, however, *M* and *L* are no longer perpendicular (see Figure 4.43*B*). The acute angle between *L* and *M* after deformation is 45°. Angular shear of *L* with respect to *M* is +45° (Figure 4.43*C*). The shear strain, $\gamma = \tan \psi$, is 1.00.

THE STRAIN EQUATIONS

Ramsay (1967) presents a step-by-step derivation of the fundamental equations that express the values of quadratic elongation (λ) and shear strain (γ) for any line in a deformed body, like *L* or *M* in Figure 4.43. Quadratic elongation and shear strain are expressed as functions of λ_1, λ_3, and the angle (θ_d) that a line makes with the *X* axis in the deformed state. Maximum and minimum quadratic elongations, λ_1, and λ_3, correspond to λ_X and λ_Z, respectively.

$$\lambda' = \frac{\lambda_1' + \lambda_3'}{2} - \frac{\lambda_3' - \lambda_1'}{2} \cos 2\theta_d \quad (4.7)$$

$$\frac{\gamma}{\lambda} = \frac{\lambda_3' - \lambda_1'}{2} \sin 2\theta_d \quad (4.8)$$

where

$$\lambda' = 1/\lambda,\ \lambda_1' = 1/\lambda_1,\ \text{and}\ \lambda_3' = 1/\lambda_3.$$

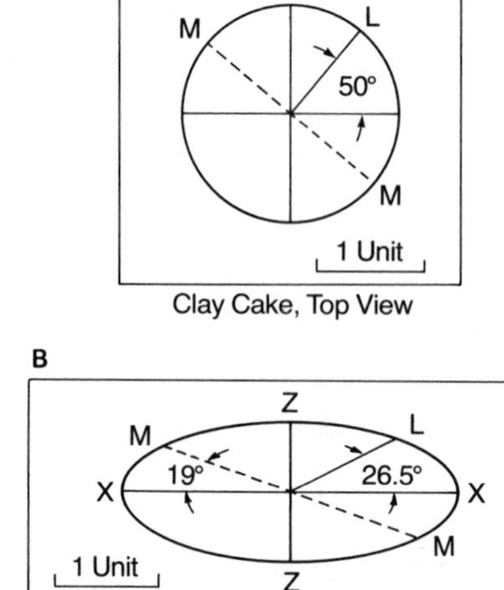

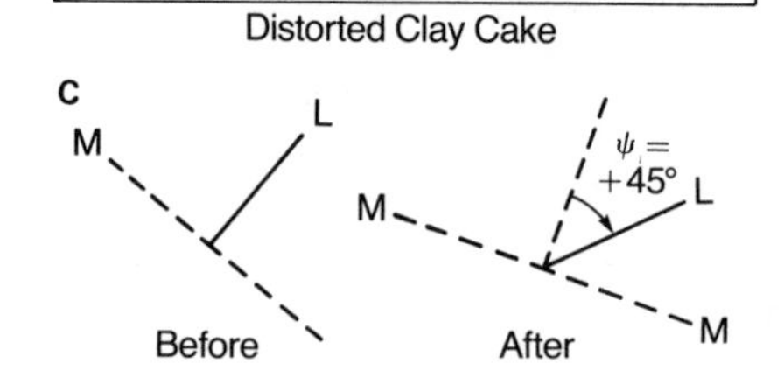

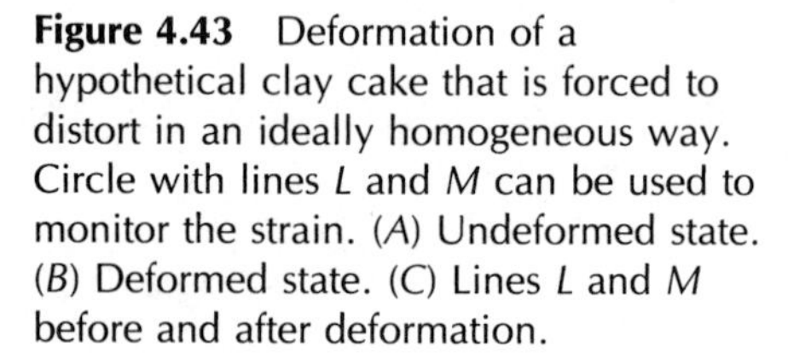

Figure 4.43 Deformation of a hypothetical clay cake that is forced to distort in an ideally homogeneous way. Circle with lines *L* and *M* can be used to monitor the strain. (*A*) Undeformed state. (*B*) Deformed state. (*C*) Lines *L* and *M* before and after deformation.

The strain equations can be solved readily with the aid of a calculator if values of λ_1, λ_3, and θ_d are known. To calculate λ and γ for line *L* in Figure 4.43*B*, the values of $\lambda_1 = \lambda_X$ and $\lambda_3 = \lambda_Z$ must be computed. Since the value of the original reference circle is 1.0 unit, the semiaxis lengths of the strain ellipse (see Figure 4.43*B*) are $1 + e_X$ and $1 + e_Z$, respectively. Using Equation 4.4,

$$\lambda_X = \lambda_1 = (1 + e_X)^2 = 2.40$$

The measured length of $1 + e_Z$ is 0.65. Thus,

$$\lambda_Z = \lambda_3 = (1 + e)^2 = 0.42$$

The measured value of θ_d is 26°.

Using Equations 4.7 and 4.8, the quadratic elongation (λ) and shear strain (λ) for line *L* are calculated as follows.

$$\lambda' = \frac{\frac{1}{2.4} + \frac{1}{0.42}}{2} - \frac{\frac{1}{0.42} - \frac{1}{2.4}}{2} \cos 52° = 0.79$$

$$\lambda = \frac{1}{\lambda'} = 1.26$$

Furthermore,

$$\frac{\gamma}{\lambda} = \frac{\frac{1}{0.42} - \frac{1}{2.4}}{2} \sin 52° = 0.774$$

$$\gamma = \frac{\gamma}{\lambda} \lambda = 0.774(1.26) = 0.975$$

$$\psi = \arctan 0.975 = 44°$$

Another strain equation presented by Ramsay (1967) is the basis for establishing the internal angle of rotation (α) that a line endures during distortion,

$$\tan \theta_d = \tan \theta \left(\frac{\lambda_3}{\lambda_1}\right)^{1/2} \tag{4.9}$$

where,

θ_d = angle between X and the line in deformed state
θ = angle between X and the line in undeformed state

Equation 4.9 reveals that the internal angle of rotation depends on only two factors: the initial orientation that the line in question makes with the X axis, and the ratio of principal quadratic elongation values.

We can try out Equation 4.9 on line L. From Figures 4.43*A* and 4.43*B* we know that $\theta_d = 26.5°$ and $\theta = 50°$. And we have already determined that $\lambda_1 = 2.4$ and $\lambda_3 = 0.42$. Substituting these values into Equation 4.9,

$$\tan 26.5° = \tan 50° \left(\frac{0.42}{2.4}\right)^{1/2}$$

$$0.4986 = 1.1918\ (0.1750)^{1/2}$$

$$0.4986 = 0.4986$$

CALCULATING THE VARIATIONS IN STRAIN

In order to see the power of the strain equations (Equations 4.7, 4.8, and 4.9) and the relations they depict, let us try them out in an applied, visual way. Figure 4.44*A* shows the lengths and orientations of a number of lines (lines *a* through *s*) that are plotted at 10° intervals in the experimentally deformed clay cake of Figure 4.43*B*. The original orientations and lengths of lines *a* through *s* are shown in Figure 4.44*B*. Figure 4.44*C* isolates lines *a* through *j* and illustrates how each changes in length and orientation as a result of the deformation.

As we have seen, change in length can be computed through Equation 4.7, shear strain can be calculated using Equation 4.8, and internal rotation can be calculated through the use of Equation 4.9. Indeed, the strain equations allow the strain of lines to be tracked in the manner summarized in Figure 4.44, especially Figure 4.44*C*.

The computations that were carried out to construct Figure 4.44 are summarized in Tables 4.1, 4.2, 4.3, and 4.4. The data of Table 4.1 permit the original orientations of lines *a* through *s* to be constructed (see Figure 4.44*B*). The data shown in Tables 4.2 and 4.3 permit quadratic elongation and shear strain, respectively, to be determined. Table 4.4 summarizes the change in orientation and length of each line, *a* through *s*.

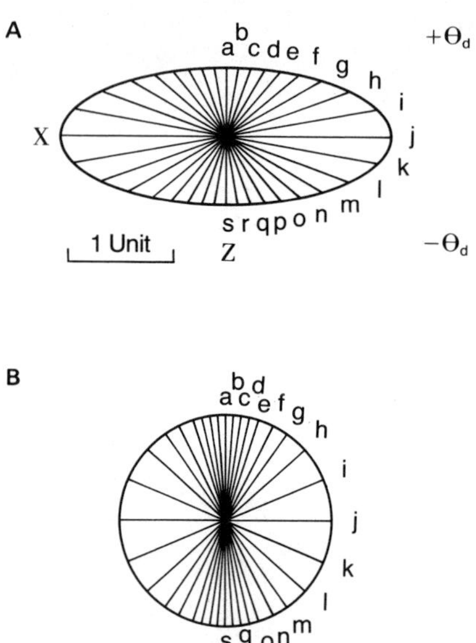

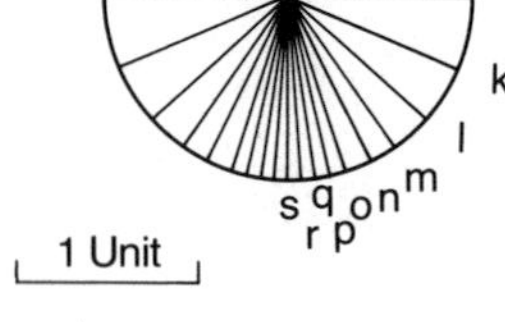

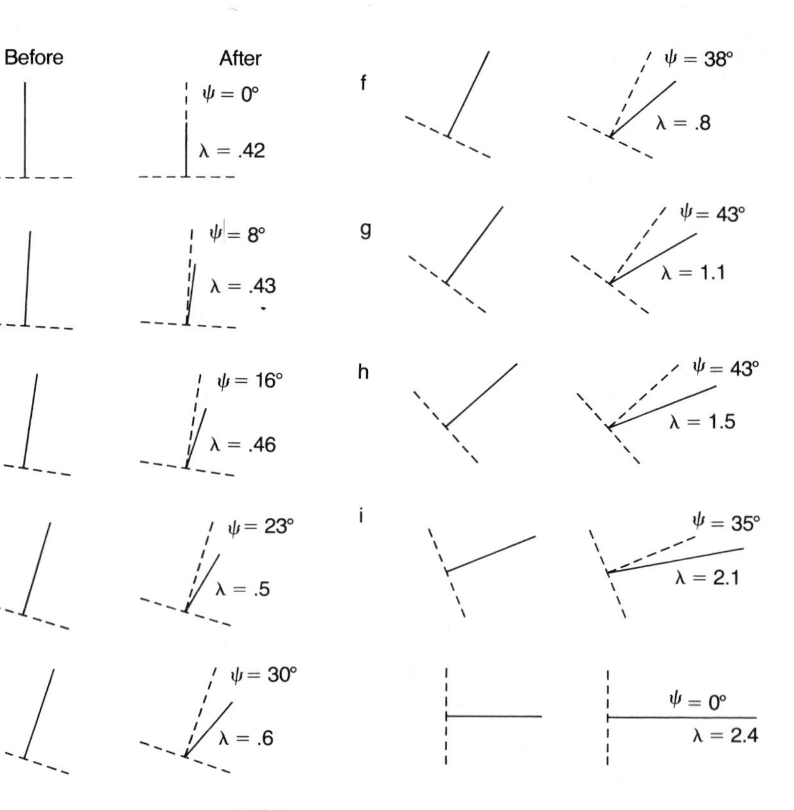

Figure 4.44 (*A*) Clay cake in the deformed state, inscribed with lines *a* through *s* at 10° intervals. (*B*) Orientation of lines *a* through *s*, as the lines would look if the strain were removed. (*C*) Changes in length and orientation as a result of the distortion of lines *a* through *s*.

Table 4.1
Calculation of the Original Orientations (θ) of Lines a–s (see Fig. 4.44)

θ = undeformed state; θ_d = deformed state

Line	θ_d	*tan* θ_d	*tan* θ	θ
a	90°	∞	∞	90°
b	80°	5.6713	13.558	86°
c	70°	2.7475	6.5682	81°
d	60°	1.7321	4.1407	76°
e	50°	1.1918	2.8490	71°
f	40°	.8391	2.0060	63°
g	30°	.5774	1.3802	54°
h	20°	.3640	.8701	41°
i	10°	.1763	.4215	22°
j	0°	.0000	.0000	0°
k	−10°	−.1763	−.4215	−22°
l	−20°	−.3640	−.8701	−41°
m	−30°	−.5774	−1.3802	−54°
n	−40°	−.8391	−2.0060	−63°
o	−50°	−1.1918	−2.8490	−71°
p	−60°	−1.7321	−4.1407	−76°
q	−70°	−2.7475	−6.5682	−81°
r	−80°	−5.6713	−13.558	−86°
s	−90°	−∞	−∞	−90°

Basic calculations: $\tan \theta_d = \tan \theta \left(\frac{\lambda_3}{\lambda_1}\right)^{1/2}$

$$\tan \theta = \frac{\tan \theta_d}{(\lambda_3/\lambda_1)^{1/2}}$$

where $(\lambda_3/\lambda_1)^{1/2} = \frac{0.42}{2.40} = .4183$

Table 4.2
Calculation of Quadratic Elongation (λ) for Lines *a*–*s* (See Figure 4.44)

State of Strain:

$e_1 = .55 \qquad \lambda_1 = 2.40 \qquad \lambda_1' = \frac{1}{\lambda_1} = .4167$

$e_3 = -.35 \qquad \lambda_3 = 0.42 \qquad \lambda_3' = \frac{1}{\lambda_3} = 2.3810$

Line	θ_d	cos $2\theta_d$	λ'	λ
a	90°	−1.000	2.3811	.42
b	80°	−.940	2.3222	.43
c	70°	−.766	2.1531	.46
d	60°	−.500	1.8900	.5
e	50°	−.174	1.5698	.6
f	40°	.174	1.2280	.8
g	30°	.500	.9078	1.1
h	20°	.766	.6465	1.5
i	10°	.940	.4756	2.1
j	0°	1.000	.4167	2.4
k	−10°	.940	.4756	2.1
l	−20°	.766	.6465	1.5
m	−30°	.500	.9078	1.1
n	−40°	.174	1.2280	.8
o	−50°	−.174	1.5698	.6
p	−60°	−.500	1.8900	.5
q	−70°	−.766	2.1513	.46
r	−80°	−.940	2.3222	.43
s	−90°	−1.000	2.3811	.42

Basic Calculations:

$$\lambda' = \frac{\lambda_1' + \lambda_3'}{2} - \frac{\lambda_3' - \lambda_1'}{2} \cos 2\theta_d \qquad \lambda = \frac{1}{\lambda'}$$

$$\frac{\lambda_1' + \lambda_3'}{2} = 1.3989, \text{ and } \frac{\lambda_3' - \lambda_1'}{2} = .9822$$

Table 4.3

Calculation of the Shear Strain (γ) and Angular Shear (ψ) for Lines *a–s* (see Figure 4.44)

State of Strain: $e_1 = .55$ $\lambda_1 = 2.40$ $\lambda'_1 = .4167$
$e_3 = -.35$ $\lambda_3 = 0.42$ $\lambda'_3 = 2.3810$

Line	θ_d	sin $2\theta_d$	γ/λ	γ	ψ
a	90°	.0000	.0000	0.00	0°
b	80°	.3420	.3359	.14	8°
c	70°	.6428	.6314	.29	16°
d	60°	.8660	.8506	.42	23°
e	50°	.9848	.9673	.58	30°
f	40°	.9848	.9673	.77	38°
g	30°	.8660	.8506	.94	43°
h	20°	.6428	.6314	.95	43°
i	10°	.3420	.3359	.70	35°
j	0°	.0000	.0000	0.00	0°
k	−10°	−.3420	−.3359	− .70	−35°
l	−20°	−.6428	−.6314	− .95	−43°
m	−30°	−.8660	−.8506	− .94	−43°
n	−40°	−.9848	−.9673	− .77	−38°
o	−50°	−.9848	−.9673	− .58	−30°
p	−60°	−.8660	−.8506	− .42	−23°
q	−70°	−.6428	−.6314	− .29	−16°
r	−80°	−.3420	−.3359	− .14	− 8°
s	−90°	.0000	.0000	0.00	0°

Basic Calculations:

$$\gamma/\lambda = \frac{\lambda'_3 - \lambda'_1}{2} \sin 2\theta_d$$

$$\gamma = \frac{\gamma}{\lambda}\lambda$$

$$\psi = \arctan \gamma$$

$$\frac{\lambda'_3 - \lambda'_1}{2} = .9822$$

Table 4.4

Summary of Internal Rotation and Changes in Length of Lines for Lines *a–s* (see Figure 4.44)

Line	*Internal Rotation* (α)	*Change in Line Length* (Δl)
a	0°	−.35 units
b	6° c*	−.34
c	11° c	−.32
d	16° c	−.29
e	21° c	−.22
f	23° c	−.10
g	24° c	+.05
h	21° c	+.22
i	12° c	+.45
j	0° c	+.55
k	12° cc*	+.45
l	21° cc	+.22
m	24° cc	+.05
n	23° cc	+.10
o	21° cc	−.22
p	16° cc	−.29
q	11° cc	−.32
r	6° cc	−.34
s	0° cc	−.35

* c = clockwise
cc = counter clockwise

The calculations and graphical displays serve to underscore the fundamental properties of homogeneous strain. Namely:

1. The principal axes of strain are directions of maximum (X) and minimum (Z) quadratic elongation and zero shear strain.
2. Within any body that has undergone a pure distortion, there are always two directions along which there has been neither lengthening or shortening; lines oriented in these directions are characterized by quadratic elongation values of 1.0.
3. Within any homogeneously distorted body there are two directions marked by maximum shear strain.
4. Quadratic elongation and shear strain values increase and decrease systematically according to direction in a deformed body; specific values of quadratic elongation and shear strain depend on the magnitudes of λ_1, λ_3, and θ_d.

THE MOHR CIRCLE STRAIN DIAGRAM

The fundamental strain equations (4.7 and 4.8) can be learned and applied more readily when the geometry reflected in these equations is understood. The strain equations were gathered into their present form by Otto Mohr (1882), who recognized that as equations of a circle they could be conveniently drawn graphically. The **Mohr circle strain diagram**, the graphical construction of the strain equations, presents the systematic variations in quadratic elongation and shear strain in a way that is both practical and versatile.

Let us construct a Mohr diagram to represent the same state of strain as that displayed by the hypothetically deformed clay cake (see Figure 4.44*B*): $\lambda_1 = 2.4$, $\lambda_3 = 0.42$. Another rendition of the deformation experiment is presented in Figure 4.45*A*, this one containing a single reference line, *A*, that in the deformed state is line A_d. Line A_d makes an angle of $\theta_d = 15°$ with the *X* axis of the strain ellipse.

Our goals are to construct a Mohr circle strain diagram on the basis of this information and to determine the values of quadratic elongation (λ) and angular shear (ψ) for line A_d. The Mohr circle strain diagram is plotted in x–y space such that values of λ', λ'_1, and λ'_3 are plotted and measured on the x axis (Figure 4.45*B*). Here λ'_1 and λ'_3 are the reciprocal values of the maximum and minimum quadratic elongations. A circle centered on the x axis and drawn through the values of λ'_1 and λ'_3 is the Mohr circle proper. Its circumference can be thought of as the locus of hundreds of points, two of which are λ'_1 and λ'_3. The x–y coordinates of each of these points on the circumference of the Mohr circle are paired values of λ' and γ/λ. These values permit quadratic elongation and shear strain to be calculated for lines of every orientation in the deformed body shown in Figure 4.44*B*.

For example, λ'_1 has (x, y) coordinates of (0.42, 0.0), corresponding to the reciprocal of quadratic elongation for lines parallel to the direction of maximum extension (*X*). Lines thus oriented are marked by 0.0 shear strain. Similarly, λ'_3 has (x, y) coordinates of (2.4, 0.0), the x value being the reciprocal of quadratic elongation for lines parallel to the axis of minimum extension (*Z*). The y value of zero again reflects 0.0 shear strain for lines thus oriented.

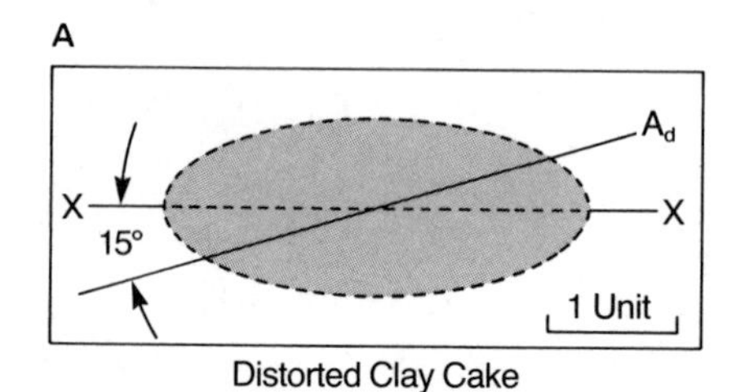

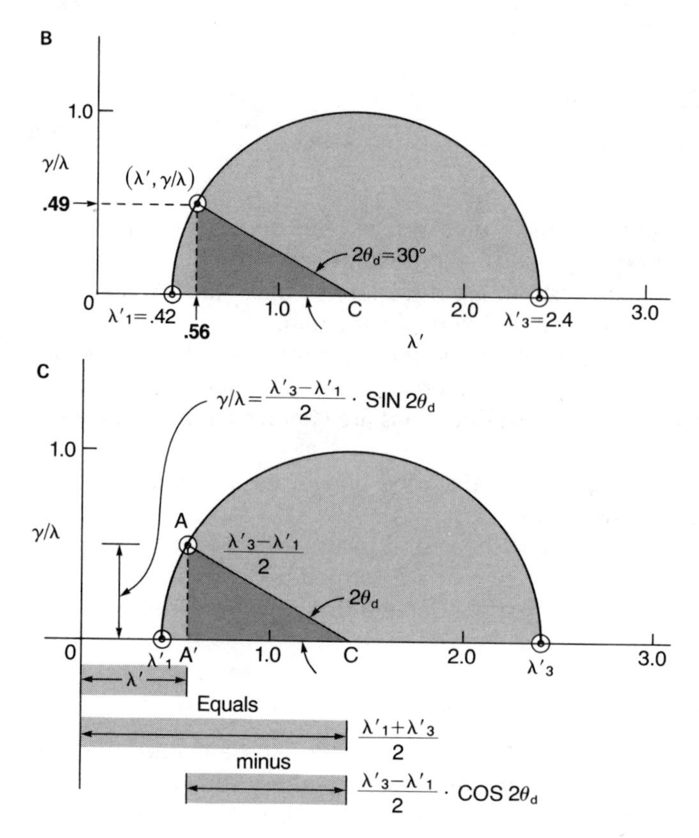

Figure 4.45 (*A*) Distortion of line *A* to line A_d. (*B*) Mohr circle strain diagram. (*C*) Mohr circle strain diagram, labeled to show the relation of the geometry of the diagram to the components of the basic strain equations.

The paired values of $(\lambda', \gamma/\lambda)$ for line A_d lie at a point somewhere on the perimeter of the Mohr circle, but where? Line A_d lies 15° **counterclockwise** from the axis of maximum extension (X) (see Figure 4.45*A*). The location of the point on the Mohr circle representing strain values for this line can be found by plotting a radius **clockwise** from λ_1' by an angle of $2\theta_d$ or 30°. The point where the radius intersects the perimeter of the circle has (x, y) values that correspond to $(\lambda', \gamma/\lambda)$. For line A_d,

$$\lambda' = 0.56$$

$$\frac{\gamma}{\lambda} = 0.49$$

Thus,

$$\lambda = \frac{1}{\lambda'} = \frac{1}{0.56} = 1.78$$

$$\gamma = \frac{\gamma}{\lambda}\lambda = 0.49(1.78) = 0.87$$

Since,

$$\gamma = \tan \psi$$

$$\psi = \arctan \gamma = 41°$$

A close-up view of the Mohr circle strain diagram (Figure 4.45C) helps to explain the relation between the geometry of the construction and the elements of the fundamental strain equations. We want to be certain that the λ' value determined for A_d in the diagram really conforms to Equation 4.7, which is,

$$\lambda' = \frac{\lambda_1' + \lambda_3'}{2} - \frac{\lambda_3' - \lambda_1'}{2}\cos 2\theta_d$$

The first component of the equation,

$$\frac{\lambda_1' + \lambda_3'}{2}$$

is the x value of the center (C) of the Mohr circle, and this value is equal to the length of OC in Figure 4.45C. The second component of the equation,

$$\frac{\lambda_3' + \lambda_1'}{2}$$

is the length of the radius of the Mohr circle. Its value is equal to the length of radii, like line CA (Figure 4.45C). The third component of the equation,

$$\cos 2\theta_d$$

is equal to the quotient CA'/CA. Substituting into Equation 4.7,

$$\lambda' = \frac{\lambda_1' + \lambda_3'}{2} - \frac{\lambda_3' - \lambda_1'}{2}\cos 2\theta_d$$

$$\lambda' = OC - CA\frac{CA'}{CA}$$

$$\lambda' = OC - CA' \text{ (see Figure 4.45C)}$$

Examining Equation 4.8 in the same fashion (Figure 4.45C),

$$\sin 2\theta_d = \frac{AA'}{CA}$$

$$\frac{\lambda_3' - \lambda_1'}{2} = CA$$

$$\frac{\lambda_3' - \lambda_1'}{2} \sin 2\theta_d = CA\frac{AA'}{CA} = AA'$$

To gain confidence in the Mohr diagram and its use, feel free to solve for the λ' and γ/λ values for any of the lines shown in Figure 4.44*B*, but this time extract the values directly from the Mohr strain circle (see Figure 4.45*B*). For example, the λ' and γ/λ values for line $C(\theta_d = +70°)$ are found by plotting a radius at an angle of $2\theta_d = +140°$ clockwise from the *x* axis, and then noting the $(\lambda', \gamma/\lambda)$ coordinates of the point where this radius pierces the circumference. The $(\lambda', \gamma/\lambda)$ values can be compared to the calculated values presented in Tables 4.2 and 4.3.

THE STRAIN ELLIPSOID AND PLANE STRAIN

The three-dimensional counterpart of the strain ellipse is called the **strain ellipsoid**. It pictures how the shape of an imaginary spherical reference object would be changed as a result of distortion. It is defined by axes of **greatest** (*X*) **intermediate** (*Y*), and **least** (*Z*) strain (Figure 4.46). The *X*, *Y*, and *Z* axes are mutually perpendicular and are parallel to directions of zero angular shear.

It is generally assumed, but it is not always true, that extension (e) parallel to *Y* equals 0.0, and if stretching in the *X* direction is perfectly compensated by shortening in the *Z* direction, there is no change in volume of the deformed body. This special state of strain is known as **plane strain**.

THE STRAIN ELLIPSOID AND ITS APPLICATION

Pioneering applied strain analysis was carried out by Ernst Cloos (1947) of Johns Hopkins University. He painstakingly analyzed tiny ellipsoids derived from originally near-spherical primary objects (ooids) in Cambrian and Or-

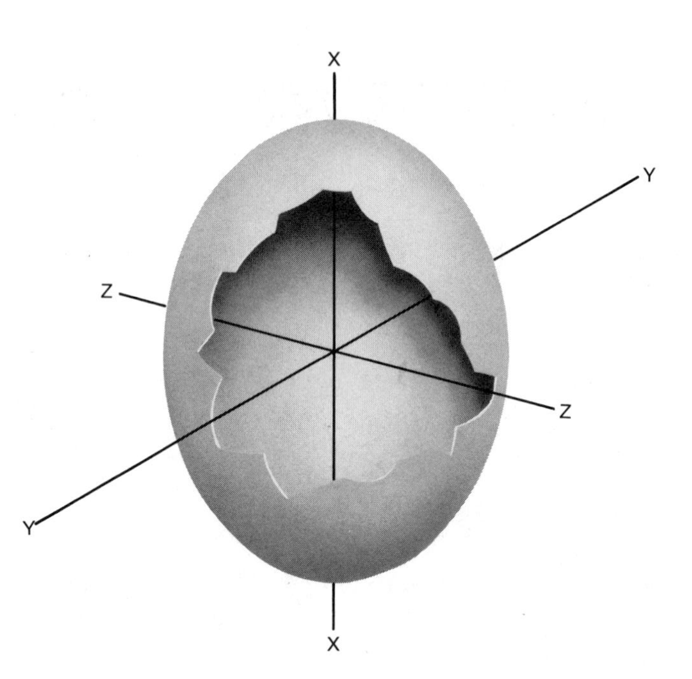

Figure 4.46 The strain ellipsoid: *X* is the axis of greatest strain, *Y* is the axis of intermediate strain, *Z* is the axis of least strain.

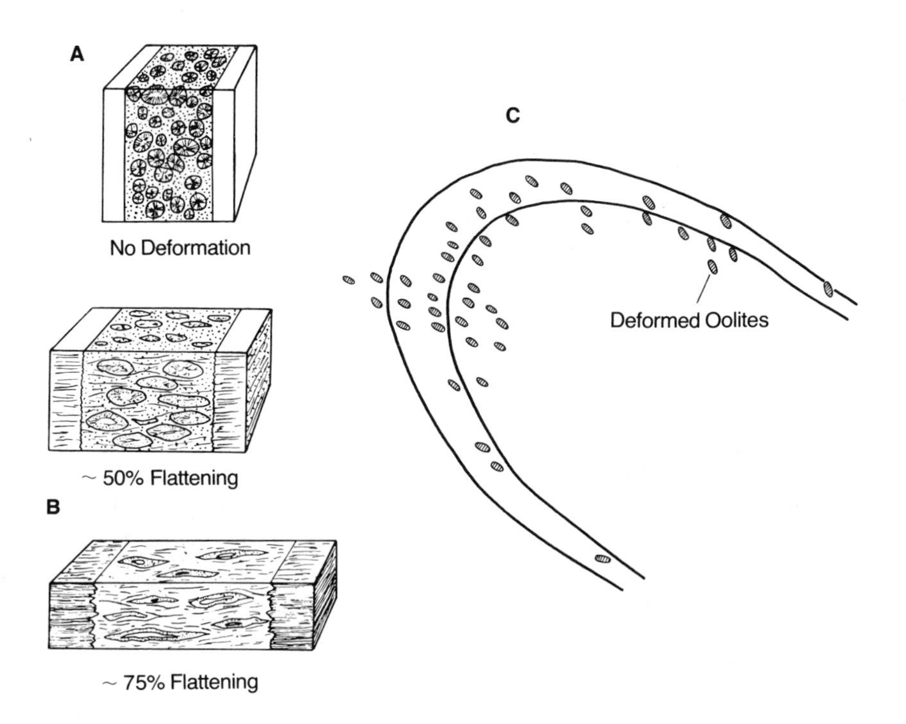

Figure 4.47 (*A*) Undeformed ooids in rocks of South Mountain, Maryland and Pennsylvania. (*B*) Ooids captured in different progressive stages of distortion. (*C*) Schematic diagram showing variation in shape and orientation of ooids as a function of position on the South Mountain fold. [From *Structural Geology of Folded Rocks* by E. T. H. Whitten, after Cloos (1947). Originally published by Rand-McNally and Company, Skokie, Illinois, copyright ©1966. Published with permission of John Wiley & Sons, Inc., New York.]

dovician carbonates exposed in the South Mountain fold in Maryland and Pennsylvania. His study is a classic in modern strain analysis.

The fold that Ernst Cloos examined is a large anticline of the Appalachian Mountain system. The reference materials for his strain analysis, the ooids, are small spherical or slightly ellipsoidal objects not uncommonly found in limestones. The ooids are generally calcareous and have grown concentrically and/or radially outward from centers of nucleation (Figure 4.47*A*). The average diameter of the ooids in rocks in the South Mountain region is less than 1 mm. Cloos reported that the diameter of ooids ranges from 0.33 to 1.2 mm. After discovering that the ooids had been deformed into ellipsoids within the South Mountain fold (Figure 4.47*B*), Cloos clearly recognized that these primary elements could be harnessed to conduct detailed strain analysis. And detailed it was! He collected 227 oriented specimens of ooid-bearing limestone. Measurements of all visible structures like cleavage and bedding and jointing were made at each station where the specimens were collected. With the aid of a microscope in the field, Cloos made preliminary measurements and assessments of the strain expressed by the ooids. In the laboratory he prepared 404 oriented thin sections of the ooid-bearing rocks, and these thin sections provided a view of the details of size, shape, and orientation of the strain-reference features.

In calculating strain, Cloos assumed that each ooid, before distortion, had been a perfect sphere. He also assumed that distortion of the ooids was accomplished without change in volume. In effect, each perfectly spherical ooid was assumed to have had an original volume defined by

$$V_s = \frac{4}{3}\pi r^3 \qquad (4.10)$$

where r = radius.

Each sphere was transformed through distortion into an ellipsoid. The volume of an ellipsoid is described by Equation 4.11.

$$V_e = \frac{4}{3} abc \qquad (4.11)$$

where,

$$a = \text{long semiaxis}$$

$$b = \text{intermediate semiaxis}$$

$$c = \text{short semiaxis}$$

Given the assumption of no change in volume,

$$V_s = V_e$$

and

$$r = \sqrt[3]{abc}$$

Thus, by measuring the length of the axes of the deformed ellipsoidal ooids, Cloos gathered data that allowed the radius of the original sphere to be determined. For example, he reported that at locality #300, the length of the long axis of the ellipsoid ($2a$) measured 8.45 mm, the length of the short axis ($2c$) measured 5.06 mm, and the length of the intermediate axis ($2b$) measured 6.74 mm. Thus,

$$a = \frac{8.45 \text{ mm}}{2} \sim 4.2 \text{ mm}$$

$$b = \frac{6.47 \text{ mm}}{2} \sim 3.4 \text{ mm}$$

$$c = \frac{5.06 \text{ mm}}{2} \sim 2.5 \text{ mm}$$

$$r = \sqrt[3]{4.2 \text{ mm} \times 3.4 \text{ mm} \times 2.5 \text{ mm}} \sim 3.3 \text{ mm}$$

Knowing the length of the radius of the initial sphere from which the ellipsoid was derived, and knowing the length of the axes of the deformed ellipsoids, the values of extension and quadratic elongation can be computed. To calculate extension (e) in the direction of axis a,

$$e_a = e_1 = \frac{l_f - l_o}{l_o} = \frac{4.2 - 3.3}{3.4} = 0.24$$

Percentage lengthening parallel to line a is,

$$0.24 \times 100\% = 24\%$$

Quadratic elongation for all lines parallel to line a is,

$$\lambda_a = \lambda_1 = (1 + e_1)^2 = (1.24)^2 = 1.53$$

By similar calculations, we write

$$e_2 = 0.02, \ \lambda_2 = 1.04 \text{ (negligible increase in length)}$$

$$e_3 = -0.23, \ \lambda_3 = 0.59 \text{ (23\% shortening)}$$

Some of the results of Cloos's strain findings are presented in Figure 4.47C, an idealized structure profile of the South Mountain fold showing the variation in orientation of the plane of flattening across the structure.

SOME PRACTICAL CONSTRAINTS IN STRAIN ANALYSIS

Not all deformed rocks contain objects, like ooids, that can be used directly to calculate the strain that is bound up in the rock. If the starting facts are meager, the quantitative strain picture that can be drawn is fragmentary. Complete analysis is thwarted by inability to determine the original size, shape, dimensions, and angular relations of primary objects in the deformed rock. Commonly the best we can do on the basis of incomplete and/or inadequate data is to prepare partial descriptions of the strain. Describing only part of the strain picture is like drawing only part of a strain ellipse: the orientations of the axes of the strain ellipse might be known, but not size or shape; the shape of the strain ellipse might be known, but not orientation or size; the size of the strain ellipse might be known, but not orientation. Guidelines exist regarding how to maximize information gained from objects like deformed pebbles or fossils (Ramsay, 1967; Ragan, 1973; Means, 1976).

Another constraint imposed by the real world is the hetereogeneous nature of strain. Naturally strained rocks typically depart from the rules of homogeneous deformation. We deal with this problem by subdividing regions of study into small domains within which strain is statistically homogeneous.

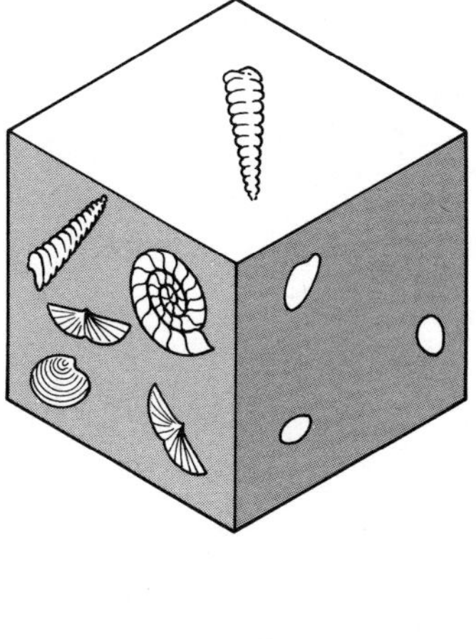

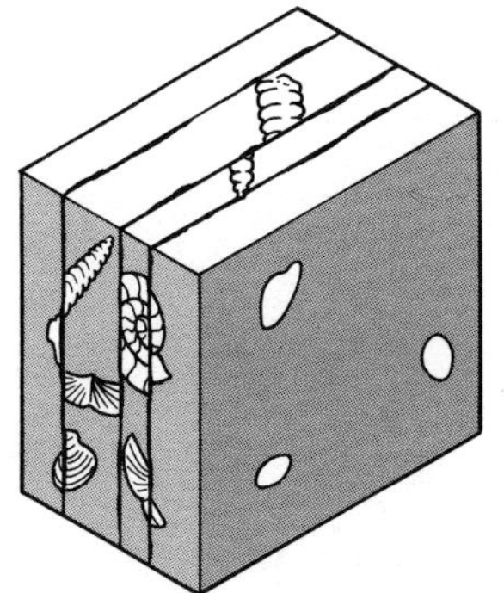

Figure 4.48 Negative dilation (i.e., loss of volume) accommodated by pressure–solution along cleavage. (Artwork by R. W. Krantz.)

DILATIONAL CHANGES

We have assumed in all examples presented so far that the strain has been a pure distortion without change in volume. This assumption has eased our entry into strain theory, but it would be misleading not to underscore the reality of ***changes in volume*** during strain. Distortion can be accompanied by increases or by decreases in volume. Easiest to visualize is distortion that opens up the rock, producing extensional cracks that are filled in by vein materials like quartz or calcite. The net effect is an increase in the volume of the distorted rock. Conversely, a rock body may undergo a decrease in volume during distortion. One of the stunning recent discoveries in structural geology is that rock can dissolve away in areas of strong compression (Figure 4.48). Part of the rock is removed to facilitate and achieve shortening even greater than would be attainable through folding and faulting alone. Material goes into solution and, perhaps, is redeposited in parts of the distorted body that are undergoing an increase in volume.

Considered two dimensionally, increases and decreases in the area of objects can be readily assessed on the basis of two-dimensional strain relationships.

$$\sqrt{\lambda_1 \lambda_3} = 1.0 = \text{no change in area} \qquad \text{(Figure 4.49}A, B\text{)}$$

$$\sqrt{\lambda_1 \lambda_3} \leqslant 1.0 = \text{decrease in area} \qquad \text{(Figure 4.49}A, C\text{)}$$

$$\sqrt{\lambda_1 \lambda_3} \geqslant 1.0 = \text{increase in area} \qquad \text{(Figure 4.49}A, D\text{)}$$

A

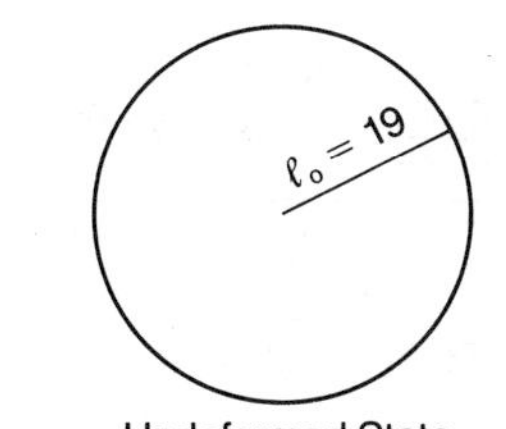

C

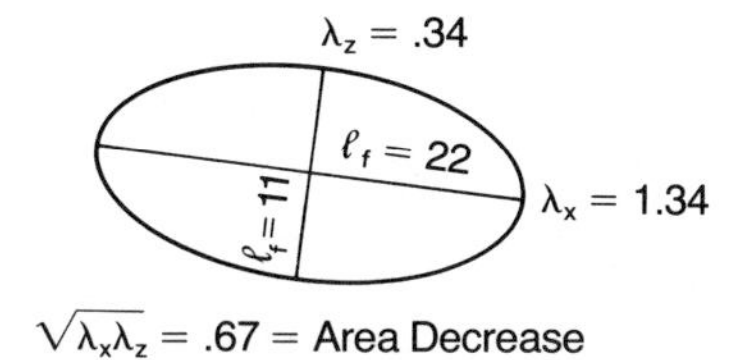

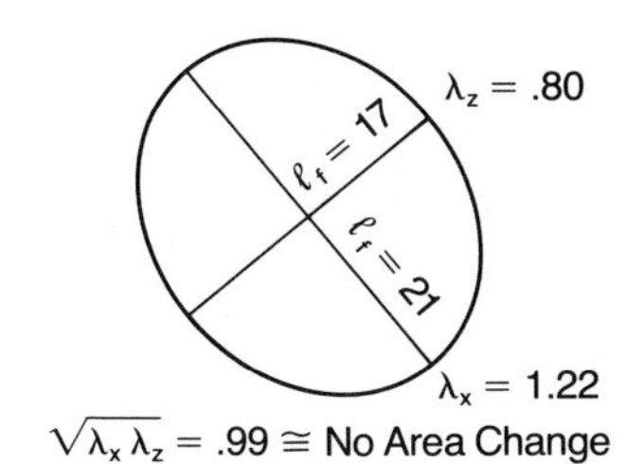

D

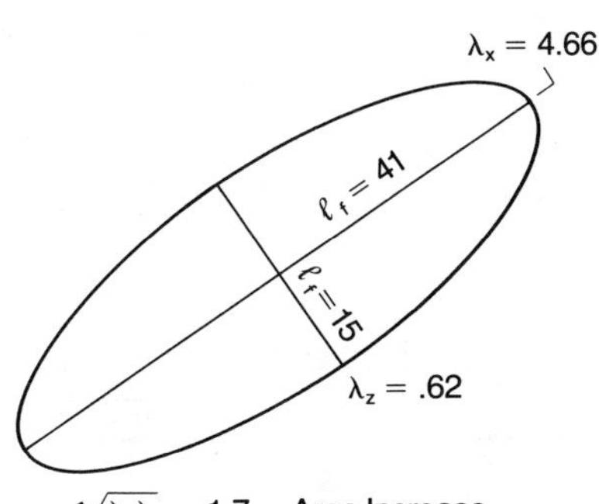

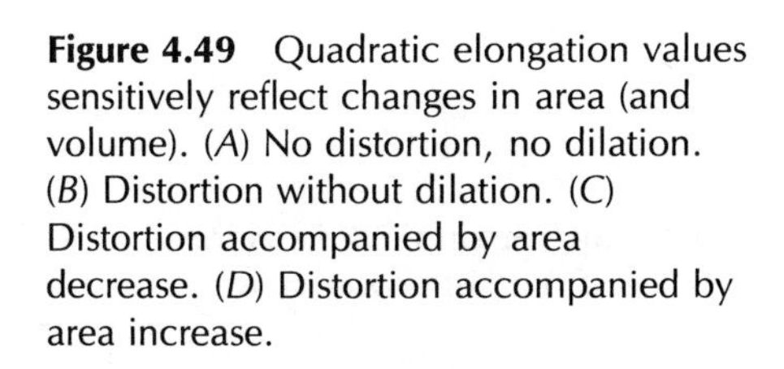

Figure 4.49 Quadratic elongation values sensitively reflect changes in area (and volume). (*A*) No distortion, no dilation. (*B*) Distortion without dilation. (*C*) Distortion accompanied by area decrease. (*D*) Distortion accompanied by area increase.

The different levels of dilation that may accompany distortions in rocks can be visualized through John Ramsay's **strain field diagram** (see Ramsay, 1967, p. 112). Using this diagram, the physical structural significance of the full range of combinations of quadratic elongation values becomes more meaningful (Figure 4.50).

For ease of reference, I have given names to the components of these diagrams. The **field of expansion** is reserved for strain characterized by quadratic elongation values (λ_1 and λ_3) greater than 1.0. It is a field marked by distortion accompanied by an increase in area ($\sqrt{\lambda_1 \lambda_3} > 1.0$). Physically the rock stretches in two directions. The **field of contraction,** in contrast, is marked by quadratic elongation values less than 1.0. Distortion is accompanied by a decrease in area ($\sqrt{\lambda_1 \lambda_3} < 1.0$) and the rock is shortened in two mutually perpendicular directions. In the **field of compensation,** λ_1 is greater than 1.0 and λ_3 is less than 1.0. **Plane strain** occurs in this field ($\sqrt{\lambda_1 \lambda_3} = 1.0$), but this is a special case in a field where increases or decreases in area are the rule.

The boundary line between the field of contraction and the field of compensation is the **field of linear shortening**, symbolizing strain marked by shortening in one direction, with no strain at right angles to the direction of shortening ($\lambda_1 = 1.0$, $\lambda_3 < 1.0$). The boundary line between the fields of compensation and expansion is the **field of linear stretching**. It describes strain values marked by $\lambda_1 > 1.0$ and $\lambda_3 = 1.0$. It represents deformation marked by extension in one direction only, with no strain at right angles to the direction of extension. Finally, the lines of stretching and shortening intersect at a point, the **field of no strain**, where λ_1 and λ_3 both are 1.0. We use the strain field diagram time and again to classify structures according to strain significance.

ROTATIONAL AND NONROTATIONAL STRAIN

In addition to dilational effects, there is yet more complexity to strain. It is the full realization that during distortion the axes of the strain ellipse, which portrays the strain at each stage of deformation, are usually not fixed in orientation. Rather, the strain axes commonly rotate! Structural geologists emphasize this reality through the use of card deck models (e.g., see Ragan, 1969, 1973). As we have seen, if a circle is imprinted on the side of a deck of computer cards, and the cards of the deck are homogeneously sheared

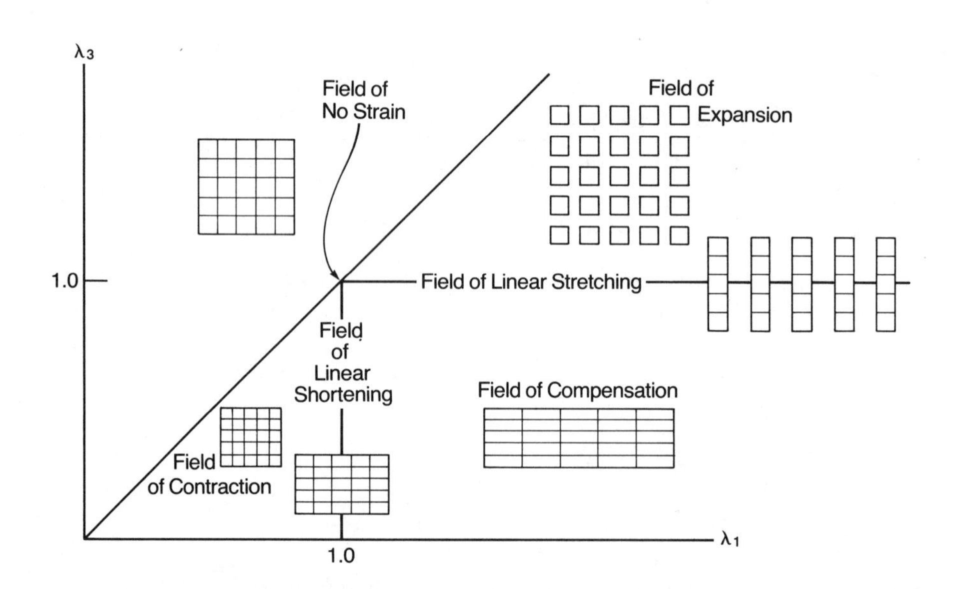

Figure 4.50 Strain field diagram. [Modified from Ramsay (1967).]

with respect to one another, the circle will be transformed progressively into an ellipse. If we view the size and orientation of the ellipse at different stages of the deformation (Figure 4.51*A*), we see that the axes of the strain ellipse rotate in space. This is an **external rotation** (**β**), to be distinguished from **internal rotation** (**α**), in which lines rotate relative to the principal axes of the strain ellipse.

As has been emphasized, part of the uniqueness of the principal axes of the strain ellipse is that these are the only mutually perpendicular lines in the distorted body that were perpendicular before deformation. These lines, in their positions **before deformation**, are referred to as the **principal axes of strain** (Ramsay, 1967). When the principal axes of the strain ellipse do not coincide in orientation with the principal axes of strain, the deformation is described as a **rotational strain** (Figure 4.51*A*). But when they do coincide, the deformation is called a **nonrotational strain** (Figure 4.51*B*). The distinction between rotational and nonrotational strain is thus based on whether the directions of maximum and minimum extension remain fixed in orientation during distortion or whether they change in orientation during the course of distortion. **Simple shear** is one kind of rotational strain, distinguished by its constant volume, homogeneous, plane strain nature. **Pure shear**, on the other hand, is a homogeneous nonrotational strain.

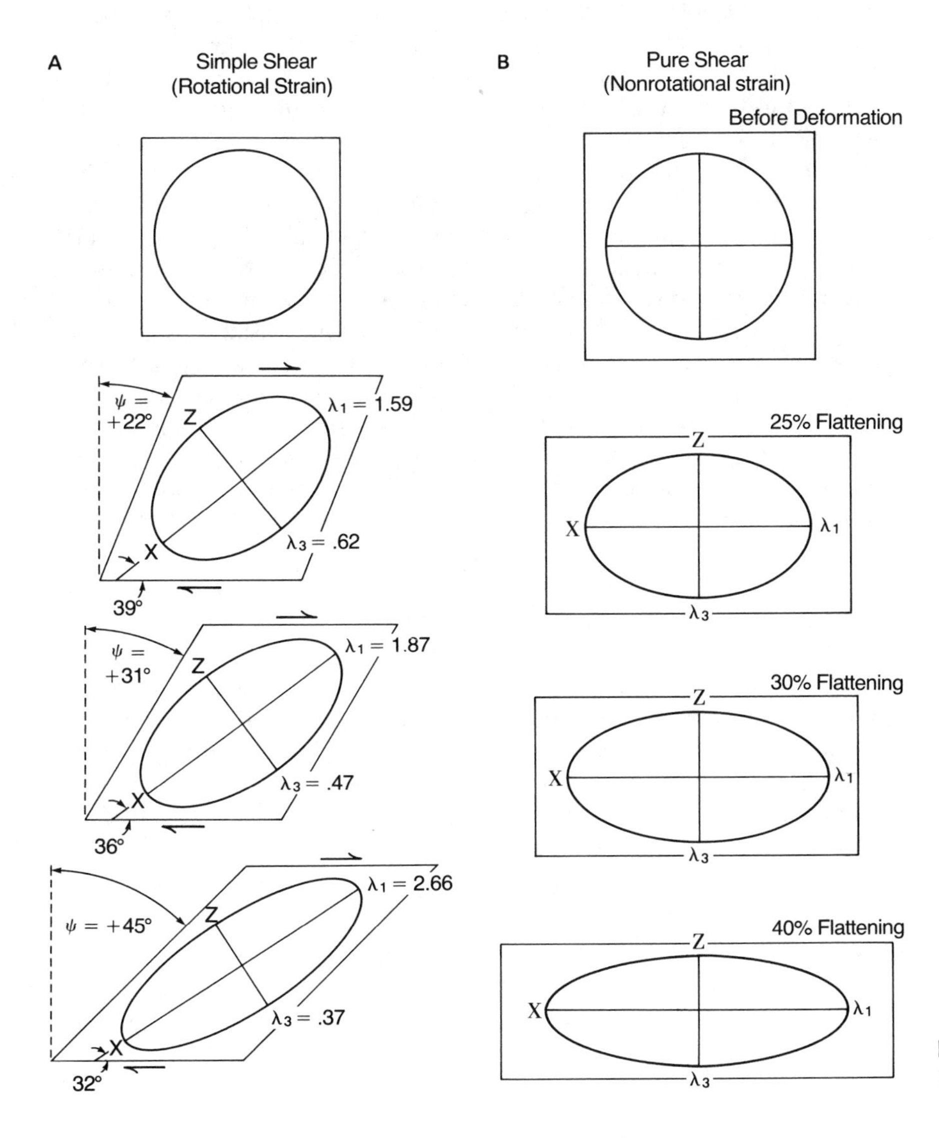

Figure 4.51 (*A*) Rotational strain. (*B*) Nonrotational strain.

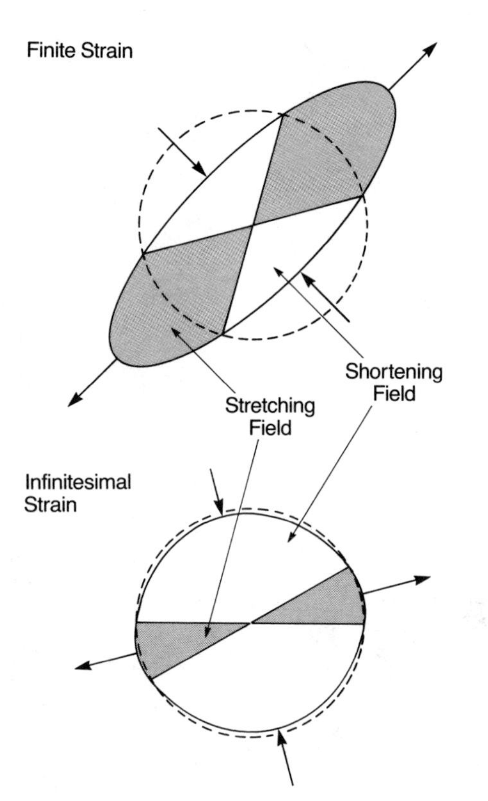

Figure 4.52 The relationship of finite and infinitesimal strain at any stage during deformation. (From *Folding and Fracturing of Rocks* by J. G. Ramsay. Published with permission of McGraw-Hill Book Company, New York, copyright ©1967.)

As structural geologists analyzing strain in outcrop, we almost never can judge whether deformation was accomplished by rotational or nonrotational strain. We can describe only the nonrotational part of the strain, as if distortion took place by nonrotational pure shear. But we should not assume that the deformation was indeed nonrotational. In fact, most geologists agree that rotational distortion by simple shear is the most important process in the shaping of natural structures.

FINITE AND INFINITESIMAL STRAIN

Strain may be appreciated more fully when it is viewed in the context of a **progressive deformation** through time. The geologic deformation that we see in outcrops and/or thin sections of rocks does not accrue overnight. Instead it is generally the product of sustained or intermittent movements over a long span of time. At each instant of progressive deformation, a rock absorbs a measure of strain that can be described as an **infinitesimal strain**, a tiny increment of strain that is added to the total **finite strain** that has already accumulated (Figure 4.52). In effect, the finite strain record is the product of superimposed infinitesimal strains.

We will learn that the ways in which incremental, infinitesimal strains are added have a profound influence on the physical and geometric nature of structures that develop in a distorted body of rock. Whether a given line (or layer) undergoes stretching or shortening at any instant of time depends on the orientation of the infinitesimal strain ellipse with respect to the line (or layer) in question. Whether a given line (or layer) undergoes total finite stretching or shortening depends on the nature of superposition of the infinitesimal strains through time.

Figure 4.53 is an image of progressive deformation, viewed regionally. It pictures in a greatly simplified, somewhat hypothetical way a superposition of incremental distortions through time. The objects of the distortion are southeastern Arizona regional rock assemblages, ranging from Precambrian to Tertiary (Figure 4.53, left column). The incremental distortions are products of the numerous deformational episodes that have affected Arizona (Figure 4.53, top). The state of strain reflected by each regional rock assemblage changes progressively as a response to the superposed deformations. The total finite strain accommodated by each assemblage (Figure 4.53, right column) reflects the number and nature of incremental strains, and the order in which the incremental strains were added.

SUMMARY

Kinematics is a three-ring circus when applied to matters of structural geology. Geometrically and mathematically we are required to keep track simultaneously of rigid body translations, rigid body rotations, distortions (rotational and nonrotational), and dilations. Fortunately, in many regions, nature has given us materials and relationships with which to work. It is up to us to learn the theory and methods well enough to use what nature has provided.

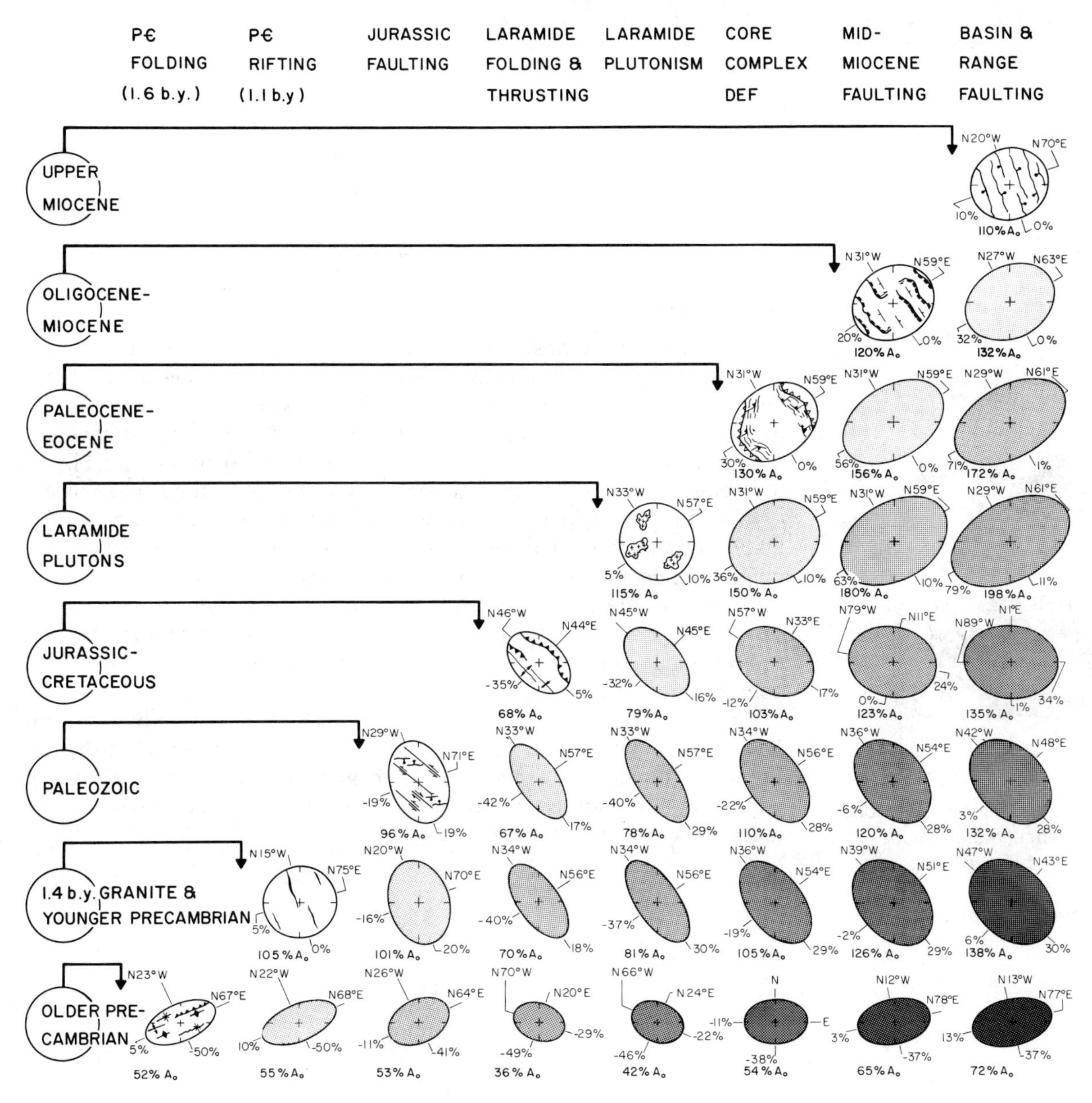

Figure 4.53 Progressive strain diagram portraying regional deformation in southeastern Arizona. Undeformed state of rock assemblages of different ages represented by circles in left column. Regional deformation to which the rock assemblages were subjected are shown at top. Ellipses show state of strain of each assemblage after each deformation. Fine print shows directions of greatest and least strain, percentages of stretching (+) and shortening (-), and changes in the original surface area (A_o) of each assemblage. [From Davis (1981), fig. 18, p. 166. Published with permission of Arizona Geological Society.

DYNAMIC ANALYSIS chapter 5

FORCE AND STRESS

DEFINITIONS

Figure 5.1 Football, the contact sport. (*A*) High stress concentration about to be sustained at contact between elbow and chest. (*B*) Contact over large surface area results in low stress concentrations. (Artwork by D. A. Fischer.)

Translations, rotations, and distortions are responses of rocks to stresses and forces. **Force** is classically defined as that which changes, or tends to change, the state of rest or state of motion of a body. This definition begins to take on personal meaning when the body is that of some professional quarterback. When such a body is subjected to the force produced by the accelerating mass of a huge opponent, its state of rest (in the pocket) or state of motion (while rolling out) may be significantly changed (Figure 5.1). The body is usually translated **and** rotated. Under a great pileup of tacklers they might even undergo slight negative dilation (decrease in volume). In not-so-rare instances, the bodies, or parts of the bodies, may endure the permanent strain of distortion, involving breaking or stretching or ripping. Whether such strain is induced depends on two factors: the strength of the body and the concentration of the stress(es).

We can think of **stress** as that which tends to deform a body of material (Jaeger and Cook, 1976). Stress will permanently deform a body only if the strength of the body is exceeded. The magnitude of **stress (σ)** is not simply a function of the **force** (F) from which it was derived, but it relates as well to the area on which the force acts,

$$\sigma = \frac{F}{A} \tag{5.1}$$

where

σ = stress

F = force

A = area

Stress is commonly expressed in units of **pounds per square inch (psi)**, or alternatively in units of **kilograms per square centimeter (kg/cm^2): 1 kg/cm^2 = 14.1 psi; 1 psi = 0.07 kg/cm^2.**

If a 270-lb (122 kg) lineman is "in flight" and descends in such a way that his accelerating mass is brought to bear on the chest of a quarterback through the tip of the lineman's elbow, the stress concentration will be crushingly high (Figure 5.1*A*). But if the "body" force of the 270-lb lineman is distributed rather evenly across the tackled quarterback, stress concentrations at any one point are low and the relationship is almost cozy (Figure 5.1*B*).

SIMPLE STRESS CALCULATIONS

Other nongeologic examples of the interrelation of force and stress abound. Consider the strategy of rescuing a skater who has just fallen through ice in a pond or lake (Figure 5.2). The skater broke through thin ice

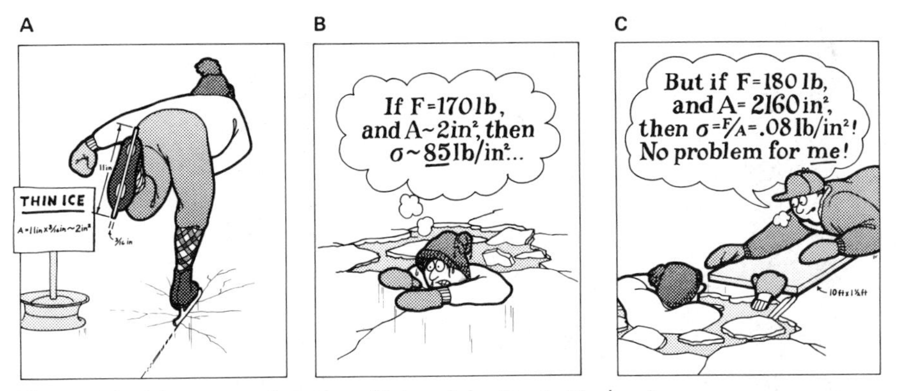

Figure 5.2 The stress of skating. (Artwork by D. A. Fischer.)

because the strength of the ice was exceeded by high stress values. The high stress values were produced by the concentration of the force of the 170-lb (77 kg) skater onto the thin blades of the skates. Assuming that the 170-lb force was not focused on the tips of the skates, but instead was distributed uniformly along one of the blades, the area of the contact between ice and blades was only about 2 in.^{2} (5.1 cm^2) (Figure 5.2*A*). The stress level produced at the ice-blade interface must have been:

$$\sigma = \frac{F}{A} = \frac{170 \text{ lb}}{2 \text{ in}^2} = 85 \text{ lb/in}^2 = 85 \text{ psi } (6 \text{ kg/cm}^2)$$

(see Figure 5.2*B*)

In contrast, the 170-lb rescuer, informed in ways of dynamic analysis, moves in close to the victim while lying on a 10 × 1.5 ft (3 × 0.45 m) plank (Figure 5.2*C*). The total load of rescuer and plank is about 180 lb (127 kg). The area of the plank is 2160 in.^{2} (5486 cm^2). The force of the rescuer is distributed in such a way that stress concentration at any point under the plank is uniformly small.

$$\sigma = \frac{F}{A}$$

$$\sigma = \frac{180 \text{ lb}}{2160 \text{ in}^2}$$

$$\sigma = 0.08 \text{ psi } (0.005 \text{ kg/cm}^2)$$

The rescue mission is a success because the breaking strength of the ice beneath the rescuer's plank is much greater than 0.08 psi.

CONCEPT OF STRESS ANALYSIS

The goal of dynamic analysis in structural geology is to interpret stresses, to describe the nature of the forces from which the stresses are (or were) derived, and to define the relationships between stress and strain. The description of stress and force is mathematical. Mathematical parameters and equations are necessary to bring forces and stresses to life.

The study of stress and force is primarily a study of their magnitudes and orientations, and such analysis can be carried out independently of any consideration of the rocks they affect. In contrast, interpreting the relationship of stress to strain involves describing forces and stresses as well as the mechanical responses of rocks to stress as a function of strength. A given rock may respond to deformation as an elastic solid in one environment and as a viscous fluid in some other environment.

Figure 5.3 Deformation of experimentally deformed cylinders of rocks. Structure is reflected in the thin-walled copper jacket that envelops each rock specimen. *Left to right*: fault in slate, conjugate faults in sandstone, ductile fault in limestone. [From Donath (1970a). Published with permission of National Association of Geology Teachers, Inc.]

MEASUREMENT OF FORCE, STRESS, AND STRAIN

VALUE OF EXPERIMENTATION

Dynamic analysis is as abstract as any of the topics and concepts presented in structural geology. To clarify the physical reality of dynamic analysis, we explore the principles through experiments in rock deformation. In deformational experiments, the origins of forces and stresses lie ultimately in the machinery used to squeeze or stretch the rocks. In nature, however, the origin of deforming forces and stresses can often be linked to the machinery of plate tectonics and plate tectonic movements (Chapter 6). Plate boundaries are commonly the sites of anomalously highly concentrated stresses.

The nature and origin of structural deformation of rocks becomes clearer through images and experiences in experimental deformation. Consider placing an undeformed cylindrical core of fine-grained sandstone into a thick-walled steel pressure vessel, squeezing the rock by hydraulic loading, and removing a conspicuously deformed specimen bearing structural characteristics identical to those of faulted rocks seen in natural outcrops (Figure 5.3). This experimental process provides a cause-and-effect glimpse of dynamics that few experiences could match. Furthermore, to perform the experiment successfully it is necessary to learn how to regulate force and stress, and how to describe the strain and strength of rocks through appropriate parameters.

SAMPLE PREPARATION

The experiment we undertake is compressional deformation of a cylinder of rock confined under pressure in a steel pressure vessel. To begin, a **core** of fine-grained, texturally homogeneous rock is extracted from a rock specimen through the use of a drill press and a diamond-drill coring device. The ends of the cylinder of rock are beveled to smooth, planar parallel surfaces on a grinding wheel. If the ends were not ground flat, small irregularities of rock projecting from the ends would focus high stress concentrations, like those focused by the tips of skates on ice. After the specimen has been prepared, a micrometer is used to measure length (l_o) and diameter (d_o) of the specimen, preferably to three significant figures. The specimen is then placed in a **jacket** of a thin-walled cylinder of copper or some other material of negligible strength. The jacket serves to seal the rock from whatever fluid (e.g., kerosene) occupies the pressure vessel. The fluid, during the test, is pressurized to produce an environment of appropriate confinement. The protection that a jacket affords assures that corrosive action or high fluid pressures do not contribute to the **failure** (i.e., faulting) of the specimen. Once all this is done, the jacketed specimen is fitted with an anvil at its base and an upper piston specimen holder at its top (Figure 5.4). These are both made of stainless steel and equipped with O-rings to prevent entry of fluids into the jacket from above or below.

The specimen and its trimmings are screwed into the **pressure vessel**, a steel vessel of sufficient wall thickness and inherent strength to resist fracturing under conditions of very high pressure. When the specimen is thus placed, simulation of natural structural deformation can proceed.

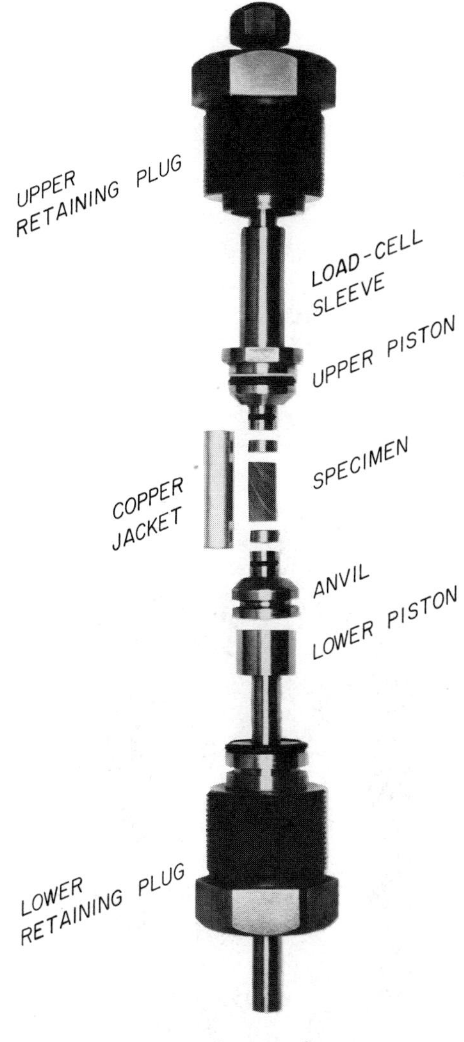

Figure 5.4 Internal parts of pressure vessel, showing the relation of the cylindrical rock specimen (and copper jacket) to pistons and anvil. [From Donath (1970a). Published with permission of National Association of Geology Teachers, Inc.]

PRESSURES, STRESSES, AND LOADS

The immediate object of the experimental deformation is twofold: (1) to subject the specimen to burial pressure conditions appropriate to the experiment, and (2) to compress and shorten the specimen **axially** (i.e., end-on) until it fails. How do we do this?

A useful instructional device to teach principles of force, stress, and rock strength is the **Donath triaxial deformation apparatus** (Figure 5.5). F. A. Donath (1970) designed and built this experimental rig in such a way that it is safe, portable, and easy to operate. The apparatus accepts small cylindrical specimens, with diameters of 0.5 in. (1.3 cm) and lengths of 1 in. (2.5

Figure 5.5 The Donath apparatus.

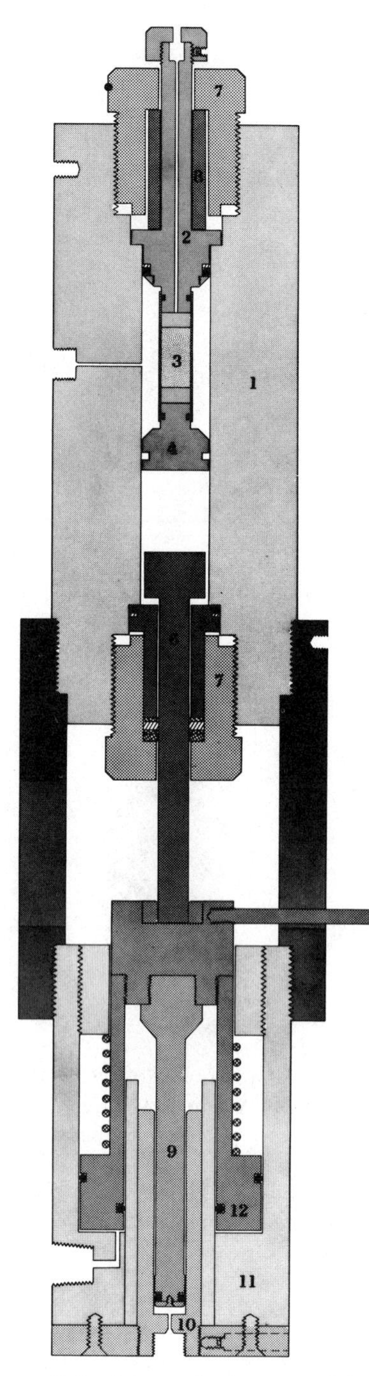

Figure 5.6 Schematic drawing of the vessel–press assembly in the Donath apparatus: 1, pressure vessel; 2, upper piston and seal; 3, specimen; 4, anvil; 5, gland and lower seal; 6, lower piston; 7, retaining plug; 8, load cell; 9, equalization piston; 10, equalization cylinder; 11, ram body; 12, ram piston; 13, collar. [From Donath (1970a). Published with permission of National Association of Geology Teachers, Inc.]

cm). Burial pressure conditions are imposed on test specimens by pumping kerosene into the specimen chamber. This produces an equal, all-sided **confining pressure** on the specimen. In nature, confining pressure at any specified level is derived from the weight of water occupying pore spaces in the overlying column of rock (**hydrostatic stress**) and the weight of the rock column itself (**lithostatic stress**). Confining pressure is thus derived from **body forces** stemming from the weight or load of a "body" of rock.

In experiments using the Donath apparatus, the level of confining pressure can be read directly in pounds per square inch (psi) from a calibrated gauge. The operational limit is 30,000 psi (2109.3 kg/cm^2). Recognizing that 1 psi (0.07 kg/cm^2) is about equal to 1 ft (30.5 cm) of burial, the raising of confining pressure to 30,000 psi simulates burying the rock specimen to a depth of 30,000 ft (9144 m), about 5.7 mi (9.1 km). Pressure *in* **bars** at this level is 30,000 psi/14.7 psi = 2040 bars, or about 2 kb. When I ask my students how far we should raise the confining pressure at the start of an experiment, the standard response has become "All the way"—all the way to the red warning line (Figure 5.5). Raising the level of the confining pressure by even small amounts, and certainly "all the way," results in a small reduction in volume of a specimen. The specimen undergoes negative dilation. Hydrostatic stress alone, even at the highest levels, cannot produce distortion. The equal all-sided compressive stresses simply reduce the size of the specimen uniformly.

When the appropriate level of confining pressure is established, a test specimen can be shortened by applying a vertical axial **load**. Load is force, and for the Donath apparatus load is calibrated in units of pounds. Load is applied to the specimen by manually pumping hydraulic jack fluid into the vessel–press assembly. As pressure of the jack fluid increases, a ram forces a piston to advance slowly up until it butts against the anvil fixed to the base of the specimen (Figure 5.6). This is a critical moment in the test. When the piston makes contact with the anvil, the stresses are transmitted through the anvil into the specimen. Moreover, the stresses are transmitted through the specimen into the specimen holder and to the steel cap that seals the top of the thick-walled pressure vessel. The steel cap can be thought of as a very strong steel spring, which deforms elastically when subjected to stress.

By knowing the relationship of stress magnitude to shortening of the steel cap, it is possible to calibrate the exact amount of force being applied by the piston at each stage of the experiment. In practice electronic signals describing the shortening of the steel cap are transmitted from a **load cell** in the cap to an *X–Y* recorder. The signals are "interpreted" by the recording device in terms of load, in pounds.

COMPUTING AXIAL STRESS

Although both confining pressure and force can be read directly from appropriate gauges and/or the *X–Y* recorder during the course of experiments, the amount of **axial stress** that is felt by the ends of the specimen cannot be displayed directly. Instead, this **stress (σ)** must be computed on the basis of the applied **force** (F) and the **cross-sectional area** (A) of the specimen,

$$\sigma = \frac{F}{A}$$

where

$$A = \pi r^2$$

Assume that the force at some stage of a deformational experiment is 4000 lb (1814 kg) and that the force is brought to bear on a cylindrical specimen whose radius is 0.243 in. (6.17 mm). Stress on the specimen can be calculated as follows.

$$\sigma = \frac{F}{A} = \frac{4000 \text{ lb}}{0.185 \text{ in.}^2} = 21{,}621 \text{ psi } (1513 \text{ kg/cm}^2)$$

MEASURING SHORTENING

Changes in the length of the test specimen are derived from measuring displacements of the piston. Progressive upward movement of the piston toward the anvil, which is fixed to the base of the specimen, is displayed on the *X–Y* recorder by pen movement horizontal and to the right (Figure 5.7).

The instant that the piston makes contact with the anvil, the pressure gauge that monitors force jumps from zero to some small positive value. As seen on the *X–Y* recorder (Figure 5.7), this point of contact, known as the **seating position**, is marked by a jump of the pen upward, recording the fact that the piston has encountered resistance. Any further movement of the piston into the specimen chamber is a record of the degree to which the specimen shortens. If rocks were perfectly unyielding, the piston could be raised no further, no matter how much force were applied to the anvil. However, rocks are deformable, and thus added increments of force always result in shortening. Shortening is revealed on the *X–Y* plotter in the form of displacement of the pen to the right of the seating position (Figure 5.7).

Shortening of the core of rock at any stage of an experiment can be described in terms of extension (e). Suppose that original specimen length (l_o) was 0.905 in. (23 mm) and that shortening was merely 0.008 in. (0.2 mm).

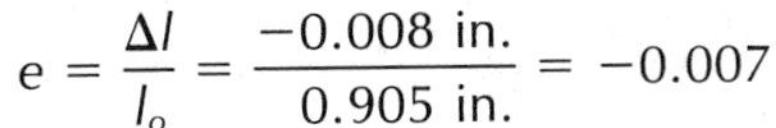

$$e = \frac{\Delta l}{l_o} = \frac{-0.008 \text{ in.}}{0.905 \text{ in.}} = -0.007$$

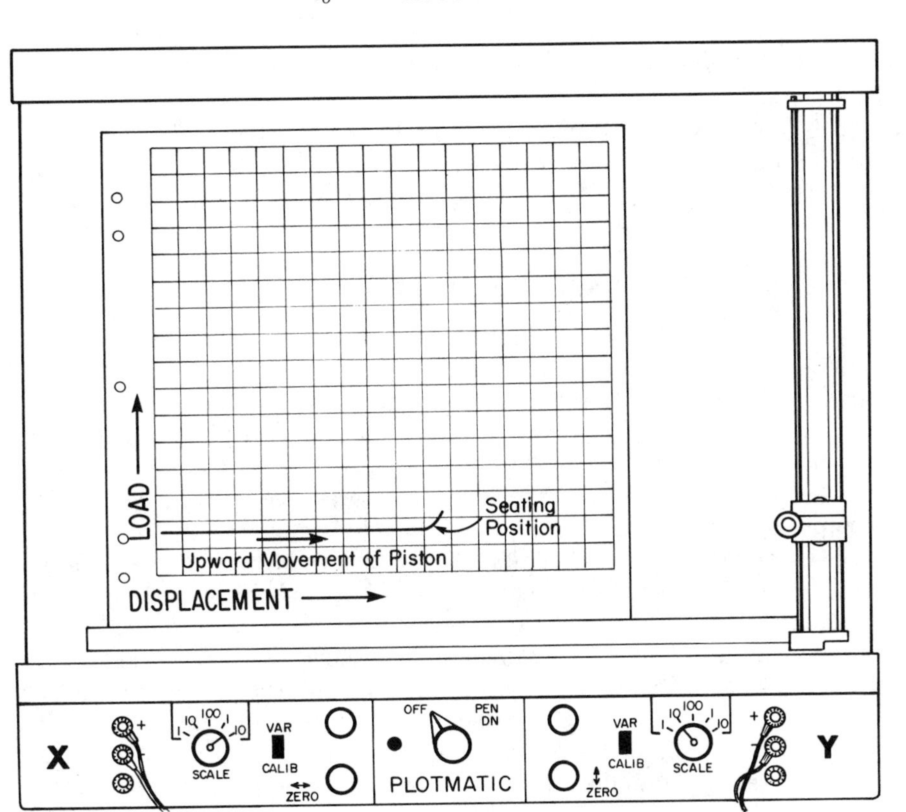

Figure 5.7 An X–Y recorder. Load is recorded in pounds on the y axis of the recorder. Displacement is recorded in thousandths of inches on the x axis. Pen path shows upward movement of piston and eventual contact of piston with anvil attached to the base of the specimen. Point of contact is known as the seating position.

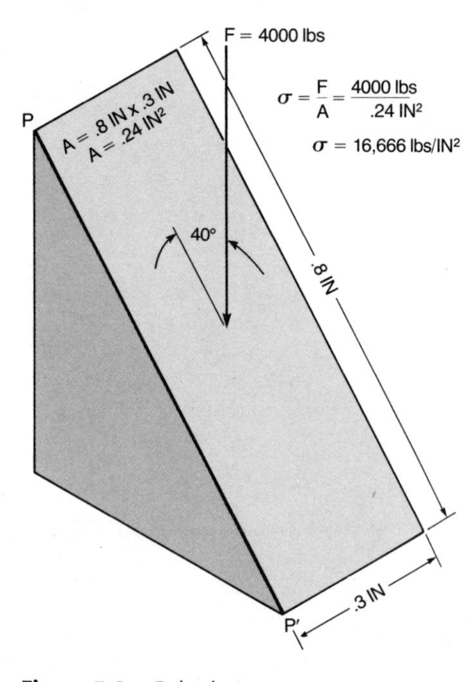

Figure 5.8 Calculating a stress vector (σ) on the basis of force (*F*) and area (*A*).

STRESS VECTORS, NORMAL STRESSES, AND SHEAR STRESSES

THE PROBLEM

The level of stress that operates on the ends and the flanks of a test specimen is relatively easy to calculate. The axial stress has a magnitude of *F*/*A*, with an orientation parallel to the vertical direction of loading. As a first approximation, this stress vector penetrates the rock internally, maintaining its magnitude and orientation throughout. The confining pressure acts horizontally and is of constant magnitude in all horizontal directions in the interior of the test specimen.

Assuming that the values of axial stress and confining pressure are known, or can be calculated, our task is to compute the level of stress that acts in each and every direction within the test specimen, not simply in the vertical and horizontal directions. Stated another way, we must learn to calculate the level of stress acting on planes of all possible orientations within the cylinder of rock. When we master this, we can begin to understand why planes of certain orientations in a given stress field are more vulnerable than others to fault translation.

COMPUTING STRESS VECTORS

Consider the two-dimensional portrayal of force *F*, which is inclined at an angle of 40° to some plane whose trace is *PP'* (Figure 5.8). Let us assume that *PP'* is the trace of a plane whose edge in the third dimension is 0.3 in. (0.76 cm) long. The length of *PP'* is 0.8 in. (2.03 cm). The area (*A*) of the plane is:

$$A = 0.8 \text{ in. } (0.3 \text{ in.}) = 0.24 \text{ in.}^2 \ (0.6 \text{ cm}^2)$$

The magnitude of the force is set at 4000 lb. The stress vector (σ) on *PP'* derived from force *F* is:

$$\sigma = \frac{F}{A} = \frac{4000 \text{ lb}}{0.24 \text{ in.}^2} = 16{,}000 \text{ psi } (1125 \text{ kg/cm}^2)$$

This calculation might seem identical to that described for calculating axial stress on the ends of a test specimen, but note that it represents the more useful general case of force that is oblique to the plane for which the stress vector is being calculated.

RESOLVING NORMAL STRESS AND SHEAR STRESS

To understand the peculiarities of stress as related to the mechanics of faulting, it is necessary to learn to resolve any stress vector (σ) into **normal stress (σ_N)** and **shear stress (σ_S)** components. The normal stress component of a stress vector acts perpendicular to the plane for which the stress vector has been calculated (Figure 5.9A). The shear stress component operates parallel to the plane for which the stress vector has been calculated. Normal **compressive stresses** tend to inhibit sliding along a plane; normal **tensional stresses** tend to separate rocks along a plane. **Shear stresses** tend to promote sliding. Faulting takes place selectively along surfaces for which there is an optimum ratio of the magnitudes of shear stress and normal stress.

Resolving a stress vector into normal and shear stress components is an easy job. It can be done graphically or numerically. The graphical construction involves "mapping" the orientation and length of the stress vector with respect to the trace of the plane on which it acts (Figure 5.9B). The map requires a scale. From the tail of the arrow representing the stress vector, a

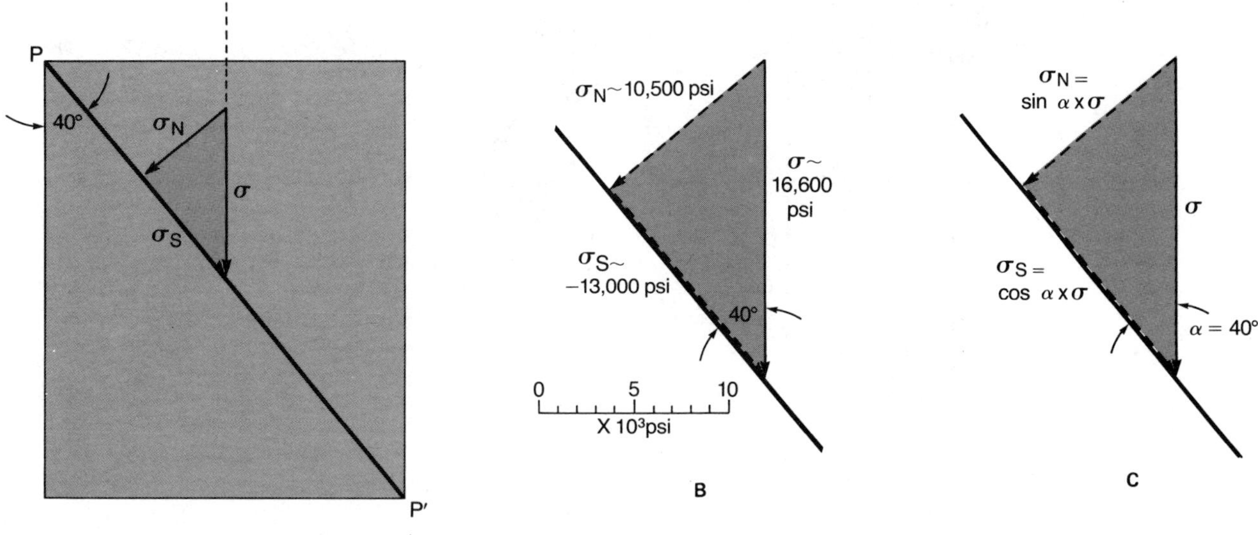

Figure 5.9 (*A*) Normal stress (σ_N) and shear stress (σ_S) components of a stress vector (σ). (*B*) Graphical determination of the normal stress (σ_N) and shear stress (σ_S) components of a stress vector (σ). (C) Calculating the components of a stress vector trigonometrically.

normal stress vector (σ_N) is constructed perpendicular to the plane on which the stress vector operates. Its value (magnitude) can be measured directly using the scale of the map. The shear stress component (σ_S) is constructed as a vector drawn from the tip of the normal stress vector (σ_N) to the tip of the stress vector (σ), and thus parallel to the plane. Its value can be directly measured.

The numerical solution for computing the magnitudes of the normal and shear stress components is trigonometric (Figure 5.9C). If alpha (α) is the angle between the stress vector (σ) and the plane on which it acts, normal stress (σ_N) can be computed as follows.

$$\sigma_N = \sigma \sin \alpha$$

The magnitude of the shear stress (σ_S) can be determined in similar fashion.

$$\sigma_S = \sigma \cos \alpha$$

Normal stresses are considered to be positive if they are compressive, negative if they are tensile (Jaeger and Cook, 1976). Shear stresses are labeled positive or negative on the basis of their sense of shear. Left-handed shear stresses are here considered to be positive, whereas right-handed shear stresses are negative.

EXAMPLE OF STRESS COMPUTATIONS

Let us compute the stress vector, normal stress, and shear stress for a plane of some specified orientation within a test specimen. Assume that the axial stress operating vertically is 15,000 psi (1055 kg/cm^2) and that the confining pressure operating horizontally is to 5000 psi (351 kg/cm^2) (Figure 5.10*A*). For this calculation we label axial stress as σ_Y and confining pressure as σ_X. They operate parallel to the *y* and *x* axes, respectively, of a Cartesian coordinate system about which the test specimen is arbitrarily suspended. Our calculations are based on the evaluation of stress at individual tiny points within the test specimen. In this way the infinitesimal volumes of material within which the planes of interest lie are considered to be weightless, contributing no body forces.

Let us first calculate the stress vector for the plane through point *O* that makes an angle (θ) of 65° relative to the *x* axis (Figure 5.10*A*). To calculate

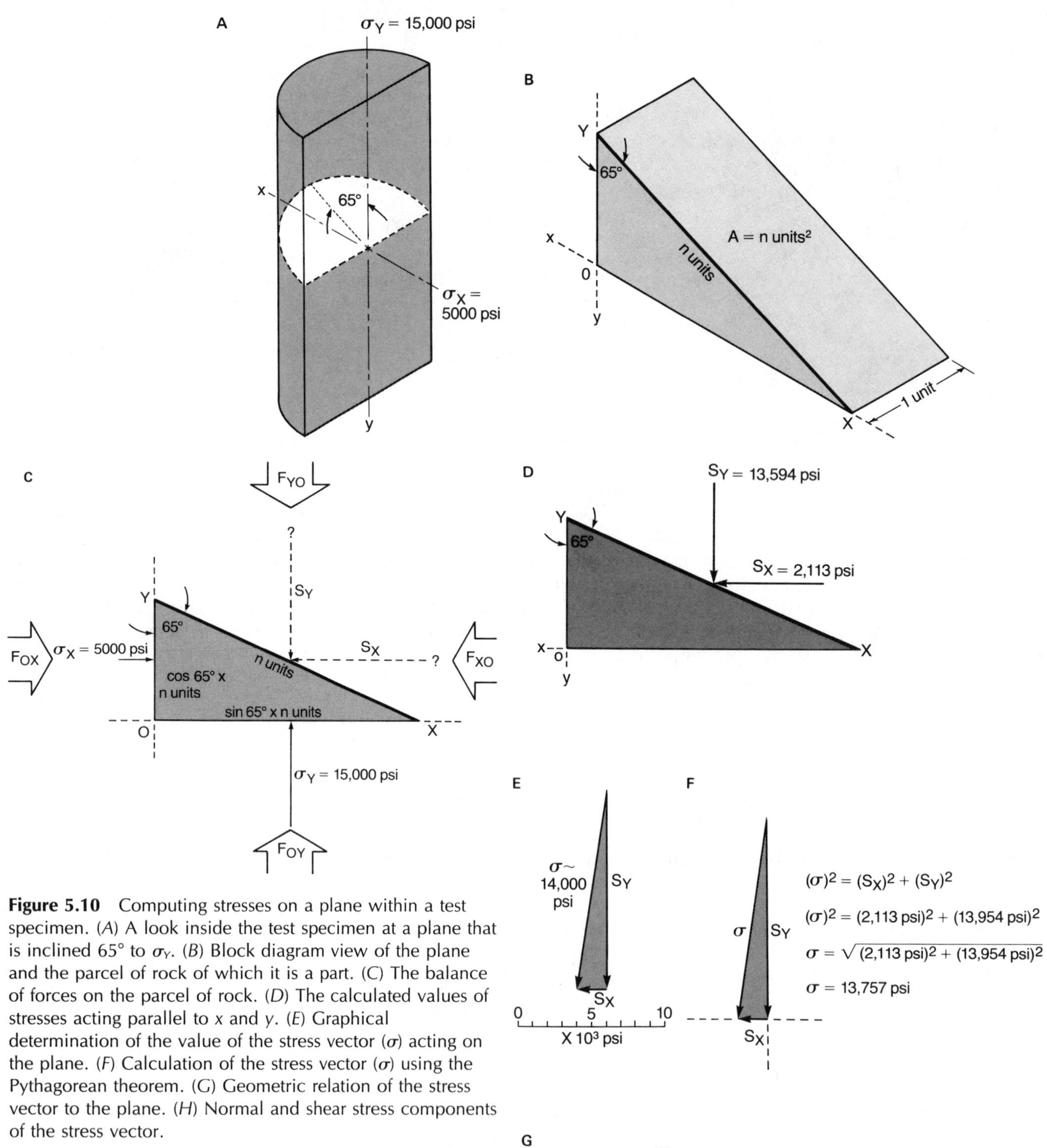

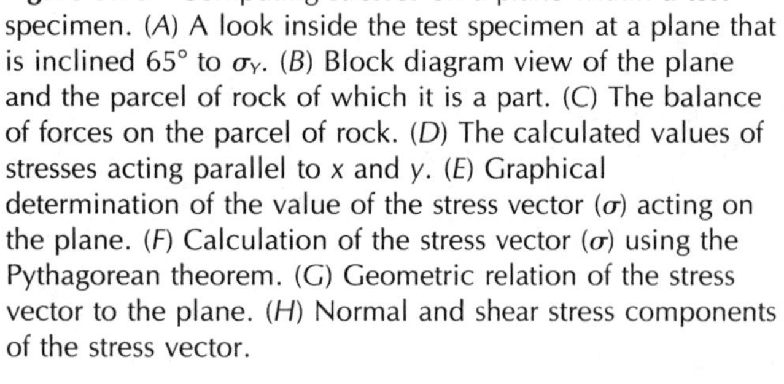

Figure 5.10 Computing stresses on a plane within a test specimen. (*A*) A look inside the test specimen at a plane that is inclined 65° to σ_Y. (*B*) Block diagram view of the plane and the parcel of rock of which it is a part. (*C*) The balance of forces on the parcel of rock. (*D*) The calculated values of stresses acting parallel to *x* and *y*. (*E*) Graphical determination of the value of the stress vector (σ) acting on the plane. (*F*) Calculation of the stress vector (σ) using the Pythagorean theorem. (*G*) Geometric relation of the stress vector to the plane. (*H*) Normal and shear stress components of the stress vector.

the stress vector (σ) for this plane, consider the plane to be part of a small parcel of rock that is triangular in cross section (Figure 5.10*B*). The hypotenuse (*XY*) of this rock triangle represents the trace of the plane whose stress vector we plan to calculate. Its length is n units. The legs of the triangle represent directions (*OX* and *OY*), which are parallel to the stresses σ_X and σ_Y. We assume that the third dimension of plane *XY* is 1 unit long (Figure 5.10*B*). Therefore the area of the plane is $n \times 1$ square units, or simply n units^2. The area of the plane whose trace is *OX* is n sin 65° units^2, and the area of the plane whose trace is *OY* is n cos 65° units^2, (e.g., n cos 65° in.^{2}).

Our approach in calculating the stress vector on *XY* is to first determine the magnitudes of the stress components acting on *XY* that operate parallel to the *x* and *y* axes. These are denoted S_X and S_Y (Figure 5.10*C*). If we can determine the values of these, we can proceed to calculate the full value of the stress vector whose components are S_x and S_Y. Following Means (1976), we set up equations from which S_X and S_Y can be calculated, assuming that the forces acting horizontally and vertically on this small chunk of rock are perfectly balanced. Otherwise the rock, in disequilibrium, would be in a constant state of rotation.

To balance forces acting horizontally we first compute the force acting in the direction of *OX* (Figure 5.10*C*).

$$F = \sigma A$$

$$F_{OX} = \sigma_X A$$

$$F_{OX} = 5000 \text{ psi } (\cos 65°)(n \text{ units})^2 = 5000\, n\, (\cos 65°) \text{ lb}$$

This force is balanced by the horizontal forces (F_{XO}) acting on the plane *XY* in the direction of *XO*.

$$F_{XO} = S_X A = S_X(n \text{ in.}^2)$$

Thus the equation that describes a horizontal balancing of force is:

$$F_{OX} = F_{XO}$$

$$5000 \text{ psi } (\cos 65°)\ (n \text{ units}^2) = S_X\ (n \text{ in.}^2)$$

Solving,

$$S_X = 5000 \text{ psi } (\cos 65°) = 2113 \text{ psi } (148 \text{ kg/cm}^2)$$

The value of S_Y can be calculated by balancing forces vertically (Figure 5.10*C*). The force (F_{OY}), operating parallel to *OY*, is:

$$F_{OY} = \sigma_Y\ (\sin 65°)\ (n \text{ units}^2)$$

This force is balanced by the vertical force (F_{YO}) acting in the direction of *YO*. The force F_{YO} is equal to the stress (S_Y) acting on plane *XY* in the direction of the *y* axis.

$$F_{YO} = S_Y n$$

Therefore, the equation that describes a vertical balancing of forces has the following form.

$$F_{OY} = F_{YO}$$

$$\sigma_Y\ (\sin 65°)\ (n \text{ units}^2) = S_Y\ (n \text{ units}^2)$$

$$15{,}000 \text{ psi } (\sin 65°)\ (n \text{ units}^2) = S_Y\ (n \text{ units}^2)$$

Consequently,

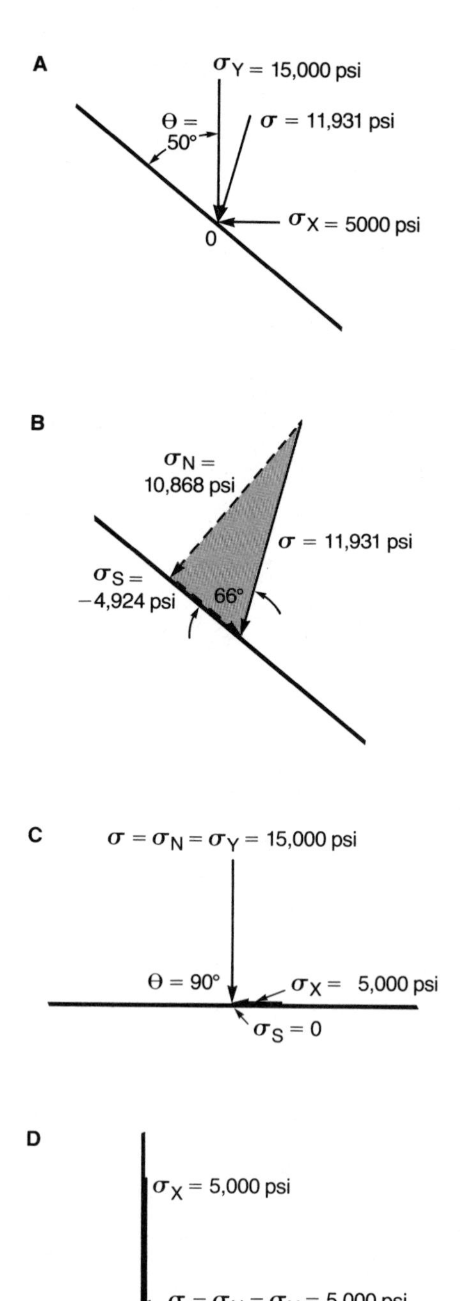

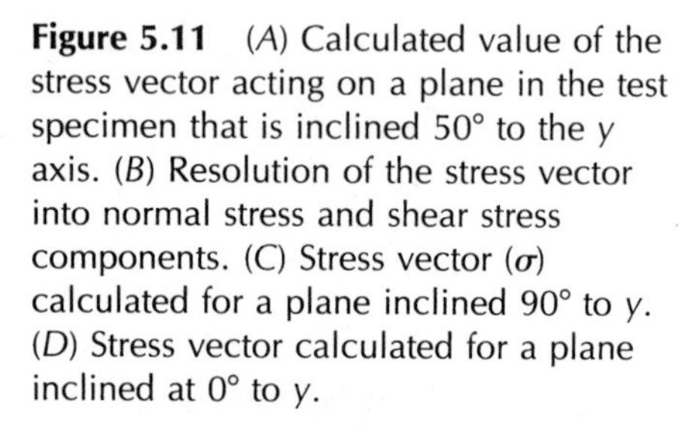

Figure 5.11 (*A*) Calculated value of the stress vector acting on a plane in the test specimen that is inclined 50° to the *y* axis. (*B*) Resolution of the stress vector into normal stress and shear stress components. (*C*) Stress vector (σ) calculated for a plane inclined 90° to *y*. (*D*) Stress vector calculated for a plane inclined at 0° to *y*.

$$S_Y = 15{,}000 \text{ psi } (\sin 65°) = 13{,}594 \text{ psi } (956 \text{ kg/cm}^2)$$

Having calculated the values of S_X and S_Y (Figure 5.10*D*), let us determine the stress vector (σ) for which these are components. This can be carried out graphically or numerically. Graphically, we map the vectors S_X and S_Y as shown in Figure 5.10*E*. Then we construct the stress vector (σ) from the tail of S_Y to the tip of S_X. Measuring its length according to the scale of the map, we find that σ has a value of about 14,000 psi (984 kg/cm²). The stress vector makes an angle of $\theta = 10°$ with the *y* axis. It is thus inclined at approximately 80°.

A more accurate way to determine the magnitude and orientation of the stress vector is through the Pythagorean theorem (Figure 5.10*F*).

$$(\sigma)^2 = (S_X)^2 + (S_Y)^2$$

$$(\sigma)^2 = (2113 \text{ psi})^2 + (13{,}954 \text{ psi})^2$$

$$\sigma = 13{,}757 \text{ psi } (967 \text{ kg/cm}^2)$$

The angle that the vector makes with respect to *y* can be found trigonometrically.

$$\sin \beta = \frac{S_x}{\sigma} = \frac{2113 \text{ psi}}{13{,}757 \text{ psi}} = 0.1536$$

$$\beta = \arcsin 0.1536 = 9°$$

As emphasized by Means (1976), stress vectors are not generally perpendicular to the plane for which they have been calculated. For example, the stress vector (σ) we just calculated is not orthogonal to plane *XY*, but instead departs from 90° by 16° (Figure 5.10*G*). Thus, in almost all cases, a given stress vector can be resolved into normal and shear stress components of some finite value (i.e., nonzero). The stress vector (σ) on plane *XY* can be resolved into a normal stress (σ_N) whose value is 13,224 psi (930 kg/cm²) and a shear stress (σ_S) whose value is −3791 psi (−266 kg/cm²) (Figure 5.10*H*).

COMPUTING TOTAL STATE OF STRESS

Stress vectors, normal stresses, and shear stresses can be calculated for planes of *all* orientations that pass through point O in the test specimen. For example, the magnitude of the stress vector to a plane that makes an angle (θ) of 50° with the *y* axis is 11,931 psi (839 kg/cm²) (Figure 5.11*A*). Inclined 66° to the plane, it can be resolved into normal stress (σ_N) and shear stress (σ_S) components of 10,868 psi (764 kg/cm²) and −4924 psi (−346 kg/cm²), respectively (Figure 5.11*B*). Stress vectors to planes inclined at $\theta = 0°$ and $\theta = 90°$ are special cases. The plane marked by an orientation of $\theta = 90°$ is horizontal, parallel to *x* (Figure 5.11*C*). The stress vector calculated for this plane is perpendicular to it and has the same magnitude as σ_Y, namely 15,000 psi. This stress vector has no shear stress component. In this special case, the stress vector is purely a normal stress. Similarly, the stress vector to a plane whose orientation is defined by $\theta = 0°$ is a normal stress whose value is 5000 psi (Figure 5.11*D*).

Using a calculator, or better yet a computer program that summarizes the mathematics of these calculations, it is possible to determine the orientations and absolute values of stress vectors, normal stresses, and shear stresses for a host of planes inclined at 5° intervals through point *O*. I have done this, keeping the stress conditions set at $\sigma_Y = 15{,}000$ psi and $\sigma_X = 5000$ psi. Results are shown in Figure 5.12. In studying the array of values in

σ_Y = 15,000 psi, σ_X = 5000 psi

Θ	σ	σ_S	σ_N	α	β
0	5000	0	5000	0	90
5	5149	868	5076	9	75
10	5570	1710	5301	17	62
15	6196	2500	5669	23	51
20	6956	3213	6169	27	42
25	7792	3830	6786	29	35
30	8660	4330	7500	30	30
35	9528	4698	8289	29	25
40	10374	4924	9131	28	21
45	11180	5000	10000	26	18
50	11931	4924	10868	24	15
55	12617	4698	11710	21	13
60	13228	4330	12500	19	10
65	13757	3830	13213	16	8
70	14198	3213	13830	13	6
75	14546	2500	14330	9	5
80	14797	1710	14698	6	3
85	14949	868	14924	3	1
90	15000	0	15000	0	0

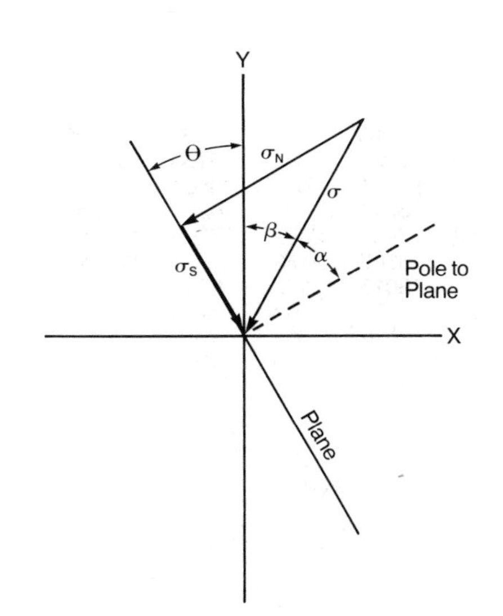

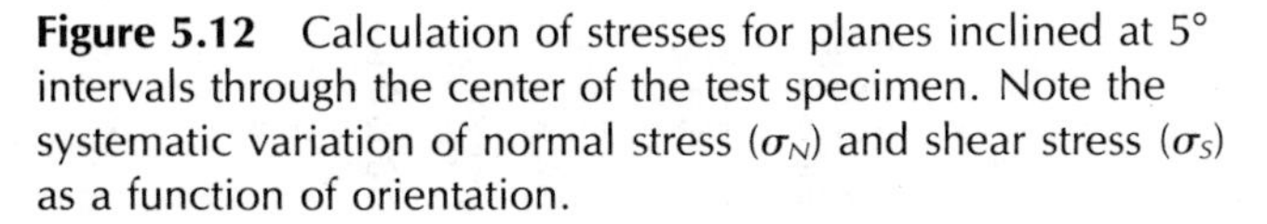

Figure 5.12 Calculation of stresses for planes inclined at 5° intervals through the center of the test specimen. Note the systematic variation of normal stress (σ_N) and shear stress (σ_S) as a function of orientation.

Figure 5.12, it becomes clear that normal stress and shear stress magnitudes vary systematically as a function of orientation: shear stress is zero for stress vectors oriented parallel to the x and y axes (θ = 0° and 90°, respectively). Shear stress (σ_S) steadily increases from 0 psi to a maximum of 5000 psi as θ increases from 0° to 45°, decreasing again to 0° between θ = 45° and θ = 90°.

THE STRESS ELLIPSE AND THE STRESS ELLIPSOID

The stress data posted in Figure 5.12 are incredibly systematic. This can be appreciated by plotting all the stress vectors such that their tails meet at a common point, the tiny point *O* containing the planes for which these vectors were computed. When the stress vectors are plotted to scale in this fashion, an elliptical form is generated, called the **stress ellipse** (Figure 5.13). The ellipse vividly pictures the orderly distribution of stress about a point in a body.

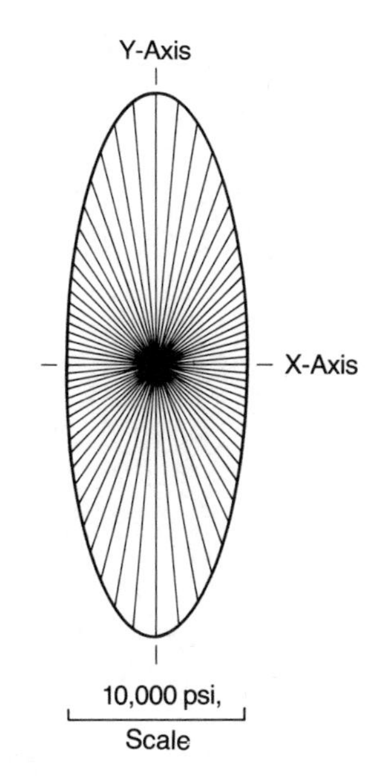

Figure 5.13 Stress vectors that are plotted according to orientation yield a stress ellipse.

The stress ellipse is a useful device for describing the state of stress at any point within a body of rock. The axes of the stress ellipse are called the **principal stress directions** (Figure 5.14). They are mutually perpendicular. The long axis of the ellipse is the **axis of greatest principal stress (σ_1)**; the short axis is the **axis of least principal stress (σ_3)**. These axes define the stress ellipse. Of all the values of normal stress (σ_N) computed for stress vectors operating about some point, the maximum normal stress component is derived from the stress vector oriented parallel to the greatest principal stress direction (σ_1). Indeed this stress vector is strictly a normal stress. It has no shear stress component. Similarly, the direction of least principal stress (σ_3) is marked by zero shear stress. The stress operating parallel to σ_3 is a normal stress whose magnitude is the lowest of all the normal stress values computed for stress vectors operating about the point.

The form of the stress ellipse changes according to the absolute and relative values of the principal stresses. Some examples are shown in Figure 5.15.

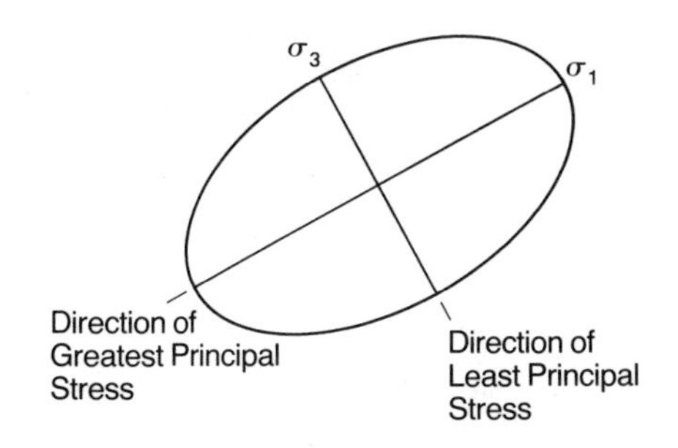

Figure 5.14 The stress ellipse.

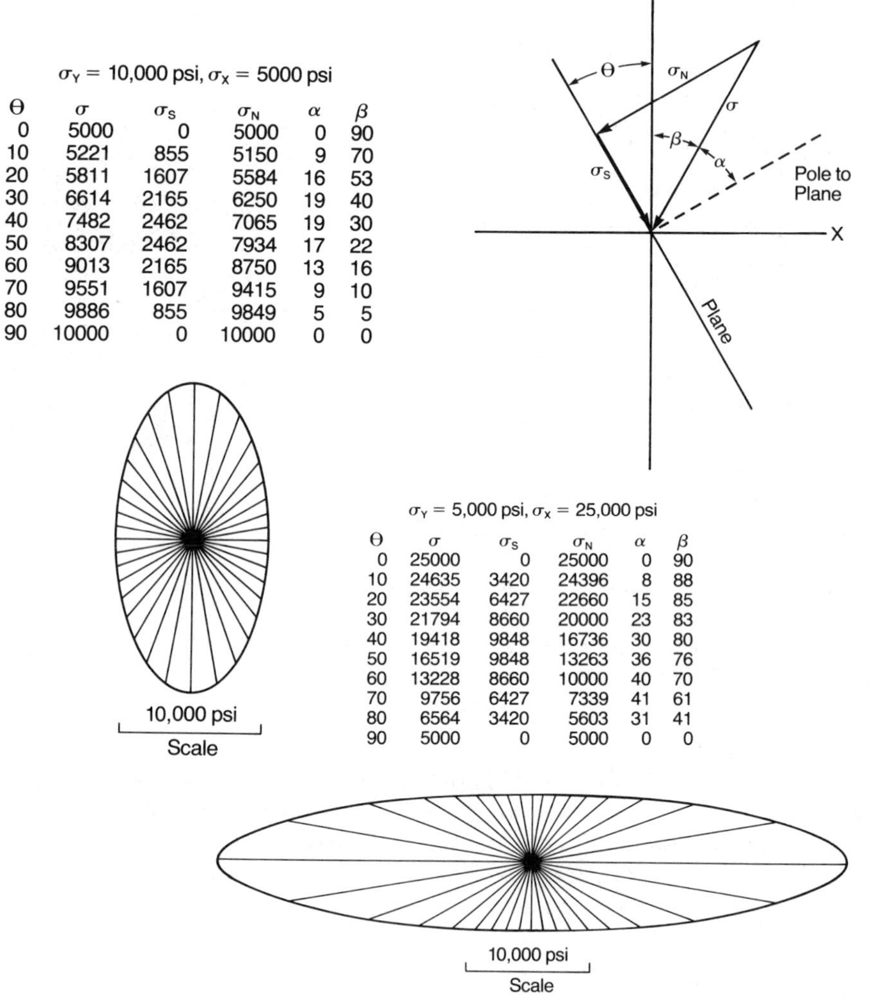

σ_Y = 10,000 psi, σ_X = 5000 psi

Θ	σ	σ_S	σ_N	α	β
0	5000	0	5000	0	90
10	5221	855	5150	9	70
20	5811	1607	5584	16	53
30	6614	2165	6250	19	40
40	7482	2462	7065	19	30
50	8307	2462	7934	17	22
60	9013	2165	8750	13	16
70	9551	1607	9415	9	10
80	9886	855	9849	5	5
90	10000	0	10000	0	0

σ_Y = 5,000 psi, σ_X = 25,000 psi

Θ	σ	σ_S	σ_N	α	β
0	25000	0	25000	0	90
10	24635	3420	24396	8	88
20	23554	6427	22660	15	85
30	21794	8660	20000	23	83
40	19418	9848	16736	30	80
50	16519	9848	13263	36	76
60	13228	8660	10000	40	70
70	9756	6427	7339	41	61
80	6564	3420	5603	31	41
90	5000	0	5000	0	0

Figure 5.15 The shapes of stress ellipses reflect the principal stress conditions. Angular relationships are shown in Figure 5.12.

In three-dimensional dynamic analysis it is the **stress ellipsoid** that provides the description of the state of stress at a point (Figure 5.16). In addition to the axes of greatest and least principal stress, the stress ellipsoid is characterized by an **axis of intermediate principal stress (σ_2)**, which is oriented perpendicular to the plane of σ_1 and σ_3. Like σ_1 and σ_3, σ_2 is a direction of zero shear stress.

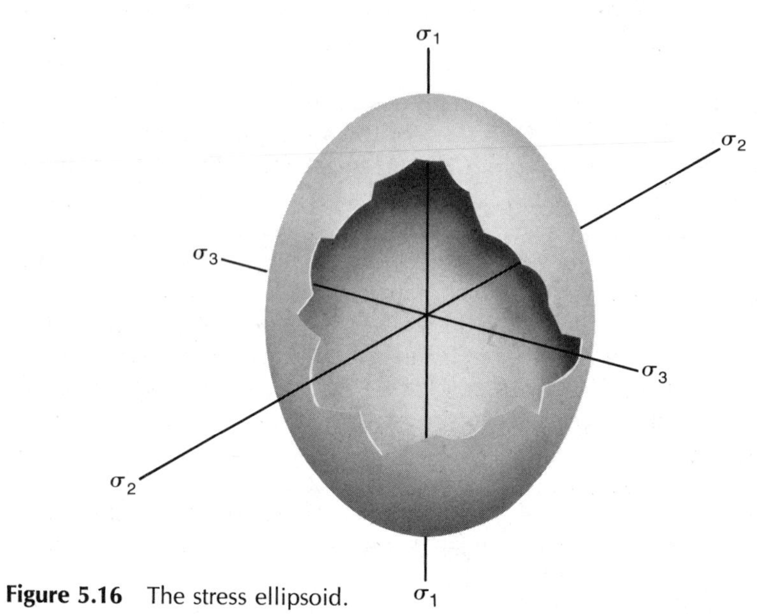

Figure 5.16 The stress ellipsoid.

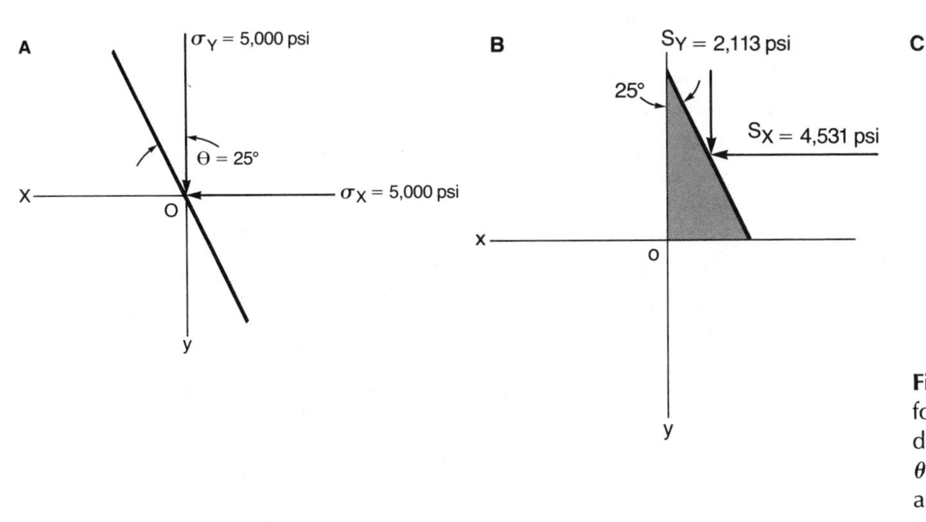

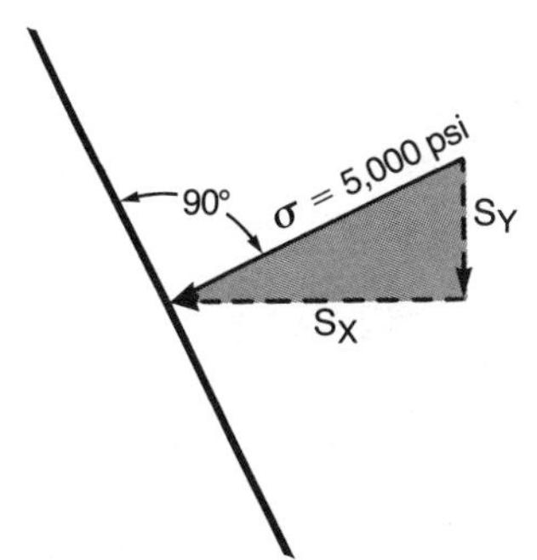

Figure 5.17 (*A*) Cross section of a plane for which the stress vector is to be determined. The plane is oriented at $\theta = 25°$. (*B*) The computed values of S_X and S_Y. (C) Determination of the stress vector (σ) based on values of S_X and S_Y.

THE SPECIAL CASE OF HYDROSTATIC STRESS

When we calculate stress vectors about a point within a hydrostatic stress field, we discover that all the stress vectors have the same value. Moreover, we find that each stress vector is oriented perpendicular to the plane for which it was calculated. Thus, the stress vectors are all normal stresses; they have no shear stress components.

By way of example, let us determine the state of stress on a plane inclined at $\theta = 25°$ within a rock subjected to an axial stress (σ_Y) of 5000 psi and a confining pressure (σ_X) of 5000 psi (Figure 5.17*A*). By the balancing of forces we find that $S_X = 4531$ psi (319 kg/cm^2) and $S_Y = 2113$ psi (Figure 5.17*B*).When we compute the stress vector for which S_X and S_Y are components, we find that S has a magnitude of 5000 psi and a direction that is perpendicular to the plane (Figure 5.17C). For this stress vector, $\sigma_N = 5000$ psi and $\sigma_S = 0$ psi! When, as before, we calculate stress vectors for planes at intervals of 5° (Figure 5.18), we discover that each combination of S_X and S_Y values always yields a 5000-psi stress vector oriented perpendicular to the plane for which it is calculated. It is no wonder that hydrostatic stresses cannot induce distortion of rocks.

σ_Y = 5,000 psi, σ_X = 5000 psi

θ	σ	σ_S	σ_N	α	β
0	5000	0	5000	0	90
5	5000	0	5000	0	85
10	5000	0	5000	0	80
15	5000	0	5000	0	75
20	5000	0	5000	0	70
25	5000	0	5000	0	65
30	5000	0	5000	0	60
35	5000	0	5000	0	55
40	5000	0	5000	0	50
45	5000	0	5000	0	45
50	5000	0	5000	0	40
55	5000	0	5000	0	35
60	5000	0	5000	0	30
65	5000	0	5000	0	25
70	5000	0	5000	0	20
75	5000	0	5000	0	15
80	5000	0	5000	0	10
85	5000	0	5000	0	5
90	5000	0	5000	0	0

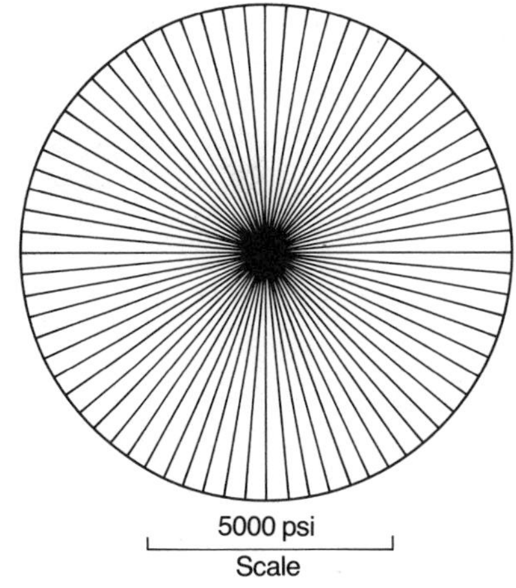

Figure 5.18 The tabulated stress vector values for planes arranged at 5° intervals in a rock suspended within a hydrostatic stress field.

THE STRESS EQUATIONS

A high goal of structural geology is to understand the stress conditions that trigger faulting. Whether a plane of a certain orientation becomes a fault largely depends on the absolute and relative magnitudes of the normal and shear stresses that operate on the plane. Normal stress that is high compared to shear stress tends to inhibit movement on a given surface. Shear stress that is high compared to normal stress tends to promote fault movement. It becomes of interest to somehow calculate the values of normal stress and shear stress that trigger faulting along a given surface.

Suppose during experimental deformation a rock is faulted along a plane inclined $\theta = 25°$ to the y axis. How do we determine the levels of normal and shear stress that were achieved on the fault plane just before failure? These values cannot be read from a chart or gauge, but must be calculated, as above. Fortunately, the magnitude of normal and shear stress can be calculated readily using **stress equations** derived in standard engineering and structural geological texts (e.g., Ramsay, 1967; Jaeger and Cook, 1976; Means, 1976). The form of the stress equations is identical to that of strain (Equations 4.7 and 4.8. They are written as follows.

$$\sigma_N = \frac{\sigma_1 + \sigma_3}{2} - \frac{\sigma_1 - \sigma_3}{2} \cos 2\theta \tag{5.3}$$

$$\sigma_S = \frac{\sigma_1 - \sigma_3}{2} \sin 2\theta \tag{5.4}$$

In illustrating the use of these equations, let us continue to focus on a state of stress characterized by $\sigma_1 = 15{,}000$ psi and $\sigma_3 = 5000$ psi. Our goal is to calculate σ_N and σ_S for a plane that makes an angle $\theta = 30°$ with respect to σ_1.

$$\sigma_N = \frac{15{,}000 \text{ psi} + 5000 \text{ psi}}{2} - \frac{15{,}000 \text{ psi} - 5000 \text{ psi}}{2} \cos 60°$$

$$\sigma_N = 10{,}000 \text{ psi} - 5000 \text{ psi} (\cos 60°)$$

$$\sigma_N = 7499 \text{ psi } (304 \text{ kg/cm}^2) \text{ (see Figure 5.12)}$$

Similarly,

$$\sigma_S = \frac{15{,}000 \text{ psi} - 5000 \text{ psi}}{2} \sin 60°$$

$$\sigma_S = 5000 \text{ psi } (\sin 60°)$$

$$\sigma_S = 4330 \text{ psi, } 304 \text{ kg/cm}^2\text{; (see Figure 5.12)}$$

For a condition of hydrostatic stress such that $\sigma_1 = \sigma_3 = 5000$ psi, we can use the stress equations to calculate σ_N and σ_S for a plane inclined at $\theta = 30°$ to σ_1.

$$\sigma_N = \frac{5000 \text{ psi} + 5000 \text{ psi}}{2} - \frac{5000 \text{ psi} - 5000 \text{ psi}}{2} \cos 60° = 5000 \text{ psi}$$

$$\sigma_S = \frac{5000 \text{ psi} - 5000 \text{ psi}}{2} \sin 60° = 0 \text{ psi}$$

THE MOHR STRESS DIAGRAM

Just as the Mohr strain diagram provides a picture of strain variations within a body (Chapter 4), the Mohr stress diagram gives us a very useful display of the stress equations. The equations describe a circular locus of paired values (σ_N, σ_S) of the normal and shear stresses that operate on planes of any and all orientations within a given body that has been subjected to known values of σ_1 and σ_3. Using the Mohr stress diagram, it is possible to identify a plane of any orientation relative to σ_1, and then to read the values of normal stress (σ_N) and shear stress (σ_S) acting on the plane.

The construction of the Mohr stress circle proceeds as follows. Principal normal stress values (σ_1 and σ_3) are plotted on the *x* axis of the diagram (Figure 5.19*A*). A circle is drawn through these points such that (σ_1–σ_3) constitutes the circle's diameter. In the example under study where $\sigma_1 = 15{,}000$ psi and $\sigma_3 = 5000$ psi, all the paired values of σ_N and σ_S as listed in Figure 5.12 exist as points on the perimeter of the circle. To define σ_N and σ_S for a specific plane (e.g., a plane oriented at $\theta = 30°$ with respect to σ_1), a radius is constructed at an angle of 2θ, or 60°, measured clockwise from the *x* axis (Figure 5.19*A*). Where this radius intersects the perimeter of the circle, a point is established whose *x*, *y* coordinates are the (σ_N, σ_S) values for the plane in question: $\sigma_N = 7500$ psi (527 kg/cm^2) and $\sigma_S = 4330$ psi. The

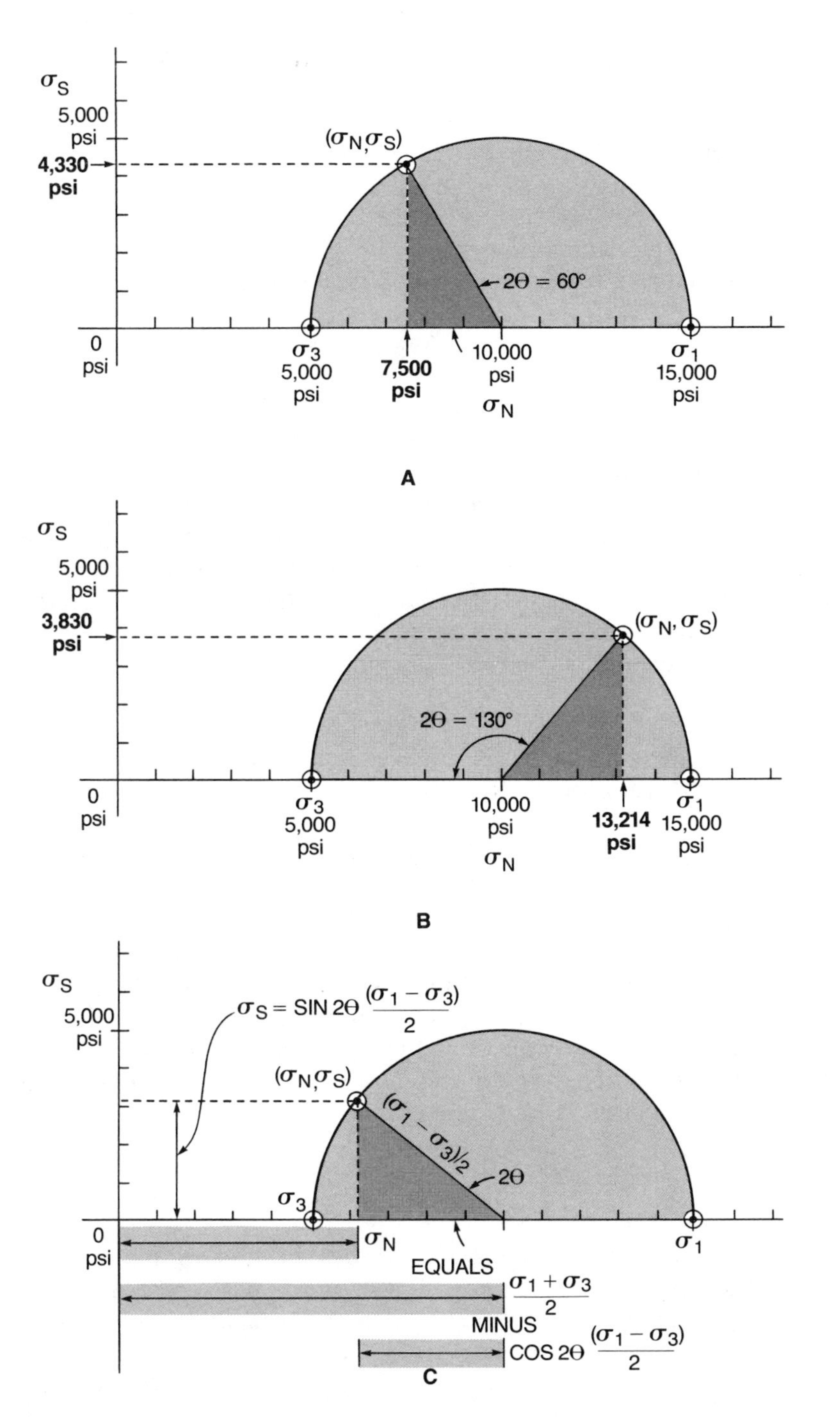

Figure 5.19 (*A*) Mohr stress diagram, constructed for the stress condition $\sigma_1 = 15{,}000$ psi, $\sigma_3 = 5000$ psi. Normal stress (σ_N) and shear stress (σ_S) values are shown for a plane that is oriented at $\theta = 30°$. (*B*) Normal stress (σ_N) and shear stress (σ_S) values are shown for a plane that is oriented at $\theta = 65°$. (C) Relation of Mohr stress circle geometry to the fundamental stress equations.

values of σ_N and σ_S for a plane oriented 65° to σ_1 are found by constructing a radius at an angle of $2\theta = 130°$, again measured clockwise from the *x* axis (Figure 5.19*B*). The *x, y* coordinates of the point of intersection of this radius with the perimeter of the circle are σ_N, σ_S = 13,214, 3830 psi. The correspondence of the stress equations to the geometry of the elements of the Mohr diagram is displayed in Figure 5.19C.

THE STANDARD COMPRESSION TEST

With this background in mind, let us return to the Donath apparatus and carry out a compressional deformation experiment from start to finish. Our purposes are to better understand the stress conditions of faulting and to

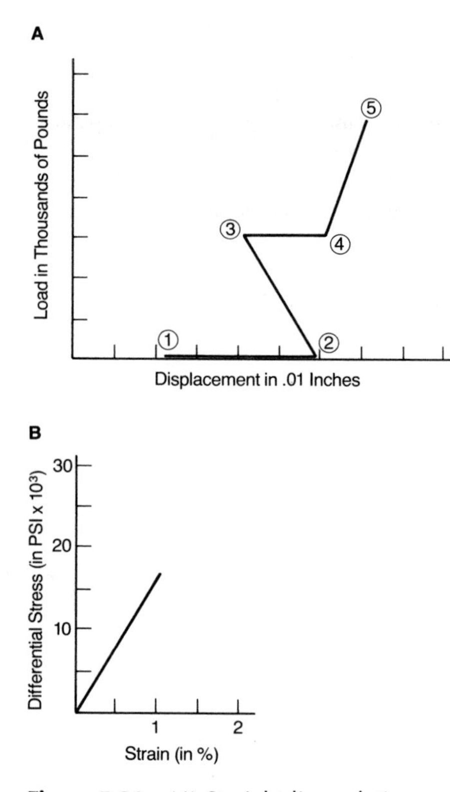

Figure 5.20 (*A*) Straight-line relation between load and displacement. Seat specimen (1-2); raise confining pressure (2-3); reseat specimen (3-4); deform specimen (4-5). (*B*) Transformation of load–displacement curve to stress–strain diagram.

study how a rock specimen responds to compression-induced shortening. We shorten a cylinder of limestone under conditions of 4000 psi (280 kg/cm^2) confining pressure. The initial length (l_o) of the specimen is 1.000 in. and the radius (r) is 0.250 in. (0.635 cm). When the specimen is subjected to 4000-psi confining pressure, the length of the specimen decreases by some tiny amount, approximately 0.001 in. (0.025 mm), because of negative dilation. The decrease in radius is negligible. Thus we calculate the cross-sectional area (A) of the specimen on the basis of the initial radius.

$$A = \pi r^2 = 3.1416\ (0.250\ \text{in.})^2 = 0.1961\ \text{in.}^2\ (0.4981\ \text{cm})^2$$

When force is applied by the piston to the base of the anvil, compressional stresses are transmitted through the rock, and the cylinder of limestone begins to shorten. In the early stage of this compression, pen movement on the *X–Y* recorder reveals that force and displacement are directly proportional. The **load–displacement curve** is a straight line (Figure 5.20*A*).

The load–displacement curve can be transformed into a **stress–strain diagram**, the standard display of experimental tests. Stress–strain diagrams plot differential stress against percentage strain (Figure 5.20*B*). **Differential stress (σ_d)**, plotted on the *y* axis of the stress–strain curve, is the difference between the greatest (σ_1) and least (σ_3) stresses:

$$\sigma_d = \sigma_1 - \sigma_3$$

where σ_d = differential stress. **Percentage strain**, plotted on the *x* axis of the stress–strain curve, is percentage shortening:

$$\text{Percentage shortening} = \frac{\Delta l}{l_o}(100)$$

The straight-line nature of the stress–strain diagram reveals that the limestone specimen is behaving elastically, like a spring. The elastic behavior is a dilational spring-action phenomenon, derived from the rock's ability to recover tiny nonrigid body changes in atomic spacings in crystal lattices of its mineral components (Ramsay, 1967). The equation of the straight line describing the proportional relationship of stress to strain for elastic bodies is **Hooke's law** (Figure 5.20*B*):

$$\sigma = Ee \tag{5.4}$$

where σ = stress, E = Young's modulus, and e = strain.

The value of *E*, **Young's modulus**, describes the slope of the straight-line stress–strain curve. Even under the same conditions of deformation, the value of *E* will vary from rock to rock, reflecting natural differences in the resistance of rock to elastic deformation. Thus the slope of a straight-line stress–strain curve is a measure of the **stiffness** of the rock. Typical values of Young's modulus are presented in Table 5.1: the higher the values, the stiffer the rock.

In the context of our experimental work, Young's modulus (*E*) can be thought of as an **elastic modulus** that describes how much stress is required to achieve a given amount of length-parallel elastic shortening of a core of rock. A second elastic modulus, known as **Poisson's ratio**, describes the degree to which a core of rock bulges as it shortens. Poisson's ratio, symbolized by the Greek letter nu, **(ν)**, describes the ratio of **lateral strain** to **longitudinal strain**:

$$\nu = \frac{e_{\text{lat}}}{e_{\text{long}}} \tag{5.5}$$

Values of Poisson's ratio appear in Table 5.2.

Table 5.1
Typical Values of Young's Modulus (E)

Rock	E ($\times 10^6$ psi)
Westerly Granite	8.1
Cheshire Quartzite	11.4
Karroo Diabase	12.2
Tennessee Marble	6.9
Witwatersrand Shale	9.8
Solenhofen Limestone	7.7

Table 5.2
Typical Values of Poisson's Ratio (ν)

Rock Type	
Limestone, fine grained	0.25
Aplite	0.20
Limestone, porous	0.18
Limestone, oolitic	0.18
Limestone, chalcedonic	0.18
Limestone, medium grained	0.17
Limestone, stylolitic	0.11
Granite	0.11
Shale, quartzose	0.08
Graywacke, coarse grained	0.05
Diorite	0.05
Granite, altered	0.04
Graywacke, fine grained	0.04
Shale, calcareous	0.02
Schist, biotite	0.01

Poisson's ratio meant very little to me, other than a Greek letter, until one day in undergraduate structural lab when my students and I placed a core of granite in a deformation press and squeezed it. The core was 3 in. (0.76 cm) in diameter and about 5 in. (12.7 cm) long. We compressed it in an unconfined state (i.e., in open air), using a "soil testing" ram. We wired the core with strain gauges to monitor changes in length (longitudinal strain) and width (lateral strain). The wires were connected to an X–Y plotter so that we could measure tiny magnitudes of elastic strain during the deformation. As we loaded the specimen with force, the rock began to shorten elastically, but to our surprise the shortening was not compensated by any increase in specimen diameter. The stress did not produce the expected lateral bulging, the "barreling," which we had anticipated. Volume decreased and stress somehow was being stored . . . that is, until the rock exploded with a blast like the sound of a shotgun. The largest rock fragment we found in the lab after the explosion was only pea sized. We were grateful to have been protected by a Plexiglass shield. I will never forget the significance of **very low** values of Poisson's ratio. Rock bursts in deep underground mines commonly represent stress releases in rocks of low Poisson's ratio.

Returning to the compression test on limestone, we find that during the elastic stage of deformation it is possible to remove the load (force) and observe that the limestone core springs back to its original length, almost instantly. As load is decreased to 0.0 lb, the pen on the X–Y recorder may not identically retrace the route of its former ascent. Instead, it may loop back to its original starting position, reflecting a brief time lag in the **recovery** (elimination) of all the bound-up longitudinal strain (Figure 5.21). Such

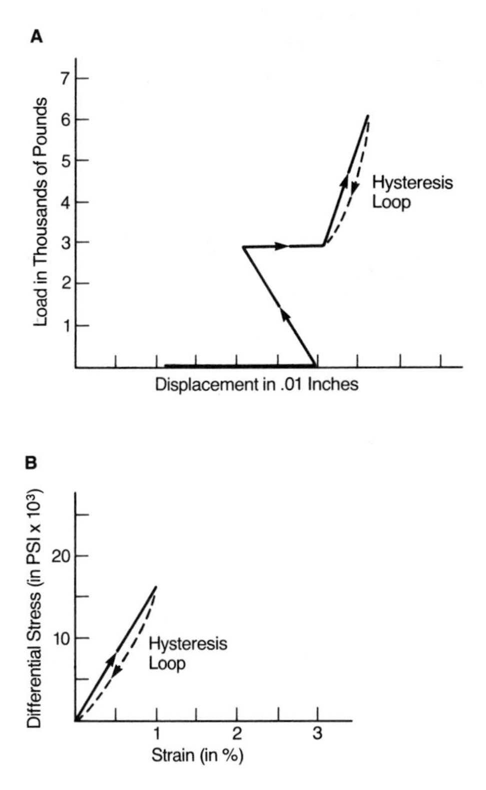

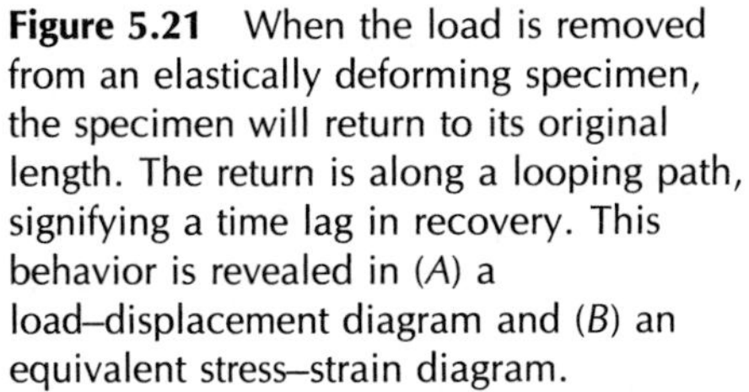

Figure 5.21 When the load is removed from an elastically deforming specimen, the specimen will return to its original length. The return is along a looping path, signifying a time lag in recovery. This behavior is revealed in (*A*) a load–displacement diagram and (*B*) an equivalent stress–strain diagram.

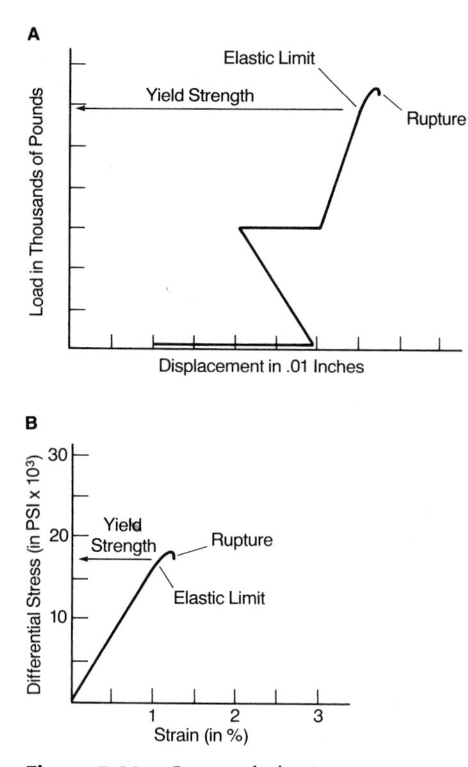

Figure 5.22 Onset of plastic deformation, then rupture, as shown in (*A*) a load–displacement diagram and (*B*) an equivalent stress–strain diagram.

retardation is known as **hysteresis**, and the loop displayed on the graph is sometimes called a **hysteresis loop**.

If instead of dropping the level of stress, it is continually raised, the limestone may eventually begin to deform plastically. In essence, the range of elastic behavior is surpassed, and permanent strain in no way recoverable begins to accumulate in the rock specimen. **Plastic deformation** produces a permanent change in shape of a solid without failure by rupture. **Ductile materials** are capable of undergoing significant plastic deformation.

Anyone who has tried to repair a toy metal slinky that has been stretched too far knows about the nonrecoverability of plastic deformation. The life expectancy of a Slinky is always short. (My experience suggests that the half-life of a Slinky is about 4 hrs.) In its all too brief youth a Slinky can smoothly descend any flight of stairs. Then without warning, it becomes internally entangled. Rescue efforts typically lead to stretching segments of the metal beyond its elastic range, at which point the metal becomes permanently plastically deformed. No efforts, however great, can undo the damage.

The onset of plastic deformation is obvious during the course of an experiment, for the load–displacement curve departs from its straight-line elastic mode and begins to bend to form a convex-upward curve (Figure 5.22). The decrease in slope of the curve signifies that proportionally less stress is required to produce a given amount of shortening of the limestone. The point of departure from elastic behavior to plastic behavior is called the **elastic limit**. Its value, known as **yield strength**, is measured in stress. Below its yield strength, a rock behaves as an elastic solid. Above the elastic limit, the rock begins to flow. If limestone behaved perfectly plastically when axially compressed under conditions of 4500-psi (281 kg/cm^2) confining pressure, it would flow continuously without rupture so long as the axial stress were applied. However, limestone under conditions of low confining pressure is so brittle that it usually ruptures by faulting almost as soon as plastic deformation begins (Figure 5.22). Indeed, the limestone may even experience a **true brittle failure** by sudden fracturing below the elastic limit in the straight-line range of the stress–strain curve.

Faulting of limestone under low confining-pressure conditions is punctuated usually by a muffled "pop" within the pressure vessel, an event that is always marked by a sudden drop in stress level and spectator cheering. The fault movement shortens the specimen in such a way that axial stress is at least momentarily relieved. In some experiments we have conducted, the axial load drops so fast that the needle in the pressure gauge plummets to 0 psi and wraps its tip around the metal peg on which it usually gently rests. However, the drop in pressure is ordinarily modest and marked on the load-displacement curve by a short fishhooklike bend (Figure 5.22*A*). If desired, axial force can again be applied, but the level of the force usually does not rise to its former magnitude because even small increments of stress create slippage on the fault surface that now cuts the specimen (Figure 5.23).

After the limestone breaks, the test usually is terminated. The confining pressure is bled off and so is the axial load. Before the limestone core is removed from the pressure vessel, the net change in length of the specimen is measured on the basis of the seating position of the piston, before and after deformation. For this test, the shortening was merely 0.010 in. (0.025 cm). This is **nonrecoverable permanent strain**. All initial elastic strain is recovered when confining pressure and axial load are removed.

Total permanent strain is calculated as follows.

$$e = \frac{\Delta l}{l_o} = \frac{-0.010 \text{ in.}}{1.000 \text{ in.}} = -0.01, \text{ or } 1\% \text{ shortening}$$

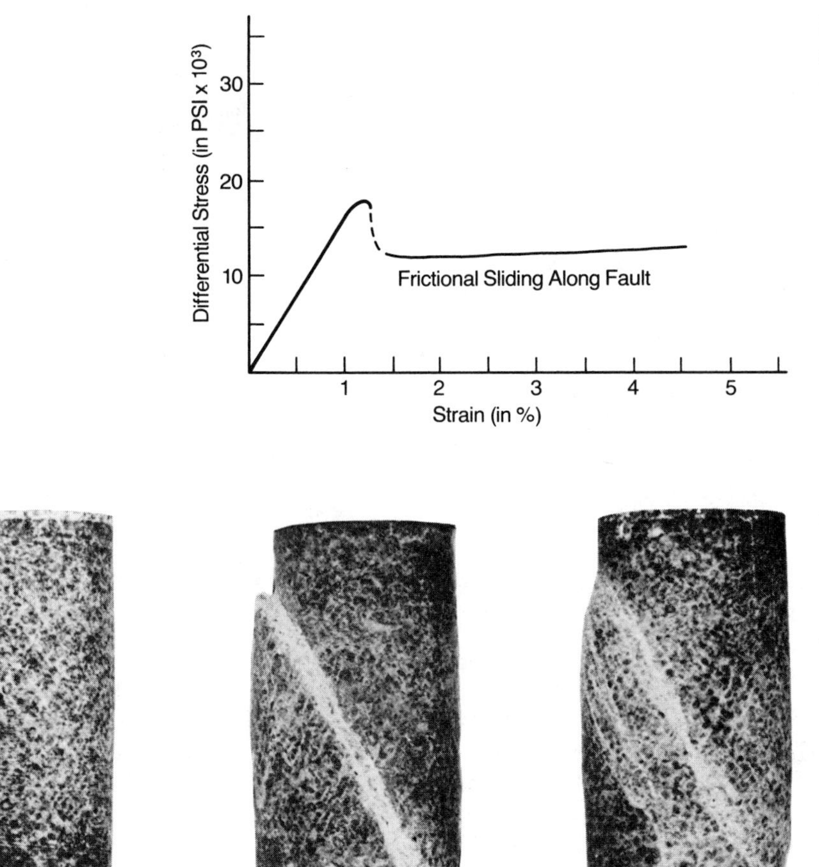

Figure 5.23 Shortening of the test specimen by frictional sliding along the fault surface.

Figure 5.24 Faulted specimens of limestone. [Reprinted by permission, "Some Information Squeezed Out of Rock," by F. A. Donath, *American Scientist*, v. 58, fig. 7, pp. 54–72, (1970b).]

When the specimen is extracted from the pressure vessel, the imprint of the fault trace is clearly visible on the surface of the copper jacket that surrounds the specimen. The angle (θ) that the trace of the fault makes with the long axis of the cylinder can be measured with a protractor. For compression tests carried out at low levels of confining pressure, θ is commonly around 25°. Upon removal of the copper jacket, the physical characteristics of the faulted limestone specimen can be examined (Figure 5.24).

COMPRESSION TESTS AT HIGHER CONFINING PRESSURES

Testing the strength of rocks in the laboratory is like eating potato chips. It is impossible to complete a test without starting on another. It is impossible to squeeze rocks without asking, "What if . . . ?" The compression test we just carried out provided data regarding the strength of limestone under conditions of 4000-psi confining pressure, room temperature, and very rapid loading. But what if we raise the confining pressure to 15,000 psi? How would the limestone then respond?

Under confining-pressure conditions of 15,000 psi, it is found that the yield strength (elastic limit) is now much higher than before (curve *A*, Figure 5.25). Above this elevated yield strength, the limestone begins deforming plastically, as reflected in the bowing of the stress–strain curve.

Before taking the axial stress to higher levels, it is instructive to bleed off all of the axial load, shortly after the limestone has begun to deform plastically (curve *B*, Figure 5.25). When this is done, it becomes evident that shortening due to elastic deformation is quickly recovered, but shortening

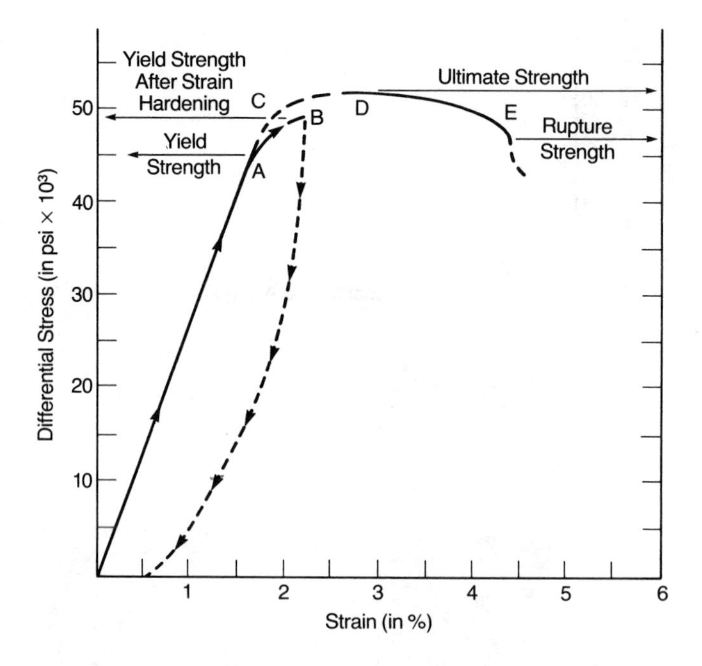

Figure 5.25 Stress–strain diagram for limestone subjected to deformation under confining pressure condition of 15,000 psi (1055 kg/cm^2) Point *A*, onset of plastic deformation. Point *B*, removal of axial load. Note the nonrecoverable plastic deformation. Curve *C*, the second loading. Curve *D*, Plastic deformation. Point *E*, Rupture.

due to plastic deformation is nonrecoverable. Plastic deformation is bound up in the strained rock permanently. When force is again applied to the specimen, the limestone once again behaves as an elastic solid, but this time the elastic limit is higher than the original yield strength first established in the experiment (curve *C*, Figure 5.25). The yield strength of the rock increases because the original fabric of the rock was modified slightly by the experience of plastic deformation. The limestone is said to have undergone **strain hardening,** thus raising its yield strength.

But let us now deal this limestone its final blow. As more and more force is applied, the specimen is "pushed" toward failure. The stress–strain curve displays plastic deformation of the limestone in the form of a smooth, gently sloping curve (*D* in Figure 5.25). Unlike the test carried out at 4000-psi confining pressure, faulting does not immediately begin after a small amount of plastic deformation. Instead, the limestone seems to be able to endure a surprising amount of plastic flow. Eventually, the limestone becomes so weak that the curve begins to descend to the right, steeply, signifying accelerated plastic deformation and impending doom by rupture. Rupture by faulting finally occurs, and stress drops abruptly (curve *E*, Figure 5.25).

The stress- strain curve beautifully records the details of this deformational experience (Figure 5.25). Above the yield strength the curve is convex upward, recording the gradual dissipation of rock strength during plastic deformation. The zenith of the curve corresponds to the **ultimate strength** of the limestone under the experimental conditions. The stress level where faulting occurs is the **rupture strength**. Ultimate strength and rupture strength, like yield strength, are both measured in terms of stress (e.g., psi or kg/cm^2).

The results of this test reveal that limestone becomes stronger at higher levels of confining pressure. This holds true regardless of the parameter that is used to describe the strength: yield strength, ultimate strength, and rupture strength values are all higher for the test carried out at 15,000 psi compared to 4000 psi. Moreover, the limestone undergoes greater plastic deformation at higher levels of confining pressure, assuming temperature and rate of loading and other such factors are held constant. The angle (θ) that the trace of the fault makes with the long axis of the core is seen to be greater, about 30°. The specimen of limestone that had faulted under confining-pressure

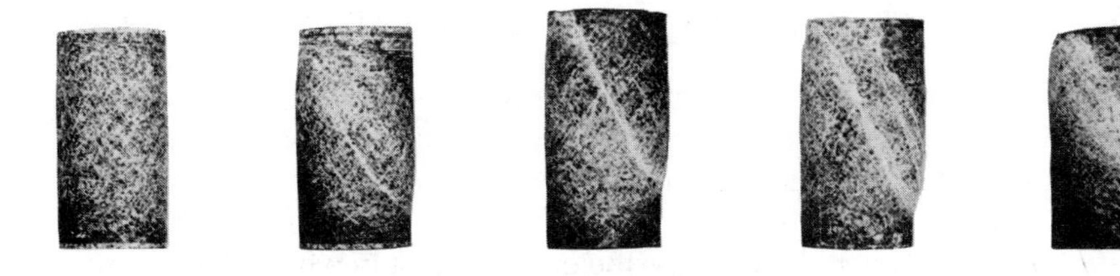

Figure 5.26 Specimens of limestone deformed to approximately the same total strain (about 15%) at different confining pressures. Increased confining pressure causes a transition in deformational mode from brittle to ductile faulting. [Reprinted by permission, "Some Information Squeezed Out of Rock," by F. A. Donath, *American Scientist,* v. 58, p. 54–72, (1970b).]

conditions of 4000 psi was seen to be very fragile, somewhat powdery along the fault, and broken by a single fracture. In contrast, the specimen subjected to rupture at 15,000 psi is seen to be more cohesive, slightly barrel shaped, and affected by a relatively wide zone of distributed fault surfaces, one of which accommodated the ultimate rupture (Figure 5.26). Under the higher confining-pressure conditions, a greater volume of the rock was affected by the deformation, reflecting greater ductility.

When compression of the limestone is carried out at even greater confining pressure, for example, "all the way" at 30,000 psi (2109 kg/cm^2), the limestone responds with even greater strength and ductility (Figure 5.26). This is evident in Table 5.3, which summarizes the conditions, measurements, and results of compression tests on limestone at confining-pressure levels of 4000, 15,000, and 30,000 psi. And when these data are all plotted on a Mohr stress diagram (Figure 5.27), it becomes even more obvious that proportionately greater levels of differential stress (σ_1–σ_3) are necessary to cause rupture when confining pressure (σ_3) is raised. Composite Mohr diagrams of this type become the basis for understanding the stress conditions under which rocks fault (Chapter 9, "Faults").

Table 5.3
Summary of Compressional Testing of Limestone under Confining Pressure Conditions of 4000, 15,000, and 30,000 psi (281, 1055, and 2109 kg/cm^2)

	Specimen #1	*Specimen #2*	*Specimen #3*
Confining Pressure	4,000 psi	15,000 psi	30,000 psi
Differential Stress at Failure	18,000 psi	46,000 psi	80,000 psi
σ_1 at Failure	22,000 psi	61,000 psi	110,000 psi
σ_3 at Failure	4,000 psi	15,000 psi	30,000 psi
Angle (θ) Between Fault and σ_1	25°	30°	33°

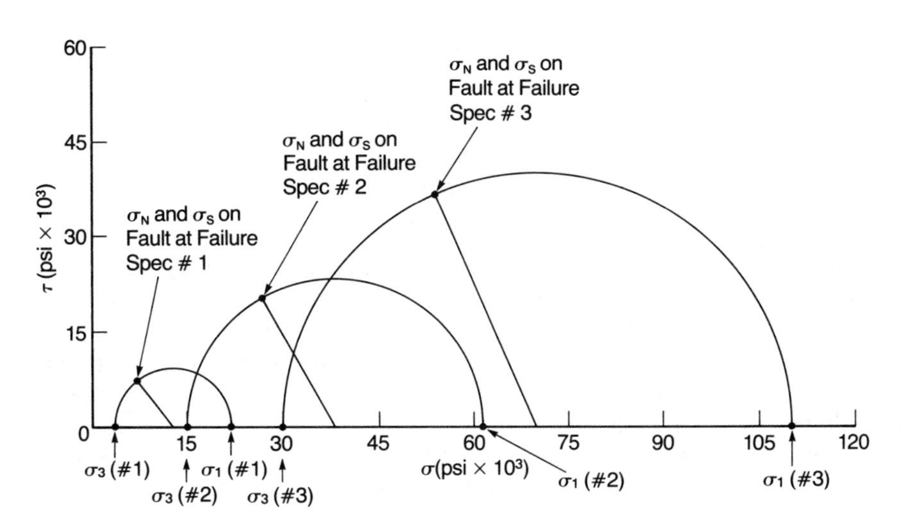

Figure 5.27 Mohr stress diagram summarizing the experimental testing of the limestone.

ROCK STRENGTH—WHAT DOES IT MEAN?

One of the chief conclusions that can be drawn from experimental compression tests is that measurements of rock strength are almost meaningless unless the conditions under which the deformation was achieved are known. Moreover, the terms "brittle" and "ductile" as descriptions of the behavior of specific rocks mean very little unless the environment in which the rock was strained is described. Let us consider the influence of the factors that contribute to the environment of deformation and thus the strength of homogeneous rocks.

LITHOLOGY

It is possible to arrange the common rock lithologies in order of increasing strength for specific conditions of confining pressure and rate of loading (strain rate). Stress- strain diagrams provide the basis for the rankings. It is understood that the rankings are only approximate, being strongly influenced by the composition, texture, and general condition of the rocks that happened to serve as "representative" test specimens for each lithology. Morever, it is understood that the nature and orientation(s) of mechanical hetereogeneity, that is, **anisotropy**, resulting from fractures, layers, foliations, and the like, profoundly influence rock strength.

Table 5.4 shows a ranking of lithologies according to strength based on stress–strain diagrams summarizing compressional tests of rocks at room temperature under low levels of confining pressure. Rocks like salt, anhydrite, shale, and mudstone are seen to be weak (and ductile) compared to rocks of intermediate strength like limestone or calcite-cemented sandstone. Quartzite, granite, and quartz-cemented sandstone are brittle and very strong by comparison.

Table 5.4
General Ranking of Lithology according to Strength, Based on Tests at Room Temperature and Low Confining Pressure

Strongest	Quartzite
	Granite
	Quartz-Cemented Sandstone
	Basalt
	Limestone
	Calcite-Cemented Sandstone
	Schist
	Marble
	Shale/Mudstone
	Anhydrite
Weakest	Salt

In a sequence of different lithologies, the rocks that are likely to behave in the most ductile manner when subjected to stress are commonly referred to as the most **incompetent**. Rocks that are likely to deform in a brittle manner, with no obvious ductile deformation, are described as the most **competent**. "Competency" and "incompetency" are relative terms. The ordering of lithologies by competency may commonly change if the conditions of deformation are changed. Changes in confining pressure, temperature, and strain rate may affect different rocks in different ways.

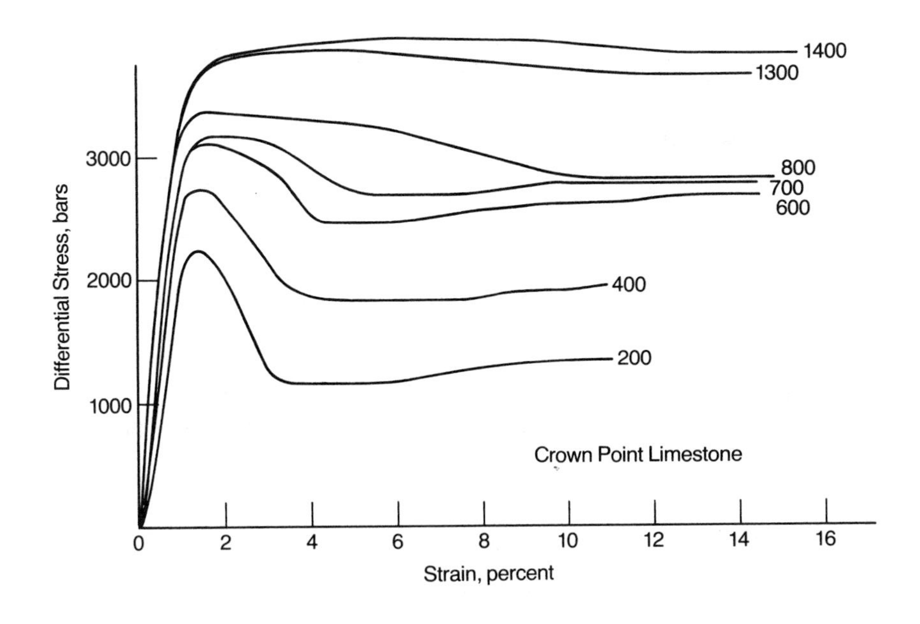

Figure 5.28 Stress–strain diagrams for limestone deformed at a variety of confining pressures. Tests conducted at room temperature. Both strength and plasticity increase with greater confining pressure. [Reprinted by permission, "Some Information Squeezed Out of Rock," by F. A. Donath, *American Scientist*, v. 58, p. 54–72, (1970b).]

CONFINING PRESSURE AND FLUID PRESSURE

We have seen that increasing the confining pressure on a rock specimen in a compression test has the effect of increasing the strength and ductility of the rock. This has been firmly documented, through experimental testing (e.g., see Handin and Hager, 1957). For any given lithology, the yield strength, ultimate strength, rupture strength, and ductility attain greater and greater values with increasing confining pressure (Figure 5.28).

The effect of confining pressure can be partially or completely offset by the presence of elevated **fluid pressure** in the rock (or test specimen) undergoing deformation. In natural sedimentary basins of accumulation, water that is entrapped in sediments during deposition may be pressurized during the course of subsidence, burial, and loading by overlying, younger sediments. As a result, a stress level of fluid pressure may be achieved that exceeds the expected hydrostatic stress level based on depth estimates alone (Hubbert and Rubey, 1959). Elevated hydrostatic fluid–pressure conditions counteract the effects of confining pressure on strength and ductility. A measure of the net effect of confining pressure and fluid pressure is **effective stress**. Effective stress equals confining pressure minus fluid pressure. If effective stress is high, a given rock will be relatively strong and ductile; if effective stress is low, the rock will display less strength and ductility. A deeply buried but highly pressurized sedimentary rock can respond to stress as if the rock were deforming in a relatively low confining-pressure environment.

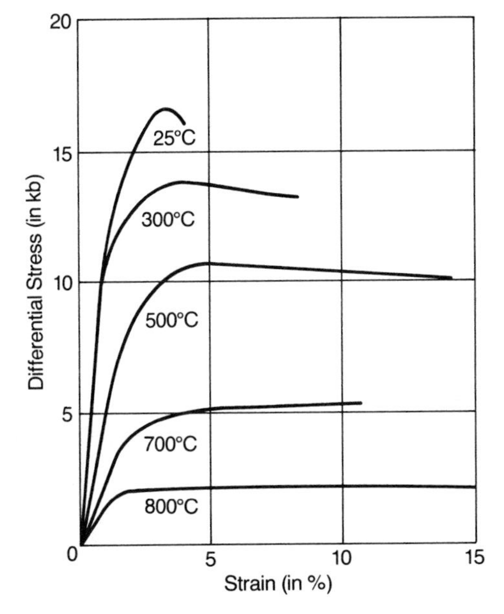

TEMPERATURE

An increase in the temperature of a rock generally depresses yield strength, enhances ductility, and lowers ultimate strength. Some rocks are more responsive to the effects of temperature than others. Igneous rocks are less affected by modest increases in temperature than sedimentary rocks (Ramsay, 1967); they are more at home in high-temperature environments. Figure 5.29 presents a typical example of the profound influence of temperature on strength and ductility. If heated sufficiently, rocks may deform in a viscous fashion and thus undergo very large permanent strains without ever rupturing or losing cohesion. **Viscous** materials are, in effect, liquids that flow when subjected to any differential stress, no matter how weak.

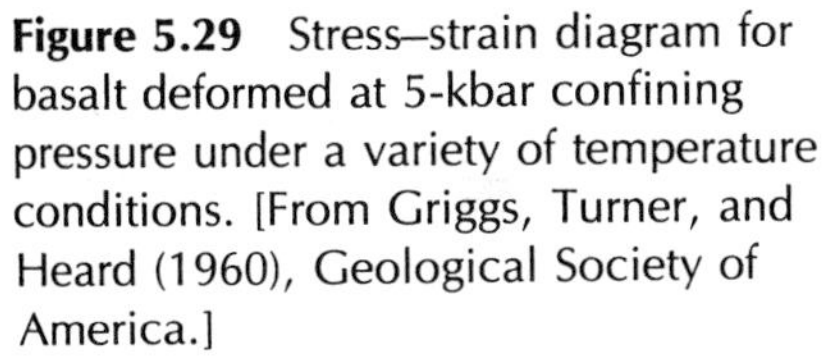
Figure 5.29 Stress–strain diagram for basalt deformed at 5-kbar confining pressure under a variety of temperature conditions. [From Griggs, Turner, and Heard (1960), Geological Society of America.]

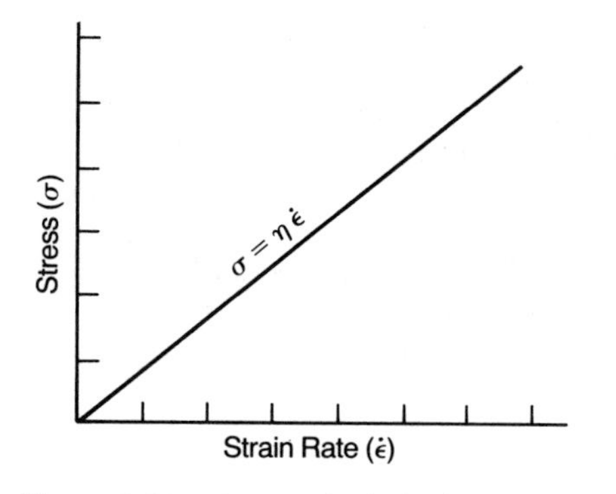

Figure 5.30 Portrayal of ideal viscous behavior on plot of stress (σ) versus strain rate ($\dot{\epsilon}$).

Unlike materials that deform plastically, truly viscous materials possess no fundamental strength threshold, which otherwise would have to be overcome to induce flow. Elevated temperatures promote viscous deformation and cause rocks to flow sluggishly.

For ideally viscous materials, there is a straight-line, linear relationship between stress and **rate of strain** (Figure 5.30);

$$\sigma = \eta\dot{\epsilon} \tag{5.6}$$

where σ = differential stress, η = viscosity, and $\dot{\epsilon}$ = strain rate. **Viscosity** is a measure of resistance to flow, just as Young's modulus can be thought of as a measure of resistance to elastic distortion. Viscosity is measured in **poises**, (i.e., dyne seconds per square meter). The greater the viscosity, the greater the internal friction of the fluid. Viscosities for common fluids are presented in Table 5.5.

Table 5.5
Viscosities of Common Fluids

Most Viscous	The Earth's Mantle	10^{23} poises
	Salt	10^{17}
	Rhyolite Lava	10^{9}
	Roofing Tar	10^{7}
	Basalt Lava	10^{3}
	Corn Syrup	10^{2}
	Castor Oil	10^{1}
	Heavy Machine Oil	6
	Olive Oil	.8
	Turpentine	.01
Least Viscous	Water at 30°C	.008

Very seldom do upper crustal rocks behave in ideally viscous fashion. Even though converted to viscous fluids by greatly elevated temperature conditions, most rocks retain the capacity of behaving elastically under relatively rapid strain rates. One of the best examples of a material that possesses this dual capacity of responding both elastically and viscously is Silly Putty. It flows continuously under conditions of very small differential stress. Yet, if subjected to rapidly applied, relatively high stresses, it will behave elastically and even break. Many rocks that have undergone deformation in relatively high-temperature environments have responded like Silly Putty, in an **elasticoviscous** manner.

STRAIN RATE

Rock strength as measured in deformation experiments is partly a function of the rate at which the stress level is raised. A rock can be forced to deform plastically at comparatively low levels of stress if the rate of loading is relatively low. Heard (1963) helped to quantitatively verify the validity of this principle through deformational experiments in which conditions were identical in every way except for the rate of deformation (Figure 5.31).

In testing the influence of "time" on rock strength using the Donath apparatus, we can compare values of yield strength, ultimate strength, and rupture strength as a function of the **rate of loading**. But normal practice is to use a deformational apparatus wherein the **rate of strain** is held constant. The test specimen is held in such a way that regardless of stress level, the specimen can shorten (or lengthen) only according to some fixed rate. When

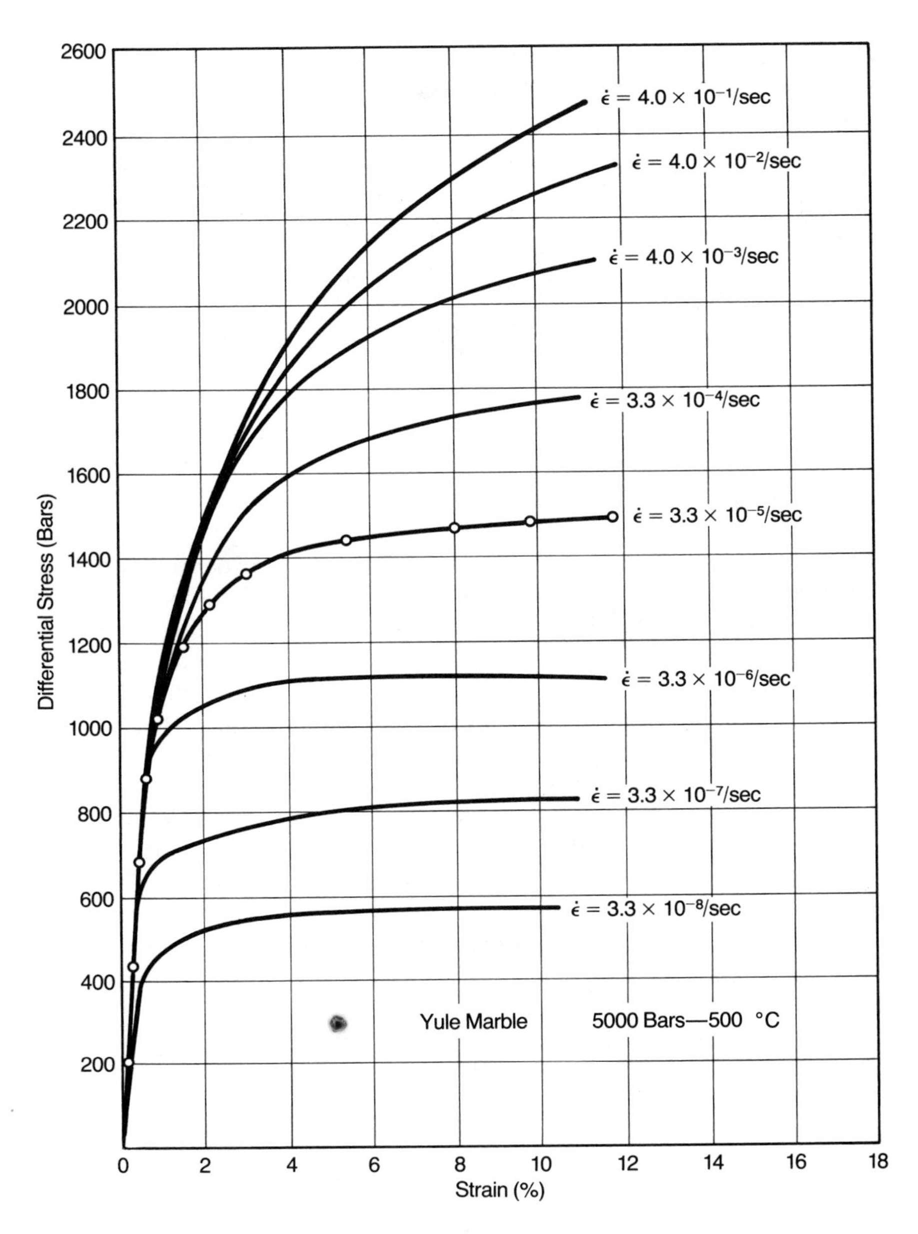

Figure 5.31 Stress–strain diagram for Yule marble deformed at different strain rate ($\dot{\epsilon}$) conditions. The higher the strain rate, the stronger the rock. [After Heard (1963). Copyright ©1963 by the University of Chicago. All rights reserved.]

the strain rate is relatively low, the amount of stress required to produce plastic deformation and ultimate failure is smaller than that for experiments in which strain rate is held at relatively high values.

The observable decrease in rock strength as a function of long-sustained stress is not too surprising. It can be thought of as a fatigue that sets in with time. Stress fractures in athletes are a human manifestation of the same phenomenon. Consider the marathon runner who trained 10 to 17 mi (16 to 27 km) per day for weeks, ran a competitive marathon, and then, after still more hard training, ran a second competitive marathon just two weeks later. The fracture of his femur was an expression of the capacity of materials to fail at very low differential stress levels. Such failure happens when stress is permitted to operate, even intermittently, for long periods of time (R. Mark Blew, M.D., Personal Communication, 1981). Perhaps it is timely to confess that my gate no longer holds, in spite of the nailed brace (see Figure 2.5*D*). Fatigue due to repeated low-stress usage by my boys fetching bikes and basketballs rendered it useless. But look at it now, pinned with 92 bolts (Figure 5.32)!

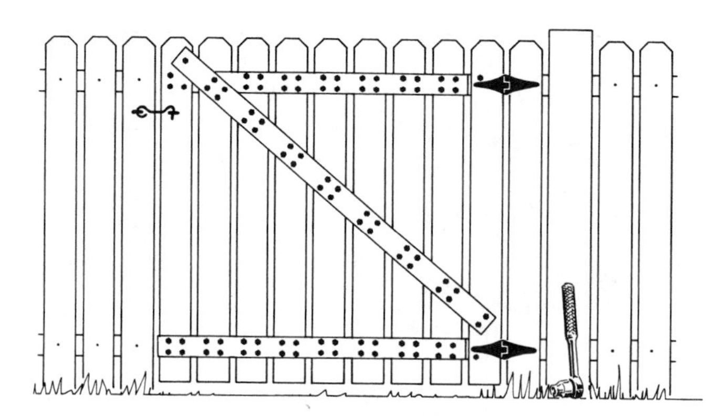

Figure 5.32 Fighting the weakness that comes with time and use. The ultimate (?) solution to strengthening the gate.

Engineers and geologists have carefully studied the time-dependent strain produced under conditions of low differential stress. The name given to this kind of strain is **creep**. Creep is the strain produced in experiments of long duration under differential stress conditions that are well below the rupture strength of the rock. Results of the tests produce consistent plots, like those shown in Figure 5.33. As soon as the load is applied, the rock experiences an immediate elastic deformation. This is followed by three distinctive kinds of mechanical response, called **primary, secondary, and tertiary creep**. Primary creep is a slightly delayed elastic deformation in which there is a general decrease in the amount of strain with time. This is followed by secondary, steady-state creep, a plastic deformation in which strain and time of loading are linearly related. Finally the rock, still under constant load, dramatically fatigues and discloses an accelerating rate of strain. This leads to failure by rupture. Tertiary creep approximates viscous deformation.

The initial elastic strain and primary creep are due to initial loading and are not particularly time dependent. The amount of strain accommodated by the rock during secondary and tertiary creep is strictly a function of the time that the load is sustained. The greater the amount of time that the rock has been subjected to some tiny stress differential, the greater the strain that is sustained by the rock.

TIME

It is natural that geologists (i.e., scientists who have 4.5 billion years to work with) focus on time dependency as one of the most important and perhaps least understood of all the independent variables that influence what we call the strength of rocks. We are not in a position to routinely run laboratory deformational experiments that are decades long, let alone centuries or millennia in duration. Without data bearing on the strength of rocks for such time intervals, specific strength values for rocks tested under short-term loading have limited analytical value in interpreting and reconstructing natural geologic deformations. Only the general qualitative results are transferable.

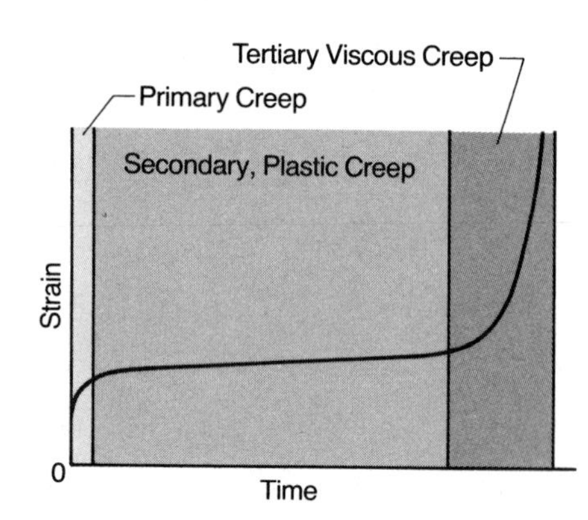

Figure 5.33 Strain versus time diagram, showing the fundamental classes of creep: primary, secondary, and tertiary.

S. W. Carey (1953), a brilliant structural geologist and tectonist, came to question whether rocks possess any **fundamental strength**. He reasoned that if rocks convert to plastic and viscous substances under conditions of long-duration loading, eventually they must fail by flow and then by rupture following accelerated tertiary creep. Carey made the scientific community aware of a state of matter that is not encompassed in the terms "solid," "liquid," and "gas." The state is that of a **rheid**:

a substance whose temperature is below the melting point, and whose deformation by viscous flow during the time of the experiment is at least three orders of magnitude (i.e., 1000×) greater than the elastic deformation under the given conditions. (From "The Rheid Concept in Geotectonics" by S. W. Carey, p. 71. Published with permission of Geological Society of Australia, Inc., copyright © 1953.)

Figure 5.34 Glacial ice, a rheid in action. Ice front of Grasshopper Glacier, Montana, shows contorted laminae caused by differential movement in the ice. (Photograph by T. S. Lovering. Courtesy of United States Geological Survey.)

Under small values of differential stress, ice behaves as a rheid after only two weeks. The ice of a valley glacier "creeping" down a mountain flank expresses a rheid in action (Figure 5.34). Salt, under small values of differential stress, behaves as a rheid after only 10 years. Salt glaciers move down mountain flanks in the Middle East and salt domes ascend like pillars through strata beneath the Gulf of Mexico. Where penetrated by mines in Texas, salt domes display folds and flow patterns that are characteristic of viscous deformation.

Salt and ice behave as rheids, but what about "real rocks"? Marble is a real rock, one that we usually think of as rigid. Yet look at the marble bench in Figure 5.35. It is behaving as a rheid after times that are short even by human standards. How is it for granite? How is it for the granite tombstone beyond the marble bench (Figure 5.35)? If granite, below its melting point, were subjected to the force of gravity for 5, or 10, or 15 billion years, would it flow like butter?

MICROSTRUCTURAL BEHAVIOR OF ROCKS

LOOKING INWARD

Sagging marble benches and barrel-shaped test specimens compel us to examine how deformation is achieved at the **grain** and **subgrain** scales of observation within rocks. What actually takes place at the microscopic and submicroscopic scales that enables rocks to change size and shape?

Figure 5.35 Marble bench bent downward by its own weight and that of the occasional occupant. The marble rheid is located in cemetery north of Soldiers' Home, Washington, D. C. (Photograph taken in 1925 by W. T. Lee. Courtesy of United States Geological Survey.)

There are a variety of **deformational mechanisms** that permit rocks and minerals to yield to stress. Insight regarding fundamental deformational mechanisms has been derived from controlled laboratory experimentation, followed by microscopic and submicroscopic examination of the experimentally manufactured **microstructures**. Knowledge gained in this way makes it possible to understand better the significance of microscopic and submicroscopic characteristics of naturally deformed rocks.

Tullis and Schmid (1982) nicely summarized and explained the dominant microstructural deformational mechanisms. These mechanisms include microcracking, cataclastic flow, dislocation glide, mechanical twinning, dislocation creep, and pressure solution. Each deformational mechanism seems to be favored by a particular range of conditions of confining pressure, temperature, and fluid pressure, and the role of each is influenced as well by mineralogy, texture, and strain rate. One of the goals of modern research in microstructures is to establish quantitatively the specific range of physical/chemical conditions under which each deformational mechanism can operate.

MICROCRACKING AND CATACLASTIC FLOW

Microcracking is a fundamental mechanism for accomplishing brittle deformation. Individual grains and crystals within a rock become fractured and crackled at the microscopic and submicroscopic scales, thus allowing a tiny domain of rock to change size and/or shape. When hairlike faults produced under brittle conditions are examined under high magnification, they are seen to be the loci of concentrations of microcracks. The fractures grow by steplike joining of microcracks (Jaeger and Cook, 1976).

When deformation is achieved at moderately high confining pressure, microcracking tends to become more pervasive, more distributed, commonly forming relatively wide bands of **cataclastic flow**. The flow is achieved by **frictional sliding** along interconnected microcracks. This leads to strain and an overall grain-size reduction. Tullis and Schmid (1982) have emphasized that cataclastic flow is very sensitive to the influences of confining pressure and fluid pressure. The ease of frictional sliding is enhanced when normal stresses acting on the linked microcracks are reduced. As we learned earlier, the level of normal stress on any fracture surface is reduced by an increase in fluid pressure or a decrease in confining pressure.

DISLOCATION GLIDE AND MECHANICAL TWINNING

Unlike cataclastic flow, **dislocation glide** and **mechanical twinning** are deformational mechanisms that are sensitive to and controlled by crystallographic planes of weakness. Dislocation glide and mechanical twinning both involve translation along discrete crystallographic surfaces.

Whether a given **intragranular** crystallographic plane will be activated to accommodate movement depends not only on the atomic bonding, but also on the magnitude of the stress vector on the plane and the ratio of the values of normal stress (σ_N) and shear stress (σ_S). The magnitude of stress required to initiate slip along individual surfaces is related to the orientation of the slip surface relative to the orientation of the stress.

Dislocation gliding is kinematically described in terms of the surface of slip (the crystallographic plane) and the direction and sense of direction of slip. The **slip systems** are defined by displacement vectors operating within crystallographic planes. The crystallographic planes that accommodate the slip may be susceptible to movement by virtue of original **crystal defects**

Figure 5.36 Slip systems do the impossible. (Artwork by D. A. Fischer.)

and/or anomously high densities of atoms (Hobbs, Means, and Williams, 1976).

Dislocation gliding is not simply a rigid body translation. Rather, dislocation gliding is achieved through a combination of both rigid and nonrigid body translation. The imaginative example that Hobbs, Means, and Williams (1976) used to illustrate this concept is the movement of a rug across a hardwood floor. The image is rendered here in cartoon form (Figure 5.36). Picture a carpet so large and heavy and so laden with furniture that a line of people pulling on it from one side cannot budge it. To move (translate) the carpet, it is necessary to curl the rug into a small anticlinal fold along one edge, and to systematically propagate the fold across the entire length of the rug. By doing this, the rug ultimately can be translated a short distance across the floor—but without removing the furniture. This becomes a simple, though time-consuming job, because *only a small part of the entire slip plane is active at any one time*. The point (or line) that separates the part of the rug that has slipped from that which has not is known as a **dislocation**. In the language of microfabric studies, the dislocation is propagated (Hobbs, Means, and Williams, 1976).

At any stage in the translation of the rug across the floor, the rug can be thought of as composed of two rigid body sheets separated by a nonrigid body segment (the fold) (Figure 5.37*A*). As a dislocation is propagated through a crystal (Figure 5.37*B*), the part of the crystal that is translated is likewise composed of two rigid body segments separated by a narrow nonrigid domain within which the crystal lattice has been distorted. Where the dislocation emerges at the end of the crystal, a small offset is produced. This offset, together with hundreds of others like it, produces the "plastic" distortion.

Mechanical twinning is similar to dislocation glide except that it occurs in only certain minerals, and within those minerals only along certain specific crystallographic planes. Moreover, the amount of slip is limited to a certain fixed amount in a certain fixed direction. The net effect of mechanical

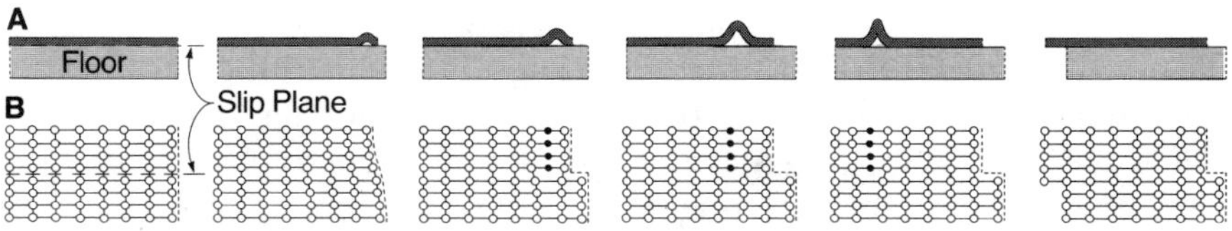

Figure 5.37 (*A*) Rigid and nonrigid segments of the rug. (*B*) Rigid and nonrigid segments of a crystal during the propagation of a dislocation. (Rendering of crystal structure reprinted with permission from *Metamorphic Textures* by A. H. Spry. Published with permission of Pergamon Press, Ltd., Oxford, copyright ©1969.)

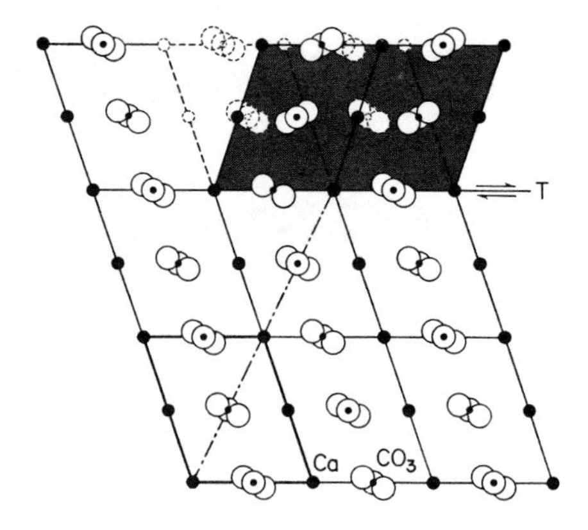

Figure 5.38 The twinning of calcite. (From *Structural Analysis of Metamorphic Tectonites* by F. J. Turner and L. E. Weiss. Published with permission of McGraw-Hill Book Company, New York, copyright ©1963.)

twinning is a constant magnitude of strain and the development of a symmetrical, "twinned" crystal structure (Figure 5.38) (Turner and Weiss, 1963; Tullis and Schmid, 1982). In contrast to microcracking and cataclastic flow, dislocation gliding and mechanical twinning are not particularly sensitive to the influence of confining pressure. It is the strength of atomic bonding that has to be overcome to produce slip, not the resistence to sliding friction. Since atomic bonds are easier to break at higher temperatures, dislocation gliding and mechanical twinning are deformation mechanisms that are favored by increases in temperature.

DISLOCATION CREEP

Dislocation creep is a deformation mechanism that permits grain-scale plasticity (**crystal plasticity**) at relatively high temperatures. It is like dislocation gliding in that dislocations are propagated along favored crystallographic planes. However, **recovery** and/or **recrystallization** take place at the same time, thus preventing the dislocations to pile up and interfere to the point that **work hardening** sets in. Recovery involves the "climb" of dislocations out of their primary slip planes, in this way avoiding some of the intragranular obstacles to gliding. Recrystallization involves the growth of new, strain-free grains at the expense of grains that had become strained by the dislocation processes. Through the combination of dislocation gliding, recovery, and recrystallization, dislocation creep emerges as a steady-state deformation mechanism that can accommodate a high level of crystal–plastic deformation at elevated temperatures (Hobbs, Means, and Williams, 1976; Tullis and Schmid, 1982).

PRESSURE SOLUTION

Pressure solution is a deformation mechanism that permits the shape of a grain, or an aggregate of grains, to be modified by selective, systematic dissolution and reprecipitation. Within an environment of marked differential stress, grains of rock in the presence of intergranular fluid can become corroded along parts of the grain boundaries that are subjected to the greatest principal compressive stress (σ_1). Solid grain material in contact with the intergranular fluid goes into solution. In fact, the fluid becomes supersaturated with respect to fluids along other parts of the grain boundary where the directed stress is less intense. A gradient is thus established that favors reprecipitation of the dissolved solids in locations impinged by the least principal compressive stress (σ_3) (Tullis and Schmid, 1982). In this way, individual grains undergo a shape transformation that reflects the differential stress environment.

Under exceptional circumstances the geometry of the growth of redeposited minerals provides a legible record of progressive deformation through time, including changes in the orientation of the least principal compressive stress. The theme of pressure solution is emphasized repeatedly in the chapters that follow. It is a deformational mechanism that seems to be manifested in a wide variety of rocks and in a broad range of temperature–pressure conditions.

LOOKING AHEAD

Dynamic analysis gives us a basis for understanding the deformational characteristics of rocks and minerals. Shifting our scale of observation, we now turn to plates and their movements, to better appreciate the natural origin(s) of the stresses that generate deformation in the Earth's crust.

chapter 6 PLATE TECTONICS

Plates and Their Boundaries

INTRODUCTION

Plate tectonics provides a useful backdrop for understanding the significance of regional structures. It is a basis for understanding the dynamic circumstances that give rise to deformational movements. Plate interactions create rock-forming environments, which in turn give rise to the fundamental, original properties of regional rock assemblages. Furthermore, plate motions, both during and after the construction of regional rock assemblages, generate the stresses that impart to rocks their chief deformational characteristics.

PLATE TECTONICS AND OROGENIC BELTS

GENERAL NATURE OF OROGENIC BELTS

Orogenic belts (or simply **orogens**) are long, broad, generally linear to arcuate belts in the Earth's crust where extreme mechanical deformation and/or thermal activity are concentrated (Figure 6.1). Major regional structures abound in orogens, and these reflect systematic distortion of the crust in which the structures are found. **Mountain systems** are a physiographic expression of orogenic belts, but the presence of mountains is not integral to our view of an orogen. Ancient orogens, still recognizable as sites of regional distortion, are beveled to flatlands in the interior of continents. And of the presently forming orogens, the structurally interesting parts may not lie in the mountains, but instead may be 10, 50, or even 700 km below the Earth's surface. In this perspective, mountains, if they exist at all, are just the roofline of an orogen.

Figure 6.1 LANDSAT image of part of the Appalachian orogen. Resistant strata define the traces of folds.

Within orogenic belts, rocks are characteristically shifted vertically out of their original positions in the crust. The vertical shifts produce **structural relief**, the kinematic analog of **topographic relief**. It is not just the topographic relief of the Andean orogen of South America that stimulates our curiosity about mountain building: it is ascending a 17,500-ft (5334 m) peak and seeing abundant marine fossils, the remains of animals that had thrived below sea level in shallow ocean waters millions of years before. The fossil-bearing strata are far out of place, challenging us to explain the 17,500+-ft shift.

Figure 6.2 Mount Cook, New Zealand. What a laboratory! (Photograph by W. B. Bull.)

Although all regions of the Earth's outer skin are targets of detailed structural analysis, orogenic belts have captured most of the attention. This is not surprising, given the magnitude of distortion that commonly is locked up in the structure of orogens. Furthermore, geologists are attracted to young orogenic belts simply because their mountainous expressions often make such beautiful and inspiring natural laboratories (Figure 6.2).

ORIGINS OF OROGENS

The generally accepted view among scientists today is that orogenic belts evolve through the interference of slowly moving rigid **plates** composed of **lithosphere**. Lithosphere is made of continental and/or oceanic crust as well as uppermost mantle material. It can be thought of as the Earth's mechanically competent outer rind, which sluggishly moves on a part of the mantle that is capable of flowing continuously.

By studying the present configuration of plates, we learn that orogenic belts mainly form along **plate margins** at or near **plate boundaries**. It is along plate boundaries that plates interfere. The breadth of an orogen reflects the degree to which plate margins are internally distorted by plate interference. Even though most distortion is concentrated in boundary regions between plates, some regional structures form well within the interior of plates, apparently through transmission of stresses for great distances from plate boundaries. **Intraplate deformation** challenges our perceptions of mountain building and our understanding of stress propagation.

The breadth of distortion that characterizes a given orogen relates to the magnitude of stresses deployed at or near plate boundaries and the persistence of plate interference through time. Breadth of distortion also reflects the fundamental strength of plates and the degree to which plates are thermally softened by elevated temperatures accompanying igneous intrusion and metamorphism.

DESCRIPTION OF PLATES

ELUSIVE NATURE OF PLATES

The plates were hard to find. They were discovered in the 1960s, and even now there are disagreements concerning how many plates exist, and where, exactly, all the plate boundaries are located. Individual plates are not made up simply of continents or ocean basins. If this were the case, the existence of plates probably would have been discovered long ago. We now know that individual plates most commonly encompass parts of continents and parts of ocean basins. Depending on the locations of the plate boundaries, an individual continent may occupy one or more plates; the same is true of ocean basins.

Since continents are generally just parts of plates, detailed studies of continental geology did not lead, and may never have led, to the discovery of the plate concept. It was exploration of the ocean basins that allowed the sum of the parts to emerge into full view.

Another factor—this one a peculiarity of the Earth's mechanical properties with depth—further prevented early recognition of the existence of plates. In trying to explain **continental drift**, geologists emphasized the mechanical importance of the seismic boundary between the crust and mantle, the so-called **Mohorovicic discontinuity** (the M discontinuity, or simply the Moho). Continents of crust were envisioned as moving on top of continuously deforming mantle. Ironically, the base of the plates does not coincide with the Moho. Rather, the lower boundary of the Earth's lithospheric plates lies in the mantle of the Earth, well below the Moho.

In short, the boundaries of individual plates do not exist where we might have expected them to be. Boundaries between continents and ocean basins generally are not plate boundaries, nor is the interface between the Earth's mantle and the Earth's crust. And since plates move so slowly by human standards, they never attracted much attention.

LITHOSPHERE AND ASTHENOSPHERE, CRUST AND MANTLE

The plates are composed of **lithosphere**—crust and upper mantle material that is **rigid** in the sense that it is capable of withstanding low levels of differential stress indefinitely without flowing (Figure 6.3). The upper reaches of individual lithospheric plates are crust, both continental and oceanic. **Oceanic crust** is relatively thin, ranging from about 4 to 9 km in thickness, and is composed predominantly of rocks that are relatively high in density (avg $\rho = 2.9$). **Continental crust** is relatively thick, ranging from approximately 25 to 70 km, and composed of materials of relatively low density (avg $\rho = 2.7$).

The **Moho**, marking the base of the crust and the top of the mantle, lies within lithospheric plates (Figure 6.3). In fact, it normally lies at a high structural level within lithosphere. The position of the Moho can be identified on the basis of a seismic–velocity discontinuity. It is thought that the Moho marks a lithological transition from gabbros to underlying ultramafic rocks.

Lithosphere is thickest under continents (Figure 6.3). The relatively thick, convex-downward bulges of lithosphere under continental terranes may serve as kinds of viscous anchors that may halt, or slow down, plate motions (Chapman and Pollack, 1977). Even so, the thickest lithosphere, when

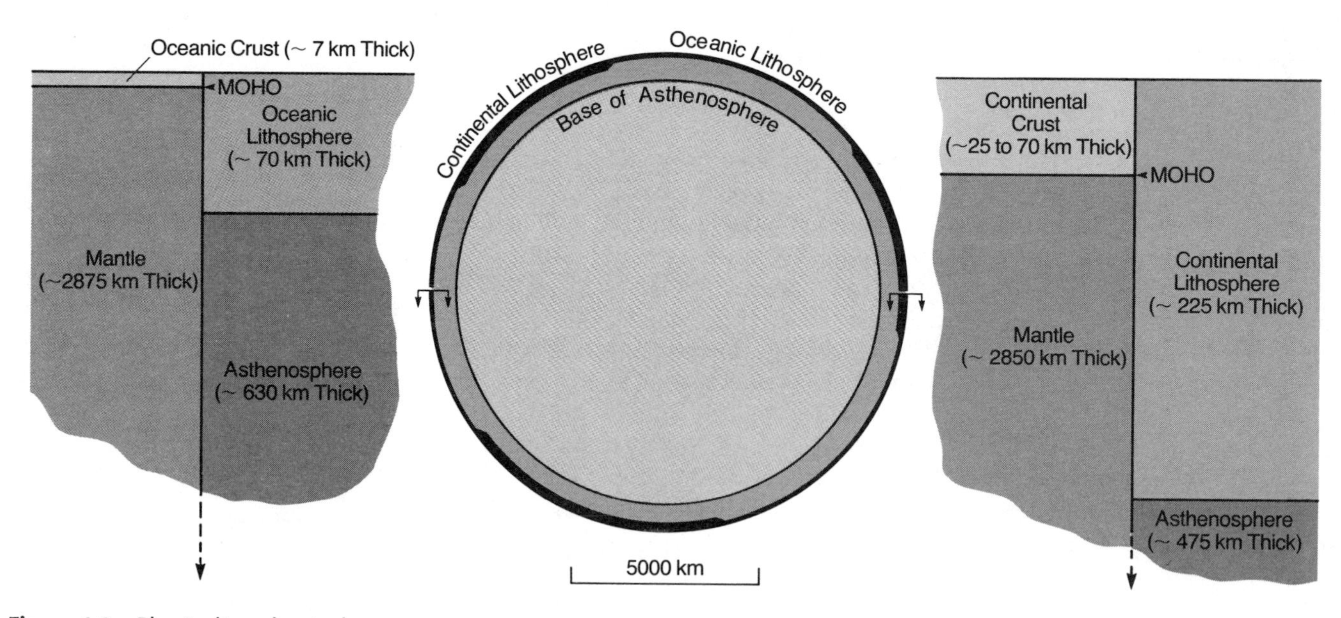

Figure 6.3 Physical/mechanical components of the crust and mantle.

viewed at the scale of the Earth as a whole, is mighty thin. Lithospheric plates are like contact lenses on the surface of an eye: exceedingly thin and gently curved. Like contact lenses, the plates move and slip readily.

The lithosphere rides on **asthenosphere** (Figure 6.3), upper mantle material that is capable of flowing continuously, even under the lowest levels of differential stress. Asthenospheric mantle is a rheid. Like Silly Putty, it is not only capable of flowing, but if loaded rapidly, it is also capable of fracturing. The ability of the asthenosphere to fracture is proved by earthquake shocks that emanate from it.

Depth of the structural transition between lithosphere and asthenosphere is vague and impossible to define precisely. The top of the asthenosphere is generally placed at the top of the **low-velocity zone** in the upper mantle, where seismic waves undergo a significant drop in velocity (Figure 6.3). Under the ocean basins, the depth to this discontinuity is approximately 75 km. Under continental crust, however, it lies at an average depth of approximately 225 km, but its position is truly variable.

The base of the asthenosphere lies at a depth of approximately 700 km. The low-velocity zone resides in the upper quarter of the asthenosphere. The 700-km depth to the base of the asthenosphere coincides with the depth level of the very deepest earthquakes.

PLATE MOTIONS

The lithosphere is not a flawless outer shell. Rather, it is fragmented into many discrete plates that move relative to one another (Figure 6.4). The motions are fundamentally rigid body motions involving combinations of translation and rotation. Only the plate margins experience significant non-rigid body deformation. Plate motions can be described in terms of convergence, divergence, and strike–slip (Figure 6.4). Velocities range from 1 to 20 cm/yr. Plate movements may combine in many ways, depending on the kinds of plate interaction that must be accommodated.

Convergence is marked by an actual relative movement (slip) that brings adjacent plates into one another. Plates in convergence are in constant

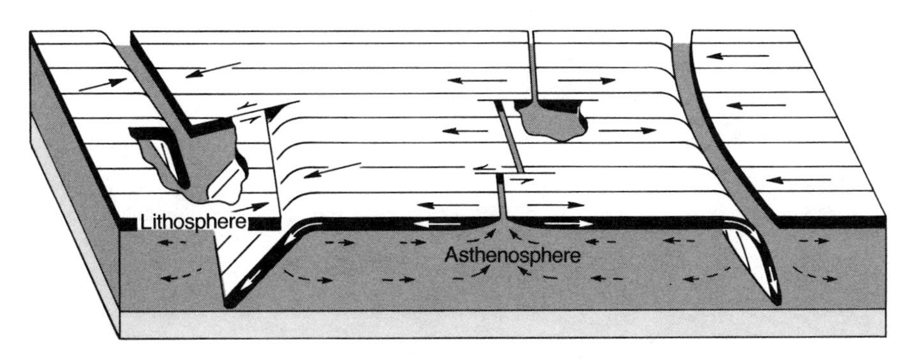

Figure 6.4 Convergence, divergence, and strike–slip of lithospheric plates. (From B. Isacks, J. Oliver, and L. R. Sykes, *Journal of Geophysical Research,* v. 73, fig. 1, p. 5857, copyright ©1968 by American Geophysical Union.)

competition for space. A common response to the space problem is the structural descent of one plate beneath the other. In effect, rock is swallowed to greater depth (Bally and Snelson, 1980). This tectonic process is known as **subduction**. An alternative response of plates to the space problem of convergence is **collision** unaccompanied by significant subduction of lithospheric material. Plate collision is like a slow-motion head-on collision between two cars whose brakes have locked on a slippery highway. Converging plates in collision can be thought of as equally buoyant. Still, two plates cannot occupy the same space and so regional-scale non rigidbody shortening must occur. The zone of contact between plates that have collided and shortened in this manner is known as a **suture**, or **suture zone**.

Divergent movement is marked by the movement of plates away from one another (Figure 6.4). The actual relative movement may be perpendicular or oblique to the boundary between adjacent plates. Without compensation, some void or opening would surely develop between diverging plates. But "would-be" voids are simultaneously and continuously filled in by upwelling igneous intrusions. When solidified, the intrusions and freshly made volcanic and sedimentary accumulations constitute new additions to the lithosphere.

Strike–slip horizontal shifting of one plate past another constitutes another movement scheme (Figure 6.4). The plates scrape past one another horizontally. Steeply dipping fault zones and shear zones absorb the mechanical effects of stresses generated during the frictional movements.

GEOLOGIC PROPERTIES OF PLATE BOUNDARIES

SEISMIC EXPRESSION

Plate boundaries are the edges of plates, the contacts between adjacent plates. Plate boundaries in modern settings are marked by intensive earthquake activity. In fact, the historic record of earthquake activity around the globe is nearly a tracing of present-day plate boundaries (Figure 6.5). Depth levels of earthquakes at or near plate boundaries vary from very shallow to very deep, depending on the nature of the plate boundary and the mechanical conditions of the plates brought into contact.

Ancient plate boundaries are seismically silent. To recognize plate boundaries in the geologic record, we must learn to identify the distinctive structures and regional rock assemblages that typify plate boundaries in modern settings.

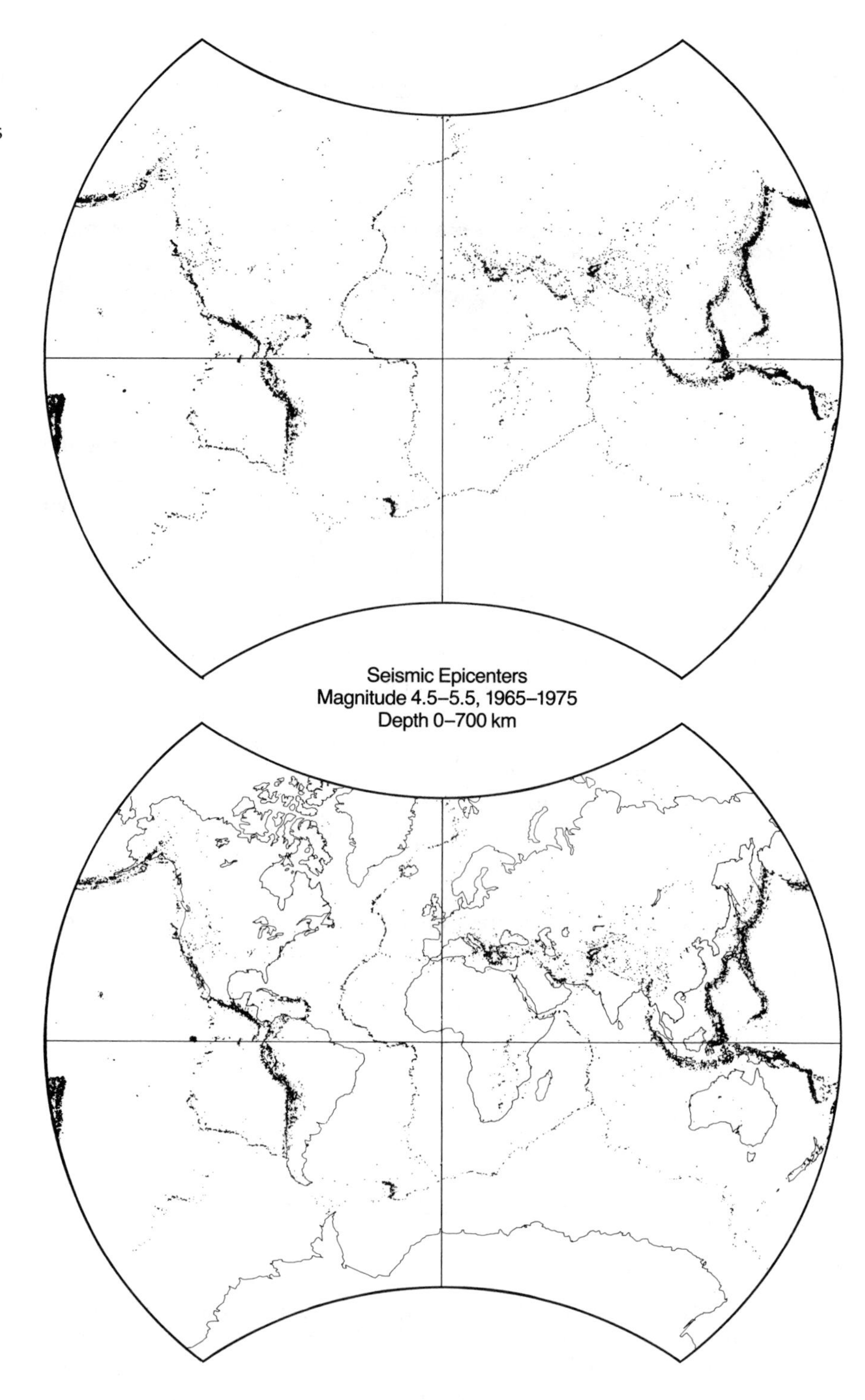

Figure 6.5 Earthquake epicenters are a tracing of plate boundaries. [From Lowman (1981). Courtesy of Goddard Space Flight Center, National Aeronautics and Space Administration.]

DIVERGENT BOUNDARIES

OCEAN RIDGE MOUNTAIN SYSTEM. The present-day divergent plate boundaries reside mainly in **intraoceanic** settings, although there are some important exceptions. Their main physiographic expression is in the form of a world-encircling **ocean ridge mountain system** (Figure 6.6).

Discovery of the ridge system as a world-encircling entity is attributed to the late Bruce Heezen of Lamont–Doherty Geological Observatory (Heezen,

1960). He and his colleagues pioneered the detailed mapping of the ocean ridge system, discovering it to be a network of ridges marked by long linear segments, broken by faults oriented transversely to the ridges. The ocean ridge mountain system, exposed to partial view in places like Iceland, is about 40,000 km long and averages 2000 km in breadth. It displays a topographic relief of approximately 2000 m. Rocks within the system are highly faulted, composed mainly of basaltic volcanics with interbedded oceanic sediments and sills, dikes, and thick-layered intrusions of mafic igneous rocks. The axis of the ocean ridge mountain system marks a major plate boundary, one along which adjacent plates move away from one another.

TRIPLE JUNCTIONS. Ocean ridge segments locally fork into other ridge segments, in a wishbone pattern. The intersection of three such ridges is one kind of **triple junction**, a place where three plates come in contact. The geography of the ridges can be indexed with respect to **ridge–ridge–ridge triple junctions** (Figure 6.6). For example, in the South Atlantic, the Mid-Atlantic Ridge, the Southwest Indian Ocean Ridge, and the America–Antarctica Ridge radiate from a common triple junction. In the Indian Ocean, a ridge–ridge–ridge triple junction marks the intersection of the Carlsberg Ridge, the Southwest Indian Ocean Ridge, and the Southeast Indian Ocean Ridge. In the eastern Pacific, the Galapagos Ridge branches from the East Pacific Rise, forming yet another ridge–ridge–ridge triple junction.

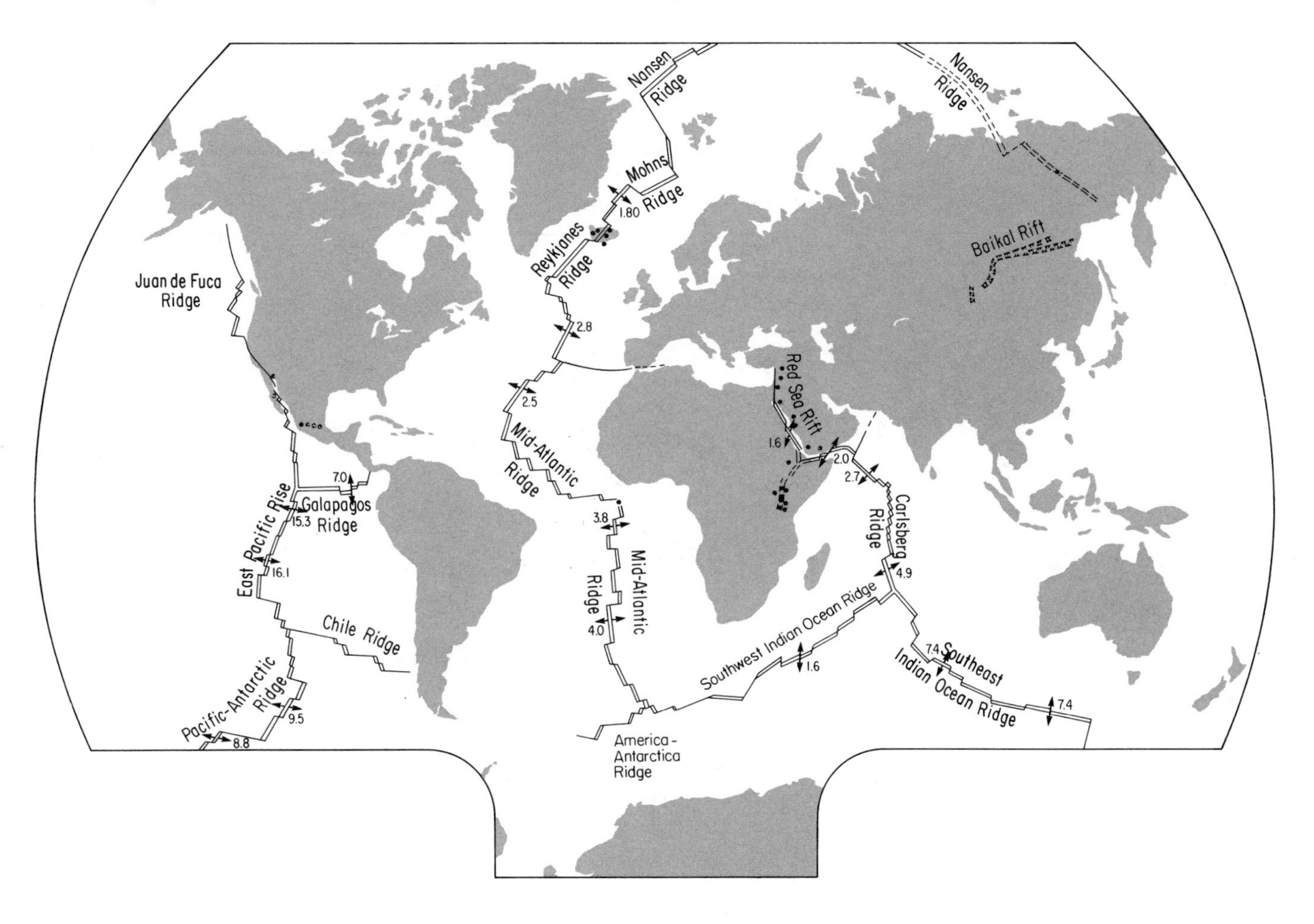

Figure 6.6 Ocean ridge systems of the world. Directions and magnitudes of spreading shown as well. Magnitudes of spreading are given in total spreading rate in centimeters per year. [From Lowman (1981). Courtesy of Goddard Space Flight Center, National Aeronautics and Space Administration.]

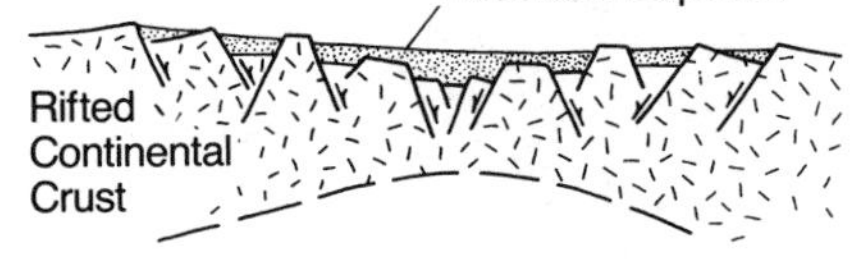

Figure 6.7 Rift basin(s) and rift basin deposits.

PLATE DIVERGENCE IN CONTINENTAL SETTINGS. Divergent rift systems cut through continents in a number of places. One of the best examples is the Red Sea. In such settings extensional faults accommodate continental crustal pull-apart. Down-dropped blocks become sites of continental basins that are filled by clastic debris derived from adjacent high-standing blocks (Figure 6.7). The sedimentary deposits that form in such environments are called **rift basin deposits**. They are typically elongate, wedge-shaped masses of poorly to moderately well-sorted debris that has been rapidly dumped. The coarse clastic debris commonly interfingers with lake deposits, including evaporites, which are intermittently formed in closed-basin depressions. The Red Sea rift valley alone is 1000 km long and averages 200 km in width. Deep basins, which opened to accommodate divergent movement of continental lithosphere, invited precipitation of evaporites over a width of 100 km, with thicknesses of up to 7 km (Goodwin, 1976).

RIDGES AND RIFTS. The ocean ridge system is very broad and is marked by ridge-symmetrical topographic relief (Figure 6.8). The dynamically active part of the ridge is very restricted, lying along the axis of the system. Within a zone no wider than 30 km, new oceanic lithosphere is created by the combination of intrusion, extrusion, and extensional faulting. Ongoing dynamic activity is expressed partly by **shallow-focus earthquakes**. The shallowness of fault-induced earthquakes reflects thin lithosphere and shallow depth to asthenosphere.

The actual divergent plate boundary along the axis of the ocean ridge system is marked by a **rift valley**, ranging commonly 2 to 10 km in width and displaying hundreds of meters of topographic relief. The very center of the rift valley generally features a central high, an edifice of volcanic accumulations (Figure 6.8). Rocks and structures within the axial rift valley express the structural processes involved in plate divergence. Cross sections of the axial rift vary in detail, but all disclose the presence of extensionally faulted volcanic and intrusive igneous rocks of mafic composition (Figure 6.8).

OPHIOLITE. **Ophiolite** is the general name given to the rock associations produced in axial rift valleys at divergent plate boundaries in intraoceanic

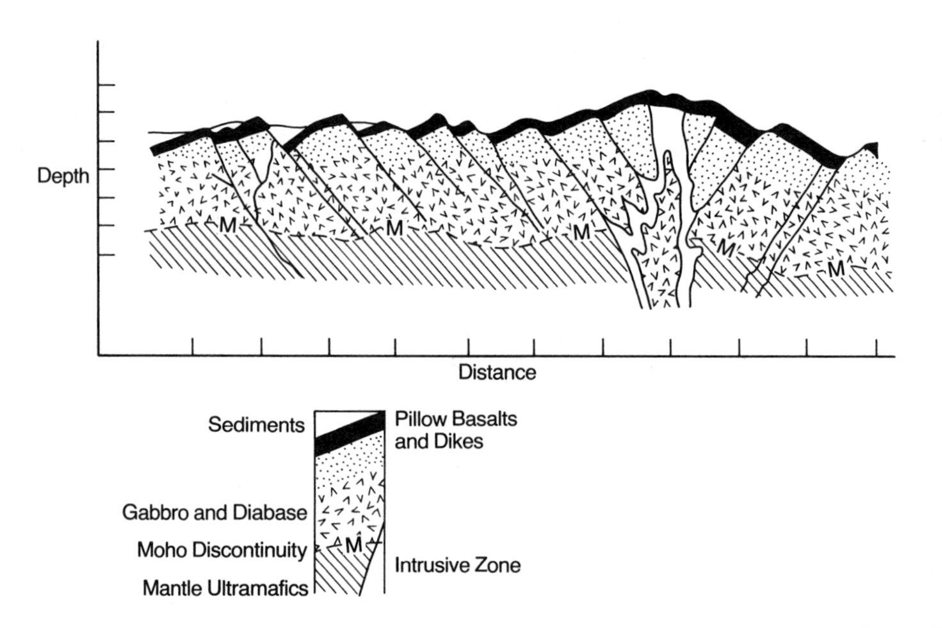

Figure 6.8 Vertically exaggerated schematic rendering of extensional processes at work in an ocean ridge system. [From Anderson and Noltimier (1973), v. 34, fig. 5, p. 144. Published with permission of *Geophysical Journal of the Royal Astronomical Society*.].

settings. A typical ophiolitic sequence is considered to consist, from top to bottom, of cherty or limey sediment; pillow basalt locally intruded by dikes and sills of diabase; metamorphosed gabbro and still more diabase dikes; massive and layered gabbros and some serpentinized peridotite; and underlying olivine periodotite mantle material (Dickinson, 1980) (Figure 6.8). The gabbros and ultramafic rocks are deformed from place to place by plastic deformation along subhorizontal zones of shear (Nicholas and Le Pichon, 1980). A typical ophiolite sequence forming in axial rift settings is 5 to 10 km thick.

SEAFLOOR SPREADING. Insight regarding interpretation of the kinematic and dynamic significance of ocean ridge systems stems from the revolutionary thinking of Bob Dietz and Harry Hess. Harnessing their extraordinary energies to rather limited, but adequate, geological and geophysical data, Dietz (1961) and Hess (1962) both reached the conclusion that oceanic ridges are sites where new lithospheric ocean floor is generated. Their model of **seafloor spreading** radically transformed geological thinking about Earth dynamics. Their ideas allow us to recognize that the extensional faults, the vertical dikes, the intrusions of gabbro, and the horizontal shearing within ophiolite sequences all contribute to the process of spreading lithospheric plates away from a divergent boundary (Figure 6.9). The ophiolite sequence constructed along a ridge axis is welded to the edges of both plates that meet at the plate boundary. There it congeals to become the youngest lithosphere within the ocean ridge orogen. Simultaneously and continuously this young lithosphere is split by faulting and intrusion and is translated laterally from the ridge axis, thus making room for yet younger lithosphere.

Vine and Matthews (1963) showed that geomagnetic field reversals are symmetrically imprinted on the seafloor lithosphere, which spreads from ocean ridge plate boundaries (Figure 6.9). Age data from many sources demonstrate that magnetic stripes at successively greater distances from ridge crests are increasingly older. Rates of seafloor spreading can be calculated on the basis of the age of ocean floor in combination with the distance

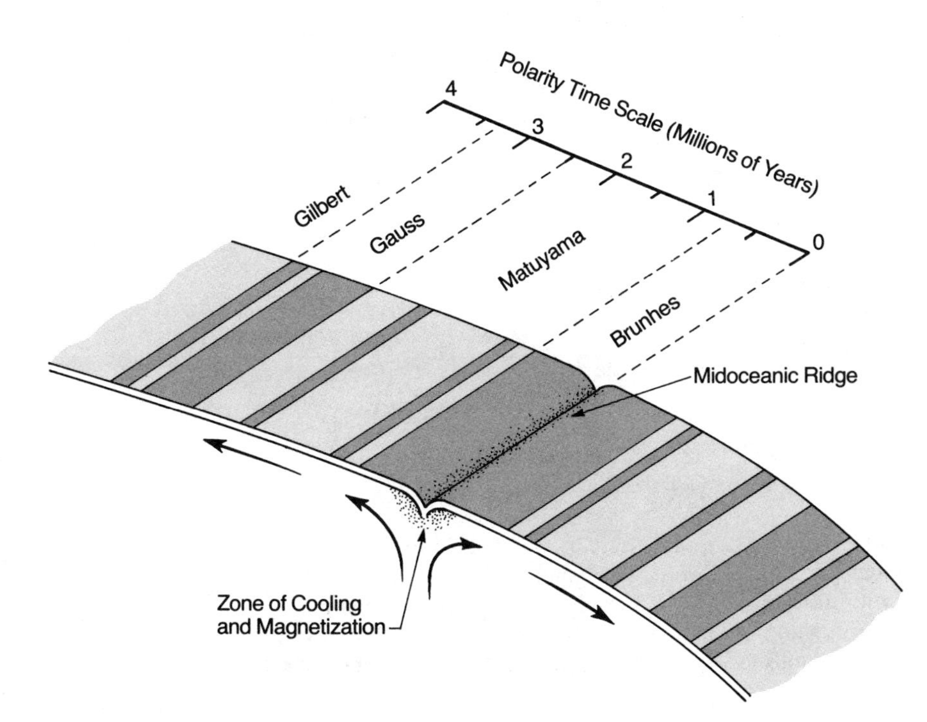

Figure 6.9 Seafloor spreading. Addition of oceanic lithosphere to the edges of plates. Geomagnetic reversals are imprinted on oceanic lithosphere and symmetrically disposed about ridge crests. (From "Reversals of Earth's Magnetic Field," by A. V. Cox, G. B. Dalrymple, and R. R. Doell, copyright ©1967 by Scientific American, Inc. All rights reserved.)

of the dated ocean floor from the ridge crest (Isacks, Oliver, and Sykes, 1968).

Rates of divergent movement along the axis of the ridge system may seem slow, ranging from 1 to 20 cm/yr. Yet in geological terms the rates of spreading are astonishingly fast. If lithosphere moves at a rate of 10 cm/yr away from a ridge axis, it will cover 100 km in 1 m.y., 1000 km in 10 m.y. An ophiolitic sequence formed at one "point" in time will be split and separated by a distance of 2000 km in just 10 m.y., assuming a spreading rate of 10 cm/yr.

GEOLOGIC DEVELOPMENT OF PASSIVE CONTINENTAL MARGINS. The natural evolution of a divergent plate boundary is recorded not only in oceanic crustal rocks, but also in rock assemblages deposited along the margin of an original supercontinent, split and spread apart by the seafloor spreading process. Initial rifting and normal faulting of an original supercontinent is attended by stretching and subsidence of the lithosphere (Figure 6.10*A*) (Dickinson, 1980; Le Pichon and Sibuet, 1981). Distance from the Earth's surface to the top of the asthenosphere is steadily reduced as stretching proceeds. Eventually basaltic magma intrudes into and extrudes through the deformed lithosphere. Ophiolitic lithosphere is thus emplaced in the developing zone of separation between adjacent, diverging plates (Figure

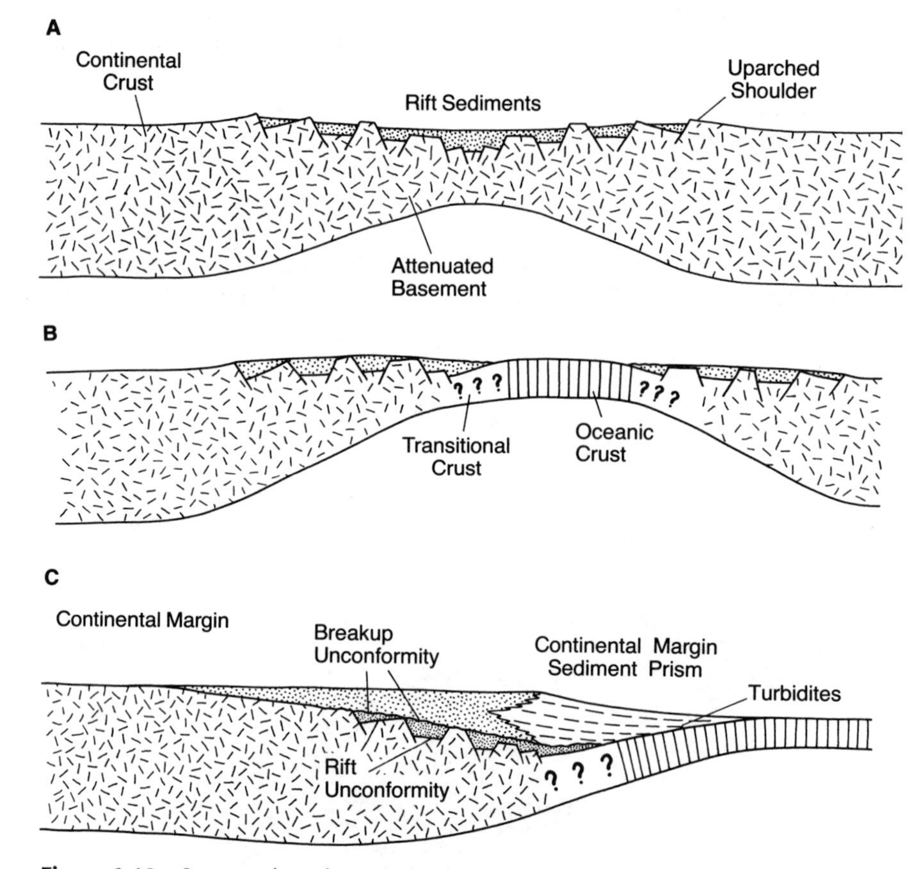

Figure 6.10 Structural evolution of a passive margin. (*A*) Stretching and subsidence. (*B*) Emplacement of new oceanic crust; buoyant uplift. (*C*) Subsidence of the passive margin and formation of continental margin prism. [From Dickinson (1980). Published with permission of Geological Association of Canada.]

6.10*B*). Oceanic crust is welded directly onto continental crustal rocks. The mixture of original continental crust and the added oceanic component at the continental edge produces a hybrid **transitional** crust, which becomes nestled between *bona fide* continental crust and young oceanic crust. The thermal input is so great during this initial rifting stage that the margin of the rifted continent is buoyantly uplifted.

With sustained seafloor spreading, the continental edge moves further and further from the **spreading center** represented by the midoceanic ridge. The edge of the continent is a **passive margin** in that stresses are no longer deforming it. Structures and rock associations of the rifting phase are frozen into rocks of the continental margin, and they are passively conveyed laterally along with the rest of the plate of which it is a part.

Systematic massive subsidence of the passive continental margin takes place as the attached ophiolitic rocks cool off (Dietz, 1963; Dietz and Holden, 1966; Dewey and Bird, 1971). Downflexing of the passive continental margin due to subsidence makes room for deposition of a **continental margin sedimentary prism** (also known as a **miogeoclinal prism**), which builds outward from the continental interior (Figure 6.10C). The basal part of the miogeoclinal prism contains conglomerates, coarse sandstones, and even some volcanics, all reflecting the waning rifting stage. These are overlain in great thickness by quartz-rich sandstones derived from the ancient continental margin. With time, a **carbonate platform** builds seaward from the continental margin, **prograding** over the basal sands. Such carbonate sequences of limestone and dolomite, along with interbedded shales and siltstones, record transgression of the sea onto the subsided continental margin (Stewart and Suczek, 1977).

The top of the continental margin sedimentary prism determines the mean elevation of the **continental shelf**, whereas the **continental slope** is built at the oceanward limit of deposition of the continent-derived sediment prism (see Figure 6.10C). Sediments pirated from the sediment prism are transported down the continental slope as turbidites, extending the sediment wedge even further. The **turbidites** interfinger seaward with thin deep-water sediments that rest on ocean crust.

THE CONTINENT INTERIOR, IN PASSING. Continentward from the passive margin lies the continental interior, or **craton**, the stable nucleus of a continent. Cratons typically are distinguished by thin, sheetlike accumulations of **platform sediments**. Flat-lying platform deposits, largely carbonate with interbedded fine-grained clastics, reflect infrequent transgressions of shallow marine waters onto the craton. The transgressions themselves may be related to changes in rates of seafloor spreading. Nestled within the continental interior and resting on stable basement of ancient crystalline rocks, platform deposits are insulated from internal strain. Although not deformed appreciably, platform sediments of continental interiors are typically warped into broad regional structures, usually basins and domes. Whatever the origin of these structures, they influence depositional patterns of the platform strata.

In all continent-interior regions of the Earth, the platform sediments locally have been stripped to expose Precambrian basement of the stable cratons (Figure 6.11). The basement terranes are known as **shield provinces** because their regional topographic form is that of a Roman shield, as it would look lying on the ground, broadly convex upward (Stokes, Judson, and Piccard, 1978). Within the shields are plutonic and metamorphic rocks, the deformed roots of ancient orogens. The rocks have been distorted and crystallized into a very rigid substratum that gives cratons their stability.

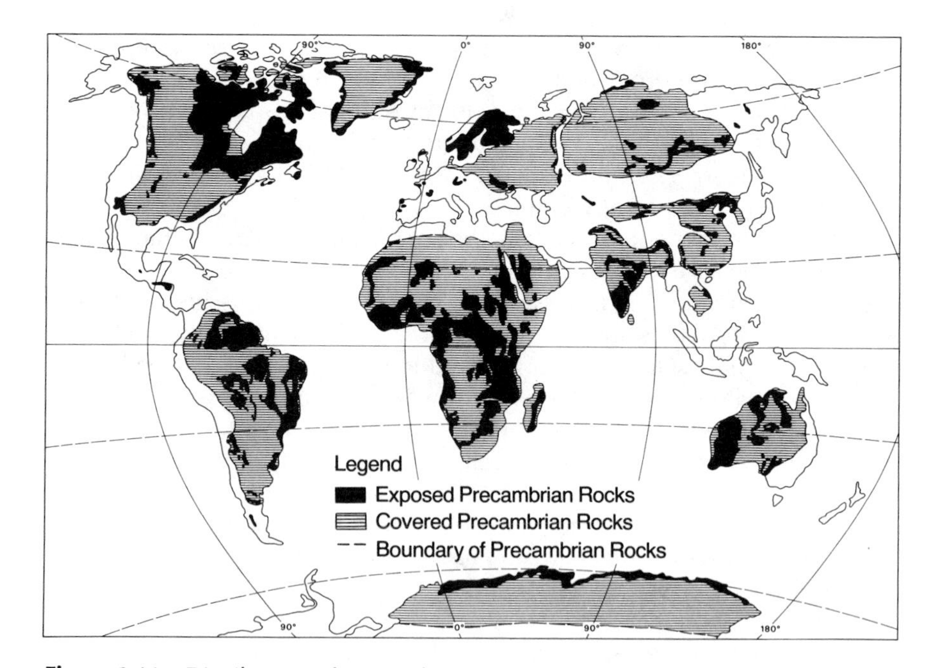

Figure 6.11 Distribution of Precambrian basement in the world's continents. (From "Giant Impacting and the Development of Continental Crust," by A. M. Goodwin *in* B. F. Windley (ed.), *The Early History of the Earth*. Published with permission of John Wiley and Sons, Ltd., Chichester, England, copyright ©1976.)

STRIKE–SLIP BOUNDARIES

INTRAOCEANIC TRANSCURRENT FAULTS. **Transcurrent faults**, discordant to the trend of ocean ridge segments, are an integral part of the ocean ridge system. The topographic expression of these faults is boldly revealed in maps of the seafloor of the world's oceans (see Figure 6.6). The faults appear to offset ridge segments within the otherwise continuous ocean ridge system. Shallow-focus earthquakes emanate from parts of these faults. The seismic activity expresses faulting and frictional sliding of thin lithosphere.

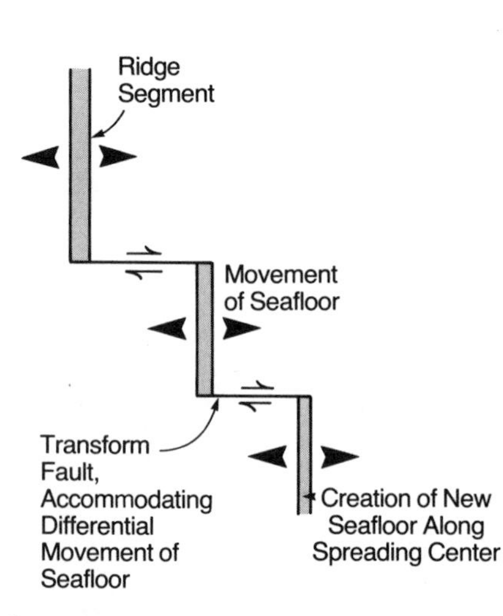

Figure 6.12 Transform fault linkage of one spreading center to another, of one part of the movement system to another.

TRANSFORM FAULTS. It was J. Tuzo Wilson, distinguished professor of the University of Toronto, who unlocked the geometric and kinematic significance of transcurrent intraoceanic faults. The stepping-stone for Wilson's interpretation was the model of seafloor spreading as set forth by Dietz (1961) and Hess (1962). Wilson (1965) reasoned that the function of transcurrent intraoceanic faults was to connect the end of one ridge segment (spreading center) to another, enabling the entire movement system along the spreading axis to be integrated (Figure 6.12). Because the faults "transform" the motion from one ridge to another, Wilson named these faults **transform faults**. Transform faults are *bona fide* plate boundaries that serve as zones of strike–slip accommodation between opposite-traveling domains of seafloor.

THE KINEMATIC CONTRADICTION OF TRANSFORMS. Rather than considering transform faults to be strike–slip faults that systematically displaced a once-continuous ocean ridge, Wilson perceived that the repeatedly offset nature of ridges was inherited from initial breakup of lithosphere. He argued that the original configuration of ridge crests and transform faults probably was not characterized by the mutually perpendicular ridge–transform patterns displayed by most mature ocean ridge systems today. Instead, formerly irregular, nonsystematic configurations of ridge crests and transform faults gradually adjusted to the kinematics of spreading.

Wilson's model has stood the test of time. Isacks, Oliver, and Sykes (1968) were able to prove that **first motions** on earthquakes along transform faults display the exact sense of movement predicted by Wilson. Their investigations showed that movements are indeed strike–slip and that the relative displacements conform to the relative motions of seafloor spreading along ridges joined by the transforms. Like Wilson (1965), Isacks, Oliver, and Sykes (1968) emphasized that the displacement vector for a given transform fault is opposite to the sense of separation that would be required to explain the "offset" of the ridge crest (Figure 6.12).

Additional insight regarding the inner workings of transform faults was provided by Sykes (1967). He proved that the only seismically active part of a ridge-to-ridge transform fault is the interior segment between the ends of ridge crests connected by the transform. This is the only segment of a transform that separates *oppositely–moving* domains of seafloor (Figure 6.12). The exterior segments of transform faults are fossil records of former interior transform fault segments, which now lie between domains of seafloor moving in the *same* direction at the same rate. No frictional interference arises, and thus these parts of the transform faults are dead. If one were to figuratively "stand" in the active part of a transform fault zone, it would be like standing on the narrow medial strip of a Los Angeles highway at 5:25 *P.M.*, watching traffic slowly go by on each side in opposite directions (Figure 6.13). The brushing of the flanks of oppositely moving cars, as well as shouts of irate drivers, would create the "seismic" noise.

Transform faults that connect offset ridge segments are the most abundant and perhaps the easiest to understand kinematically. However, two other fundamental types of transform fault exist as well, and these too were recognized by Wilson (1965). One links the end of a ridge with the end of a **trench**, the surface expression of a subduction zone; the second connects the ends of two subduction zones. Like **ridge-to-ridge** transform faults, **ridge-to-trench** and **trench-to-trench** transform faults exist because they serve to accommodate differential strike–slip movement of adjacent plates. Figure

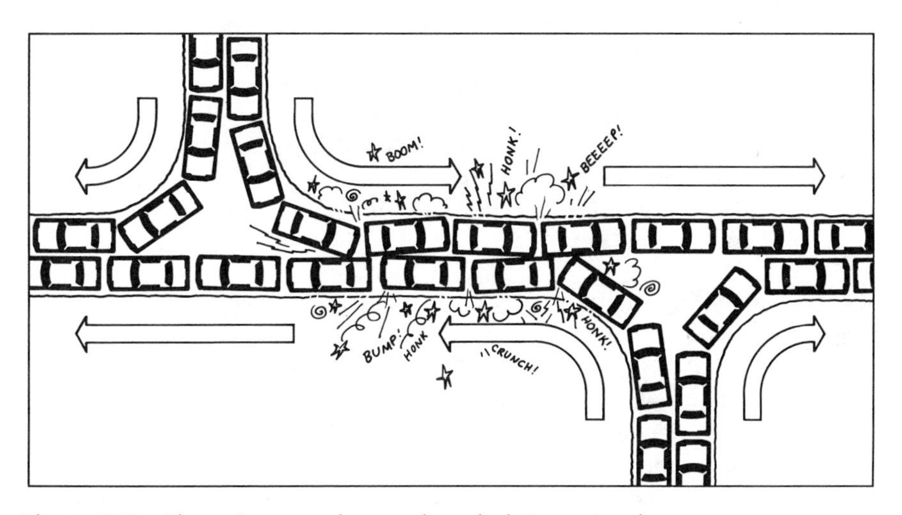

Figure 6.13 The noisy part of a transform fault is restricted to the interior segment, which lies between the ends of spreading centers connected by the fault. (Artwork by D. A. Fischer.)

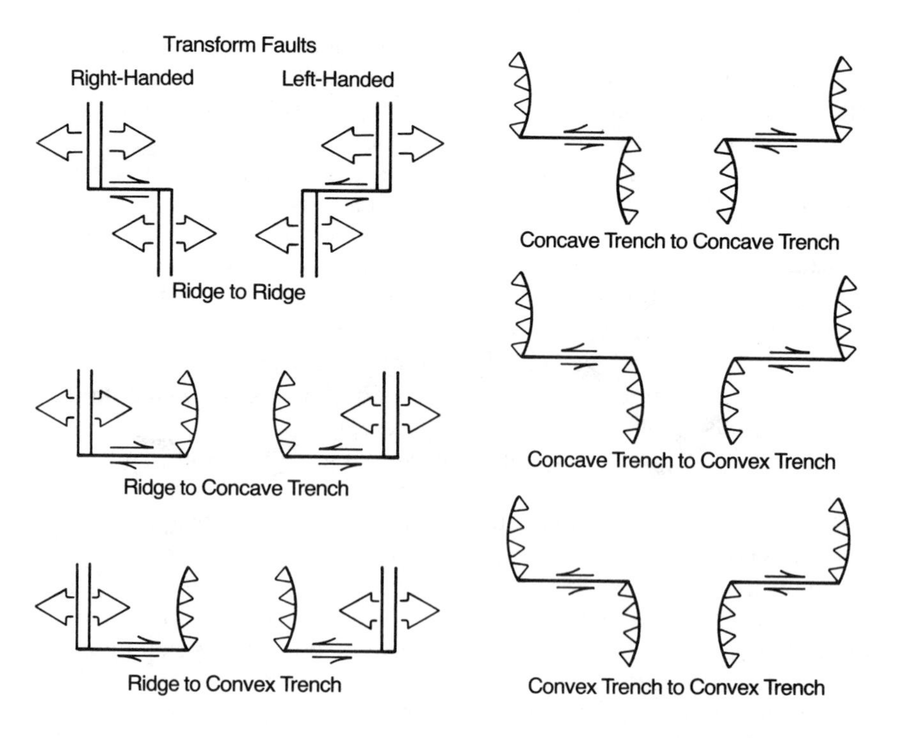

Figure 6.14 The classes of transform faults. Teeth are on overriding plate. [From Wilson (1965). Reprinted by permission from *Nature*, v. 207, pp. 343–347, copyright ©1965 by Macmillan Journals Ltd., London.]

6.14 presents the full spectrum of transform faults and the movements they accommodate.

Not all transform faults are intraoceanic. Tanya Atwater in 1970 stunned the geologic community with her interpretation that the San Andreas fault is a transform fault that slashes through continental crust. The San Andreas fault is indeed a ridge-to-ridge transform, which connects the East Pacific Rise and the Juan de Fuca Ridge (see Figure 6.6). It has accommodated major strike–slip faulting.

CONVERGENT BOUNDARIES

GENERAL PICTURE. Convergent boundaries are found in intraoceanic and continental margin settings where lithospheric plates press against each other. In some cases convergent boundaries are marked by the slow descent of one plate beneath the other, an underthrusting or **subduction** of the denser, heavier plate. In effect, one of the plates is swallowed down a subduction zone into the Earth's mantle. In other cases convergent boundaries are marked by **collision**, and the **accretion** of rocks from one plate onto the leading edge of the other.

ISLAND ARC MOUNTAIN SYSTEMS. **Island arc mountain systems** occur at major plate boundaries in intraoceanic settings where plates move into one another. At the boundary proper, one plate underthrusts the other (Figure 6.15). Partial melting of the underthrusted oceanic lithosphere and the conveyed ocean floor sediments generates magmas that are intruded and extruded. Island complexes called **volcanic arcs** are built, and these arcs shed debris that is eventually deposited as sediments in adjacent basins.

One of the best examples of a region of current island arc mountain building is in the western Pacific, from the Aleutian arc southward through a belt of archipelagos that includes the Japan, Mariana, Philippine, and New

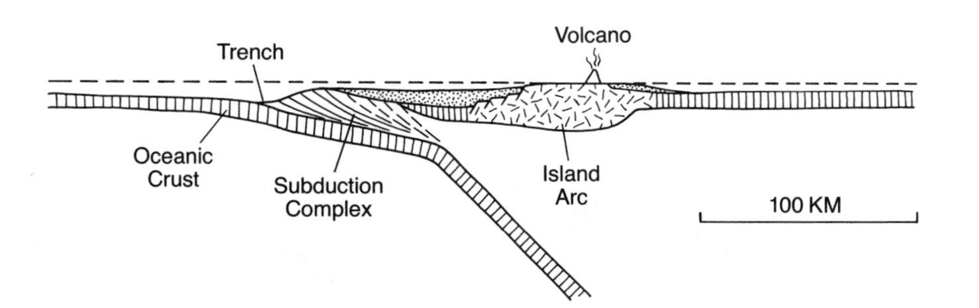

Figure 6.15 Intraoceanic subduction. (From W. R. Dickinson, *American Geophysical Union Maurice Ewing Series 1*, fig. 1.A, p. 34, copyright ©1977 by American Geophysical Union.)

Hebrides Islands. The topographic relief of individual island arc mountain systems is staggering, in some places exceeding 10,000 m (33,000 ft).

CONTINENTAL MARGIN MOUNTAIN SYSTEMS. **Continental margin mountain systems**, expressions of continental margin orogens, are liberally distributed in all of the continents of the Earth (Figure 6.16). They differ widely in specific character and in time(s) of formation, but they all display a **regional mechanical strain** expressed in the presence of structures like folds, faults, fractures, and cleavages. Thermal effects are almost always present in the form of volcanic, plutonic, and/or metamorphic rocks. Active uplift and deformation in very young continental margin orogens, like the Himalayan and Andean systems, are displayed in the physiographic forms whose bold topographic relief is so overpowering.

Continental margin orogens for the most part have formed through plate convergence. If positioned at the leading edge of a plate, rocks in a continental crustal mass are vulnerable to strain-producing movements that arise from convergent movement(s) into adjacent plates. For example, the South American Cordillera, or Andean mountain system, has resulted from stresses generated during interference of the lithospheric plate that South America occupies and the lithospheric plate occupied by oceanic crust of the southern Pacific. Underthrusting and partial melting of the oceanic plate has contributed to outpourings of volcanic materials within the Andean region. Plate interference has led to shortening of the continental rocks in westernmost South America by folding and thrust faulting.

The orogenic influence of plate interactions is manifest in the Alpine, Apennine, Carpathian, and Himalayan ranges as well (Figure 6.16). These mountains comprise an east–west-trending system that has resulted from convergence and continental collision of the plates that Africa and India occupy against that of Europe and Asia. The ancient Tethys Sea was closed (repeatedly) by the northerly relative movement of Africa, thrusting oceanic crust and sediment onto the European platform. The damage is almost immeasurable.

BENIOFF ZONES AND TRENCHES. The descent of one plate beneath another during subduction creates frictional interference that is disclosed seismically by severe earthquake activity. Earthquake foci describe an inclined zone, the inferred **subduction zone**, which projects as deep as 700 km. These so-called **Benioff zones**, named after the scientist who first recognized them (Benioff, 1954), are believed to mark the sites of convergent plate boundaries at depth.

Figure 6.16 Worldwide distribution of mountain systems in continental settings.

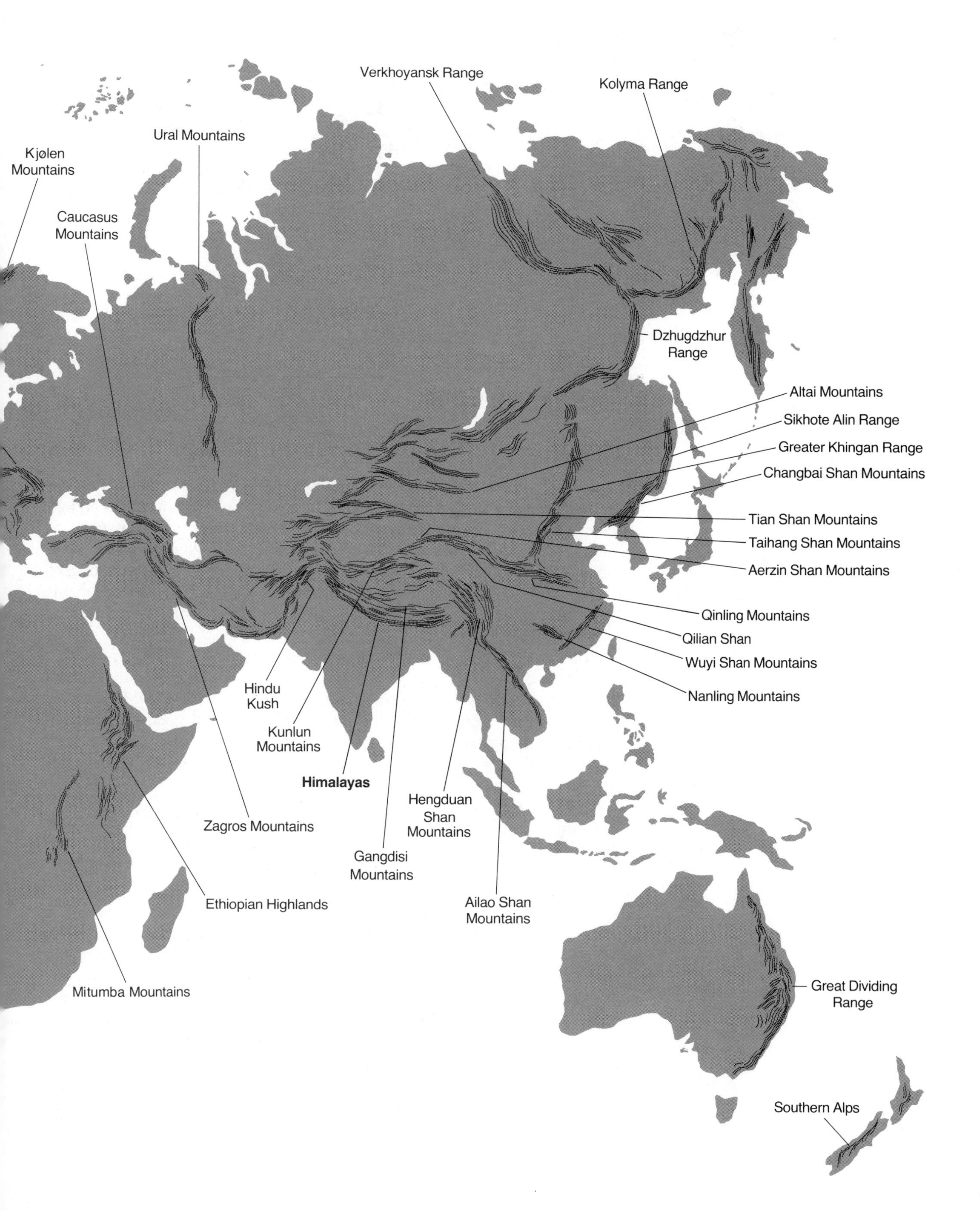
Verkhoyansk Range
Kolyma Range
Ural Mountains
Kjølen Mountains
Caucasus Mountains
Dzhugdzhur Range
Altai Mountains
Sikhote Alin Range
Greater Khingan Range
Changbai Shan Mountains
Tian Shan Mountains
Taihang Shan Mountains
Aerzin Shan Mountains
Qinling Mountains
Qilian Shan
Wuyi Shan Mountains
Nanling Mountains
Hindu Kush
Kunlun Mountains
Himalayas
Hengduan Shan Mountains
Zagros Mountains
Gangdisi Mountains
Ailao Shan Mountains
Ethiopian Highlands
Mitumba Mountains
Great Dividing Range
Southern Alps

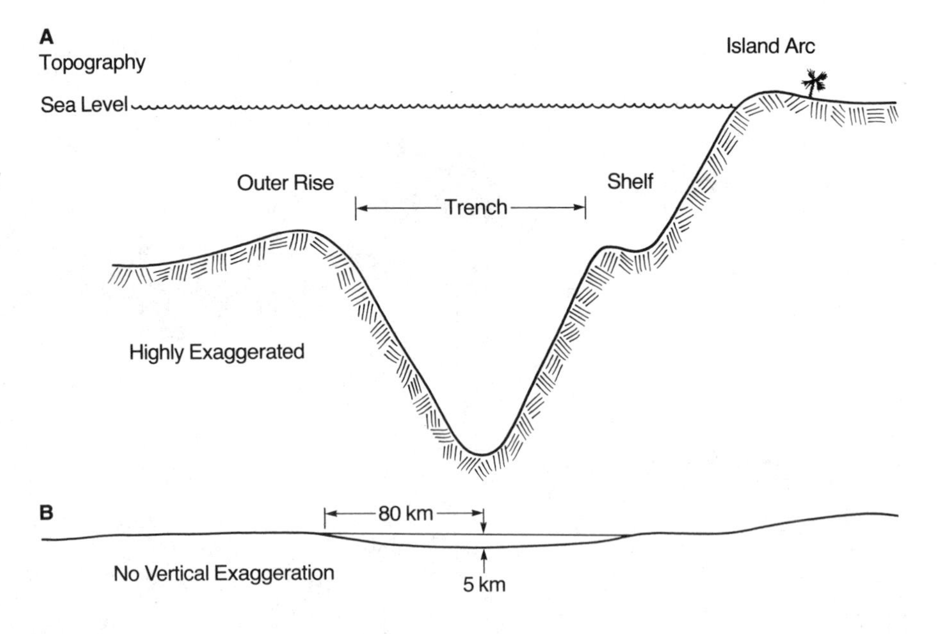

Figure 6.17 (*A*) Profile view of a typical trench, with vertical exaggeration. (*B*) Profile view of a typical trench, without vertical exaggeration. [From Melosh and Raefsky (1980), v. 60, fig. 1, p. 344. Published with permission of *Geophysical Journal of the Royal Astronomical Society*.]

An oceanic **trench** is created along the surface trace of a Benioff zone (see Figure 6.15). Both in intraoceanic and continental margin settings, trenches mark the positions at which the slab to be underthrust begins to flex and go under. Trenches are deep, averaging 8 km with water depths as great as 12 km. The average breadth of a trench is about 100 km. Cross-sectional topographic profiles of trenches reveal slightly asymmetrical forms (Figure 6.17*A*), with the steepest trench wall, the **inner wall**, facing the plate that is being subducted. The **outer wall** of a trench may rise gently away from the trench for distances of 100 to 200 km.

Profiles of trench topography, like that shown in Figure 6.17*A*, are normally constructed with significant vertical exaggeration to identify features that would be subtly expressed, at best, on ordinary profiles. Profiles drawn without vertical exaggeration and viewed in a regional perspective reveal that trenches are actually very slight depressions (Figure 6.17*B*). Such profiles understate the extraordinary mechanical processes taking place beneath the trenches.

Trenches of the world marking sites of present-day subduction are shown in Figure 6.18. One manifestation of the dynamics of these sites of subduction is the presence of active volcanoes. An even clearer signal of the dynamic state of subduction zones is the earthquake activity along them.

SUBDUCTION. It is amazing what junk is jammed into subduction zones at the sites of trenches. Deep-water sediments and slabs of underlying ophiolitic crust are conveyed into the sites of subduction zones. Topographic features, like submarine **volcanic plateaus** and **seamounts**, ride the moving lithospheric slab and are likewise brought into trenches. **Turbidity currents** triggered by earthquake shocks and composed of high-density slurries of mud and sediment flow into trenches under the influence of gravity. These too contribute to the mass that must be swallowed. Most extravagant of all, the steady movement of plates may eventually feed the leading edge of a continent into a trench. The buoyancy of continental masses precludes significant subduction. When a continent is fed into a subduction zone, the subduction zone may become inoperative, sutured by collisional tectonics. Alternatively, the dip direction (**polarity**) of the subduction zone may reverse

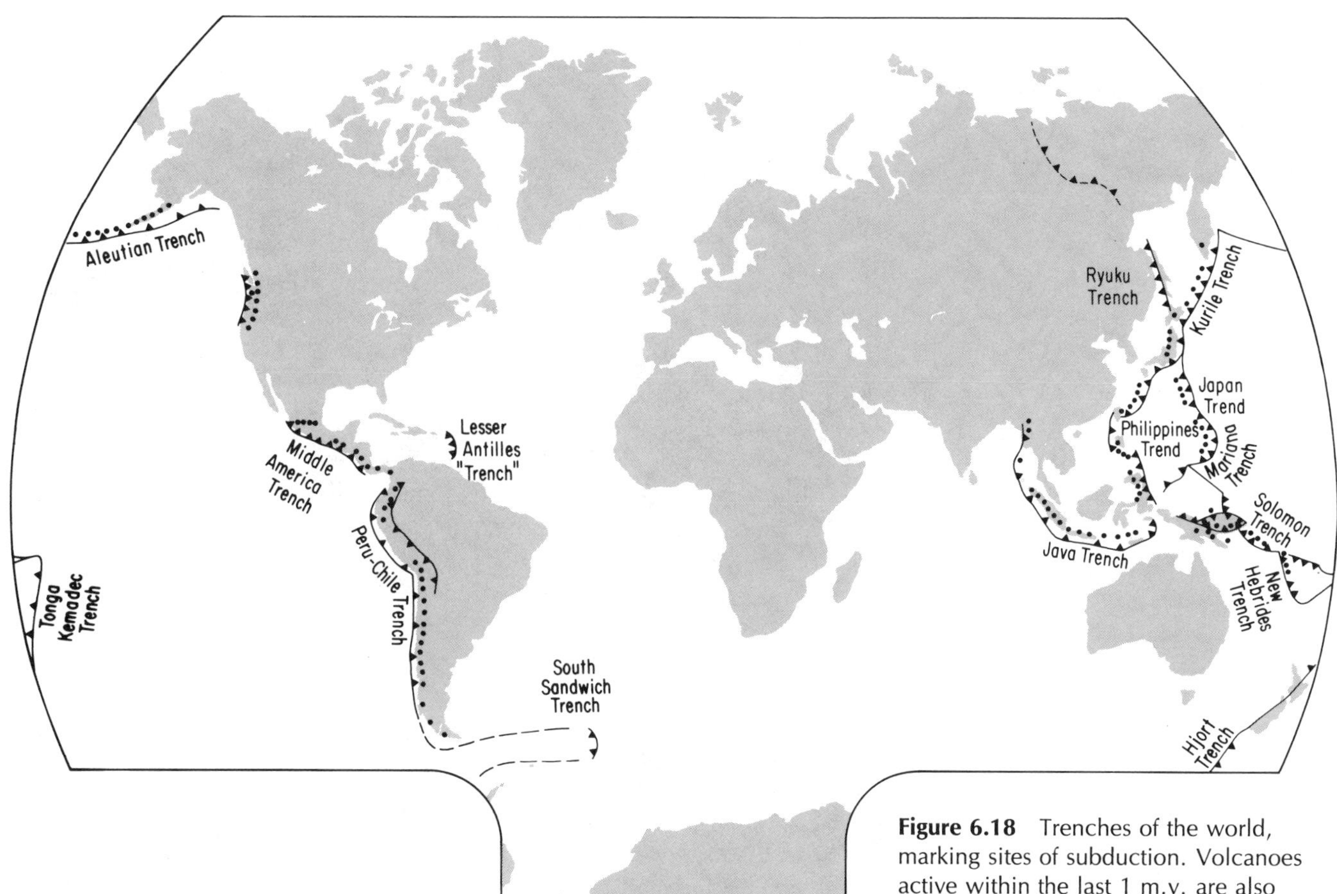

Figure 6.18 Trenches of the world, marking sites of subduction. Volcanoes active within the last 1 m.y. are also shown. [From Lowman (1981). Courtesy of Goddard Space Flight Center, National Aeronautics and Space Administration.]

and the former overriding plate may become the subducted plate. In any event, the leading edge of the continent becomes an **active margin** of deformation.

MELANGES. Normally during subduction, uppermost sediments and seafloor brought into the trench are partly scraped off and plastered or accreted to the inner wall of the trench (Figure 6.19). The result is an **accretionary melange wedge** of low-density ($\rho = 2.3$), water-saturated materials distorted by the chaotic structure of fault imbrication and plastic folding (Figure 6.20). The topographic expression of the melange wedge forms a lip, which dams

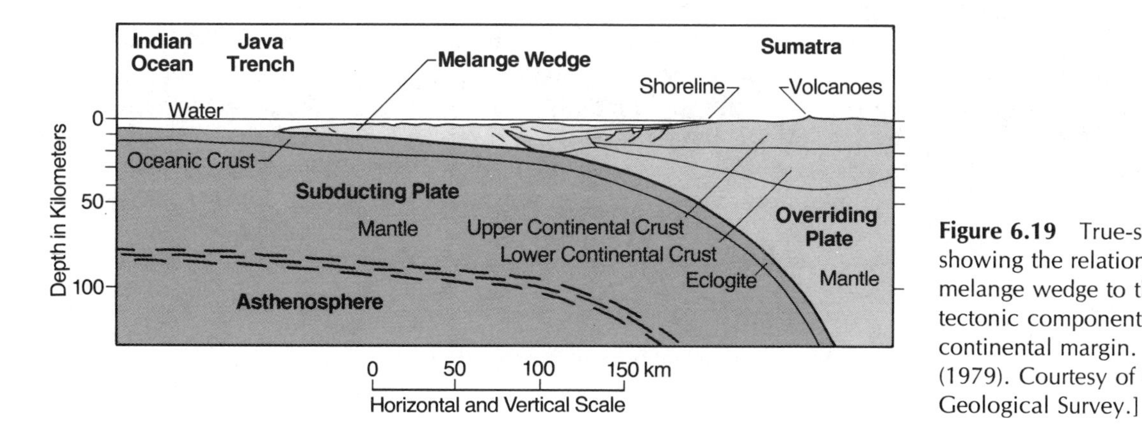

Figure 6.19 True-scale profile view showing the relations of an accretionary melange wedge to the fundamental tectonic components of an active continental margin. [From Hamilton (1979). Courtesy of United States Geological Survey.]

Figure 6.20 Folds and penetrative foliation in melange exposed in the San Juan Islands, Washington. (Photograph by G. H. Davis.)

basins of accumulation of material that would otherwise be transported into the trench from the overriding plate (see Figure 6.19).

The term **melange** means different things to different people. Its most notable characteristic is what it does not display—coherent internal layering and coherent internal order (Darrel Cowan, Personal communication, 1982). Exposures consist of blocks of competent, resistant rocks protruding from poorly exposed, nonresistant matrix material. Internal structure is so complex that it is nearly impossible to map. The constant dilemma of workers in interpreting the formation of melange lie in evaluating the relative roles of soft-sediment deformation and postconsolidation deformation, and evaluating the extent of tectonic versus gravitational dismemberment and emplacement.

Figure 6.21 Tectonic inclusions in melange, Kodiak Islands, Alaska. Largest inclusion is 2 m long. [From Moore (1978). Published with permission of Geological Society of America and the author.]

Melanges typically contain resistant **blocks** of **tectonic inclusions** of diverse origin that rest within a sheared, foliated "shaly" matrix (Figure 6.21). Blocks of several hundred meters in dimension are common. **Exotic blocks**, composed of lithologies that are foreign to the setting, are also sprinkled within melange. A common component of melange is **radiolarian ribbon cherts**, which are typically disrupted by folds, faults, and penetrative cleavages.

Mixing of the various rocks within a melange can be achieved in a number of ways, both sedimentary and tectonic (Goodwin, 1976). For a given deposit, some of the details of mechanical mixing will never be known. Melanges of sedimentary origin (**olistostromes**) may form by gravity sliding and slumping of blocks of largely sedimentary rocks down the inner slopes of a trench, where they are incorporated into muddy sediments. Melanges of a tectonic origin may form by the movement of a subducting plate past the trench inner wall, in simple shear fashion. The simple shear would tend to flatten the rocks in one direction and stretch them in another. Some of the flattening and stretching is achieved by plastic distortion of water-rich sediments. Much of it, however, is accomplished by the action of subparallel, **imbricate** faults that shuffle and stack the deforming sequence.

Sediments on the outer slope of a trench are pristine and undeformed, consisting of **abyssal plain sedimentation**, commonly in the form of thin, undeformed, sheetlike muds and siliceous oozes. These sediments are drawn ever slowly into the trench, where eventually they are transformed into melange.

NEVER-NEVER LAND: SUBDUCTION AND/OR ACCRETION. What goes on beneath a trench within a Benioff zone is not clearly known. But there is much speculation. Beneath a trench, the downgoing slab descends at an angle determined by factors such as the rate of convergence and the relative densities of the adjacent converging plates (Luyendyk, 1976). The deepest trench in the world's oceans, the Marianas Trench, is linked to a Benioff zone that dips fully 90°. It is receiving, through subduction, some of the oldest ocean floor of the Pacific (150 m.y). In contrast, the Mexico Trench, which is swallowing oceanic lithosphere 20 m.y. in age, is the mouth of a subduction zone that is inclined merely 15° to 20° (England and Wortel, 1980). In general, relatively old oceanic lithosphere may be subducted with relative ease because the ophiolitic rocks are cold and dense. Old, cold, relatively heavy lithosphere sinks like a rock. Conversely, young, warm, relatively light oceanic lithosphere may possess an inherent buoyancy that resists subduction, thus creating strong horizontal compressive stresses along the zone of contact (England and Wortel, 1980). For these reasons and others, the dip of Benioff zones beneath trenches is highly variable. Descent velocities of subducted slabs are likewise variable, ranging from 2 to 10 cm/yr.

Subduction is not a perfect disposal system. Although subduction appears to have been responsible for destroying all coherent terranes of oceanic crust older than about 150 m.y., there is growing evidence that convergent plate boundaries have been unable in some cases to swallow up, or keep down, major masses of rock. For example, in much of Alaska and western North America, continental margin mountain systems are dominated by the deformed remains of Paleozoic–Mesozoic ocean ridges, volcanic islands, and deep-sea sediments. These appear to have been plastered against the western edge of North America along with other scraps and pieces of oceanic floor (Cox, Dalrymple, and Doell, 1967; Coney and others, 1981). The mapped distribution of the **accreted terranes** that have been added to North America and Central America is enormous. They comprise half to two thirds of the Western Cordillera of North America (Figure 6.22). The intensity of deformation within some of the accreted terranes is staggering, whereas others are not greatly deformed at all.

Peter Coney coined the term "**suspect terrane**" for regional rock assemblages that may be exotic to the home continent in which they now reside. Some suspect terranes are composed of rocks that have been transported thousands of kilometers from their place of origin. The accretion process amounts to international kidnapping. Rocks that started out within or atop one plate are forced to come aboard some other plate at a convergent boundary. The mechanical transfer takes place through collision, aborted subduction, and/or strike–slip shearing. What a structural challenge suspect terranes represent! What a never-never land!

HIGH-PRESSURE METAMORPHISM. Frictional resistance of downward movement of a subducted slab not only generates deep-focus earthquakes, it also leads to high-pressure metamorphism at shallow depth. Characteristic **blueschist metamorphic assemblages** are formed at the expense of partially subducted melange sediments and underlying, ophiolitic oceanic crust (Miyashiro, 1973). The low-temperature character of blueschist metamorphic assemblages is attributed to the downsinking of relatively cool rocks into relatively hot regions (Ernst, 1975). The high-pressure metamorphism takes place at depths of 20 to 35 km. Tremendous vertical faulting is required to lift these rocks to the level of surface exposures. The presence of blueschist facies metamorphic rocks at the surface is yet another testimony of the great structural relief bound up in orogenic belts.

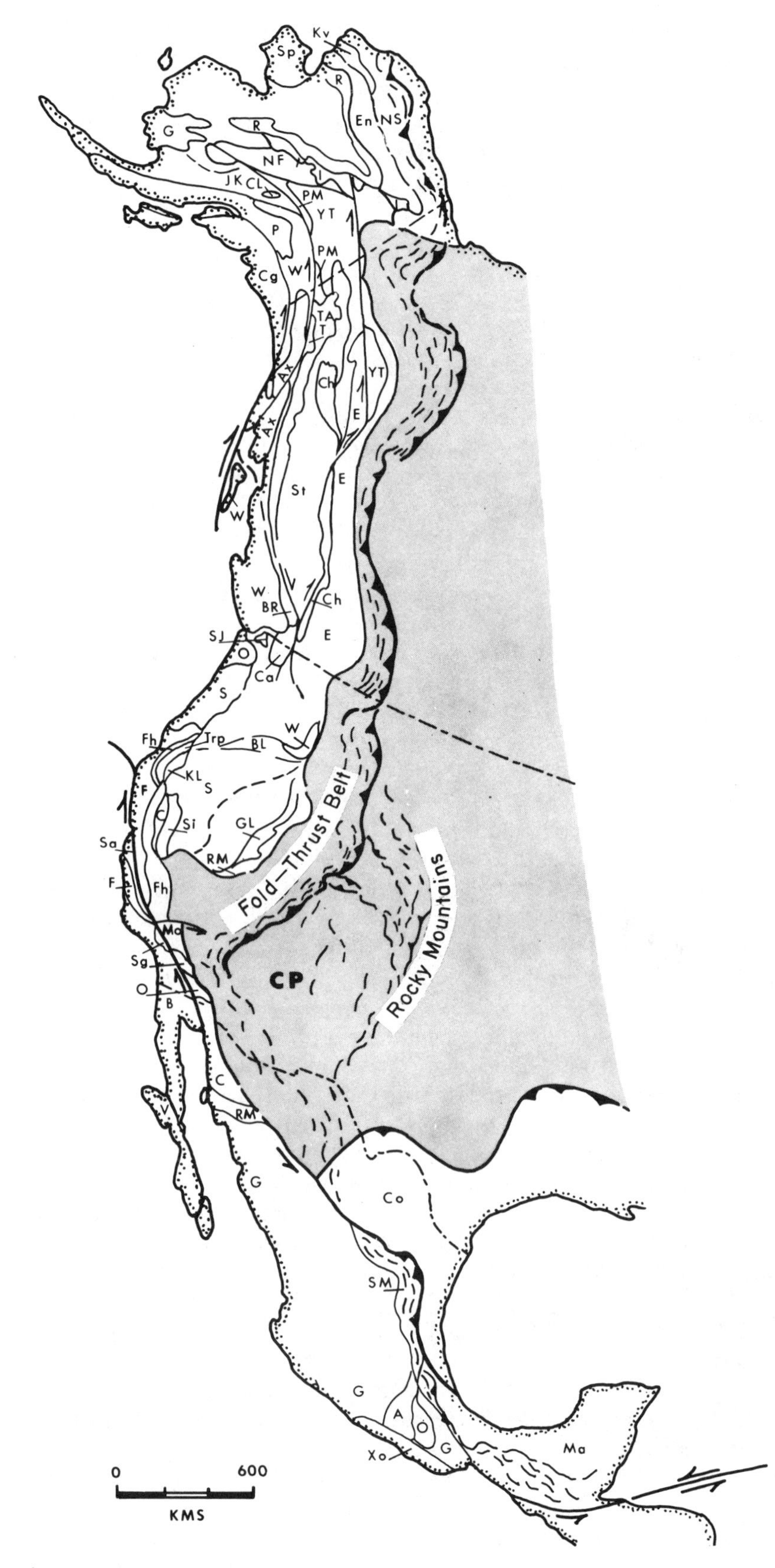

Figure 6.22 Accreted terranes of western North America. [From Coney and others (1981). Courtesy of United States Geological Survey.]

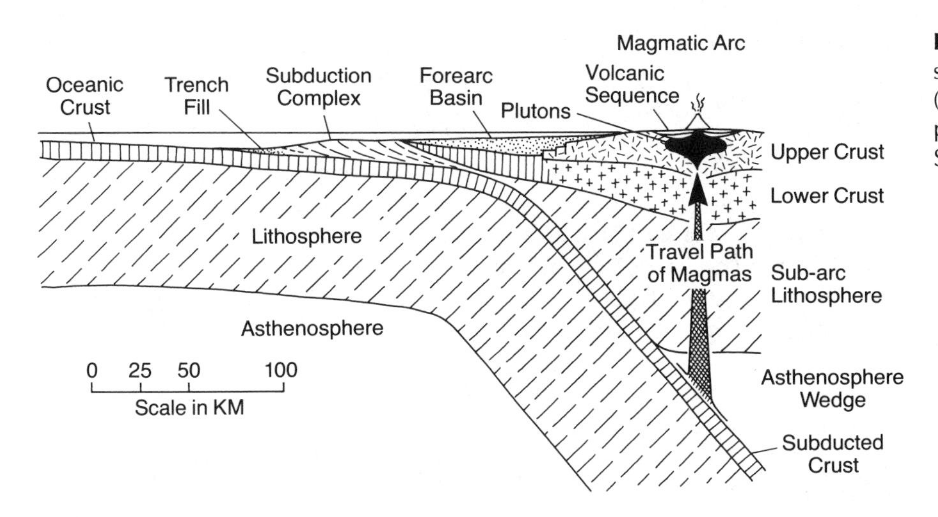

Figure 6.23 Generation of magma during subduction. [From Dickinson and Payne (1981), cover illustration. Published with permission of Arizona Geological Society.]

THE MAGMATIC ARC. Ultimately a subducted slab descends to the point at which partial melting begins to take place and magmas buoyantly rise from depth. Under the influence of gravity, these high-temperature, low-density magmas move up, into, and through the overriding plate (Figure 6.23). The products of intrusion and extrusion contribute to the formation of a **magmatic arc**, the uppermost reaches of which comprise a **volcanic arc**.

Partial melting to form magma begins when the descending slab penetrates asthenosphere, and this occurs on average at a depth of about 100 to 150 km. The angle of inclination of the Benioff zone and the depth to the asthenosphere combine to determine the location where the magmatic arc will be constructed.

The magmas generated within zones of partial melting beneath continental crust rise through the crust to form large intrusive igneous bodies of dominantly quartz monzonitic to granitic composition (Figure 6.23). The flooding of the crust by magma produces a belt of low-pressure regional and/or contact metamorphism. Geothermal gradients in belts of low-pressure metamorphism are on the order of 25°/km, contrasted with the average 10°/km gradient that characterizes sites of high-pressure metamorphism (Miyashiro, 1974).

VOLCANIC ARCS. Volcanic arcs are produced near sites of subduction. They constitute enormous, heterogeneous rock assemblages. In intraoceanic settings, magmatism creates the island arcs that are so classically displayed in the western Pacific. The arcuate pattern of the islands, as viewed in plan, has long been an object of question and curiosity. In all probability the arc shape is derived from the intersection of two curved surfaces, one the inclined Benioff zone, the other the top of the asthenosphere (Dickinson, 1980). Magma ascending vertically from this curved line of intersection gives rise to an arcuate volcanic chain.

The volcanic arcs in intraoceanic settings commonly range in composition from tholeiitic basalts through calc-alkaline basalts and andesites. Pyroclastic deposits composed of gas-charged particulate matter are most abundant, with lava flows occurring mainly near the eruptive centers. Aprons of clastic debris run off the composite volcanoes of the islands, to be deposited in adjacent deep basins as volcaniclastic turbidites, marine tuffs, and submarine ash deposits.

Volcanic arcs are produced in great abundance at active continental margins where subduction has been long-lived. The circum-Pacific Ring of Fire is the most often cited example. Magma extruded in such settings is of intermediate composition because of contamination of the primary magmas during ascent through the continental crust. Classically, the volcanic products are andesitic.

Volcanic edifices in arcs of continental margin orogens are large **composite volcanoes** featuring interlayered flows and pyroclastic deposits. In addition there are **cinder cones**, formed almost exclusively of pyroclastic material and occasional **volcanic domes** of rhyolitic materials. Some of the most impressive products are **ash flows**, otherwise known as welded tuffs or ignimbrites. Ash flows form through processes that involve the high-velocity movement of hot, glowing, gas-charged avalanches of particulate material, notably glass shards and pumice fragments in a very fine-grained matrix (Smith, 1960a, b). Single sources have produced flow fields larger than 12,000 mi^2 (19,400 km^2), containing more than 500 mi^3 (800 km^3) of material. Single flows may travel 70 mi (113 km) from their source.

The volume of material discharged from volcanic centers is sometimes so great that the huge, largely evacuated magma chambers collapse inward and cause a foundering of the volcanic edifice. The result is a collapse **caldera**, kilometers in diameter. An outstanding treatise on the formation of a caldera is Howell Williams's report on the geology of Crater Lake National Park (Williams, 1942). In it he describes the geological history of Mount Mazama, the former occupant of the site of Crater Lake. The volcano blew out so much magma in the form of pyroclastic material and flows that its foundation gave way. As a result, a volcanic mountain, estimated to be 6000 ft (1820 m) high and 17 mi^3 (27 km^3) in volume, vanished into the collapsed foundation. This event occurred only 5000 years ago.

SEDIMENTARY BASIN DEPOSITS. Volcanic arcs are vulnerable to rapid erosion. Transport of enormous quantities of eroded debris spill toward opposite sides of the arc complex. They may be deposited in a variety of basins, the names of which are commonly referenced with respect to the position of the magmatic arc (Figure 6.24). **Forearc basins**, **interarc basins**, and **backarc basins** form within or immediately adjacent to the active continental margin. They form in intraoceanic arc settings as well. The sediments accumulate in thick wedge-shaped to prism-shaped deposits whose specific forms reflect **structural and depositional environment**.

Deposits of forearc basin settings include fluvatile–deltaic deposits, continental shelf depositional prisms, and deep-water turbidity current deposits and submarine fans (Dickinson and Seely, 1979). Interarc basins typically are wedge-shaped to lens-shaped deposits of very locally derived clastic material. Deposition is largely subaerial. Some of the interarc deposits are simply alluvial fans and lahars (volcanic mud flows) interlayered with volcanic flows and ash deposits of the arc. Others are basinal accumulations at sites of down-dropped blocks.

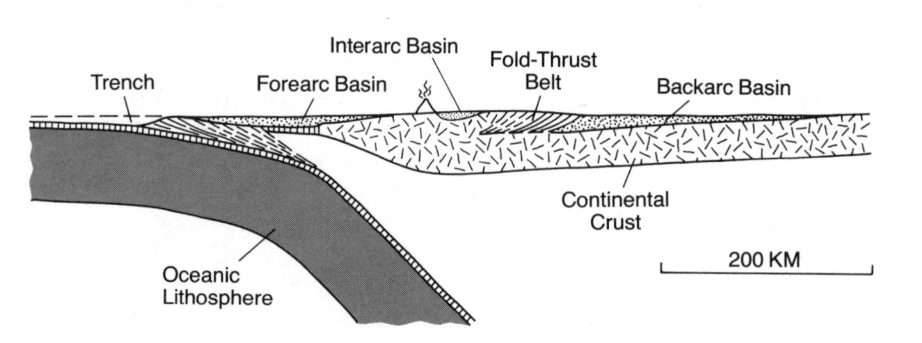

Figure 6.24 Sedimentary basins at the site of an active continental margin. (From W. R. Dickinson, *American Geophysical Union Maurice Ewing Series 1*, fig. 1B, p. 34, copyright ©1977 by American Geophysical Union.)

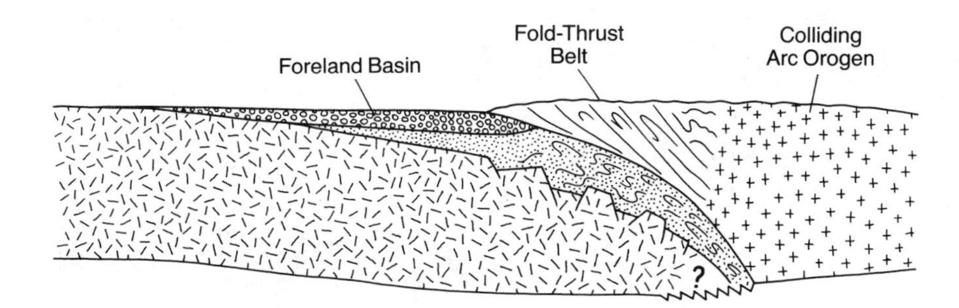

Figure 6.25 Foreland basins, a response to bending of the lithosphere under the load of thrusted and folded sediments. [From Dickinson (1980). Published with permission of Geological Association of Canada.]

Tectonic stresses generated by plate convergence and/or magmatic processes within the arc can lead to stretching and/or shortening of the framework behind the arc. Extensional faulting may open up rifted basins that become backarc basins of accumulation of poorly sorted deposits of locally derived sedimentary and volcanic detritus. **Backarc spreading** is especially common in island arc intraoceanic settings (Karig, 1971).

Each basinal deposit is highly vulnerable to mechanical distortion by folding and faulting. Convergence accompanied by subduction and/or collision can distort the freshly deposited sediments and volcanics. Stresses that issue from plate boundaries can even be transmitted beyond the active continental margin into the **foreland** of the continental interior (Figure 6.25). In some cases, foreland basins develop between the active continental margin and *bona fide* craton. Foreland basins form by downbending of the lithosphere, as a response to the load created by the vertical stacking and thickening of rocks that is accomplished by thrust faulting and folding. Such downbending of lithosphere has been compared to the bending down of the leaf springs in a car. Sedimentary wedges and prisms derived from highlands built by faulting are shed into foreland basins. Thousands of meters of sediments may accumulate to form the deposits, which coarsen and thicken toward the plate margin, toward the source of dynamic disturbance.

Plate Kinematics

PRESENT-DAY PLATE CONFIGURATION

Divergent, convergent, and strike–slip boundaries are linked together in a world-encircling structural system. Together they combine to form the boundaries of the plates of the Earth. Soon after the significance of oceanic ridges, Benioff zones, and transform faults was perceived, Xavier Le Pichon (1968) recognized that the Earth could be subdivided into discrete fundamental plates. Refinement of his original model has led to our present view of the lithospheric plates of the world, as presented in Figure 6.26.

The largest of the plates are the **North American, South American, Eurasian, African, Antarctic, Indian–Australian, Pacific**, and **Nazca** plates. The plate edges are geometrically intricate and involve limitless combinations of ridge, trench, and transform boundaries. The plates are generally composed both of oceanic and continental terranes.

Relatively small, lesser plates, complete the fundamental system of lithospheric plates of the Earth (Figure 6.26). The **Juan de Fuca**, **Cocos**, and **Scotia** plates each consist of young ophiolitic seafloor spreading from ridges and guided by transforms into nearby subduction zones. The **Arabian** plate is

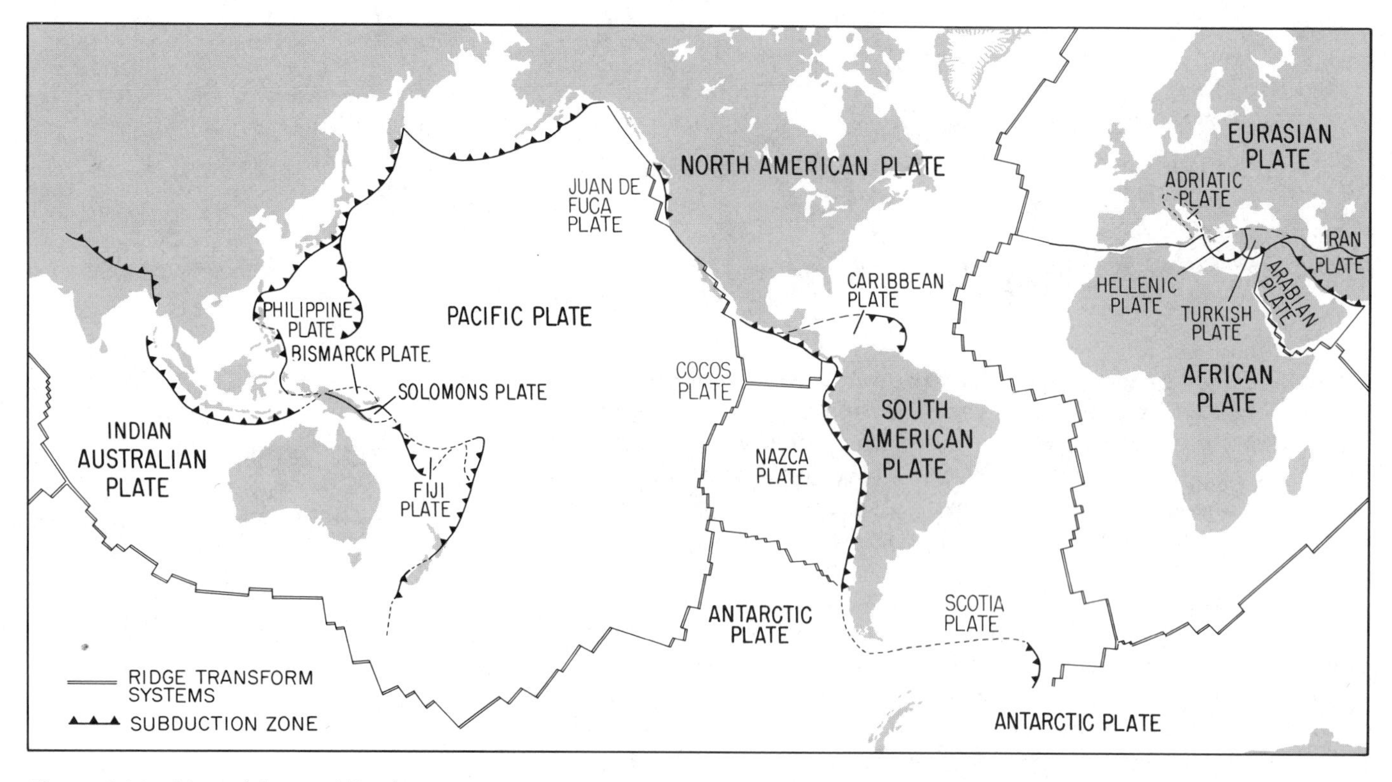

Figure 6.26 Map of the world's plates. (From "Plate Tectonics" by J. F. Dewey, copyright ©1972 by Scientific American, Inc. All rights reserved.)

composed of continental lithosphere diverging from the Red Sea rift, guided both by oceanic and continental transforms into the suture zone where continental crust of the Eurasian and Arabian plates collide. The **Philippine, Bismarck, Solomons,** and **Fiji** plates in the western Pacific occupy marginal seas bounded by combinations of trenches and transform faults.

Knowledge of the geography of the plates of the Earth is helpful, if not essential, to understanding the tectonic setting of specific regions of interest and study. Plates and their boundaries are necessarily represented on world maps as static features. But both the plates and their boundaries are continually on the move. How they move is the subject of plate kinematics.

PLATE CONFIGURATIONS IN THE PAST

The movement of plates through time is fascinating. Reconstructions of the positions of continents and the plates they occupy impart a sense of awe regarding the degree to which our "stable foundations" are active. We see in plate reconstructions, such as those by Bally and Snelson (1980), the fundamental plate movements that took place in the Mesozoic and Cenozoic eras (Figure 6.27).

The breakup of the supercontinent of Pangaea began with rifting beween North America and Africa in the mid-Jurassic period, about 180 m.y. ago (Figure 6.27). The rifting initiated seafloor spreading and the formation of the northern Atlantic Ocean and the Gulf of Mexico. At about 138 m.y.b.p., in the early Cretaceous, South America and Africa began to split apart, forming the beginnings of the South Atlantic. At approximately the same time a former continental terrane consisting of India, Antarctica, and Australia began to diverge from the east flank of what is now Africa. In the late Cretaceous, 90 m.y. ago, Greenland and North America separated, as did India and Madagascar. Fifty m.y.b.p., Australia and Antarctica separated from each other, while India continued on its flight toward Eurasia, with which it

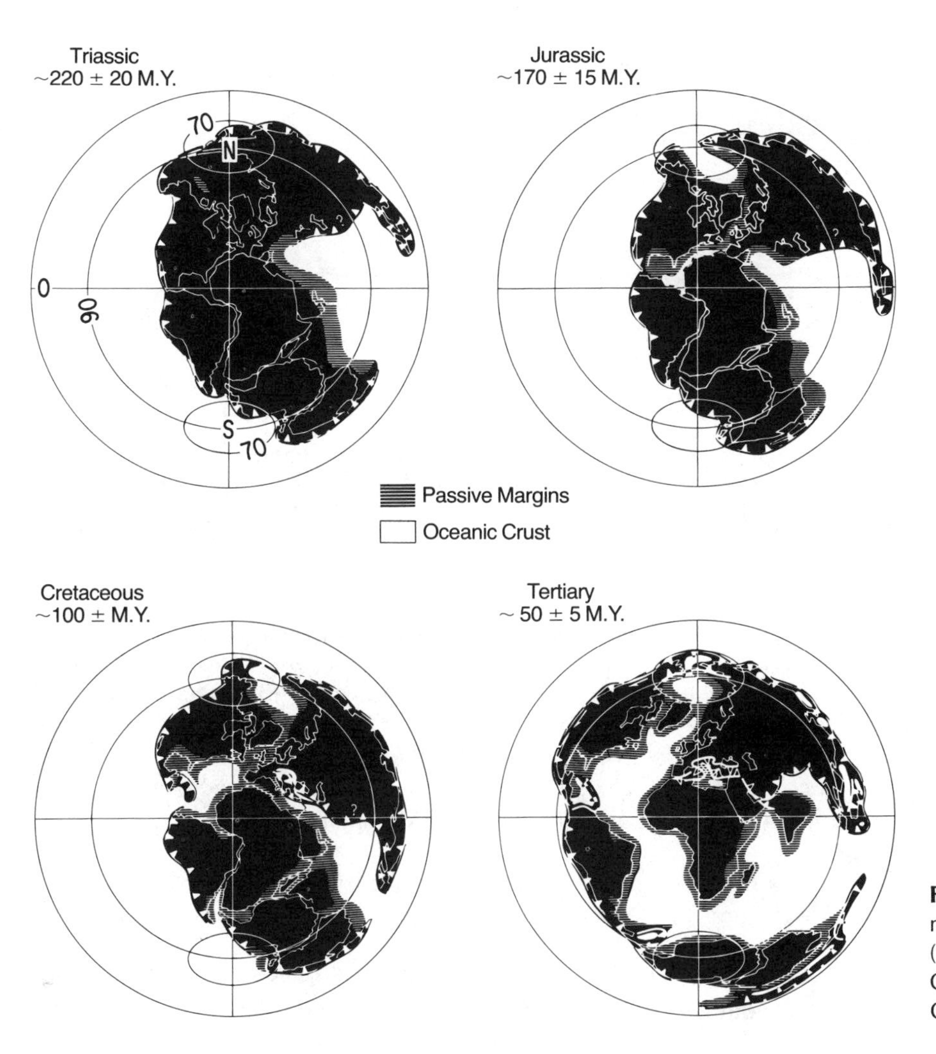

Figure 6.27 The drama of past plate movements. [From Bally and Snelson (1980). Published with permission of Canadian Society of Petroleum Geologists.]

finally began to collide 40 m.y.b.p. In upper Miocene, 10 m.y.b.p., the Red Sea and the Gulf of Aden opened. The Gulf of California opened 5 m.y. ago when the western part of Mexico pulled away at the Baja peninsula.

Before the theory of plate tectonics was accepted, all these movements and shifts of continents were described as **continental drift**. But more than just the continents were moving. The continents were simply passengers on lithospheric ships moving on mantle-level asthenosphere. What an unpredictable drama! What will the major scenes be like in the next 200 million years?

CROSS-SECTIONAL VIEWS OF PLATE INTERACTIONS

Map-view reconstructions, like those by Bally and Snelson (1980), provide one kind of insight into plate kinematics. A complementary perspective can be derived from cross-sectional views of plate interactions. Dewey and Bird (1971), in a classical paper describing plate tectonic configurations, provided geologists with a clearer picture of the impact of seafloor spreading and plate tectonics on the fashioning of continental geology through time. In essence they constructed schematic world-view cross sections that display how the existing plates interact (Figure 6.28).

The South American plate is shown diverging westward from the Mid-Atlantic Ridge (Figure 6.28). The continent of South America is positioned on

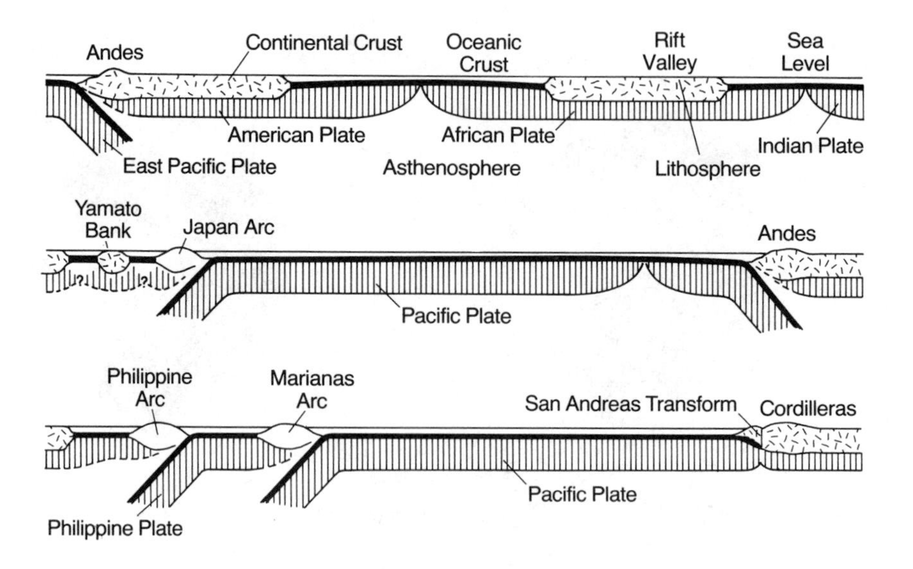

Figure 6.28 Worldwide cross sections of plate interactions. (From J. F. Dewey and J. M. Bird, *Journal of Geophysical Research*, fig. 2, p. 2627, copyright ©1970 by American Geophysical Union.)

the plate such that its eastern edge is a passive margin. In contrast, the western edge of South America is an active margin beneath which the Pacific Ocean floor of the Nazca plate is continually being subducted. The African continent resides in the center of the African plate. Both its eastern and western edges are passive margins. Seaward from the coasts of Africa, ophiolitic oceanic crust is welded to the continental crust. The oceanic crust has spread from the Carlsberg Ridge and the Mid-Atlantic Ridge in symmetrical fashion.

The Pacific plate spreads westward from the East Pacific Rise over a broad expanse (Figure 6.28). Near Asia it descends into the trench complexes of the western Pacific. Volcanic arcs continue to build above the subducted lithosphere. Off western South America, the Pacific plate is subducted beneath the South American plate. Off western North America, the Pacific plate comes in direct contact with the North American plate. The San Andreas transform fault marks the boundary. As a transform linking the East Pacific Rise and the Juan de Fuca Ridge, the San Andreas fault accommodates strike–slip faulting between the North American and Pacific plates.

Such examples give color to some of the many possible plate tectonic interactions that can take place. John Dewey, more than anyone, has made it clear that the kinematic possibilities are so broad ranging that it will remain very difficult to reconstruct the details of past movements.

RELATIVE VERSUS ABSOLUTE MOTIONS

Visualizing and calculating the movements of plates and plate boundaries is an exercise in determining **relative motions**. We met this concept earlier in Chapter 4. Slip is defined as the actual relative displacement between points that were once in contact. Slip can be evaluated, even though we seldom can determine the **absolute movement**. Let me give an example.

Suppose that I load my triaxial deformation apparatus into a van and proceed to deform a core of sandstone while driving 95 mph along Interstate 10 between Tucson and Tempe. Throughout the trip, the specimen is subjected to slow, steady, uniform loading. In Tempe, I remove the faulted sandstone from the pressure vessel and show it to structural geologist Donal Ragan. Even though Dr. Ragan is a first-class structural geologist, he would not be able to determine on the basis of the sample alone the absolute movements that had taken place during the deformation experiment. He

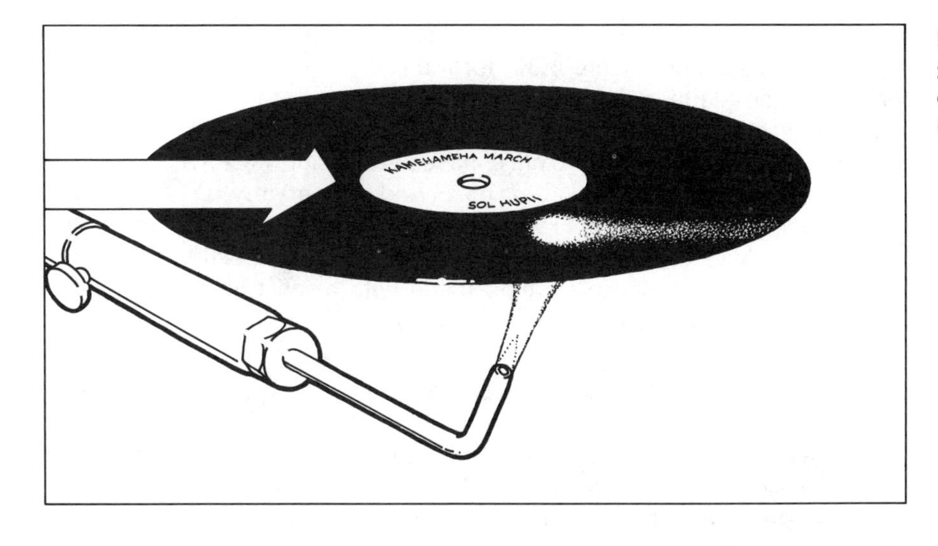

Figure 6.29 The Hawaiian Islands were generated by the translation of lithosphere of the Pacific plate over a fixed hot spot. (Artwork by D. A. Fischer.)

would only be able to measure the **actual relative displacement** on the order of 2 mm or so. There would be no clue that during the experiment the hanging wall of the faulted specimen had moved, in absolute terms, along a displacement vector of 106 mi (170 km) in a N33°W sense of direction. So it is with plates. Without some form of extra special information, we describe the motion of a plate in relation to some other plate, not within an absolute framework.

What kind of extra special information is required? If Dr. Ragan had been given, in addition to the faulted rock specimen, the 12 speeding tickets I received during my Tucson-to-Tempe trip, he would have been able to evaluate the absolute movements that the rock specimen experienced. Each speeding ticket records when and where I was stopped. He could have used this information to map out absolute movements.

Even though plates can never be accused of speeding, nature sometimes provides some bits of special information to help pin down their absolute motions. Jason Morgan of Princeton University proposed that fixed **hot spots** in the mantle may constitute absolute points of reference for describing plate motions (Morgan, 1971, 1972). A hot spot is regarded as a **thermal plume** of uncertain origin that sears the overriding lithosphere. The path of movement is thus recorded in the form of a linear scar of volcanoes and/or igneous intrusions in the lithosphere. The mapped locations of the igneous rocks, combined with radiometric age dates for same, yield the time-and-place data necessary to unravel absolute motions. One image that comes to mind in visualizing hot spots is one of slowly passing a 78-rpm record over a blowtorch, thus steadily melting and blistering the record along the line of movement (Figure 6.29).

A likely geological record of the passing of a plate over a hot spot is the Hawaiian Islands–Emperor Seamounts chain in the Pacific Ocean (Figure 6.30). Oriented west–northwest, the Hawaiian Islands systematically increase in age from east to west. The island of Hawaii, at the eastern end of the chain, lies directly above or just northwest of the hot spot and thus still contains active volcanoes. In contrast, the Midway Islands, at the western end of the chain, are composed of 40-m.y.-old volcanic rocks. The Midway Islands formed when that part of the lithosphere was located where Hawaii is located today. Knowing that the distance between Hawaii and the Midway Islands is 3000 km, the actual relative movement of the Pacific plate with respect to the fixed hot spot in asthenosphere can be determined. The average velocity is 3000 km/40 m.y. (7.5 cm/yr).

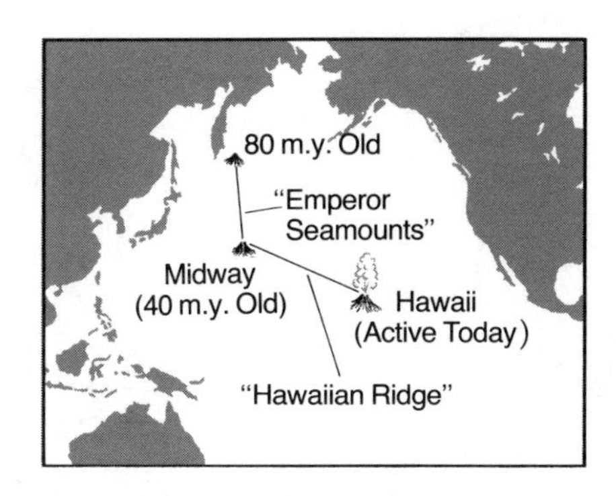

Figure 6.30 Map showing the Hawaiian Islands–Emperor Seamounts chain, an inferred hot-spot track.

Northward from Midway Islands, the Emperor Seamounts trend along a north–northwest line (Figure 6.30) for a distance of 3000 km, where they encounter the Aleutian Trench. Volcanic rocks of the seamounts increase in age from 40 to 80 m.y. from south to north along the chain. We conclude from these facts that the Pacific plate moved north–northwest with respect to a fixed hot spot in the mantle during the period 80 to 40 m.y.b.p. Combining the Emperor Seamounts and Hawaiian Islands data, Morgan postulated that 40 m.y.b.p. was a moment of dramatic shift in the direction of relative movement of the Pacific plate lithosphere with respect to asthenospheric mantle beneath the plate.

MIGRATION OF PLATE BOUNDARIES

Relative motions are the basis for evaluating the migration of plate boundaries. Consider, for example, a segment of the Mid-Atlantic Ridge at latitude 25°S, where oceanic floor of the South American plate and the African plate diverge at a **total spreading rate** of 4 cm/yr (Figure 6.31*A*). The **half-spreading rate** of 2-cm/yr describes the movement of the South American

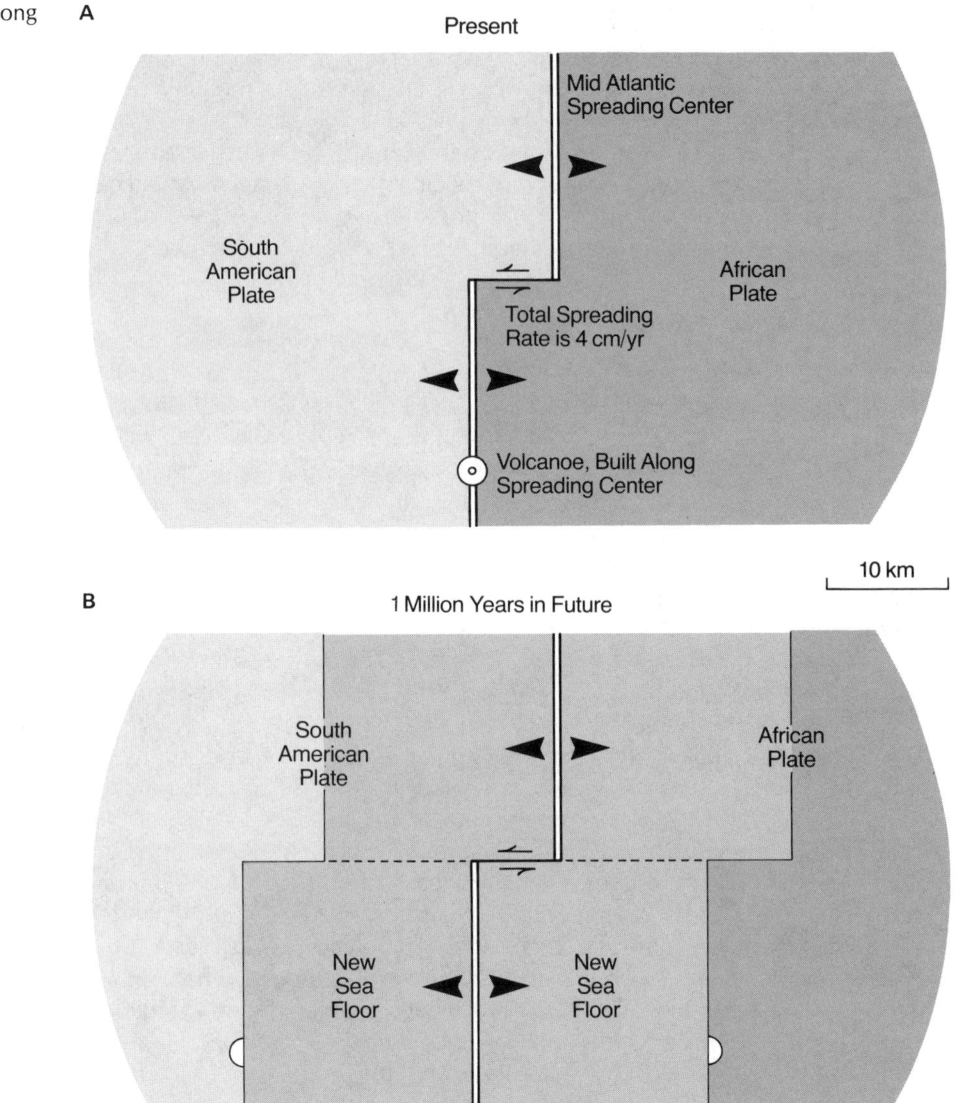

Figure 6.31 Actual relative motion along the Mid-Atlantic Ridge in the South Atlantic. (*A*) Present configuration. (*B*) Configuration 1 m.y. hence.

plate **relative** to the Mid-Atlantic Ridge, imagining the ridge as fixed. Alternatively, the 2-cm/yr velocity might be viewed as the motion of the African plate relative to the Mid-Atlantic Ridge. If a volcanic island that had been constructed along the ridge were "split" by seafloor spreading along the plate boundary, each half would migrate laterally away from the ridge at 2 cm/yr. In 1 m.y., the two halves of the original volcano would be 40 km apart (Figure 6.31*B*).

Let us now change our perspective in viewing this system. Let us fix the westernmost edge of the African plate as it exists today, and then describe the movement of both the Mid-Atlantic Ridge and the South American plate relative to the African plate. To visualize this perspective, consider a "ridge-shaped" tape measure (not available at most hardware stores) that is capable of feeding out steel tape from both sides (Figure 6.32*A*). Consider the emerging tape as new seafloor lithosphere that is fed out at the rate of 2 cm/yr (Figure 6.32*B*). Relative to a fixed African plate, the ridge moves westward at a rate of 2 cm/yr. However, a point in the interior of the South American plate will move at twice that velocity, namely 4 cm/yr, westward from the African plate. In 1 m.y., the ridge will have moved 20 km westward from the present western edge of the African plate. A point on the eastern edge of the South American plate today will move 40 km westward during the next million years, as measured with respect to the African plate.

Relative movements are calculated for convergent plate boundaries as

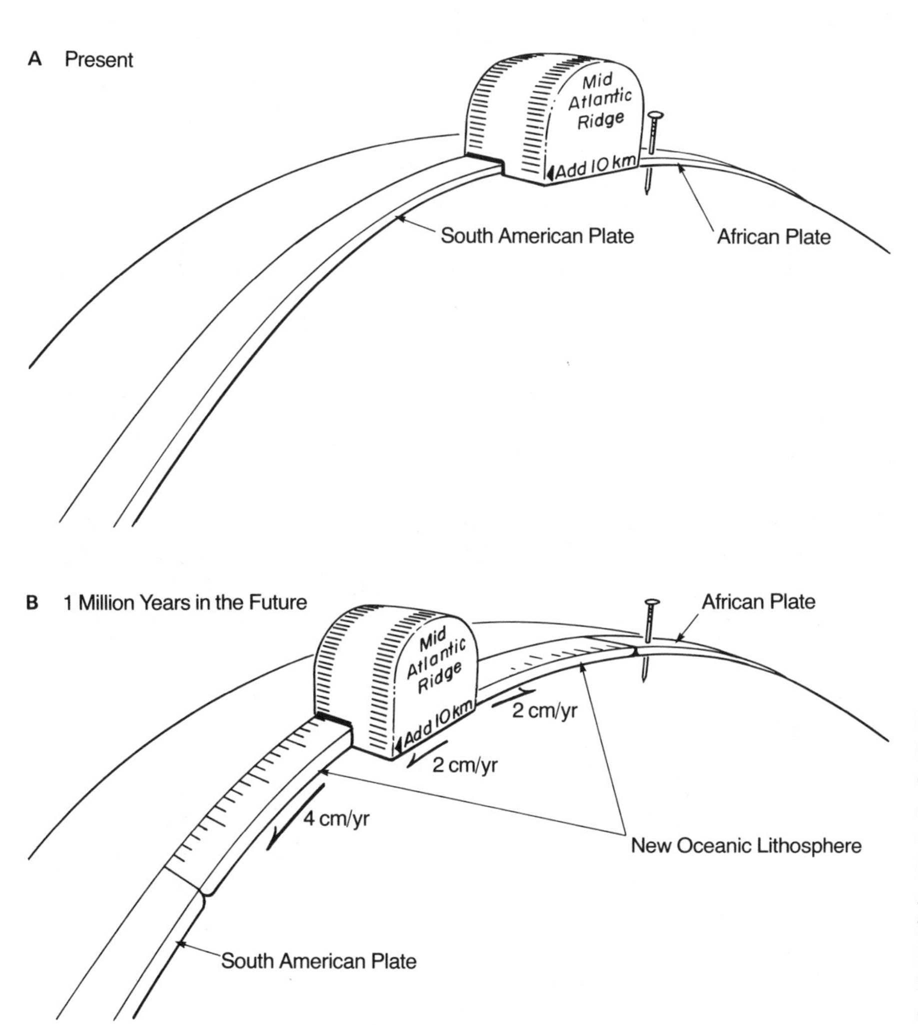

Figure 6.32 Tape measure imagery of relative motions. (*A*) Present configuration of the African and South American plates and the Mid-Atlantic Ridge. (*B*) Configuration 1 m.y. hence. New oceanic lithosphere has been added to the South American and African plates. With respect to a *fixed* African plate, the Mid-Atlantic Ridge moves away at 2 cm/yr and the South American plate moves away at 4 cm/yr.

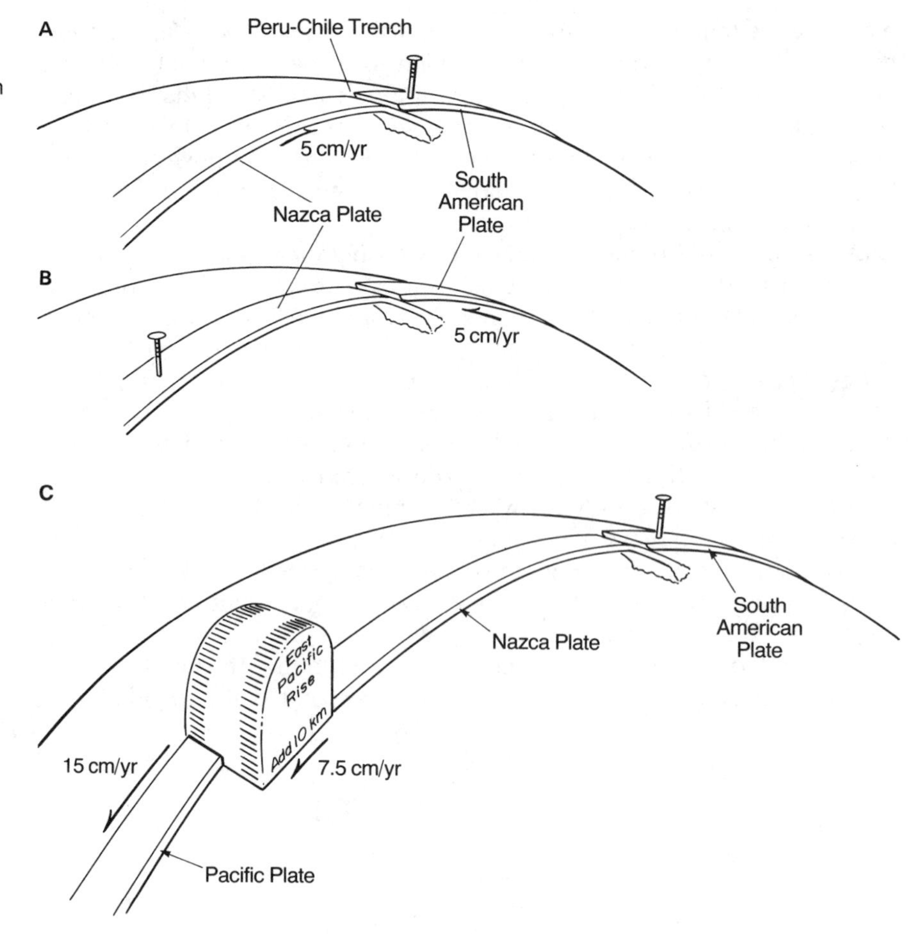

Figure 6.33 (*A*) Relative motion of the Nazca plate with respect to the South American plate (fixed). (*B*) Relative motion of the South American plate with respect to the Nazca plate (fixed). (*C*) Relative motion of the East Pacific Rise and the Pacific plate with respect to the South American plate (fixed).

well, like the tectonically active western continental margin of South America (see Figure 6.26). Pacific seafloor of the Nazca plate is underthrusting the South American plate at an inferred rate of 5 cm/yr. This means that if the Peru–Chile Trench is considered to be fixed in space, a point on the Nazca plate moves toward the trench, eastward, at the rate of 5 cm/yr (Figure 6.33*A*). Another way to look at this relative motion is to hold the Nazca plate fixed and visualize the Peru–Chile Trench moving westward over the Nazca plate at the rate of 5 cm/yr (Figure 6.33*B*). Will the trench overtake the East Pacific Rise? This depends on how rapidly the rise moves westward from the western edge of the Nazca plate. Since the present rate of spreading along that part of the rise is about 7.5 cm/yr, it appears that the rise will outdistance the trench and that the breadth of the Nazca plate will actually expand (Figure 6.33*C*).

EVOLUTION OF TRANSFORMS

With relative motion in mind, we can take the next step in plate kinematics: computing the changes in distance between ridges and/or trenches that are linked by a common transform fault zone. The basis for this kind of analysis was presented by J. Tuzo Wilson (1965). The goal of the exercise is to determine if a given transform fault configuration changes through time and, if so, in what way.

Figure 6.34*A* is a map of the geometry of a ridge-to-ridge transform. New seafloor is being generated at the ridge at a half-spreading rate of 5 cm/year. The ridge segments on opposite sides of the transform fault migrate eastward

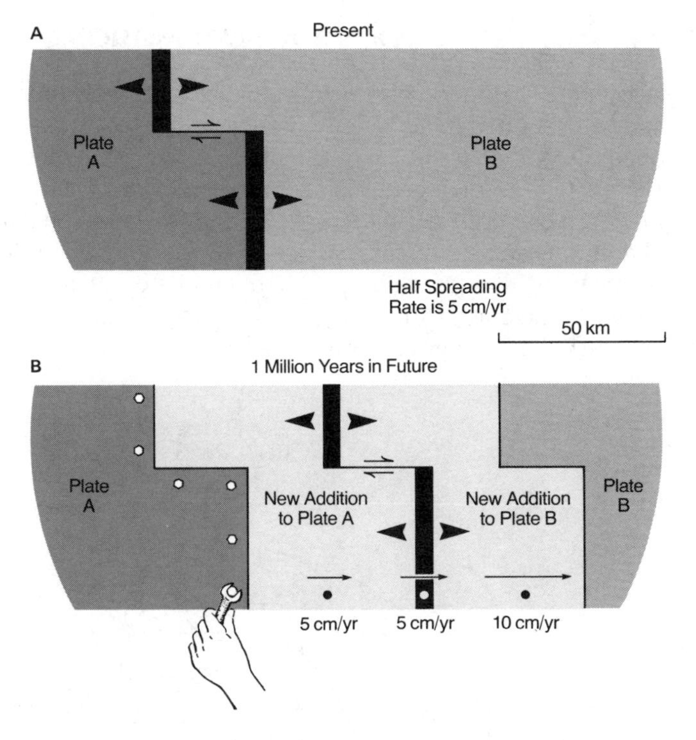

Figure 6.34 (*A*) Ridge-to-ridge transform system. (*B*) Configuration of system remains the same 1 m.y. hence. (From Wilson (1965). Reprinted by permission of *Nature,* v. 207, pp. 343–347, copyright © 1965 by MacMillan Journals Ltd., London.)

at a rate of 5 cm/yr with respect to plate *A*, which is assumed to be fixed. In 1 m.y., each ridge segment shifts eastward a distance of 50 km (Figure 6.34*B*). Even though the total translation is significant, the configuration of the ridge-to-ridge transform does not change.

Such a static configuration does not hold for the ridge-to-trench transform shown in Figure 6.35*A*. The trench in this example is the mouth of an eastward-dipping subduction zone. Holding plate *A* fixed and assuming a spreading rate of 5 cm/yr, the ridge migrates eastward at a rate of 5 cm/yr (Figure 6.35*B*). Plate *B* moves away from the ridge at a rate of 5 cm/yr, and at the subduction zone underthrusts a fixed plate *A* at the rate of 10 cm/yr. The net effect is a steady decrease in the distance between the ridge and the trench, and thus a steady decrease in the length of the ridge–trench transform. The rate of movement of the ridge toward the trench (assumed to be fixed) is 5 cm/yr (Figure 6.35*B*). If the subduction zone dipped westerly, there would have been no change in the configuration between the ridge and the trench. The evolution of ridge–trench transform systems depends critically on the dip directions of the subduction zones that are connected.

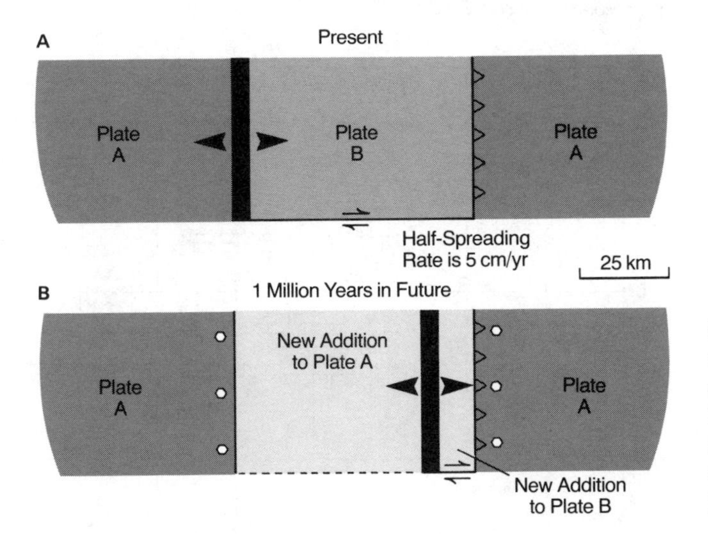

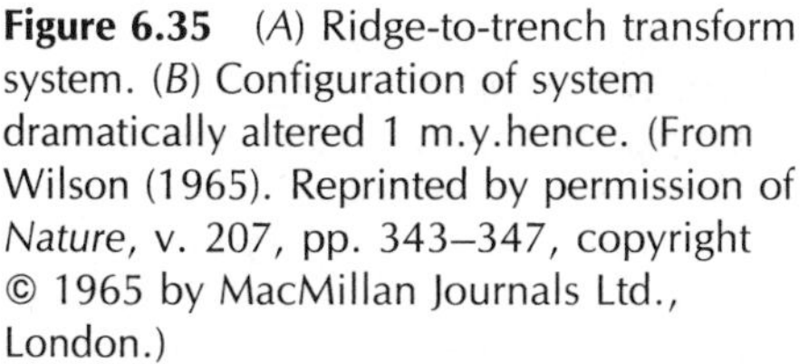
Figure 6.35 (*A*) Ridge-to-trench transform system. (*B*) Configuration of system dramatically altered 1 m.y.hence. (From Wilson (1965). Reprinted by permission of *Nature*, v. 207, pp. 343–347, copyright © 1965 by MacMillan Journals Ltd., London.)

POLES AND PLATE MOTIONS

The relative motion between two plates that meet along a spreading center can be described quite handily. Principles for the constructions were developed by Morgan (1968) and Le Pichon (1968). The rules are elegant in their simplicity. The divergent motions between two plates at any point along a ridge can be described with respect to a pole of rotation (Figure 6.36). A pole of rotation can be thought of as an imaginary point on the surface of the Earth representing the surface projection of an axis of rotation passing through the center of the Earth. Adjacent plates at a spreading center can be thought of as pie-shaped wedges whose tips attach to the pole.

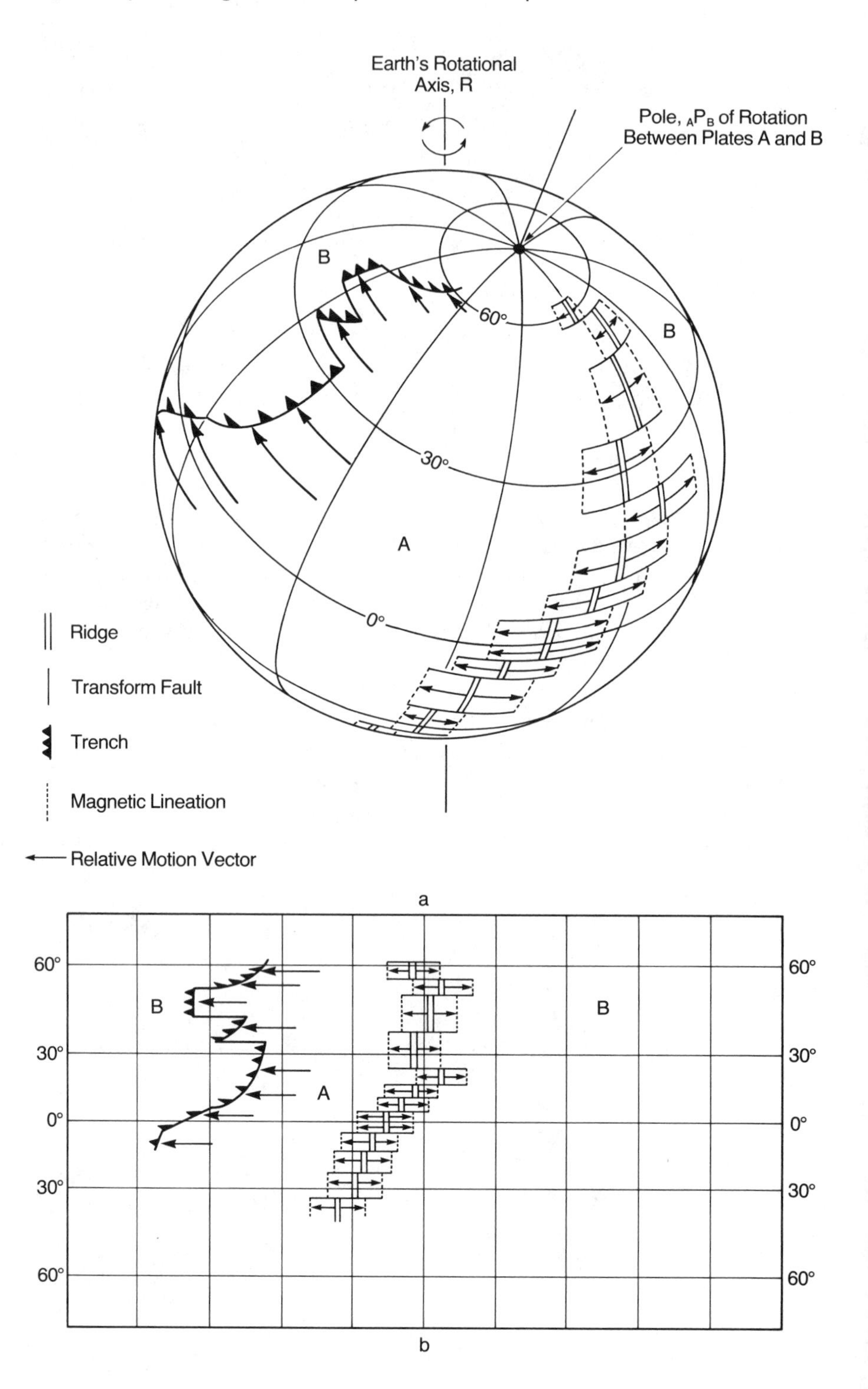

Figure 6.36 Spherical geometric description of the relative movement between two plates. (After *The New View of the Earth* by S. Uyeda, copyright ©1978 by W. H. Freeman and Company, San Francisco. All rights reserved.)

The location of the pole of rotation for any two diverging plates is found by constructing great circles perpendicular to transform fault traces that link the ridge segments (Figure 6.36). The transform faults are small-circle traces of path movement about the pole of rotation whose location is being sought. Great circles drawn perpendicular to the transform fault traces will intersect ideally at the pole of rotation. Defining the exact location of the pole becomes a statistical matter. Small circles drawn about the pole of rotation represent the "tracks" of movement. By combining these kinematics with age dates of seafloor whose sample locations are known, the **average linear velocity** of the relative movement of seafloor from a ridge can be determined. Maximum rates of spreading are found farthest from the pole of rotation; minimum rates of spreading are closest to the pole of rotation. Coupling linear velocity with direction of movement becomes the basis for establishing displacement vectors of movement for any given time period.

It is instructive to portray this geometry stereographically, on an equal-area net. The stereographic projection is used as an **upper hemisphere projection** such that constructions reflect what would be seen on the surface of a globe. Using stereographic projection, it is possible to construct a hypothetical ridge system, complete with ridge-to-ridge transform faults (Figure 6.37*A*), and to evaluate what the ocean ridge system would look like at some time in the future, say 15 m.y. hence. The pole of rotation that describes the relative motion between any two diverging plates, like plates A and B in Figure 6.37*A*, is positioned at the *N* point of the projection. Small circles of the projection describe appropriate orientations of transform faults that would help accommodate the spreading. Ocean ridge segments, in contrast, trace out on great circles.

To evaluate how the plate configuration shown in Figure 6.37*A* would look 15 m.y. in the future, it is necessary to represent the spreading rate not as linear velocity (in km/m.y.), but rather as **angular velocity**. Angular velocity is expressed in degrees per million years. In this example, we assume that the angular velocity (ω) is 1°/m.y.

In representing the spreading stereographically, the individual ridge segments as well as plate B are moved with respect to a fixed plate A (Figure

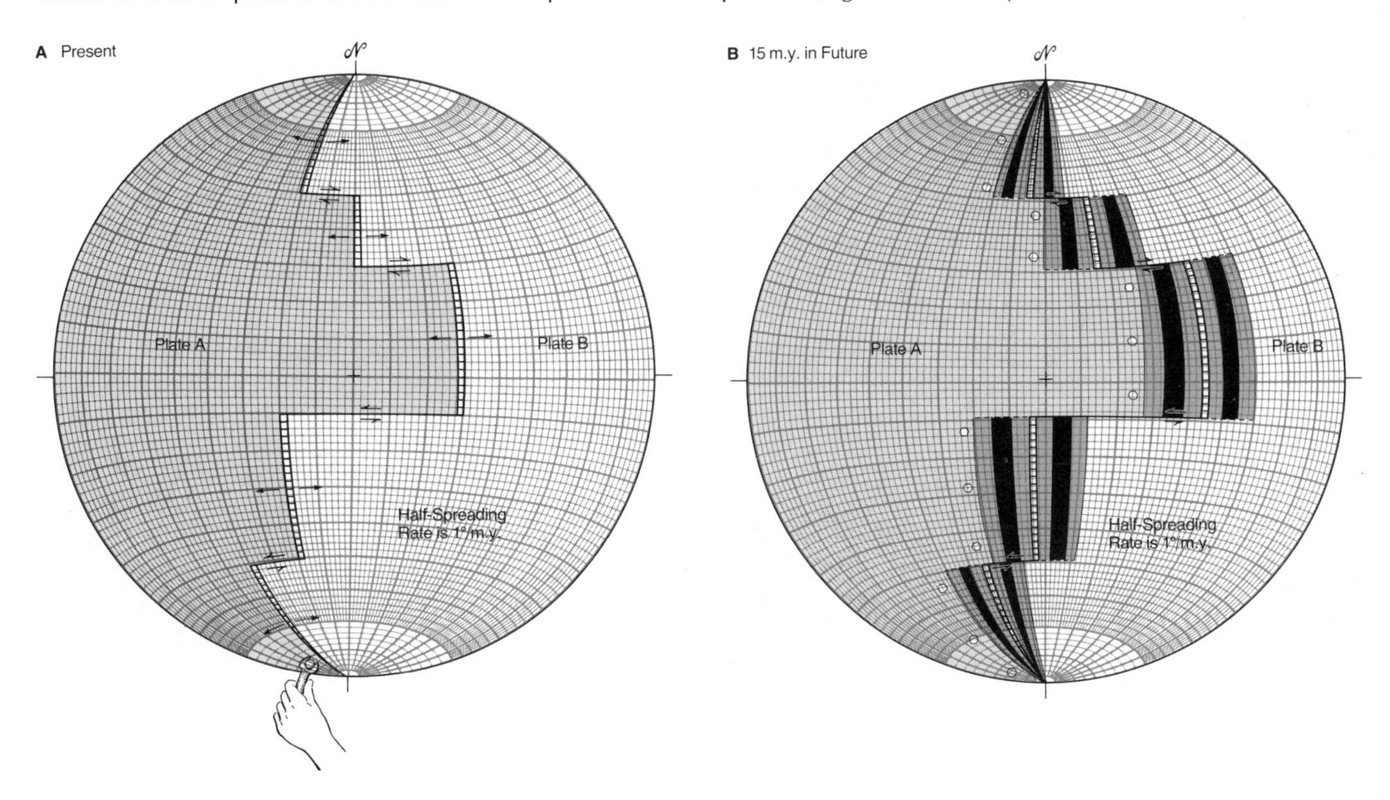

Figure 6.37 Stereographic display of the relative motion between plates. (*A*) Starting configuration of ridge segments and transform faults. (*B*) Configuration 15 m.y. later, assuming a half-spreading rate of 1°/m.y.

6.37*B*). To add interesting detail to the construction, newly generated seafloor is shaded dark or light, depending on the assumed polarity of the Earth's magnetic field during the time of formation of new oceanic crust. For convenience and simplicity, Figure 6.37*B* is drawn such that the polarity of magnetization reverses each 5 m.y. Points in plate B move with respect to A at an angular velocity of 2°/m.y., and the ridge moves eastward from the present-day eastern boundary of plate A at an angular velocity of 1°/m.y. The configuration of magnetic stripes and plate boundaries issues systematically from the movements, with the final pattern frozen in lithosphere in the manner shown in Figure 6.37*B*.

VECTOR DIAGRAMS AND PLATE MOVEMENTS

When we look at the plate configuration map of the world (see Figure 6.26), we are reminded instantly that the computing of plate interactions between just two plates is only a small part of the analysis. It becomes necessary to consider multiple plates and the relative motions between them. **Triple junctions**, points at which three plates come together, become the focus of study. As mentioned earlier, some triple junctions are **ridge–ridge–ridge triple junctions**, like the triple-junction contact between the South American, African, and Antarctic plates (see Figure 6.26). The zone of contact between the Nazca, Antarctic, and Pacific plates is a **ridge–ridge–transform triple junction.** The North American, Pacific, and Juan de Fuca plates meet at a trench–transform–transform triple junction (see Figure 6.26).

The challenge of evaluating triple junctions is twofold: (1) determining the relative motions between any two of the plates that meet at a triple junction, and (2) evaluating the movement of the triple junction and the changes in the configuration of plate boundaries through time. The first of these considerations is most important to our needs. The second is beyond the scope of this presentation, but can be explored in a very challenging and important article by McKenzie and Morgan (1969).

It is possible to measure the relative motion between any two plates that are in contact at a triple junction, provided information is available regarding the relative motions along two of the three boundaries. Facts regarding linear velocity and direction of relative movement are represented as vectors, and **vector circuit diagrams** are constructed to determine relative motion along the third boundary. Vector circuit diagrams are not unlike those we have already constructed in calculating forces and stresses (Chapter 5).

Consider the ridge–ridge–ridge triple junction that brings the Indian–Australian, African, and Antarctic plates into contact (Figure 6.38*A*). The ridges that meet at this triple junction in the Indian Ocean are the Carlsberg Ridge, whose half-spreading rate is 2.5 cm/yr along a N60°E line; the Southeast Indian Ocean Ridge, whose half-spreading rate is 3.7 cm/yr along a N30°E line; and the Southwest Indian Ocean Ridge, whose half-spreading rate and direction we would like to determine. These starting facts describe the movement of the Antarctic plate relative to the Indian–Australian plate, and the Indian–Australian plate relative to the African plate. We want to determine what the movement of the African plate is relative to the Antarctic plate.

To solve this problem, a vector circuit diagram is constructed. The strategy is to map the speed and direction of movement of the Indian–Australian plate with respect to a fixed Antarctic plate, to map the speed and direction of the African plate with respect to a fixed Indian–Australian plate, and then to measure directly from the map the movement of the Antarctic plate with

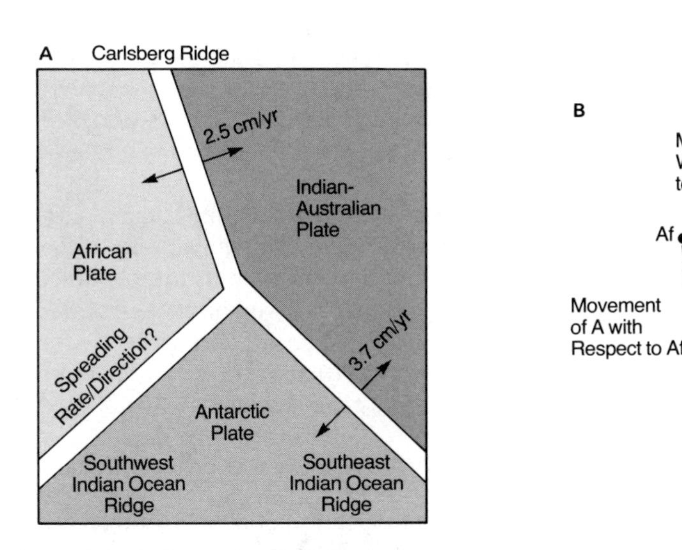

Figure 6.38 Relative motion among plates at the ridge-to-ridge triple junction between the Indian–Australian, African, and Antarctic plates. (*A*) Configuration and half-spreading rates. (*B*) Vector circuit diagram to determine the relative velocity of the Antarctic plate with respect to the African plate.

respect to a *fixed* African plate. Using this approach, the circuit of movement is counterclockwise around the triple junction. (Using a different strategy, a clockwise circuit could be employed.)

The Antarctic plate is fixed in vector space by plotting the point labeled *A* in Figure 6.38*B*. Points in the Indian–Australian plate move according to the total spreading rate (2 × 3.7 cm/yr) from the Antarctic plate along an azimuth of N30°E. After a reasonable map scale has been chosen for the vector diagram, a vector is constructed whose scale length is 7.4 cm and whose trend is N30°E. An arrowhead is placed at the tip of the vector, denoting the direction sense of the movement. At the tip of the arrowhead, the label "I–A," for Indian–Australian plate, is marked. A second vector is constructed, which describes the movement of the African plate relative to the Indian–Australian plate. Its trend is S60°W. Its scaled length of 5 cm is the total spreading rate. Its southwestward tip is marked by an arrowhead labeled "Af" for African plate. Thus having fixed the location of the African plate in the vector circuit diagram, the relative movement of the Antarctic plate relative to the African plate can be directly measured. Using the map scale, it is found to be 3.9 cm/yr, signifying a half-spreading rate of 1.9 cm/yr.

Let us work through a second example. Consider the transform–transform–trench triple junction that marks the common point between the Juan de Fuca, Pacific, and North American plates at the Mendocino triple junction (Figure 6.39*A*). The rate of right-slip transform faulting between the North American and Pacific plates along the San Andreas fault is about 5.6 cm/yr. The rate of right-slip transform faulting between the Juan de Fuca and Pacific plates is 5.8 cm/yr. We would like to determine the nature of the relative motion between the Juan de Fuca and North American plates where they converge along the subduction zone that separates them.

The proper vector circuit can be constructed by moving clockwise from the North American plate about the triple junction. Mark first on the vector diagram the point "NA," symbolizing the North American plate fixed in vector space (Figure 6.39*B*). The movement of the Pacific plate with respect to the North American plate is 5.6 cm/yr. This rate is plotted along a direction that is N30°W, parallel to the trace of the San Andreas fault. The tip of the vector arrow is marked "P," for Pacific plate. Keeping the Pacific plate fixed, we next plot the movement of the Juan de Fuca plate relative to it along the transform fault boundary (the Mendocino fracture zone) (Figure 6.39*B*). Its rate is 5.6 cm/yr. Its direction, parallel to the trace of the transform, is S80°E. The tip of the arrow is marked "JdF" for Juan de Fuca. With these two

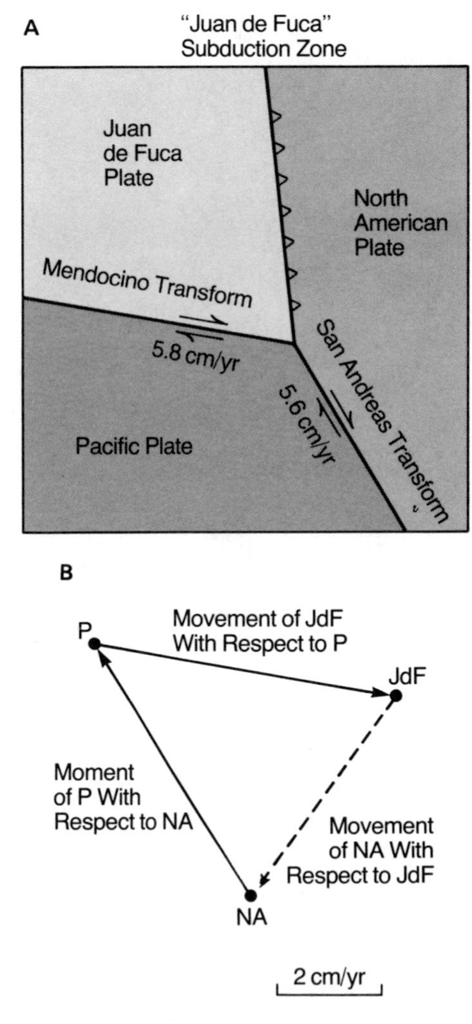

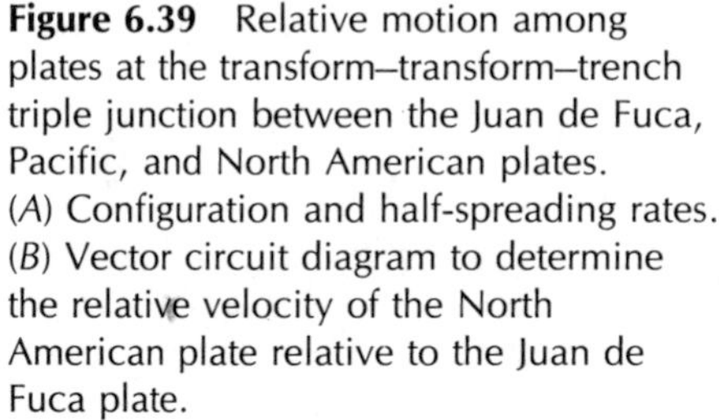

Figure 6.39 Relative motion among plates at the transform–transform–trench triple junction between the Juan de Fuca, Pacific, and North American plates. (*A*) Configuration and half-spreading rates. (*B*) Vector circuit diagram to determine the relative velocity of the North American plate relative to the Juan de Fuca plate.

vectors drawn, the movement of the North American plate relative to the Juan de Fuca plate can be directly measured. Its magnitude is 4.7 cm/yr, and its direction is S35°W. Thus, the vector diagram predicts that seafloor of the Juan de Fuca plate is being subducted under North America at a rate of 4.7 cm/yr.

The construction of vector circuit diagrams requires accurate data on rates and directions of movement. Such data are abundantly available for ridge boundaries, less available for transform boundaries, and difficult to extract from trenches. At trenches, the record that would otherwise provide information on relative motion is swallowed up. Consequently, in practice, subduction rates are derived secondarily from construction of vector diagrams using data describing motions at transform and ridge boundaries.

PLATE MOVEMENTS AND REGIONAL DEFORMATION

Peter Coney (1978, 1981) made a pioneering contribution in relating plate motions to regional deformations. On the basis of hot spots and magnetic anomalies, he reconstructed the position of North America with respect to "Pacific" plates to the west for specified time slots between the Mesozoic era and the present. For each time slot he calculated the relative velocity vector between western North America and the immediately adjacent "Pacific" plate, in the hopes of identifying a correspondence between the relative velocity vectors and the patterns of deformation in western North America. He struck paydirt within the time frame of 80 to 40 m.y., the time of the Laramide orogeny. Coney was able to show that the relative velocity between western North America and oceanic plates to the west was anomalously high (namely, 14 cm/yr) during the interval 80 m.y.–40 m.y. This compares with an average relative velocity of about 7 cm/yr from the early Mesozoic era to the present day. Moreover, Coney's calculations revealed that the 14-cm/yr relative velocity vector was oriented almost perfectly perpendicular to the boundary between the North American plate and the so-called Farallon plate to the west. Plate motion was purely convergent! This too was unique for the interval early Mesozoic–present. Coney concluded that strong, head-on plate convergence generated the stresses that created the fold-and-thrust belts and Rocky Mountain uplifts that characterize the Laramide deformation pattern of western North America.

CONCLUSIONS

Plate tectonics is fundamental to understanding the formation of structures. The geometry and kinematic evolution of structures of the Earth becomes more understandable in the light of plate tectonics, as does the dynamic origin of stresses and forces. In addition, as we have learned, plate motions predetermine the nature of regional rock assemblages and the primary, original structures that develop within them.

II STRUCTURES

chapter 7 CONTACTS

INTRODUCTION

One of the requirements of structural geology is the ability to cross over from one geologic environment and/or plate tectonic setting to another, prepared to carry out detailed structural analysis of any rock system encountered. Geologic opportunities draw us into sedimentary, igneous, and/or metamorphic settings at a moment's notice, opportunities like exploring for oil or metals, mapping geologic quadrangles, analyzing folds, investigating groundwater potential, evaluating geologic hazards, designing underground mines, and—like the geologist astronaut Harrison Schmitt—taking a field trip to the Moon. Most would agree that geologic experience in a rich variety of terranes can stimulate extraordinary personal scientific growth.

It is useful to anticipate in advance the kinds of **contact** one might encounter in different geologic settings and to learn the diagnostic signatures for recognizing each. It is also beneficial to learn to identify **primary structures** and **primary structural relationships** that develop before lithification while the rocks are forming, to be able to distinguish their interpretive messages from those derived from **secondary structures** and **secondary structural relationships** that form after lithification. Contacts are the subject of this chapter. Primary structure is the subject of the chapter that follows.

PERFECT FIT EVERY TIME

Rock bodies are the building blocks of the crust of the Earth. They come in all sizes, shapes, and strengths. They are modules, of sorts, that make up the whole. Through processes of deposition and/or deformation, they generally fit together perfectly along tight contacts. A wonderful nongeological image of the ultimate in perfect fit is seen in the mortarless contacts between limestone blocks in walls of the Inca architectural marvel known as Saqsaywaman, Cuzco, Peru (Figure 7.1).

In nature, rocks may be brought into contact through deposition, intrusion, faulting, and shearing. The fundamental contacts that form are normal depositional contacts, unconformities, intrusive contacts, faults, and ductile shear zones. The ways in which rock bodies fit together is deduced from geologic mapping, supplemented wherever possible by drilling and geophysical data. Map symbols for contacts are shown in Table 7.1.

Table 7.1
Map Symbols for Contacts

Symbol	Description
 52	Normal Depositional Contact, Including Unconformities, Showing Dip
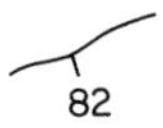 82	Intrusive Contact, Showing Dip (intrusive contacts may be igneous or sedimentary)
 32	Fault Contact, Showing Dip
 46	Major Fault, Showing Dip
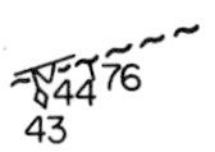 44 76 43	Shear Zone, Showing Strike & Dip of Zone; Strike & Dip of Internal Foliation; & Trend & Plunge of Internal Lineation
 14 15 14 13	Major Shear Zone, Showing Strike & Dip of Zone; Strike & Dip of Internal Foliation; & Trend & Plunge of Internal Lineation

NORMAL DEPOSITIONAL CONTACTS

During normal depositional process, sedimentary layers and/or volcanic layers are deposited on each other **conformably**, forming a sequence of parallel to subparallel beds (Figure 7.2). Conformable deposition of one layer on another takes place during continuous deposition, or during separate events so close in time that the age difference(s) between the younger

Figure 7.1 Limestone blocks in Inca wall fit together perfectly along mortarless contacts. Windowlike opening is about 1 m high. Like many geologic contacts, these man-made contacts are difficult to interpret: What processes were required to bring the limestone blocks into perfect contact? (Photograph by L. A. Lepry.)

and older layers cannot be detected with our existing timepieces. Contacts formed between layers in a conformable sequence are referred to as **normal depositional contacts**. Normal depositional contacts are usually planar to slightly irregular in form. Rock layers astride normal depositional contacts are distinctive because of properties such as color, composition, lithology, texture, and thickness.

Normal depositional contacts can occur as **intraformational contacts** within formally designated rock formations, or they may be **interformational contacts** that mark the interface between two formations. Intraformational contacts are not usually shown on geologic maps unless they delineate an especially distinctive and useful marker bed.

In structurally complicated areas, the recognition of contacts as normal depositional contacts is sometimes achieved in a roundabout way—by noting the absence of the characteristic physical signatures of unconformities, intrusive contacts, faults, and shear zones.

UNCONFORMITIES

GENERAL NATURE

An **unconformity** is a depositional contact between two rocks of measurably different ages. Unconformities are special because they are surfaces that represent globs of time during which either there was no deposition of

Figure 7.2 Normal depositional contacts within rhyolite ignimbrite in the barranca country of Sinaloa, Mexico. Contacts are beautifully exposed in this region, the most deeply dissected part of the Sierra Madre Occidental. The volcanic rocks are part of a magmatic arc, approximately 400 × 1500 km in surface area. The volcanics of the arc erupted almost entirely during Miocene time. (Photograph by D. J. Lynch.)

rock-forming material, or there was erosional removal of the entire rock record for the interval of time represented by the unconformity.

TYPES OF UNCONFORMITY

Unconformities are divided into three major classes: nonconformities, angular unconformities, and disconformities (Figure 7.3). **Nonconformities** are depositional surfaces separating younger sedimentary or volcanic rocks from underlying older crystalline rocks (i.e., metamorphic and/or intrusive igneous rocks). An **angular unconformity** is an unconformity that separates sedimentary and/or volcanic strata whose attitudes are discordant to each other. The term "angular" expresses the fact that layers above and below angular unconformities are not parallel. A **disconformity** is an unconformity separating strata that are parallel to each other.

Nonconformities are common in foreland and active continental margin settings. They are also the fundamental contacts that separate platform sediments from underlying basement in stable continental interiors. Nonconformities reflect a geologic history starting with intrusion and/or metamorphism at relatively deep levels in the crust, followed by vertical uplift, erosional uncovering of the crystalline rocks, and finally, deposition of sediments and/or volcanics onto the erosional surface. Nonconformities include the **great unconformities** that separate Precambrian crystalline basement rocks from much younger overlying sedimentary or volcanic strata (Figure 7.4*A*). But they also include the unconformities separating post-Precambrian crystalline rocks from overlying, younger, sedimentary and/or volcanic strata (Figure 7.4*B*). Standing on such a nonconformity is something special, particularly if the discontinuity between cover and crystalline rock represents 2 or 3 b.y. of Earth time.

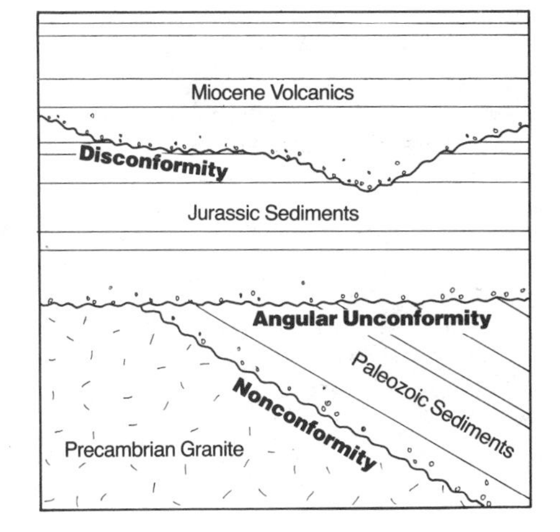

Figure 7.3 Schematic portrayal of the three kinds of unconformity: nonconformity, angular unconformity, and disconformity.

Classical angular unconformities are horizontal depositional surfaces separating young horizontal strata above from old steeply dipping strata below. A Grand Canyon example features tilted 1 b.y. old Precambrian sediments unconformably overlain by 500 m.y. old, flat-lying Cambrian sediments

Figure 7.4 (*A*) Nonconformity between flat-lying Cambrian Tapeats Sandstone (Єt) and underlying Precambrian crystalline rocks (PЄ), as exposed in Granite Gorge of the Grand Canyon. (Photograph by N. W. Carkhuft. Courtesy of United States Geological Survey.) (*B*) Nonconformity between mid-Tertiary andesitic volcanic rocks (black) and underlying early Tertiary Gunnery Range Granite, southwestern Arizona. (Not all contacts are plain as black and white.) (Photograph by D. J. Lynch.)

(Figure 7.5). The geometrical properties of angular unconformities can be interpreted as follows: deposition of older beds; rotation of the layers to shallow, moderate, steep, or overturned dips by folding and/or faulting; beveling of the upturned ends of such beds by erosion; and deposition of the younger beds. Histories like this are commonplace within active continental margin settings.

Spectacular angular unconformities are sometimes displayed in single outcrops or cliff exposures. In contrast, subtle angular unconformities marked by very slight angular discordance can be "seen" only through careful regional mapping. Consider a regional assemblage of sedimentary rocks tilted at an angle of just 4° (Figure 7.6*A*). If this regional rock unit is beveled by a 1°-dipping erosional surface, the resulting 3° of angular discordance at the unconformity usually would not be discernible at the outcrop scale of view (Figure 7.6*B*). However, the angular discordance would be conspicuous when seen in the perspective of tens of kilometers of viewing

Figure 7.5 Angular unconformity between flat-lying Paleozoic strata and underlying tilted upper Precambrian strata, exposed in a section of wall in the Grand Canyon of the Colorado River. [From Powell (1875). Courtesy of United States Geological Survey.]

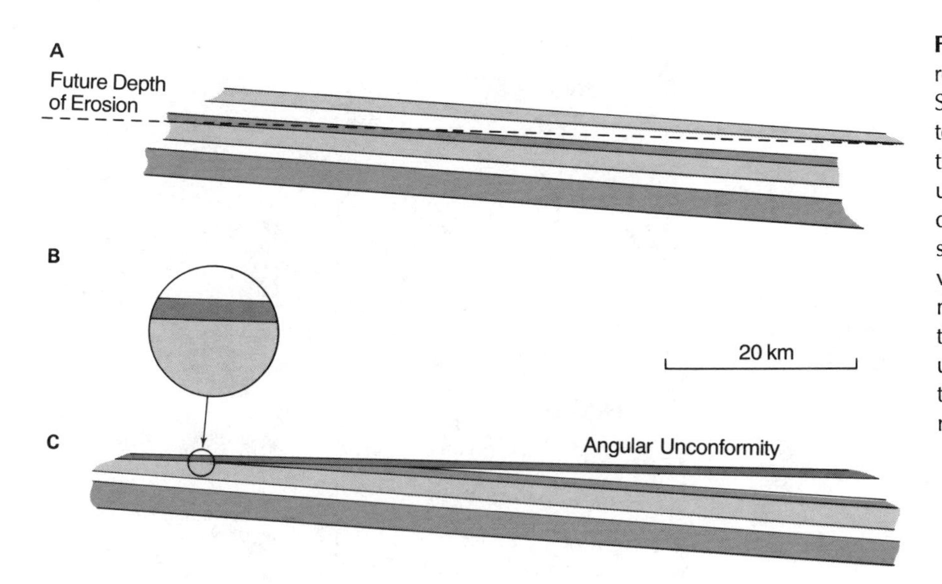

Figure 7.6 Development of a subtle regional angular unconformity. (*A*) Sedimentary sequence tilted at 4°, about to be beveled by erosion to the depth of the 1°-dipping reference line. (*B*) Angular unconformity looks like normal depositional contact when only a tiny segment of the regional relationship is viewed. (*C*) Regional view shows that a number of the 4°-dipping units beneath the unconformity pinch out at the unconformity, thus drawing attention to the existence of the subtle unconformable relationship.

distance (Figure 7.6*C*). Units below the unconformity would pinch out against it regionally. The younger strata would **overlap** the older strata. Such unconformable relationships are common in foreland terranes and in parts of miogeoclinal prisms.

The physical presence of a disconformity may be hard to detect. Recognition requires complete knowledge of the ages of beds within the sequence of strata that contains the disconformity. In the Paleozoic geologic column of the Grand Canyon, Silurian rocks are completely missing and thus Devonian strata, and more often Mississippian strata, rest disconformably on the Upper Cambrian Muav Limestone.

The scenario of interpretation of disconformities involves deposition of older strata in a horizontal or subhorizontal attitude, nondeposition or continued deposition of strata followed by erosional removal of same, and deposition of younger strata on the surface of nondeposition or erosion. Platform deposits of stable craton interiors are riddled with disconformities, as are parts of miogeoclinal prisms.

OUTCROP CHARACTERISTICS

Nonconformities and intrusive igneous contacts can sometimes be confused. Angular unconformities and fault contacts likewise can be misinterpreted. Disconformities can be completely overlooked because they lack geometric expression. Chances for misinterpretation are enhanced by poor outcrop exposure and complex structural overprinting. In their raw unblemished state, however, unconformities possess a number of outcrop-scale physical and geometric properties that aid in their identification.

The strata above an unconformity commonly feature a **basal conglomerate**, normally composed of clasts of the rock below the unconformity. Figure 7.7*A* pictures a nonconformity separating Precambrian granite from Cambrian Bolsa Quartzite near the sleepy town of Dos Cabezas in southeastern Arizona. The basal conglomerate immediately overlying the unconformity is spectacular, composed of cobbles of resistant quartz and quartzite from the Precambrian foundation (Figure 7.7*B*). Basal conglomerates advertise erosional intervals. The basal conglomerate may range in coarseness from a thin granule conglomerate to a thick boulder conglomerate.

Figure 7.7 (*A*) Nonconformity between basal conglomerate of the Cambrian Bolsa Quartzite overlying 1.4 b.y. old Precambrian granite. (*B*) Dash-line representation of the exact trace of the nonconformity. (*C*) Close-up view of the basal conglomerate. (Photographs by G. H. Davis.)

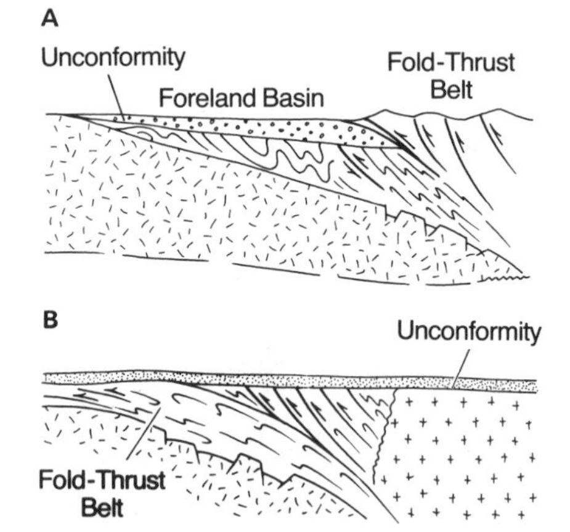

Figure 7.8 (*A*) Folding and thrusting took place after the deposition of the youngest strata beneath the unconformity, and before the deposition of the oldest strata above the unconformity. (*B*) Accretion followed the deposition of the youngest rock deformed in the suture zone, and preceded the time of deposition of the oldest sediment that covers the suture zone unconformably.

Surfaces of unconformity locally possess **topographic relief** that can be recognized as the product of ancient erosion, perhaps including **channeling** (see disconformity in Figure 7.3). Under ideal conditions, fossil soil profiles, called **paleosols**, are preserved along unconformities. These may be baked where overlain by lava flows. Physical characteristics like boulder conglomerates, channels, and paleosols, together with regional map relationships of rock distributions, give clues to the presence of unconformable relationships.

TECTONIC SIGNIFICANCE

Unconformities have important tectonic implications. An angular unconformity separating folded and thrusted strata below from a foreland basin deposit above places constraints on the timing of the crustal contraction (Figure 7.8*A*). The timing of folding and thrusting is bracketed by the age of the youngest rock below the unconformity and the oldest rock above it. Where parts of island arc mountain systems have been accreted to active continental margins (Figure 7.8*B*), the timing of the accretion may be bracketed by the age of the youngest rocks accreted in the suture zone and the age of the oldest rocks resting in angular unconformity on the sutured mass.

Buttress unconformities commonly form in regions of tectonic unrest where faulting produces scarps and fault-bounded basins (Figure 7.9). Rapid deposition of clastic materials results in complete filling of the fault-bounded basins. Unconformities that emerge in the geological column from such a sequence of events are marked by unusually high topographic relief. Furthermore, the basinal deposits directly above the unconformity may show abrupt facies changes. Ideal settings for the formation of buttress unconformities are rift basins and interarc basins.

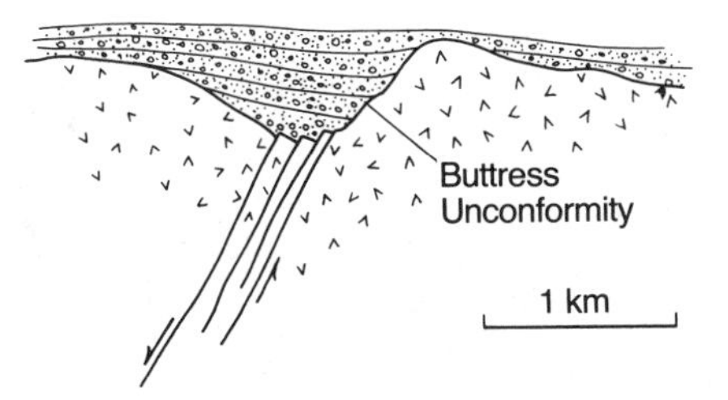

Figure 7.9 Buttress unconformity. Faulting transforms a flat landscape into one marked by high topographic relief and steep slopes. Coarse clastic materials are rapidly deposited in topographic "lows." Freshly deposited sedimentary units laterally abut against the older bedrock beneath hill slopes and mountain slopes of the former topography.

Important unconformities are now being recognized offshore in the subsurface geology of passive continental margins. On the Atlantic seaboard of the eastern United States, seismic reflection profiles reveal preservation of some of the structural–geologic movements that accompanied rifting of North America from Europe by seafloor spreading. In particular, deep parts of the profiles record extensional faults and fault blocks whose geometry corresponds to products of rifting (Figure 7.10). Basins created by the faulting are filled by sediments whose thicknesses vary according to the fault block geometry. Salt evaporites, which formed in the basins, are seen to have risen buoyantly as domes and pillars of salt. The top of the entire assemblage of faults and rift basins is truncated abruptly by a subhorizontal surface of unconformity, above which are concordant layers of younger sediment. Such an unconformity has come to be known as a **breakup unconformity**. Timing of the breakup of the supercontinent Pangaea is constrained by ages of rocks above and below the unconformity.

GEOMETRY

Unconformities can retain their original orientation, or they can be rotated and/or distorted as a result of later movements. The names of unconformities remain the same regardless of the absolute orientation of the unconformity.

To prepare accurate descriptions of the structural geometry of unconformities, it is necessary to compare and contrast relative orientations of rocks above and below the unconformity, and to measure its orientation. This is usually easy to do at the outcrop scale. The respective orientations of bedding and the contact of unconformity can simply be measured with a compass. However, unconformities are regional features, and orientation data collected at a single outcrop describe only a small part of the total geometric picture. To determine the regional attitude of an unconformity, it is beneficial to represent the surface of unconformity in the form of a structure contour map. Petroleum exploration geologists prepare such maps routinely.

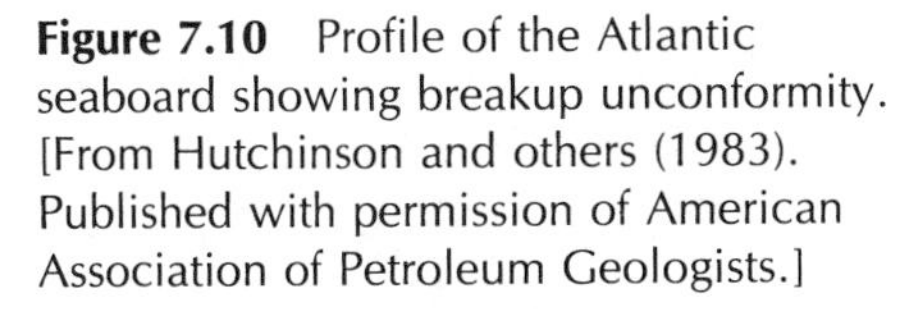

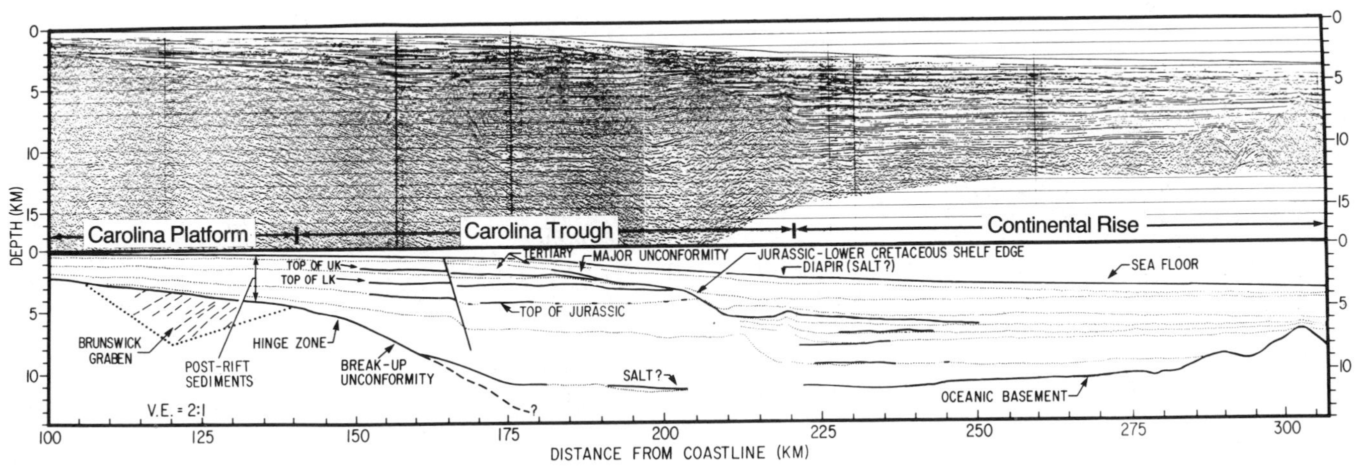

Figure 7.10 Profile of the Atlantic seaboard showing breakup unconformity. [From Hutchinson and others (1983). Published with permission of American Association of Petroleum Geologists.]

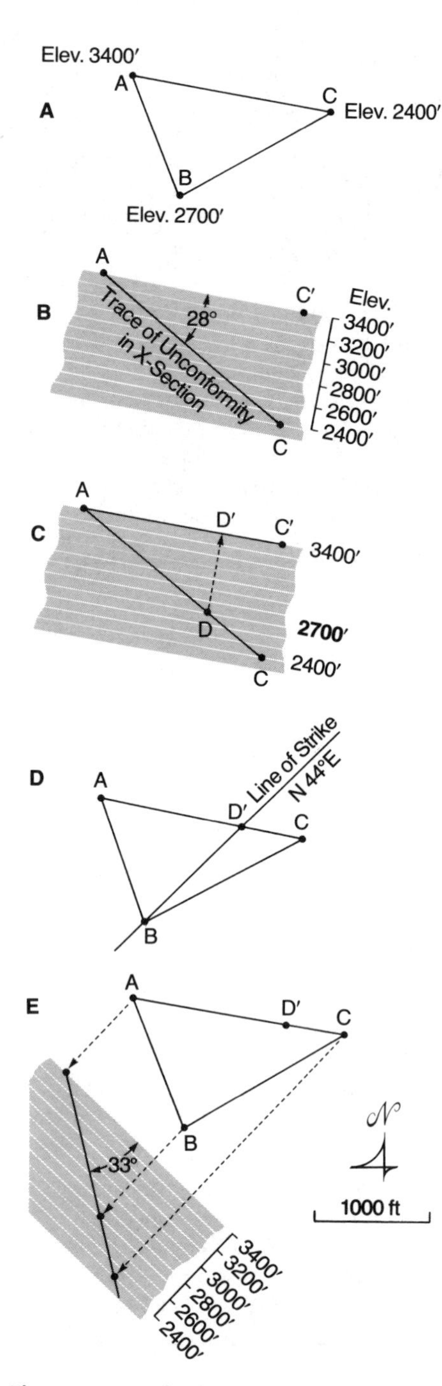

Figure 7.11 Three-point problem. (*A*) Starting data: locations and elevations of three points on a common plane, in this case a planar unconformity. (*B*) Trace of unconformity as seen in cross section through points *A* and *C*. (*C*) Identification of point *D* on the trace of unconformity when elevation is the same as point *B*. Identification of vertical projection of point *D* to the line of section *AC*. (*D*) Line of strike connects points *B* and *D*! (*E*) Determination of true dip of unconformity.

In some studies, it may be appropriate to assume that a specific unconformity is planar within a given area. If such an assumption is reasonable, the average strike and dip of the unconformity over a relatively large area may be determined through an orthographic construction known as a **three-point problem** (Figure 7.11). Solving a three-point problem requires elevation control for at least three points *that lie on a common plane,* in this case the surface of unconformity. The elevation control may be derived from topographic maps used in conjunction with geologic maps and/or subsurface drilling information. The three control points used in three-point constructions should define a triangular array of relatively widely spaced points, which mark different elevations.

Consider points *A*, *B*, and *C*, at elevations of 3400, 2700, and 2400 ft (1030, 818, and 727 m), respectively (Figure 7.11*A*). Each point lies on the angular unconformity whose attitude is sought. Point *B* is intermediate in elevation between points *A* and *C*. The trace of the unconformity as seen in a cross section passing through points *A* and *C* is constructed in vertical profile (Figure 7.11*B*). It is inclined at an apparent dip of 28° toward *C*. Somewhere along its trace is a point whose elevation is the same as *B*. The location of this point (*D*) is found by constructing a horizontal reference plane whose elevation is the same as that of *B* (Figure 7.11*C*). The intersection of the trace of the unconformity with the reference plane is projected vertically to *D′*, which represents the vertical projection of the point on the unconformity whose elevation is 2700 ft. By connecting *B* and *D′* (i.e., points of equal elevation on the angular unconformity), the line of strike is defined (Figure 7.11*D*). Its trend is N44°E.

To determine the dip of the unconformity, it is necessary to construct a structural profile at right angles to the line of strike, and to project the trace of the unconformity into this plane. Such a profile is shown in Figure 7.11*E*, with elevations projected into it from control points *A*, *B*, and *C*. The inclination of the unconformity in this special profile is a measure of true dip, namely 33°SE.

Three-point constructions are not restricted to determining the strike and dip of unconformities. The three-point method can be used effectively to establish the strike and dip of any extensive planar structure, like a bed, formation, fault, or intrusive contact.

INTRUSIVE CONTACTS

TYPES OF INTRUSION

Intrusive contacts form where viscous liquids or rheids move past the country rocks through which they flow. The term **country rock** refers to the rock assemblage, whatever its nature, that receives the intruder. The intrusive contact proper is the interface between country rock and the **intrusive body**.

"Intrusive contact" generally brings to mind the **igneous intrusion** of a hot viscous fluid (i.e., **magma**) into country rock. However intrusions may be igneous *or* sedimentary. The term **diapir** is used to describe any body that has been able to flow in the fluid or rheid state and to intrude surrounding country rock.

Igneous magmatic intrusions are responsible for fashioning the most commonly encountered intrusive contacts in the Earth's crust (Figure 7.12*A*). Country rock intruded by magma may include not only sedimentary, volcanic, and metamorphic rocks, but intrusive igneous rocks as well. Igneous magma forms a rock body that is younger than the country rock it intrudes.

Sedimentary intrusions are of two types: **soft-sediment intrusions** and **salt diapirs**. Soft-sediment intrusions involve the squeezing and/or buoyant rise of buried but yet-unconsolidated water-rich muds and sands into adjacent or overlying country rock (Figure 7.12*B*). The mud and sand intrusions typically originate from within **source beds** marked by **high fluid pressure**. Salt diapirs are domes, pillars, and walls of salt that buoyantly rise as rheids from thick beds of evaporites into overlying sedimentary country rock (Figure 7.12*C*). Sedimentary intrusions are generally made of rocks that are older than those they penetrate.

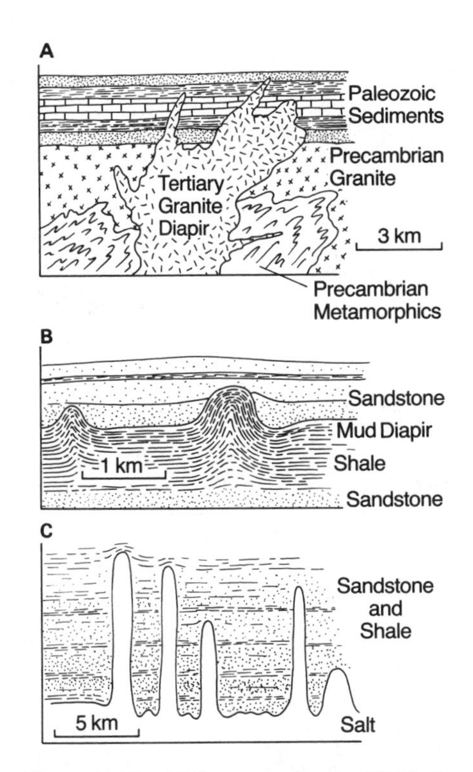

Figure 7.12 (*A*) Igneous diapir. (*B*) Shale (mud) diapir. (*C*) Salt diapirs.

SOME CHARACTERISTICS OF INTRUSION

Intrusive contacts may be recognized on the basis of a number of features. Intrusions in general, whether they be of igneous or sedimentary origin, commonly send **apophyses** of irregular tonguelike injections into country rock (Figure 7.13). **Dikes** and **sills** may intrude the country rock as well. **Dikes** are tabular injections that cross-cut layering or foliation in the surrounding wall rock (Figure 7.14). **Dikelets** are small dikes (Figure 7.15). **Sills** are sheetlike intrusions that invade country rock parallel to layering or foliation (Figure 7.16). Tabular intrusions of mudstones and sandstones are referred to as **clastic dikes** (Figure 7.17), or **clastic sills**, depending on their relation to layering in the country rock they intrude. Igneous dikes sometimes occur as **dike swarms** of subparallel, usually vertical members (Figure 7.18), or as **radial dikes** that spray outward in all directions from volcanic centers of eruption. **Ring dikes**, concentric dike-filled fractures, play an important role in accommodating the foundering or collapse of volcanic edifices.

Pieces of fractured country rock, known as **inclusions** or **xenoliths**, may become detached from the wall of contact during intrusion to become incorporated within the main igneous or sedimentary intrusion or within apophyses branching from it. Some inclusions are in fact composed of older,

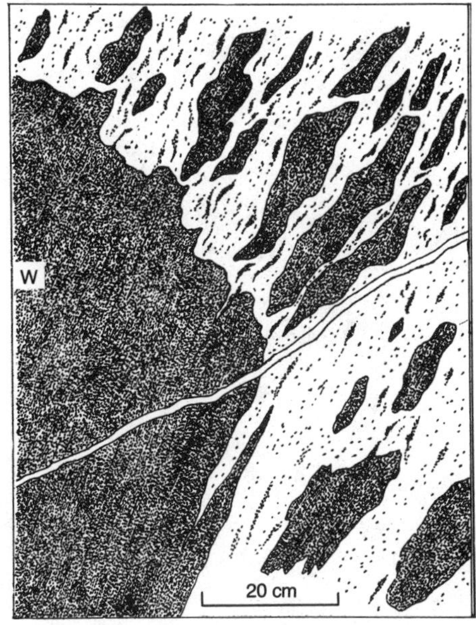

Figure 7.13 Tonguelike apophyses of quartz monzonite (white) invade faintly foliated diorite (dark), Sierra Nevada, California. (Field sketch by E. B. Mayo.)

Figure 7.14 Great dike running north from West Spanish Peak, Colorado. The face of the dike shows cast of horizontal bedding planes of enclosing Eocene strata. (Photograph by G. W. Stose. Courtesy of United States Geological Survey.)

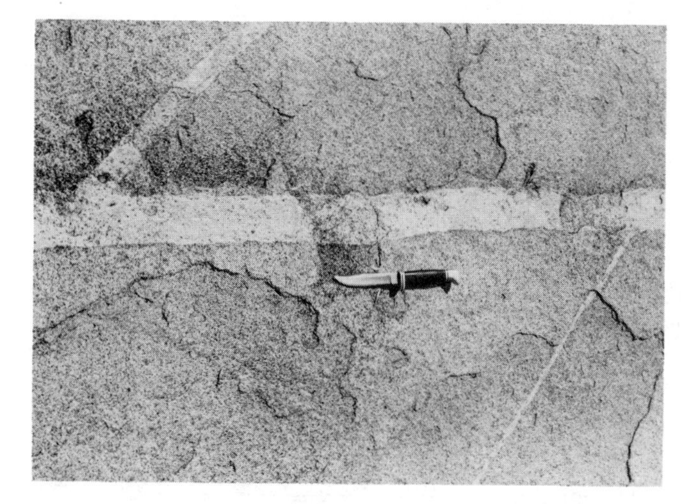

Figure 7.15 Aplite dikelets in granite. (Photograph by G. H. Davis.)

Figure 7.16 Aplite sills in foliated granite. Note that the sills are parallel to the elongate dark inclusions in the granite. (Photograph by G. H. Davis.)

Figure 7.17 (*A*) Clastic dikelet of Jurassic mudstone (dark) cutting through Permian limestone. (Photograph by G. H. Davis.) (*B*) Clastic dikes cut through horizontal strata in the Badlands of South Dakota. (Photograph by R. W. Krantz.)

congealed parts of the intruding mass. Inclusions come in all sizes and shapes (Figure 7.19). The injection of apophyses, dikes, and sills into country rock and the presence of inclusions of country rock within the intrusion are clues that shed light on relative age relationships.

Figure 7.18 Dike swarm of pegmatites intruding quartz diorite near Mount Lemmon in the Santa Catalina Mountains, southeastern Arizona. Geologist is Evans B. Mayo. (Photograph by G. H. Davis.)

Figure 7.19 Some examples of inclusions in igneous rocks. (*A*) Dark mafic inclusions in granite. (Photograph by G. H. Davis.) (*B*) Angular inclusions in granite. (Photograph by E. B. Mayo.)

THERMAL EFFECTS OF IGNEOUS INTRUSIONS

The thermal effects of igneous intrusion add signatures for contact recognition that are absent along the margins of sedimentary intrusions. Country rocks invaded by magma respond to heat by recrystallization and metamorphism. Sedimentary rocks, unaccustomed to hot environments, are particularly vulnerable to thermal alteration. Igneous intrusions that penetrate country rocks at high, relatively cool levels in the crust may impart to the wall rocks a local **contact metamorphism** that produces a restricted halo or **aureole** of metamorphism in wall rocks along the contact. The width of the aureole, which may range from meters to several kilometers, is a function of the intensity of the thermal gradient and the favorability of the wall rocks to recrystallization. Ore deposits are commonly found in such settings. The intrusive magma itself may quickly cool upon contact with the wall rock, forming a **chill zone** of very fine-grained igneous rock at the border of the igneous body.

At relatively deep levels in the Earth's crust, the emplacement of magma may contribute to regional metamorphism of the country rock. Chill zones ordinarily are not present in such environments, nor are discrete contact

Figure 7.20 Dark roof pendant of metamorphosed sedimentary rock surrounded by granite of the Sierra Nevada. (Photograph by E. B. Mayo.)

metamorphic aureoles. Nevertheless, the intrusive bodies commonly send aphophyses and dikes into wall rocks, and inclusions may become bound up in the magma.

It is not at all uncommon in igneous intrusive environments to map large rootless masses of metamorphosed country rock that project down and into the intrusive body (Figure 7.20). **Roof pendants**, as they are called, provide an unusual stop-action glimpse of the ascent of magma into country rock. Likewise, **screens** or **septa** are large tabular walls of country rock that serve to separate two intrusive igneous bodies, or the lobes of a common body.

PROBLEMATIC INTRUSIVE CONTACTS

In arc terranes, magmas may intrude their own volcanic pile. Where this has happened, recognition of intrusive contacts may be very difficult, especially where the intrusive rock is relatively fined grained, altered, poorly exposed, and compositionally similar to its volcanic counterpart. The problem is compounded if intrusion is concordant with volcanic layering, in which case sills may be very difficult to distinguish from lava flows. Ideally, a sill should be characterized by *chill zones* along both its upper and lower contacts. In contrast, volcanic flows ought to display clear signs of having moved across a former topographic surface. Messy arc terranes require detailed geologic mapping coupled with careful radiometric isotopic analysis.

In some cases the contact between two intrusive igneous rocks is difficult to interpret because of contradictory field relationships. The mapped array of apophyses and inclusions may suggest that each igneous body has intruded the other. Field relationships of this sort could suggest that both magmatic bodies are of the same age or, alternatively, that the thermal effects of a young igneous intrusion resulted in the partial melting and mobilization of part of the older igneous rock. Contacts of this type are head-scratchers. To evaluate them properly requires geologic mapping, isotopic analysis, detailed petrographic study, and a fundamental knowledge of igneous petrology.

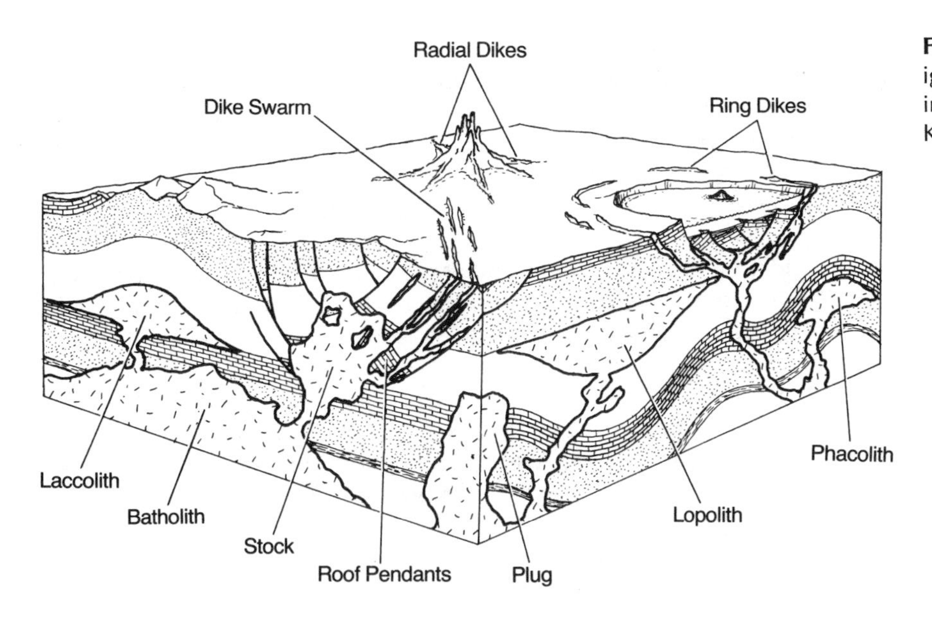

Figure 7.21 Index to names of some igneous intrusive bodies and igneous intrusive structures. (Artwork by R. W. Krantz.)

SHAPES OF INTRUSIONS

INCENTIVE FOR ANALYZING SHAPES OF INTRUSIONS. Geologic mapping coupled with subsurface exploration has demonstrated that the geometry of intrusive contacts is commonly much different from the geometry of contacts of deposition, faulting, and shearing. Inasmuch as the shape of an intrusive contact reflects the shape of the body of which it is a part, knowledge of the common three-dimensional forms of intrusive bodies is very helpful in the interpretation of contacts. Analysis of the form(s) of intrusive contacts is of more than academic interest. Given the common occurrence of mineral deposits along igneous intrusive contacts and the trapping of oil and gas near sedimentary intrusive contacts, it is economically rewarding to learn to construct reasonable interpretations of the shapes and positions of intrusive contacts in the subsurface.

SHAPES OF IGNEOUS INTRUSIONS. An established vocabulary aids in describing the common forms of igneous intrusions (Figure 7.21). **Pluton** is the name given to a relatively large body of intrusive igneous rock, exclusive of simple dikes and sills. Plutons whose contacts are generally parallel to the layering and/or foliation of country rock are known as **concordant intrusions**. Plutons whose contacts cut across and thus truncate the layering and/or foliation of country rock are considered to be **discordant**. Most plutons are in part concordant and in part discordant to the internal structure of the country rock that surrounds them.

The terms **plug** and **stock** are commonly used in reference to a largely discordant pluton whose exact form is either unknown or is inferred to be a steep-sided cylinder or a moderately steep-sided, bulbous dome or cone (Figure 7.22). The distinction between "plug" and "stock" is mainly one of size: plugs are generally less than approximately 0.25 km^2 in area as seen in map view; stocks are larger, but less than about 40 km^2 in map view. Plutons greater than about 40 km^2 in map-view area are called **batholiths**. Although locally discordant to the internal structure of country rock, batholiths are viewed regionally as enormous horizontal, sheetlike intrusions that are broadly concordant to the structure of the crust they intrude (Hamilton and Myers, 1967). Enormous batholithic complexes occupy the western margins

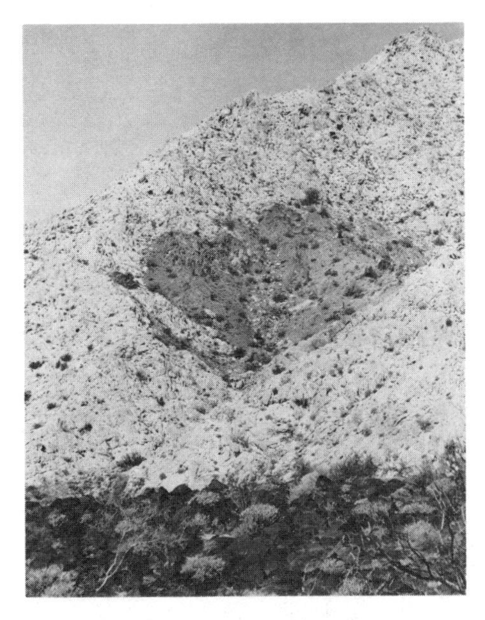

Figure 7.22 Plug of hornblende andesite (dark) intruded into granite (white) in southwestern Arizona. (Photograph by D. J. Lynch.)

of North and South America as fossil magmatic arcs. They include the Coast Range batholith of Canada, the Sierra and Southern California batholiths of western United States, and the Andean batholith complex of South America.

Mapping of contacts has revealed that there are large concordant intrusive bodies that are not strictly sill-like in form (see Figure 7.21). Some, called **laccoliths**, have planar soles and convex tops. The magma from which they formed was capable of doming the overlying strata. In contrast, **lopoliths** are large concordant intrusive bodies whose upper contacts are essentially planar, but whose lower contacts are concave. **Phacoliths** are concordant intrusions which are horseshoelike in cross section, for they occupy the crests of folds.

The Colorado Plateau of the western United States hosts a number of laccolithic intrusions, including the Henry Mountains, Navajo Mountain, and the Carrizo dome (Hunt, Averitt, and Miller, 1953). Lopoliths are rare, but where found they are typically composed of mafic igneous rocks. The Sudbury complex of Ontario (International Nickel Company, 1946; Hawley, 1962) and the Bushveld complex of South Africa (Hall, 1932, Willemse, 1959) are both associated with lopolithic intrusions. Together these lopoliths host rich reserves of iron, nickel, chromite, and platinum.

The names of the full range of intrusive bodies, when read in sequence, have quite a ring: batholith, laccolith, lopolith, and phacolith, not to mention harpolith and bysmalith. Best of all is **cactolith**.

> A quasi-horizontal chonolith composed of anastomosing ductoliths, whose distal ends curl like a harpolith, thin like a sphenolith, or bulge discordantly like an akmolith or ethmolith. (From "Geology and Geography of the Henry Mountain Region, Utah" by C. B. Hunt, P. Averitt, and R. L. Miller, 1953, p. 150. Courtesy of United States Geological Survey.)

Only very careful mapping and analysis of igneous intrusive contacts will permit such a body, if it exists at all, to be discovered.

SHAPES OF SALT AND SHALE DIAPIRS. The form of sedimentary intrusions, other than clastic dikes and sills, range from steep-walled, blunt-topped pillars and spines to round-topped domes and ridgelike walls. Contacts are smooth and regular, lacking the apophyses and irregular projections into country rock that typify many igneous intrusions.

Salt and shale intrusions are the most common of the large sedimentary intrusions. More than oddities, they occur by the hundreds in regions such as the Gulf Coast basin of the southern United States, and they contribute to the trapping of oil and gas in country rock sediments that have been domed and/or pierced by the intrusions (Halbouty, 1969).

Salt and shale intrusions are normally described as **diapirs**, domelike, pipelike, or pluglike intrusions that ascend hundreds to thousands of meters from their source bed(s) into and/or through the overlying sedimentary cover. "Diapir" means to "to pierce" (Dennis, 1967), and indeed many salt and shale **piercement diapirs** literally pierce sediments of the cover, as if the diapir had risen through a cylindrical hole (Figure 7.23*A*). Diapirs that dome but do not actually puncture the overburden are called **nonpiercement diapirs** (Figure 7.23*B*).

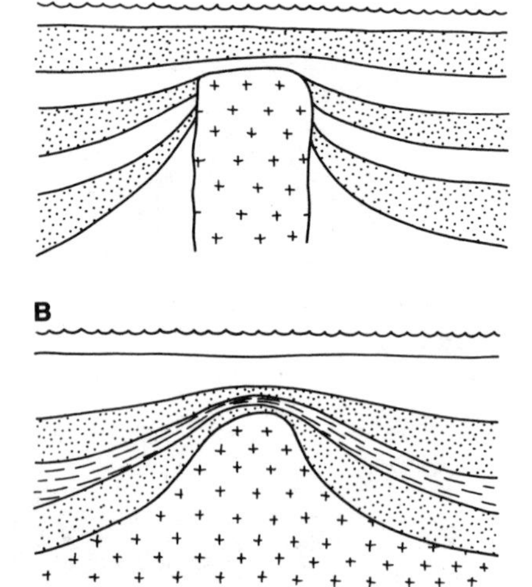

Figure 7.23 (*A*) Piercement and (*B*) nonpiercement diapirs. [From Bishop (1978). Published with permission of American Association of Petroleum Geologists.]

Salt and shale diapirism is favored by a depositional history that yields thick layers of salt and mud overlain by even thicker accumulations of sand and/or sand–mud mixtures. Sedimentary columns of this type are distinguished by **density inversion**, produced by the presence of relatively low-density salt or uncompacted mud (ρ = 2.2 or less) beneath higher density sand and sand–mud mixtures (ρ = 2.4–2.7). Density inversions of this type may lead to the buoyant rise of the lower density materials. In fact the fundamental properties of diapirs have been successfully replicated in the laboratory through scale-model experiments (Figure 7.24). These experiments feature the disturbance of stratified soft materials (like clays and oils and asphalts) arranged in an order of density inversion (Nettleton, 1934; Parker and McDowell, 1955; Ramberg, 1967).

Salt diapirs in the Gulf Coast are commonly circular in map-view expression, averaging 3 km in diameter, and never larger than 30 km in diameter (Woodbury and others, 1973; Bishop, 1978). Many of the salt diapirs have risen 10 km through the Cenozoic overburden to very shallow levels (Figure 7.25). Some diapirs have surfaced landward from the Gulf, where their map-view expression can be directly observed. The Grand Saline salt dome, located 90 km east of Dallas, Texas, is a case in point (Balk, 1949; Muehlberger and Claybaugh, 1968). The surface expression of the Grand Saline salt dome is interesting to structural geologists who are fond of symmetry relationships. Axial, circular symmetry is expressed in many ways. The surface projection of the dome is a 2-km diameter circular topographic low marked by gas seeps and saline springs and marshes. Low hills controlled by Eocene bedrock encircle the topographic depression, and these in turn control a circular and radial drainage system centered on the dome. The oldest bedrock exposed at the surface is displaced 2000 m above its normal stratigraphic position! Radial faults cut the formations, and these are marked by displacements that are greatest nearest the center of the dome.

Mining of the salt of the Grand Saline dome has resulted in a carved network of tunnels, 200 m below the surface, and the walls and ceilings of the tunnels provide a unique look at the interior of a salt dome. The axial symmetry of the system is dramatically expressed in vertically plunging folds whose patterns look like folds that form in a handkerchief when it is pulled through a small ring (Balk, 1949) (Figure 7.26). The folds are conspicuously displayed in the ceilings of huge mined-out rooms. Folded sequences consist of white salt interlayered with anhydrite, which appears dark gray to black. Wavelengths of the largest folds are greater than 30 m; the smallest folds are microscopic.

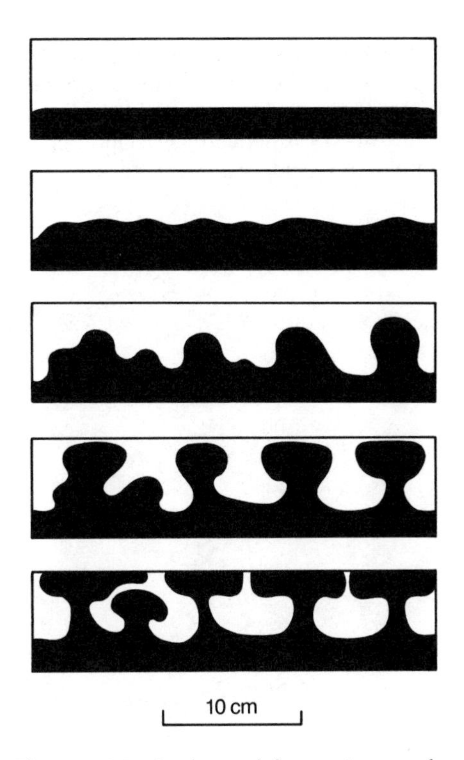

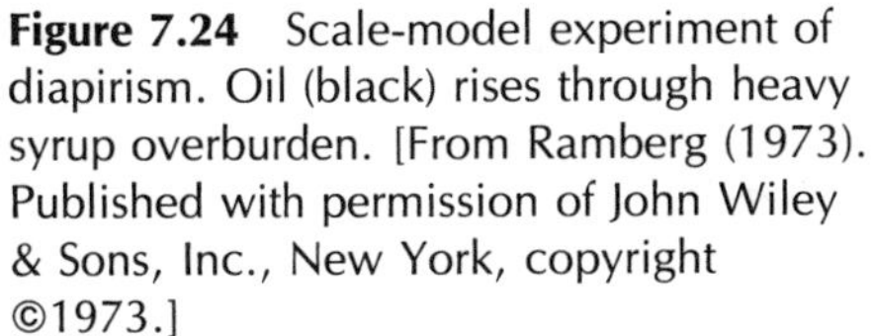

Figure 7.24 Scale-model experiment of diapirism. Oil (black) rises through heavy syrup overburden. [From Ramberg (1973). Published with permission of John Wiley & Sons, Inc., New York, copyright ©1973.]

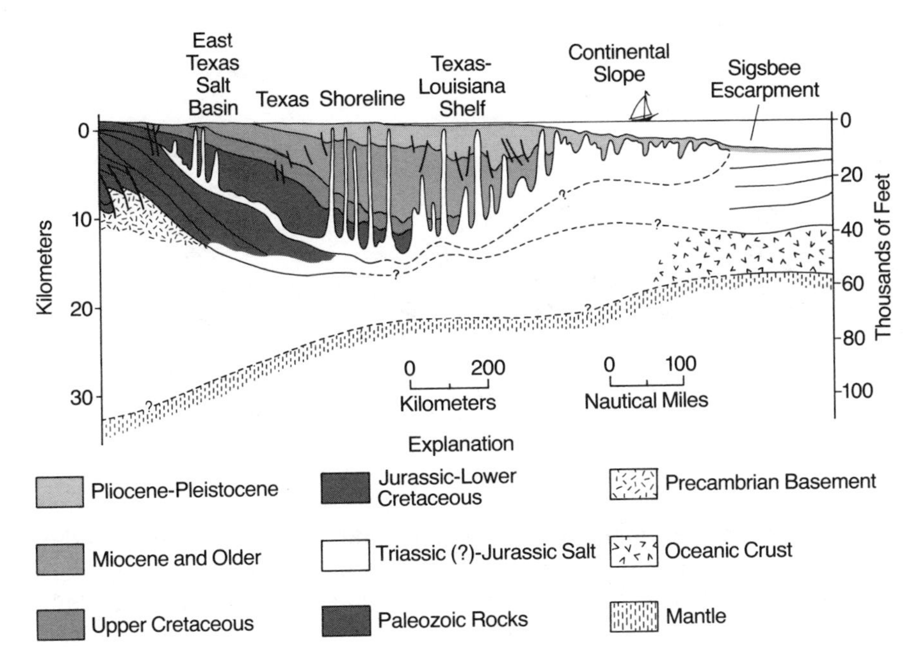

Figure 7.25 Cross-sectional profile of salt domes and salt spines in the Gulf Coast region. [From Martin (1978). Published with permission of American Association of Petroleum Geologists.]

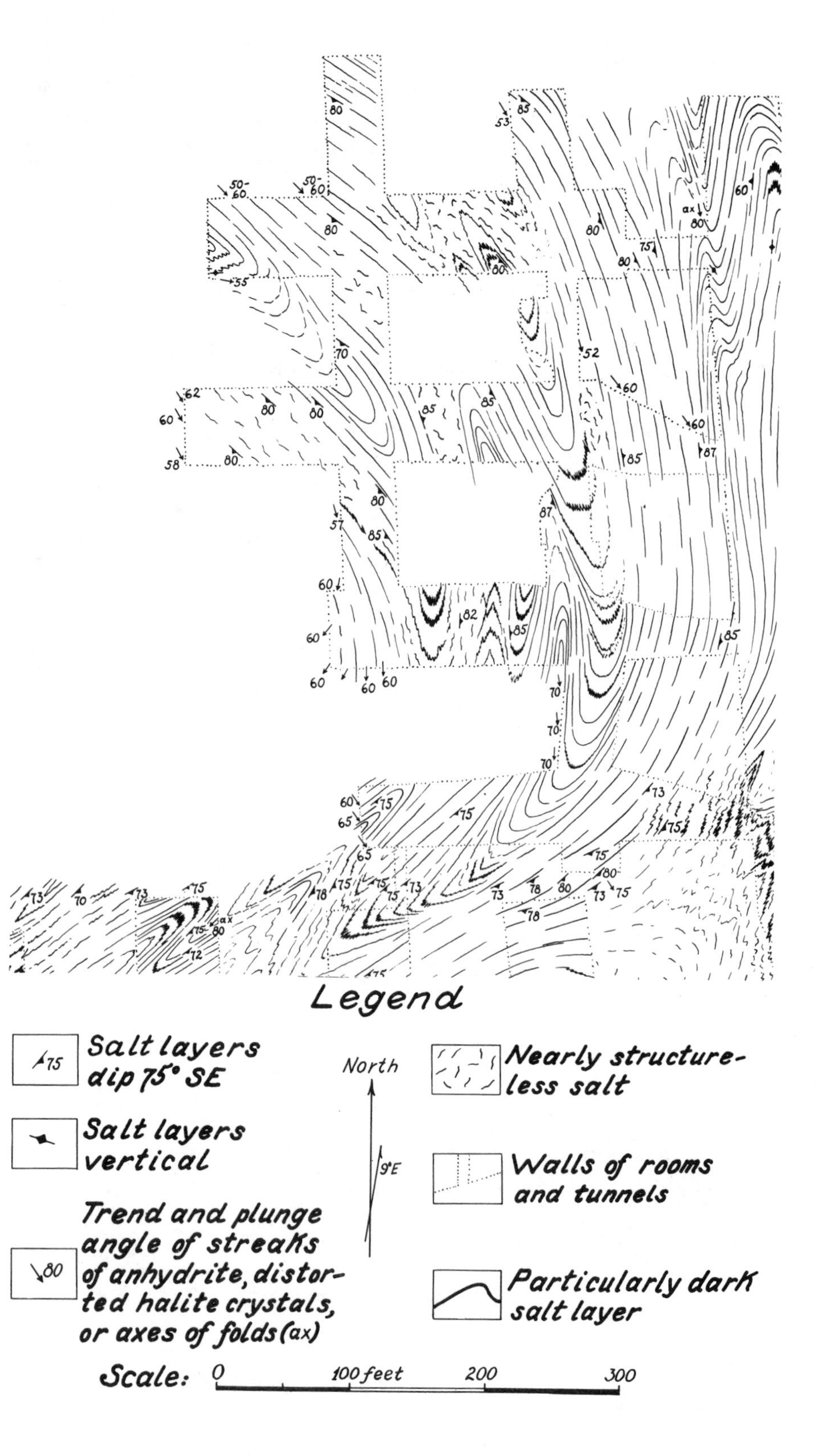

Figure 7.26 Geologic map of part of the subsurface of the Grand Saline salt dome, Texas. [From Balk (1949). Published with permission of American Association of Petroleum Geologists.]

Balk pictured the Grand Saline as a dome 5 km high whose contacts dip outward at 70°. Its top is blunt and circular, about 400 to 800 m in diameter. He estimated that the volume of the mass is 22 km^2. Balk's remarks regarding origin fit in beautifully with our emphasis on detailed structural analysis.

The large-scale movement of salt through the dome is, of course, the combined result of the motion of immense numbers of individual crystals. . . . the movements of small [translations] were systematically coordinated to the large-scale motions of the salt mass through the dome. The preferred orientation of the anhydrite has its exact counterpart in the alignment of prismatic phenocrysts in igneous plugs parallel with the axis of the plug; the subvertical alignment of pebble axes, prismatic minerals, and metacrysts along the borders of plutons and diapirs; and the preferred orientation of prismatic suspensions parallel with the axis of a pipe in which a fluid is flowing. (From "Structure of Grand Saline Salt Dome, Van Zandt County, Texas" by R. Balk, 1949, p. 1818. Published with permission of American Association of Petroleum Geologists.)

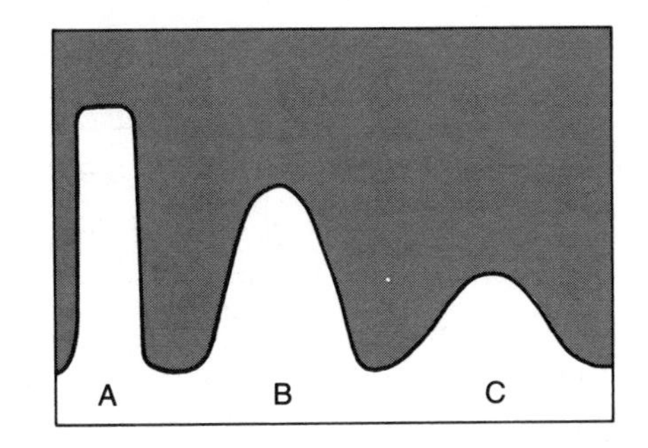

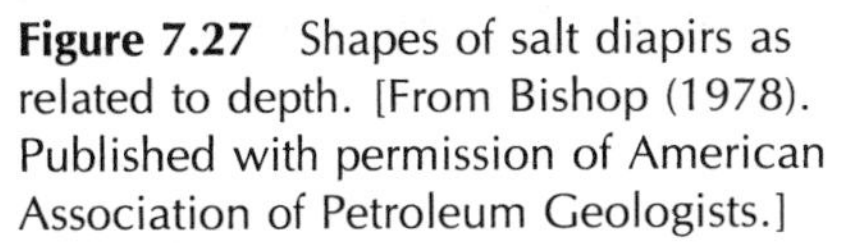

Figure 7.27 Shapes of salt diapirs as related to depth. [From Bishop (1978). Published with permission of American Association of Petroleum Geologists.]

It is now well documented that salt diapirs vary in shape as a function of depth (Bishop, 1978). Shallow ones (no deeper than 2 km), like the Grand Saline, are marked by steep sides and flat tops, almost like cylinders. Shallow to moderately deep diapirs tend to display rounded tops and less steeply inclined flanks. And the deep salt diapirs have round, domelike tops and relatively gently sloping sides (Figure 7.27). It would seem that at deep levels salt diapirs tend to expand laterally as they move upward, but at shallow levels their drive is channeled wholly vertically in the form of spines or needles of constant diameter.

Rates of sediment accumulation, and thus rates of loading, influence the nature of the map configuration of salt diapirs (Bishop, 1978). Where sedimentation rates are slow (ca. 100 m/m.y.), salt diapirs occur as small, discrete, isolated, nonpiercement diapirs. Intermediate rates of sedimentation (ca. 300 m/m.y.) generate larger, more uniformly distributed nonpiercement and piercement diapirs. Rapid sedimentation (ca. 500 m/m.y.) leads to the growth of large piercement diapirs, which display preferred, fault-controlled (?) alignments as well as networks of salt walls.

Shale diapirs tend to be pluglike masses, no taller than 1 km. A typical example is the Laward diapir in southeastern Texas (Figure 7.28). Composed of Eocene and Oligocene shale, the diapir pierces the lower part of Oligocene sandstone. Flat-topped and marked by steeply dipping flanks, it stands 365 m high and about 700 m wide (Bishop, 1977).

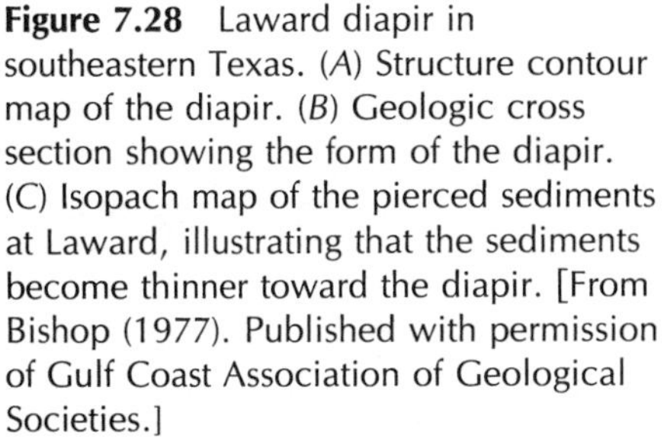

Figure 7.28 Laward diapir in southeastern Texas. (*A*) Structure contour map of the diapir. (*B*) Geologic cross section showing the form of the diapir. (*C*) Isopach map of the pierced sediments at Laward, illustrating that the sediments become thinner toward the diapir. [From Bishop (1977). Published with permission of Gulf Coast Association of Geological Societies.]

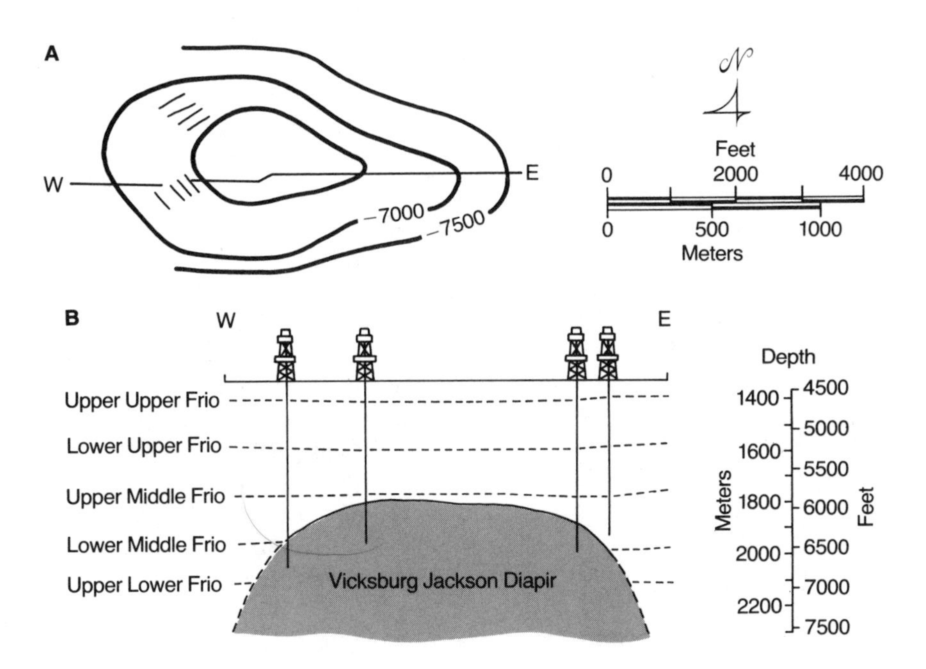

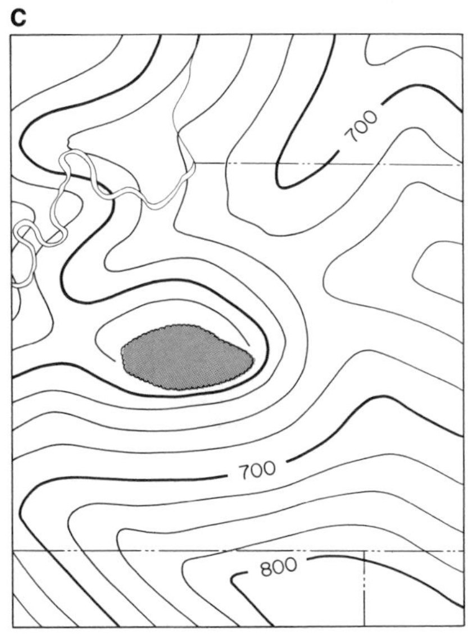

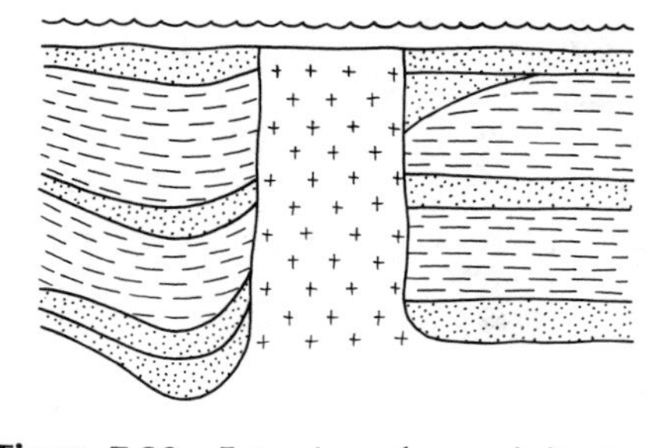

Figure 7.29 Extrusion of a mud diapir. [From Bishop (1978). Published with permission of American Association of Petroleum Geologists.]

Originally it was surmised that the rise of piercement diapirs of shale was a wholly intrusive process, one in which the head of the diapir punctured and shouldered aside overburden caught in its way. Now it is known that some layers of the pierced overburden of many shale diapirs never blanketed the diapirs. Rather many shale diapirs intrude upward through their sand or sand–mud overburden, and then **extrude** onto the basin floor, eventually to be covered by more sediment (Figure 7.29). This reality alone underscores both the difficulty and importance of interpreting contact relationships properly.

GULF COAST DIAPIRISM

It is no wonder that density inversions arose in the sedimentary column of the Gulf Coast region. First of all, the Gulf Coast region was a site of thick accumulations of mid-Jurassic salt (the Louann salt) in rift basins related to the opening of the Gulf of Mexico (Martin, 1978; Humphris, 1979). Second, during the Cenozoic era the Gulf Coast basin received tens of thousands of meters of sand and mud, which resulted in the **progradation** of the continental shelf outward from the continental margin, as it existed in earliest Cenozoic times (Antoine and others, 1974). Cenozoic sedimentary accumulations off the Louisiana coast are estimated to be 15,000 m! The Jurassic salt deposits thus became covered by an ever-increasing load of sediments. And thick Cenozoic mud layers deposited from time to time during the course of Gulf Coast sedimentation typically were covered by the progradational deposition of sand and sand–mud layers.

Loading of salt by sediments gave rise to salt diapirism in the Gulf Coast region. Nettleton (1934) demonstrated that salt may be forced to buoyantly rise as a response to differential stress produced by uneven vertical loading of a salt layer at depth. If the weights of two different columns of sediment above a salt layer are unequal, the salt will flow upward as a rheid into the overlying sediment column, to balance the forces. The rheidity of salt is about 10 years. As some measure of the capacity of salt to flow rapidly in terms of human history, it is amazing to learn that

> in salt mines in the Austrian Salzburg district, old tunnels were discovered in which bodies of ancient Celtic miners were found with their mining tools, embedded in salt. (From "Structure of Grand Saline Salt Dome, Van Zandt County, Texas" by R. Balk, 1949, p. 1823. Published with permission of American Association of Petroleum Geologists.)

Salt diapirism is typically long-lived, because the magnitude of the density inversions increases as the sand and sand–mud overburden becomes increasingly compacted. In the Gulf Coast region, salt diapirism fed from the Louann salt layer(s) continues to be active today, a testimony to the endurance and persistance of rheid flow, given the proper geological conditions.

Shale diapirism in the Gulf Coast region also is generated by sediment loading, in this case the loading of water-rich mud. Unlike salt diapirs, shale diapirs have a short life expectancy because their upward rise slows and ceases as the source shale bed compacts. Density contrast between salt and overburden sediments increases as the overburden sediments become more and more compacted. In contrast, any initial density contrast that may have existed between water-rich mud and overlying sand is eliminated as water is squeezed out of the mud during compaction. Shale diapirs, then, never grow into spines or needles.

Figure 7.30 And the sea became dry land. Faulting during the Alaskan earthquake of 1964 resulted in significant vertical displacement. The northwest block (left) of the Hanning Bay fault was raised 4 to 5 m, relative to the southeast block. The white coating on the exposed seafloor bottom consists of the bleached remains of calcareous algae and bryozoans. The height of the fault scarp near the ponded water is 4 m. [Photograph by G. Plafker. From Plafker (1965). Courtesy of United States Geological Survey.]

FAULT CONTACTS

PRESENT-DAY FAULTING

Fault contacts are easy to identify in areas where faulting is presently active. Modern faults produce **offset** of both natural and man-made objects (Figure 7.30). Needless to say, the offset of features like roads, streams, and beaches simplifies the search for fault contacts. The offset of bedrock marker beds flags the locations of faults as well (Figure 7.31).

Figure 7.31 Fault offset of marker layer in highly deformed melange in the San Juan Islands, Washington. Margi Rusmore is the geologist. (Photograph by G. H. Davis.)

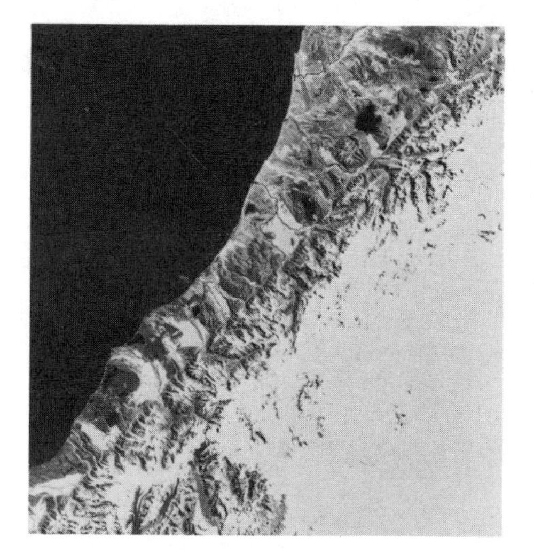

Figure 7.32 LANDSAT view of physiographic expression of the Alpine fault, New Zealand.

Present-day faults, and faults that were active in the recent past, commonly have splendid physiographic expression. Long, straight to curved **lineaments** defined by aligned topographic features serve to mark the locations of fault traces (Figure 7.32). Lineaments can be recognized on aerial photographs and topographic maps, although ground-checking is required to confirm the interpretation that a specific lineament is indeed the trace of a fault.

Fault scarps form where fault movements have produced or are producing vertical displacements. Sometimes the height of a scarp accurately reflects the true vertical component of slip on the fault (Figure 7.33*A*). Usually however, scarps along the surface trace of a fault are *not* a direct reflection of either the inclination of the fault surface or the magnitude of fault displacement. Original fault inclination and scarp height is usually reduced by erosion. Scarps called **fault-line scarps** usually give a false impression of the actual vertical component of displacement along a fault. These are scarps, located along or near the trace of a fault, marked by a topographic relief that

Figure 7.33 (*A*) Red Canyon fault scarp in the Montana earthquake area, as it looked in August 1959. Scarp height reflects the actual fault displacement. (Photograph by J. R. Stacy. Courtesy of United States Geological Survey.) (*B*) Fault-line scarp near Lake Mead. The topographic relief along the scarp does not reflect slip. Rather, it reflects differential erosion along the fault interface between resistant volcanic rock and nonresistant volcanic rock. Apparent relative displacement in the rock on which Ernie Anderson stands actually moved *upward* relative to the resistant volcanic rock. (Photograph by G. H. Davis.)

Figure 7.34 Triangular facets along the front of the Wasatch mountain block near Provo, Utah. (Photograph by C. Glass, G. Brogan, and L. Cluff. Courtesy of Woodward-Clyde.)

simply reflects the fortunes of erosion (Figure 7.33*B*). Commonly they display the relative resistance of the rocks brought into contact by faulting. Only under fortuitous circumstances will a fault-line scarp reflect accurately a magnitude and sense of apparent displacement that actually conforms to that of true slip.

Former planar fault scarps may be dissected by erosion to yield **triangular facets**, physiographic signatures that call attention to the location(s) of fault contacts. An outstanding display of triangular facets may be seen along the Wasatch Front in northern Utah (Figure 7.34).

PHYSICAL PROPERTIES

Fault contacts can be recognized at the outcrop scale by virtue of an array of characteristic physical properties, not all of which will be assembled in any one place. Commonly one should expect to find a discrete fracture break or discontinuity, a **fracture surface** (Figure 7.35) whose orientation and position satisfactorily describe the interface between the rocks brought into contact. Alternatively, a **fault zone** may form instead of a discrete fault surface. Fault zones consist of numerous closely spaced fault surfaces, commonly separating masses of broken rock. Minor folds may be found within fault zones. Whether a discrete fault surface or a fault zone is produced by faulting depends on many factors, none of which can ever be precisely

Figure 7.35 Reasonably discrete fracture surface separates rough-weathering Wasatch formation (Eocene) on left from smooth weathering Straight Cliffs formation (Cretaceous) on right. Paunsaugunt fault near Bryce Canyon, Utah. (Photograph by G. H. Davis.)

Figure 7.36 (*A*) Students measuring the orientation of a polished, striated fault surface near Patagonia, Arizona. (*B*) Slickensided fault surface in northwestern Guatemala. Alston Boyd III is the geologist. (Photographs by G. H. Davis.)

known for a given fault. Rock strength, strain rate, the physical environment of deformation, and duration of faulting are all important. Fault surfaces may be finely polished to **slickensided surfaces** as a result of the differential movement. Some surfaces are mirrorlike (Figure 7.36). Other fault surfaces are not slickensided at all, either due to the host rocks' inability to take a polish or because the former polished surface had been blemished by weathering and erosion.

Fault surfaces commonly display **striations** that reflect fault movement (Figure 7.37). Striations are most noticeable on slickensided surfaces, but they are by no means restricted to them. Some striations appear to be actual scratches in the faulted rock, produced by differential rigid body movement (Figure 7.38). In contrast to scratches, some striations are actually lineations produced by crystallization of minerals. The striae in quartz shown in Figure 7.37 may have formed this way. Large fault zones may display surfaces which contain deeply furrowed, slickensided **grooves** (see Figures 1.15 and 3.19). Some of these surfaces look exactly like bedrock surfaces that have been polished and grooved by glacial flow.

The wall rocks of faults respond to faulting in ways that depend both on the environment of deformation and lithology. Under conditions of brittle faulting, wall rocks become anomalously highly fractured by **joints** and **microfaults**. Faulted rocks may be converted into **breccias** made up of

Figure 7.37 Striations on a fault surface. Fine-grained sandstone (gray) displays faint striations. The bold striae are displayed by a quartz vein (white). Platform on which dime rests is a chattermark.

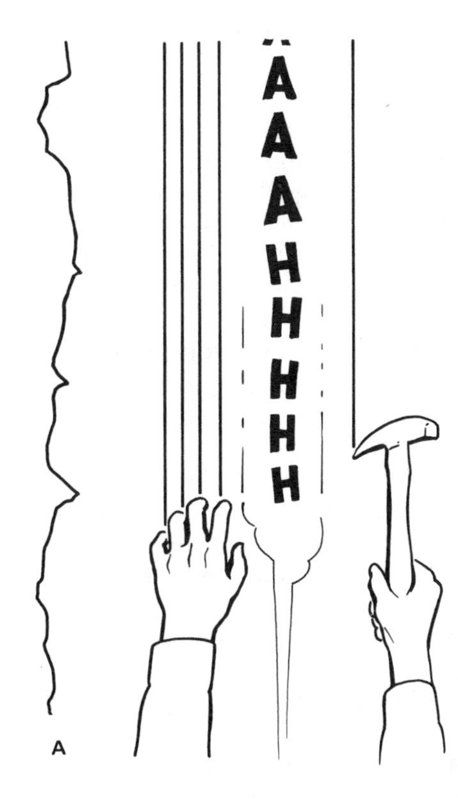

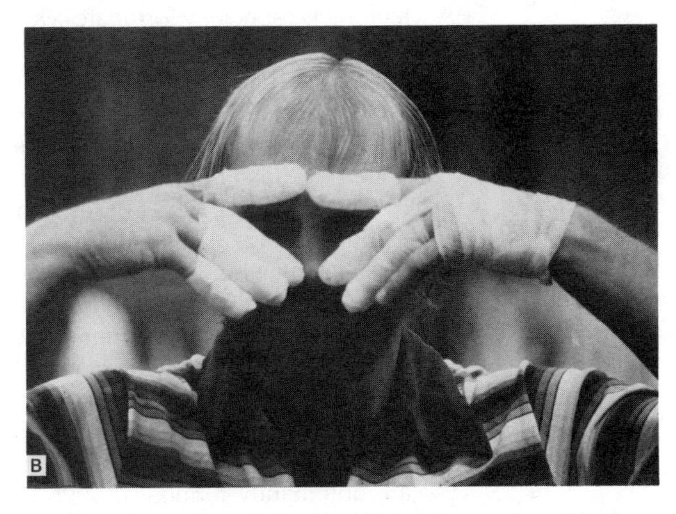

Figure 7.38 (*A*) Striae produced by differential rigid body movement. You would be rigid too! (*B*) Geologist Steve Lingrey left such striae on the Paunsaugunt fault, at the locality shown in Figure 7.35.

angular fragments of wall rock set in a finer grained matrix of crushed material (Figure 7.39). Ordinarily the angular fragments that comprise breccias move with respect to one another during faulting. Dilation and volume increase are characteristic of the process, and this implies that breccias form in environments of relatively low confining pressure. Voids representing

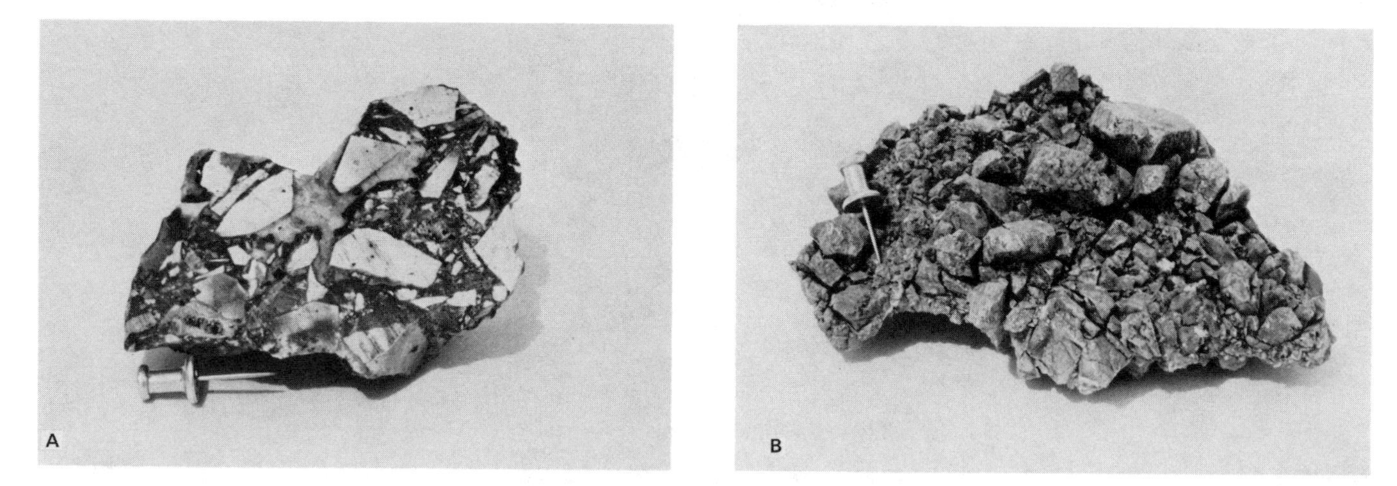

Figure 7.39 Fault breccias, inside (*A*) and out (*B*). (Photographs by G. Kew.)

Figure 7.40 (*A*) Fractured, veined Paleozoic limestone just above a flat fault. (*B*) Close-up of cockscomb structure in the limestone. [From Davis and others (1979). Published with permission of Geological Society of America and the authors.]

pull-apart and dilation may not be completely filled by the finer grained matrix at the time of faulting. The void space may later be partially or completely filled by precipitation of minerals through circulation of ground water and/or hydrothermal solutions. **Cockscomb** structures characterized by crustification of minerals around the lining of a void are a typical product of partial open-space filling (Figure 7.40). Formation of open space during faulting is especially fortunate when mineralizing solutions have filled voids with precious or base metals.

Wall rocks that are weak and incompetent respond in a plastic manner to faulting and are commonly converted to **gouge** along the fault contact (Figure 7.41). Gouge is a very fine grained, clayey crushed rock. In its dry state it feels like talcum powder, but admittedly a poor grade of powder because it retains bits and pieces of resistant grains that did not completely succumb to the crushing processes. When wet, gouge crops out as a sticky clay. Zones of gouge may be wispy and thin, or alternatively they may be meters wide.

Under conditions of high stress concentration, high confining pressure, and moderately high temperature, wall rocks along a fault contact can be converted into **cataclastic** and **mylonitic rocks**. Cataclastic rocks owe their texture to pervasive crushing or **comminution** of mineral grains achieved through penetrative frictional sliding along networks of microcracks. Although mylonitic rocks partly reflect the influence of cataclastic flow, they are dominantly formed through crystal–plastic dislocation creep. In other words, the grain-size reduction in mylonites is mainly due to the **dynamic recrystallization** of original grains that deformed by dislocation creep (Tullis and Schmid, 1982), not simply to comminution accompanying cataclastic flow. Mylonites occupy zones of profound laminar flow, especially zones of simple shear.

A whole family of unusual and interesting cataclastic rocks and mylonitic rocks has been described, and serious study of such rocks requires microscopic petrography. Higgins (1971) classified cataclastic and mylonitic rocks according to a number of criteria, especially grain size, presence or absence of foliation, and degree of recrystallization. For our purposes the terms **microbreccia** and **mylonite** are especially useful. Microbreccias are cata-

Figure 7.41 Fault gouge (white material beneath fracture surface) formed along low-dipping, nearly horizontal fault in the Rincon Mountains, Arizona. (Photograph by G. H. Davis.)

clastic rocks in the form of well-indurated, massive rocks that have been crushed to very fine grain size through cataclastic flow. Microbrecciated granites, for example, are fine-grained rocks whose feldspar crystals are broken into tiny angular chips. Mylonitic rocks, on the other hand, are intensely deformed rocks that flowed and dynamically recrystallized in the solid state, producing a foliation or **fluxion structure** and, commonly, lineation as well. Mylonites display a planar parallel arrangement of broken mineral grains and shear surfaces (Figure 7.42*A*). Viewed microscopically, the mylonites display beautiful lensoidal fabrics. Quartz typically looks like it has flowed like butter, whereas the feldspars obviously have behaved brittlely, breaking into chips, which sometimes become oriented within the fluxion structure. **Ultramylonites** (Figure 7.42*B*) are mylonites taken to the edge of recognition, comminuted and recrystallized to the point that fluxion structure is hardly recognizable in handspecimen view.

A somewhat rare, but intriguing component of certain major fault zones, particularly those that have faulted deep reaches of the Earth's crust, is pseudotachylite (Figure 7.43). Tachylite is basalt glass, but **pseudotachylite** is not a volcanic product. It is a **fault melt** that can be made from many different starting materials through faulting. Pseudotachylite may be restricted to discrete fault surfaces, or it may depart from the main fault surface to occupy spiderlike networks of gash fractures. They are in essence fossil records of deep earthquakes (Sibson, 1980). To form pseudotachylite there must be powerful shock under very high-pressure conditions and an instantaneously imposed strain. Pseudotachylite is found in and around some grand fault zones, like those mapped in deep structural levels of the Alpine Mountain system.

There are other characteristics that allow fault contacts to be recognized, but these involve signatures other than those that describe the nature of the fault surface(s) and the nature of the immediately adjacent wall rock. As will be seen in Chapter 9, the displacement patterns produced by faulting are a dead giveaway.

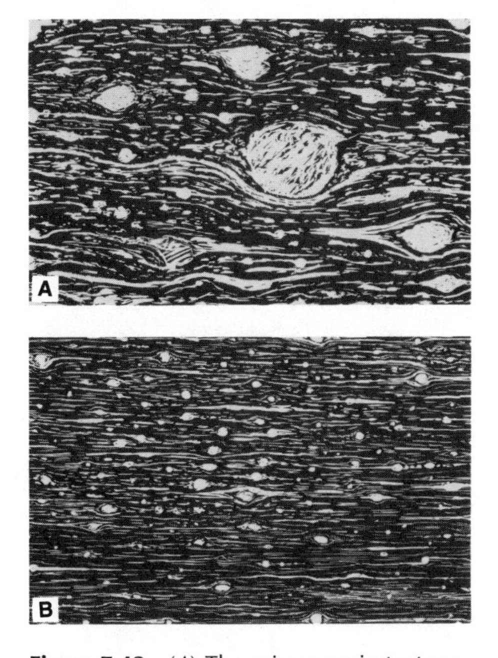

Figure 7.42 (*A*) The microscopic texture of mylonite. The white represents fragmented grains (porphyroclasts) and breccia streaks. The black represents finely laminated material. (*B*) The microscopic texture of ultramylonite. [From Higgins (1971). Courtesy of United States Geological Survey.]

Figure 7.43 Photomicrograph of fault glass, pseudotachylite. Glass (black) contains microbreccia fragments. The banded laminae are ultramylonite. (Reprinted with permission from *Journal of Structural Geology*, v. 2, R. H. Sibson, "Transient Discontinuities in Ductile Shear Zones." Pergamon Press, Ltd., Oxford, copyright ©1980.)

Figure 7.44 Ductile shear zone. [From Davis (1980). Published with permission of Geological Society of America and the author.]

DUCTILE SHEAR ZONES

Ductile shear zones produce contacts entirely different from contacts marking ordinary faults (Figure 7.44). Unlike ordinary brittle fault surfaces, ductile shear zones commonly do not display any physical break. Instead, differential translation of rock bodies, separated by the shear zone, is achieved entirely by ductile flow (Figure 7.45). Markers pass through ductile shear zones without necessarily losing their continuity, but the effects of the shearing are reflected in distortion of the markers and in the transformation of the original rocks to sheared rocks like mylonites. Distortion is expressed both in shape and orientation changes.

Ductile shear zone contacts are most commonly formed in igneous or metamorphic environments, where elevated temperature and/or confining pressure renders one or both wall rocks ductile. However, ductile shear zones can also form during soft-sediment deformation of unconsolidated sands and muds. In such environments, the ductile response is due to the water-rich nature of the sediments.

Geologic mapping has revealed that ductile shear zones are tabular to curved, centimeters to kilometers thick, and marked by penetrative foliation(s). If shearing is purely ductile, no cohesion is lost at the interface between the two rock bodies that are brought into contact. In fact, the absence of a conspicuous fracture surface tends to disguise the fact that a given shear zone can really accommodate tens of meters to kilometers of displacement. Where early ductile shearing culminates in late-stage brittle faulting, the shear zone is marked by a discrete fault surface or fault zone bordered by wall rocks that are plastically deformed. Ramsay (1980) calls these structures **brittle–ductile shear zones**.

Shear zones can accommodate either shortening or stretching of crustal rocks. Diagrams drawn by Ramsay (1980) illustrate two of the possibilities (Figure 7.46). Older rock can be brought up against younger, higher level rock during crustal shortening (Figure 7.46*A*); relatively young, high-level rock can be sheared down onto deeper, older rocks during crustal extension (Figure 7.46*B*).

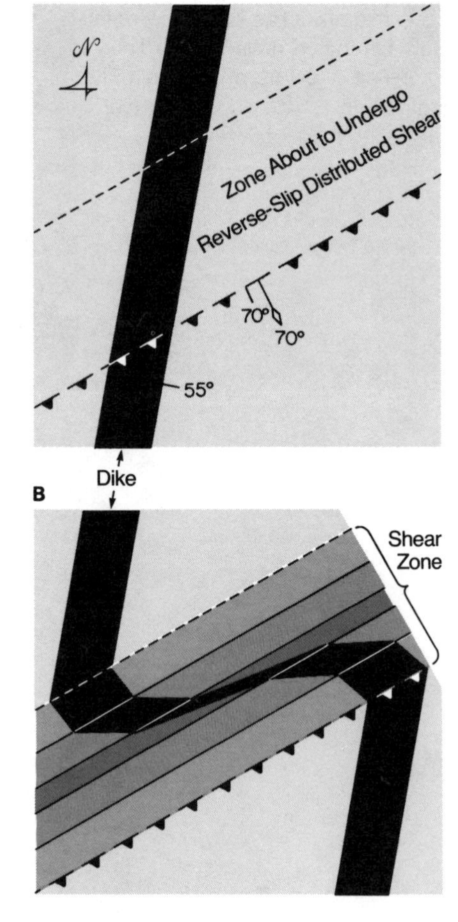

Figure 7.45 (*A*) Map view of dike about to be deformed by ductile shearing. (*B*) Distortion of dike achieved by ductile shearing. Contacts of shear zones separate distorted rock from undistorted rock.

Properties of ductile shear zone contacts are predictable from the fundamentals of strain analysis (Ramsay, 1980). If a body of rock is cut by a shear zone of some uniform thickness, the original bedrock caught within the zone of shear will be distorted such that it lengthens in one direction and shortens in another (Figure 7.47*A*). The direction of stretching corresponds to the

plane of flattening in the shear zone, a plane whose physical manifestation is a penetrative foliation, inclined at an acute angle to the slip (or shear) planes. The greater the degree of simple shear, the more the foliation rotates toward parallelism with the borders of the shear zone (Figure 7.47*B*). At the borders proper, primary features in the host rock outside the shear zone may be traced into their distorted counterparts within the zone of shear.

If the shear zone expands in thickness during the deformation, perhaps as a result of strain hardening, a broader zone of original country rock will become distorted (Figure 7.47*C*). Foliation in the interior part of the zone, having formed first, will rotate the most during progressive shearing. Relatively newly formed foliation near the exterior of the shear zone will rotate least. Differential rotation of foliation creates the systematically curved **sigmoidal** foliation pattern that characterizes most shear zone contacts.

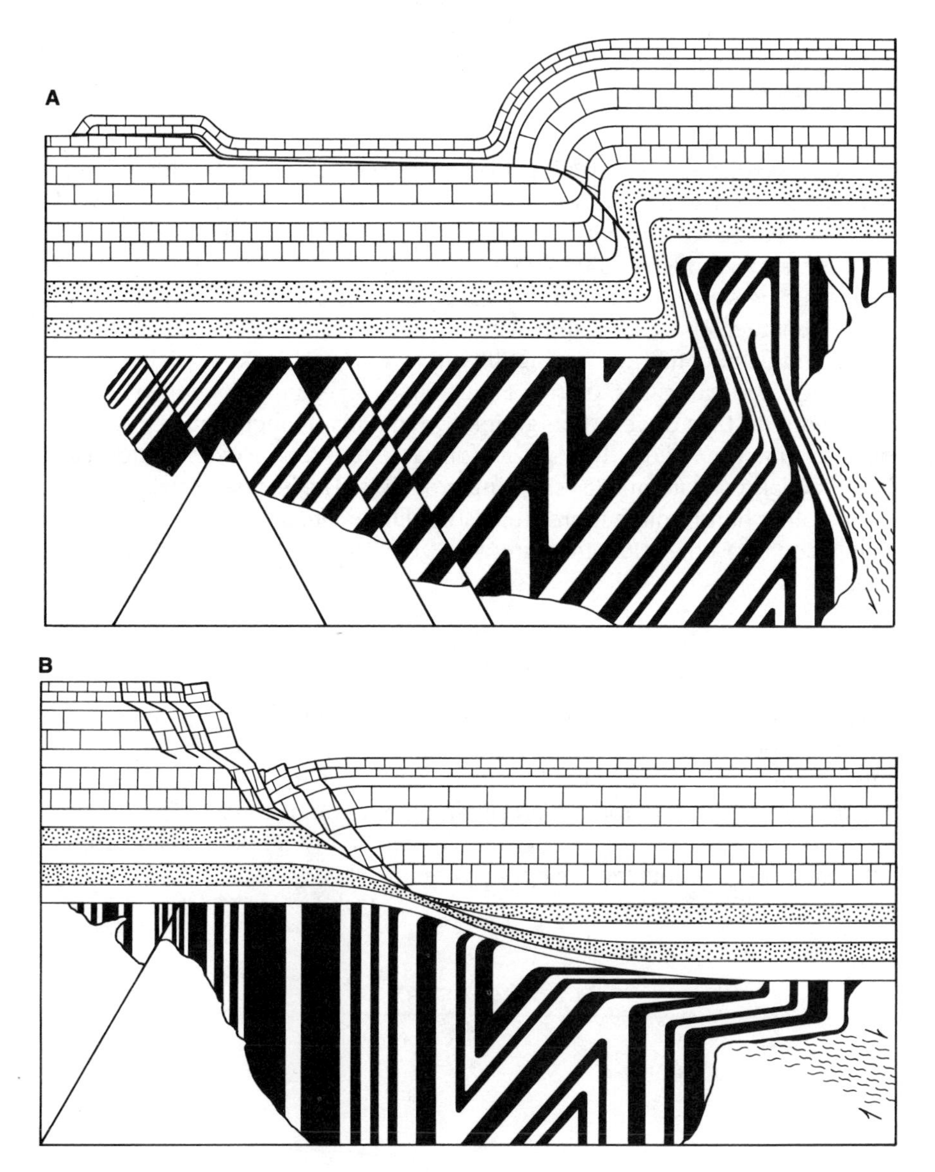

Figure 7.46 (*A*) Ductile shearing that accomplishes layer-parallel shortening. (*B*) Ductile shearing that accommodates layer-parallel stretching. (Reprinted with permission from *Journal of Structural Geology*, v. 2, J. G. Ramsay, "Shear Zone Geometry: A Review. Pergamon Press, Ltd., Oxford, copyright ©1980.)

A

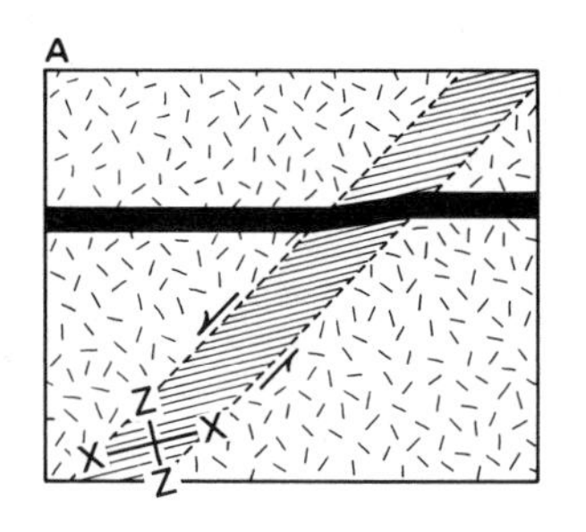

B

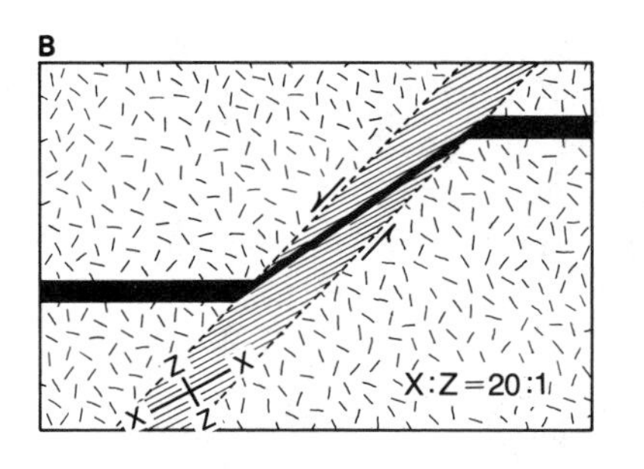

C

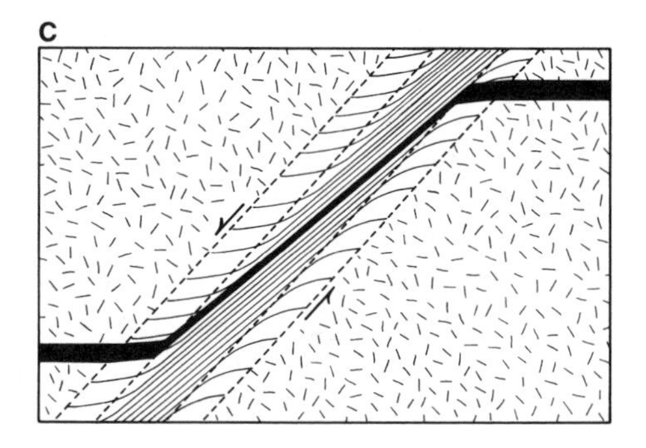

Figure 7.47 Kinematic development of a shear zone. (*A*) Shearing results in simultaneous stretching and flattening. (*B*) Progressive deformation by shearing results in rotation of plane of flattening toward parallelism with shear zone walls.(C) Expansion of thickness of shear zone and consequent differential rotation of plane of flattening into sigmoidal form.

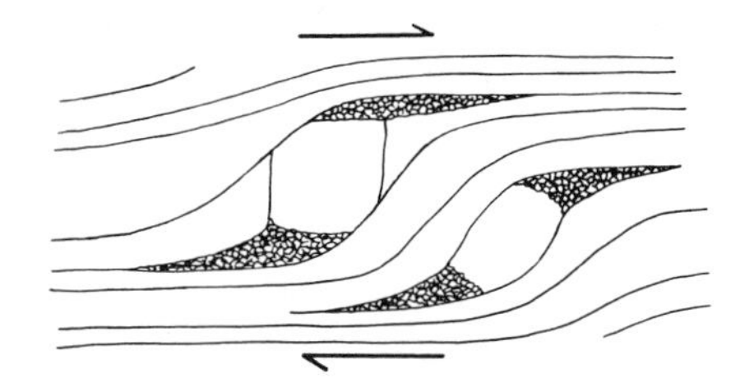

Figure 7.48 Microstructural indicator of the sense of shear within a shear zone. This asymmetric augen structure is marked by "tails" composed of fine-grained material of the same composition as the host porphyroclast. [From Simpson and Schmid (1983). Published with permission of Geological Society of America.]

The sense of shear within a ductile shear zone can be determined if offset markers are exposed. However, in the absence of conspicuously offset markers it is necessary to examine microstructures for clues to sense of shear. Simpson and Schmid (1983) have provided some microstructural criteria for deducing the sense of movement in sheared rocks. Microstructural clues to movement are observed in thin sections cut perpendicular to the flattening plane and parallel to the shearing direction. One of the most reliable micro-movement indicators is shown in Figure 7.48, an example of augen feldspar adorned with **tails** of very fine-grained, dynamically recrystallized feldspar derived from the augen. The soft, deformable tails tend to rotate effortlessly into the plane of flattening as shearing proceeds. The less deformable augen, on the other hand, rotate more sluggishly. Differential rotation of augen and tails leads to a characteristic sigmoidal geometry that reveals the sense of shear.

SUMMARY

Contacts are simple to distinguish in theory, but where structures are complex, deformations multiple, and exposures poor, recognizing the nature of contacts can be a very difficult and challenging assignment. *Of all the missions of geologic mapping, proper interpretation of contacts is of greatest importance*. Accurate reconstruction of geologic history depends on correct interpretation of contacts.

chapter 8 PRIMARY STRUCTURE

NATURE OF PRIMARY STRUCTURES

Primary structure is structure that originates during the formation of rocks, either through depositional processes or deformational processes. Primary structure encompasses a great variety of features; common examples are bedding, volcanic layering, cross-stratification, ripple marks, soft-sediment folds, growth faults, and columnar jointing. Primary structure in sedimentary rocks forms before lithification. In volcanic and intrusive igneous rocks, primary structure forms during the flow and late-stage congealing of magma, before and during final crystallization. Metamorphic rocks, strictly speaking, do not possess primary structure, for metamorphic rocks are made secondarily at the expense of preexisting sedimentary, igneous, or metamorphic rocks. The form and fabric of metamorphic rocks are to a great extent influenced by the original lithology and texture of the **protolith**, the rock that is transformed by the metamorphism.

USEFULNESS OF PRIMARY STRUCTURE

GUIDE TO STRAIN

The evaluation of primary structure is useful in many ways. Primary structures, where found deformed, can be valuable guides to internal strain. The internal strain disclosed by deformed primary structures may reflect **primary and/or secondary strain**. Reduction spots are primary structures, and we have already seen how they can be used to monitor ***secondary*** strain. The small, bleached-white spheres in red siltstones and shales are transformed into miniature strain ellipsoids when distorted by deformation during folding (Figure 8.1). In similar fashion, certain primary structures record ***primary*** strain. The bubblelike **orbicules** in dike rock shown in Figure 8.2 were originally spherical primary structures in the magma. They became distorted into ellipsoids during flow of the magma before it congealed.

DISTINGUISHING UP FROM DOWN

Many primary structures in sedimentary and volcanic rocks can be used as guides for determining whether rock layers are rightside up or upside down. For example, cross-beds often converge into parallelism with the lower, older bedding surface and are truncated along the upper, younger bedding surface (Figure 8.3). Structures like these have a **polarity of form** that can be used to distinguish the original upper surface of a layer, even if the layer is upside down. If the upper surface can be identified as such, the relative ages of the beds in a stratigraphic surface can be interpreted.

Telling "up" from "down" within a rock sequence constitutes the **determination of facing**. The **face** of a layer is its upper surface, like the cheese

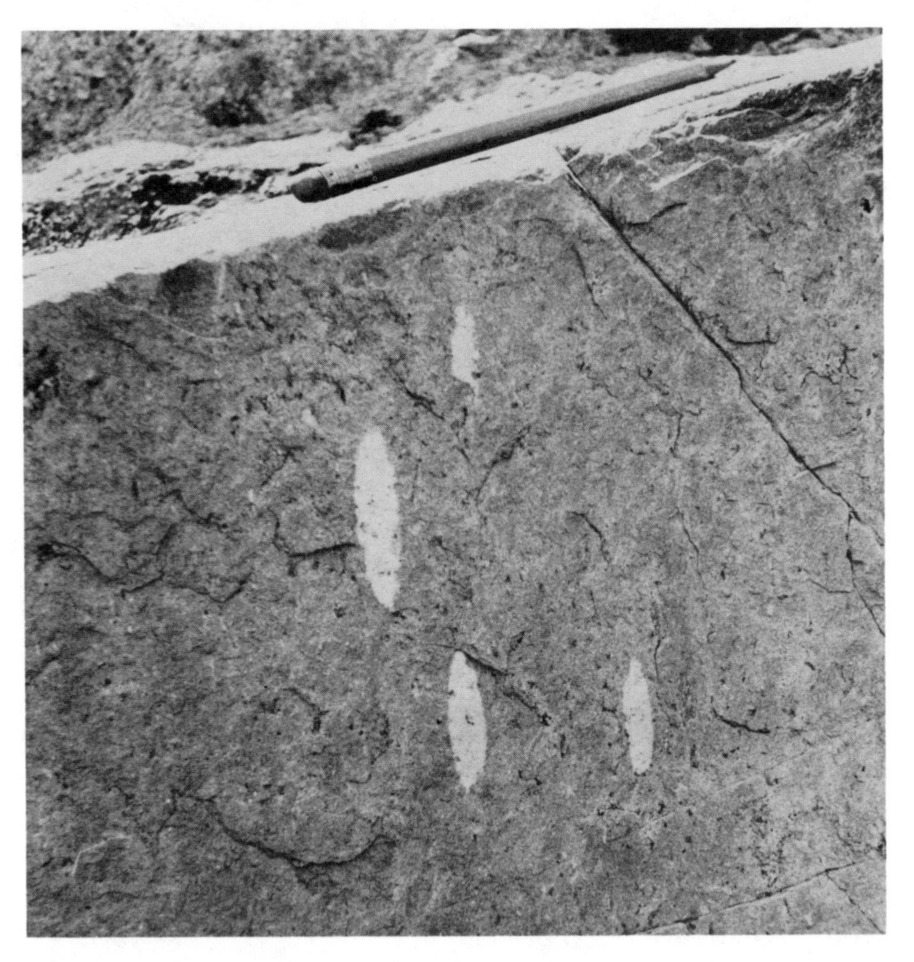

Figure 8.1 Deformed reduction spots. The distortion reflects a secondary strain. (Photograph by O. T. Tobisch.)

Figure 8.2 (*A*) These orbicules in dike rock from Colorado were spherical originally. (*B*) But before the magma congealed into rock, they were distorted to ellipsoids in the plane of flow of the magma. (Photograph by G. Kew.)

Figure 8.3 Cross-bedding in Navajo Sandstone near Checkerboard Mesa, Zion National Park. Geologist is Drew Davis. Truncation of cross-strata along upper bedding surface where Drew is sitting reveals that the Navajo sandstone is rightside up. (Photograph by Drew's Dad.)

side of an open-face sandwich. Some rocks have facing indicators as conspicuous as cheese on bread, and these permit the relative age progression within a sequence to be determined at a glance. Other signatures are much more subtle but nonetheless useful. Knowledge of facing within sequences of partially exposed, folded sedimentary rocks is absolutely essential to making accurate reconstructions of fold geometry.

CLUES TO TRANSPORT DIRECTION(S)

Primary structures are useful in yet another way. They sometimes display geometric properties that can provide guides to kinematic movements that took place during the formation of rocks (Figure 8.4). As kinematic indicators, some primary structures reflect depositional movements; some reflect deformational movements. Kinematic interpretation can focus on such diverse themes as directions of stream transport, long-shore drift along barrier islands, wind directions in ancient dune fields, flow direction of lava, internal creep within a rising diapir, slip direction of a slump or landslide. Many primary structures have a form that permits the **direction** of movements to be determined. Some primary structures are exceptionally useful because they display a polarity of form that allows even the **sense of direction** of movement to be interpreted.

The hero we look to in the use of primary structures as transport indicators is Henry Clifton Sorby, the remarkably creative nineteenth-century scientist whose contributions cross all the earth sciences and beyond (Folk, 1965). Sorby (1859) measured 20,000 primary structures, mainly cross-bedding and lineation, to reconstruct ancient depositional patterns in England.

[Sorby] had first noticed current-structures in sandstones in 1847 [age 19!] while sheltering from the rain in a rock quarry. At the time he had been engaged in observing flow characteristics and sand deposition in a river and was *at once* struck with the application of his research to ancient rocks. (From "Henry Clifton Sorby (1826–1908), the Founder of Petrography" by R. L. Folk, 1965, p. 44. Published with permission of National Association of Geology Teachers.)

Figure 8.4 Current direction indicators on the underside of an overturned sandstone bed in Borneo. The linear grain is parallel to current direction. (Photograph by W. B. Hamilton. Courtesy of United States Geological Survey.)

SEDIMENTARY STRUCTURES OF DEPOSITIONAL ORIGIN

BEDDING

Bedding, ranging from massive layering to delicate lamination, is a fundamental primary structure in sedimentary rocks. Distinctive because of color, texture, composition, and resistance to erosion, bedding imparts to sedimentary rocks their fundamental architecture (Figure 8.5). E. S. Hills said it best.

> It is the sum of its lithological features that characterizes a bed, with the implication that it was laid down under a particular set of conditions, either uniform or systematically varying. (From *Elements of Structural Geology* by E. S. Hills, p. 7. Published with permission of Methuen & Co., Ltd., London,) copyright ©1972.

Figure 8.5 (*A*) Flat-lying bedding in sedimentary rocks in Monument Valley, southernmost Utah. (Photograph by G. H. Davis.) (*B*) Expression of bedding in gently dipping Miocene mudstones, Kaikoua Peninsula, South Island, New Zealand. (Photograph by W. B. Bull.)

The presence of bedding allows the internal structural geometry of sequences of deformed strata to be measured and described.

In the perspective of regional structural analysis, it is useful to combine bedded formations of sedimentary rocks into unconformity-bound **sequences** (Sloss, 1963) that have undergone a common depositional history, perhaps within a uniform or systematically changing plate tectonic setting. Unconformities that mark the tops and bottoms of regional sequences have structural significance, whether it be signaling **orogenic** events of folding, faulting, and igneous intrusion, or reflecting broad, regional **epeirogenic** warping. Sedimentary sequences constitute primary structural systems, the size, shape, and internal structure of which exert influence on mechanical response to secondary deformation. If the form and internal structure of two regional sequences differ, their mechanical responses to deformation will also differ, even if the conditions of deformation are the same for each. Thus we should expect to see contrasts in the mechanical behavior of different sequences, like continent interior platform deposits, miogeoclinal continental margin prisms, wedge-shaped rifted basin deposits, forearc basin deposits, and accretionary wedge melanges.

CROSS-STRATIFICATION

Cross-bedding is a common primary structural element found within clastic sedimentary rocks, especially siltstones and sandstones (Figure 8.6). Cross-stratification is characterized by bedding or lamination oriented at an angle to the bedding surfaces that mark the top and bottom of the cross-stratified unit. In their most useful and diagnostic form, the **cross-beds** or **cross-laminations** are tangential to the lower bedding surface and they are sharply truncated along the upper surface. Cross-stratification is inclined in the general direction of sediment transport and current flow. Where cross-stratification is essentially planar, the down-dip direction represents the sense and direction of sediment transport by water or wind. Where they are scoop shaped, the axis of the scoop plunges in the direction and sense of current flow. Truncation of the upper reaches of cross-beds is achieved by erosion shortly after deposition.

The geometric form of cross-stratification imparts a polarity to rock layers that can be used to determine both facing and current direction. As an

Figure 8.6 Cross-bedding. Truncated layers of older beach rock are capped by a thin sandstone layer that follows the slope of the existing beach, Bikini Atoll on north side of Uorikku Island, Marshall Islands. (Photograph by K. O. Emery. Courtesy of United States Geological Survey.)

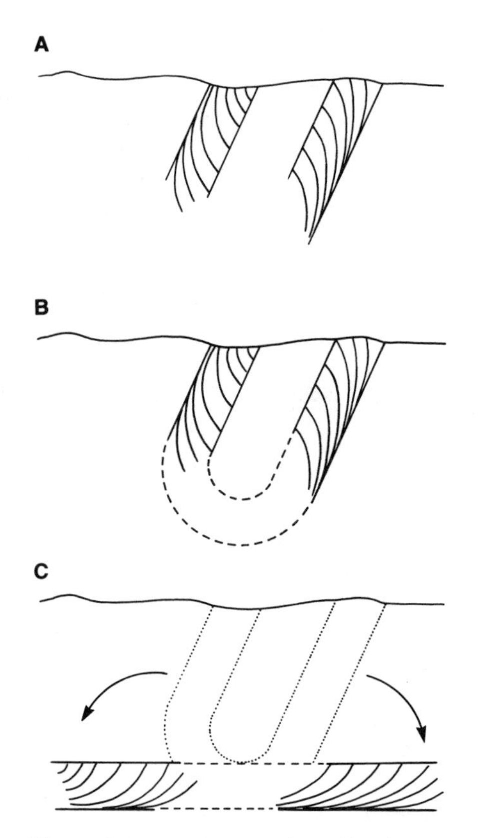

Figure 8.7 (*A*) Structural profile showing moderately steeply dipping cross-stratified sandstone beds. The geometry of the cross-beds discloses that some layers are rightside up and others are upside-down. (*B*) Interpretation. (*C*) Reconstruction of paleocurrent direction by unfolding the folds.

example, consider the structural profile of cross-bedded sandstones shown in Figure 8.7*A*. All the layers dip by the same amount in the same direction, but some are rightside up while others are upside down. Were it not for facing indications derived from cross-beds, the overturning might go unnoticed, and so might the folding that is responsible for the overturning (Figure 8.7*B*). Upon unfolding of the limbs of the folds (Figure 8.7*C*), it is possible to use the polarity of the cross-beds to deduce the sense of paleocurrent transport.

Cross-stratification forms in many different physical environments, including dunes, rivers, beaches, shallow seas, and deltas. Thus, cross-beds are of widespread use as indicators of facing and current direction.

RIPPLE MARKS

Ripple marks are repeated wave forms of sand, silt, and mud that are created in shallow water because of the action of currents, waves, swells, and the play of wind on the surface of the water. **Oscillation ripple marks** have a symmetrical, concave form that reveals facing within a sedimentary sequence (Figure 8.8*A*). **Current ripple marks** are asymmetrical, imparting a polarity to the primary structure by which current direction can be determined (Figure 8.8*B*). Local paleocurrent direction is interpreted to be oriented at right angles to the crests of current ripples, in the sense of inclination of the steeper, shorter face.

PARTING LINEATION

Parting lineation is a subtle primary structure that commonly occurs in siltstone and sandstone. Expressed as a faint linear grain on bedding surfaces, it records the current direction at the time that the sand and silt were being deposited (Figure 8.9). The physical expression of parting lineation is

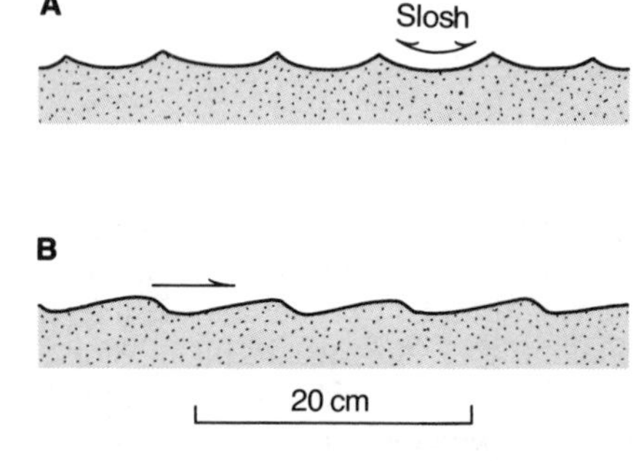

Figure 8.8 (*A*) Oscillation ripple marks. (*B*) Current ripple marks.

Figure 8.9 Parting lineation in the plane of cross-stratification in the Navajo Sandstone, Zion National Park. Knife is aligned parallel to the lineation. The parting lineation records the wind direction during an instant of time in the Jurassic when the Navajo sand was being formed in a great desert dune field.

due to the subparallel alignment of the longest dimensions of silt or sand grains within bedding laminae. The grains are aligned in the direction of paleocurrent (Conybeare and Crook, 1968). Whereas both cross-bedding and current ripple marks reflect direction and sense of current flow, parting lineation records only direction.

GRADED BEDDING

Some sandstones and conglomerates are marked by zones of **graded bedding**, stratigraphic intervals on the order of centimeters or meters within which grain or clast size decreases systematically upward. For example, a pebble conglomerate at one stratigraphic level may grade upward through granule conglomerate to coarse sand (Figure 8.10). Since the grain size becomes finer upward, graded bedding constitutes a useful facing indicator. The presence of graded bedding reflects the settling out of suspended materials from what had been rapidly flowing, sediment-laden current that experienced a sudden decrease in velocity of flow.

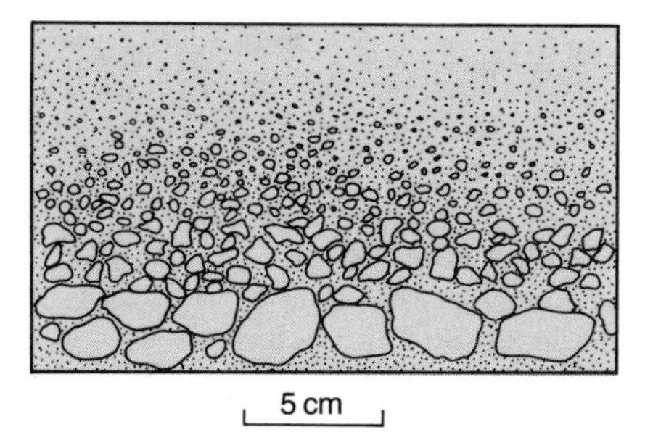

Figure 8.10 Graded bedding, another indication of facing. Clast and/or grain size decreases upward within the unit, in the direction of the youngest part of the bed. (Photograph by G. H. Davis.)

Turbidity current deposits commonly display graded bedding. Turbidity currents move sand, silt, and mud as high-density slurries. They move sediment into trench, forearc basin, and continental slope settings, and into deep lakes as well. When a turbidity current reaches the foot of the declivity down which it flows, it spreads out on flat plains, loses velocity, and drops sediment load in a natural, size-sorted way.

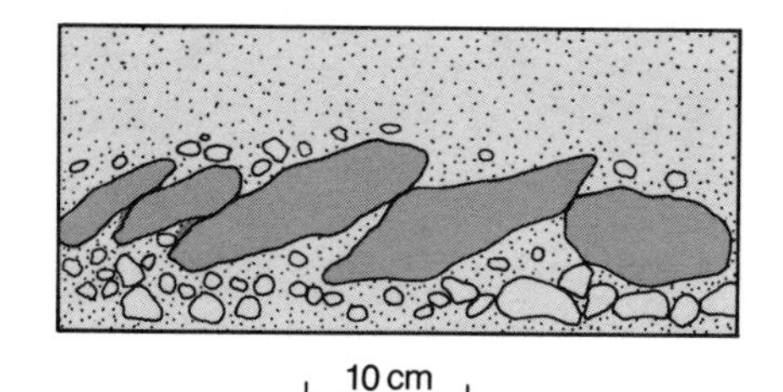

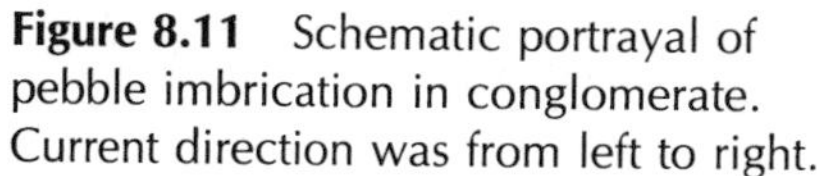

Figure 8.11 Schematic portrayal of pebble imbrication in conglomerate. Current direction was from left to right.

PEBBLE IMBRICATION

Pebble imbrication is marked by shingled overlap of disc-shaped to flat pebbles (Figure 8.11). As in modern stream or shoreline settings, discoidal and platy pebbles tend to dip in the upcurrent direction. Pebbles thus oriented resist being flipped by the current. Given the stability that upcurrent dip provides, the pebbles become shingled. The resulting pebble imbrication is valuable in interpreting current direction and direction of source area for conglomeratic sedimentary sequences, like those that form in rift, interarc, and foreland basins.

SOLE MARKS

Sole marks are so named because they are best preserved on the underside or sole of a given bed. They occur preferentially along bedding surfaces between mud below and sand or silt above. An unblemished muddy seafloor bottom is vulnerable to etching by scouring due to currents and to the movement of "sticks and stones" across the surface. The etched forms may be preserved if the cut and scoured depressions are filled by sand (or silt) during burial of the mud. Upon lithification, sand casts reflect the sizes and shapes of the original bottom markings. The markings become exposed to view on the underside of a sandstone (or siltstone) bed where erosion has removed the underlying shale layer, the former muddy bottom sediment on which the sand (or silt) was deposited (see Figure 8.4).

The forms of sole marks are interesting indeed! **Groove marks** are relatively long, linear casts produced by the scratching and plowing of current-propelled objects across the soft surface of mud (Figure 8.12). Their orientations yield directions of local paleocurrent. **Flute casts** are footprint-shaped features that taper from wide (size EEE) to narrow (size AAA) from toe to heel. Formed by current scouring, they splay in the downcurrent direction. The heel of a flute cast is thickest, and this reflects greatest depth of scouring in the upcurrent direction.

Figure 8.12 A look at the underside of a steeply inclined bed of Pennsylvanian sandstone in the Marathon Basin, Texas. The sole of the bed displays a nice array of grooves and flute casts. Current direction responsible for deposition of the sandstone was from right to left. (Photograph by E. F. McBride.)

SEDIMENTARY STRUCTURES OF DEFORMATIONAL ORIGIN

SOFT-SEDIMENT BEHAVIOR

Certain deformational structures may develop in sediments before the sediments are hardened into rock. The nature of the structures provides insight regarding the mechanical consistency of the materials before lithification, and the environment within which the sediments were deposited and buried. Familiarity with the nature of soft-sediment structures is necessary to bypass confusion with secondary structures. However, distinguishing between primary and secondary structures in some environments may be hopeless and/or arbitrary.

Plastic, soft-sediment deformation is enhanced by the water content of freshly deposited muds, silts, and clays. The physics of this behavior is an important focus of study for geologists and engineers. Water-rich sediments commonly respond to stress in a **hydroplastic** manner (Elliot, 1965). Hydroplastic sediments can be distorted easily because individual grains of sediment, cushioned all around by water, are free to move independently of surrounding grains. Fortunately, bedding laminae and other primary depositional features are often preserved, and these give expression to the distortional effects. However, when sediments that are especially rich in water are distorted, they behave as quasi-liquids and all internal structure is lost to view (Coneybeare and Crook, 1968).

Figure 8.13 Primary fold in the Green River Shale (Eocene) in western Colorado. The fold is intraformational, overlain and underlain by undistorted laminae of rock. (Photograph by G. H. Davis.)

Primary sedimentary structures of deformational origin are wonderful examples of nonrigid body deformation. At one extreme the strain is purely dilational: for example, the compaction of sediments due to loading during burial results in layer thinning and volume reduction. At the other extreme, the primary strain may be purely distortional. Distortional soft-sediment deformation commonly affects only the uppermost bed(s) of a freshly deposited sequence within a basin. The wet sediments may slump, flow, fold, or fault where inadequately supported or where triggered to do so by earthquake shocks. Resulting structures are confined to a very narrow stratigraphic interval overlain and underlain by strata that are not themselves internally deformed (Figures 8.13 and 8.14).

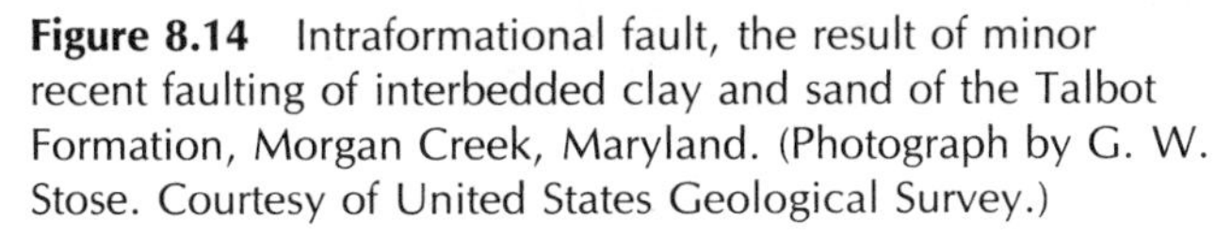

Figure 8.14 Intraformational fault, the result of minor recent faulting of interbedded clay and sand of the Talbot Formation, Morgan Creek, Maryland. (Photograph by G. W. Stose. Courtesy of United States Geological Survey.)

MUD CRACKS

When the wet mud of a river bottom or lake bed or beach is permitted to dry out, it will characteristically become fragmented into polygonal blocks that are bounded by **mud cracks** (Figure 8.15). The cracking accommodates the progressive volume reduction that must take place as water in the drying mud evaporates. Because the amount of shrinkage is constant in all horizontal directions within the plane of the mud layer, the mud crack pattern tends to be random.

Mud cracks, when filled by sand, may be preserved in the geologic record. Where found in ancient sediments, they may be used to interpret depositional environments. Furthermore, since they commonly taper downward, mud cracks become instruments for interpreting facing of the sequence of rocks in which they are found.

Figure 8.15 Mud cracks in the Willcox dry lake (playa), southeastern Arizona. (Photograph by B. J. Young.)

DIFFERENTIAL COMPACTION

Compaction of a layer of sediment during burial is a natural response to vertical force created by the load of a steadily increasing thickness of younger sediments. Compacting a sediment is like placing a brick on a wet sponge. Water is squeezed out, pore space is reduced, density increases, and the layer is thinned. The degree to which a layer can be thinned varies markedly with lithology and texture. Under the same conditions of loading, the range of **compaction strain** can vary from 0 to 70% among different lithologies (Chilingarian and Wolf, 1975). The percentage compaction strain can be calculated according to standard strain parameters. For example,

$$e_c = \frac{t_f - t_o}{t_o}(100)$$

where e_c = compaction strain,
t_f = final thickness of layer,
and t_o = original thickness of layer.

Lateral variations in thickness, load, and compactability of sediments give rise both to regional and outcrop-scale structures. The principles are illustrated in Figure 8.16. Several situations are depicted. Location A features a layer of constant thickness whose compactability increases laterally from 40 to 60% as a consequence of variations in lithology. Between locations A and B there is a stratigraphic interfingering of two lithologies, one whose compactability is 60%, the other whose compactability is only 20%. Location C features a stratigraphic layer of 50% compactability that has been deposited nonconformably against and over a perfectly rigid mass of igneous rock. After burial and compaction of these layers, differential changes in thickness produce an interesting array of structures. A simple **homocline** forms at A, wherein all the strata dip by the same amount in the same direction. **Fishtail structure** develops between A and B. The ragged-edge interfingering expresses a form quite similar to structures mapped and mined in the anthracite coal fields of eastern Pennsylvania where highly compactable peat beds were once interlayered with sand beds more resistant to compaction. A **compaction fold** forms at location C. Such folds have nothing to do with tectonic stresses. They are due instead to differential responses to load-induced compaction (Nevin and Sherrill, 1929).

EARTH CRACKS

In the desert regions of the American Southwest, the role of differential compaction is being accelerated by the impact of man. Rapid withdrawal of groundwater from unconsolidated basin-fill alluvium has led to compaction of the alluvium at rates much greater than normal. This can lead to **differential subsidence** between areas underlain by the compacted alluvium versus alluvial deposits that have not yet been compacted. The subsidence is accommodated by long-continuous fractures marked by small, almost negli-

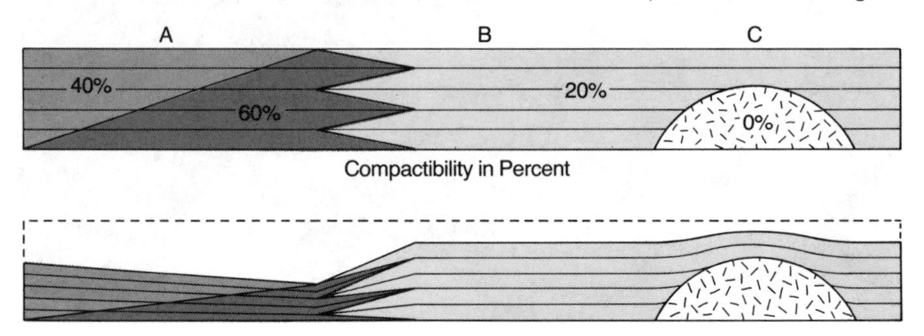

Figure 8.16 Schematic illustration of differential compaction.

Figure 8.17 Earth crack in the desert floor near Picacho, Arizona, between Tucson and Phoenix. Creosote bushes are about 1 m high. (Photograph by T. L. Holzer. Courtesy of United States Geologic Survey.)

gible, dilational and/or shear movements. The surface expression of many of these cracks is foreboding (Figure 8.17): open, fissurelike **earth cracks** as wide as ±3 m and as deep as ±5 m. The fissurelike nature of these cracks forms by **piping**, the percolation of water along the initial, narrow fracture, resulting in caving and the formation of tunnels through which the unconsolidated alluvium is removed. Some earth cracks can be traced for distances of kilometers, through subdivisions and across highways and railroad tracks.

BOUDINS AND PINCH-AND-SWELL STRUCTURE

Where unconsolidated layers of strongly contrasting strength properties are permitted to stretch as they compact, **pinch-and-swell** structures and **boudins** can form (Figure 8.18). Pinch-and-swell structure is characterized by a gentle pinching or necking of the relatively stiff layers within the multilayer sequence. Boudins are formed when the stiff layers neck down radically and/or rupture completely to accommodate layer-parallel lengthening. Whether boudins or pinch-and-swell structures form within a sequence of layers depends largely on the **ductility contrast** between the layers that are subjected to layer-parallel stretching (Figure 8.18).

Boudins sometimes look like links of sausage when viewed in cross section, hence the name (*boudin* is French for sausage). The process of forming boudins is called **boudinage**. Where stretching takes place in the plane of layering such that the sediment is pulled apart in two mutually perpendicular directions, another edible-sounding structure forms: **chocolate tablet structure**. It looks like the top surface of a Hershey bar. Pinch-and-swell structures, boudins, and chocolate tablets are by no means restricted to soft-sediment deformation. As we will see, they are commonplace in deformed sequences of sedimentary and metamorphic rocks where folding and shearing have distorted layers characterized by strong ductility contrast.

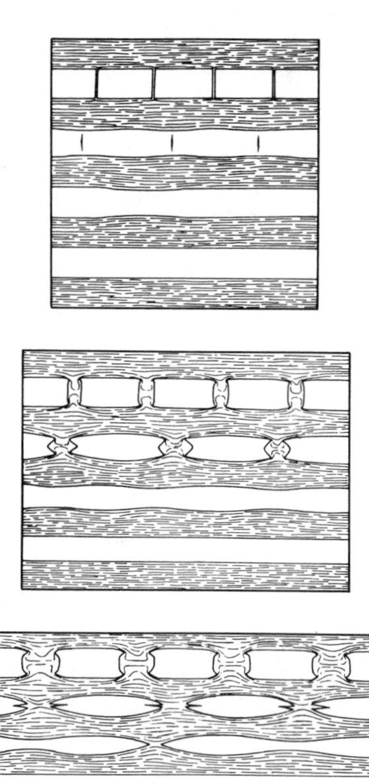

Figure 8.18 Formation of pinch-and-swell structure and boudins by layer-parallel stretching. Ductility contrast between layers determines the extent to which the stiffer layers pinch, neck, and/or break. (From *Folding and Fracturing of Rocks* by J. G. Ramsay. Published with permission of McGraw-Hill Book Company, New York, copyright ©1967.)

PENECONTEMPORANEOUS FOLDS

Penecontemporaneous folds can form during deformation of unconsolidated hydroplastic sediments. Such folds are typically **intraformational**; that is, they are restricted to a given bed or a given sequence of beds (Figure 8.19*A*). The restriction of penecontemporaneous folds to narrow stratigraphic intervals is consistent with folding on the floor of a depositional basin or at very shallow depths of burial.

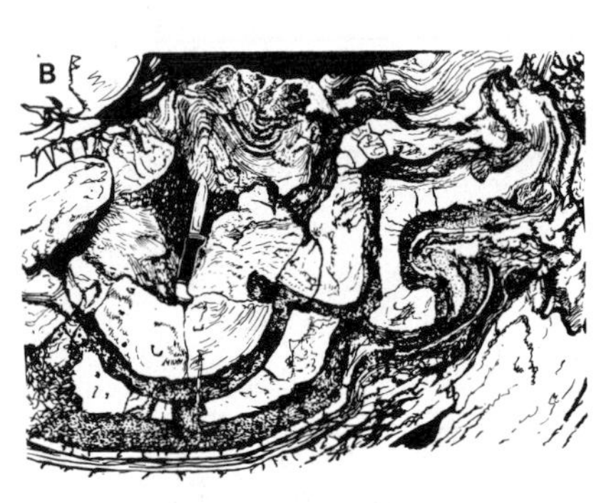

Figure 8.19 (*A*) Intraformational recumbent folds in Pleistocene Flora Formation near Saskatoon, Saskatchewan. [From Hendry and Stauffer (1977). Published with permission of Geological Society of America.] (*B*) Disharmonic folding. Layers of sediment were not yet consolidated at the time they were formed. [From Davis and others (1979). Published with permission of Geological Society of America and the authors.]

Most penecontemporaneous folds are simply irregularly contorted beds whose fold properties are **disharmonic**. Disharmonic folds have forms that do not propagate upward or downward in any systematic, predictable manner (Figure 8.19*B*). They commonly show radical changes in layer thickness, a reflection of mobility of the mass of material thus distorted.

Gravity plays an important role in the development of intraformational folds. Where layers of unconsolidated muds and sands are perched precariously in part of a basin where topographic slope or primary initial dip is oversteepened, they may slump and fold when triggered by earthquake shock, rapid sediment loading, or creep.

PENECONTEMPORANEOUS FAULTS AND DUCTILE SHEAR ZONES

In some environments the cohesion of yet-unconsolidated materials is such that hydroplastic flow is minimized. Instead, soft-sediment deformation may be accomplished by **penecontemporaneous faulting** and **ductile shearing**. The faults and shear zones serve to extend or shorten the not-yet-consolidated layers. During shearing, soft material of one layer can be smeared onto the faulted face of a stiffer layer.

GROWTH FAULTS

Growth faults are an especially important class of faults. Although often they are large regional structures, growth faults can be considered to be primary deformational structures because they operate during sedimentation. Growth faults form characteristically, but not exclusively, in unconsolidated sediments freshly deposited in basins that are actively growing in breadth and depth. Faulted depressions in the floor of the depositional basin become filled in by accumulating sediments. The structural record for this is clear: within a given layer cut by a growth fault, sediment thickness is disproportionately great on the down-dropped side (Figure 8.20*A*). In fact, thickness of sediment in the downthrown block may be 10 times the thickness of sediment in the relatively upthrown block (Edwards, 1981).

Relatively down-dropped blocks of growth faults commonly display **rollover anticlines**, also known as **reverse drag** (Figure 8.20*B*). Rollover anticlines result from the slumping of strata toward the fault as a means of eliminating would-be open space created by movement on curved fault surfaces (Figure 8.20*C*). The strain effect of rollover anticlines can also be achieved by **collapse** featuring the down-dropping of fault-bounded blocks (Figure 8.20*D*).

The Gulf of Mexico contains a growth fault system whose characteristics are especially well known because of intensive petroleum exploration (Figure 8.21) (Hardin and Hardin, 1961; Murray, 1961; Ocamb, 1961; Halbouty, 1969; Martin, 1978). The faults have been mapped in the subsurface both offshore and onshore. The growth faults dip, on average, from 40° to 60°, and they are distinguished by radical increases in thicknesses of beds on the relatively down-dropped blocks (Edwards, 1981). Commonly, the faults curve and shallow out with depth. Fault dip is generally basinward, and the strike directions of individual faults are generally parallel to the depositional strike of Tertiary units in the Gulf Coast basin (Bruce, 1972) (Figure 8.21).

The origin of the growth fault system in the Gulf Coast region seems to be related to a combination of favorable dynamic circumstances. For one thing, the Louann salt source bed, which marks the base of the enormous sediment package of the Gulf Coast depositional system, is flowing basinward as a rheid in response to vertical loading by sediments (see Figure 7.25). The Sigsbee escarpment is an imposing salt scarp formed where the salt, driven by gravity, has moved tens of kilometers over the youngest Gulf Coast sediments. Some of the growth faulting may reflect extension of the Cenozoic sediment package as it glides basinward on the salt.

Specific local controls for the development of the growth faults relate to the sediment distribution of muds and sands, and the rheological properties of unconsolidated muds and sands (Bruce, 1972; Edwards, 1981). Bruce (1972) has shown that the **regional contemporaneous faults** coincide with the basinward flanks of huge, buried elongate masses of shale. The **shale masses**, as they are called, are tens of kilometers long, up to 40 km wide, and

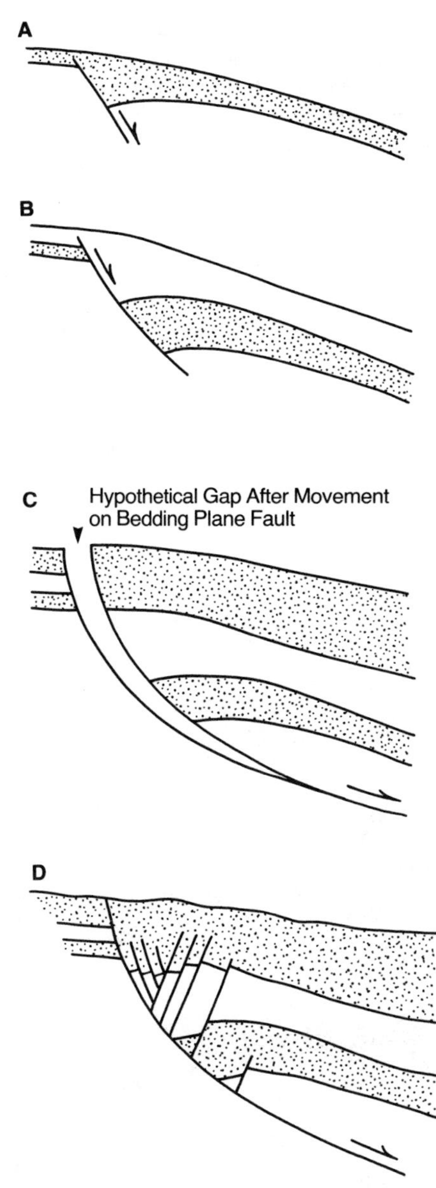

Figure 8.20 (*A*) Characteristic form and geometry of a growth fault. (*B*) Rollover anticline, reverse drag of the hanging wall strata. (*C*) Movement along a listric growth fault creates the potential for fissurelike open space. Hypothetical open space is eliminated by (*D*) the formation of a rollover anticline and/or collapse along antithetic faults. [From Bruce (1972). Published with permission of Gulf Coast Association of Geological Societies.]

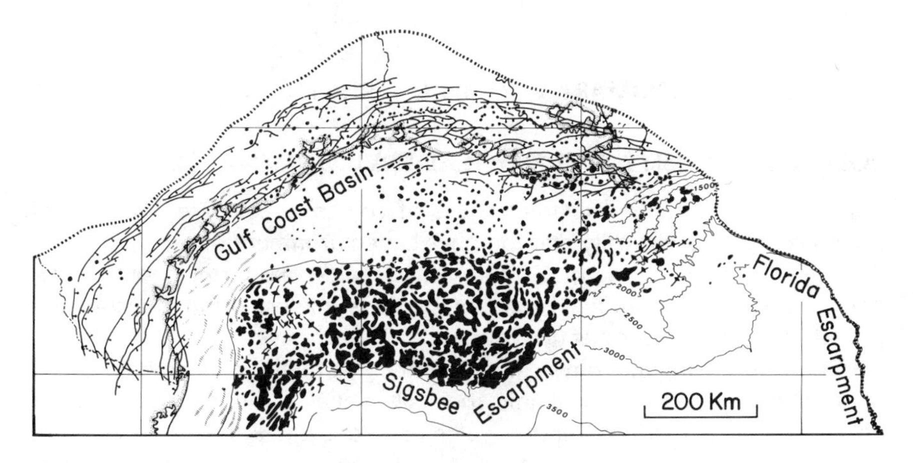

Figure 8.21 Map of the distribution of growth faults and salt domes (black) in the northern and eastern Gulf of Mexico region. The growth faults cut Cenozoic deposits in lower coastal plain and continental shelf regions. [From Martin (1978). Published with permission of American Association of Petroleum Geologists.]

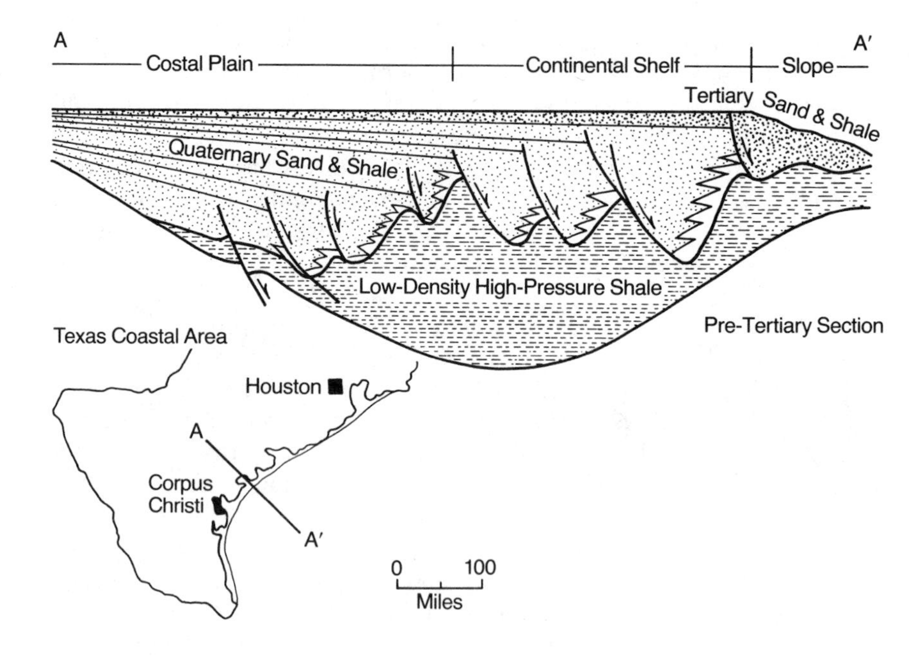

Figure 8.22 Relation of Gulf Coast growth faults to pressurized shale masses. [From Bruce (1972). Published with permission of Gulf Coast Association of Geological Societies.]

3 km thick. They are oriented subparallel to the coast. Undercompacted and characterized by high fluid pressure, the shale masses serve to control the locations of the growth faults. Specifically, the faults separate compactible sand sequences from the tops and/or flanks of the pressurized shale masses (Figure 8.22). In similar fashion, Edwards (1981) has emphasized that the major zone of growth faulting in south Texas coincides with the location where huge Paleocene–Eocene sand deltas prograde over unstable basin muds at the shelf margin. Thus the faulting may reflect, in part, a kind of differential compaction of interfingering sediment facies (Carver, 1968).

Ernst Cloos (1968) tackled the growth fault system of the Gulf of Mexico experimentally. He was able to reproduce the dominant physical and geometric characteristics of the system through well-planned experiments using sand, clay, and paraffin. Cloos concluded that the Gulf Coast sediment package has undergone, and continues to undergo, gravity-induced down-to-the-basin creep.

PRIMARY VOLCANIC STRUCTURES

ROLES OF VISCOSITY AND PHYSICAL ENVIRONMENT

The primary structural characteristics of volcanic rocks are largely controlled by magma viscosity and the physical environment of extrusion. This theme is the cornerstone of Gordon Macdonald's masterful book, *Volcanoes*, a treat for any student of geology who is interested in understanding primary volcanic structures (Macdonald, 1972).

It is not surprising that physical environment strongly influences the kinds of structure that form in volcanic rocks. It makes a difference whether a volcanic flow is extruded under water or under air. Rates of cooling are markedly different in the two settings. The steepness and roughness of topography is also bound to influence external and internal flow movements, including relative speed. The forms of volcanic centers and volcanic products can be quite diverse (Figure 8.23)!

Magma viscosity is resistance of magma to flow. Chemical composition has greatest influence on magma viscosity, although temperature, gas con-

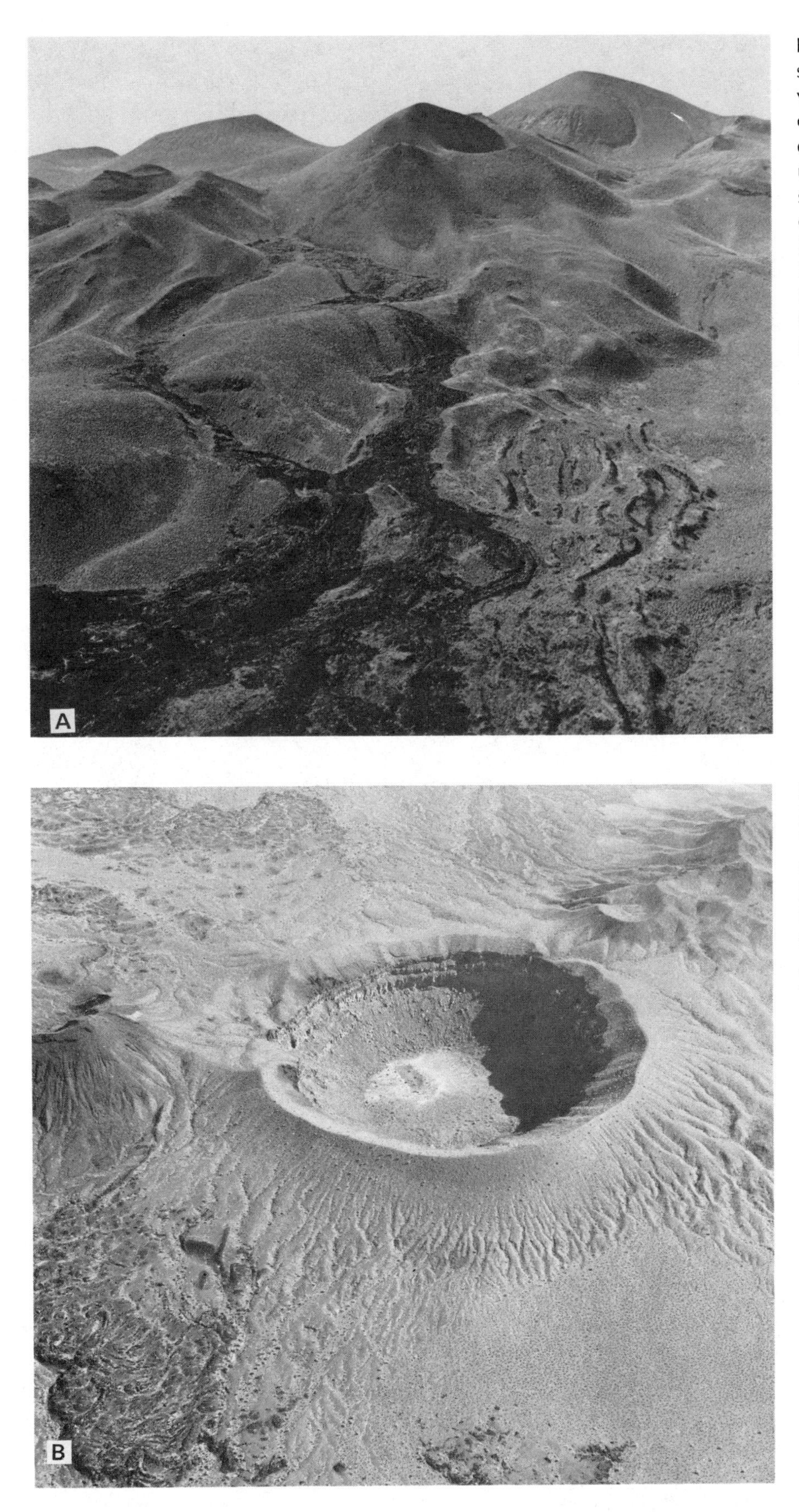

Figure 8.23A Carnegie Cone near the summit of Volcan Santa Clara, Pinacate volcanic field, Sonora, Mexico. Both cinder and flowing lava were produced during eruption. Most of the cinder piled up to form a cone around the vent, but some traveled far enough to mantle the countryside, covering earlier lava flows. Lava levees of earlier flows are seen as black ridges in lower right. Youngest lava flow (black rock river) issued from a fissure at the base of the cone. (Photograph by D. J. Lynch.)

Figure 8.23B Sykes Maar Crater in the Pinacate volcanic field, Sonora, Mexico. The Maar Crater was "excavated" by steam explosions produced when rising basalt magma interacted with groundwater in the unconsolidated sediments beneath the volcanic field. The walls of the crater are composed of early-formed basalt flows. (Photograph by D. J. Lynch.)

Figure continues

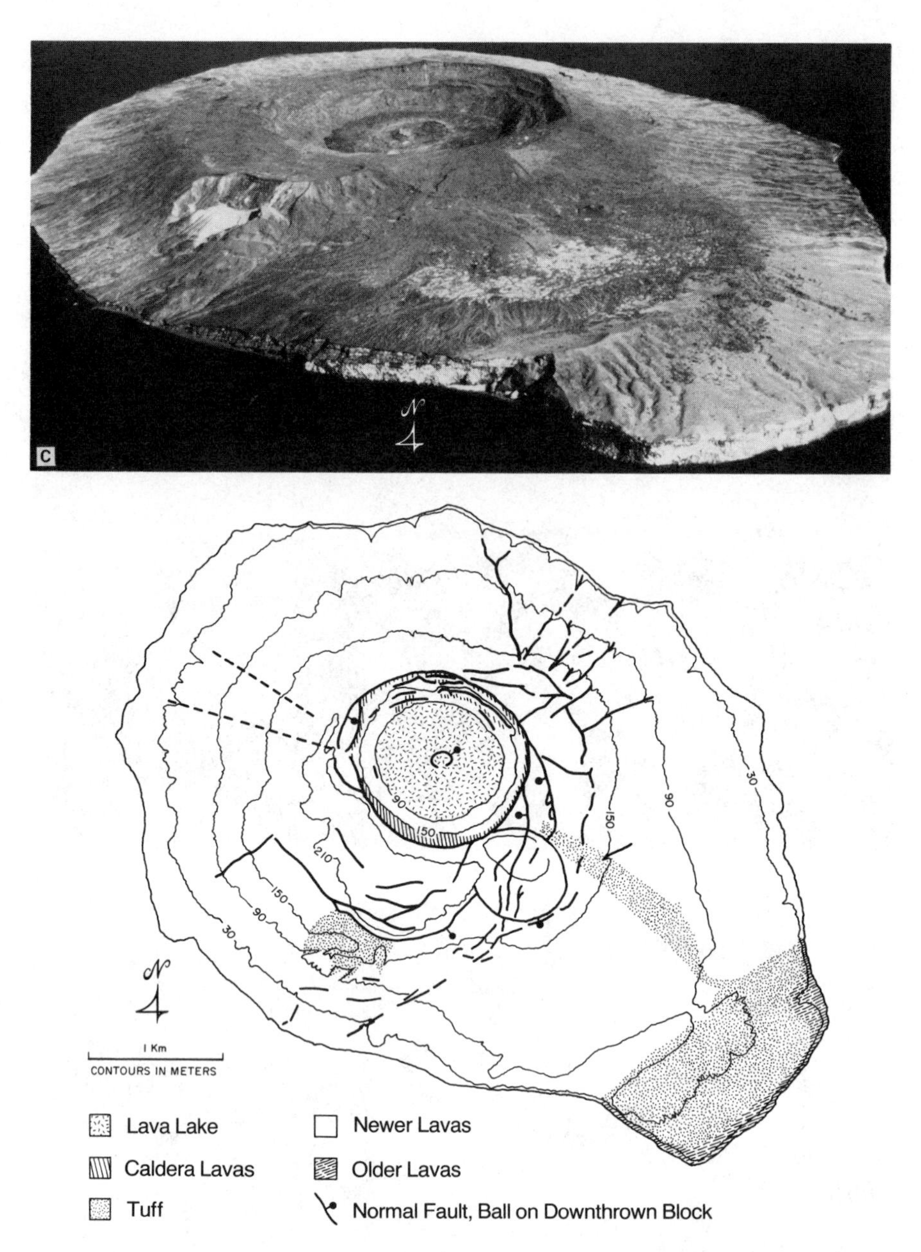

Figure 8.23C Photograph and geologic map of the Isla Tortuga, a recently formed volcanic island in the Gulf of Mexico. The volcanic island is composed mainly of basalt and vitric tuff. During its latest stage of activity, the volcano suffered caldera collapse, accompanied by extrusion and the formation of a lava lake. Collapse was accommodated by the concentric rings of inward-dipping faults. [From Batiza (1978). Published with permission of Geological Society of America and the author.]

tent, and amount of solid load are important factors as well (Macdonald, 1972). Mafic, basaltic magmas tend to be of relatively low viscosity (ca. 5×10^4 poises). Basalt magma is thus very fluid. Felsic, rhyolitic magmas are of relatively high viscosity (ca. 5×10^{12} poises), and thus tend to be very stiff. The difference in fluidity of basaltic versus rhyolitic magma partly accounts for the dominantly intrusive habit of granites and the dominantly volcanic habit of basalt.

FLOW STRUCTURE

Flow structures preserved in volcanic lava commonly provide clues to movement plan and facing. Of all flow structures in basalt, **ropy lava** is the most expressive (Figure 8.24). It forms in relatively low-viscosity, **pahoehoe** basalt. Ropy lava results from local lava currents that drag fold the plastic skin of a flow, contorting it into a series of nested arcs, which tend to be convex in the sense of current flow. The internal complexity of eddy patterns and other disharmonic movements are usually difficult to reconcile with

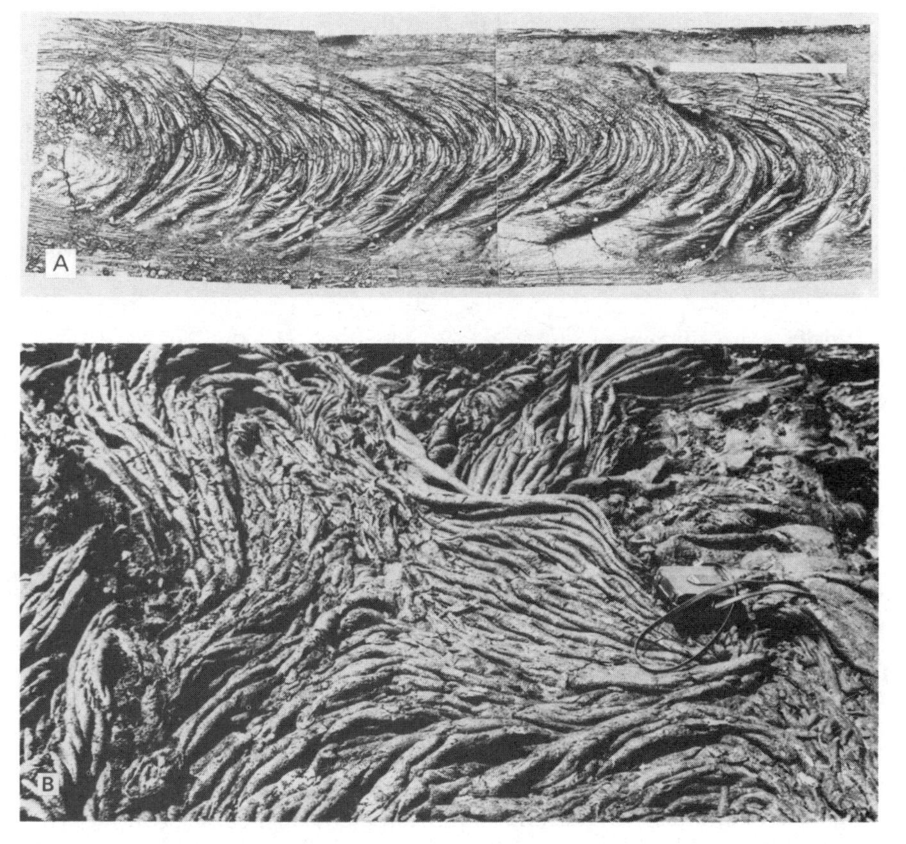

Figure 8.24 Ropy pahoehoe structure, Kilauea Volcano, Hawaii. Bar scale represents 2 m. [From Fink (1980). Published with permission of Geological Society of America and the author.] (*B*) Close-up view of ropy pahoehoe in the 1919 flow within Kilauea Caldera, Hawaii. (Photograph by G. A. Macdonald. Courtesy of United States Geological Survey.)

overall movement plan of the flow as a whole. Like the image of foam layers on the surface of a Canadian pool, the pattern of ropy lava frozen into the lava is a stop-action record of just a single part of a large flow at a single instant in time. The record may or may not validly represent the movement of the lava flow as a whole.

To better sense the mechanical consistency of the skin that forms on pahoehoe lava, and to gain greater appreciation of Gordon Macdonald's depth of experience as a volcanologist, read these words.

> Pahoehoe flows quickly cool to a semisolid state, but the crust thus formed remains somewhat plastic. It is this crust that may become wrinkled and twisted to form the ropy structure. It is possible to cross a flow on this plastic crust while the flow itself is still moving. The crust sags and bends under one's weight, like "rubber" ice on a partly frozen pond, but does not break. Similarly, one can jump on the top of small "toes" along the edge of an active pahoehoe flow and cause the liquid lava to spurt from the end of the toe, without breaking through the top, as long as the toe remains filled with liquid. (From *Volcanoes* by G. A. Macdonald, p. 73. Published with permission of Prentice-Hall, Inc., Englewood Cliffs, New Jersey, copyright ©1972.)

Within basaltic lava flows there are a number of structures that reflect larger scale flow movements. Hollow, abandoned **lava tubes** are preserved in the depths of flows, the remains of feeders that once funneled lava to a downslope destination. Differential stresses that build as a result of surges of new lava may create **squeeze-ups** in the form of dikelike intrusions, which squirt upward from the molten interior of a flow into the upper and outer chilled margins. The frozen skin of lava flows is commonly deformed by surface folding into **pressure ridges**. Viewed from high altitude, pressure ridge patterns resemble ropy lava (Figure 8.25).

Figure 8.25 Pressure ridges in rhyolite obsidian flow, Big Glass Mountains, Medicine Lake Highlands Volcano, California. Bar scale is 500 m. [From Fink (1980). Published with permission of Geological Society of America and the author.]

Where basaltic lava is extruded underwater, or where lava pours into the sea, **pillow lavas** instantly form (Figure 8.26*A*). "Pillows" are generally flattened, with rounded tops. Where a pile has been formed, the bottoms of the pillows conform in shape to the rounded tops of those beneath them, so that in vertical cross section they show lobes projecting downward between the intersecting curved tops of the underlying pillows (Figure 8.26*B*). Given their cross-sectional forms, pillows are valuable as facing indicators, even within metamorphosed basalts (i.e., greenstones) of Precambrian age. Mid-oceanic ridges are great pillow factories.

Vesicles are swarms of tiny holes in volcanic rock. They are the frozen record of gas bubbles in lava. Vesicles are versatile elements for use in structure analysis. Their concentration at the tops of lava flows make them valuable facing indicators. In addition, their ellipsoidal forms reflect late-stage distortional flow movements just before crystallization. The strain record, in turn, provides clues to reconstructing the internal movement plan within flows. Thus, under ideal conditions, vesicles may provide help in assessing facing, primary strain, and movement plan. In addition, vesicles offer open space for the precipitation of minerals from circulating hydrothermal fluids. The filled vesicles are called **amygdules**. The most famous amygdules occur in Precambrian basalt in the Keweenawan Peninsula of Michigan. They are locally filled with native copper!

A. Waters (1960) has described rather unusual primary structures in the flood basalts of the Columbia River Plateau. Known as **spiracles**, these structures are large gas chimneys that have channeled gas and vapor upward from steaming water or burning organic material beneath the flow. The vapors rise through the base of the flow and actually chill the interior of the

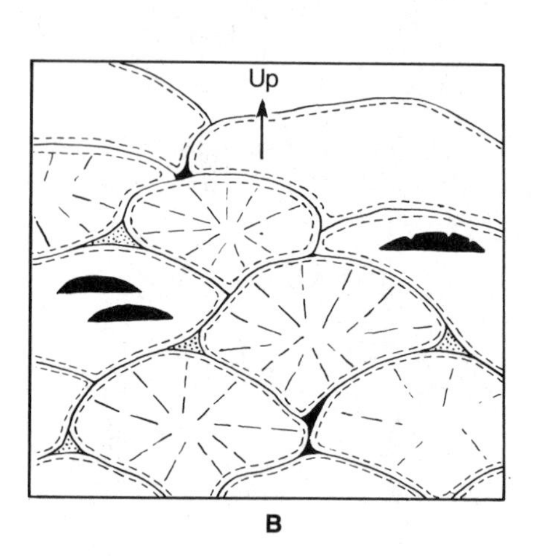

Figure 8.26 (*A*) Pillow lava at Alicastello, Sicily. (Photograph by R. Greeley.) (*B*) The characteristic rightside-up form of pillows. [From Macdonald (1967). Published with permission of Wiley-Interscience, Inc., New York, copyright © 1967.]

hotter basalt, fashioning an internal cavity. The chimneys range from centimeters to meters in diameter. The smallest are called **pipe vesicles** and **pipe cylinders**. Upon continued movement of the flow after the initial formation of these primary volcanic structures, the tops of spiracles, pipe vesicles, and pipe cylinders each can bend as a result of simple shear movements. Bending is in the direction of movement, disclosing sense of flow.

Rhyolitic to rhyodacitic flows are of such high viscosity that the frictional resistance to flow produces internal simple shear. Simple shear in turn distorts the lava and produces, among other structures, **asymmetric intraformational folds**. Progressive deformation with continued flow can result in the refolding of earlier formed folds (Figure 8.27). The orientations of asymmetric intraformational folds can be a useful guide to direction of flow. Resolving the direction(s) of movement of a flow may in turn place constraints on the location of the vent(s) from which the lava issued.

Figure 8.27 Fold within a fold in rhyodacite specimen collected in the Chiricahua Mountains, Arizona. (Photograph by G. Kew.)

COLUMNAR JOINTING

In few places is the comparison of architecture and structural geology so rich as in descriptions of the columnar-jointed basalts (Figure 8.28). Composite basalt flows are commonly described in terms of an upper colonnade, an entablature, and a lower colonnade. In architectural terms, "colonnade" refers to a series of columns at regular intervals; and the "entablature" is the upper section of a wall supported on columns or pilasters. The columns of basalt flow architecture are produced by **columnar jointing**, a fracturing that accommodates negative dilation during final congealing and shrinkage of a flow. Columns tend to be polygonal (Figure 8.29) in the same way that mud cracks resulting from shrinkage of mud are polygonal. Columns ranging from 1 m to 20 m in height are found in the lower colonnade of Columbia River Plateau basalts. They are vertical and beautifully developed. Within the entablature, columns are much narrower then those in the lower colonnade, and they display a greater range of orientations. The upper colonnade is relatively thin, containing thick columns that lack perfection of form (Waters, 1960; Schmincke, 1967).

Figure 8.28 Architectural display of columnar jointing in the lower colonnade of the prehistoric Mauna flow exposed at the Boiling Pots near Hilo, Hawaii. Faint horizontal striae on the columns are a record (an "etching") of crack advance within the cooling basalt. [From Ryan and Sammis (1978). Published with permission of Geological Society of America and the authors.]

Figure 8.29 Columnar joints in basalt, exposed at San Miguel Regla, Hidalgo, Mexico. (Photograph by C. Fries. Courtesy of United States Geological Survey.)

Waters (1960) demonstrated that the orientations of columns in the entablature section of flood basalts sometimes reveal the sense of direction of lava flow. Columns are locally seen to plunge toward of the rear of the flow of which they are a part.

COMPACTION STRAIN IN ASH-FLOW TUFFS

The products of hot, gas-charged ignimbrite eruptions are particularly widespread in the Western Cordillera of the United States and Mexico. They form volcanic piles thousands of meters thick and covering tens of thousands of square kilometers (see Figure 7.2). Most are mid-Miocene in age. The cause of their eruption is not yet understood, but the event could have been triggered by deep-seated rifting of a magmatic arc.

One of the distinctive and peculiar characteristics of ash flow tuffs is the presence of **welded** textures of primary origin. Welding reflects the compaction strain of a hot, heavy, chemically active body. Incipient welding is marked by the sticking together of glassy volcanic fragments at their points of contact (Figure 8.30*A, B*). Complete welding achieves cohesion of the full surfaces of adjacent glass fragments (Figure 8.30*C, D*). Complete welding is accompanied by the elimination of all pore spaces (including vesicles) and the distortional flattening of glass shards and pumice fragments. The result of welding is **eutaxitic structure**, a foliation defined by aligned flattened shards and pumice fragments (Figure 8.31). At the base of some ash flow sheets the flattened pumice is ellipsoidal in the plane of foliation. Such elongation reflects kinematics in the form of late-stage movements of the flow as a whole (Smith, 1960a, b).

Ash flow sheets resulting from a single eruption, or from quick successive eruptions, form **simple cooling units** (Smith, 1960b). Compaction strain is imparted to the sheet(s) during uninterrupted episodes of cooling. A simple cooling unit consists of internal zones that vary in their degree of welding. The bottoms and tops are represented by nonwelded rocks. Inward from top

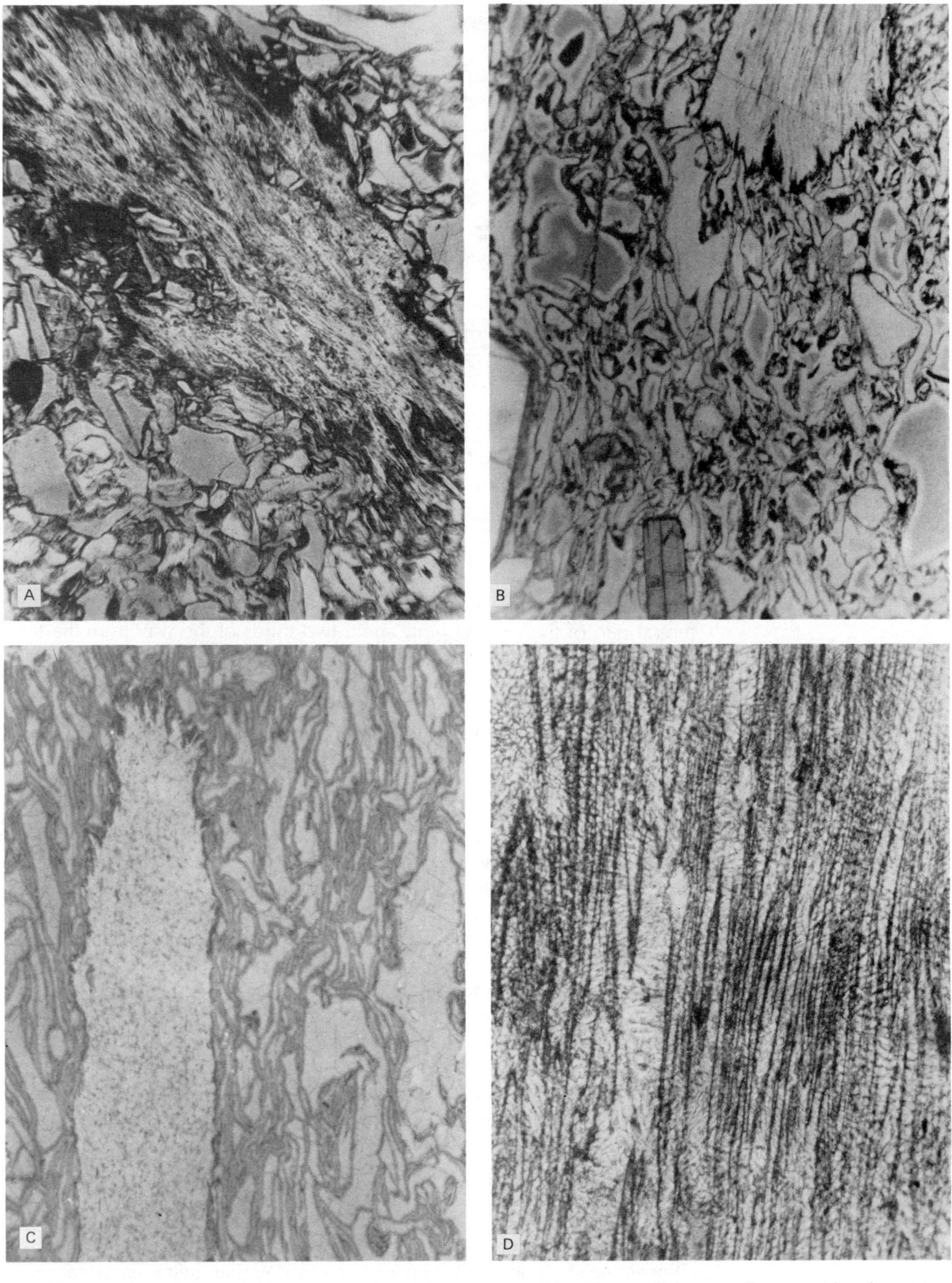

Figure 8.30 Photomicrographs of welded ash flow tuff illustrating the progressive loss of the internal structure of pumice by welding. (*A*) Least welded. (*B*) More welded. (*C*) Most welded. (Photographs by R. L. Smith.) (*D*) Extreme compaction probably accompanied by stretching due to slight mass flowage. Also shown is superposed devitrification in the form of a subtle grain that crosses the compaction fabric. [From Ross and Smith (1960). Courtesy of United States Geological Survey.]

Figure 8.31 Eutaxitic structure within a Tertiary ash flow sheet in Nevada. Lower right half of outcrop is composed of densely welded tuff. Partings in the tuff follow concentrations of pumice lapilli which have been preferentially weathered out. (Photograph by P. W. Lipman. Courtesy of United States Geological Survey.)

and bottom, the rocks become incipiently welded to partially welded. In the central part of the cooling unit, about two thirds of the way from the top, there is a zone of dense welding, which appears as a dark obsidianlike glass.

The internal strain reflected in welded textures can be quantitatively measured. Ragan and Sheridan (1972) evaluated the internal strain of the Bishop tuff, a Pleistocene ash flow tuff in eastern California. Their studies indicate that the compaction process moves from a fluid stage through a viscous stage. In the early fluid stage, pumice fragments behave as rigid particles, free to rotate as the surrounding gas-rich matrix is continuously flattened by load-induced degassing. The fragments first rotate into parallelism with the plane of flattening. When porosity reaches about 50%, the pumice fragments begin to collapse along with the matrix. When pore space is negligible, further compaction is achieved by pure flattening of the pumice fragments.

Ragan and Sheridan (1972) were able to measure strain in a number of ways. Deformed gas bubbles (i.e., vesicles) provided one measure of strain, but use of this primary structure was limited to zones of incipient or partial welding. The use of **glass shards** as strain devices was more helpful to these investigators. Glass shards may be thought of as the rinds of closely packed bubbles in vesicular parts of ash flow sheets. If the bubbles are spherical and uniform in diameter, the rinds between bubbles form branching glass shards that everywhere intersect at angles of about 120°. Ragan and Sheridan measured departures from the original 120° angles as a measure of the internal distortion within welded parts of the cooling units. They then converted the angular relationships to ratios of principal strains.

Ragan and Sheridan eventually discovered an easier way to measure compaction strain in the Bishop tuff. They found that their measurements of distortion could be correlated with **bulk density** of the samples of volcanic rocks: the greater the density, the greater the degree of welding. It is not often that two distinctly different methods of strain analysis complement each other so well.

PRIMARY PLUTONIC STRUCTURES

GRANITE TECTONICS

Discovery of the enchantment of the primary internal structures of granitic rocks was pioneered by Hans Cloos (1925). Through excruciatingly detailed mapping and petrographic analysis of granitic bodies in Europe, Cloos came to realize that primary flow and fracture elements within granitic bodies tend to be systematically coordinated. Moreover, he recognized that the flow and fracture geometry in plutons reflects kinematics of intrusion. From the work of Cloos we learn that magmas should be thought of as viscously deforming bodies that undergo changes in shape and/or size during the course of buoyant ascent through country rocks. Under some circumstances, every part of an intrusive igneous magma may feel the effects of the distortional movements, preserving the nature of the movements in the geometry of internal primary structures. Salt dome diapirism gives us a model for visualizing this. But unlike salt diapirs, granitic diapirs are capable of fracturing during the late stages of their emplacement, and even the fracture geometry may express kinematic movement of the pluton, albeit the wanning stages.

English-speaking geologists are indebted to Robert Balk (1937), Ernst Cloos (1946), and Evans Mayo (1941) for communicating the basis of Cloos's methods in **granite tectonics** and for extending the research to magmatic settings in North America.

FLOW STRUCTURES

Flow structures form relatively early in the process of emplacement of magma, and they continue to form until the mass is stiff. Although primary fractures form late, they are in part synchronous with the formation of some flow structures. The image that best illustrates the compatibility of flow and fracture is the frontispiece of Robert Balk's (1937) memoir on the *Structural Geology of Igneous Rocks* (Figure 8.32). It is an image familiar to us by now . . . foam layers within a mountain pool. The foam layers form a broad arc in the plane of the surface of the water. At the crest of the arc are discrete fractures and fissures that cut the flow layers. The fractures developed through stretching of the foam layers on the convex part of the arc. If water and foam layers are capable of such fracturing, so too must be viscous granitic magma in its late stages of flow.

Alignments of **crystals** often reflect flow kinematics in a pluton. The most commonly occurring crystals that are useful in orientation analysis are needlelike crystals of hornblende, tabular crystals of feldspar, and platy mica. Aligned elongate minerals, like hornblende and/or feldspar, can produce **lineation** (Figure 8.33). **Mineral lineation** is generally oriented parallel to local flow direction.

Aligned tabular and/or platy minerals commonly define a **flow foliation** in igneous bodies. Foliations in igneous rocks are typically produced by parallelism of mica sheets and/or parallelism of the broad 010 faces of feldspar crystals. The alignment reflects laminar flow of the intruding pluton (Figure 8.34).

There are a number of possible preferred orientations of mineral arrangements, and all require that the distinction between foliation (planar alignment) and lineation (linear alignment) be grasped. For example, orthoclase feldspar phenocrysts can be oriented in such a way that the configuration produces both lineation and foliation (Figure 8.35*A*). This is achieved by a

Figure 8.32 "Cracks" in surface of a mountain pool. [From *Structural Behaviour of Igneous Rocks* by R. Balk (1937), Geological Society of America.]

parallelism of the long axis of each of the crystals combined with a parallelism of 010 crystal faces as well. Alternatively, if the 010 faces of orthoclase feldspar crystals are in parallel alignment, but such that the long axes of the feldspar crystals are randomly oriented, the rock possesses foliation but not lineation (Figure 8.35*B*). To determine whether a given igneous rock possesses foliation and/or lineation, it is necessary to study the orientations of crystals as exposed on different surfaces of the rock body. The surfaces may be the sides of a handspecimen, the flanks and top of an outcrop, the walls of a quarry, or thin sections of different orientations.

Hans and Ernst Cloos carefully documented the orientation of crystals within a pluton of sanidine trachyte (Cloos and Cloos, 1927). The plug they studied may have been a feeder vent for a volcano. Known as Drachenfels on the Rhein, the intrusive plug displays foliation defined by parallelism of 010 faces of large sanidine crystals (Figure 8.36). The structure map of the body reveals flat foliation at the top of the body, and steep to overturned dips of foliation on the flanks. Hans and Ernst Cloos interpreted the overall form of the body on the basis of the internal foliation, concluding that it is cylindrical, marked by a flat, blunt top and steep flanks. They reproduced the geometry of the Drachenfels by forcing plastic gypsum containing originally randomly oriented shale fragments into a pipelike cylinder. The Drachenfels structure was produced, in essence, near the top of the pipe.

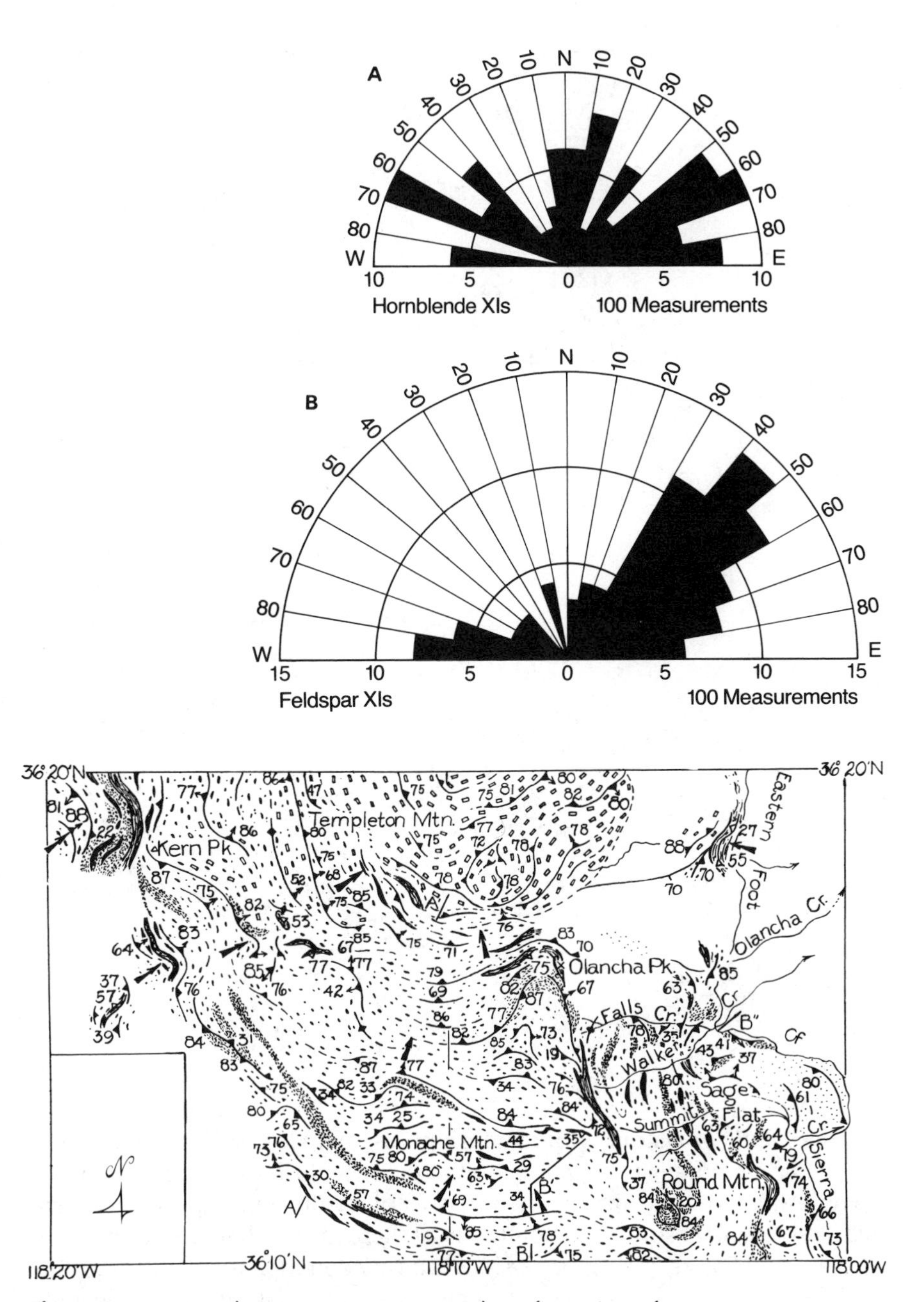

Figure 8.34 Map showing orientations and configuration of flow foliation in granitic rocks in the Temple Mountain area of the Sierra Nevada. (Mapping by E. B. Mayo.)

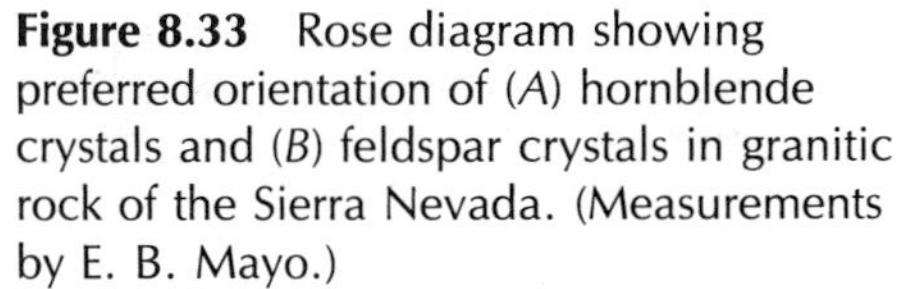

Figure 8.33 Rose diagram showing preferred orientation of (*A*) hornblende crystals and (*B*) feldspar crystals in granitic rock of the Sierra Nevada. (Measurements by E. B. Mayo.)

Riedel (1929) carefully reproduced the experiment, an experiment that underscores the distortion of a viscous body that is forced to occupy a restricted space (Figure 8.37).

Foliations and lineations in granitic rocks are not always conspicuous. Many granitic rocks display very subtle foliations and/or lineations whose physical reason for existence is not at all obvious. Microscopic study is useful in determining the petrographic basis for the expression of foliation and lineation. Foliation and lineation may reflect any number of textural configurations, including aligned feldspars, elongate or discoidal quartz, and close-spaced fractures or cleavages. Sometimes you must become a "true believer" to recognize the subtle foliations and lineations that your friends and colleagues claim to see.

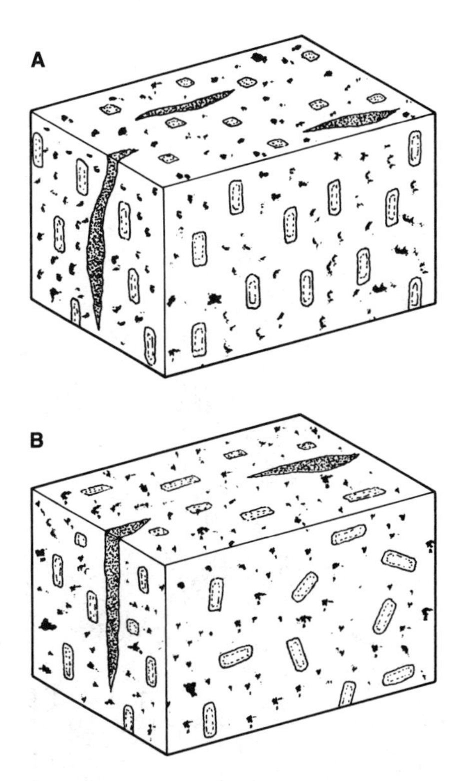

Figure 8.35 (*A*) Lineation and foliation in same rock. (*B*) Foliation and "fake" lineation. In this case, the feldspar crystals do not possess a true parallel alignment. Instead they are randomly oriented within the foliation. Note that the top views of *A* and *B* are nearly identical.

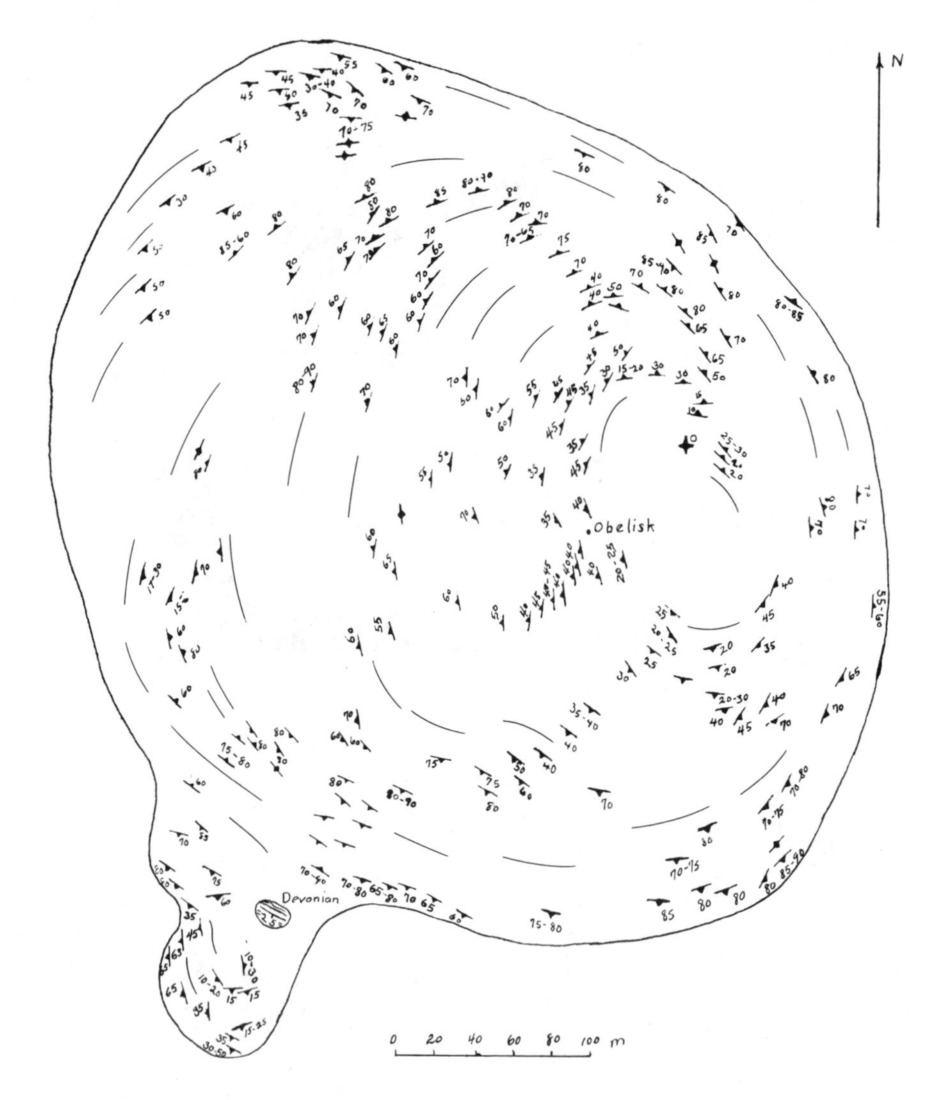

Figure 8.36 Map of foliation in the Drachenfels volcano. [From Cloos and Cloos (1927).]

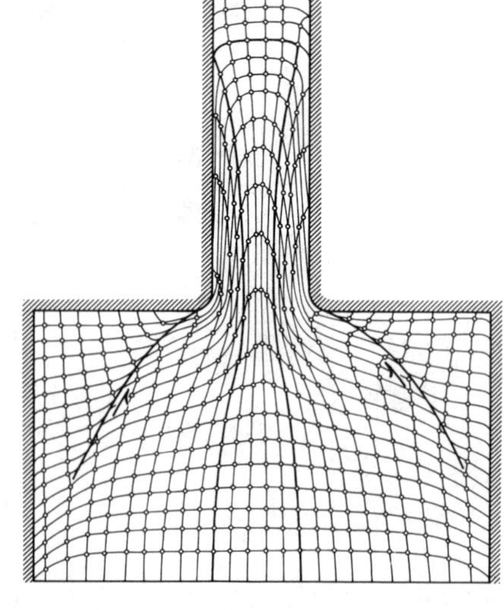

Figure 8.37 Experimental replication of the internal structure of the Drachenfels. Original gridwork is profoundly distorted by the flow of the viscous materials that are intruded upward. [From Riedel (1929).]

Alignments of xenoliths and inclusions also help to disclose the geometry of primary internal structure in granitic rocks. **Inclusions** are disc-shaped to spindle-shaped dark, fine-grained masses of ferromagnesium minerals that may represent absorbed xenoliths or perhaps mineral segregation within magma (see Figure 7.19*A*). Inclusions are very common in granitic magmas, and their orientations can provide a glimpse of internal structure even when the rock matrix does not disclose foliation or lineation. Disclike xenoliths generally are oriented parallel with flow foliation defined by crystal alignments. The parallelism of rod-shaped xenoliths may impart flow lineation.

Another primary plutonic element that may contribute to foliation goes by the odd-sounding name of **schlieren** (Figure 8.38). Schlieren are composed of the fundamental minerals of the igneous body they occupy, but the minerals are concentrated in ways that give rise to wispy flow layers. Fuller appreciation of the structure and the term is enhanced by reading a footnote from Balk's 1937 memoir.

> Schliere is an old German word which denotes a flaw in glass, a zone of abnormal composition, which disturbs the optical properties of the piece. The term is also used in describing the crudely stratified composition of air above heated ground. (From "Structural Behaviour of Igneous Rocks" by R. Balk, 1937, p. 15. Courtesy of Geological Society of America.)

Figure 8.38 Schlieren in the Sierra Nevada. Geologist is Dan Lynch. (Photograph by D. Trent.)

MODELING INTRUSION AND FLOW

Ramberg's classical centrifuge models represent a wonderland of consideration of the kinematics and dynamics of the emplacement of plutons by flowage (Figure 8.39). Large geological bodies are inherently much weaker than small bodies (Hubbert, 1937). Thus to experimentally simulate Earth structure in a realistic way, Ramberg chose to use very soft, weak materials. This is the practice followed in all serious quantitative experimental modeling of regionally deformed belts. Ramberg made an important step beyond his peers by scaling down "time" as well. He succeeded in doing this by perturbing the gravitational constant: by mounting the weak materials to be deformed in a centrifuge and whirling them rapidly. Because his experiments typically feature density inversion achieved by placing the lowest density materials at the base of the layered models and the heavier, stronger material above, the rapid whirling creates gravity-induced buoyant rising. The forms and internal structures of the rising masses match what we know of the form and internal structure of natural diapirs, both igneous and sedimentary.

STATISTICAL REPRESENTATION OF FLOW STRUCTURE

The presence and intensity of development of flow foliation and lineation in plutonic rocks is gauged by constructing stereographic pole-density diagrams and/or rose diagrams. Meaningful diagrams require a minimum of 50 measurements, and preferably many more. Long axes of crystals and rod-shaped xenoliths or inclusions are plotted stereographically as points in terms of trend and plunge. Strike and dip of platy minerals, clots, disc-shaped xenoliths, and schlieren are plotted stereographically as poles. Where exposures only permit two-dimensional appraisal of orientations, rose diagrams are used to present the geometric relationships. On these diagrams the trends of structures are posted in class intervals of 5° or 10°, almost like an ordinary histogram. The orientation of the surface of the

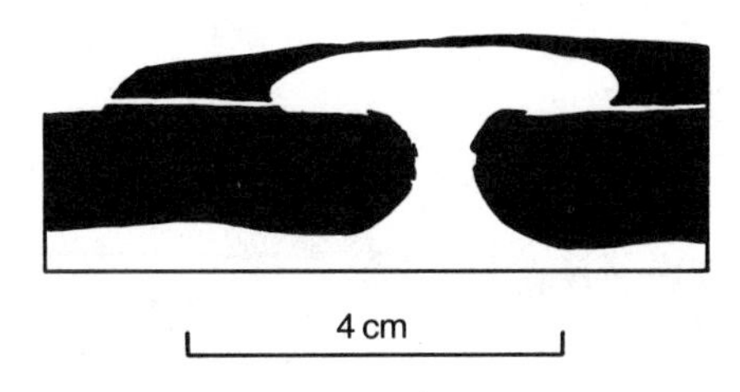

Figure 8.39 Ramberg experiment simulating diapirism of igneous plutons. The diapir is composed of silicone putty. [From Ramberg (1973). Published with permission of John Wiley & Sons, Inc., New York, copyright ©1973.]

exposure from which the measurements were taken should be noted on all rose diagrams. Ideally, both pole-density diagrams and rose diagrams are displayed along with the results of detailed structural geologic mapping, of the quality exemplified by the work of Cloos, Balk, and Mayo.

LATE-STAGE FRACTURING

Primary fractures evolve in plutons as a natural response to the **thermodynamics** of intrusion. The ***primary*** nature of fractures formed before final crystallization is disclosed in a number of ways: the geometrical coordination of the fracture system with primary flow foliation and lineation, the filling of fractures by aplite and pegmatite dikes genetically related to the magma, and the presence of veins and/or alteration assemblages that form by late-stage hydrothermal solutions circulating within the still-hot body.

It is desirable to try to distinguish primary fractures related to the thermodynamics and kinematics of intrusion from fractures created by regional stresses. Making such a distinction is not always possible. Regional stresses that operate long after the crystallization of a particular pluton may fracture the pluton, just as they can fracture any other rock body. Moreover, as we see in Chapter 10 ("Joints"), when a pluton is intruded into a regional setting marked by active differential stresses, primary fractures may form that are systematically oriented with respect to the regional stresses, not to the geometric and kinematic order of intrusion.

Plutons that contain a primary flow structure lend themselves to fracture analysis, for the lineation and/or foliation can be "played" geometrically against the fracture pattern to see if a kinematic coordination exists. On this basis, Hans Cloos recognized fundamental types of primary fractures: **cross fractures, longitudinal fractures, stretching surfaces, marginal fissures**, and **marginal thrusts**. The configuration of these primary fractures is represented in the idealized block diagram of the Strehlen, a German granite body analyzed by Cloos (1922) (Figure 8.40). The fractures are shown in relationship to primary foliation and lineation within the body.

Cross joints are long, planar fractures, evenly spaced and often coated with minerals, which are oriented perpendicular to lineation. They accommodated an elongation parallel to lineation. Cloos concluded that flow lineation commonly expresses stretching of the congealing skin of granite. Both lineation and cross joints are particularly well developed near the roof of a pluton in the zone where stretching is enhanced by surges of magma from below.

Longitudinal fractures, of less certain strain significance than cross fractures, are steeply dipping and strike perfectly parallel to the trend of linea-

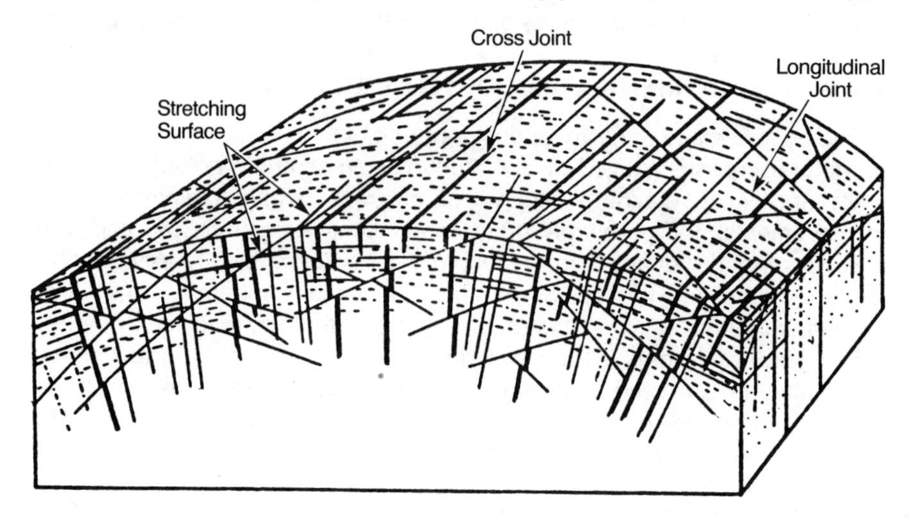

Figure 8.40 Block diagram showing ideal primary fracture pattern in a granitic pluton. [From Cloos (1922).]

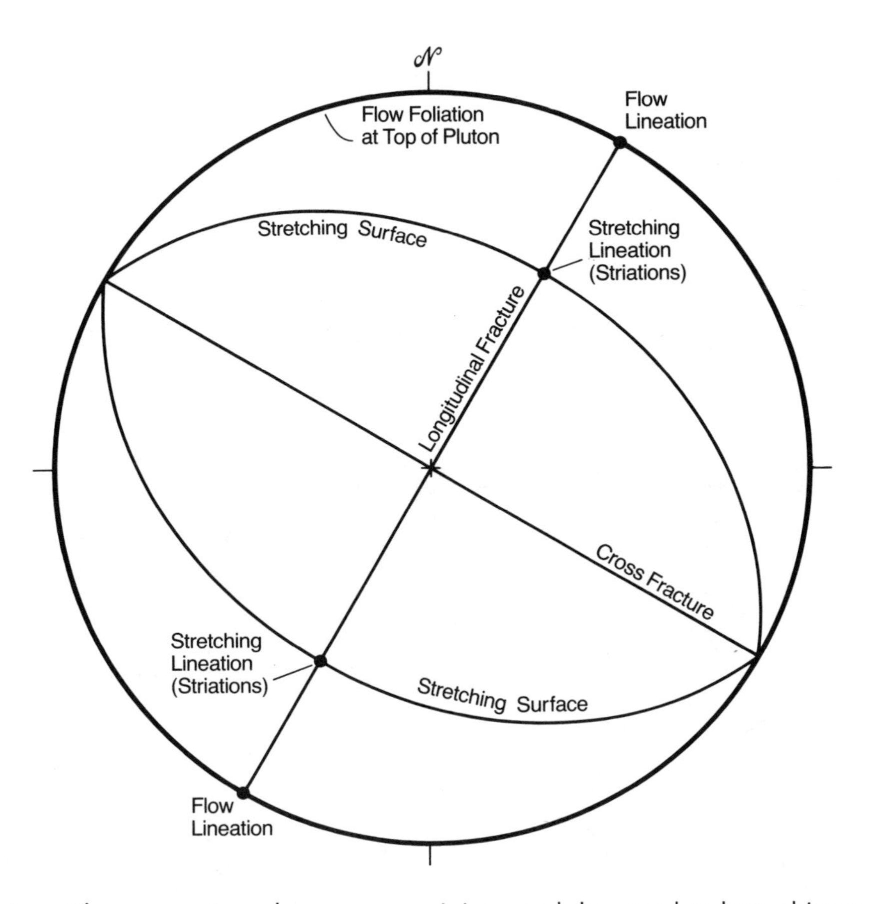

Figure 8.41 Stereographic projection showing an idealized configuration of cross fractures, longitudinal fractures, and stretching surfaces at the top of a foliated, lineated pluton.

tion. They are not as planar as cross joints, and they tend to have thinner mineral coatings.

Stretching surfaces occur in the upper reaches of a pluton. They are low-dipping fault surfaces marked by striations that trend in the direction of lineation. Where offset can be determined through visible displacement of dikes, schlieren, clots, xenoliths, or veins, the faults are seen to be extensional, accommodating lengthening of the top of the intruding mass. Formation of stretching surfaces may be thought of as a pure shear. Although the stretching surfaces may initially dip moderately steeply, the rotation accompanying flattening during stretching results in a progressive decrease in the angle of the dip.

Uncommonly, at the margins of the flanks of an intrusion, there are **marginal fissures** filled by dikes. The fissures dip at moderate angles into the pluton, and they affect both the pluton and the country rock. The orientation and the open nature of the fissures are consistent with dilation in the direction of maximum elongation, which is controlled by the amount of simple shear movement of the pluton with respect to wall rock. Expansion of the mushrooming top of a pluton can convert the marginal fissures into **marginal thrusts**, along which the granitic body moves out and over the country rock.

The stereographic portrayal of an ideal primary fracture system in a granitic pluton is systematic and clearly coordinated with primary flow structures (Figure 8.41).

FINAL COMMENT

The full range of primary structures is limitless. Our coverage in this chapter barely scratches the surface. Appreciation of primary structure requires an intimate knowledge of sedimentary and igneous processes, as well as an understanding of the physics of viscous behavior.

chapter 9 FAULTS

Recognition and Classification

SOME DEFINITIONS

Faults are fractures along which there is visible offset by shearing (Figure 9.1). Faults can occur as single discrete breaks, but where the rock that is faulted is weak or where faulting is especially intense, no discrete break may be evident. What forms instead is a **fault zone** composed of countless subparallel and interconnecting closely spaced fault surfaces. Faults and shear zones range in length and magnitude of offset from small structures visible in handspecimens to long spectacular crustal breaks extending hundreds of kilometers and accommodating displacements of tens to hundreds of kilometers.

Some joints are faults, **microfaults** to be sure, but they generally go unrecognized as such because the tiny offset caused by shear cannot always be seen. Large fractures that have sustained major dilational opening (by 20 cm or more) are known as **fissures.** They are not considered to be faults. Fissures form at the surface of the Earth where processes like crustal stretching or subsidence locally pull the earth apart (Figure 9.2). Where faulting takes place at deep levels under conditions of elevated temperature and/or confining pressure, **shear zones** may develop.

Figure 9.1 Fault in metamorphic rock in the Tortolita Mountains, southern Arizona. [Tracing by D. O'Day of photograph by G. H. Davis. From Davis (1980). Published with permission of Geological Society of America and the author.]

Figure 9.2 Fissure cutting the Hiliana Pali Road at Kalanaokuaiki Pali, Kilauea Volcano, Hawaii. The crack formed on December 25, 1965. Initially only 3 ft (1 m) wide, the crack soon enlarged to 8 ft (2.4 m) as a result of slumping along the edges. Note columnar jointing on far side of road. (Photograph by R. S. Fiske. Courtesy of United States Geological Survey.)

RECOGNIZING FAULTS

GEOLOGIC MAP RELATIONS

Faults are relatively easy to recognize in continental regions of moderate to excellent exposure. Systematic geologic mapping has proved to be an extremely effective method for locating faults, particularly where the faults cut a geologic column of sedimentary and/or volcanic rocks whose stratigraphy is well known. Geologic maps reveal the plan-view expression(s) of fault patterns. Faults are identified and tracked on the basis of mapped patterns that disclose **truncation and offset** of one or several bedrock units.

Truncation and offset can take many forms, depending on the orientations of the faults, the movements on them, and the orientations of the rock layers that are cut and displaced (Figure 9.3). Truncation and offset usually result in an apparent horizontal shifting of mapped bedrock units.

Where faults trend parallel to the strike of bedding, translation does not produce the simple but conspicuous horizontal shifting of layers. Instead, the presence of the fault must be recognized on the basis of **inconsistent stratigraphic patterns**. The rock formation shown in Figure 9.4 is composed of 10 distinctive units (numbered in the figure) and is cut by two faults that strike parallel to bedding. These dip such that they cut across bedding at steep angles. The presence of the two faults might not be immediately apparent in the geometry of the map pattern, but knowledge of the stratigraphy of the formation permits the faults to be recognized by **repetition and omission of strata**. If, during a traverse across the map area (Figure 9.4), we move up-section from unit 1 to unit 3, only to find ourselves crossing back unexpectedly into a repeated section of the same units, we would recognize that we had crossed a fault. And if during our continued traverse we were to cross directly from unit 3 into unit 8, the omission of units 4 to 7 would alert us to the fact that we had crossed a second fault. Unconformities can also account for omission of strata, so be careful.

Figure 9.3 Truncation and offset at the outcrop scale. (Photograph by G. H. Davis.)

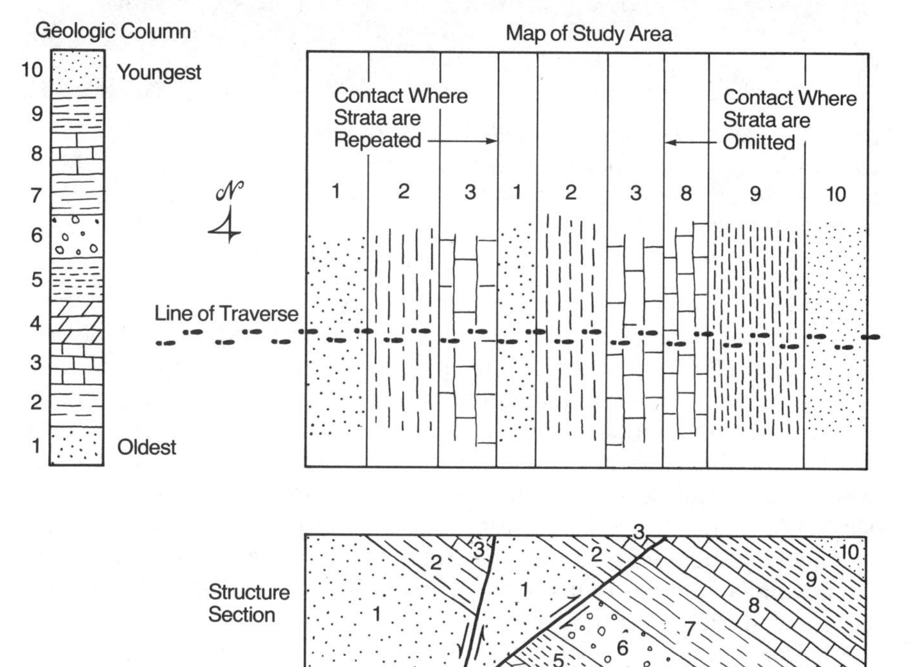

Figure 9.4 Identification of locations of faults on the basis of inconsistent stratigraphic patterns. One fault is marked by repetition of strata, the other by omission of strata.

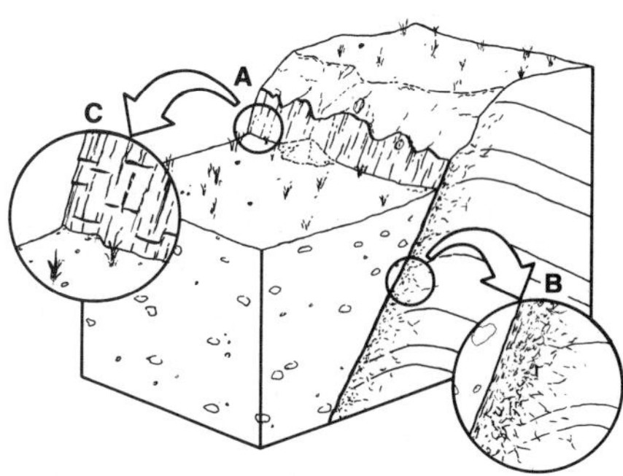

Figure 9.5 Schematic diagram showing some physiographic and physical characteristics of faulting. (*A*) Fault scarp. (*B*) "Fault rocks," like gouge, breccia, and mylonite. (*C*) Striations and chatter marks. (Artwork by R. W. Krantz.)

PHYSICAL FEATURES

Many of the physical signatures that are valuable in establishing the presence of faults were discussed in Chapter 7 in the context of distinguishing fault contacts from unconformities, intrusive contacts, and ductile shear zones. Physiographic features like straight topographic lineaments, fault scarps, triangular facets, and fault-line scarps reveal the locations of faults, especially recent faults (Figure 9.5*A*). Gouge, breccia, microbreccia, and mylonite are fault rocks that can signal faulting and shearing as well (Figure 9.5*B*). Many fault zones are marked by alteration, especially silicification, and by the presence of ore minerals. The old prospectors were excellent fault finders! It is rare to come upon a fault zone in the southwestern United States

Figure 9.6 Striations and chatter marks on fault surface cutting siltstones in the Tucson Mountains, Arizona. Chatter marks are best expressed in the quartz vein coatings (white). Steep faces of the chatter marks are directed to the right. (Photograph by G. H. Davis.)

that is not marked by a pit, shaft, or "diggings" of one kind or another. Alteration and mineralization are favored by the increased permeability that results from closely spaced fractures in fault zones.

Slickensides, striations, and grooves are direct products of frictional sliding on faults. Unfortunately, these surface markings are not present on all faults, either because they never formed or because they were removed by weathering at the level of surface exposures. Limestones seldom display slickensides and striations because carbonate, even in semiarid regions, is extremely vulnerable to dissolution weathering. However, siliceous veins occupying fault zones in limestone commonly reveal well-developed striations and slickensided surfaces.

Chatter marks commonly are formed on the surfaces of faults (Figure 9.5C). These small, asymmetrical, steplike features are typically oriented perpendicular to striations (Figure 9.6). Their origin is related to differential shearing along fault surfaces. "Topographic relief" on chatter marks is usually less than 5 mm. Some investigators have compared the formation of chatter marks to the shaping of roches moutonnées in glacially scoured areas. **Roches moutonnées** are sculptured polished bedrock hills that feature a relatively short, steep face sloping in the direction of the ice movement, and a long gentle declivity dipping oppositely to the direction of ice flow. Sense of slip deduced on the basis of asymmetry has been related to chatter marks in a parallel way. Experimental evidence contradicts this interpretation as a general rule of thumb: the asymmetry of chatter marks does not appear to be a dependable indicator of displacement sense. Nonetheless, the presence of chatter marks is helpful in confirming that a particular surface is actually a fault.

GEOPHYSICAL SURVEYING

Faults are difficult and costly to locate in continental areas of very poor rock exposure, in the deep subsurface, and in ocean basins. Nonetheless, major faults are routinely discovered by explorationists through geophysical methods, especially seismic, gravity, and magnetic surveying. Abrupt contrasts in the geophysical signatures of rocks at depth can signal the sharp truncation of bedrock at locations of faults. Fault patterns are mapped on the basis of the geophysical data.

Geophysical exploration of fault structures in the ocean floor completely changed the way in which Earth dynamics is viewed. Mapping the distribution of magnetic patterns in seafloor sediments and volcanics revealed the presence of transform faults, by far the most important fault network on the face of the globe (Chapter 6). Had it not been for discovery of these faults and interpretation of their significance, the revolution called plate tectonics would have been delayed.

At the present time, deep seismic-reflection probing of the crust and upper mantle of the Earth has triggered a revolution in exploring the roots of regional fault systems in continental regions. Although petroleum explorationists have been engaged in **seismic-reflection** studies for decades, advanced computer technology now permits fuller exploitation of the method. The Consortium for Continental Reflection Profiling (COCORP) is using the technique to probe structural relationships at depths as great as 70 km.

In carrying out seismic-reflection profiling, sound waves are set up in the subsurface by shaking the ground surface with large truck-mounted vibrators. The waves speed to depth, where they eventually are reflected by discontinuities marking sharp contrasts in density between rocks above and below (Figure 9.7). Since faults often bring rocks of different lithology and density into contact, they are among the discontinuities that are capable of causing sound waves to be reflected. Reflected sound energy radiates back to the surface, where it is collected and measured through vibration sensors known as **geophones**.

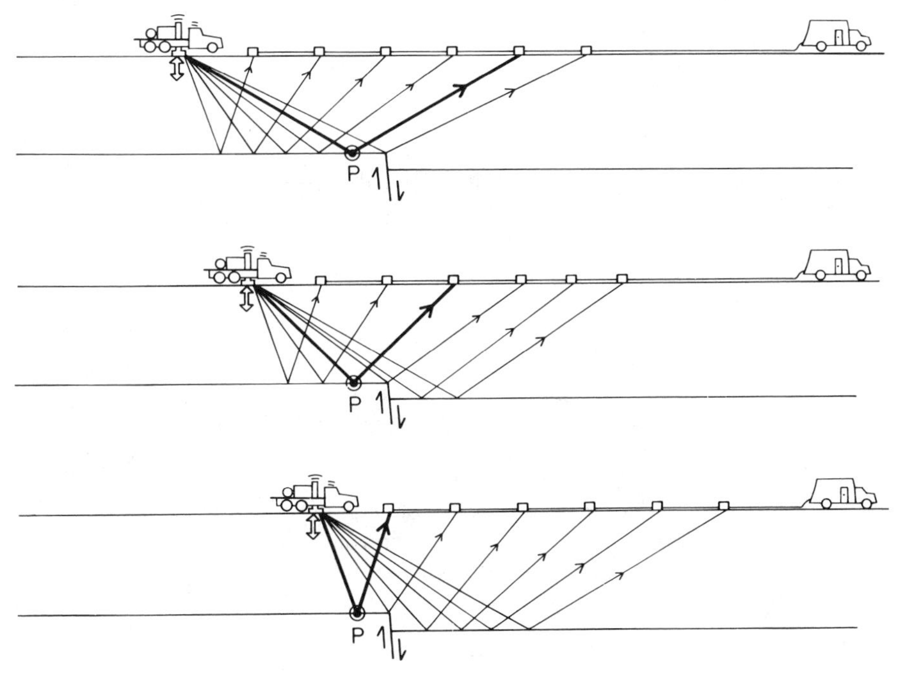

Figure 9.7 Schematic illustration of the process of seismic-reflection profiling. (From "The Southern Appalachians and the Growth of Continents" by F. A. Cook, L. D. Brown, and J. E. Oliver. Copyright ©1980 by Scientific American, Inc. All rights reserved.)

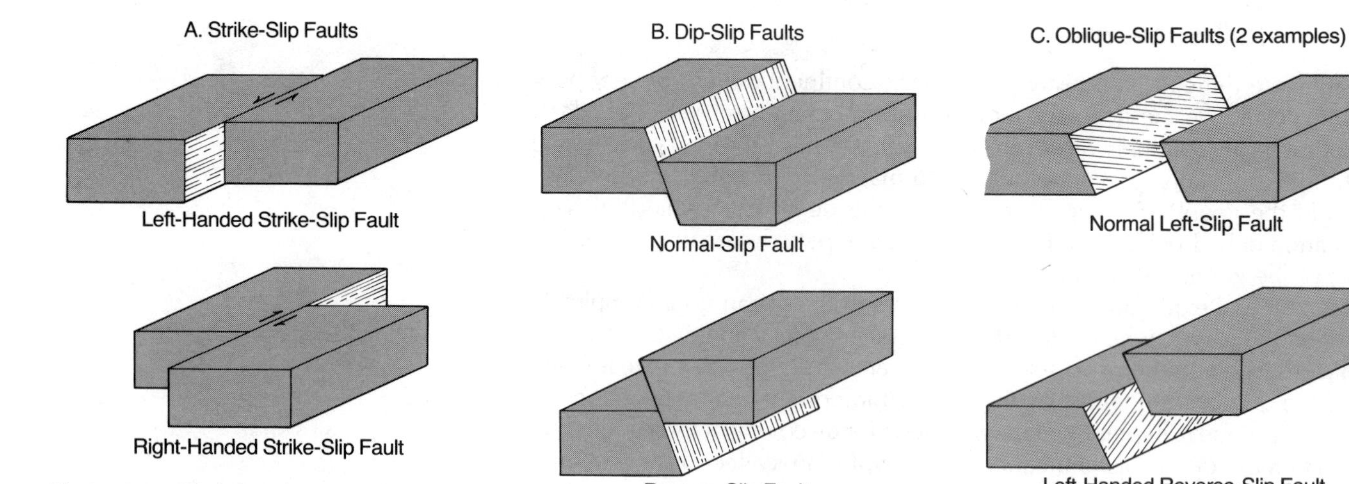

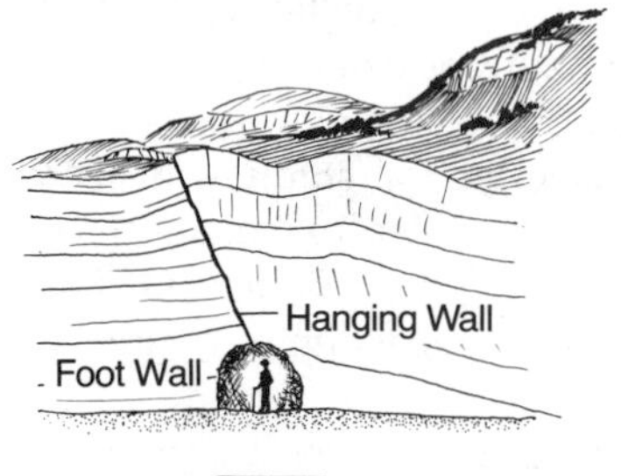

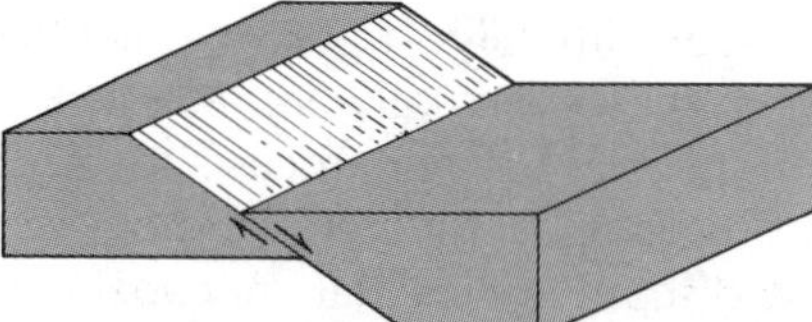

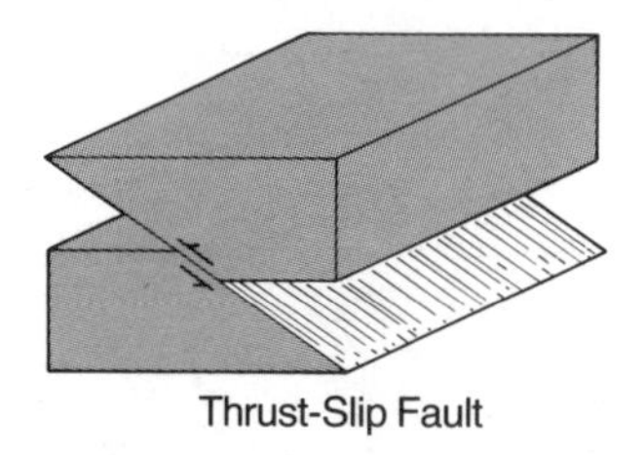

Figure 9.8 Slip classification of faulting. (*A*) Block diagrams showing left-handed and right-handed strike-slip faulting. (*B*) Block diagrams showing normal slip, reverse-slip, low-angle normal slip, and thrust-slip faulting. Prospector in tunnel pauses to think about the difference between the footwall and the hanging wall of a fault. (*C*) Examples of oblique-slip faults, including a normal left-slip fault and a left-handed reverse-slip fault. (*D*) Schematic block diagram of a rotational fault.

Interpreting sound waves reflected from the crust and mantle has been compared to interpreting the signals discharged by a bat flying about in a small room. Some of the complexity of analysis is revealed in a statement by Cook, Brown, and Oliver (1980).

> The computer analysis of the COCORP data include(s) . . . the elimination of particularly noisy data, the calculation of wave velocities, the collection of all reflection signals from each common depth point, the compensation for differences in the distance between sources and receivers, the adjustment for differences in topography and near-surface geology, and the stacking, or superposition, of the several coherent signals for each common reflection point. The result of this intensive processing is a seismic cross section that resembles a geological cross section except for the important differences that depth is represented not by distance but by travel time and that lateral variations in the

velocity of sound in the rocks can *distort* the geometry of the reflections. (From "The Southern Appalachians and the Growth of Continents" by F. A. Cook, L. D. Brown, and J. E. Oliver, p. 160.)

To build an accurate geological cross section, the velocity **distortions** must be removed, and this requires both a theoretical and a practical understanding of the idiosyncrasies of seismic reflection. Once seismic distortion has been removed, structural geological distortion remains, and this, after all, is the object of the probe.

CLASSES OF FAULTS

BASIS OF CLASSIFICATION

We have seen that transform faults are a class of faults whose movements and functions are understood only in the context of plate tectonics. Most faults in continental regions cannot be clearly and quantitatively linked to specific plate tectonic movements or configurations, present or past. Consequently, they are analyzed and understood in an entirely different way.

Faults are classified according to two criteria: slip and separation. Wherever possible, faults are named on the basis of **slip**, the **actual relative displacement** that they have accommodated. To classify faults in this manner, it is necessary to know the direction and sense of translation. If these cannot be established, the faults are classified and named on the basis of **separation**, the **apparent relative displacement**. Using separation as the basis for classification, the end product of faulting is described irrespective of the movement path along which faulting took place. Learning to distinguish slip from separation is one of the most important lessons in fault analysis (Crowell, 1959; Hill, 1959).

SLIP CLASSIFICATION

Slip on a fault is the displacement vector between two points that were coincident before faulting. **Strike–slip faults** accommodate horizontal slip between adjacent blocks (Figure 9.8*A*). They are described as **left handed** or **right handed**, depending on the sense of actual relative movement.

Dip–slip faults are marked by translation directly up or down the dip of the fault surface (Figure 9.8*B*). Movement on a dip–slip fault is described with reference to the actual relative movement of **hanging wall** *and* **footwall**. The hanging wall is the fault block toward which the fault dips. The footwall is the fault block on the underside of the fault. The terms "hanging wall" and "footwall" are derived from old mining jargon: a prospector working in a drift he has tunneled along an inclined fault finds his head close to the hanging wall, his feet on the footwall (Figure 9.8*B*).

Several kinds of dip–slip faults exist (Figure 9.8*B*). A **normal-slip fault** is one in which the hanging wall moves down with respect to the footwall. Many normal-slip faults dip at moderate to steep angles, averaging 60° or so. Normal-slip faults that dip less than 45° are referred to as **low-angle normal-slip faults**. **Thrust-slip** and **reverse-slip faults** are marked by translation of the hanging wall upward relative to the footwall. Thrust-slip faults dip less than 45°, usually around 30° or so. Reverse-slip faults dip more steeply than 45°.

Translation on **oblique-slip faults** is inclined between strike–slip and dip–slip (Figure 9.8*C*). The rake of the displacement vector lies somewhere between 0° to 90°. To describe oblique-slip faults accurately, the terms "normal," "reverse," "thrust," "right handed," and "left handed" are combined in ways that conform with the interpreted direction and sense of translation. If the main component is strike–slip, then right-handed or left-handed is used as the second modifier, preceded by normal, reverse, or thrust, depending on the nature of the dip–slip movement. If the chief component is dip–slip, then normal-slip, reverse-slip, or thrust-slip is used as the second modifier, preceded by right-handed or left-handed, depending on the nature of the strike–slip component of movement (Figure 9.8*C*).

Some faults are rotational or scissorslike (Figure 9.8*D*). The geometric complications that arise in evaluating slip on rotational faults are challenging indeed! A rotational fault changes both in its magnitude of slip and in its sense of slip along strike. Thus, a rotational fault may be normal slip along part of its length and reverse slip along another part (Figure 9.8*D*).

EVALUATION OF SLIP

RECONSTRUCTION OF FAULTED LINES. There are a number of ways to determine the direction and sense of slip on a fault. One way is to identify a line that has been offset by faulting. The illustration of this concept in Chapter 4 focused on a *geometric* line of reference, the intersection of a dike and a sandstone bed. Crowell (1959) has described other faulted lines, both geometric and physical, that are especially useful in defining slip. These include the intersection of an angular unconformity with inclined bedding, the edge of a sandstone lens, fossil shorelines marking lines of abrupt facies change, contour lines on isopach or structure contour maps, and the faulted hinge of a fold.

The principles of using faulted lines to classify faults on the basis of slip is illustrated here using the faulted hinge of a fold as a basis for reference. The hinge of a folded layer is like the crease in a folded piece of paper. In the mapped relationship shown in Figure 9.9*A*, the hinge of an overturned syncline plunges 15° S10°E. The fold hinge is offset by a fault that strikes N75°W and dips 65°SW. Displaced parts of the faulted fold hinge crop out on both sides of the fault (locations *A* and *B*, Figure 9.9*A*). Before faulting, these hinges were part of a straight, continuous hingeline in a sandstone bed. Slip on the fault is deduced by projecting each of the offset hinge segments into the plane of the fault. Once accomplished, the distance and direction between these projected lines can be measured.

The orthographic solution involves constructing vertical structural profiles oriented parallel to the trend of the fold axis (N10°W) (Figure 9.9*B*). On the footwall side, the hingeline projects downward at 15° toward the projected position of the fault in the subsurface. The projection of the hingeline pierces the fault at point *C*, elevation 2610 ft (791 m). On the hanging wall side, the hingeline projects 25° skyward from its outcrop elevation to point *D* (elevation 2645 ft; 801 m) where it pierces the upward projection of the fault plane. The relative positions and elevations of points *C* and *D* indicate that faulting was achieved by oblique slip, involving the combination of normal and left-handed slip.

The actual strike–slip and dip–slip components of translation are measured in the plane of faulting (Figure 9.9*C*). To accomplish this, points *C* and *D* are plotted in the plane of the fault at their respective elevations and locations. The strike–slip, dip–slip, and net slip components of translation

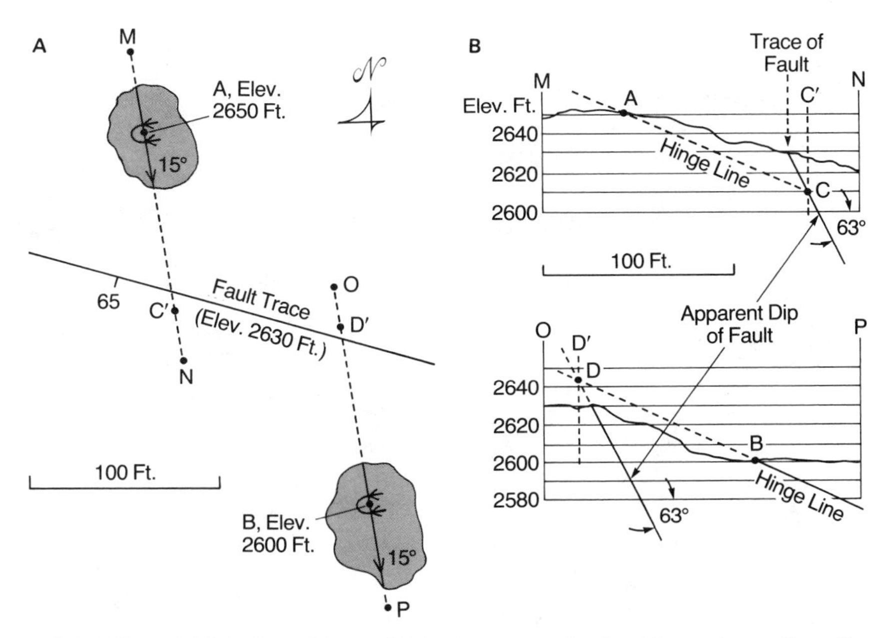

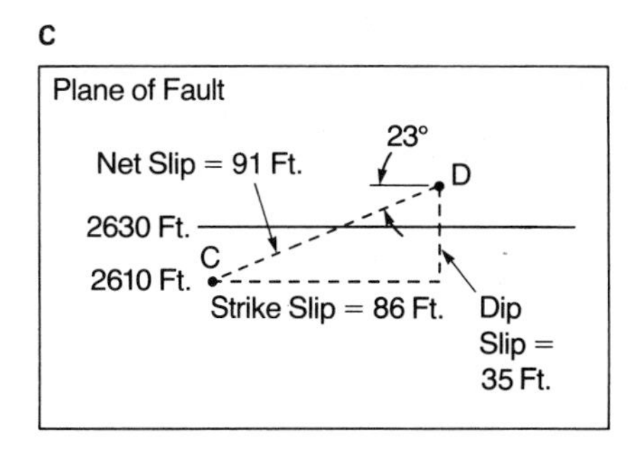

Figure 9.9 Determination of the translation along a fault on the basis of the offset of the hingeline of a fold. (*A*) Map relationships showing the fault offset of the trace of an overturned syncline. (*B*) Structure profiles showing where hingelines intersect fault surface. (*C*) Slip components as seen in the fault plane.

are 86, 35 and 91 ft (26, 11, and 28 m), respectively. Since the strike–slip component of translation exceeds the dip–slip component, the fault is named a reverse left-handed strike–slip fault. The rake of the displacement vector on the fault plane is 23°W. This corresponds to a net slip orientation of 21° N86°W, as computed stereographically.

STRIATIONS AND GROOVES. Rarely do geologic data within a study area permit such detailed constructions to be carried out. Usually only direction and sense of slip can be determined, and these are evaluated on the basis of outcrop features.

The direction of slip is most readily established through the orientations of striations and grooves. The rake of striae and grooves on a fault surface are an inscription of the direction of net slip. Faults and fault systems that accommodate single movements and/or sustained but simple movement plans commonly reveal striations of uniform orientation. These are useful in evaluating slip on faults. Consider the mapped fault relationship in Figure 9.10. The fault strikes N30°E and dips 70°SE. It separates volcanic rocks of Jurassic age on the footwall from Permian sedimentary rocks on the hanging wall. Permian strata were faulted upward with respect to the Jurassic volcanics, but along what line? Striations on the fault surface reveal the direction of slip. They rake 80°SW. This indicates oblique-slip faulting, featuring a major dip- slip component of translation. Given the orientation of the striations and the age relationships of the rocks on hanging wall and footwall, the fault is classified as a left-handed reverse-slip fault.

There are difficulties in using striations as a guide to slip. Many faults undergo multiple, complex histories of movement. Slickensided surfaces can be marked by striations of broad-ranging orientations. "Scatter" in striation orientation prevents simple naming of faults according to slip criteria. As if this were not enough, there is a generally held opinion among many geologists that striations on fault surfaces reflect only the latest fault movement. If indeed faults are good erasers, heavy reliance on striations alone as a guide to the direction of fault movement may lead to oversimplified or incorrect interpretations. Nonetheless, striations and grooves remain an integral part of the descriptive record. They are useful kinematic guides to slip, provided they are treated with the care that the foregoing words of caution warrant.

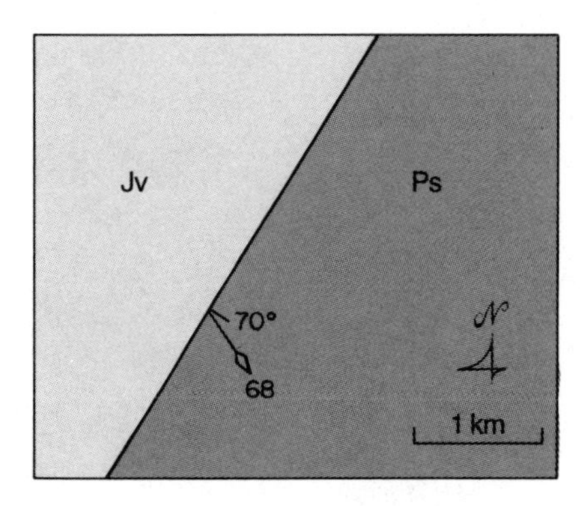

Figure 9.10 Map of fault contact between Jurassic and Permian strata.

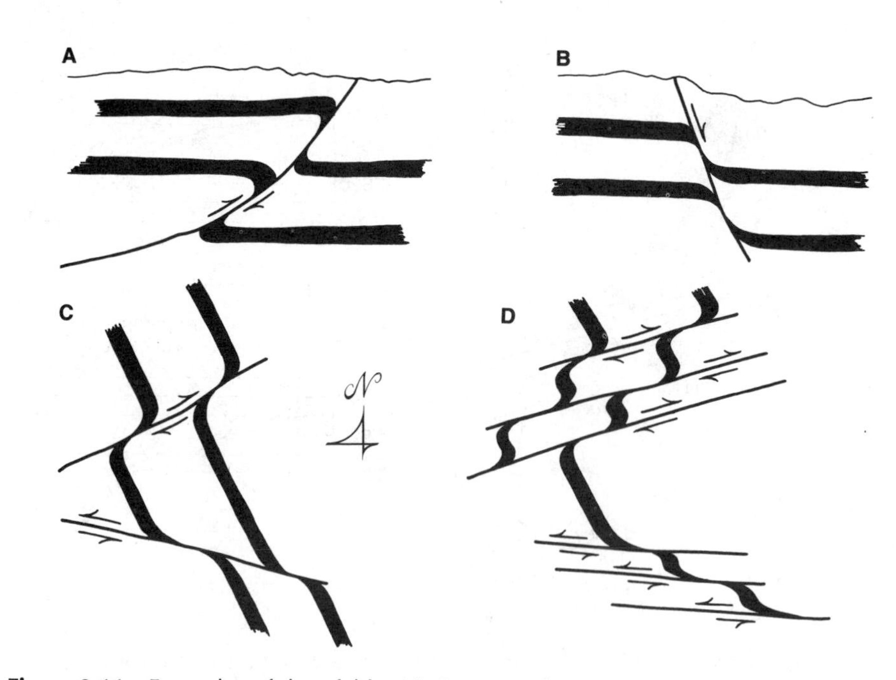

Figure 9.11 Examples of drag folds. (*A*) Strata on the hanging wall of thrust-slip and reverse-slip faults are dragged into an anticline, whereas strata on the footwall are dragged into a syncline. (*B*) Drag fold relationships along a normal slip-fault. (*C*) Patterns of drag folding along right-handed and left-handed strike-slip faults. (*D*) Sigmoidal drag folding along closed-spaced, distributed right-handed and left-handed strike-slip faults.

DRAG FOLDS. **Drag folds** are another structure that can be used to determine the direction and sense of slip during faulting. Drag folding is a distortion of bedding, or other layering, resulting from shearing of rock bodies past one another. Consider how a sequence of sedimentary rock is affected by fault movements (Figure 9.11). Strata close to the fault surface are deformed by frictional drag into folds that are convex in the direction of relative slip. The truncated ends of dragged layers point away from the sense of actual relative movement. Ideally, hanging–wall strata of a thrust-slip or reverse-slip fault are dragged into an anticline and the footwall strata are dragged into a syncline (Figure 9.11*A*). Similarly, under ideal circumstances, hanging wall strata of a normal-slip fault are dragged into a syncline, while the footwall strata are dragged into an anticline (Figure 9.11*B*).

Drag folds resulting from strike–slip faulting can be spectacular, especially where steeply dipping layering is radically folded. Right-handed and left-handed patterns of drag are distinctly different (Figure 9.11*C*). Within broad fault zones marked by spaced strike–slip faults, individual layers can become completely fault bounded. And if each end of a fault-bounded layer is curled by drag, **sigmoidal drag folds** result (Figure 9.11*D*). The layers are doubly curved. Sigmoidal drag folds formed by right-handed simple shear are gently S shaped. Those formed by left-handed simple shear resemble backward S's, the kind that kids paint on signs. Folds shaped like backward S's are referred to as Z shaped. A wonderful example of sigmoidal folding within a strike–slip fault zone is shown in Figure 9.12, a detailed structure map of the Stockton Pass fault, southern Arizona, prepared by Swan (1976).

Where normal-slip, reverse-slip, and thrust-slip faults cut and drag horizontal strata, the axes of drag folds will be oriented approximately horizontal, perpendicular to the direction of slip (see Figure 9.11*A, B*). Where pure

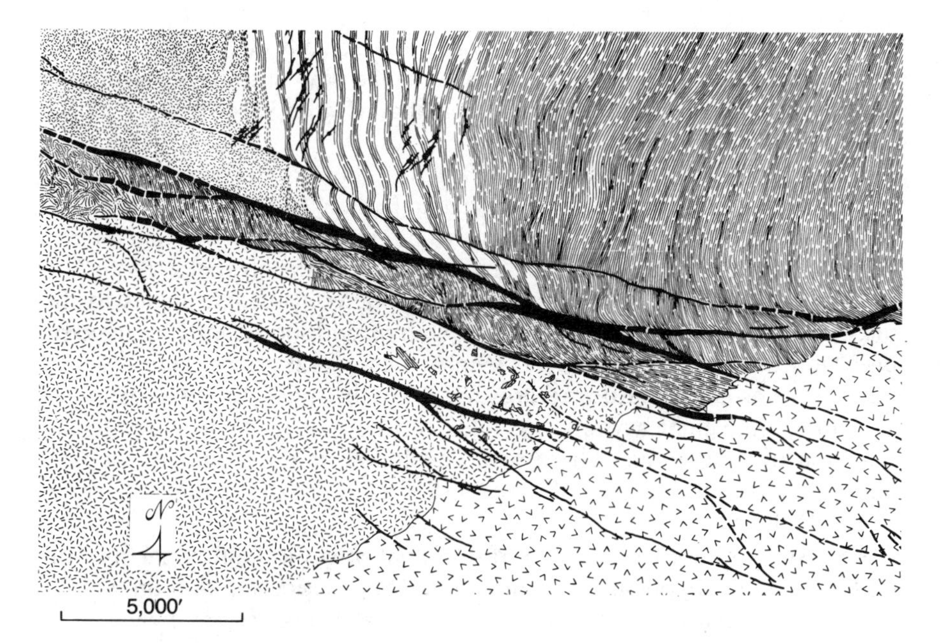

Figure 9.12 Drag folding of foliation in Precambrian gneiss, Stockton Pass, Arizona. Drag folding was caused by left-handed strike-slip faulting along the Stockton Pass fault zone. [From Swan (1976).]

strike–slip faulting operates along vertical surfaces cutting steeply dipping layers, the axes of resulting drag folds will be steep, perpendicular as before to the direction of movement (see Figure 9.11*C, D*). The orientations of striations and drag folds should be mutually perpendicular.

Tight asymmetrical drag folds commonly occur in the interior of fault zones; these are very useful in evaluating the sense of simple shear movement during faulting. The asymmetrical folds, when viewed downplunge, can be described as S shaped or Z shaped (Figure 9.13). The clockwise rotation that typifies asymmetrical, Z-shaped drag folds reflects right-handed simple shear. The counterclockwise rotation that characterizes asymmetric S shaped drag folds reflects left-handed simple shear. Learning to interpret asymmetric drag folds is an important aid in naming faults according to slip.

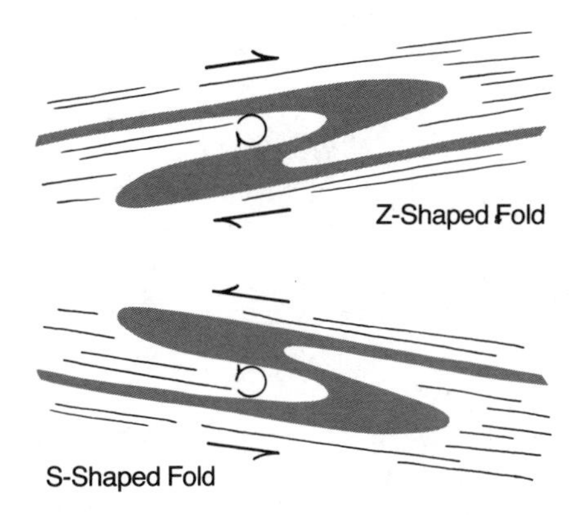

Figure 9.13 The forms of Z-shaped and S-shaped drag folds: Z-shaped fold forms result from clockwise internal rotation; S-shaped folds result from counterclockwise internal rotation.

Bold generalizations are dangerous when it comes to drag folding. Rocks do not always cooperate in the ideal ways just described. The most notorious exception to the rule is **reverse drag** on normal-slip faults, including growth faults. Originally horizontal bedding in the hanging wall of a normal-slip fault can become dragged in such a way that it actually dips toward the fault surface, thus forming a rollover anticline (Figure 9.14*A*). Hamblin (1965) first recognized this phenomenon in faulted Colorado Plateau strata. He reasoned that reverse drag may be unique to **listric** fault geometries. Listric faults are curved faults that flatten with depth (Figure 9.14*A*). Sustained movement on a listric normal-slip fault favors a pulling away of the hanging

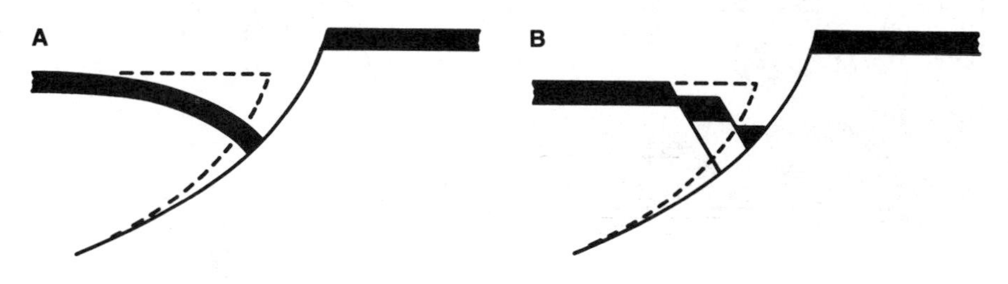

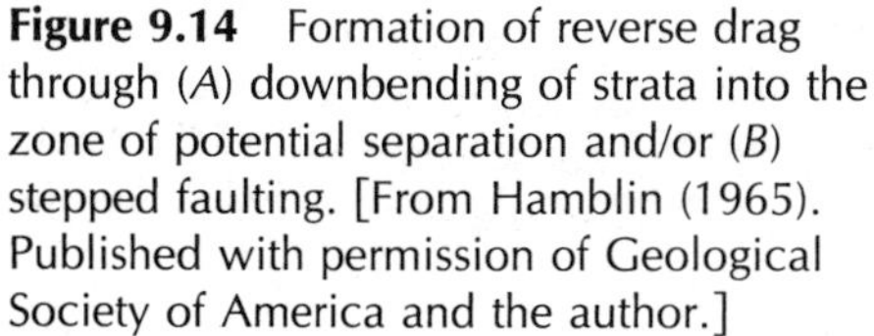

Figure 9.14 Formation of reverse drag through (*A*) downbending of strata into the zone of potential separation and/or (*B*) stepped faulting. [From Hamblin (1965). Published with permission of Geological Society of America and the author.]

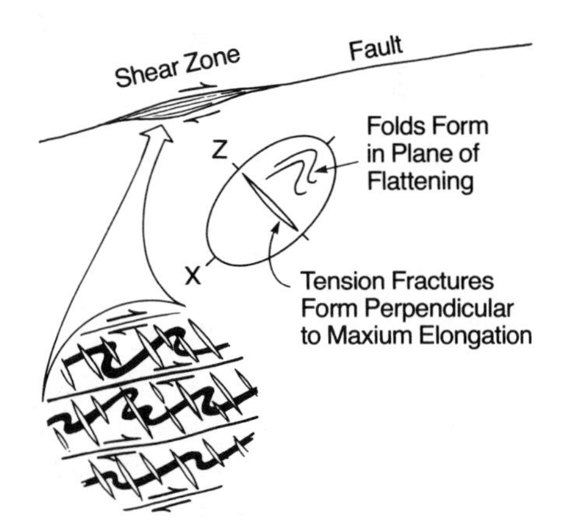

Figure 9.15 Simple shear origin of tension fractures and tight folds within a shear zone.

wall block from the footwall. Such a gap never actually develops because of such compensating mechanisms as the draglike flow of strata, or **antithetic faulting**, into the zone of potential separation (Figure 9.14*B*).

GASH FRACTURES AND TIGHT FOLDS. There is yet another way in which the direction and sense of slip due to faulting can be evaluated on the basis of minor structures. The method is a direct application of strain theory. Think of fault zones as zones of simple shear that distort rocks. Distortion stretches the rock in one direction and shortens it at right angles to the direction of stretching. The effects of stretching and shortening produced by distortion during faulting can be seen in many fault zones (Figure 9.15). Tension veins called **gash fractures** form at right angles to the direction of stretching. And **tight asymmetric folds** can develop at right angles to the direction of shortening. Where well developed, the combination of gash fractures and tight folds virtually establishes the sense of slip that took place during faulting.

Analysis of minor structures of distortional origin in fault zones has important practical application in mining geology. Suppose a rich ore vein is abruptly intercepted by a fault and translated safely out of sight. How can its offset position be found? If striations and drag folds are present, the strategy of the search is reasonably straightforward. If they are absent, the strain within the fault zone may be the only record of the sense of slip on the fault.

Consider the steeply dipping, N15°E-striking fault zone shown in the map view in Figure 9.16*A*. Within the zone are vertical gash fractures filled with quartz. They strike approximately N70°E. Horizontal striations on the fault surfaces indicate that the fault zone accommodated strike–slip movement. However it is not known whether translation was right handed or left handed. To solve for sense of slip, assume that the gash fractures formed at right angles to distortional stretching of the rock during faulting (Figure 9.16*B*). The attitude of the gash fractures is perpendicular to the axis of maximum extension (*X*) of the strain ellipse that describes the fault-induced distortion. Its orientation in this example is compatible with right-handed strike–slip faulting.

THE SEPARATION PROBLEM

Interpreting the direction and sense of slip on faults is often complicated by deceptive patterns created by the interference of structure and topography and by the absence of minor structures that if present would have helped to define the slip path. *The offset observed along faults in outcrops, in map*

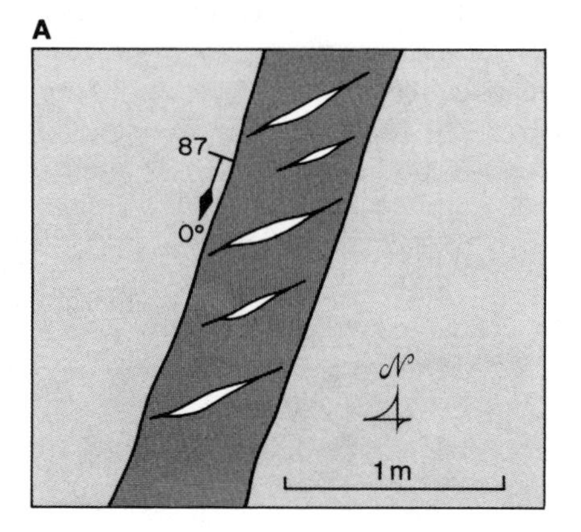

Figure 9.16 Use of gash fractures to determine direction and sense of translation on a fault. (*A*) Map showing steeply dipping fault zone and vertical gash fractures. (*B*) Geometric relationship between the orientation of gash fractures and direction of stretching allows the fault to be classified as a right-handed strike–slip fault.

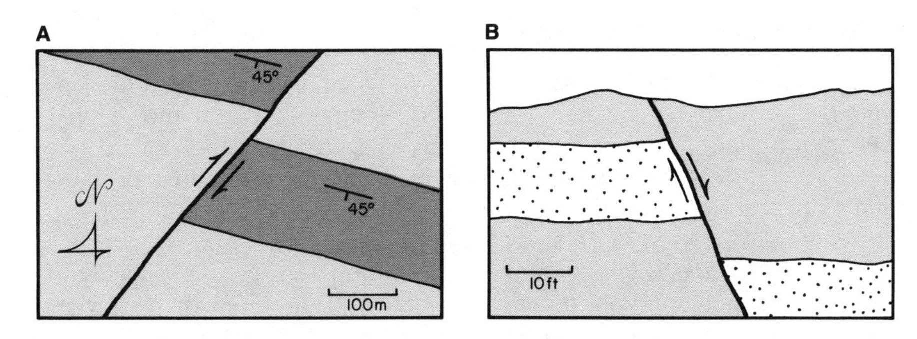

Figure 9.17 (*A*) Map view of right-lateral separation of a basalt layer. (*B*) Cross-sectional view of normal separation of a sandstone layer.

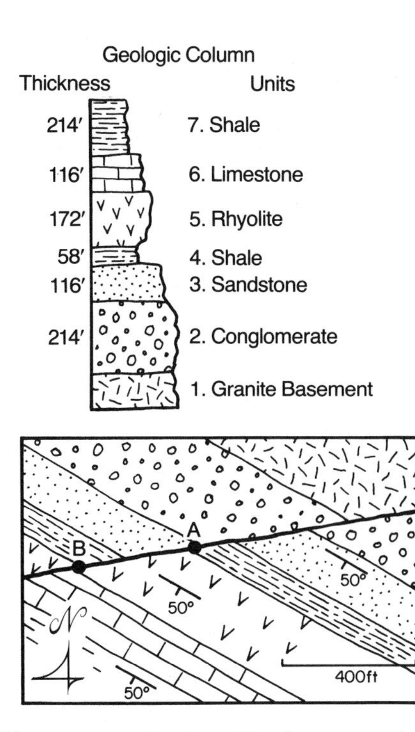

Figure 9.18 Stratigraphic throw at point *A* is equal to the combined thickness of units 3 and 4, or 174 ft (53 m). Stratigraphic throw at point *B* is simply equal to the thickness of unit 5, which is 172 ft (52 m).

patterns, and in structure profiles is separation. "Separation" refers only to the sense and magnitude of offset along faults. It is the apparent relative movement, and it is described irrespective of the actual relative displacement. The faulted basalt layer shown in map view in Figure 9.17*A* displays 200 m of right-lateral separation. The faulted sandstone layer seen in cross section in Figure 9.17*B* is marked by 20 ft (6 m) of normal separation.

Stratigraphic throw is a special measurement of separation, one that is commonly used in exploration geology as a convenient measure of the magnitude of faulting. **Throw** is the thickness of the stratigraphic interval between two beds that are brought into contact by faulting (Figure 9.18). At point *A* in Figure 9.18, faulting has brought the base of unit 3 and the top of unit 4 into juxtaposition. Throw thus equals the combined thickness of units 3 and 4, namely 174 ft (53 m). At point *B*, the base of unit 5 is faulted into contact with the base of unit 6. Throw, in this case, equals the thickness of unit 5, namely 172 ft (52 m).

Apparent relative movement seldom corresponds with actual relative movement. The separation that we view on geologic maps and in canyon walls is a product of many influences: orientation of layering; strike and dip of the fault surface; slip on the fault, including direction, sense, and magnitude of displacement; and the orientation of the exposure in which separation is viewed. Consider some simple examples: the tilted sedimentary rocks shown in Figure 9.19*A* are cut by a normal-slip fault whose magnitude of net slip is 600 ft (182 m). The fault strikes at right angles to the strike of bedding. Following erosional beveling of the faulted terrain to a common level (Figure 9.19*B*), the outcrop relationships convey the false impression that left-handed strike–slip faulting had taken place. Conversely, left-handed strike–slip faulting of a tilted sequence of rocks (Figure 9.20*A*) can produce structural relationships that in cross-sectional view beckon interpretation by normal-slip faulting (Figure 9.20*B*). The possibilities are limitless!

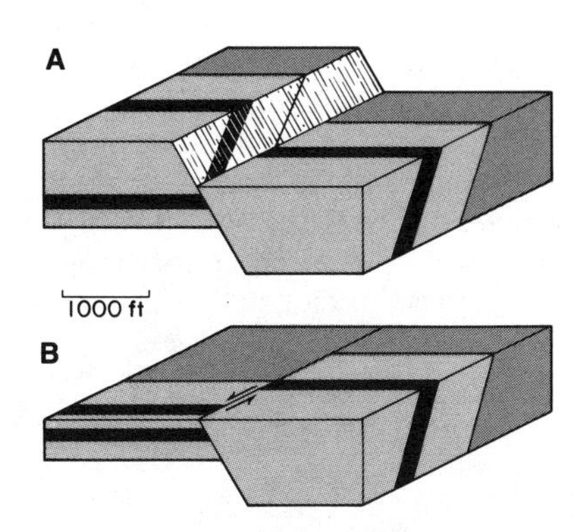

Figure 9.19 Block diagrams that underscore the difference between slip and separation. (*A*) Inclined layers are displaced along a normal-slip fault. (*B*) Erosion of the upper reaches of the footwall block creates the illusion of left-handed strike–slip faulting.

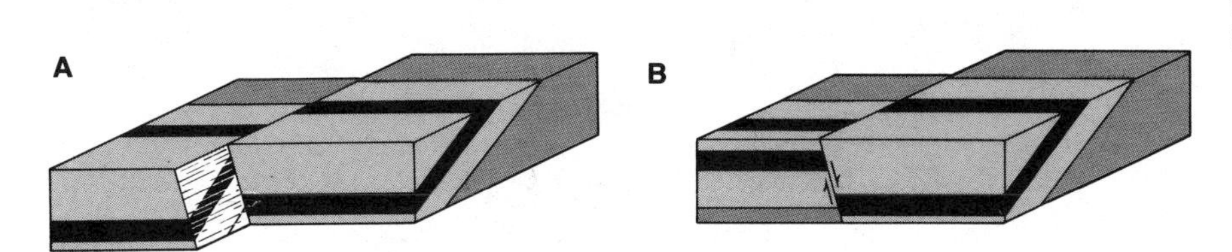

Figure 9.20 (*A*) Left-handed strike-slip faulting of inclined layers. (*B*) Erosion of the front end of the footwall block gives the impression that the faulting was normal slip.

SEPARATION CLASSIFICATION

In confronting the separation problem we learn that slip cannot be determined on the basis of apparent relative displacement. At best, knowledge of separation places modest constraints on what the translation might have been. Such limitations notwithstanding, evaluation of separation constitutes the main descriptive record for faulted rocks or faulted regions.

Separations viewed in cross-sectional exposures are described simply as normal, thrust, and reverse (Figure 9.21*A*). Normal separation is marked by offset of the hanging wall downward relative to the footwall. Thrust and reverse faults, in the separation sense, are characterized by offset of the hanging wall rocks upward relative to the footwall. Thrust faults dip less than 45° and reverse faults dip more steeply than 45°. *Separations viewed in plan view are described as left lateral or right lateral* (Figure 9.21*B*). The suffix "-handed" is reserved for proclaiming slip.

Complete descriptions of separation are based both on plan-view and cross-sectional observations of offset (Figure 9.21*C*). If the main separation is lateral, right-lateral or left-lateral is used as the second modifier, preceded by normal, reverse, or thrust, depending on the nature of the separation seen in cross-sectional view. If the chief separation is normal, reverse, or thrust, one of these terms is used as the second modifier, preceded by right-lateral or left-lateral, depending on the nature of the lateral separation observed in plan (Figure 9.21*C*).

MAP SYMBOLOGY

Slip and separation classification schemes provide a vocabulary that can be used to communicate accurately the level of our understanding of the geometric and kinematic properties of faults that we have mapped and

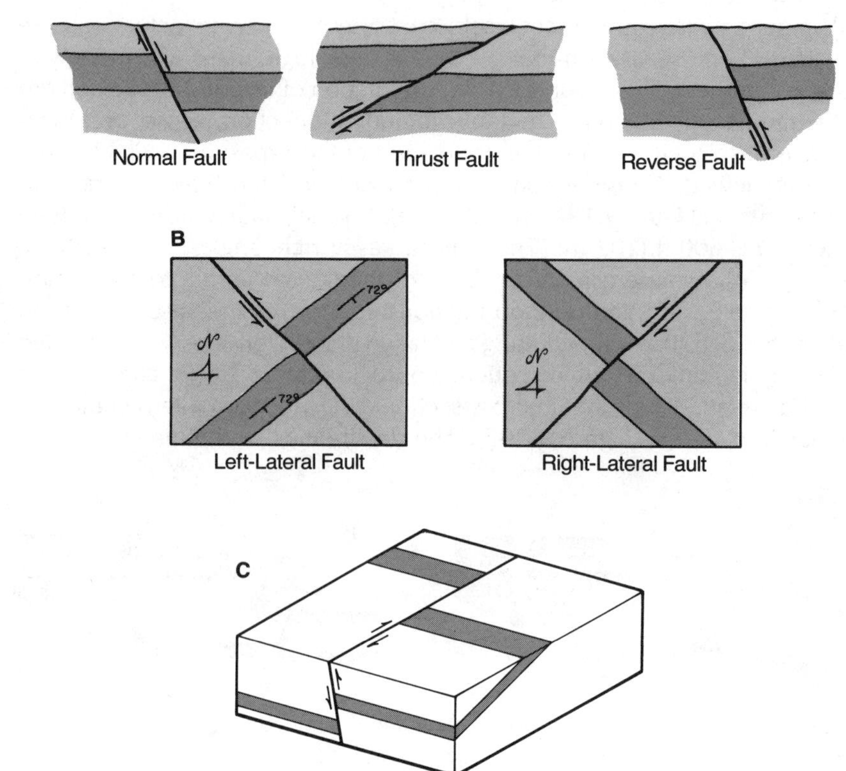

Figure 9.21 Classification of faults according to separation. (*A*) Cross-sectional views showing normal, thrust, and reverse faults. (*B*) Map views showing left-lateral and right-lateral faults. (*C*) The combination of map and cross-sectional views permits this fault to be classified as a right-lateral reverse fault.

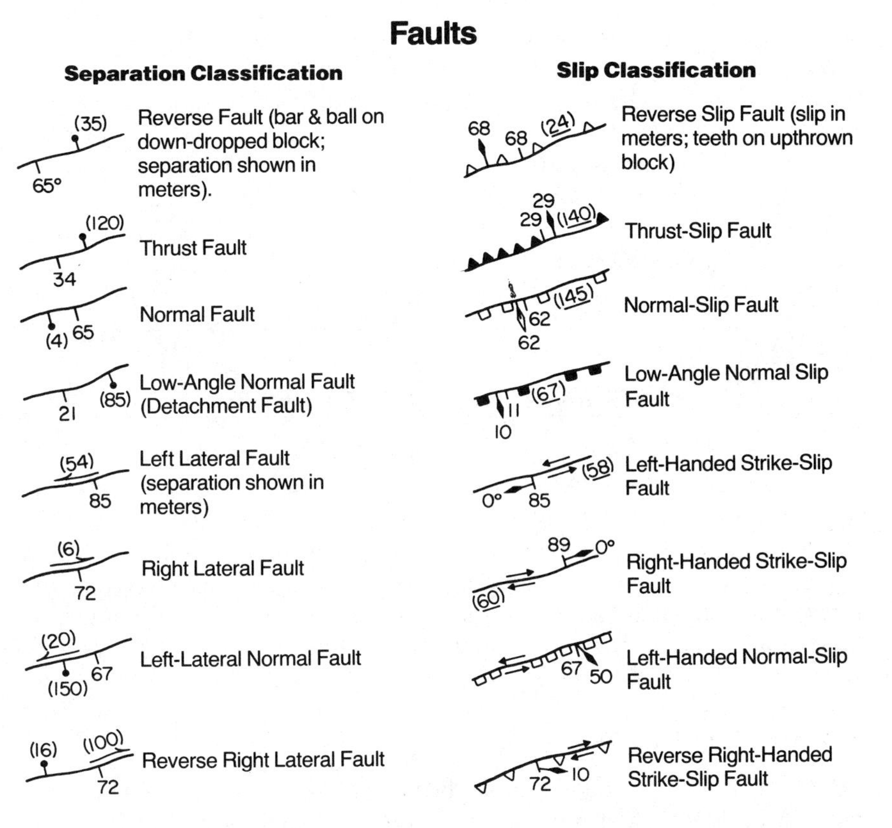

Figure 9.22 Suggested map symbols for distinguishing between separation and slip.

studied. The same distinctions can be communicated in geologic maps if attention is paid to symbology. Hill (1959) originally proposed that geologic map symbology for faults should distinguish slip and separation. Figure 9.22 contains my suggestions on how this can be done.

Faults classified according to slip are given eye-catching decorations. Such faults are worth emphasizing! "Blocks" mark the hanging wall sides of normal-slip and low-angle normal-slip faults. "Barbs" mark the hanging wall sides of thrust-slip and reverse-slip faults. Blocks and barbs on low-angle faults are darkened, whereas those representing faults dipping steeper than 45° are left open. Right-handed slip and left-handed slip are portrayed with triangle-tipped arrows.

For oblique-slip faults, barbs, blocks, and arrows are combined (Figure 9.22). If dip–slip movement is the chief component of an oblique-slip fault, the barbs and blocks are shown in their normal close spacing. Where strike–slip movement is the chief component of movement, the barbs and blocks are shown widely spaced. To these fault trace symbols are added dip direction, dip magnitude, and symbols for the minor structures that permitted slip to be interpreted. Symbols for striations, grooves, drag folds, gash fractures, and the like are posted at the locations along the fault trace where the structures were observed.

Separation symbols, in the scheme that I use, are less elaborate than those for slip (Figure 9.22). A "bar-and-ball" symbol is placed on the side of the fault that has apparently moved relatively downward. This symbol, combined with the symbol for dip direction and magnitude, permits normal, thrust, and reverse separation to be distinguished. Sense of separation, right or left lateral, is denoted by a single-tipped arrow. Bar-and-ball symbols and single-tipped arrows can be combined in ways to describe separation observed both in plan-view and cross-sectional exposures.

According to standard convention, all fault traces are shown in bold lines. They are drawn unbroken where exposures permit their exact positions to be located. They are dashed where fault location is approximated. They are dotted where the faults are concealed by alluvium or by some other cover, like municipal hospitals and reservoirs. Fault traces are queried (- ? -) where there remains some question as to whether the fault actually exists.

Kinematics

STRAIN SIGNIFICANCE OF FAULTS

Faults exist because the rocks they occupy were distorted by deforming stresses. *Normal-slip faults, thrust- and reverse-slip faults, and strike–slip faults have a common function: to stretch the crust in one direction and shorten it in another.* The directions of stretching and shortening are at right angles to one another. In examining how this is accomplished, we assume plane strain. In other words, we assume that volume is conserved during faulting and that there is neither stretching nor shortening in the third dimension.

NORMAL-SLIP FAULTS

Normal-slip faults accommodate horizontal extension and vertical shortening. Consider the flat-lying sedimentary layer 120 m long and 10 m thick shown in Figure 9.23*A*. When this layer is cut and displaced by twenty 60°-dipping normal-slip faults, the end points of the layer are shifted to positions 132 m apart (Figure 9.23*B*). The layer is said to be stretched, even though you know and I know that it was not stretched like a rubber band. The 12 m of stretching is the sum of the **gaps** between offset layers (Figure 9.23*B*). Extension and quadratic elongation for the stretched layer are easily computed.

$$e = \frac{l_f - l_o}{l_o} = \frac{132 - 120}{120} = 0.10$$

$$\text{percent lengthening} = 0.10\,(100) = 10\%$$

$$\lambda = (1 + e)^2,$$

$$\lambda = (1 + 0.1)^2 = 1.21.$$

If the normal-slip faulting had been uniformly and penetratively stretched to its 132- m final length, the thickness of the layer would have been reduced

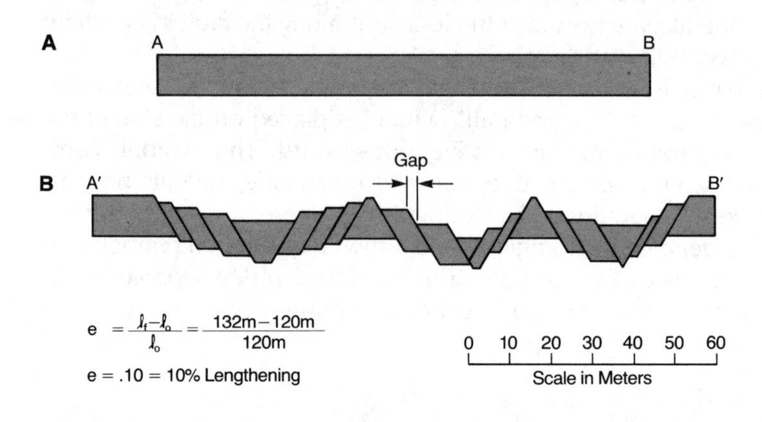

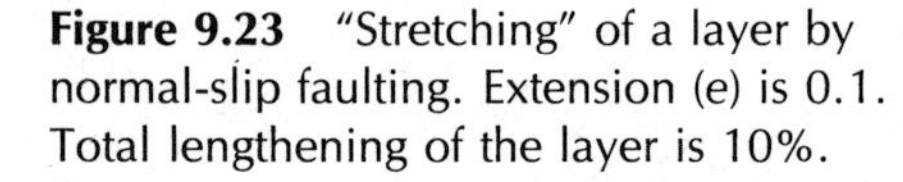

Figure 9.23 "Stretching" of a layer by normal-slip faulting. Extension (e) is 0.1. Total lengthening of the layer is 10%.

from 10 m to 9 m. The final thickness, t_f = 9 m, can be calculated by assuming that the cross-sectional area (A) of the layer remained the same before and after deformation.

$$A_o = A_f$$
$$(l_o)\,(t_o) = (l_f)\,(t_f)$$
$$(120 \text{ m})\,(10) = (132 \text{ m})\,(t_f)$$
$$t_f = \frac{1200\text{m}^2}{132\text{m}} \sim 9 \text{ m}$$

Final thickness (t_f) calculated in this way, can be called the **virtual thickness** of the layer. It is the average thickness that a faulted layer would possess if the lengthening (or shortening) of the layer had been achieved by a homogeneous penetrative flow equivalent in magnitude to the sum of the offsets along spaced, discrete faults.

THRUST- AND REVERSE-SLIP FAULTS

Thrust-slip and reverse-slip faults perform the opposite function of normal-slip faults. Thrust-slip and reverse-slip faults allow a layer of rock to be shortened horizontally and extended vertically. Figure 9.24 shows how this is achieved. A layer 188 m long and 10 m thick is cut and displaced by a combination of thrust-slip and reverse-slip faults (Figure 9.24*A*, *B*). Translation on each of the faults creates an **overlap** between offset layers that effectively shortens the initial layer. The sum of the overlaps in this example is 35 m. Final length (l_f) of the layer is 153 m. Quadratic elongation and extension can be calculated as follows.

$$\lambda = (1 + e)^2$$
$$e = \frac{\Delta l}{l_o} = \frac{-35 \text{ m}}{180 \text{ m}} = -0.19, \text{ or } 19\% \text{ shortening}$$
$$\lambda = (0.81)^2 \sim 0.65$$

Virtual thickness is computed by equating the cross-sectional areas of the layer, before and after deformation.

$$A_o = A_f$$
$$(l_o)\,(t_o) = (l_f)\,(t_f)$$
$$(188 \text{ m})\,(10 \text{ m}) = (153 \text{ m})\,(t_f)$$
$$t_f = \frac{1880 \text{ m}}{153 \text{ m}} \sim 12.3 \text{ m}$$

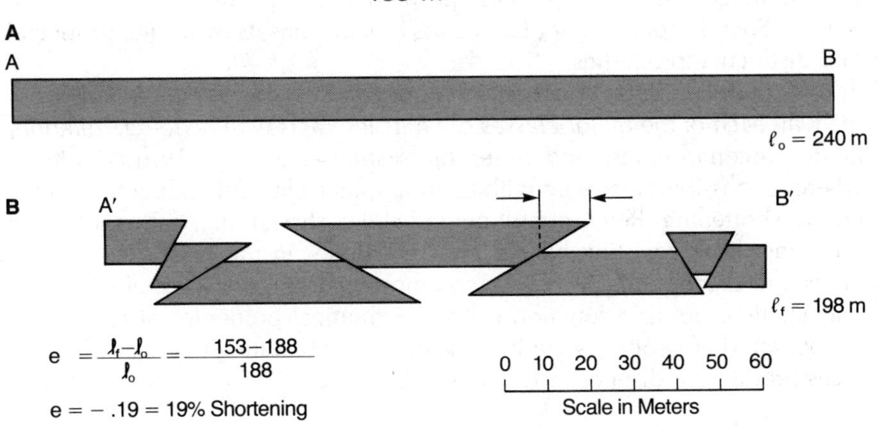

Figure 9.24 "Shortening" of a layer by thrust-slip and reverse-slip faulting. Extension (e) is .19. Shortening of the layer is 19%.

STRIKE–SLIP FAULTS

If we view the cross sections shown in Figures 9.23 and 9.24 as if they were plan-view maps of fault patterns, we see the effects of two-dimensional strain produced by strike–slip faulting. Strike–slip faulting results in stretching and shortening in the plane of horizontal layering. The faulting has no effect on thickness of the deformed layer. Although the directions of extension and shortening are mutually perpendicular, the absolute orientations of these directions depend on the relative degree of development of right-handed versus left-handed strike–slip faults. If translations along right-handed and left-handed strike–slip faults are balanced, strain is nonrotational. On the other hand, if one sense of strike–slip faulting predominates, the strain is rotational.

DISTINCTIONS AMONG FAULT CLASSES

The main classes of faults are distinctive because the absolute directions of stretching and shortening are different for each. Normal-slip faulting accommodates horizontal stretching and vertical shortening, whereas thrust-slip and reverse-slip faulting accommodate horizontal shortening and vertical stretching. With regard to strike–slip faulting, stretching and shortening both take place in the horizontal plane. Taken together, the main classes of faults are a versatile array of structures: they can distort the crust in a variety of ways, as the requirements of plate tectonic configurations and/or local stresses dictate.

The strain significance of the main classes of faults can be pictured in Ramsay strain-field diagrams. Considered in cross-sectional view (Figure 9.25*A*), thrust-slip, reverse-slip, and normal-slip faults all occupy the same field, the field of compensation. Strike–slip faulting, on the other hand, occupies the field of no strain. Considered in plan view (Figure 9.25*B*), the main classes of faults occupy different fields. Thrust-slip and reverse-slip faults lie in the field of linear shortening. Normal-slip faults occupy the line of linear stretching. And strike–slip faults occur within the field of compensation.

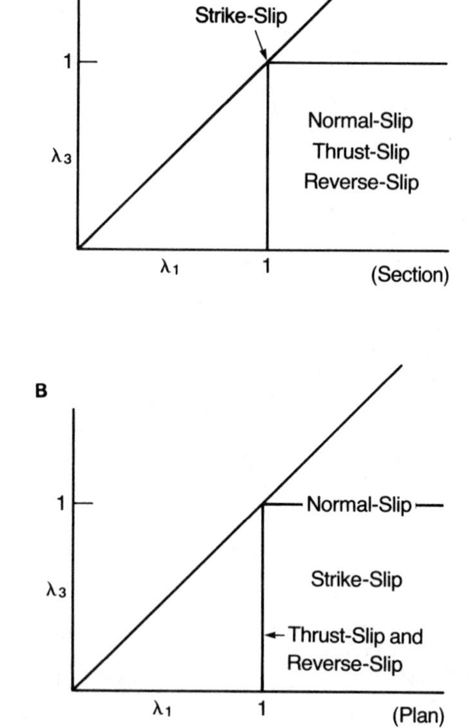

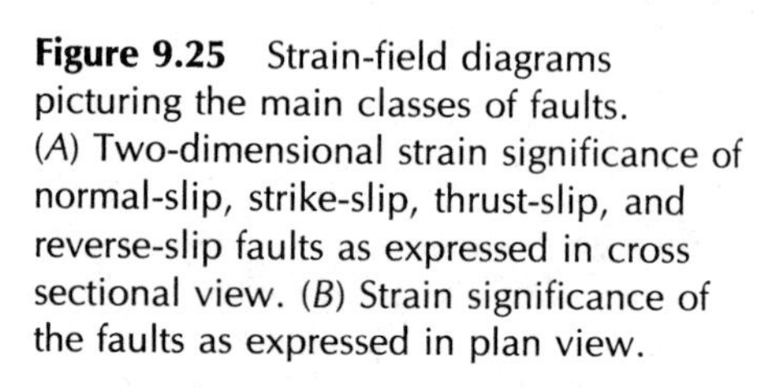

Figure 9.25 Strain-field diagrams picturing the main classes of faults. (*A*) Two-dimensional strain significance of normal-slip, strike-slip, thrust-slip, and reverse-slip faults as expressed in cross sectional view. (*B*) Strain significance of the faults as expressed in plan view.

INFLUENCE OF PREEXISTING STRUCTURES ON FAULTING

In discussing faulting as a process, it is customary to refer to thrust faulting, reverse faulting, normal faulting, and strike–slip faulting, not in a separation sense, but with the understanding that these processes give rise to thrust-slip, reverse-slip, normal-slip, and strike–slip faults. Given that the main fault classes all have the same strain function, to achieve simultaneous stretching and shortening, we might guess that each brand of faulting would produce fault systems of identical *physical* properties, irrespective of actual orientations. Such is not the case. Each class of faults has its own suite of special and distinctive properties.

Differences in physical properties among fault systems result largely from the way each of the major classes of faults interacts with bedding, foliation, faults, unconformities, and other preexisting structures. Thrust faulting, wherever possible, teams up with bedding-plane slip and folding to achieve crustal shortening. Reverse faulting and strike–slip faulting commonly take advantage of preexisting high-angle *weaknesses* in the crust, like ancient faults and shear zones, to help accommodate distortion. Normal-slip faults may shallow out as a function of the mechanical properties of formations encountered at depth. As we have seen, normal faulting along curved surfaces produces a distinctive rotation of bedding.

The influence of preexisting primary and secondary structures on faulting is profound. In studying natural fault systems we quickly learn that the Earth is not like a homogeneous test specimen used in a simple laboratory experiment. Distortion by faulting is imposed on heterogeneous rocks that are full of discontinuities and strength differences, which influence the ways in which the kinematics of faulting are achieved.

THRUST FAULTING

TECTONIC ENVIRONMENT

Systems of thrust faults form at or near the active margins of convergent plates. Melange wedges, forearc and backarc basin deposits, and miogeoclinal prisms are all especially vulnerable to thrust faulting. Whether thrusting occurs depends in part on the nature of plate tectonic configurations through time (Burchfiel and Davis, 1975). Cause-and-effect relationships are not yet understood, but velocity and direction of plate convergence, the relative buoyancy of the converging plates, and the dip of subduction zones must be significant factors.

The very best expressions of regional thrusting are in miogeoclinal prisms of sedimentary strata that have been foreshortened. The most instructive examples feature great thicknesses of sedimentary strata that were undeformed and essentially horizontal before shortening by thrust faulting and associated folding.

REGIONAL CHARACTERISTICS

The structural characteristics of thrust belts are beautifully portrayed in regional structural profiles, like those constructed for the Canadian Rockies of Alberta and British Columbia (Figure 9.26). Great expanses of strata have been deformed by thrusting in this region. Price and Mountjoy (1970) estimated that the original miogeoclinal prism in the Canadian belt was about 300 mi (480 km) wide, tapering from a thickness of 40,000 ft (12,000 m) on the west to 6000 ft (1820 m) on the east. The westernmost strata were moved at least 125 mi (200 km) to the northeast. The nonconformity between sedimentary cover and crystalline basement is preserved underneath the thrusted folded mass (Figure 9.26). It dips gently westward. The thrust-slip faults do not cut into basement, but feed into ductile sedimentary layers right above the basement/cover interface.

The eastern Idaho/western Wyoming thrust belt is similar to the Canadian belt. Royse, Warner, and Reese (1975) estimate that the miogeoclinal prism was originally about 120 mi (190 km) wide, tapering from 70,000 ft (21,000 m) on the west end to 11,000 ft (3300 m) on the east. These thicknesses include **syntectonic** sedimentary basin deposits derived from fault-bounded uplifts and dumped during the course of thrusting. Strata in the westernmost margin of the belt were moved about 50 mi (80 km) eastward to their present site. As in the Canadian belt, the nonconformity separating basement and sedimentary cover dips gently and is not broken by the thrusts.

Rocks that have been thrust great distances are considered to be **allochthonous**. They have been moved a long distance from their original site of deposition. Allochthonous rocks come to rest on **autochthonous** rocks, rocks that have retained their original location within the lithosphere of which they are a part. Regional thrust faults serve to separate rocks of an allochthon from rocks of the autochthon (Figure 9.27). **Windows** through allochthonous cover can provide a deep look into autochthonous rocks,

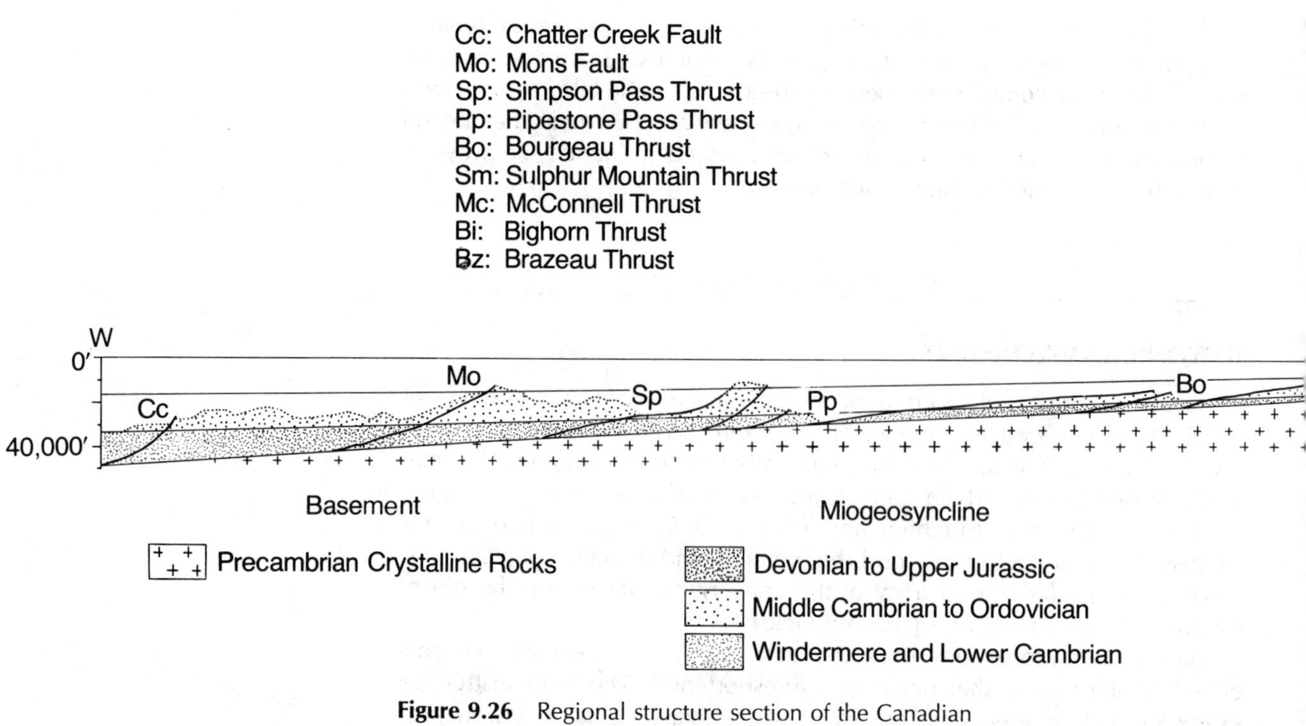

Figure 9.26 Regional structure section of the Canadian Rockies of Alberta and British Columbia, and restoration of the faulted region into its original configuration. [From Price and Mountjoy (1970). Published with permission of Geological Association of Canada.]

which are otherwise concealed. Isolated **klippen** of allochtonous rocks can disclose the former extensiveness of overthrust strata.

THIN-SKINNED DEFORMATION AND DECOLLEMENT

The contradiction of overthrusting is revealed in the structure profile of the Canadian Rockies (see Figure 9.26): strongly folded, faulted sequences of miogeoclinal strata rest atop basement that clearly has not been involved in the distortion. Dahlstrom, in reference to the Canadian Rockies belt, wrote:

> . . . there is never a balance between the length of Mesozoic and Paleozoic beds and the length of basement. The "cover" beds are always too long, which

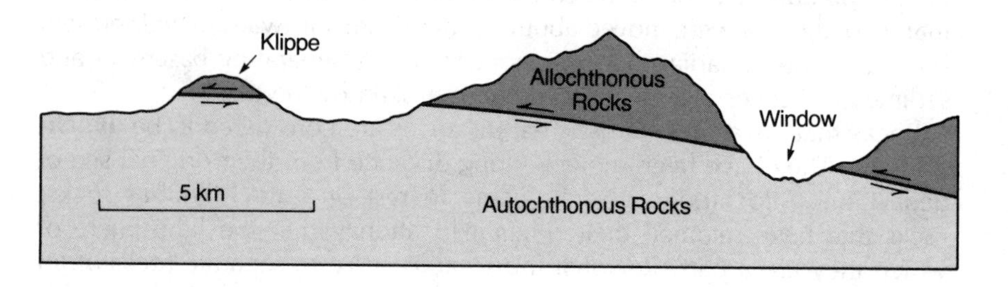

Figure 9.27 Cross-sectional view of allochthonous and autochthonous rocks, as well as windows and kippen.

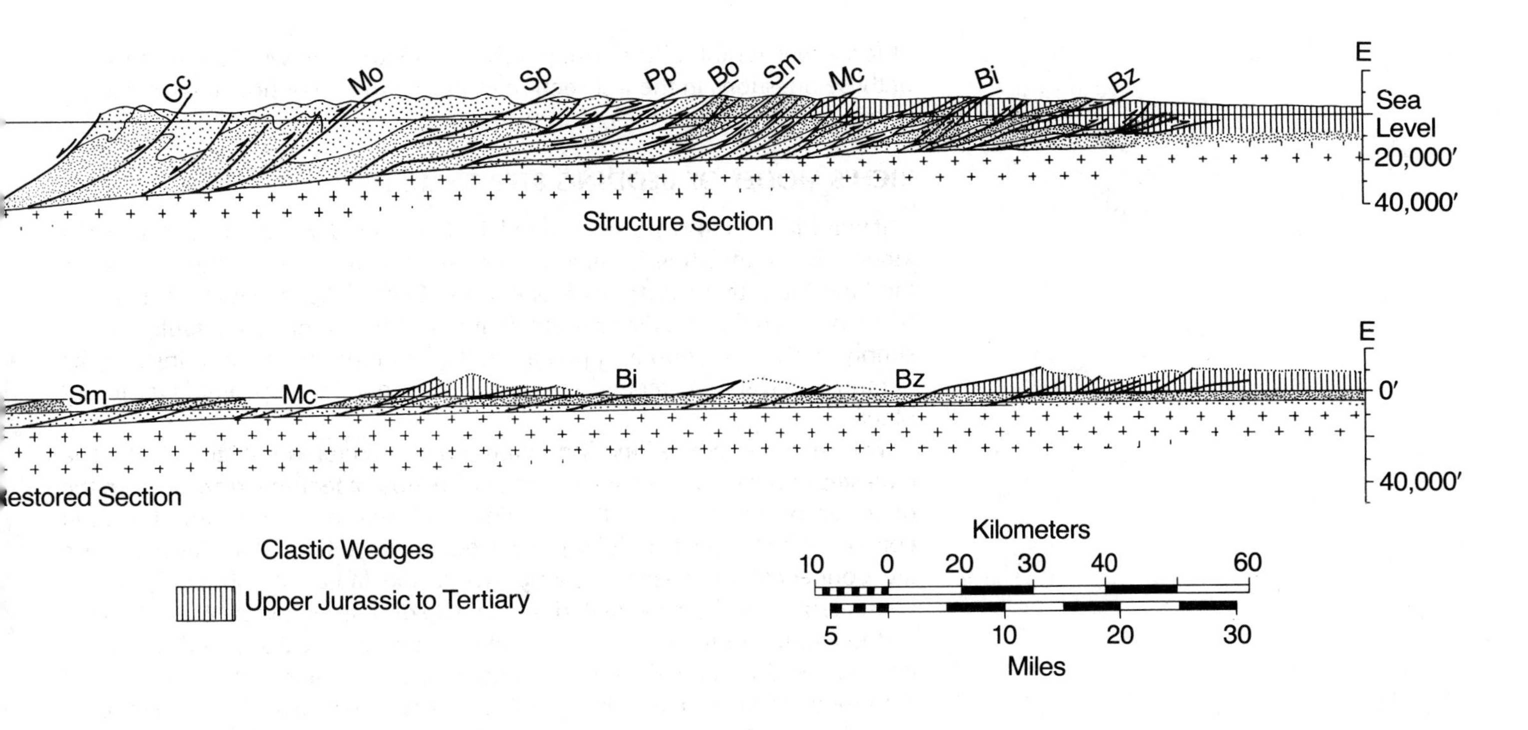

is explained in cross section by a sole fault along which the upper beds have been moved (shoved? glided?) into the cross section from the west. (From "Balanced Cross Sections" by D. C. A. Dahlstrom, 1969, p. 746. Reproduced by permission of National Research Council of Canada from *Canadian Journal of Earth Sciences*.)

"Sole fault" as used by Dahlstrom has other names as well. Some call it a **basal shearing plane** (DeSitter, 1964). The commonly used term is **decollement**, although as DeSitter points out the term refers not to a structure but to a process: "the detachment of the upper cover from its substratum." The position of decollement is normally at a discontinuity marking major ductility contrast, permitting cover to become distorted independently of what is underneath. The great nonconformity between Precambrian basement and sedimentary cover can provide such an interface of ductility contrast.

Although decollement surfaces are easy to describe, they are tough to explain. The reality of **decollement faulting** and **thin-skinned overthrusting** was first recognized by pioneer geologists of the nineteenth century mapping in the Jura Mountains and the Alps. They mapped enormous, far-traveled **nappes** of folded and faulted strata resting on crystalline rocks of an entirely different structural character. The image of thin-skinned overthrusting haunted them, as it now does us, with questions regarding the dynamics of overthrusting: If the thrusts formed by compressional shortening, why was the basement not affected? (Or was it?) If the thrusted, folded cover moved into position by gravitational gliding along the interface between basement and cover, where did it come from and how did it move *up* the dip of the nonconformity? If the deformed cover moved as a whole, what magnitude of force was required to translate it by ±150 km? We will face up to some of these questions in the last section of this chapter. For now it is instructive to examine the kinematics of thrusting.

RICH'S MODEL OF BEDDING-STEP THRUSTING

It was Rich (1934) who recognized the key to modern understanding of the kinematics of thrusting in regional terranes. While studying the geology of the Pine Mountain overthrust block in the Central Appalachian Mountains, he recognized that the Pine Mountain thrust, a major thrust-slip fault, did not simply slash up-section as a planar fault of uniform orientation. Instead, he demonstrated that it stepped up-section from one incompetent layer to another.

The geometry of the Pine Mountain thrust is pictured in Figure 9.28. The fault steps up-section westward, in the direction of **tectonic transport**, in the direction of translation of the hanging wall relative to footwall. The fault consists of two layer-parallel segments occupying ductile shale layers. These are connected by a **step** or **ramp** where the fault cuts obliquely across competent beds. Rich estimated the net slip on this fault to be 6 mi (10 km).

Recognition of the stepped or ramped nature of thrust-slip faults, like the Pine Mountain, opened the door to understanding the structural geology of deformed strata in many thrust belts. Harris (1979) and Suppe (1980a,b) have described the geometric requirements of this model beautifully. They have emphasized that the segmented nature of step thrusts forces a distinctive geometry upon layered rocks of the hanging wall. When faulted hanging wall strata are forced to move up a thrust ramp, they are flexed into an angular syncline (Figure 9.29*A*). The form of this syncline is propagated upward into the hanging wall strata. Its position remains fixed as long as the ramp thrust remains active. When faulted hanging wall strata are translated to the top of the step or ramp, to the position where the fault once again assumes a layer-parallel position, they are flexed into an anticline (Figure 9.29*B*). The anticline is angular, almost kinklike. Called **snakehead folds** (L. D. Harris, Personal communication, 1980) or fault-bend folds (Suppe, 1980 a, b), these structures grow in amplitude as faulting continues until the full thickness of hanging wall strata has reached the summit of the ramp. From that time on, the fold grows laterally, but not in amplitude, as long as translation continues (Figure 9.29*C*).

It is a tribute to Rich's geometric perception that he was able to recognize that the flat-topped anticlines and synclines in the hanging wall of the Pine Mountain thrust could be explained by a fault that steps in and out of layer-parallel orientations.

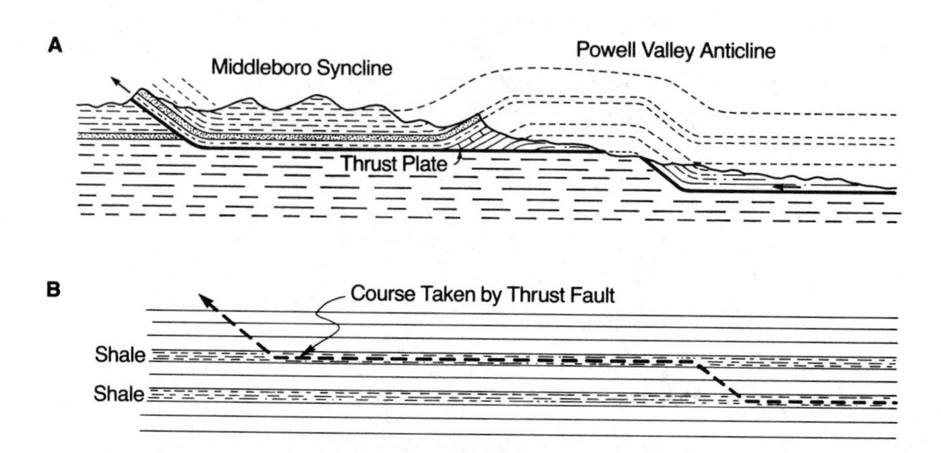

Figure 9.28 (*A*) Structure section of the Pine Mountain thrust in the Central Appalachian Mountains. (*B*) Path of fault through the stratigraphic column as seen before thrusting. [From Rich (1934). Published with permission of American Association of Petroleum Geologists.]

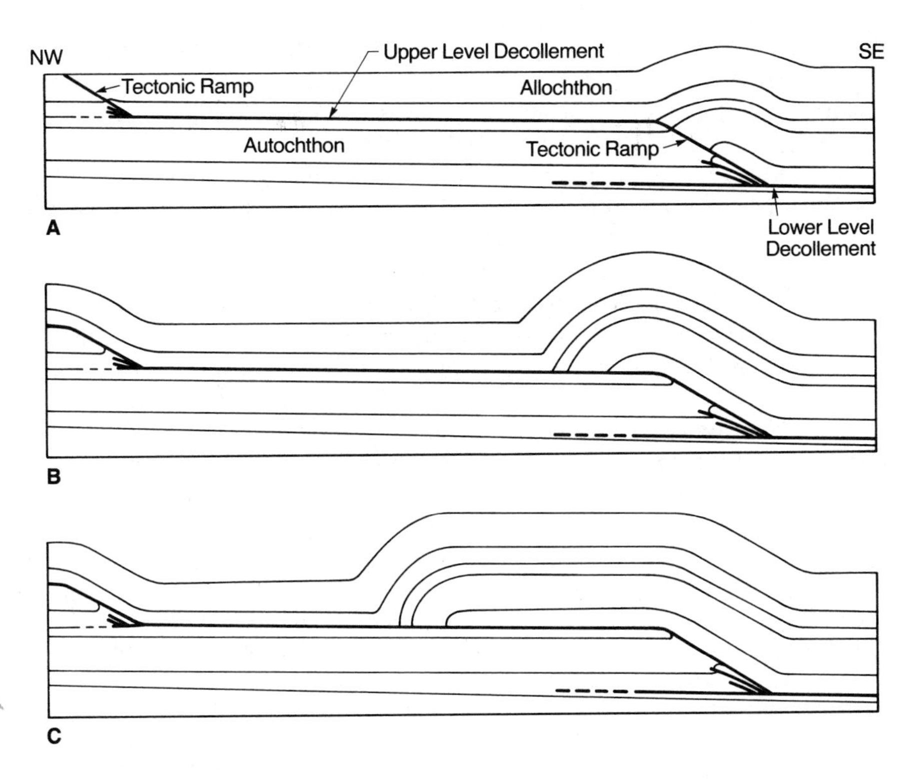

Figure 9.29 Kinematic evolution of regional bedding-plane step thrusts. Diagrams *A*, *B*, and *C* show progressive westward movement of upper plate strata. [From Harris (1979). Published with permission of Geological Society of America and the author.]

THE KINEMATIC RULES

Much of our present understanding of the geometry and kinematics of thrusting has been derived from structural geologists, like Rich, who have applied their skills in the arena of petroleum exploration. Explorationists like Bally, Gordy, and Stewart (1966), Dahlstrom (1969), and Royse, Warner, and Reese (1975) have developed an unusual appreciation for the full three-dimensional nature of thrust systems. They have achieved this understanding through insights derived from geologic mapping, subsurface drilling, and seismic reflection profiling carried out in the search for petroleum.

Thrust systems evolve in such a way that the major faults cut up-section in the direction of tectonic transport. They step from layer-parallel segments within soft, incompetent layering and cut obliquely across stiff, competent beds on route to the next favorable incompetent unit. The major thrusts split off from one another, and the resulting pattern is marked by successively younger faults in the direction of tectonic transport (Dahlstrom, 1969; Royse, Warner, and Reese, 1975).

Although the general kinematics of the thrusting process are reasonably well known, it is the interpretation of specific fold/thrust configurations in the subsurface that is fundamental to petroleum exploration. In spite of all that is known about thrusting, it remains very difficult on the basis of geologic mapping alone to confidently interpret subsurface structural details. In theory, an infinite variety of subsurface geometric configurations of beds, folds, and thrusts can be drawn to fit the surface control (geologic map patterns) and subsurface control (well data regarding depths to formations and structures).

Dahlstrom (1969) developed some kinematic rules that provide a means to test and improve interpretations of subsurface geology in thrust belts. The rules are summarized in the construction of **balanced cross sections**, structural profiles drawn to avoid violating the original cross-sectional area of the rocks that were subjected to deformation. If it can be shown that deformation did not alter bed thickness, balanced structural profiles simply require that

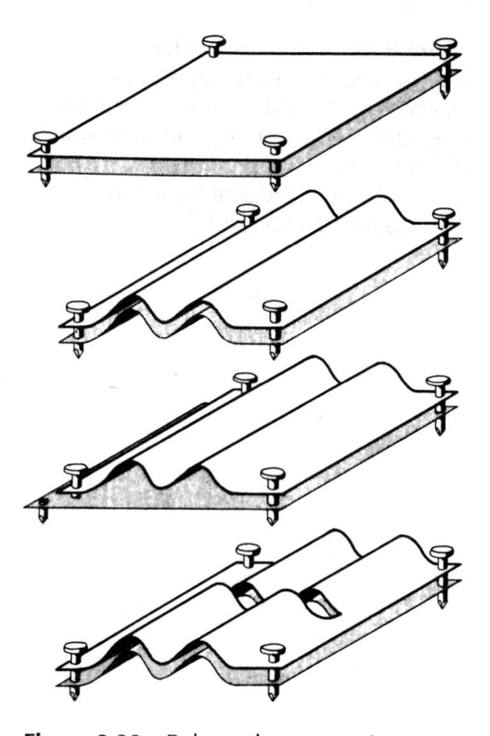

Figure 9.30 Balanced cross sections are constructed in a way to preserve the original cross-sectional areas of beds that are involved in the deformation. [From Dahlstrom (1969). Reproduced with permission of National Research Council of Canada from *Canadian Journal of Earth Sciences*.]

the thickness of each bed be held constant and that the beds shown in profile view be drawn of equal length (Figure 9.30). Dahlstrom (1969) and Royse, Warner, and Reese (1975) have emphasized that thrust faulting and folding in the Canadian Rockies and the eastern Idaho/western Wyoming thrust belt indeed took place without appreciable volume change. Furthermore, they demonstrated that bedding was neither thickened nor thinned significantly during the deformation.

The main kinematic principle emphasized by Dahlstrom (1969) in the preparation of balanced cross sections is **consistency of displacement**. Slip measured along any given fault should be uniform in sense and magnitude, assuming that the rocks are indeed rigid (Figure 9.31). If data show that the slip varies from place to place along a fault (Figure 9.31*A*), this circumstance must be explained by interpretations that involve the **interchange** of different degrees of folding and faulting to accommodate a common shortening (Figure 9.31*B*), or the **replacement** of one fault by a series of **imbricate splay faults** (Figure 9.31*C*). Dahlstrom explains:

> Thrust faulting and folding are both mechanisms for making a packet of rock shorter and thicker than it was originally, so that one could expect the two mechanisms to be interchangeable. Imbrication is the distribution of displacement from one large fault to several minor ones. (From "Balanced Cross Sections" by D. C. A. Dahlstrom, 1969, p. 746–747. Reproduced with permission of the National Research Council of Canada from *Canadian Journal of Earth Sciences*.)

Balanced cross sections must stay within the bounds of all subsurface, seismic, and geologic data on hand, and they should be constructed with an understanding and appreciation of kinematic principles. The balancing of sections provides a way to place realistic constraints on the structures that might exist at depth.

MAP-VIEW EXPRESSION OF THRUST BELTS

A final concept that Dahlstrom (1969) and others have emphasized affects our perception of plan-view renditions of thrust relationships. Displacement on a single fault may decrease along strike to zero, even though regional

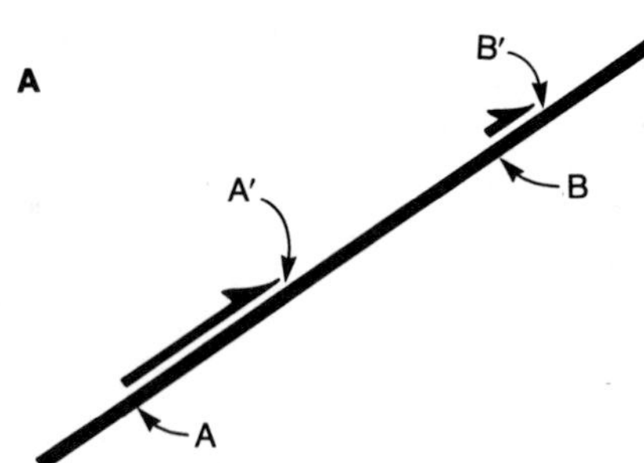

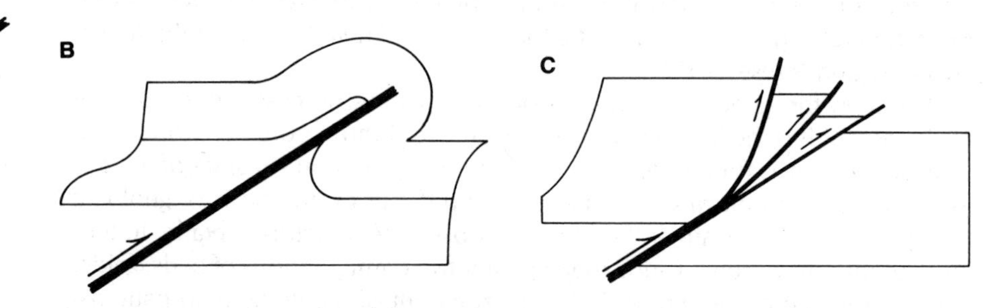

Figure 9.31 (*A*) Fault showing an inconsistency of displacement. The slip along the lower reaches of the fault is much greater than that toward the surface. The inconsistency can be accommodated by (*B*) the interchange of faulting and folding or (*C*) the replacement of the single fault by several splay faults. [From Dahlstrom (1969). Reproduced with permission of National Research Council of Canada from *Canadian Journal of Earth Sciences*.]

shortening along that distance remains strong. This apparent contradiction can be understood by recognizing that displacement (slip) can be transferred from one thrust fault to another (Figure 9.32). **Transfer zones** are the overlapped ends of faults in which decreasing slip on one fault is compensated by increasing slip on the other (Dahlstrom, 1969).

The plan-view expression of overthrust belts is commonly marked by transverse strike–slip faults (Figure 9.33). Known as **tear faults**, they form mainly because of the impossibility of translating a huge rock mass as a single unit. A miogeoclinal wedge hundreds or thousands of kilometers long, hundreds of kilometers wide, and kilometers thick simply cannot move along thrusts of unlimited length. Instead, larger masses are broken up into smaller structural units bounded by thrust faults and tear faults. Movements of the smaller structural units are orchestrated with the movements of the larger mass as a whole, including the movement of all of its parts.

Tear faults have also been described as **compartmental faults**. Recognized by Brown (1975) in the Wyoming foreland province, these serve as partitions between domains of rocks in which a common magnitude of shortening has been achieved in different ways. For example, a given compartmental fault might separate a fold-dominated compartment from another that is fault dominated (Figure 9.34). Compartmental faults are unlike classic strike–slip faults in that the sense of slip need not be uniform along the length of the fault. Moreover, compartmental faults are surprisingly short when compared to the magnitude of slip they accommodate. In this way they are somewhat like continental versions of transform faults.

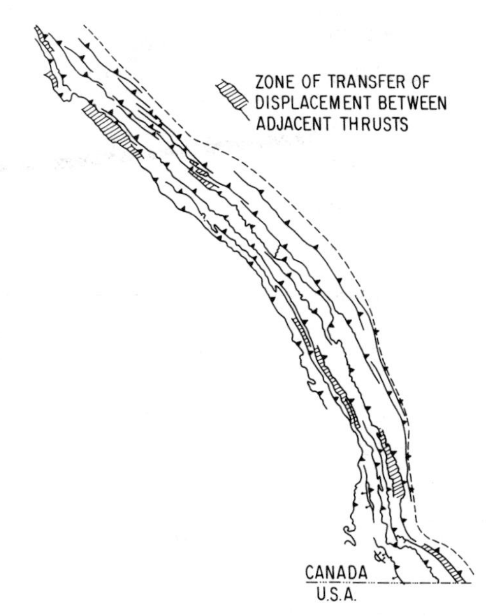

Figure 9.32 Map of regional thrust faults in the Canadian Rockies. Cross-hatched symbols identify transfer zones where decreasing translation on one "dying" fault is compensated by increasing translation on an "emerging" fault. [From Dahlstrom (1969). Reproduced with permission of National Research Council of Canada from *Canadian Journal of Earth Sciences*.]

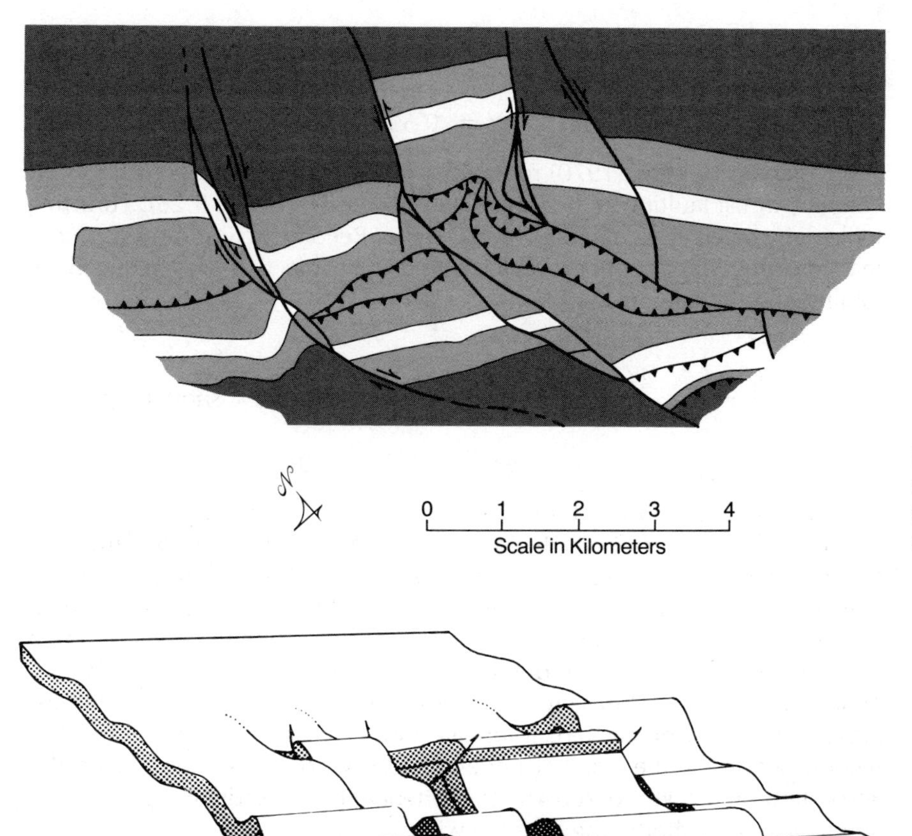

Figure 9.33 Map showing tear faults within a regional system of folds and thrust faults. [From Price, Mountjoy, and Cook (1978).]

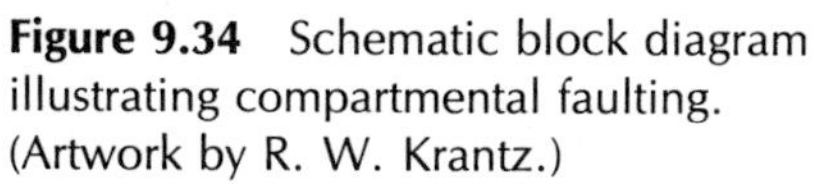

Figure 9.34 Schematic block diagram illustrating compartmental faulting. (Artwork by R. W. Krantz.)

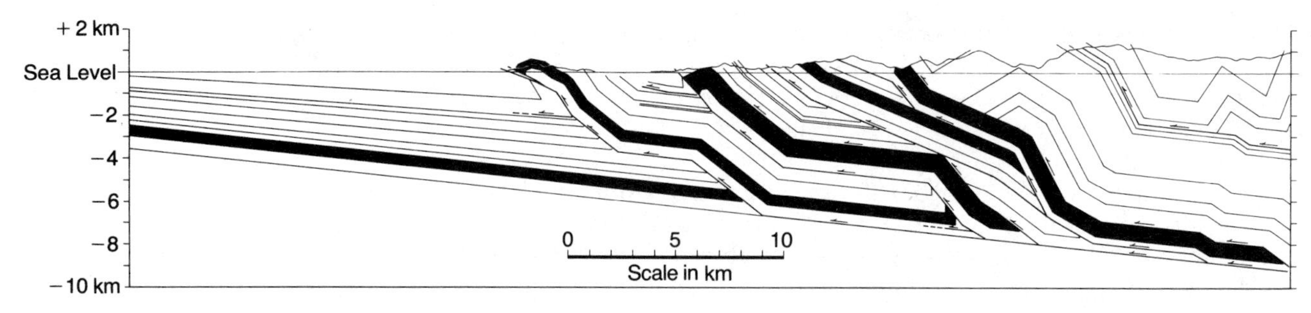

Figure 9.35 Regional structure section showing an array of stacked thrusts in Taiwan. [From Suppe (1980a). Published with permission of Geological Society of China.]

STRAIN SIGNIFICANCE OF REGIONAL THRUSTING

Strain estimates derived from balanced cross sections of thrust belts are staggering. In the eastern Idaho thrust belt, the miogeoclinal prism was shortened by about 50% because of thrusting and associated folding that took place between latest Jurassic and early Eocene. On the basis of carefully balanced geologic cross sections, Royse, Warner, and Reese (1975) computed a total shortening of 52 mi (83 km) for the region between the Wasatch Mountains of Utah and the Green River Basin of Wyoming. Their estimate of the original width of the rocks in the belt is 111 mi (178 km). These data permit extension and quadratic elongation to be calculated.

$$e = \frac{\Delta l}{l_o} = \frac{-52}{111} = -0.47 \text{ or } 47\% \text{ shortening}$$

$$\lambda = (1 + e)^2 = (0.53)^2 = 0.28$$

Price and Mountjoy (1970) reported a similar magnitude of crustal shortening by thrust faulting in the Canadian Rockies (see Figure 9.26). Thrusting took place between late Jurassic and Eocene. Reconstruction shows an original miogeoclinal sediment package 475 km wide that was reduced by folding and thrusting to 240 km.

$$e = \frac{l_f - l_o}{l_o} = \frac{240 \text{ km} - 475 \text{ km}}{475 \text{ km}} = -0.49 \text{ or } 49\% \text{ shortening}$$

$$\lambda = (1 + e)^2 = (0.51)^2 = 0.26$$

For purposes both of tectonic understanding and petroleum exploration, Suppe has been carefully analyzing the fold-and-thrust system of Taiwan in the context of ramp thrusting. Taiwan is part of a collisional orogen wherein oceanic and island arc lithosphere of the Philippine Sea plate is being thrusted over Oligocene and younger miogeoclinal sediments of the Chinese continental margin. The rate of plate convergence there is 7 cm/yr (Suppe, 1980a). Suppe's structural cross sections reveal imposing architectural **duplexes** composed of stacked, imbricate thrust packages (Figure 9.35). His reconstructions of the original undeformed state of the strata demand the absorption of 160 km to 200 km of shortening by the sediment prism. This amounts to 50% shortening over the width of the region of study. Given the present rate(s) of plate convergence, all this shortening probably has been achieved since late Pliocene!

NORMAL FAULTING

EXPERIMENTAL MODELS

The characteristics of normal faulting were elegantly modeled in simple clay deformation experiments by Hans Cloos (1936), and later by his brother Ernst Cloos (1955). The delicate, intricate fault patterns that can emerge in deformed clay that is properly prepared and deformed are astonishingly similar to natural patterns.

The clay cake to be deformed is prepared by mixing dry kaolin with water until a soft buttery consistency is achieved. The clay is spread as a layer ±4-cm thick, approximately 15 × 15 cm in size, and smoothed with a putty knife. A reference circle may be gently impressed onto the surface of the clay cake to serve as a strain marker.

Hans and Ernst Cloos recommended two different methods for deforming the clay. In the first experiment the clay cake is built on top of an elastic rubber sheet (Figure 9.36*A*). Then the rubber sheet is stretched slowly and uniformly. Since the base of the clay cake sticks to the rubber sheet, the entire clay layer experiences the effects of the stretching. As stretching commences, normal-slip faults, along with tear faults, emerge to accommodate the extension (Figure 9.36*B*). Gradually the faults become uniformly distributed throughout the clay cake.

The top of a clay cake stretched in this way reveals numerous closely spaced faults and fault scarps. They are typically oriented at a high angle to the direction of stretching. Some fault traces are straight; others are curved. Fault surfaces exposed along the fault scarps are marked by dip–slip striations.

The cross-sectional view of a clay cake stretched on a rubber sheet reveals a system of **fault-bounded blocks** (Figure 9.36*B*). The bounding faults are all

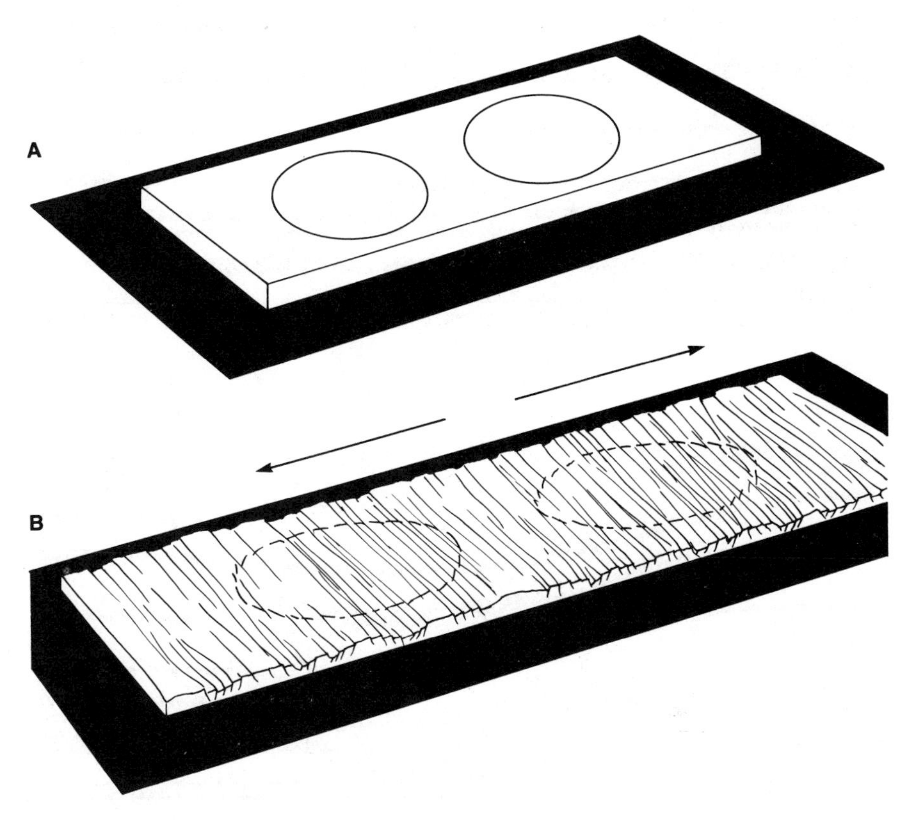

Figure 9.36 Clay cake simulation of normal faulting. (*A*) Plan view of the stretching of a clay cake on a rubber sheet. (*B*) Cross section of the clay cake reveals a series of fault-bounded horsts and grabens. [From Cloos (1955), Geological Society of America.]

normal-slip faults, dipping on the average of 60°. Sets of oppositely dipping normal-slip faults are about equally developed. The relatively uplifted blocks bounded by normal-slip faults are known as **horsts**. The depressed blocks between horsts are known as **grabens**.

When viewed from a distance, the final deformed state of the clay cake shows lengthening in the direction of stretching, and thinning (**flattening**) of the layer as a whole. Extension and quadratic elongation values can be determined by comparing original and final thicknesses and original and final lengths of the clay cake.

In the second experiment recommended by Hans and Ernst Cloos, the clay cake is placed on top of overlapping sheets of sheet metal. To achieve stretching of the clay, the ends of the sheets of metal are pulled slowly and uniformly away from each other (Figure 9.37*A*). This movement causes stretching of the clay in the region directly above the overlap of the sheet metal. The area affected by the extension increases in breadth as the sheets are pulled further and further apart. Experimental deformation of this nature results in the development of a single complex graben above the area of overlap of sheet metal (Figure 9.37*B*). The surface of the clay layer bows downward at the site of the graben, and the whole width of the graben is pervaded by normal-slip faults.

The major boundary faults of grabens formed in this way dip inward, at about 60°. In fact, almost all the faults in the clay cake tend to dip inward toward the center of the graben. There are, however, exceptions. **Antithetic faults** form close to the major boundary faults but dip in the opposite direction (Figure 9.37*B*). The role of antithetic faults is to eliminate the gap(s) that

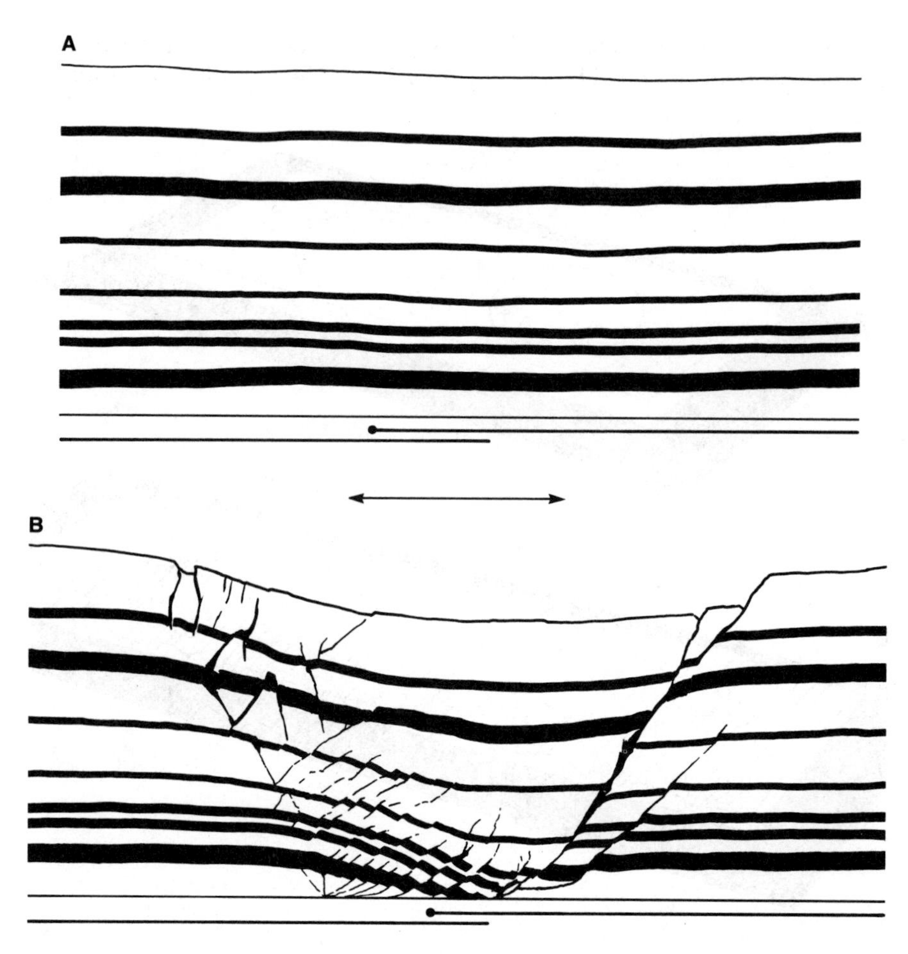

Figure 9.37 A second clay cake simulation of normal faulting: deformation of a clay cake on top of overlapping panels of sheet metal. Stretching and normal faulting are achieved by pulling apart the sheet metal panels. (*B*) Cross-sectional view of the deformed clay cake shows a series of tilted step blocks. [From Cloos (1968). Published with permission of American Association of Petroleum Geologists.]

would otherwise be produced by displacements on curved fault surfaces. Antithetic faults and reverse drag perform the same function; they fill in potential voids.

Ultimately, the surface of a clay cake stretched on sheet metal is distended into a series of **tilted step blocks** (Figure 9.37*B*). The step blocks dip toward the surface on which they are rotated. The strike of each tilted step block is about perpendicular to the direction of stretching.

The formation of faults in soft clay is yet another reminder that when attempting to replicate experimentally the structures of large regions of the Earth's crust, it is necessary to scale down not only the size, but also the strength of the materials used (Hubbert, 1937). Large crustal segments, when considered as a whole, are very weak.

TECTONIC ENVIRONMENT

Normal-slip fault systems form where crustal rocks undergo stretching and lengthening. Along with transform faults, they are the dominant fault class along midoceanic spreading centers. Where ocean ridges have stepped into continental regions, as in the Red Sea, the expression of normal faulting can be seen much more clearly.

The breakup of a continent by rifting is accomplished by normal faulting. Such is the case in the East African rift zone today and in the eastern part of the Appalachian Mountain system 200 million years ago when Europe and North America were pulling apart from one another.

Normal faulting also flourishes in regions that are marginal to strike–slip faults and continental transform faults. In these environments stretching is one manifestation of the simple shear rotational strain created by horizontal strike–slip faulting.

THE RED SEA REGION

Details of normal faulting in the Red Sea region have been described beautifully by Lowell and Genik (1972) and by Lowell and others (1975). Their descriptions and interpretations are especially vivid because as members of a petroleum exploration team they were able to build a solid three-dimensional picture on the basis of both geophysical and geological data. Their work not only provides insight into normal faulting but also illuminates the process whereby continental lithosphere becomes rifted apart.

The tectonic environment of the Red Sea region is essentially that of the Carlsberg Midoceanic Ridge, except that the ridge underlies continental crust (Figure 9.38). Like normal midoceanic spreading centers, the Red Sea region is marked by high heat flow, active seismicity, symmetrical magnetic anomaly patterns, a gravity high, and even some midocean ridge basalts.

Stretching of the Red Sea region was achieved by penetrative, distributed normal-slip faulting (Figure 9.39). Thinning by normal faulting resulted in subsidence and the formation of a complex graben system. As in the clay-on-sheet-metal experiment, subsidence was so significant that the land surface dropped below sea level. Transgression(s) of ocean waters resulted in deposition of oceanic sediments, including 5 to 7 km of evaporites. Horsts, grabens, and tilted step blocks were fashioned by distributed normal faulting of the continental crust.

Ultimately, the lithosphere was stretched and thinned to the point of rupture, permitting entrance and ascent of oceanic basalts (Figure 9.39). Oceanic crust continues to be emplaced in the innermost part of the Red Sea at a rate consistent with the spreading rate of Saudi Arabia with respect to Africa.

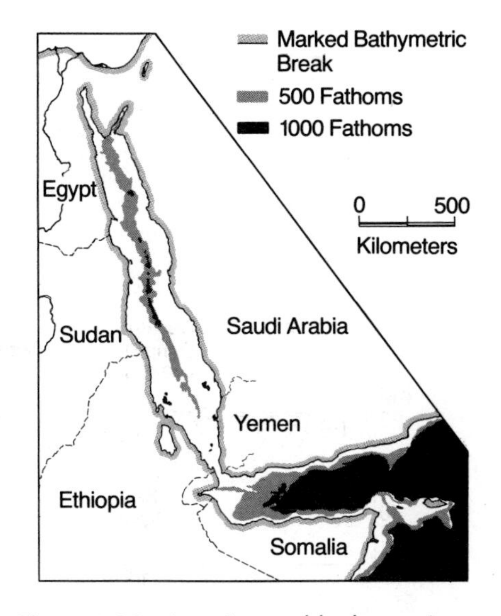

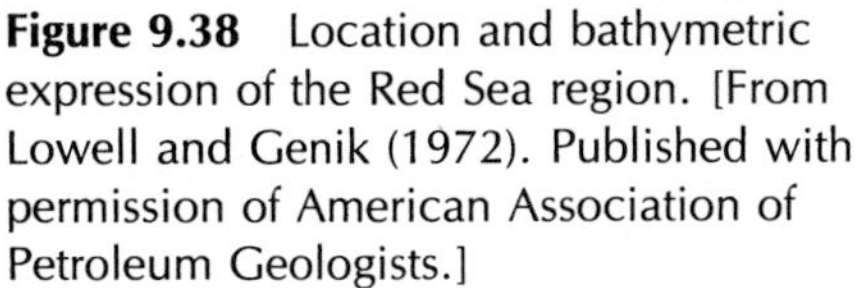
Figure 9.38 Location and bathymetric expression of the Red Sea region. [From Lowell and Genik (1972). Published with permission of American Association of Petroleum Geologists.]

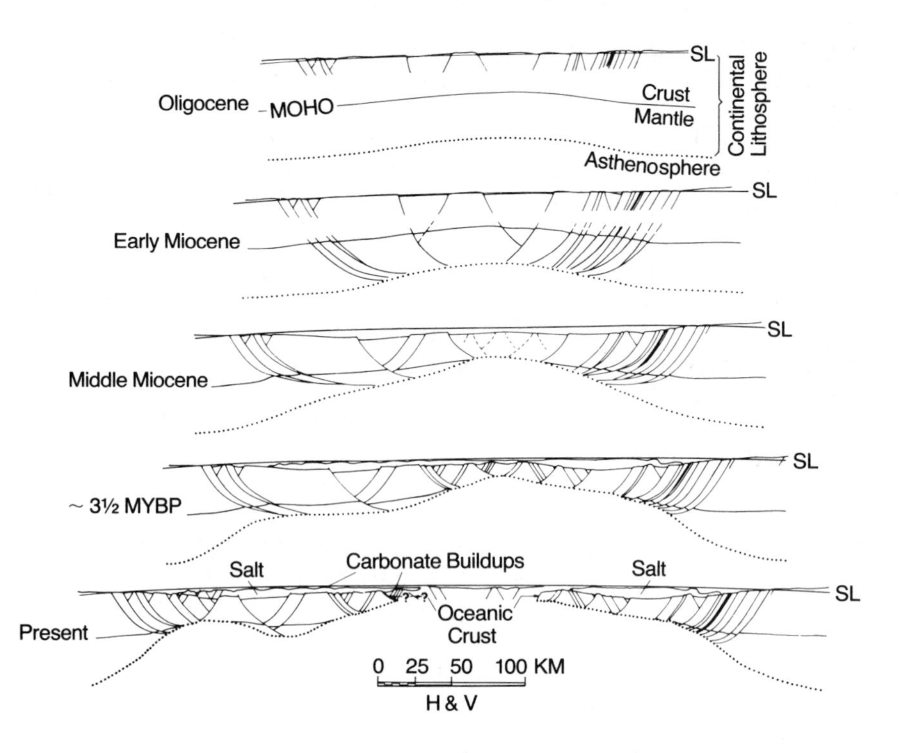

Figure 9.39 Kinematic evolution of faulting of the Red Sea region by distributed normal-slip faulting. [From Lowell and Genik (1972). Published with permission of American Association of Petroleum Geologists.]

Kinematic analysis of the faulting in the Red Sea region is full of lessons. Lowell and Genik (1972) warn that the eastern and western margins of the Red Sea were never in direct contact, and thus the evaluation of slip requires more than simply matching shorelines. Rather, accurate calculation of the displacement vector for faulting requires summing up the displacements on all the normal-slip faults in the system. Based on structure profiles, like that shown in Figure 9.40, Lowell and Genik estimated that the width of the region was initially 315 km. Normal faulting stretched it to 450 km. Moreover, there may be another 100 km of stretching bound up in plastic deformation and translations on minor faults for which data are not available. The interpretation of a 135-km increase in length is thus a conservative estimate. Even so,

$$e = \frac{\Delta l}{l_o} = \frac{135}{315} = 0.43 = 43\% \text{ stretching}$$

$$\lambda = (1 + e)^2 = (1.43)^2 \sim 2$$

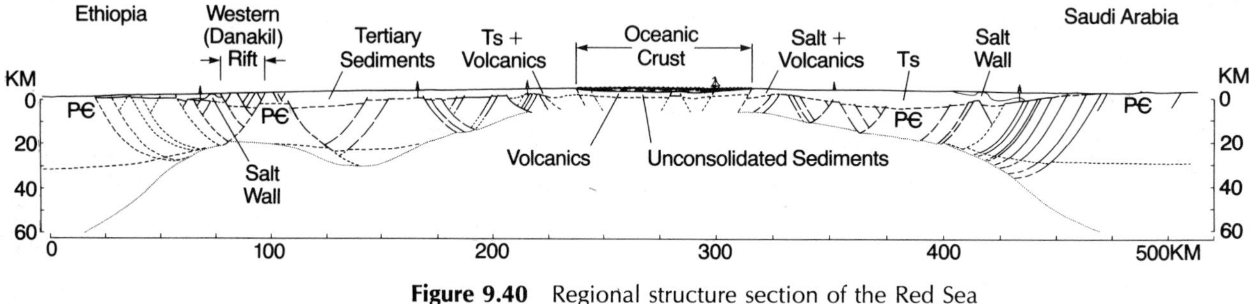

Figure 9.40 Regional structure section of the Red Sea region. [From Lowell and Genik (1972). Published with permission of American Association of Petroleum Geologists.]

Kinematic reconstruction of the Red Sea region is shown in Figure 9.41. Pull-apart is oblique, not perpendicular, to the trend of the Red Sea itself.

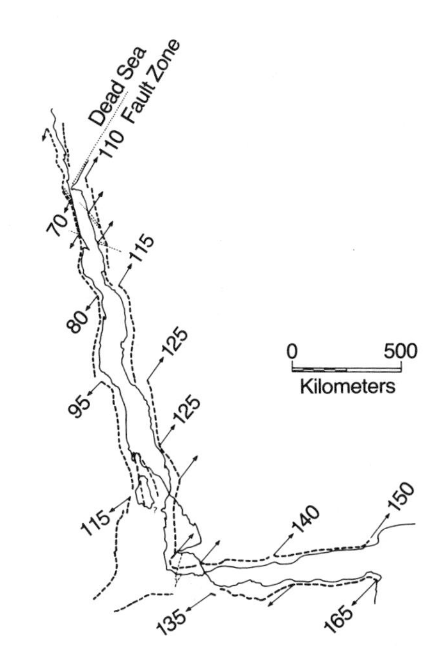

Figure 9.41 Map showing the magnitudes and directions of slip required to close the Red Sea rift. [From Lowell and Genik (1972). Published with permission of American Association of Petroleum Geologists.]

NORMAL FAULTING IN THE BASIN AND RANGE PROVINCE

HIGH-ANGLE NORMAL FAULTING. The Basin and Range province of the western United States and northern Mexico is an extraordinary region of distributed normal faulting and associated strike–slip faulting. The structural system is especially complicated in places because the extensional faulting was superimposed on a framework that had already suffered complex deformation during thrusting, intrusion, and metamorphism (Davis, 1981). Basin and Range normal faulting commenced about 20 m.y. ago and ended about 5 m.y. ago.

The most recent Basin and Range normal faulting took place between about 12 m.y. ago and 5 m.y. ago. As portrayed in Eaton's (1980) tectonic map of the Basin and Range province, the faults trend north–northeast, north–south, or north–northwest (Figure 9.42). The normal faulting serves to block out the mountains and valleys that give the Basin and Range province its distinctiveness (Figure 9.43). Fault-bounded basins and ranges are generally 25 to 70 km broad and 50 to 300 km long. The map-view expression of the ranges in the southern part of the Basin and Range was once described by United States Geological Survey geologist Clarence Dutton as "an army of caterpillars marching northward out of Mexico" (King, 1959, p. 152). In detail, the basins and ranges are seen to divide and split, forming complex physiographic and structural patterns. These patterns in turn reflect the branching and bending of the traces of normal-slip faults and associated strike–slip faults, the orientations of which are locally controlled by preexisting faults and other weaknesses.

The uniformly dipping **homoclinal** attitudes of bedding in mountain blocks of the Basin and Range province indicate that tilted step blocks are the dominant structure of the extended terrane. The major boundary faults are seldom well exposed. Their traces are typically covered by basin fill and alluvial fan gravels shed from the relatively uplifted blocks. Fortunately, there are places where such basin fill accumulations have been eroded deeply, thus exhuming major faults.

The properties of the young, high-angle normal-slip faults of the Basin and Range are nicely displayed where they have disturbed rocks of the adjacent tectonic province, the Colorado Plateau. Part of the breathtaking scenery of Utah is due to the influence of regional normal-slip faults. Three of the main fault systems are the Hurricane, the Sevier, and the Paunsaugunt (Figure 9.44). Each is composed of a family of fault surfaces that trends north–northeast for about 300 km. Colorful sedimentary and volcanic strata are dropped, west side down, along these normal-slip faults. Net slip on the Hurricane fault locally exceeds 3000 m! Slip along the Sevier and Paunsaugunt faults is 600 m or less, varying from place to place along the trace of each of these faults.

The faults of the western Colorado Plateau are parallel, but do not line up. Instead, their tips are arranged in such a way that they overlap in a systematically stepped fashion (Figure 9.44). The configuration is like transfer zones between faults in thrust belts. In all likelihood, decreasing slip along one of the major normal-slip fault zones is compensated by increasing slip on an en echelon structure.

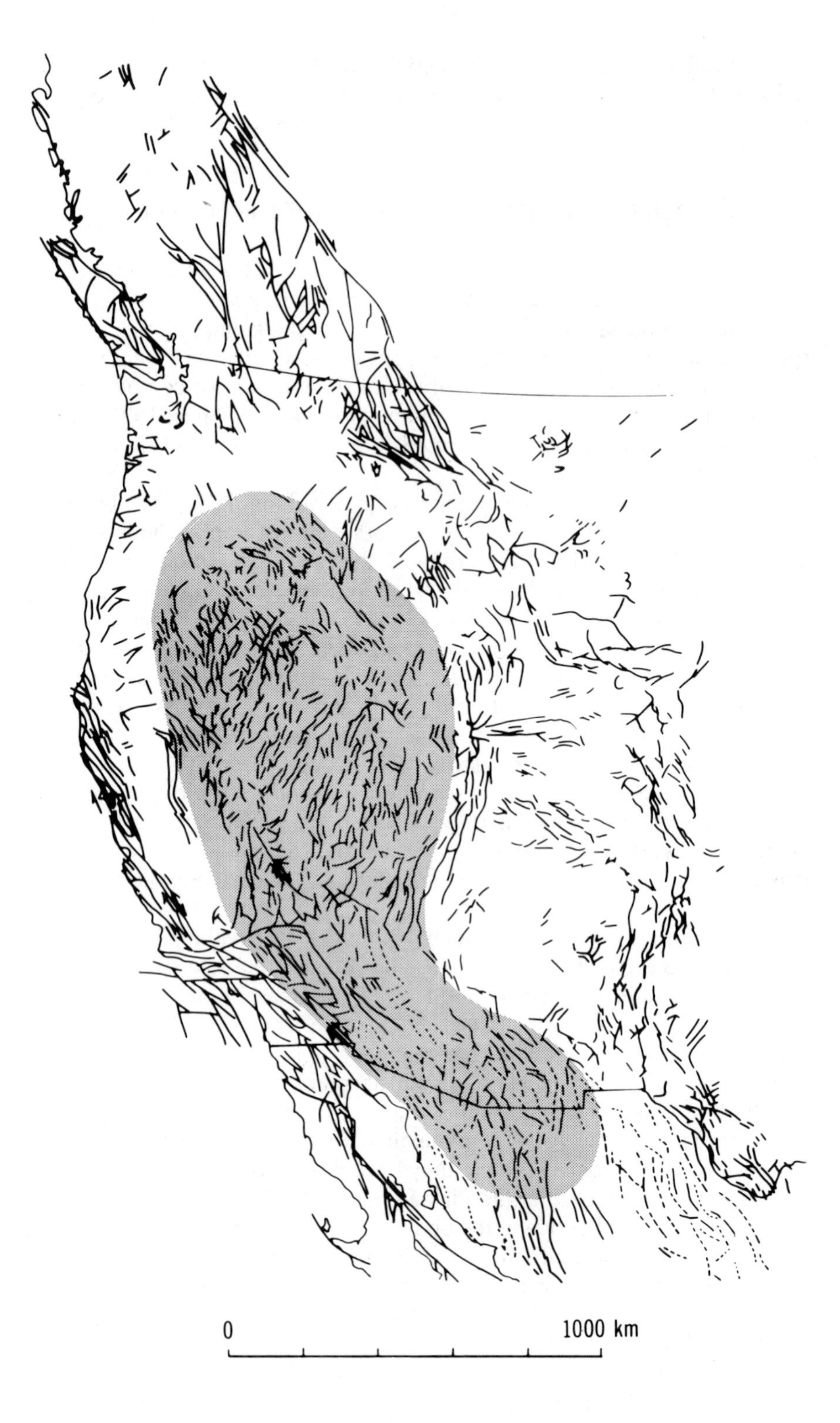

Figure 9.42 Regional structure map showing the distribution and orientations of faults in western North America, including normal faults in the Basin and Range province of the American Southwest. [Reproduced from Eaton (1980) in *Continental Tectronics,* National Academy Press, Washington, D.C.].

Figure 9.43 Typical Basin and Range physiography, Tinajas Altes Mountains, southwestern Arizona. (Photograph by W. B. Bull.)

LOW-ANGLE NORMAL FAULTING. The traditional view of the Basin and Range province is that its distinctive structural character has been fashioned exclusively by **high-angle** normal faulting. However, continued mapping and analysis is demonstrating that the Basin and Range province is pervaded by low-angle normal-slip faults that have accommodated profound crustal stretching. Many of these **detachment faults** are Oligocene–Miocene in age. They are cut and offset by bona fide Basin and Range high-angle normal faults.

In their simplest form the low-angle fault relationships consist of steeply tilted hanging wall strata resting on expansive flat to gently dipping fault surfaces (Figure 9.45). Rocks of almost any age and origin can be found in the hanging wall blocks, but those containing steeply tilted Oligocene and Miocene strata in the hanging wall best reveal the youthful age of faulting. The footwall rocks beneath the low-angle normal-slip faults represent deep levels of the geological column. They are commonly Precambrian. Within the fault zones the footwall rocks are microbrecciated and chloritically altered. The hanging wall rocks are locally shattered by fracturing.

Fault relationships of this type are found preserved in mountain ranges throughout much of the Basin and Range province. Spectacular examples of low-angle **detachment faulting** have been mapped by Anderson (1971) in the Lake Mead region (see Figure 4.21*A*), G. A. Davis and others (1980) in the Whipple Mountains of California (Figures 9.45, 9.46), and Davis (1980) in the Rincon Mountains, southeastern Arizona (Figure 9.47).

Historically, low-angle fault contacts like these had been interpreted as thrust faults produced during compressional shortening. But workers like Anderson (1971) and Armstrong (1972), demonstrated that many of these low-dipping faults are products of extensional normal faulting. Armstrong (1972) called these faults **denudational faults**—faults that place high-level relatively young rocks on deep-level relatively old rocks. Denudational faulting is the structural equivalent of erosional stripping. Indeed, low-angle normal faulting can be thought of as a tectonic stripping of cover to expose progressively deeper levels of crustal rocks.

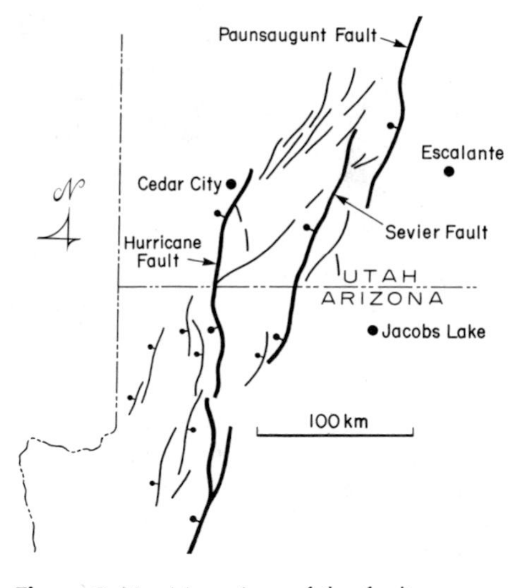

Figure 9.44 Map view of the fault pattern in the High Plateaus of western Utah.

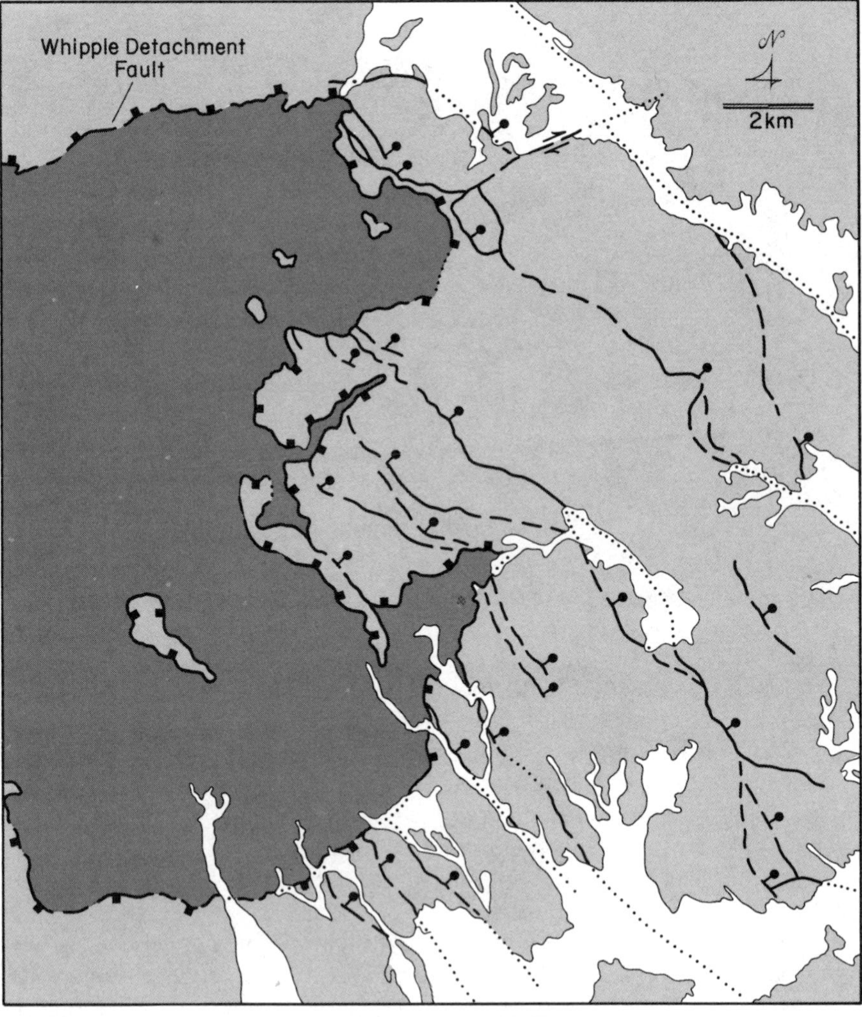

Figure 9.45 Geologic map of detachment faulting in the Whipple Mountains, southeastern California. [From Davis and others (1980). Published with permission of Geological Society of America and the authors.]

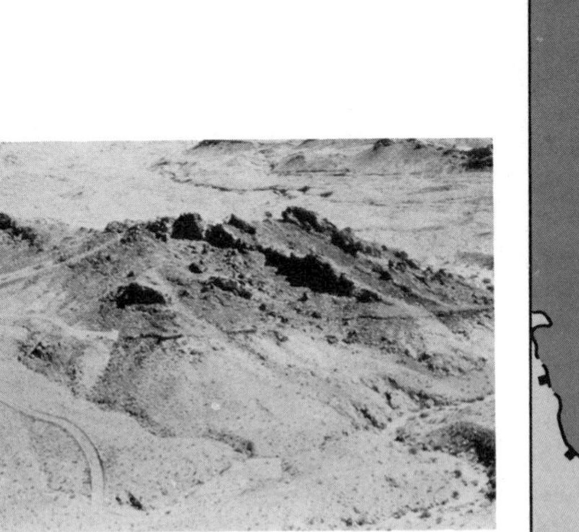

Figure 9.46 View of southwest-tilted Tertiary strata separated from underlying mylonitic gneisses by a detachment fault. [From Davis and others (1980). Published with permission of Geological Society of American and the authors.]

Figure 9.47 The Catalina fault in the Rincon Mountains separates microbreccias (dark cliff) from overlying Pennsylvanian-Permian limestones. [From Davis (1980). Published with permission of Geological Society of America and the author.]

Figure 9.48 One of the great turtleback structures in Death Valley. The fault surface accommodated major normal-slip translation. (Photograph by G. H. Davis.)

The magnitude of individual low-angle normal faults in the Basin and Range is captured in the work of Wright and Troxel (1973) in Death Valley. They demonstrated that the great **turtleback structures** of Death Valley (Figure 9.48) are enormous fault grooves or **fault mullions** preserved along the upper surface of Precambrian crystalline footwall rock. As smooth, polished mountain ridges, they plunge northwestward in the direction that hanging wall strata were translated during faulting. Remnants of the faulted, rotated hanging wall strata make up the Panamint Range, which forms the western flank of Death Valley. The thick pile of strata there is dipping moderately to steeply eastward. The original position of this section of rock was above and east of the present location of the turtlebacks. Net slip was at least 9 km.

A STRUCTURAL ANALYSIS OF NORMAL FAULTING. Proffett's (1977) study of low-angle normal faulting in the Yerington district of Nevada is very revealing. Geometric and kinematic details of the fault system emerged from exploration and development of a major porphyry copper deposit.

The site of Proffett's study was just south of Reno in west-central Nevada. Radiometric potassium–argon age dating indicates that the faulting there began in the Miocene, 17 to 18 m.y. ago. East-dipping concave-upward fault surfaces, with large troughlike grooves, accommodated normal-slip translation and accompanying rotation of hanging wall strata (Figure 9.49). During the faulting, ignimbrites of Oligocene age were back-tilted from horizontal to steep westward dips. Some rocks rotated by more than 90°! Relatively old normal-slip faults, which became inactive, were later cut and rotated to shallow dips by relatively young faults. Indeed, the youngest faults in the Yerington district are the steepest, whereas the oldest faults typically dip shallowly (Figure 9.49).

Proffett constructed detailed structural profiles of the fault relationships on the basis of geological mapping and subsurface drilling. From these he computed slip for the major faults. The largest slip on any fault was found to be 4000 m (13,000 ft). Proffett determined this value of net slip on the basis of a truncated and offset geometric line, a line formed by the pinch-out of a bed against an unconformity.

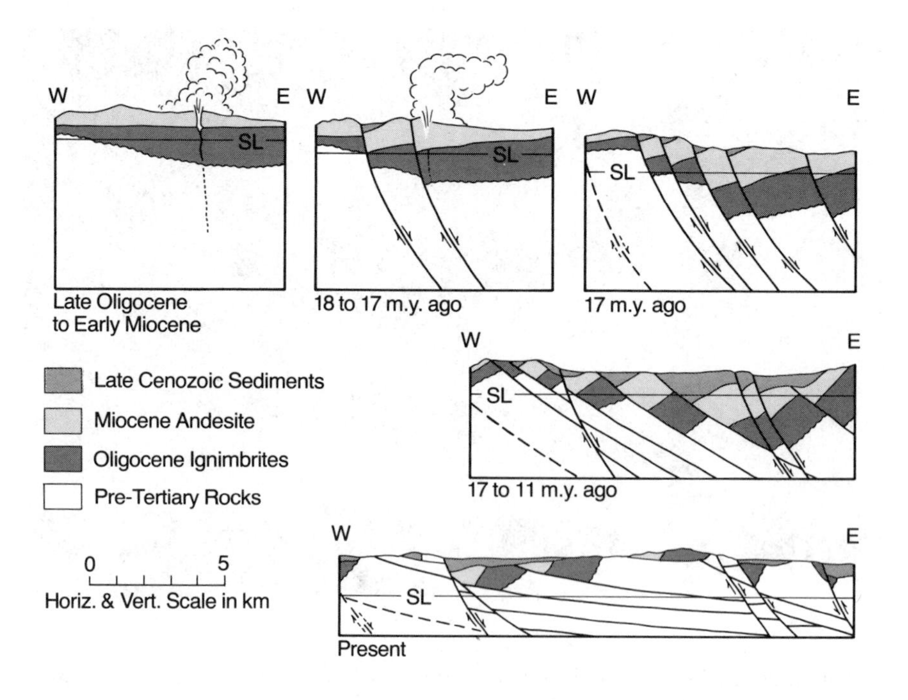

Figure 9.49 Kinematic evolution of normal faulting in the Yerington district, west-central Nevada. [From Proffett (1977). Published with permission of Geological Society of America and the author.]

The strain implications of the fault system in the Yerington district are astounding: more than 100% stretching. Based on structure profiles, Proffett estimated the original east–west width of the study site to have been 7.3 km. He interpreted the extended width, after normal faulting, to be 17.3 km.

$$e = \frac{l_f - l_o}{l_o} = \frac{17.3 \text{ km} - 7.3 \text{ km}}{7.3 \text{ km}} = 1.37 = 137\% \text{ stretching}$$

$$\lambda = (1 + e)^2 = 5.6$$

BASIN AND RANGE STRETCHING. Proffett's estimate of total stretching across the entire Basin and Range province is very conservative, much less than the 137% cited above. Based on regional cross sections, he estimated that the Basin and Range has been stretched by about 170 km in Nevada. This amounts to a 35% stretching. Other workers have suggested values closer to 100% (Hamilton, 1978).

Crustal stretching by normal faulting was compensated by thinning and flattening. Indeed, there is geophysical evidence for such compensation. The thickness of the crust in west-central Nevada is only about 25 km, compared with normal crustal thicknesses of 42 km in the Colorado Plateau and 56 km in the Sierra Nevada. Assuming a normal crustal thickness of 49 km, the thinning of the crust, due in part to normal faulting, can be determined as follows,

$$e = \frac{l_f - l_o}{l_o} = \frac{25 \text{ km} - 49 \text{ km}}{49 \text{ km}} = -0.48 = 48\% \text{ thinning}$$

In all probability, the entire thickness of the crust and part of the upper mantle in the Basin and Range were affected by the east–west stretching and crustal thinning brought about by normal faulting. We are only beginning to understand the many ways in which crustal stretching can be accommodated by normal faulting (Wernicke and Burchfield, 1982).

ORE DEPOSITS AND LOW-ANGLE NORMAL FAULTING. Recognition and understanding of the kinematics of low-angle detachment faulting in the

Basin and Range is fundamental to clear-eyed mineral exploration. An almost unparalleled discovery of a porphyry copper deposit in southeastern Arizona illustrates the point. Lowell (1968) correctly surmised that the top of a mineralized pluton, the San Manuel ore body, had been truncated and displaced by low-angle normal faulting. Based on fault analysis, he predicted that the hanging wall part of that ore body would be found 3 km to the southwest along a S50°W bearing. Drilling proved his interpretation to be correct. Structural analysis gave birth to the Kalamazoo ore body.

The discovery was *almost* unparalleled. A similar discovery had been made in the same region almost a decade earlier. The Pima porphyry copper deposit southwest of Tucson is the beheaded part of the Twin Buttes deposit, which lay to the south. Beheading was caused by low-angle normal faulting (Cooper, 1960). Translation was approximately 2 km. The southern Arizona porphyry copper deposits apparently make unusual and unusually profitable reference markers for establishing displacement vectors!

STRIKE–SLIP FAULTING

DISTRIBUTED STRIKE–SLIP FAULTING

EXPERIMENTAL MODELING. Clay model experiments illustrate the fundamentals of strike–slip faulting quite nicely, and they provide a framework for introducing terminology. Ernst Cloos (1955) described some simple experiments in which clay cakes become penetratively deformed by closely spaced, distributed strike–slip faulting. The experimental setup involves fastening a wire-mesh cloth to a hinged wooden frame that is capable of being deformed from a square to a rhomb. A 2-cm thick clay cake is constructed on top of the wire-mesh screen, and a 5-cm diameter reference circle is impressed on the surface of the clay, so that strain can be monitored during the course of ensuing deformation (Figure 9.50*A*). Deformation of the clay cake can be achieved in one of two ways: (l) pulling two opposing corners of the frame away from each other, or (2) shifting two parallel members of the wooden frame in simple shear fashion. Both motions distort the wire mesh and the clay cake such that the reference circle is transformed to an ellipse (Figure 9.50*B*). In the first case, the clay is distorted by pure shear, and the axes of the strain ellipse remain fixed in orientation throughout the course of the experiment. In the second case, the clay is distorted by simple shear rotational strain, and the axes of the strain ellipse rotate systematically and continuously during the deformation.

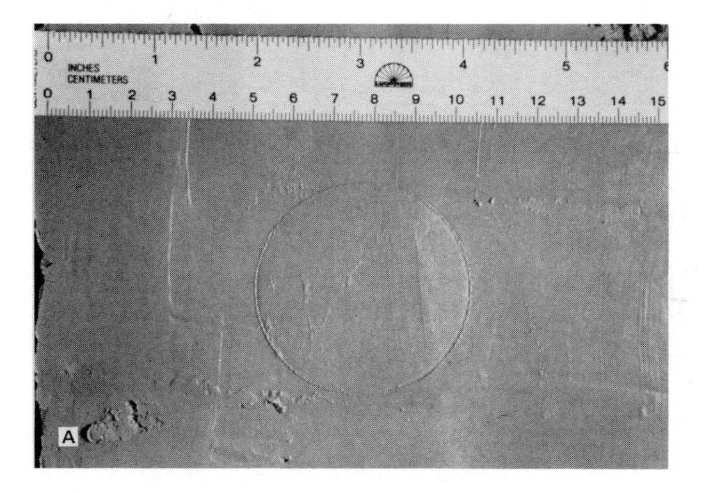

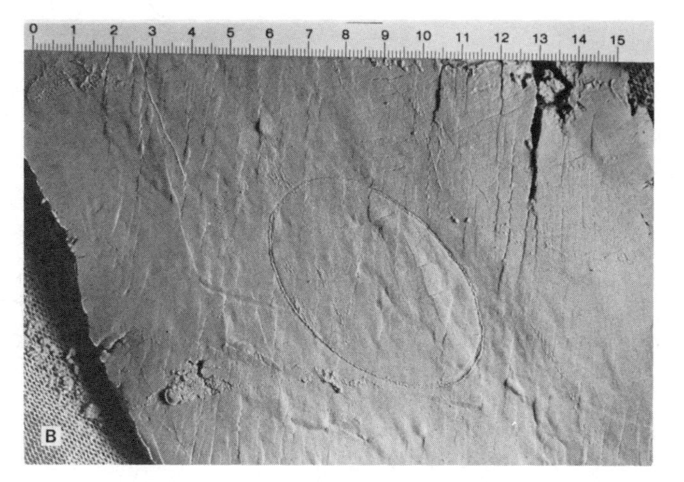

Figure 9.50 Clay-cake simulation of strike-slip faulting. (*A*) Experimental set-up consists of clay cake on a hinged wire-mesh screen. (*B*) Distortion of the clay cake.

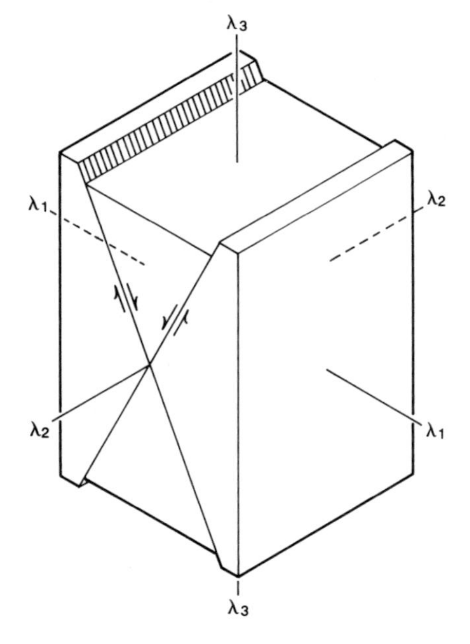

Figure 9.51 Nature of conjugate faulting, including the geometric relationship between the conjugate faults and the axes of strain.

CONJUGATE FAULTING

Distortion of the clay is largely accomplished by conjugate strike–slip faulting. The term "conjugate" means that the faults occur in two intersecting sets that are coordinated kinematically. Each set is distinctive both in orientation and in sense of shear. One set is right handed; the other is left handed (Figure 9.51).

The expression **conjugate faulting** is not used exclusively in reference to strike–slip faults. Normal-slip faults and strike–slip faults generally occur in conjugate sets as well. Conjugate faulting brings about simultaneous stretching and shortening. Triangular, wedge-shaped fault-bounded blocks move in toward one another in the direction of minimum extension. Complementary wedge-shaped fault-bounded blocks move outward to bring about an extension of the clay. The initial acute angle of intersection between the conjugate sets of strike–slip faults that emerge in the clay cake is about 60°. The line that bisects this **conjugate angle** is parallel to the axis of minimum extension of the strain ellipse (Figure 9.51).

ROTATION OF FAULTS. Conjugate strike–slip faults that form in the clay cake are not fixed in orientation through time. Instead, they rotate as deformation proceeds. As we have learned, two kinds of rotation are possible. In simple shear deformation, **external rotation** of the conjugate faults accompanies rotation of the axes of the strain ellipse (Figure 9.52*A*). If the axes of the strain ellipse rotate, so must the conjugate fractures within the ellipse. For both simple shear and pure shear deformation, an **internal rotation** of the conjugate faults results from increasing distortion (Figure 9.52*B*). The conjugate angle between the faults steadily increases with progressive distortion. If the clay is distorted to an extreme degree, the original 60° conjugate angle can expand by **flattening** to 120° or more. External and internal rotation of conjugate faults must be kept in mind when we view the end result of strike–slip faulting that took place over millions of years.

TENSION FRACTURES. Tension fractures also develop in clay cakes that are deformed in the manner described by Cloos (1955). The tension fractures form at right angles to the axis of maximum extension (X). If the surface of the clay is moistened by a thin film of water, thus breaking the surface tension of the clay, open tensional **gash fractures** emerge as the dominant structure. In the absence of a water film, the tension fractures are only modestly developed, subordinate to delicate, closely spaced **conjugate** strike–slip faults.

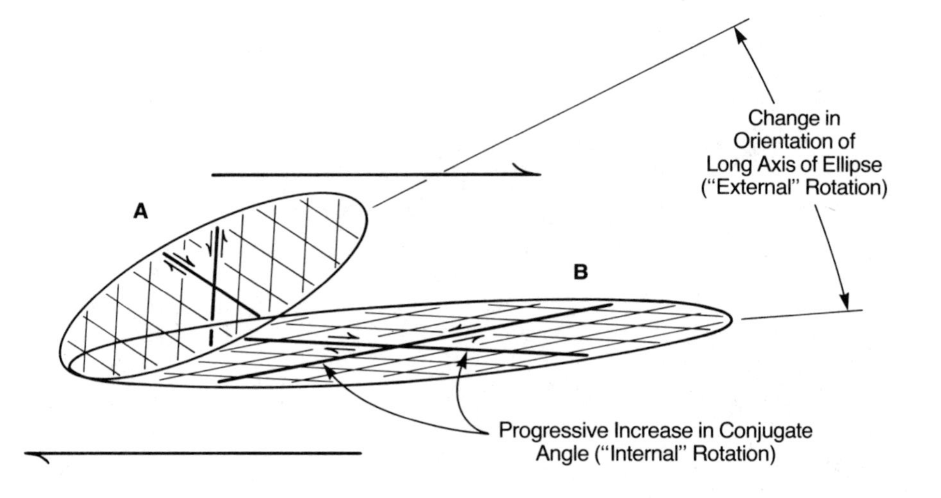

Figure 9.52 (*A*) Progressive external rotation of conjugate faults during simple shear deformation. (*B*) Internal rotation of conjugate faults during progressive distortion increases the conjugate angle between the faults.

STRUCTURAL PATTERNS ASSOCIATED WITH STRIKE–SLIP FAULTING

EXPERIMENTAL MODELING. Wilcox, Harding, and Seely (1973) designed some clay cake experiments to evaluate the structural patterns that develop in sedimentary strata above deep-seated strike–slip faults (Figure 9.53). Clay cakes are prepared on panels of sheet metal that can be moved oppositely and uniformly past one another during the course of an experiment. The line of contact between the metal panels predetermines the location and orientation of the linear zone of strike–slip faulting that develops in the overlying clay. Reference circles are impressed on the surface of the clay cakes before deformation so that strain, translation, and rotation can be evaluated (Figure 9.53*A*). Initial strike–slip movement of the metal panels results in distortion of the clay in such a way that the reference circles are transformed to ellipses (Figure 9.53*B*). Eventually the clay begins to fault within a zone parallel to the underlying metal panels (Figure 9.53*C*). When this happens, the distorted traces of the reference circles are truncated progressively and translated (Figure 9.53*D*- *F*).

SYNTHETIC AND ANTITHETIC FAULTS. Both right-handed and left-handed strike–slip faults emerge in clay deformed in this way (Wilcox, Harding, and Seely, 1973). The faults combine to form conjugate sets marked by an initial conjugate angle of intersection of about 60° (see Figure 9.53*C*). Of the two conjugate fault sets, the one whose sense of slip is identical to that of the main zone of faulting is called **synthetic**. The one whose sense of slip is opposite to that of the main zone is called **antithetic**. The synthetic faults are typically oriented at a small acute angle to the trace of the main fault zone; the antithetic faults are oriented at a very high angle to the main zone (see Figure 9.53*E*).

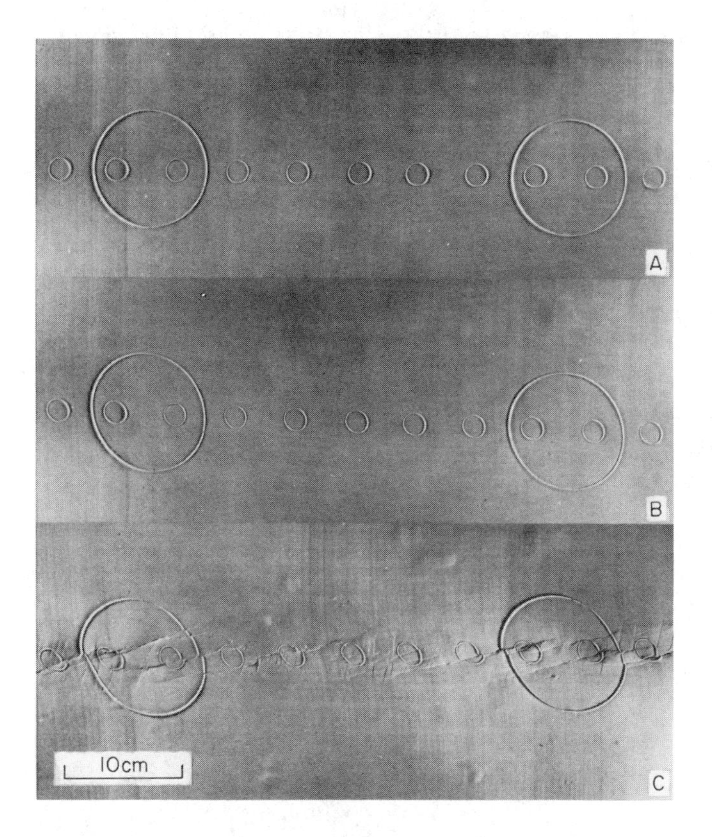

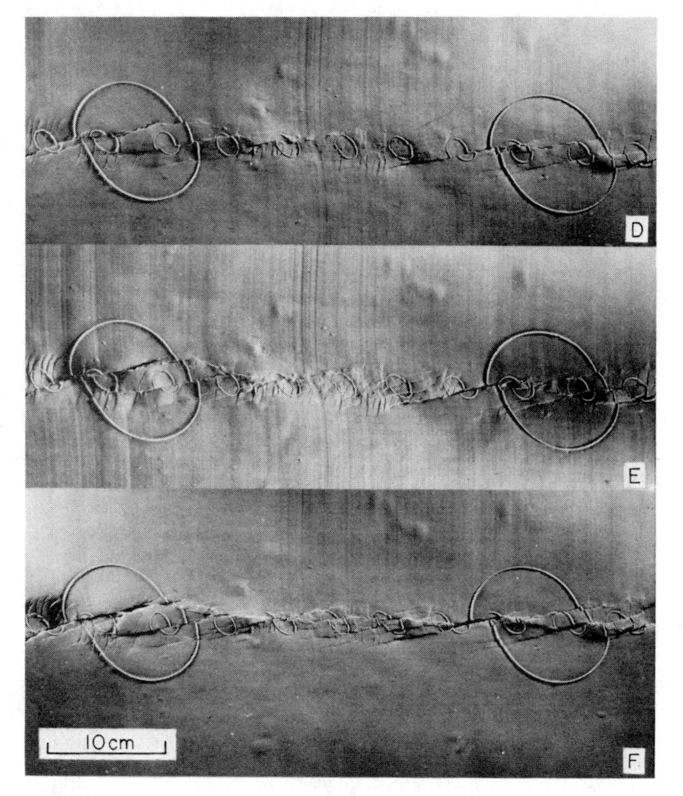

Figure 9.53 Clay cake deformation experiments simulating strike-slip faulting. Clay cake is placed on adjoining panels of sheet metal. Strike-slip faulting is achieved by shifting the panels horizontally past one another. (*A*) Starting configuration. (*B*) Initial distortion of clay. (*C*) Onset of faulting and the formation of synthetic and antithetic faults. (*D*) to (*F*) Continued faulting. Folds that develop become oriented parallel to the direction of greatest extension (*X*). [From Wilcox, Harding, and Seely (1973). Published with permission of American Association of Petroleum Geologists.]

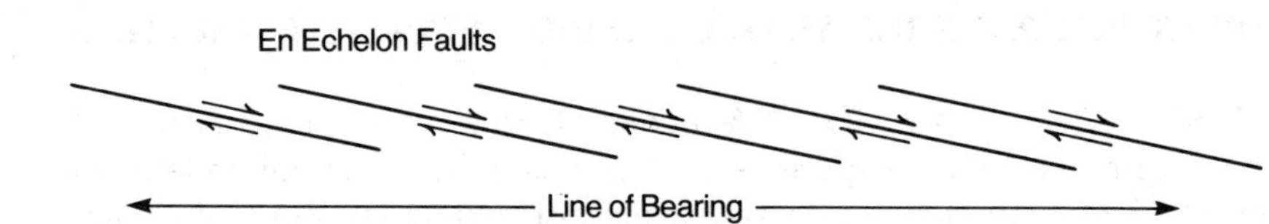

Figure 9.54 Geometric configuration of an echelon faults.

EN ECHELON PATTERNS. Faults of like sense in a strike–slip fault zone are oriented in **en echelon patterns**. For structures to be "en echelon" they must be parallel to one another and they must be arranged along a common **line of bearing** (see Figure 9.54). The trend of individual structures is oblique to the trend of the line of bearing. In the clay model experiments, the synthetic fault set is composed of en echelon elements (see Figure 9.53*C*).

FOLDS AND TENSION FRACTURES. Not only do strike–slip faults form in the course of clay cake experiments of the type carried out by Wilcox, Harding, and Seely (1973), but folds and tension fractures can develop as well. Tension fractures form perpendicular to the axis of maximum extension (*X*) and they are arranged en echelon along the trace of the main fault zone. Folds are aligned with axes parallel to the direction of maximum extension (*X*). They are also arranged en echelon to the main zone of faulting. Folds resulting from strike–slip faulting are described as right-handed or left-handed, depending on the sense of shear along the main fault zone. **Right-handed folding** means that if we were to walk the length of one of the en echelon folds, to the point at which it disappears, we would have to turn to the right to search and discover the next fold in line.

STRAIN DUE TO CURVATURE OF STRIKE–SLIP FAULTS

There are yet more peculiarities to strike–slip fault patterns. Crowell (1974) has emphasized that bends in strike–slip faults invite high concentrations of strain. The strain is distributed within distinctive suites of structures.

Movement along perfectly planar strike–slip faults results in coherent structural patterns. Wall rocks slide past each other without much interference (Figure 9.55*A*). **Branching and braiding** of faults is minimal. In contrast, if a strike–slip fault is marked by an abrupt bend, or even a gradual bend, complications arise (Figure 9.55*B*). Bordering country rocks are required to adjust to stress buildups by stretching or shortening.

Crowell (1974) has described fault curvature in terms of **double bends**. Movement of wall rock along a fault with a double-bend curve leads to convergence or divergence, depending on the sense of motion and the sense of curvature (Figure 9.55*B*). **Releasing double bends** tend to create open space. **Restraining bends** are sites of crowding. Releasing bends are Z shaped along right-handed strike–slip faults and S shaped along left-handed faults (Figure 9.55*B*). Conversely, restraining bends are S shaped along right-handed faults and Z shaped along left-handed faults.

Deformation at releasing bends is marked by extensional deformation, especially normal faulting. Fault-bounded grabens formed at releasing bends are capable of receiving thousands of meters of clastic deposits. In mineral deposit settings, releasing bends are likely sites of open-space filling and ore deposition. Restraining bends are marked by shortening. Shortening is achieved by thrusting and folding. A common response to such crustal shortening is vertical uplift of the thickened block.

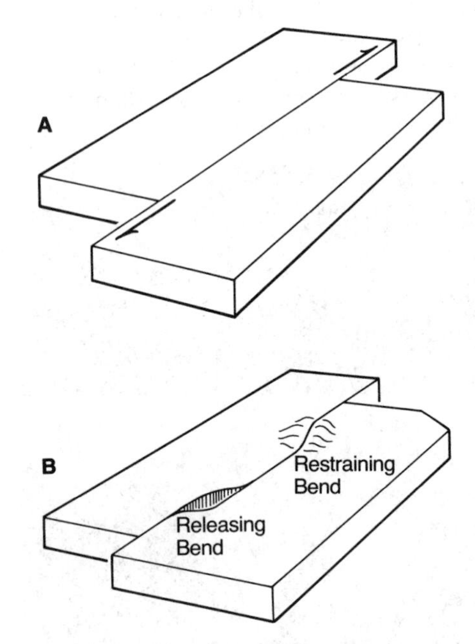

Figure 9.55 (*A*) Strike-slip movement along perfectly planar faults produces neither gaps or overlaps. (*B*) Strike-slip movement along irregularly curved faults produces gaps at releasing bends and crowding at restraining bends. [From Crowell (1974). Published with permission of Society of Economic Paleontologists and Mineralogists.]

Fault-bounded **wedges** of rock form during the natural evolution of a double bend. Resistance at a restraining bend may become so large that a newly formed fault cuts around the restraint (Figure 9.56), thus isolating a wedge of country rock. Wedges can be large or small. Some associated with the San Andreas system are 100 km long. A wedge, during continued fault movement, can rotate such that one tip subsides to form a basin while the other tip rises to produce an uplift. The uplift becomes a source area for clastic debris, which inevitably is deposited in the nearest tectonic depression.

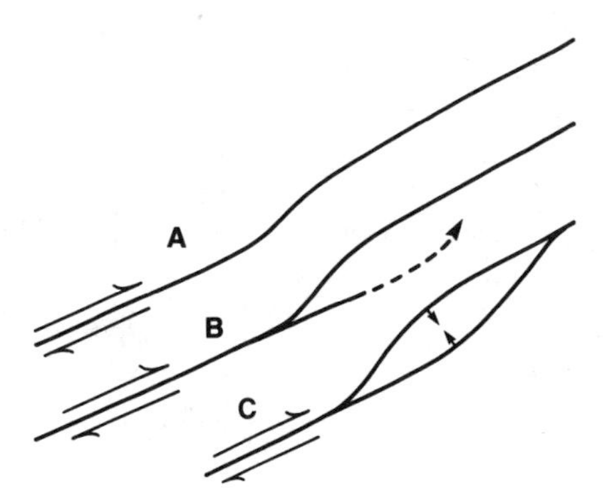

Figure 9.56 (*A*) to (*C*) The progressive development of fault-bounded structural wedges at restraining bends of strike-slip faults. [From Crowell (1974). Published with permission of Society of Economic Paleontologists and Mineralogists.]

TECTONIC SETTING

We have already discussed strike–slip faults of a transform nature that operate in oceanic settings (Chapter 6). They comprise a fundamental class of faults whose geometric and kinematic properties are quite distinctive. Major strike–slip faults in continental settings are "landlubber" counterparts to the midoceanic transform faults. Like oceanic transform faults, most of the truly large strike–slip faults in continental terranes are fundamental plate boundaries. The San Andreas fault in California is a transform fault boundary between the North American plate and the Pacific plate (see Figure 6.26). The Motagua fault in Guatemala is a transform fault boundary between the Caribbean and the North American plates (Figure 9.57). The Alpine fault in New Zealand constitutes a transform fault boundary between the Pacific and India plates (see Figure 6.26). These faults, and some others like them, are thousands of kilometers long and have accommodated hundreds of kilometers of strike–slip displacement. Although strike–slip faults occur as compartmental and tear faults within systems of thrust, reverse, and normal faulting, these cannot compare in size, slip, and overall importance to those that currently are, or formerly were, major plate boundaries.

STRIKE–SLIP FAULTING IN GUATEMALA

Guatemala has been ripped by strike–slip faulting. Signals of tectonic unrest in the country are abundant: long, straight to gently curved lineaments; stream valleys offset by recent strike–slip faulting; marine Pliocene strata perched at elevations as high as 14,500 ft (4400 m); earthquake shocks; smoking volcanoes; crater lakes fed by hot springs; and imposing topographic relief. Slopes in northwestern Guatemala are locally so steep that cornfields must be climbed on hands and knees.

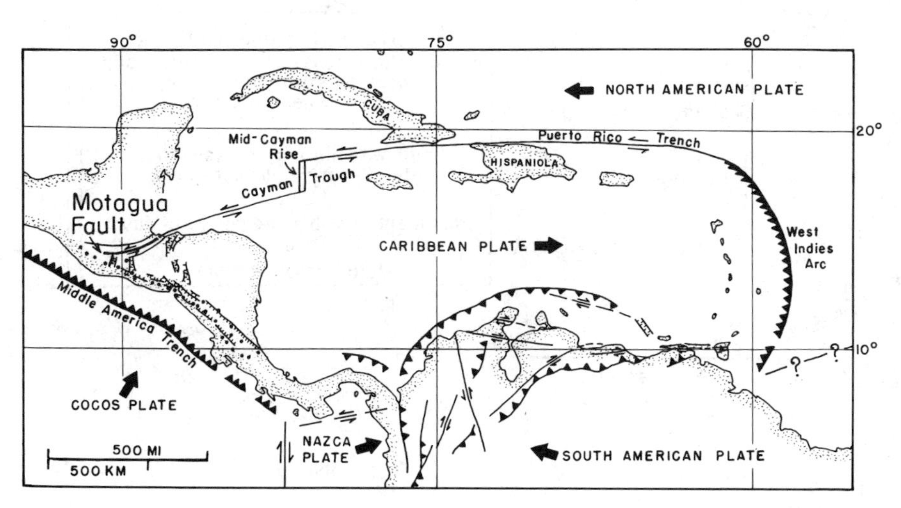

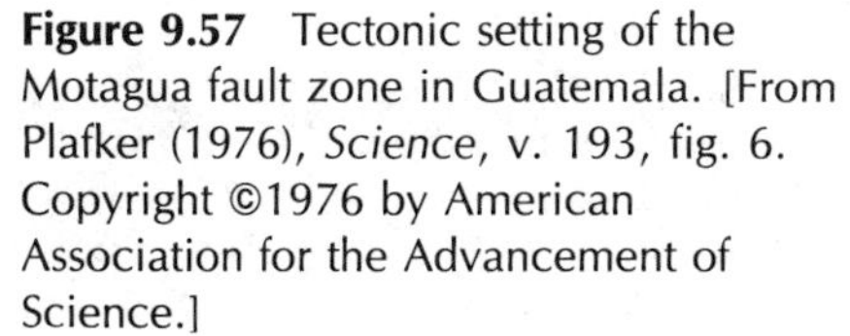

Figure 9.57 Tectonic setting of the Motagua fault zone in Guatemala. [From Plafker (1976), *Science*, v. 193, fig. 6. Copyright ©1976 by American Association for the Advancement of Science.]

Physiographic and geologic indicators of tectonic unrest are academic compared to the events of February 4, 1976, when sudden strike–slip movement along the Motagua fault resulted in more than 74,000 injuries and the loss of 23,000 lives. Fully one fifth of the country's population were victims of the tragic earthquake. George Plafker of the United States Geological Survey reported that the Guatemalan earthquake produced more surface faulting than any earthquake event in the Western Hemisphere since the San Francisco earthquake of 1906 (Plafker, 1976).

The structures resulting from movement along the Motagua fault during the 1976 earthquake bear a remarkable resemblance to those produced experimentally. According to Plafker (1976), 230 km of the Motagua fault zone was activated during the earthquake (Figure 9.58). Movement was left-handed strike–slip, with an average slip of 108 cm (Figure 9.59). The maximum net slip observed, calculated on the basis of an offset road, was 340 cm. From place to place along the fault zone, dip–slip movement was observed, but nowhere was it especially systematic in terms of sense of displacement. The minor structures that formed along the fault include synthetic left-slip faults oriented subparallel to the fault trace, en echelon tension fractures trending at a high angle to the fault trace, and left-handed en echelon fold structures.

Figure 9.58 Geologic map showing the trace of the Motagua fault, as well as measured horizontal slip (circled numbers in centimeters) along the fault during the 1976 earthquake. [From Plafker (1976), *Science*, v. 193. fig. 1. Copyright ©1976 by American Association for the Advancement of Science.]

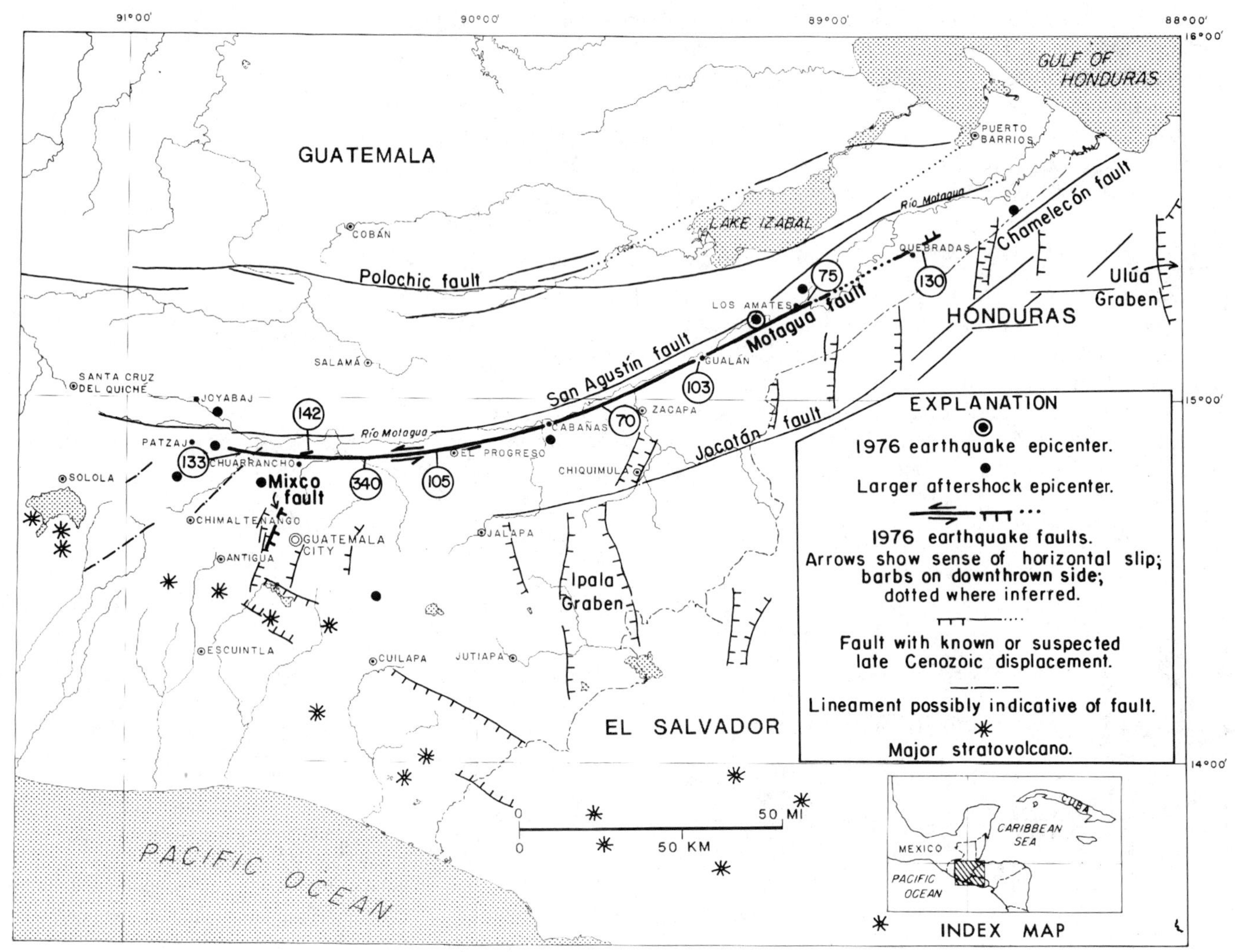

Figure 9.59 Left-lateral offset of a line of trees by left-handed slip along the Motagua fault. [From Plafker (1976), *Science*, v. 193, fig. 4. Copyright ©1976 by American Association for the Advancement of Science.]

Strike–slip faulting in Guatemala is nothing new. The structural grain of the country bears a clear record of long sustained distortion by strike–slip faulting and associated folding (Anderson and others, 1973).

THE ALPINE FAULT IN NEW ZEALAND

As part of the boundary between the Pacific plate and the Indian plate, the Alpine fault connects two opposite-dipping subduction zones (see Figure 6.16 and Figure 7.32). The fault cuts across the South Island of New Zealand for a trace length of 600 km. Right-handed en echelon folds are aligned along its trace, and restraining bends have attracted significant thrusting and folding.

Within the past 40 m.y., the Alpine fault has accommodated approximately 460 km of right-handed strike–slip movement (Allis, 1981). But during the past 10 to 15 m.y. slip along the Alpine fault has not been purely strike–slip. Instead, relative plate motions have created a strong east–west convergence across it. Differential uplift (10 mm/yr!) by compression-induced reverse and thrust faulting is changing the face of South Island.

THE SAN ANDREAS FAULT

BACKGROUND. The San Andreas fault has been studied as carefully as any fault in the world, in part because of its threat to huge metropolitan centers in California. The San Andreas fault was not always recognized as a strike–slip fault. The first serious proposal that the San Andreas fault had accommodated really significant strike–slip displacement was made in 1953 by Hill and Dibblee. They saw in the geologic record abundant evidence for right-handed strike–slip displacement.

J. Tuzo Wilson (1965) originally perceived the San Andreas system as a transform fault linking the East Pacific Rise to the Juan de Fuca Ridge (see Figure 6.26), but it was Tanya Atwater's exquisite interpretation of the evolution of the San Andreas fault system that truly made the fault understandable (Atwater, 1970). Her interpretation washed plate tectonics to

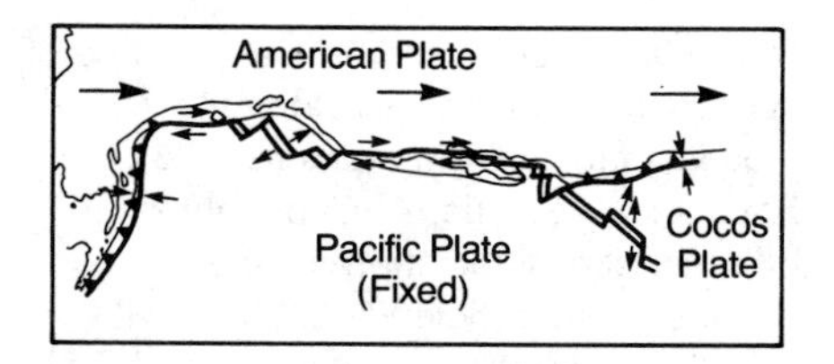

Figure 9.60 Map of the trace of the San Andreas fault as portrayed on a Mercator projection. [From Atwater (1970). Published with permission of Geological Society of America and the author.]

shore. Until the time of her work, most geologists regarded seafloor spreading and plate tectonics as an ocean-bound concept, only vaguely applicable to understanding structural deformation in rocks within continental settings. By interpreting the structural history of western California in the direct context of plate interactions, Atwater demonstrated just how encompassing the revolution called plate tectonics would be.

ATWATER'S MODEL. The role of the San Andreas fault as a plate boundary between the North American and Pacific plates can best be appreciated when the fault is shown in a special map projection, drawn about the pole of relative motion between the Pacific and North American plates (Figure 9.60). As Morgan (1968) and Atwater (1970) have explained, the trace of the San Andreas fault in this projection is horizontal, occupying the trace of a small circle about the pole of relative motion of two plates. Relative motion between the Pacific and North American plates is about 6 cm/yr, right handed, and oriented parallel to the trace of the San Andreas fault system. The San Andreas fault accommodates part of the motion (approximately 4 cm/yr). The rest of the motion is distributed on other faults and in other ways (Atwater, 1970).

Direct contact between the Pacific and North American plates was achieved through a steady, systematic evolution of the plate configuration through time. Forty million years ago and earlier, the Pacific and North American plates were everywhere separated by the Farallon plate (McKenzie and Morgan, 1969) (Figure 9.61). In the early Tertiary, a trench between the Farallon and North American plates continuously consumed Farallon plate lithosphere. Subduction took place faster than new Farallon lithosphere was being created. Consequently, the entire Farallon plate was subducted along the trench. In effect, the birth of the San Andreas fault system required elimination of the Farallon plate and the oceanic ridge from which it was spawned. Once eliminated, the Pacific and North American plates were permitted to come into contact (Figure 9.61).

PHYSICAL CHARACTERISTICS OF THE SAN ANDREAS FAULT. The kinematics of development of the San Andreas fault make sense only in the context of plate tectonics. But the physical and geometrical features of the fault zone are the same as that of any strike–slip fault. In fact, the major and minor structures along it conform to much of what we learned in the context of clay cake deformation (Wilcox, Harding, and Seely, 1973). En echelon right-handed anticlines are commonplace along the fault (Crowell and Ramirez, 1979). Pull-apart grabens and tensional rifts are oriented perpendicular to the trends of the folds, perpendicular to maximum extension (X).

The Newport–Inglewood fault zone on the southwest side of the Los Angeles basin is a microcosm of structural deformation along the San Andreas (Harding, 1973; Crowell, 1974). Right-handed en echelon fold structures trend north to northeastward and display structural relief ranging from 180 to 760 m. The folded structures are offset by right-handed synthetic en echelon strike–slip faults. The folds, although faulted, are great oil traps. Oil is obtained from sediments ranging in age from mid-Miocene to Holocene.

Basins along the Newport–Inglewood fault zone trend northwest/southeast and range in length from 50 to 100 km. They too are arranged en echelon. Average depth of the basins is about 3 km. The basins are rhombic to lens shaped in plan view, defined by straight margins on the northeast and southwest sides. The straight margins are traces of en echelon synthetic right-handed strike–slip fault zones; the irregular margins are defined by normal-slip faults, oriented perpendicular to major extension.

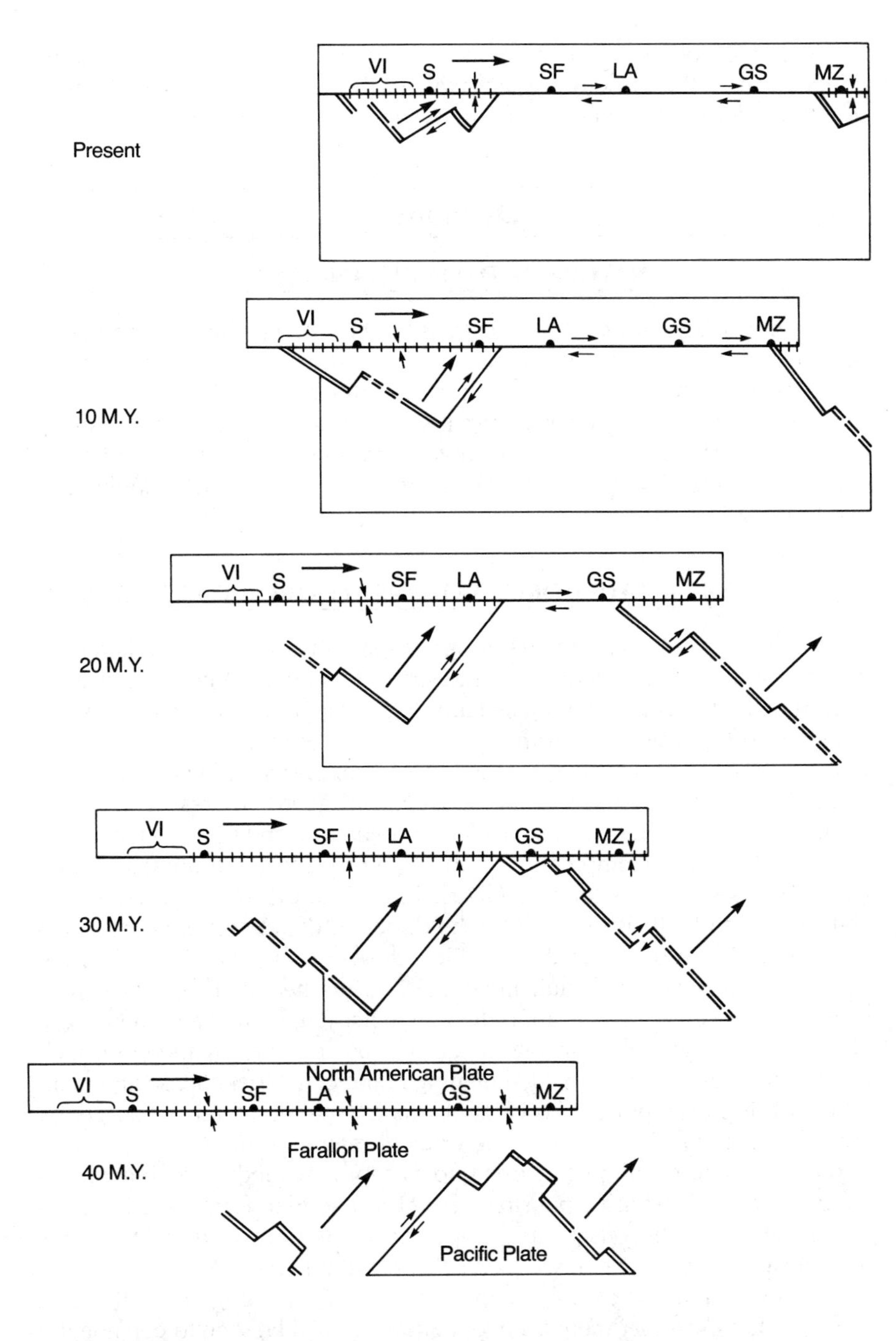

Figure 9.61 Plate tectonic evolution of the San Andreas fault. S = Seattle; SF = San Francisco; LA = Los Angeles; GS = Guymas; and MZ = Mazatlan. [From Atwater (1970). Published with permission of Geological Society of America and the author.]

RELATION OF FAULTS TO PRINCIPAL STRAIN DIRECTIONS

Conjugate faults, whether they be thrust-slip, normal-slip, or strike–slip, generally intersect in a line that is parallel to the axis of intermediate strain (*Y*) (Figure 9.62). The axes of maximum and minimum strain (*X* and *Z*) occupy a plane that is perpendicular to the axis of intermediate strain (*Y*). The axis of minimum strain (*Z*) bisects the conjugate angle between faults. The conjugate angle is normally an acute angle except where modified by flattening brought about by internal rotations.

Minor structures like gash fractures and striations can be used to help define the principal strain axes. Tension fractures are oriented perpendicular to the axis of maximum extension (*X*). Striations are parallel to the line of intersection of each fault with the *X–Z* plane.

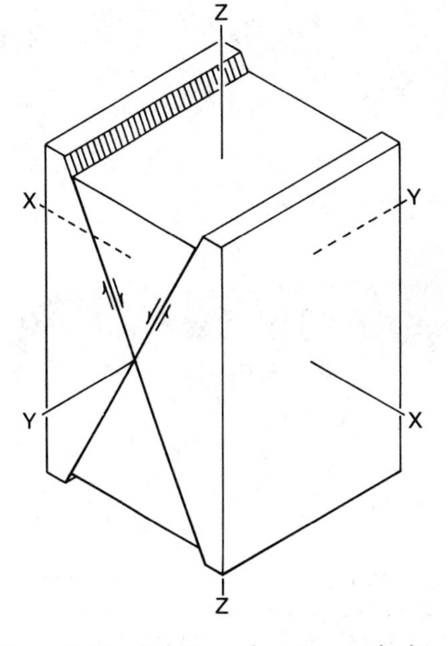

Figure 9.62 Relation of conjugate faults to the principal strain directions.

As a part of the kinematic analysis of faulting, it is useful to construct the orientations of the principal strain axes on the basis of the orientations of structural elements like faults, striations, and tension fractures.

Dynamics

NATURE OF DYNAMIC ANALYSIS

Dynamic analysis of faulting allows us to understand why faults can be so conveniently separated into normal-slip, thrust-slip, and strike–slip categories. We learn that there is a dynamic basis for the fact that, on average, strike–slip faults dip vertically, normal-slip faults dip at 60°, and thrust-slip faults dip at 30°. Dynamic analysis of faulting is concerned both with stress conditions under which rocks break and with the orientations of faults relative to stress patterns.

RELATION OF FAULT ORIENTATIONS TO STRESS DIRECTIONS

Both field relationships and deformation experiments affirm that faults are oriented systematically with respect to stress directions. One of the clearest images of the relation of conjugate faults to principal stress directions can be rendered through simple deformation of a block of moist clay. We buy 10 × 10 cm blocks of "moist clay" at the local art supply shop, remove one from its plastic wrapping, shape it into a cube, and gently impress a reference circle on its top surface. We let the clay dry overnight and squeeze it in a vice the next afternoon. Compression in the vice produces, without fail, conjugate faults, tension fractures, and striations (Figure 9.63). The conjugate angle between faults is bisected by the direction of greatest principal stress (σ_1), which is the direction that the vice closes on the clay. Typically, the initial angle between each fault and σ_1 is 30°. Thus the initial conjugate angle between the paired faults is about 60°. By persistent cranking of the handle of the vice, this initial conjugate angle of ±60° can be expanded through internal rotation accompanying flattening of the clay block. Sense and magnitude of displacement on individual faults can be monitored by observing the displacement of the trace of the original reference circle.

Tension fractures in the clay open up as gaping wounds perpendicular to the direction of least principal stress (σ_3). These tend to form early in deformation and may be offset later along the faults that develop. Striations become exposed on fault scarps in the clay, but the best views of the striations are obtained by pulling apart the clay block along one of the fault surfaces after the experiment is ended. Striations will be seen to pervade the fault surfaces. Striation orientations on a given fault are defined by the intersection of the fault surface with the σ_1/σ_3 plane.

Principal stress directions remain parallel to principal strain directions throughout the moist clay deformation experiment because the distortion is by pure shear: σ_1 is parallel to Z; σ_3 coincides with X. The systematic relationships that exist between conjugate faults, tension fractures, and principal stress directions provide a basis for interpreting paleostress directions in rocks deformed millions of years ago (Figure 9.64 and see Figure 5.3).

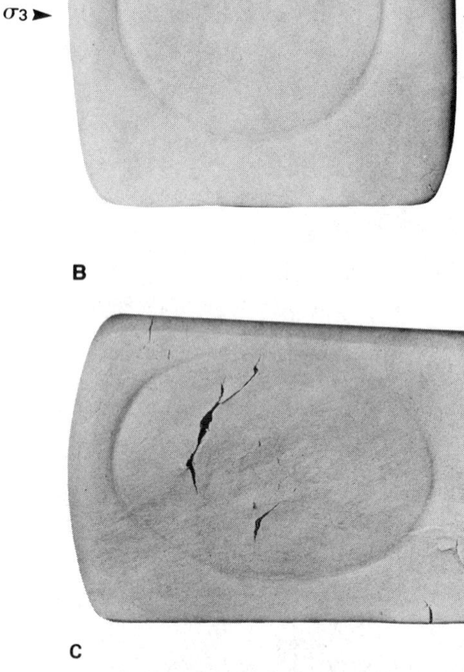

Figure 9.63 The clay block experiment. Progressive shortening of a clay block is accommodated by the development of faults and tension fractures.

MOHR ENVELOPE OF FAILURE

We learned in Chapter 5 that Mohr circle diagrams provide a convenient graphical means to compute values of shear stress and normal stress on any

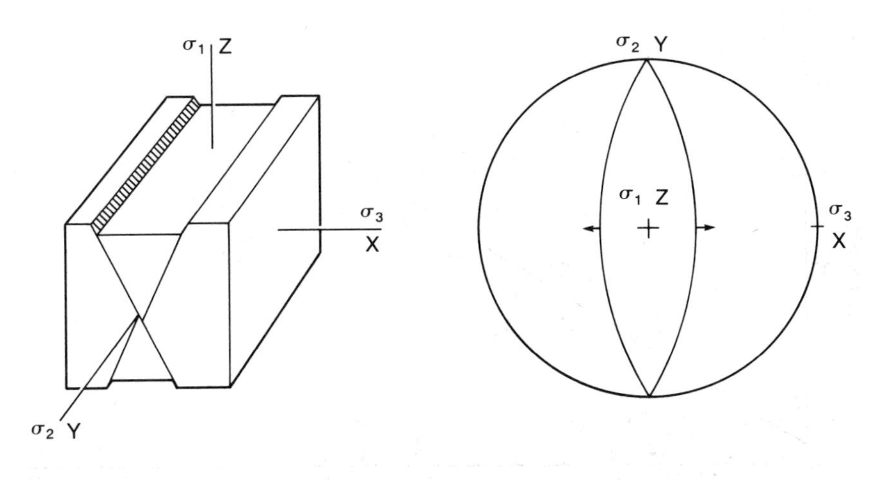

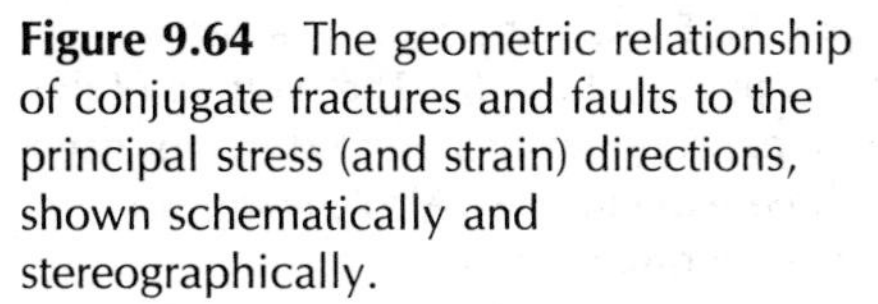
Figure 9.64 The geometric relationship of conjugate fractures and faults to the principal stress (and strain) directions, shown schematically and stereographically.

plane within a body subjected to known values of greatest and least principal stress. We now use the Mohr stress circle to map the resistance of a rock to faulting.

Suppose we want to establish the range of mechanical stability of limestone for several different confining-pressure conditions: 4000, 15,000, and 30,000 psi (292, 1096, and 2190 kg/cm^2). We proceed by subjecting three cylindrical cores of limestone to standard compression tests, taking each test specimen to failure by faulting. The load (i.e., force) at the instant of failure is recorded for each of the tests. And after each faulted specimen has been removed from the pressure vessel, the angle (θ) between the fault trace and the direction of greatest principal stress (σ_1) is measured and recorded.

To prepare a Mohr circle diagram of the test results, we need to plot the values of σ_1 and σ_3 for each specimen at failure. Stress σ_3 is equal to the value of confining pressure, and σ_1 is equal to the force (F) on the specimen divided by the cross-sectional area (A) of the specimen.

Principal stress magnitudes and θ values for the series of experiments are plotted on a common Mohr circle diagram, as shown in Figure 9.65*A*. The three circles represent the stress conditions achieved at the time of faulting during each of the experiments. To each circle is added a radius that makes an angle 2θ with the σ_N axis of the diagram; the 2θ angle is "swung" clockwise. The values of normal stress (σ_N) and shear stress (σ_S) on each fault plane at failure are reflected by the coordinates of the point marking the intersection of each circle with its radius. These are the **failure points**.

By connecting the failure points for each of the stress circles, an **envelope of failure** for the limestone is established (Figure 9.65*A*). The failure envelope represents the stress threshold that must be surpassed for faulting to occur under the conditions specified. The envelope of failure makes it possible to predict the level of differential stress ($\sigma_1 - \sigma_3$) that must be attained for faulting to occur.

Suppose we wanted to know the magnitude of the greatest principal stress that would be required to fault the limestone under confining pressure conditions of 18,500 psi (1352 kg/cm^2). Would 30,000 psi (1974 kg/cm^2) be high enough? Not a chance! A stress circle drawn through points on the σ_N axis of the Mohr diagram corresponding to $\sigma_3 = 18{,}500$ psi and $\sigma_1 = 30{,}000$ psi is not large enough to intersect the envelope of failure (Figure 9.65*B*). The stress circle resides entirely within the field of stability. For the differential stress ($\sigma_1 - \sigma_3$) to be great enough to cause failure, σ_1 must be raised much higher than 30,000 psi. Indeed, if σ_1 is raised to the level of 71,000 psi (5183 kg/cm^2), while confining pressure is held constant, the stress circle becomes tangential to the envelope of failure at failure point σ_N, $\sigma_S = 30{,}600$ psi

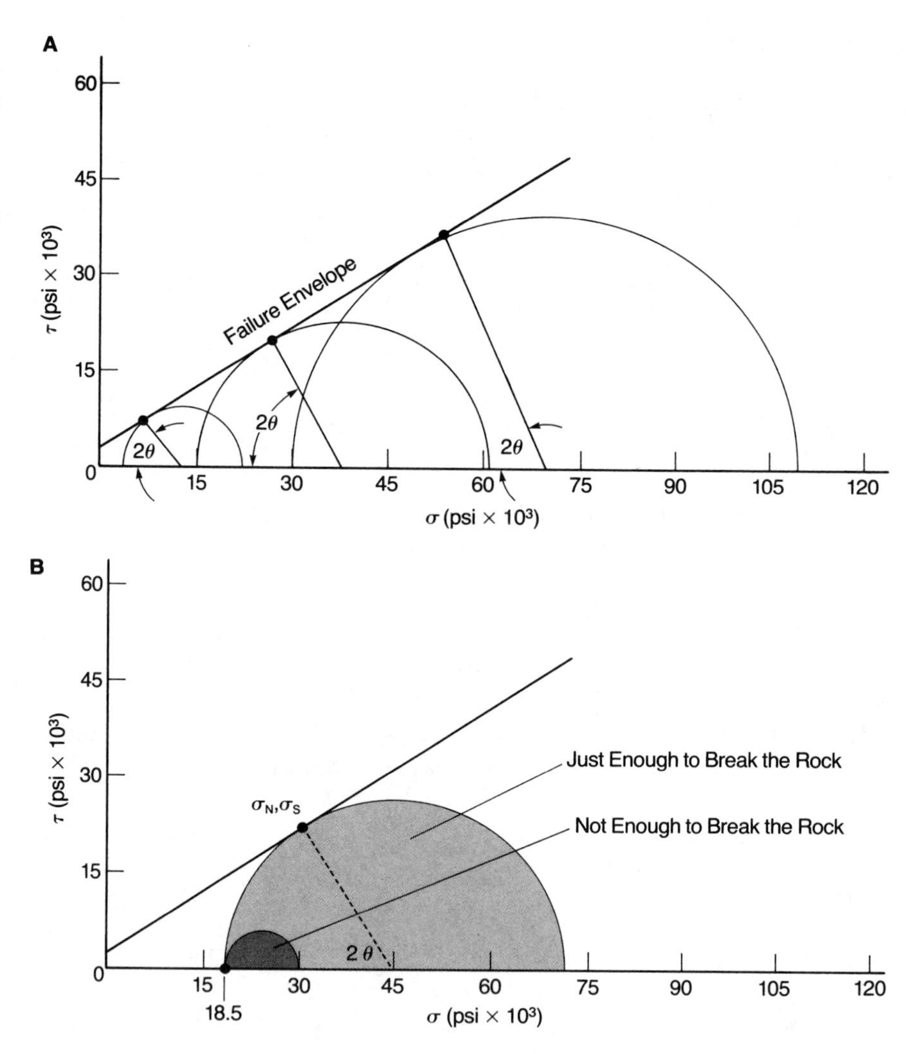

Figure 9.65 Construction of a Mohr envelope of failure. (*A*) The plotting of the principal stress values (σ_1 and σ_3) for each of three experiments. Also shown, for each experiment, is the angular relationship of faulting to the direction σ_1. The envelope of failure passes through the "failure points" representing each of three experiments. (*B*) Use of the failure envelope as a guide to determining the stress level at which the rock will fault under confining pressure of 18,500 psi (1352 kg/cm²).

(2234 kg/cm²), 22,200 psi (1621 kg/cm²) (Figure 9.65*B*). On this basis we would surmise that σ_1, when raised to 71,000 psi, would cause the specimen to fault. Moreover, the angle (θ) between the fault and σ_1 could be predicted to be approximately $2\theta/2$, or 28°.

Predicting the failure point of a rock is some of the fun of rock deformation experiments. Great shouts, screams, and moans, not to mention the clinking of nickles and dimes, have issued forth from our rock deformation lab. Guessing the value of greatest principal stress (σ_1) required to break the very first specimen in a series of tests is almost mindless speculation. But after stress data and θ values for two or three tests are posted on a Mohr diagram, the predicted stress values for failure begin to cluster within a very narrow range. Someday one-armed triaxial deformation machines will fill the casinos in Las Vegas, and when this happens, smart managers will prevent the players from constructing Mohr circle diagrams.

MOHR–COULOMB LAW OF FAILURE

Countless failure envelopes have been generated through laboratory testing. They all seem to have several things in common. They tend to be straight over much of their length, though admittedly curved in regions of especially high or especially low confining pressure. They have positive slopes that average about 30°, although the slope may range anywhere from 20° to 45°. And when projected to the σ_S axis of the Mohr diagram, they intersect at small positive values.

The characteristics of failure envelopes can be understood in the context of the **Mohr–Coulomb law of failure**, a law based on dynamic models presented by Coulomb (1773) and Mohr in 1900. The law describes the **critical shear stress level (σ_c)** which is required to break a rock.

The strength of a rock to resist faulting is derived from two sources. One is the natural **cohesive strength (σ_o)** of the rock, that is, the bonding that holds the rock together. The second is the **internal frictional resistance to faulting**. The stress required to fault a rock must be large enough to overcome cohesive strength, so that the rock may fracture. At the same time it must be large enough to overcome the rock's internal resistance to faulting, so that movement can take place along the fracture.

Cohesive strength is an intrinsic rock property. For a given rock it does not vary appreciably, even when the conditions of deformation are altered significantly. Granites, basalts, and quartzites are more tightly bonded than calcite-cemented sandstones.

Internal frictional resistance to faulting is *not* an intrinsic rock property. The level of internal frictional resistance to faulting varies from rock to rock, both as a function of the **coefficient of internal friction (tan ϕ)** and the level of **normal stress (σ_N)** acting on the potential plane of faulting. To understand this, consider the two variables that influence how easy (or how hard) it is to push a sled on a cold winter afternoon. These are (1) the coefficient of **sliding friction** between the runners and the hard-packed snow over which the sled is required to move; and (2) the weight of the person on the sled (Figure 9.66). The coefficient of sliding friction is *analogous* to the coefficient of internal friction: sleds on ice are faster than sleds on cinders. The weight of the person on a sled is analogous to the level of normal stress acting on the discontinuity between runners and ice. It is easier to give a push to a small child on a sled than to try to budge the overweight dad who has not been able to give up his old Flexible Flyer.

The coefficient of sliding friction is a constant for each pair of materials brought into contact. Likewise, the coefficient of internal friction is constant for each rock type. Normal stress, on the other hand, is not a constant. Rather it varies with factors like the number of riders on the sled, or, geologically speaking, the level of confining pressure in the environment of deformation.

Figure 9.66 The relationship, if any, of fault dynamics to sled riding. (Artwork by D. A. Fischer.)

With this introduction in mind, let us look at the specific phrasing of the Mohr–Coulomb law of failure. It is stated as an equation,

$$\sigma_c = \sigma_0 + \tan \phi(\sigma_N) \quad (9.1)$$

where,

σ_c = critical shear stress required for faulting

σ_0 = cohesive strength

ϕ = angle of internal friction

$\tan \phi$ = coefficient of internal friction

σ_N = normal stress

The Mohr–Coulomb law of failure begins to mean something when we see its geometric expression in the Mohr diagram. Figure 9.67 shows, once again, the envelope of failure for limestone. A shear stress (σ_S) of approximately 22,200 psi is required to fault the limestone when placed under a confining pressure of 18,500 psi. In terms of the Mohr–Coulomb law of failure, this shear stress value is σ_c, the critical shear stress necessary for faulting to occur. Part of its magnitude is cohesive strength (σ_o), expressed in units of stress. The value of σ_o can be read directly from the Mohr diagram as the *y* intercept of the envelope of failure (Figure 9.67). The rest of σ_c is the stress required to overcome internal frictional resistance to faulting. This component is labeled σ_f in Figure 9.67. The value of σ_f can be expressed in terms of the normal stress (σ_N) acting on the fault plane, and the angle of the internal friction (ϕ), which is the slope of the envelope of failure.

$$\tan \phi = \frac{\sigma_f}{\sigma_N}$$

$$\sigma_f = \tan \phi \ (\sigma_N)$$

Thus,

$$\sigma_c = \sigma_0 + \tan \phi(\sigma_N) \qquad \text{(Figure 9.67).}$$

We thus learn that the stress level at which rocks will fault is strongly influenced by ϕ, the angle of internal friction. The angle of internal friction for most rocks lies between 25° and 35°. Consequently, the coefficient of internal friction, $\tan \phi$, commonly ranges from 0.466 to 0.700. If we can

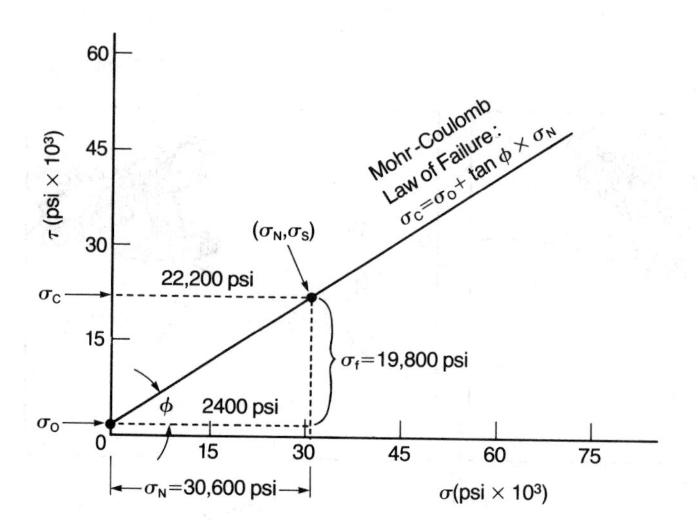

Figure 9.67 Expression of the Mohr-Coulomb law of failure in terms of the components of a Mohr diagram.

assume that cohesive strength is a very small part of the critical shear stress required for faulting, most faults form when shear stress on the plane of failure reaches a level that is slightly more than 50% of the normal stress acting on the surface. The exact optimum ratio of shear stress to normal stress will vary from rock to rock.

The angle of internal friction (ϕ) determines the angle (θ) between the fault surface and the direction of greatest principal stress (σ_1). Based on the geometry of the Mohr stress diagram (Figure 9.67), we have

$$\phi = 90 - 2\theta$$

Thus,

$$2\theta = 90 - \phi$$

$$\phi = \frac{90 - \phi}{2}$$

Since most rocks in nature possess an angle of internal friction of about 30°, the value of θ for most fault relationships is also 30°.

HUBBERT'S SANDBOX EXPERIMENT

Further appreciation of the Mohr–Coulomb law of failure can be gained in a simple but elegant experiment described by Hubbert (1951). The experiment allows us to develop a working knowledge of constructing failure envelopes and applying the results to fault analysis. Sand is substituted for rock, but otherwise all conditions hold.

FAILURE ENVELOPE FOR SAND

The first step is to construct an envelope of failure for sand. Two small, lightweight wooden or metal frames are stacked one on top of the other (Figure 9.68). The surfaces of the frames should be very smooth so that the frames can effortlessly slide past each other. The interior of the double-frame box is filled with medium-grain sand. A piece of masonite is cut to fit snugly into the interior of the upper frame. When positioned, it rests entirely on sand.

To construct an envelope of failure for sand, we determine the level of shear force that is required to cause differential movement of the upper and lower frames under a variety of conditions of normal force. For the frames to move with respect to each other, the sand must actually fault.

Normal force (F_N) is applied to the masonite by placing rock samples of known weight on top of the masonite (Figure 9.68*A*). The rocks are weighed as the experiment proceeds, and the values (in ounces or grams, pounds or kilograms) are written directly on them using a marking pen. Once a given amount of normal force is loaded onto the masonite, the **shear force (F_S)** required to induce faulting of the sand is determined. This is achieved by pulling on a "spring" scale attached to an eyescrew on the upper frame. At the instant the upper frame begins to move, the load registered on the spring scale is read and recorded. This load constitutes the critical shear force (F_S) required for faulting.

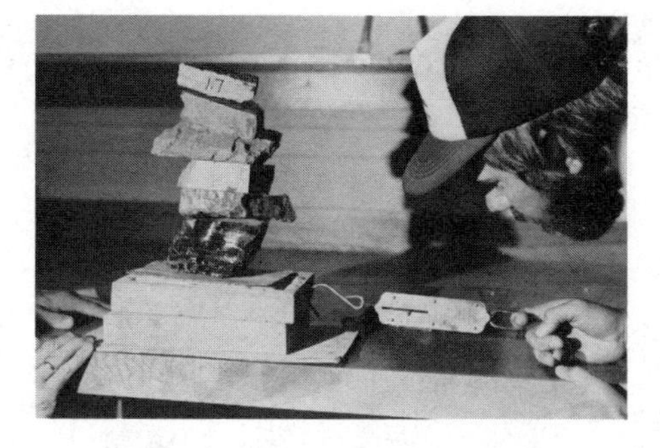

Figure 9.68 Construction of the failure envelope for sand. Experimental setup consists of two wooden frames, sand, a piece of masonite, rocks of known weight, a spring scale, and a baseball cap. The experiment itself involves measuring the amount of shear force required to move the upper frame for a number of given conditions of normal force. Andrew Arnold shows the technique. (Photograph by R. W. Krantz.)

Each combination of normal force (F_N) and shear force (F_S) is plotted on the *x* and *y* axes of a Mohr diagram, respectively (Figure 9.69). With careful attention to weighing and plotting, an amazingly straight-lined envelope of failure can be fit to the half-dozen or so failure points. Because sand is

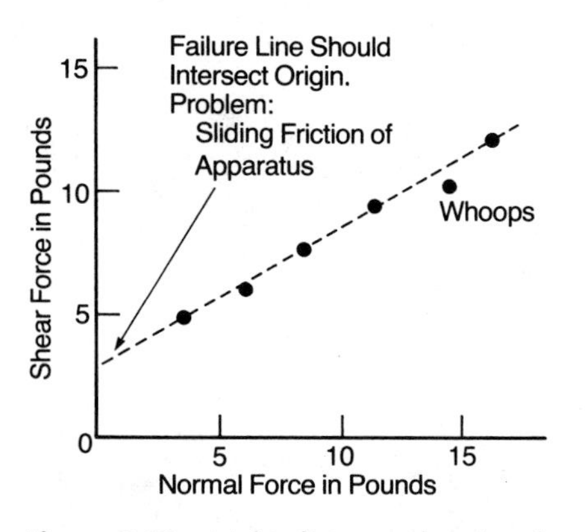

Figure 9.69 Mohr diagram showing the envelope of failure for sand based on paired measurements of shear force (F_s) and normal force (F_n).

cohesionless, the failure envelope ought to pass through the origin of the x–y coordinate system. In point of fact, the intercept typically lies just above the origin because of the effect of sliding friction between the frames.

Hubbert found that the failure envelope for loose sand displays an angle of internal friction (ϕ) of 30°. For compacted sand the angle of internal friction is 35°. It would be interesting to determine the angle of internal friction for other cohesionless materials, like breakfast cereals. Calculating the internal friction of materials like Grape Nuts, Cheerios, Rice Crispies, and Wheaties might reveal the degree to which differences in ϕ values arise from differences in size and/or shape and/or density of constituent particles.

THE SANDBOX

The second part of Hubbert's experiment involves generating normal-slip and thrust-slip faults in sand. A sandbox of the type shown in Figure 9.70 is filled with layers of sand separated by marker horizons of white or colored dry powdered clay. A vertical wooden or metal partition serves to separate two compartments, a smaller one on the left in which normal-slip faults form, and a larger one on the right in which thrust-slip faults form. Defor-

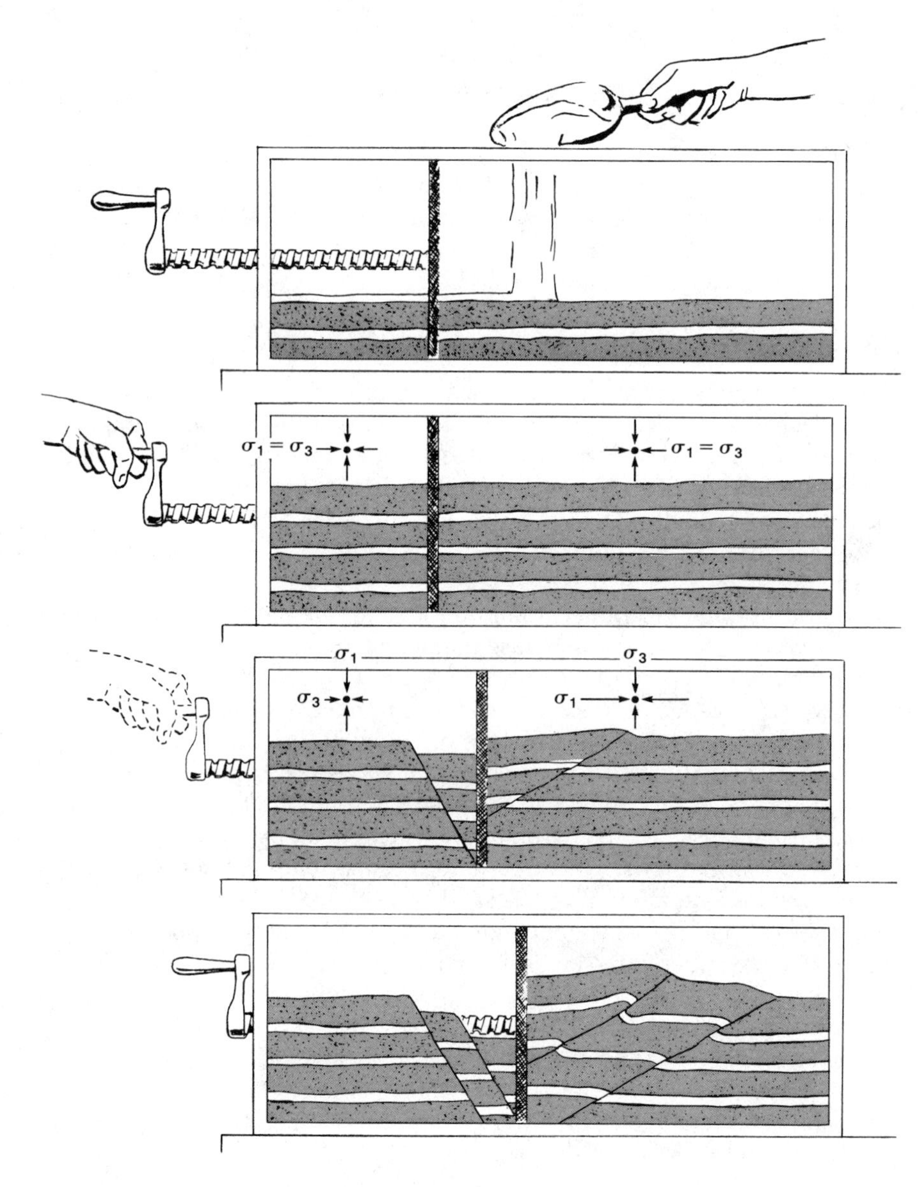

Figure 9.70 The famous sandbox experiment.

mation of the sand simply requires moving the partition to the right by means of a manually driven threaded-screw arrangement. The instant that the partition moves, a normal-slip fault develops in the left-hand compartment (Figure 9.70). Its dip is typically 60°, the complement of the angle of internal friction for sand. As the partition is forced to move further and further to the right, the first-formed normal-slip fault increases in displacement. Other normal-slip faults form as well. As a result of the normal faulting, the upper surface of the sand in the left-hand compartment develops a fault scarp topography.

Compressional shortening of sand in the right-hand compartment eventually forces the development of a thrust-slip fault. The sand first arches slightly and then is cut by a thrust-slip fault, which cuts up-section at an angle of approximately 30°. No steps or ramps or layer-parallel segments evolve because the sand is unlayered, loose, and homogeneous. After a certain amount of translation has been accommodated by the thrusting, a second thrust-slip fault develops, and this fault typically forms beneath the first, cutting up to the surface beyond it in the direction of tectonic transport. Translation on the first thrust ceases the instant that translation is initiated on the second. Steady translation on the second thrust results in folding of the first. Sometimes **back-limb**, antithetic thrusts develop that dip oppositely to the main thrust-slip faults.

The structural relationships produced in the sandbox experiment conform to what we have learned about dynamic analysis of faulting. Before movement of the partition in the sandbox, both compartments of sand are marked by a state of lithostatic stress in which stress is equal and all sided (see Figure 9.70). Movement of the partition to the right relieves horizontal stress in the left compartment. At the same time, horizontal stress intensifies in the right-hand compartment. Thus, very early in the experiment, different states of stress evolve in the two compartments. In the left compartment, σ_1 is vertical and σ_3 is horizontal. In the right compartment, σ_1 is horizontal and σ_3 is vertical. Given these principal stress directions and our knowledge of the angle of internal friction (ϕ) for sand, it is possible to predict the orientations of the faults that must develop in the sandbox experiment. Faults form in loose sand and at angles of $\theta = 30°$ to the direction of greatest principal stress. Since σ_1 is vertical in the left compartment during ongoing deformation of the sand, the faults that form there *must* dip 60°. Conversely, since σ_1 is horizontal in the right-hand compartment during deformation, the faults that form there *must* dip 30°.

MOHR DIAGRAM VIEW OF THE SANDBOX EXPERIMENT

The buildup of differential stress during the sandbox experiment can be pictured by means of a Mohr diagram. We start with a figure that shows only the envelope of failure and the lithostatic state of stress that exists in the sand, before deformation. The lithostatic stress state is one in which $\sigma_1 = \sigma_3$; it is represented on the Mohr diagram by a single point on the σ_N axis. As soon as the partition begins to move, the lithostatic state of stress is altered to one of differential stress. In the left-hand compartment σ_3 becomes weaker and weaker, but σ_1 remains constant (Figure 9.71*A*). Because of the progressive decrease in the value of σ_3, differential stress eventually becomes large enough to "break" the sand. The differential stress at failure is represented by values of σ_1 and σ_3 that define a circle that just touches the envelope of failure for sand.

In the right-hand compartment, the steady increase in horizontal compressive stress is represented by a steady increase in the value of σ_1 (Figure

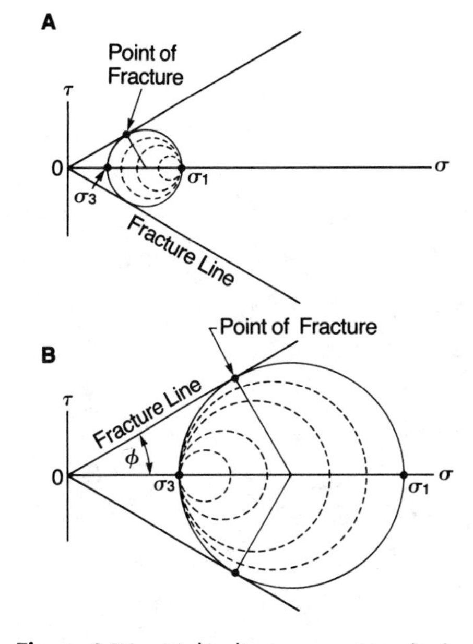
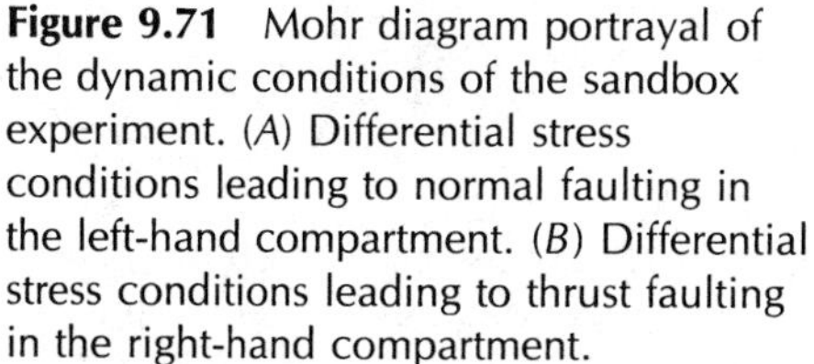

Figure 9.71 Mohr diagram portrayal of the dynamic conditions of the sandbox experiment. (*A*) Differential stress conditions leading to normal faulting in the left-hand compartment. (*B*) Differential stress conditions leading to thrust faulting in the right-hand compartment.

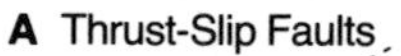

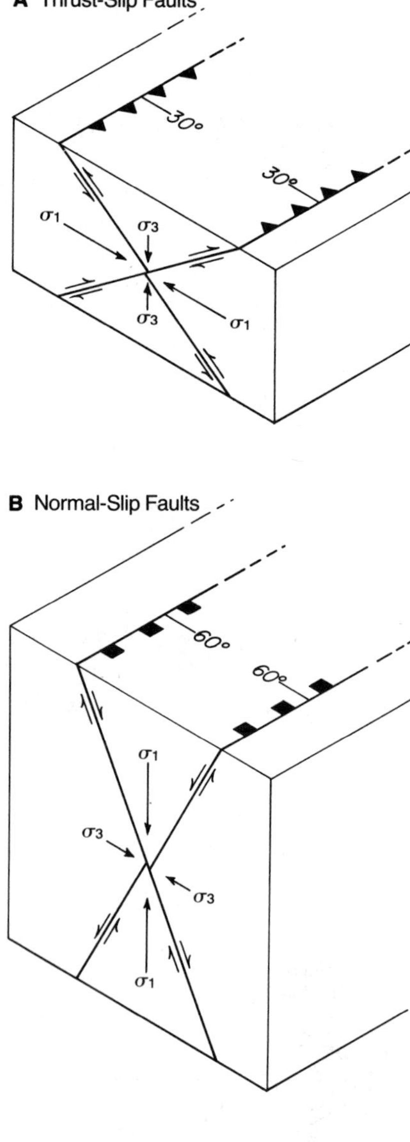

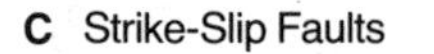

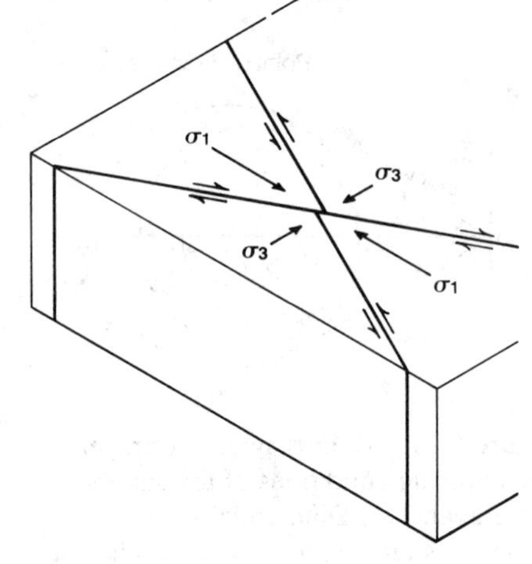

Figure 9.72 The relation of (*A*) thrust-slip, (*B*) normal-slip, and (*C*) strike-slip faults to the principal stress directions.

9.71*B*); σ_3 remains fixed at the original lithostatic stress level. Thrust faulting occurs when differential stress reaches such a level that a circle drawn through values of σ_1 and σ_3 touches the envelope of failure for sand.

The Mohr diagram representing faulting in sand reveals that sand, like rock, is much stronger in compression than in tension. This is revealed by the positive slope of the envelope of failure.

ANDERSON'S INSIGHTS ON FAULTING

Anderson (1951) recognized that the properties of principal stress directions, in combination with the Mohr–Coulomb law of failure, require that only strike–slip, thrust-slip, and normal-slip faults form at or near the surface of the Earth. Considering the Earth as a perfect sphere, Anderson reasoned that the discontinuity between air and ground at any point on the Earth's surface is a plane along which shear stress is zero. Only on very windy days do pedestrians get swept off their feet by shear stress, and as great as that wind-generated stress might feel, it is not of sufficient intensity to cause faulting of bedrock.

Since the principal stress directions are directions of zero shear stress, *the surface of the Earth must be a principal plane containing two of the three principal stress directions*. The third principal stress direction is oriented perpendicular to this principal plane and thus, at any point, is perpendicular to the surface of an ideally spherical Earth. If principal stress directions are vertical or horizontal at or near the surface of the Earth, and if the angle of internal friction for most rocks is about 30°, only normal-slip, strike–slip, and thrust-slip faults should be able to form at or near the Earth's surface. Thrust-slip faults form when σ_3 is vertical (Figure 9.72*A*). Normal-slip faults form when σ_1 is vertical (Figure 9.72*B*). Strike–slip faults form when σ_2 is vertical (Figure 9.72*C*).

STEREOGRAPHIC REPRESENTATION OF FAULTS AND PRINCIPAL STRESSES

INTERPRETING FAULT ORIENTATIONS FROM STRESS DIRECTIONS

The likely orientations of faults that should form in a given stress field can be evaluated stereographically, provided the angle of internal friction is known. In interpreting likely fault orientations, we assume that rocks are homogeneous and isotropic, and that the Mohr–Coulomb law of failure holds.

Let us consider what kinds of fault should develop in a stress field where $\sigma_1 = 0°$ N20°E, $\sigma_2 = 0°$ N70°W, and σ_3 is vertical. Assume $\phi = 30°$. We know that thrust-slip faults should form simply on the basis of the vertical orientation of σ_3. The probable orientations of the thrusts can be predicted by applying what we have already learned about the formation of conjugate faults. *Conjugate faults intersect in σ_2. And the trace of each fault, when viewed in the σ_1/σ_3 plane, is oriented at an angle of $\theta = \phi$ with respect to σ_1* (see Figure 9.62).

To give these relationships geometric reality, we stereographically plot points that portray the orientations of the principal stress directions (Figure 9.73*A*). We then define the σ_1/σ_3 principal plane by fitting σ_1 and σ_3 to a common great circle (Figure 9.73*B*). Counting 30° along the great circle from σ_1, we locate two reference points (1 and 2) that represent the intersection of the fault planes with the σ_1/σ_3 plane. By fitting σ_2 and reference point 1 to a common great circle, one of the thrust-slip faults is defined (Figure 9.73*C*). And by fitting reference point 2 and σ_2 to a common great circle, the

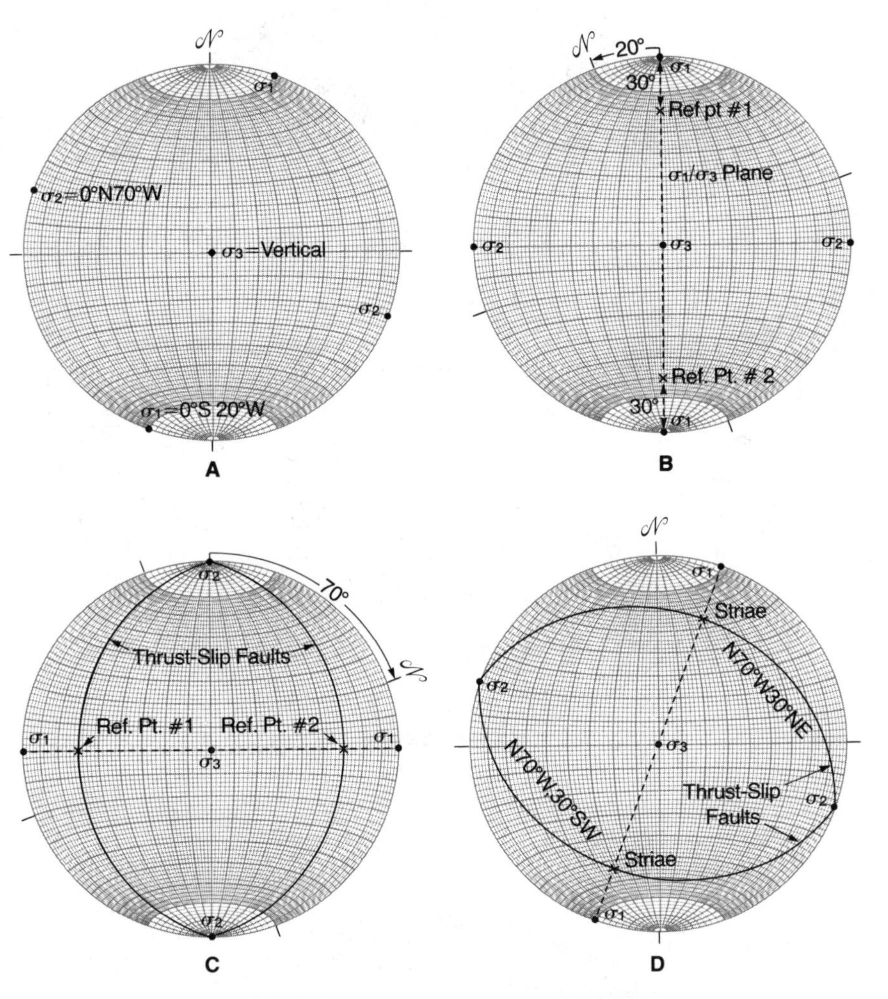

Figure 9.73 Stereographic representation of the relation of faults to the principal stress directions. (*A*) Stereographic portrayal of principal stress directions. (*B*) Identification of the σ_1/σ_3 plane, and the plotting of reference points (1 and 2) at $\theta = 30°$ from σ_1 on the σ_1/σ_3 great circle. (*C*) The orientation of one of the thrust faults is represented by a great circle that passes through reference point 1 and σ_2. The orientation of the second thrust fault is defined by the great circle that passes through reference point 2 and σ_2. (*D*) Portrayal of actual orientations of the faults and the striae that occur along them.

second thrust-slip fault of the conjugate set is established. The actual orientations of the faults can then be determined by normal stereographic procedures (Figure 9.73*D*). Comparable solutions for defining the orientations of strike–slip and normal-slip faults within appropriate stress fields are shown in Figures 9.74 and 9.75, respectively.

If one or more of the principal stress directions is inclined, the solution is more difficult to visualize, but the stereographic operations are the same. Consider the general situation wherein none of the principal stress directions are vertical or horizontal; instead all are inclined (Figure 9.76). As before, we define the σ_1/σ_3 plane by fitting σ_1 and σ_3 to a common great circle, and then we set off reference points from σ_1 along the great circle at distances corresponding to the value of ϕ. The orientations of the fault planes are each defined by one of the reference points and σ_2 (Figure 9.76).

USING FAULTS TO INTERPRET STRESS. In the course of field-oriented structural analysis, it is common to work through this problem backward. Suppose we want to interpret principal stress directions on the basis of the orientations of two conjugate faults. The faults are first plotted stereographically as great circles, as in Figure 9.77*A*. The intersection of the two great circles is taken to be the orientation of σ_2; in this example, σ_2 plunges 20° N50°W. We know that the σ_1/σ_3 plane is perpendicular to σ_2. Given the trend and plunge of σ_2, the σ_1/σ_3 plane must strike N40°E and dip 70°SE (Figure 9.77*B*). σ_1 lies somewhere along the great circle representing the σ_1/σ_3 planes, but where?

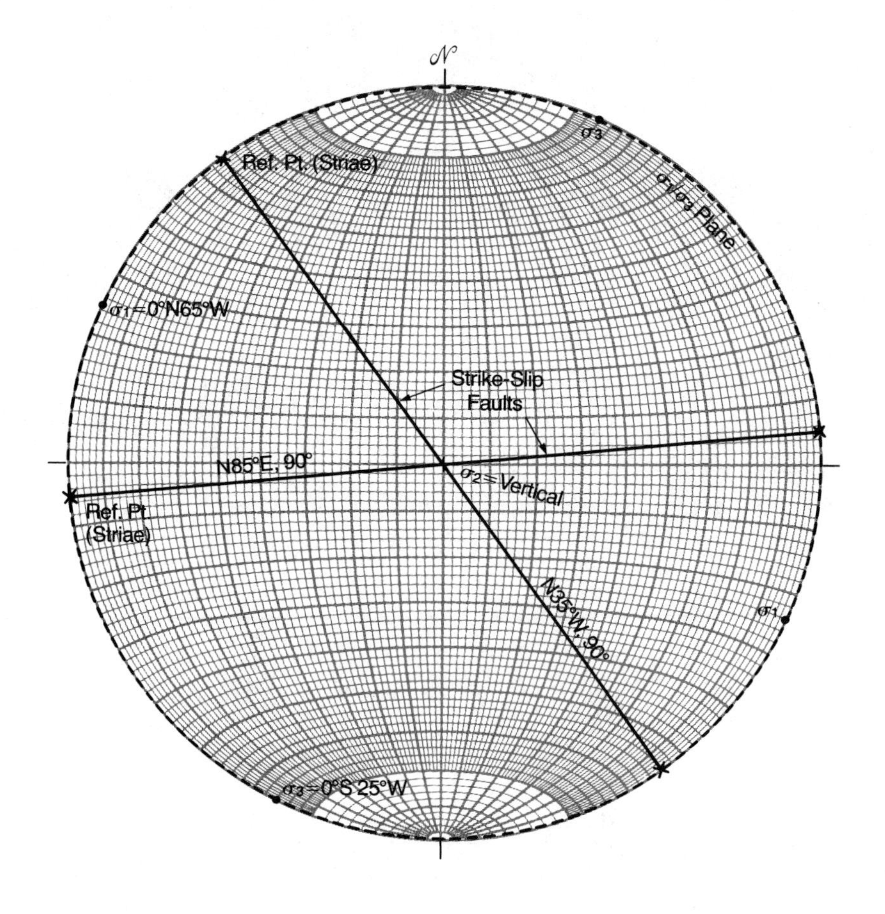

Figure 9.74 Stereographic representation of the relation of conjugate strike-slip faults to the principal stress directions.

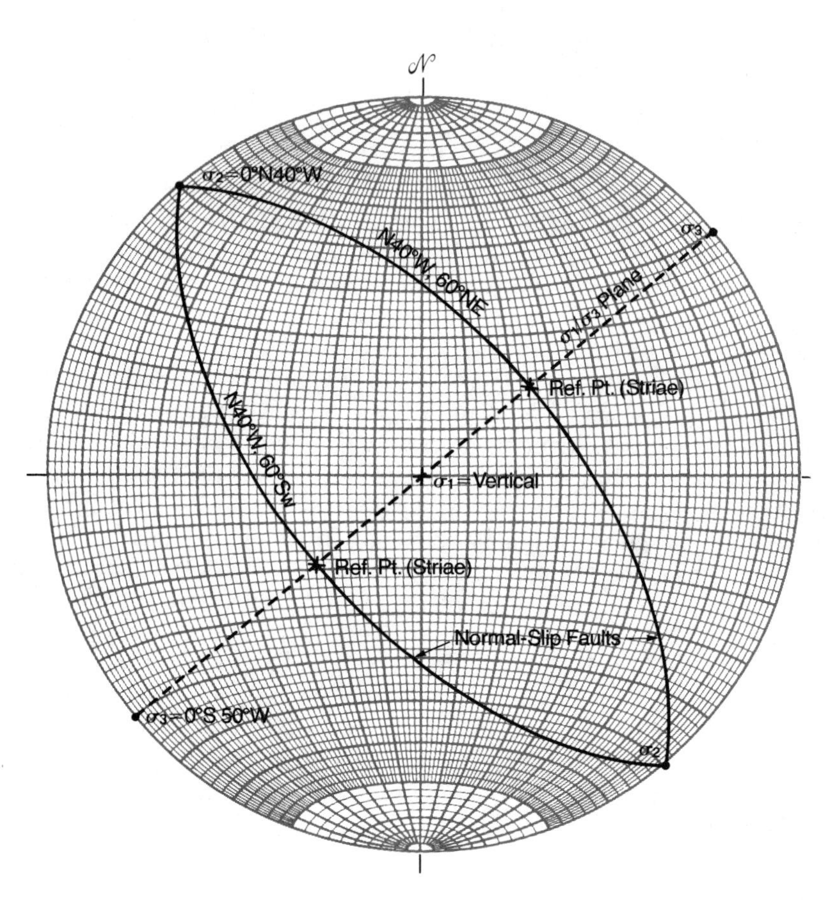

Figure 9.75 Stereographic representation of the relation of conjugate normal-slip faults to the principal stress directions.

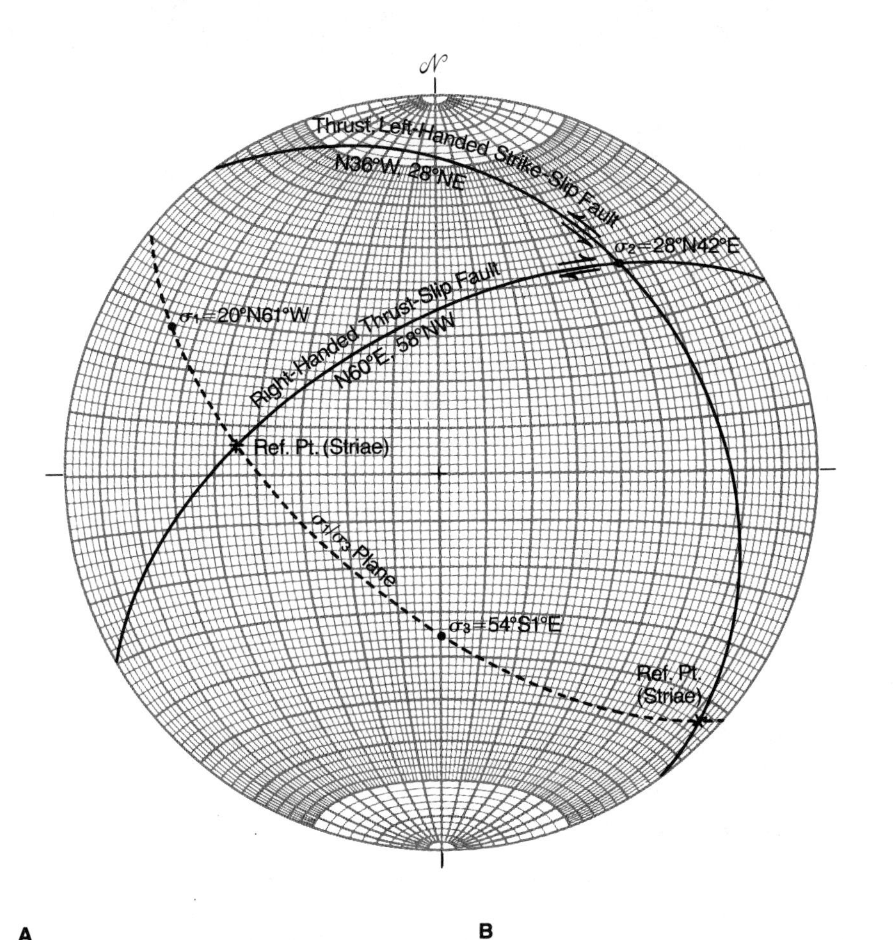

Figure 9.76 Stereographic determination of the orientations of conjugate faults that would develop in a stress system characterized by inclined principal stress directions.

A

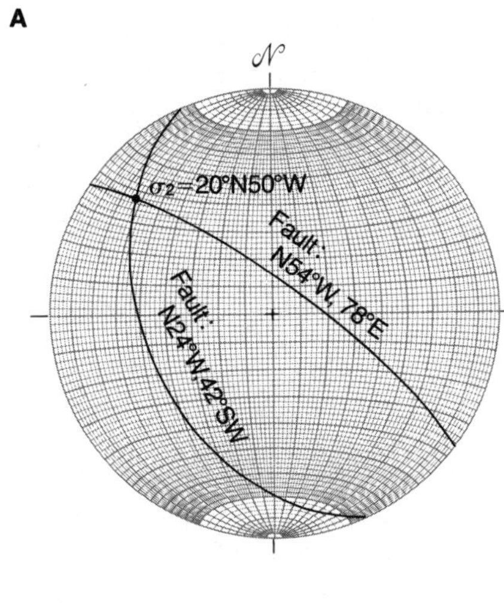

B

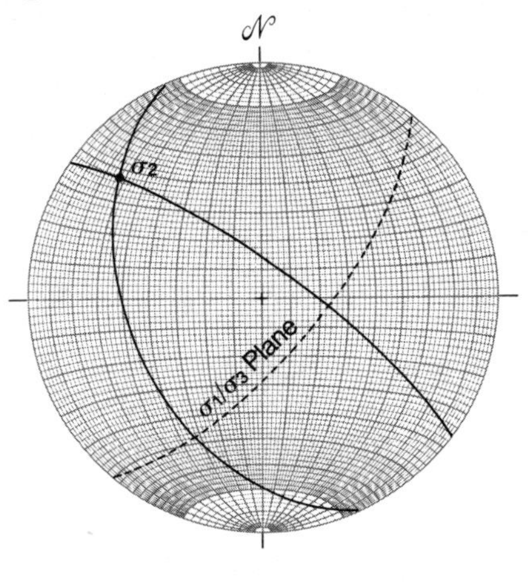

C

D

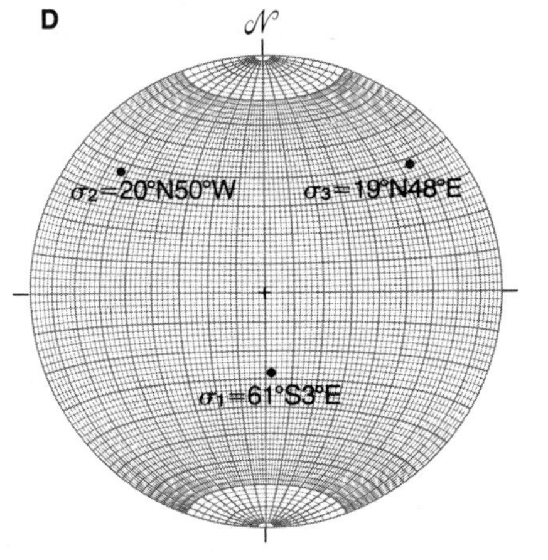

Figure 9.77 Interpretation of principal stress directions on the basis of the known orientations of conjugate faults. (*A*) Stereographic representation of the faults as great circles. The faults intersect in σ_2. (*B*) Identification of the σ_1/σ_3 plane, the great circle whose pole is σ_2. (*C*) Determination of the stereographic locations of σ_1 and σ_3. (*D*) Portrayal of the orientations of σ_1, σ_2 and σ_3, as solved.

We know from experiments that σ_1 bisects the acute angle between conjugate faults. Stereographically this means that we can locate σ_1 by bisecting the angle between the fault traces that lie in the σ_1/σ_3 plane. To do this, rotate the overlay such that the strike of the σ_1/σ_3 plane becomes aligned along the north–south line of the net (Figure 9.77*C*). Then measure the acute angle between the reference points. Since the acute angle measures 64°, the bisector (σ_1), is located 32° from each reference point along the σ_1/σ_3 plane; σ_3 is perpendicular to σ_1. The orientations of the principal stress directions are ultimately defined in terms of trend and plunge (Figure 9.77*D*).

EXCEPTIONS TO THE LAW

THE PROBLEM

Both the Mohr–Coulomb law of failure and Anderson's theory on faulting illuminate our understanding of fault relationships. However, there are a number of facts and relationships that cannot be explained in the context of the "laws" presented thus far. Here are some of the problems and some of the solutions.

GRIFFITH CRACKS

Rocks commonly fail at levels of critical shear stress that are significantly lower than the magnitudes of stress that are predicted by conventional theory. Furthermore, failure envelopes generated for many rocks are found to be parabolic instead of straight. The parabolic curves decrease in slope at higher and higher levels of normal stress. **Parabolic failure envelopes** express the fact that the conjugate angle between faults approaches 90° at high levels of confining pressure. Parabolic envelopes contradict the Mohr–Coulomb law of failure.

Observations such as these led Griffith (1924) to reexamine the Mohr–Coulomb law of failure and to evaluate the response of rocks to stress at the microscopic level. He concluded that the fracturing of rocks, even rocks as homogeneous as glass, is influenced by microscopic flaws and cracks. The cracks, now known as **Griffith cracks**, weaken the rock's resistance to faulting by permitting stress concentrations to build at the inside corners of crack aperatures. Parabolic failure envelopes are consistent with Griffith's model of the influence of microscopic flaws on the dynamics of faulting. As a consequence of Griffith's theory, engineers and geologists have modified the Mohr–Coulomb theory in attempts to describe more accurately the conditions under which rocks will fail.

ROLE OF ANISOTROPY

Directional planar weaknesses like bedding, foliation, or fractures substantially alter the response of a test specimen to stress. The level of critical shear stress required to break a rock varies as a function both of the physical nature and the orientation of the **anisotropy**. The clearest examples of this kind of behavior have been derived from experimental deformation of test specimens that contain preexisting fractures. Handin (1969) was able to show that fractures oriented at angles as high as 65° to σ_1 can be activated as faults. He also showed that the critical stress level required to cause faulting along preexisting fracture surfaces is less than that required to break an unfractured specimen of the same lithology.

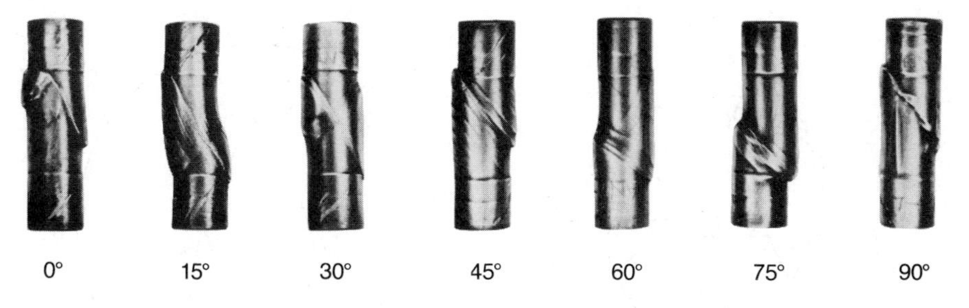

Figure 9.78 Specimens of anisotropic rock compressed at various angles to foliation. The orientation of faulting, in each case, is influenced by the angular relationship between the orientation of foliation and the direction of greatest principal stress (σ_1). [From Donath (1961), Geological Society of America.]

Conventional Mohr–Coulomb theory does not apply in a direct way to the faulting of prefractured specimens. First, fractured specimens do not actually possess cohesive strength (σ_o) along the fractures that become activated as faults. And second, the **coefficient of external sliding friction** on fracture surfaces emerges as the dominant factor in determining the level of shear stress required to produce faulting along preexisting fractures. Thus, internal friction takes a back seat to external friction.

The presence of a foliation, like schistosity, also influences the dynamics of faulting. Donath (1961) showed the degree to which the orientation of foliation influences the orientation of faults that develop in test specimens (Figure 9.78). When foliation is oriented at a very high angle to σ_1, the specimen will fault, as usual, at an angle such that $\theta \sim \phi$. But if foliation is at a closer angle to σ_1, the orientation of the fault will differ from what it would be in a homogeneous specimen. For example, if foliation is parallel to σ_1, the angle ϕ will be very low, perhaps 10° or 20°. If foliation is inclined between 25° and 45° to σ_1, the fault surface will commonly develop right along the foliation, thereby overriding whatever the dictates of the coefficient of internal friction might have been.

Rocks in the Earth are full of planar weaknesses, in the form of bedding, foliation, layering, dikes, faults, joints, and contacts. Considered at the regional scale, flaws in basement have a profound influence on the stress required for faulting to occur and on the orientation of the faults that emerge. If a relatively young faulting event takes place through reactivation of an ancient fault surface, it is unlikely that principal stress orientations responsible for the faulting can be deduced from the geometric facts available.

ROLE OF THREE-DIMENSIONAL STRAIN

Mohr–Coulomb theory predicts that faults should form in conjugate sets. And this seems to be supported by the results of triaxial deformation experiments. Yet faulted rocks in nature commonly display two conjugate pairs of faults arranged in **orthorhombic symmetry**. Donath (1962) mapped such a pattern in south-central Oregon (Figure 9.79). One way to explain the presence of four fault sets is to call on two episodes of faulting, each of which yields conjugate pairs of faults. However, it is now clear that four fault sets can indeed be generated in a single event. How is this possible?

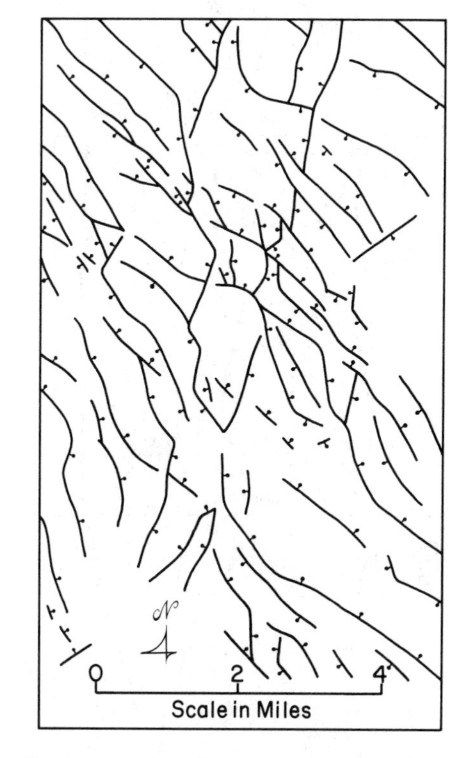

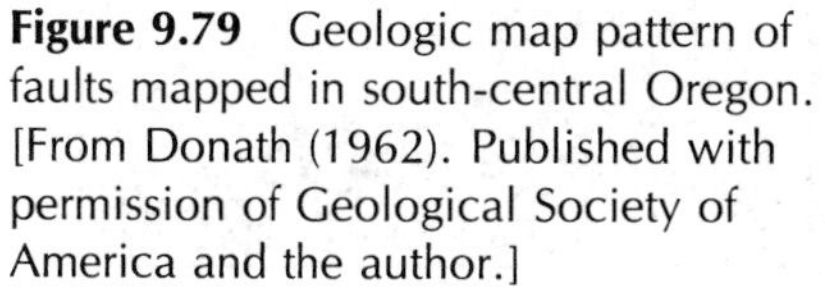

Figure 9.79 Geologic map pattern of faults mapped in south-central Oregon. [From Donath (1962). Published with permission of Geological Society of America and the author.]

Reches (1978a) has emphasized that the presence of three or four fault sets in a given region of study is the natural result of faulting within a three-dimensional strain field. The patterns seem peculiar only because rock deformation experiments traditionally have been carried out under conditions of two-dimensional stress and strain. Where specimens are allowed to shorten (or stretch) by different amounts in *three* mutually perpendicular directions, the characteristic fault pattern that emerges is one of three or more sets arranged in orthorhombic symmetry (Figure 9.80). Such a fault pattern was produced by Oertel (1965) in a clay cake subjected to stretching in a three-

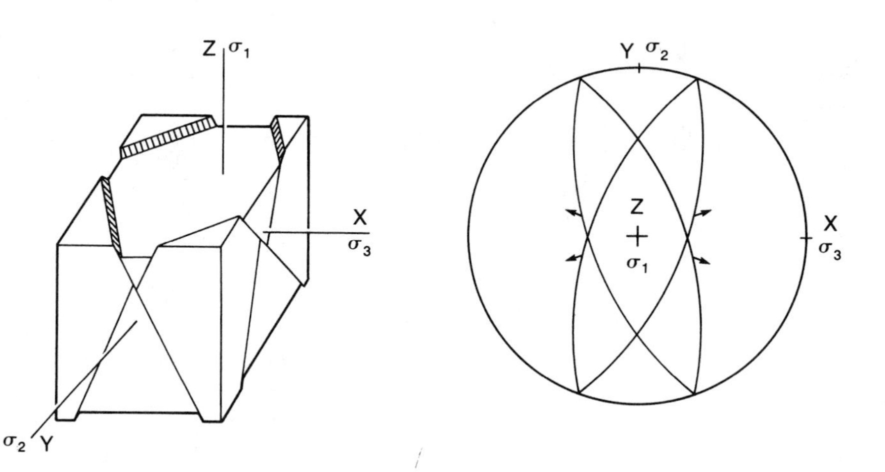

Figure 9.80 Fault sets produced in three-dimensional strain field. [From Reches (1983), *Tectonophysics*, v. 95. Published with permission of Elsevier Scientific Publishing Company, Amsterdam.]

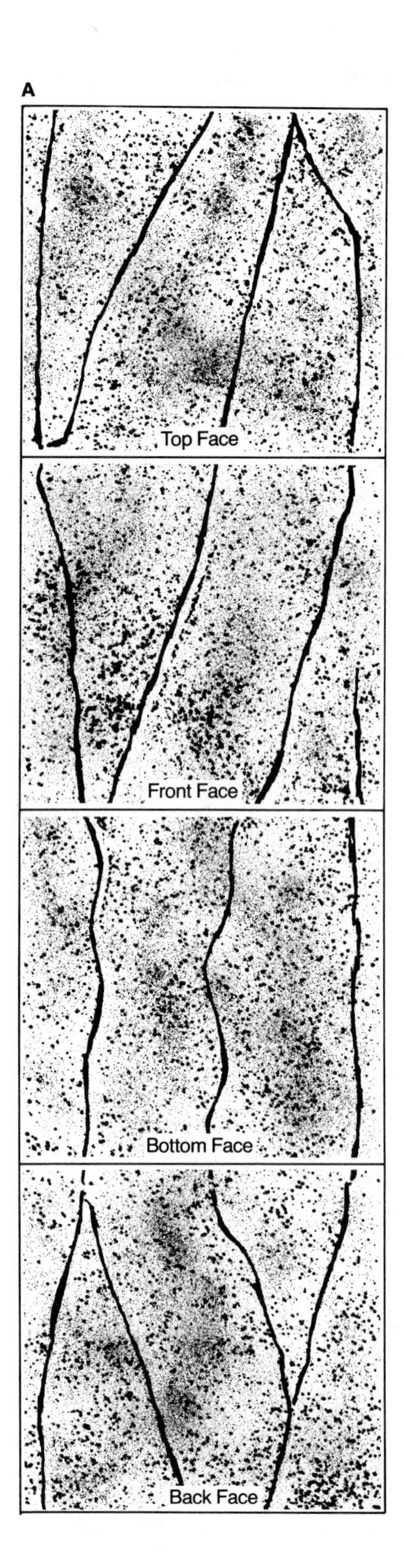

dimensional strain field. And Reches and Dieterich (1983) produced such patterns in cubes of sandstone, granite, and limestone that were subjected to compression in a three-dimensional strain field (Figure 9.81). Reches (1983) further showed that the angular relationships among fault sets formed in this way not only relate to the angle of internal friction (ϕ) of the host rock, but also to the ratio of strain along the X, Y, and Z principal strain axes.

This exception to the law has finally yielded an explanation that will make fault interpretations more manageable. The work by Reches and Dieterich puts us on notice that rocks are highly sensitive to the difference between **plane strain** and **three-dimensional strain**. We should be, too.

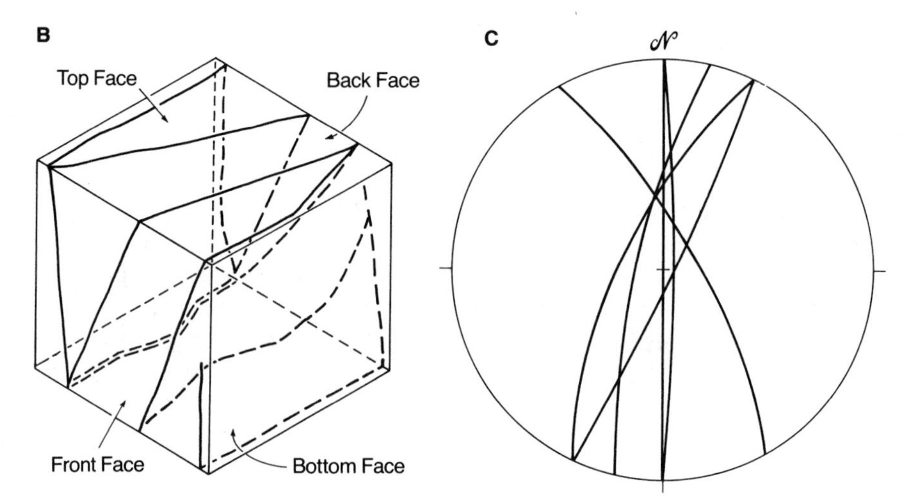

Figure 9.81 Fault pattern in cube of rock subjected to three-dimensional strain. (*A*) "Mapped" traces of faults on faces of deformed cube. (*B*) Three-dimensional portrayal of the faulted specimen. (*C*) Stereographic projection of the faults. [From Reches and Dietrich (1983), *Tectonophysics*, v. 95. Published with permission of Elsevier Scientific Publishing Company, Amsterdam.]

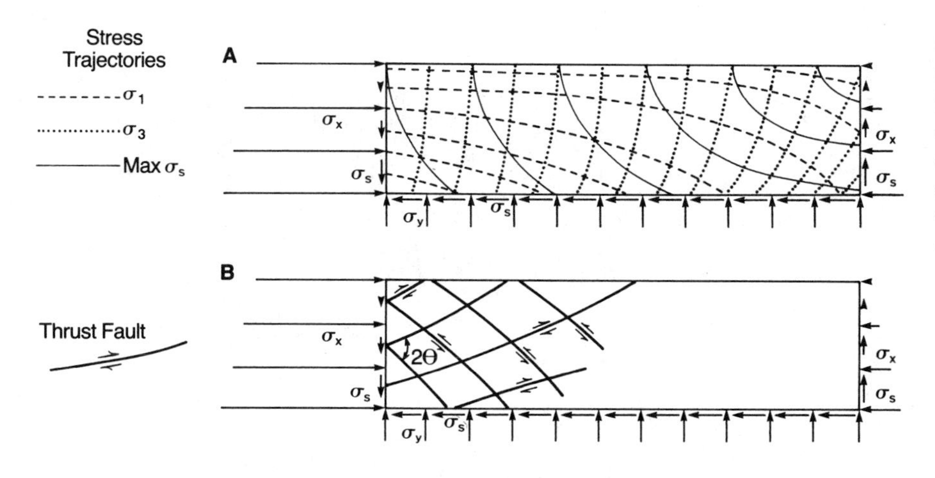

Figure 9.82 (*A*) Pattern of stress trajectories that would be produced in a block subjected to strong horizontal compressive stresses that die out laterally. (*B*) The pattern of curved faults that would emerge within the field of curved stress trajectories. [From Hafner (1951), Geological Society of America.]

THE PROBLEM OF REVERSE FAULTS

Where does reverse faulting fit into the Anderson model of faulting and Mohr–Coulomb theory? Reverse-slip faults are found in orogenic belts around the world, yet they are not featured in Anderson's fundamental classes of faults.

One obvious explanation for the origin of some reverse-slip faults is that they occupy former sites of normal faulting or strike–slip faulting. Where reverse faulting is a result of **fault reactivation**, the dip of the reverse-slip fault is inherited from a previous event.

Another explanation of reverse faulting is that principal stress directions are not necessarily vertical and horizontal at depth. Rather, the orientations of the principal stress directions may be inclined. **Stress trajectories** become inclined and/or curved as the result of changes in the state of stress both laterally and vertically. Hafner (1951) demonstrated this theoretically. He showed, for example, that the dissipation of horizontal compressive stresses can result in a strain field marked by curved stress trajectories (Figure 9.82*A*). Curved compressive stress trajectories give rise to continuously curved faults, parts of which are thrust-slip faults, but parts of which are reverse-slip faults (Figure 9.82*B*).

Hafner's work (1951) showed that compression-induced thrusts at depth may **steepen upward** into reverse-slip faults. In contrast, Sanford (1959), Stearns (1978), and Friedman and others (1976) have demonstrated experimentally that **differential vertical uplift**, involving no compression, can produce upper level thrusts that **steepen downward** into reverse-slip faults (Figure 9.83). Interpreting the dynamics of formation of reverse faulting thus requires full three-dimensional views of faults and fault systems.

The dynamic significance of reverse faulting has been hotly debated. In this regard, the Wind River uplift in Wyoming has attracted much attention. At issue is whether the uplift owes its existence to horizontal compressive stresses or to differential vertical uplift in a noncompressive environment. A major part of the problem is trying to establish whether moderately dipping faults exposed at or near the surface steepen with depth to form high-angle reverse-slip faults, or whether they shallow with depth into pure thrust-slip faults (Figure 9.84). Both options are possible from the theoretical and experimental point of view.

Seismic profiling has helped to disclose the deep three-dimensional structure of the Wind River reverse fault. Berg (1962) demonstrated that the Wind River fault, bounding the western margin of the Wind River Mountains, is a northeast-dipping thrust-slip fault (Figure 9.85). Drilling and seismic data

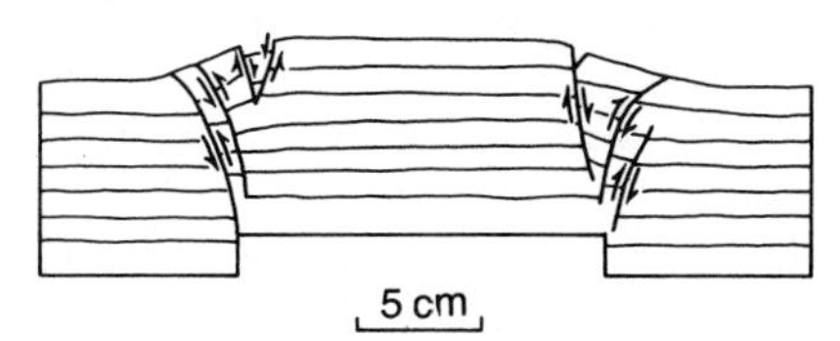

Figure 9.83 Curved reverse-slip faults produced by differential vertical uplift in deformation experiment. [From Sanford (1959), Geological Society of America.]

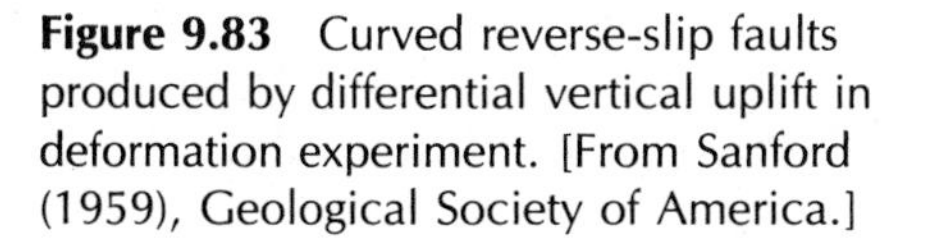

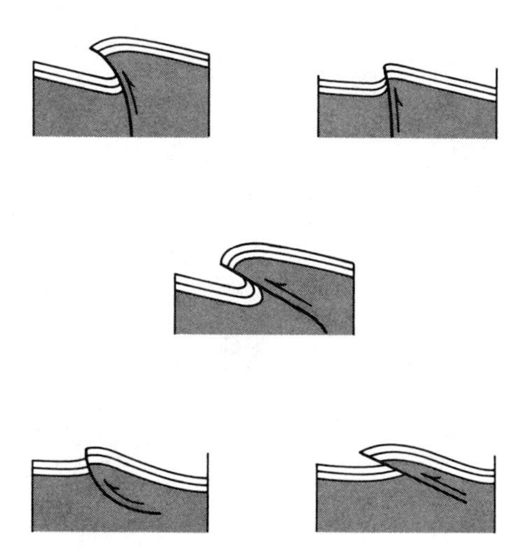

Figure 9.84 Possible configurations of the Wind River fault at depth. [From Smithson and others (1978). Published with permission of Geological Society of America and the authors.]

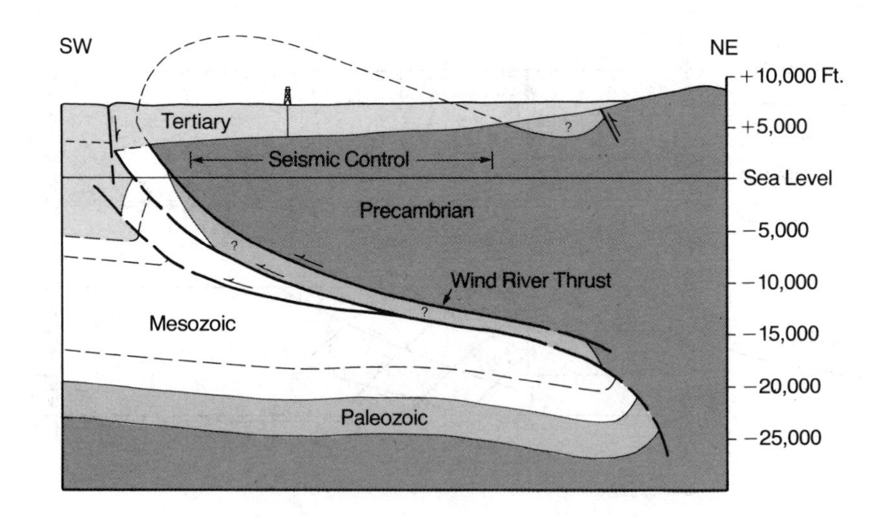

Figure 9.85 Structure section of the Wind River Range, Wyoming. [From Berg (1962). Published with permission of American Association of Petroleum Geologists.]

available to Berg led him to conclude that the Wind River thrust has a maximum vertical displacement of 40,000 ft (12,000 m) and a maximum horizontal displacement of 50,000 ft (15,000 m), and that the zone of faulting ranges in thickness from 1000 to 2500 ft (300 to 760 m). What does the thrust do at even greater depth? Does it steepen or flatten?

In an attempt to see deeper into the lithosphere, seismic-reflection profiling was undertaken by COCORP over a 160-km traverse length at the south end of the Wind River Mountains. The conclusion of the study, as reported by Smithson and others (1978), is that the Wind River thrust is visible on the profile at depths down to 24 km and that at depth the dip is approximately 30° to 35°. This interpretation, if correct, suggests that the Wind River thrusting was an accommodation to regional crustal shortening. It also would indicate that Mohr–Coulomb theory may be applicable on a scale that goes well beyond the confines of a sandbox.

MECHANICAL PARADOX OF OVERTHRUSTING

The most provocative paper ever written on the dynamics of faulting was that by Hubbert and Rubey (1959) entitled, *The Role of Fluid Pressure in Mechanics of Overthrust Faulting*. Hubbert and Rubey underscored the dominant, if not essential, role of fluid pressure in the low-angle tectonic transport of great overthrust sheets. Hubbert and Rubey calculated the approximate amount of force that would be required to translate allochthonous thrust terranes of the size that are known to exist in the Canadian Rockies, the Western Cordillera of the United States, and the Appalachians. They came to the conclusion that the calculated force, if applied to the rear of the mass, would far exceed the crushing strength of granite. In fact they concluded that the longest thrust sheet that could be pushed across a horizontal surface would be limited to 10 km or so. Considering the possibility that thrust sheets are not pushed, but rather slide down structural gradients under the influence of gravity, Hubbert and Rubey proceeded to calculate the dynamic requirements of gravitational models. The results were the same: large thrust sheets should not exist, even if gravity is the propelling mechanism.

The solution to the paradox of overthrusting involved modifying the Mohr–Coulomb law of failure. Hubbert and Rubey asked, how it is possible to decrease the magnitude of critical shear stress that is required for faulting to occur? They examined closely Equation 9.1:

$$\sigma_c = \sigma_o + \tan \phi \, (\sigma_N)$$

and attacked the variable σ_N, the normal stress acting on the plane of faulting. Hubbert and Rubey were able to show that **high fluid pressure (σ)** in rocks tends to offset the magnitude of lithostatic normal stress (σ_N) acting on the plane of faulting. The ***effective stress*** **(σ_N − σ)** acting normal to a fault surface would equal the difference between the normal stress and the fluid pressure. The more the normal stress is reduced, the lower the value of critical shear stress that is required to produce faulting. With these concepts in mind, Hubbert and Rubey (1959) modified the Mohr–Coulomb law of failure as follows:

$$\sigma_c = \sigma_o + \tan \phi(\sigma_N - \sigma) \tag{9.3}$$

Many sedimentary basins in the world, like the Gulf of Mexico, are marked by bedding-parallel zones of fluid pressure that are so highly elevated that they approach the value of the lithostatic stress of the overlying sedimentary cover. The fluid pressure essentially supports the weight of all the rock above. When effective stress approaches zero, one entire term of the Mohr–Coulomb law of failure is eliminated:

$$\tan \phi(\sigma_N - \sigma) = \tan \phi(0) = 0$$

What is left is the simple expression:

$$\sigma_c = \sigma_o$$

When this state exists, all that is required to move a thrust sheet is to break the cohesion of the rock. Once cohesion is lost, only minimal stress is necessary to keep the thrust moving. As structural geologist Ralph Kehle facetiously remarked, the problem is not getting thrusts to move, but stopping them once they get rolling.

Hubbert and Rubey (1959) demonstrated the fundamentals of the fluid pressure model in their now-famous beer can experiment. Sample preparation consists of drinking two beers, preferably out of nonaluminum cans (Figure 9.86*A*). Place *one* of the empties in the freezer (Figure 9.86*B*), and

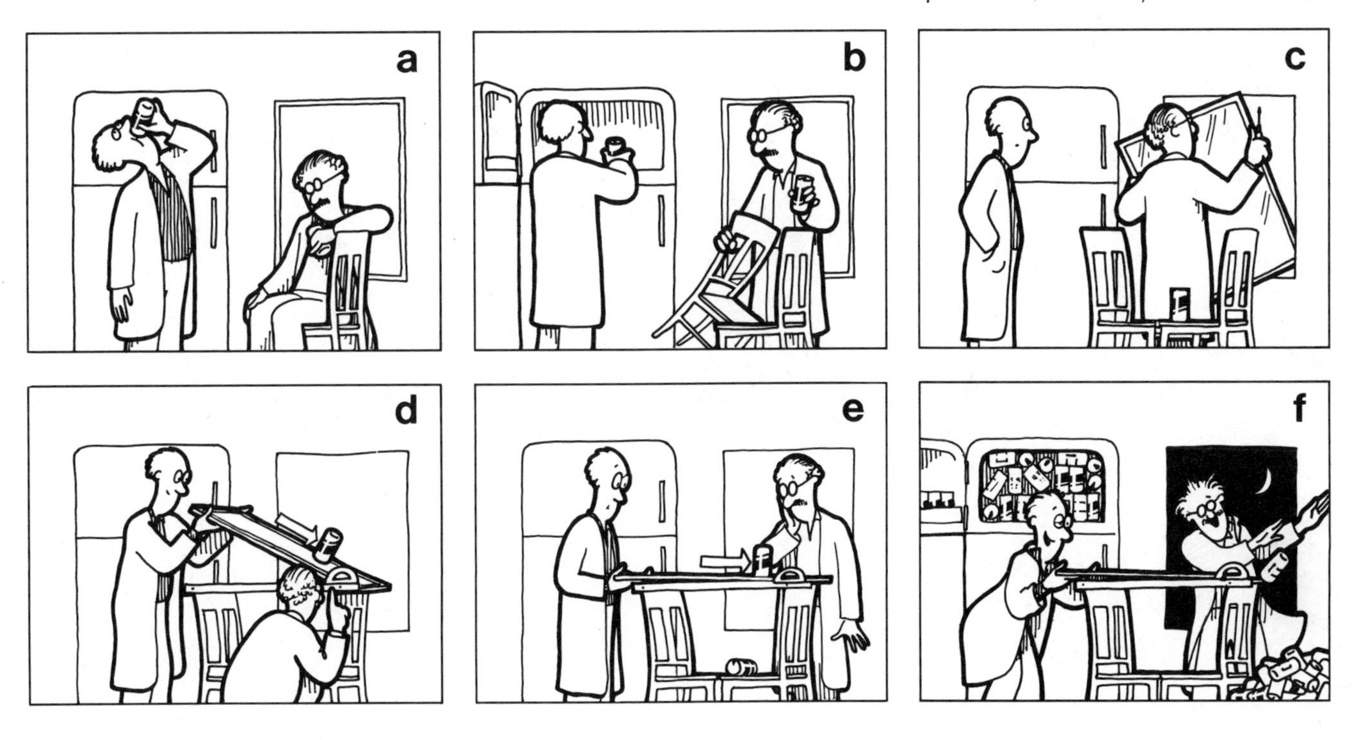

Figure 9.86 The famous beer can experiment. (Artwork by D. A. Fischer.)

remove a window from your house or apartment or lab (Figure 9.86*C*). Clean the glass with detergent, rinse, and leave it wet with a thin film of water. Place beer can 2, top down, on the pane of glass. Now lift one end of the glass to form an inclined plane, and, with protractor in hand, measure the angle at which the beer can commences movement down the plane (Figure 9.86*D*). Hubbert and Rubey report typical angles of about 17° corresponding to a coefficient of sliding friction of metal on wet glass of 0.3. After beer can 1 has been chilled in the freezer, quickly pull it out and perform the same exercise (Figure 9.86*E*). This time the beer can begins to move down the inclined plane at negligible angles of slope (~ 1°). It moves easily not because the glass is wet. Rather it moves because a fluid pressure derived from expansion of the warming air inside the can offsets the normal stress exerted by the can on the glass. Hours can be spent enjoying experiments on the role of fluid pressure in overthrusting (Figure 9.86*F*).

CONCLUDING REMARKS

The study of faulting encompasses geometric, kinematic, and dynamic analysis in the broadest possible way. If we are to understand faulting, we need to integrate field, experimental, and theoretical research. The rewards for such activities are great. Faulting is one of the most important mechanisms of rock deformation. It produces regional distortion. It traps petroleum and controls ore deposition. It provides clues to rearrangements of the Earth's architecture through time. Experimental deformation and theoretical studies help us immensely in understanding the properties of faults that we find in nature.

chapter 10 JOINTS

NATURE AND VALUE

GENERAL CHARACTERISTICS OF JOINTS

Joints are fracture surfaces along which there has been imperceptible movement (Figure 10.1). They are fractures where adjacent slabs and masses of bedrock join. Joints in rock bodies permit infinitesimal adjustments to take place as the regional units within which they are found change size and/or shape during structural movements like subsidence, uplift, thrusting, contraction, expansion, and folding. Except under unusual circumstances, bedrock never opens up or slips appreciably along joints.

Joints occur at the outcrop scale in virtually all rocks and thus comprise the most abundant structural element in the crust of the Earth. Joints commonly display extraordinarily systematic **preferred orientations** (Figure 10.2). The patterns often show a striking symmetry. But in spite of their abundance and their systematics, joints may be the least useful of structures in interpreting the stress and strain conditions of past deformational events.

HARDSHIPS IN INTERPRETING JOINTS

A number of factors detract from the usefulness of joints in kinematic and dynamic analysis. First, it is very difficult to establish the time of formation of specific joints or joint sets, and thus to know exactly when they formed relative to a given deformational event. Second, since joints are surfaces of negligible discernible movement, neither separation nor slip can be measured except in special circumstances. Third, as **cohesionless surfaces**, joints are activated and reactivated in multiple deformational events. Thus,

Figure 10.1 Joints and joint surfaces exposed in a rock quarry near Glen Echo, Maryland. (Photograph by G. K. Gilbert. Courtesy of United States Geological Survey.)

Figure 10.2 Aerial photograph of jointing in the Entrada Sandstone (Jurassic) near the campground at Arches National Park, Utah. Two prominent sets of joints come into clear, crisp resolution when the photo is viewed at a low angle to the plane of the page, parallel to the strike of the joints. (Photograph by R. Dyer.)

even where separation along a joint can be measured, the magnitude and sense of separation must be viewed as a composite of all past slip movements and not generally as the simple product of a single movement. Finally, joints can be formed in so many ways that it is commonly impossible to distinguish among the numerous possible interpretive models that can explain adequately their formation.

PRACTICAL AND AESTHETIC VALUE OF JOINTS

Although joints are often difficult to interpret, they are nonetheless very important structures. For ages, quarry workers have taken advantage of joint-controlled planes of weakness in removing building blocks of granite and limestone from bedrock. These fracture weaknesses exert profound control on weathering and erosion, and thus on fashioning landscape. Many scenic attractions owe much of their uniqueness to weathering and erosion of horizontal layers of rock systematically broken by steeply dipping joints. The rock chimneys (**hoodos**) of Bryce National Park (Utah) and Chiricahua

Figure 10.3 Jointing exerts a major control on landform development. (*A*) Jointing in flat-lying Miocene ignimbrites, Chiricahua National Monument, Arizona. (Photograph by G. H. Davis.) (*B*) Joint-pervaded landscape in Canyonlands National Park, Utah. (From G. E. McGill and A. W. Stromquist, *Journal of Geophysical Research*, v. 4, p. 4547–4563. Copyright ©1979 by American Geophysical Union.)

National Monument (Arizona), the columns of Devil's Post Pile (California), the arches of Arches National Park (Utah), and the mesa and butte country of the Colorado Plateau, especially Monument Valley (Utah and Arizona), all serve as remarkable and inviting examples (Figure 10.3).

Beyond their scenic value, joints constitute a structure of indisputable geologic and economic significance. The presence of joints invites circulation of fluids, including rain and groundwater, hydrothermal mineralizing solutions, and oil and gas. As cracks in rocks, joints can be thought of as structures that significantly contribute to the bulk porosity and permeability of rocks.

The influence of jointing on the flow of fluids through rock is especially well known to those who love to explore caves. The shapes and orientations of rooms and passageways in caves are commonly controlled by the selective solution removal of limestone along major joint trends. For example, Left Hand Tunnel of Carlsbad Caverns in New Mexico is elongate east–northeast, parallel to the predominant set of joints that cuts the limestone in that part of the caverns (Jagnow, 1979). Cross-sectional profiles of Left Hand Tunnel prepared by Jagnow reveal that tall, narrow passages are centered on prominent vertical joints that once guided the groundwater circulation.

Explorationists appreciate the benefits of circulation of fluids through jointed rocks. Petroleum geologists evaluate the nature and degree of development of joints as one guide to the reservoir quality of sedimentary formations. In fact, to increase the yield of reservoir rocks in oil and gas fields where production is waning, it is common practice to "crack" the rocks artificially, either through explosives or through the high-pressure pumping of fluids in the well(s).

The natural circulation of **hydrothermal fluids** through joints in hot rocks at depth constitutes an increasingly acknowledged source of energy, namely geothermal energy. For geothermal systems to be operational, the presence of thoroughly jointed rocks is as essential as heat and fluids.

Joints can serve as sites of deposition of metallic and nonmetallic minerals. In almost all **hydrothermal deposits**, a part of the mineralization is localized in and around joints. The minerals are deposited either through **open-space filling** of joints or through **selective replacement** of chemically favorable rocks adjacent to the joint surfaces along which hydrothermal fluids once circulated. Even where joints do not carry economically significant levels of mineralization, they may be marked by veinlets and/or alteration assemblages of distinctive silicate and sulfide minerals. Economic geologists use alteration minerals as clues to possible hidden locations of ore deposits.

ROLE OF JOINTING IN MASS WASTING

There are harmful side effects to the ease of circulation of fluids through jointed rock systems. Landsliding, slumping, and other processes of mass wasting are enhanced by the saturation of jointed rocks with rainwater and groundwater, especially in terrains of high topographic relief. The fluid pressures exerted by groundwater in cracks in rocks weaken the level of normal stress on fracture surfaces, thus enhancing the potential for slip. Where fluid-filled joints in rocks beneath a hillslope dip outward toward the free face of the hill, the steady force of gravity, in concert with fluid pressure exerted by water in the cracks, can cause sliding of earth and houses outward and downslope (Figure 10.4).

Engineers and consulting geologists address the problem of **mass wasting** not only in residential areas and municipal construction sites, but also in the designing of open-pit mines. Appropriately stable slope angles for a given open-pit mine depend on a number of factors, including joint orientation and abundance of joints. The engineering problem demands the judgment and balance required of a tightrope walker (Figure 10.5). The object is to maximize the slope angle of the pit, so that buried parts of the ore deposit can be uncovered and exploited through removal of the least amount of wasterock overburden. At the same time, the engineers must minimize the risk of slope failure by making certain that the slopes of the pit are not oversteepened, given the geological conditions that exist. Slope failure, when it occurs, results in loss of capital equipment, disruption of road and/or track systems for hauling ore, and infilling of the open pit by wasterock. Pit-slope failure can shut down marginally economic mining operations.

Figure 10.4 Block glide at Point Fermin, near Los Angeles. (Photograph by Spence Air Photos. Published with permission of National Research Council.)

A

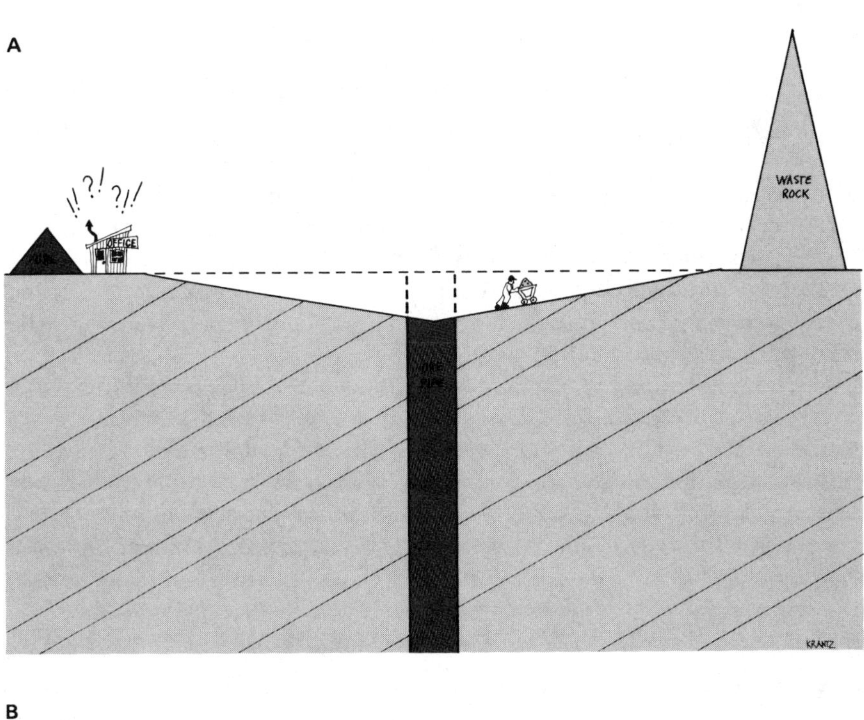

B

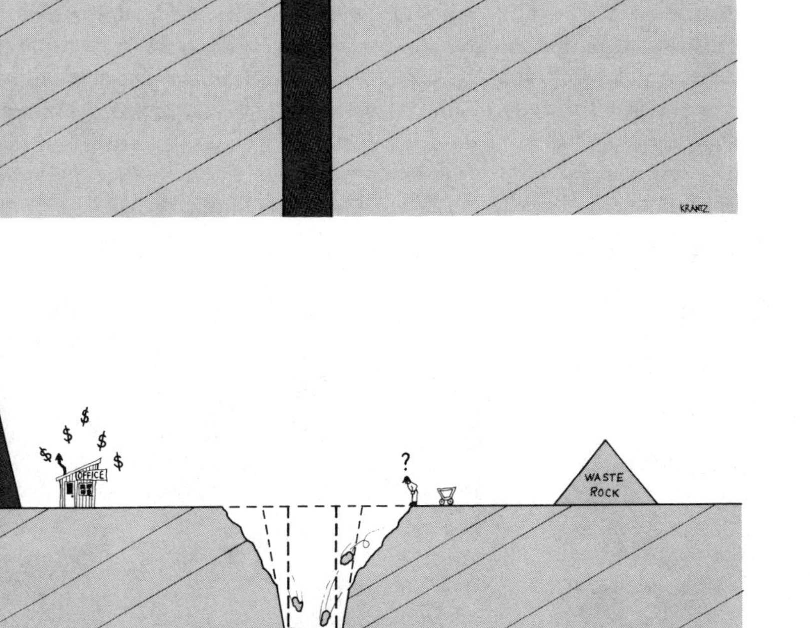

Figure 10.5 Trials and tribulations of pit-slope design. (*A*) Gently dipping pit walls are safe but uneconomical. (*B*) Steeply dipping pit walls are profitable in the short term but risky in the longer view. (Artwork by R. W. Krantz.)

DEFINITIONS AND DISTINCTIONS

The term **joint** is most commonly used in reference to relatively continuous and through-going fractures that are reasonably planar and along which there has been imperceptible movement. The infinitesimal movement that gives rise to a joint can be a shear parallel to the joint surface, a dilation perpendicular to the joint surface, a shortening perpendicular to the joint surface through actual loss of material, or some combination of these factors. In most cases the movements are microscopic and cannot be recognized in the outcrop record.

It is not uncommon to recognize offset along fractures that would otherwise be classified as joints. The offset may be a shear separation and/or a

Figure 10.6 Quartz veins. [From *The Minor Structures of Deformed Rocks: A Photographic Atlas* by L. E. Weiss. Published with permission of Springer-Verlag, New York, copyright ©1972.]

dilational separation. **Shear separation** is produced by movements parallel to the face of a joint; **dilational separation** is produced by movements perpendicular to the face of a joint. If shear separation is evident along a jointlike fracture, and if the amount of separation is very small (less than 1 cm), the fracture may be described as a **microfault**. Dilational separation along joints is normally preserved in the form of **veins** (i.e., filled fractures Figure 10.6). Where a vein is excessively wide (more than about 20 cm), the term **fissure vein** is used to emphasize the large dilational aperture. Open, *unfilled* fractures with dilational separations less than about 20 cm are called **gash fractures**. Open unfilled fractures with dilational separations greater than about 20 cm are called **fissures**.

The best developed joints are eye-catching **systematic joints** that are planar, parallel, and evenly spaced (Figure 10.7). A given outcrop area or region of study is typically marked by more than one ***set*** of subparallel, systematic joints, the existence of which is obvious and/or can be demonstrated through statistical analysis of orientation measurements.

Joints may closely resemble bedding or cleavage, especially if the joints are planar, parallel, and closely spaced. It is easy to misidentify jointing as

Figure 10.7 Some examples of systematic jointing. (*A*) Aerial view of joints in the Moab Member of the Entrada Sandstone (Jurassic), Arches National Park, Utah. The spacing of the joints ranges from 25 to 50 m. (Photograph by R. Dyer.) (*B*) Joints in Cretaceous sandstone in the Tucson Mountains, Arizona. (Photograph by G. H. Davis.) (*C*) Amazing orthogonal pattern of jointing as photographed from the top of a small juniper tree in northern New Mexico. Field notebook for scale. (Photograph by G. H. Davis.)

bedding in massively stratified rocks that lack the conspicuous color and textural variations that would otherwise call attention to the attitude of stratification. Where bedding and jointing are difficult to distinguish, it is necessary to carefully examine the bedrock for primary structures reflecting the orientation of the internal stratification.

Fracture cleavage (Figure 10.8), now more commonly and appropriately referred to as **spaced cleavage**, resembles a very closely spaced, planar, parallel jointing. Spaced cleavage is typically associated with tightly folded sedimentary and low-grade metasedimentary rocks. Spaced cleavage and jointing are distinctly different from the point of view of strain. It can often be shown that spaced cleavage surfaces are discontinuities along which significant volumes of host rock have been removed by **pressure solution** during the compression-induced shortening that accompanied the folding. Clay residues typically line the fracture. Where spaced cleavage is well developed, typical volume losses are estimated to range from 30 to 50% or more. Spaced cleavage surfaces thus are marked by "perceptible" structural movements; they do not qualify as joints even if they do resemble jointing.

Figure 10.8 Jointlike spaced cleavage in folded impure limestones and marls of the Earp formation (Pennsylvanian-Permian) at Agua Verde near Tucson, Arizona. Note anticlinal fold on right. Height of exposure is approximately 3 m. (Photograph by J. M. Crespi.)

Some joints are so irregular in form, spacing, and/or orientation that they cannot be readily combined into distinctive, through-going sets. These are **nonsystematic joints**. They display such disorder that it seems hopeless to try to understand their structural significance. Nonetheless, nonsystematic joints are part of the deformational record, reflecting some aspect of the strain history of the rocks in which they are found.

Almost all rocks are jointed at the outcrop scale. And most outcrops contain both systematic and nonsystematic joints. The number of joints that are intercepted in a given outcrop varies considerably from place to place and from rock to rock. Incompetent rocks like shale or mudstone are usually more highly jointed than competent rocks like limestone or sandstone. And relatively thin layers tend to be more closely jointed than thicker layers (Harris, Taylor, and Walper, 1960). Spacing of systematic joints in thinly laminated incompetent rocks may be so close (several millimeters to 1 cm) that the fractures are best referred to as **microjoints**. In contrast, large, through-going joint surfaces in very thick competent layers may be so widely spaced that broad stretches of a given rock exposure remain absolutely joint free.

Some rocks, especially those within fault zones, may be **shattered** along randomly oriented, closely spaced fractures. Fractures of this nature do not qualify as joints in the conventional sense because the density of fracturing is so high and because there is no sense of preferred orientation in the form of easily recognized and/or documented sets. Fractures in such shattered rocks are called fractures, not joints. The array of fractures is referred to as **shattering**.

PHYSICAL CHARACTERISTICS

Individual joints are planar to curviplanar, generally featureless surfaces that intersect the tops and flanks of outcrops as lines. Erosion and spalling of rocks along joint surfaces reveal **joint faces** (Figure 10.9). No outcrop that I know of reveals an entire joint surface that can be recognized as such. Thus the actual shape of a joint surface is conjectural, and it must be deduced from partial views. Woodworth (1896) concluded long ago that joints are elliptical. He based this interpretation on careful inspection of three-dimensional exposures of jointed bedrock. Woodworth noted that the long axis of the elliptical form of a given joint is generally parallel to the trace of bedding or layering, where present. In conventional structural studies, no

Figure 10.9 Joint face in the Pennsylvanian Bonaventure formation along the shore of Chaleur Bay, New Brunswick, Canada. The joint face is marked by an exquisite display of plumose markings. Geologist is Wayne Nesbitt. (Photograph by G. H. Davis.)

attempt is ever made to determine what the joint shape in a given rock system might be. The joints are simply treated as partial exposures of surfaces whose dimensions will never be seen in full.

Some joint faces are characterized by **plumose markings** (Figure 10.9), a structure distinguished by featherlike surface patterns. They are most commonly seen on joint surfaces in sandstones and siltstones, but they occur in all competent rocks. Plumose markings splay symmetrically from the central axis of the main joint face toward the fringes (Figure 10.10). They are physically composed of a series of tiny ridges and troughs, a microtopographic relief on the main joint face (Roberts, 1961). The ridges and troughs, and thus the plumose patterns themselves, are best seen under the low-incidence lighting conditions that occur in early morning or late afternoon. The central axis from which plumes radiate is commonly aligned parallel to the trace of bedding or rock layering.

Figure 10.10 Nearly complete joint face exposed in handspecimen of fine-grained quartzite. Plumose markings splay symmetrically from central axis. (Photograph by G. Kew.)

Normal plumose markings diverge sharply at angles of 30° to 35° from the central axis, gradually curving to angles of about 75° near the edge of each fringe. The V-shaped plumose markings on adjacent rock faces that meet at a common joint are virtually identical. The ridges and troughs on one face nestle perfectly into the troughs and ridges on the adjacent face. The perfect

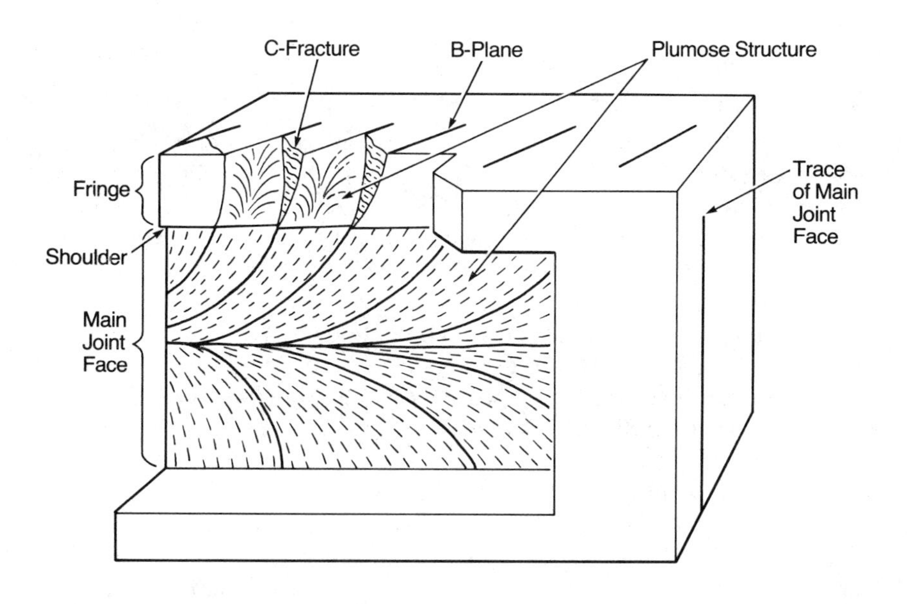

Figure 10.11 Schematic diagram showing the ideal attributes of a joint face. [From Hodgson (1961), *American Journal of Science*, v. 259.]

fit indicates that minuscule movements are sufficient to create plumose markings.

Experimental work has revealed that the presence of plumose markings signals a fracturing achieved by a rapid, near-instantaneous, snapping apart of the rock, almost in an explosive way. Syme Gash (1971) described plumose patterns as characteristic of high-energy, fast-running fractures. In fact, he pointed out that the ridges and troughs are probably the expressions of systematic microfracturing created by stress waves advancing in front of a fast-moving main fracture. The direction of propagation of the opening of individual joints is opposite to the direction in which the plumes "V." The plumose markings on a given joint "V" or converge toward the point in the rock where the fracture originated (Secor, 1965).

An ideal exposure of an ideal plumose joint consists of a smooth, planar, **main joint face** bordered by more roughly hewn **fringes** (Figure 10.11) (Hodgson, 1961). The fringes project outward from the main joint face by some small amount.

Fringes of joints are the outermost margins of a given joint surface. They are terminations of the main joint face, where the energy required for propagating the joint dissipates. Where well developed and unweathered, the fringes display a serrated appearance produced by two interfering, very closely spaced fracture sets (Roberts, 1961). One set, made up of **border planes**, forms an en echelon alignment of short, closely spaced fractures. These too may display plumose markings. The border planes (B planes) are connected to one another by curved to rough **cross planes** (C planes) that strike at right angles to the border planes. The border planes intersect the main joint face at angles of 20° to 25° (Figure 10.11).

JOINT-RELATED STRUCTURES

VEINS

Joints commonly become sites where minerals are precipitated in the form of **veins** (Figure 10.12). Precipitation of minerals in joints can be thought of as a **healing** of fractures. Mineralizing solutions invade rock bodies along joints, precipitating crystals from solution when the chemical conditions are right. Precipitation is triggered by favorable temperature and/or pressure conditions, and by the mixing of different fluids that happen to meet at the

Figure 10.12 Handspecimen of calcite-filled fractures in siltstone. (Photograph by G. Kew.)

Figure 10.13 Crustification displayed in a rock specimen from the Commonwealth Mine, Pearce, Arizona. Top and bottom margins of specimen are remnants of the wall rock for this crustified vein. Vein itself contains symmetrical bands of mineralization. Central band (white) is lacy agate bordered outward by bands of amethyst quartz (gray).

intersection of joint-controlled channelways. Quartz and calcite veins are the most common, but precious-metal and base-metal veins are the most attractive.

Veins form along joints through replacement and/or open-space filling. Most veins show conspicuous evidence of open-space filling. Dilational opening of wall rock creates offset of preexisting structures, like bedding. The actual amount of offset and the sense of offset depend on the thickness of the vein and the configuration of the vein and the offset structures (see Figure 4.18). **Replacement veins** are deposited through selective replacement of wall rock adjacent to joints. No dilational opening is required, and thus preexisting structures in the wall rock, like bedding, are not offset.

Evidence for open-space filling is preserved in **crustification** and **cockscomb textures**. Crustification is marked by the presence of distinctive, symmetrically arranged mineral bands deposited in joint spaces parallel to the surfaces of the wall rocks that are separated (Figure 10.13). The mineral bands are distinctive because of differences in color, texture, and mineralogy. In each crustified vein, the outermost symmetrically paired bands were the first to form; and the innermost compositional band in the middle of the vein represents the final product of precipitation. Crustification testifies to sustained circulation of fluids through joints. The mineral bands record the changing temperature/pressure and chemical conditions. Cockscomb texture provides a stop-action glimpse of the crustification process (Figure 10.13). The "cockscomb" appearance is produced by angular faces of crystals that project into yet-unfilled open space in the interior of the joint opening. The crystals usually grow at right angles to the wall rock surfaces.

Some vein-filled joints display **crystal fiber growths** (Figure 10.14). As the term implies, crystal fiber growths are marked by elongate, fiberlike or needlelike aggregates of fine-grained minerals, commonly quartz and calcite (Durney and Ramsay, 1973). The fibers may be straight or systematically curved. Where straight, they are not necessarily perpendicular to wall rock surfaces.

Durney and Ramsay (1973) have shown that the formation of crystal fibers in veins is due to joint dilation accompanied by simultaneous vein filling, without the development of open space (Figure 10.15). Crystals are precipitated during step-by-step **infinitesimal** dilations of the joint. The final product, the **crystal fiber vein**, reflects the total **finite strain** accomplished by the

Figure 10.14 Crystal fiber vein cutting melange at Cowhead Point, San Juan Islands, Washington. Note that the crystal fibers are not perfectly perpendicular to the walls of the vein. (Photograph by G. H. Davis.)

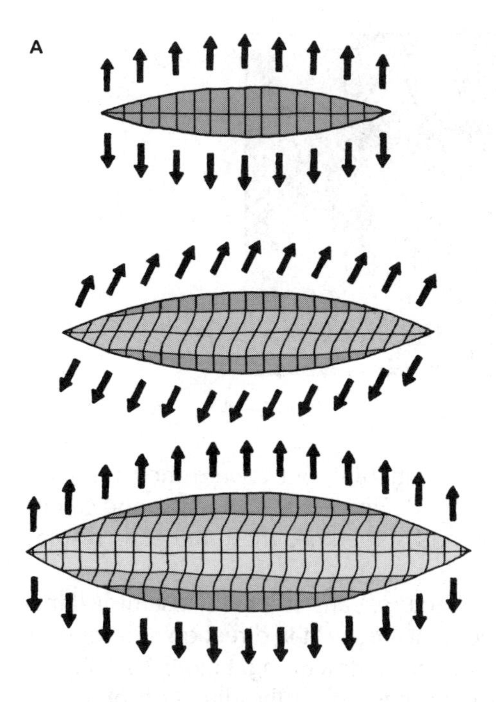

Figure 10.15 (*A*) Schematic rendering of the development of curved and kinky crystal fiber veins as a response to changes in direction of vein opening. (*B*) Curved crystal fiber vein of quartz and adularia, developed in sandy shale in Miller Mountain, Arkansas. (Photograph by A. E. J. Engel. Courtesy of United States Geological Survey.)

dilation. In ordinary veins the crystals grow perpendicular to the walls of the joint. But crystal fibers display orientations that are independent of the attitude of the vein walls. Instead, the growth of crystal fibers is ultrasensitive to changes in the direction of progressive opening of the vein. During each incremental step, the crystal fibers grow in the direction of differential displacement, which corresponds to the direction of maximum principal extension (Figure 10.15*A*). With changing directions of differential displacement, the crystal fibers take on curved forms. Changes in the direction of differential displacement can often be viewed as the natural result of progressive, rotational deformation during movements involving simple shear.

En echelon gash vein sets are marked by three or more relatively, short, sometimes stubby veins that are parallel and overlapping, and arranged in a line (the **line of bearing**) (Figure 10.16). Individual veins typically trend at 45° or less to the line of bearing of the set as a whole. Like gashes or wounds,

Figure 10.16 En echelon gash veins of quartz. (Photograph by G. H. Davis.)

Figure 10.17 Sigmoidal quartz veins. (Reprinted with permission from *Journal of Structural Geology*, v. 2, J. G. Ramsay, "Shear Zone Geometry: A Review." Pergamon Press, Ltd., Oxford, copyright ©1980.)

the veins are thickest at their centers, pinching out toward their margins. Some are **sigmoidal**, gently doubly curved (Figure 10.17). Quartz and calcite are the most common vein fillings, and sometimes these minerals display crystal fiber growth patterns.

En echelon gash veins (and gash fractures) are especially useful in kinematic analysis because their arrangements record the direction and sense of the simple shear that was responsible for the fracturing (Figure 10.18). The line of bearing of gash veins is usually parallel to the direction of simple shear. Individual fractures form as tensional openings oriented at 45° to the direction of simple shear, and perpendicular to the direction of maximum extension (Figure 10.18*A*). As they form, the fractures may be filled as veins. Sense of simple shear is recorded by the polarity of the intersection of any one vein and the line of bearing.

If the magnitude of simple shear is very small, gash veins will remain at 45° to the line of bearing. But if simple shear movements continue during a progressive deformation while the shear zone expands, the early-formed veins will rotate in accord with the principles of simple shear rotational deformation (Figure 10.18*B*). Sense of rotation is clockwise in the case of right-handed simple shear, and counterclockwise in left-handed simple shear. The rotation causes early-formed veins to be distorted by end-on shortening. In response, the veins pucker into their characteristic gashlike forms. During each incremental step of the progressive deformation, new veins may form, some of which propagate from the tips of the first-formed veins. Each *newly formed* fracture or fracture-filled vein is initially aligned at 45° to the line of bearing. But these too will rotate if deformation by simple shear continues (Figure 10.18*C*). The progressive deformation produces the sigmoidal forms so typical of gash veins and gash fractures. And if the veins are marked by crystal fiber growth, the individual crystal fibers will show curved, sigmoidal patterns as well. Rather than giving a single stop-action view of deformation, en echelon gash fracture and gash veins provide a video replay of the sequential steps in their formation.

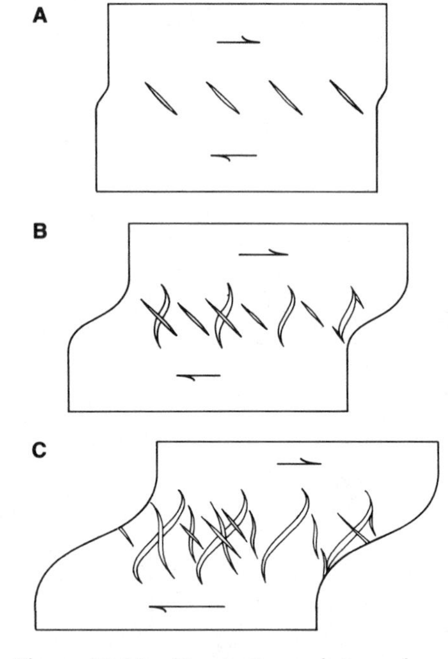

Figure 10.18 Kinematic evolution of en echelon fractures and gash veins within a zone of simple shear. (*A*) Tension fractures initially are oriented at 45° to the walls of the shear zone. (*B*) Continued simple shear results in rotation of early-formed fractures and/or veins. (*C*) Rotation of fractures and/or veins results in shortening and puckering. [From Durney and Ramsay (1973). Published with permission of John Wiley & Sons, Inc., New York, copyright ©1973.]

STRIATED SURFACES

Some fracture surfaces are like joints in that they show no discernible movement, but are like faults in that they are marked by striationlike lineations. It is convenient to refer to these as **striated surfaces**. The striated appearance can be produced by an actual mechanical abrasion due to frictional sliding, by the growth of crystal fibers during selective crystallization, or by some combination of these processes. Durney and Ramsay (1973) have explained that crystal fibers on fracture surfaces are precipitated on the leeward side of tiny ridges or bumps (Figure 10.19). The microtopography in combination with shear tends to produce open space, in the same way that open space tends to form along releasing bends of curved faults. Narrow veins of crystal fibers grow on fracture surfaces in such a way that individual crystal fibers trend parallel to the direction of differential displacement (Figure 10.19*B*); the sense of shear is disclosed by the plunge of the crystal fibers relative to the fracture surface; and the length of individual fibers records the magnitude of slip accommodated by the fracture surface during the time of crystal fiber growth.

Striated surfaces, whatever their physical nature, commonly form on fractures parallel to bedding, or on bedding surfaces proper. Striated surfaces also form on joints that discordantly cut bedding. The obvious kinematic value of striated surfaces is that they define the directions of local movements within deformed rock bodies. As will become clear in the next chapter, movements revealed by the directional properties of striated surfaces are often closely coordinated with the geometry and kinematics of folding.

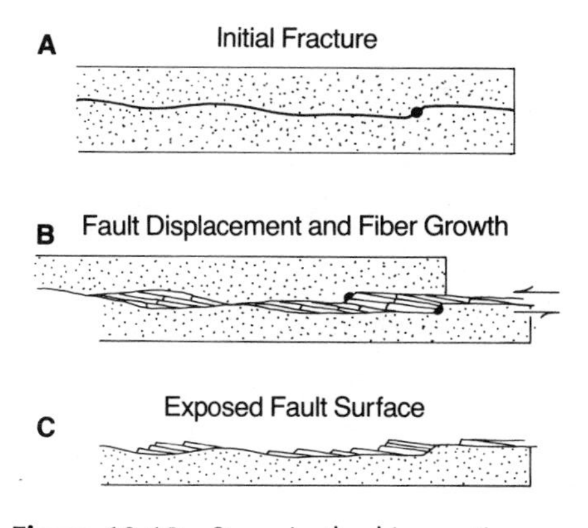

Figure 10.19 Steps in the kinematic evolution of crystal fiber lineation on a "striated" surface. (*A*) Formation of fault surface. (*B*) Fault displacement and simultaneous growth of crystal fibers in the direction of least stress. (*C*) Striated surface as exposed by weathering and erosion. [From Durney and Ramsay (1973). Published with permission of John Wiley and Sons, Inc., New York, copyright ©1973.]

STYLOLITIC JOINTS

Stylolitic joints are surfaces along which rock has been removed by pressure-induced chemical dissolution, a process generally referred to as **pressure solution** (Stockdale, 1922, 1926; Droxler and Schaer, 1979). Limestones and sandstones are especially sensitive to pressure solution. By accommodating the shortening of rock through volume loss, stylolitic surfaces perform the opposite function of gash fractures or veins (Fletcher and Pollard, 1981). Stylolitic surfaces typically form along bedding planes and preexisting joints, discontinuities that offer channelways for solution transport.

As seen in outcrop, the traces of stylolitic joints often look like tiny brain sutures, classical **stylolites** (Figure 10.20). The sutures evolve as adjacent rock walls interpenetrate one another. Teeth that form along stylolites during the dissolution process are conical or columnar, usually less than 5 or 10 mm long. The larger teeth tend to be flat-topped columns, whereas the smallest teeth are conical. **Cones** and **columns** on one side of a stylolite fit perfectly into the array of cones and columns on the opposite wall. The teeth are not necessarily perpendicular to the face of the stylolitic joint on which they occur (Choukroune, 1969). As stylolitic joints are formed, the host rock

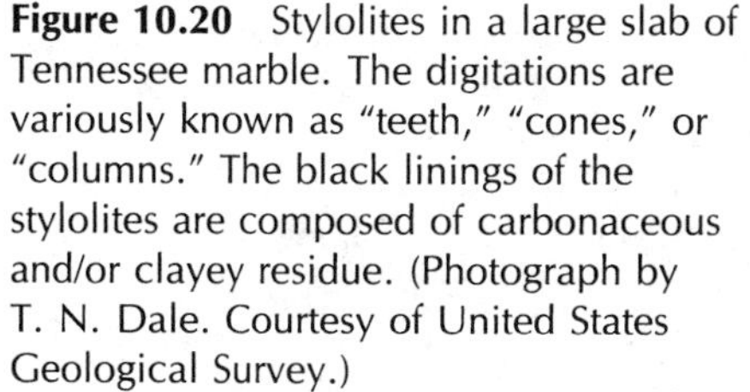

Figure 10.20 Stylolites in a large slab of Tennessee marble. The digitations are variously known as "teeth," "cones," or "columns." The black linings of the stylolites are composed of carbonaceous and/or clayey residue. (Photograph by T. N. Dale. Courtesy of United States Geological Survey.)

progressively dissolves in the direction of greatest principal stress (Blake and Roy, 1949; Geiser and Sansone, 1981; Arthaud and Mattauer, 1969). Given this sensitivity to dynamic conditions, the teeth of sutured stylolites become preferentially aligned in the direction of greatest principal stress, regardless of the overall orientation of the stylolitic joint.

Stylolitic joints are typically marked by the presence of very thin seams of clayey or carbonaceous material, commonly only a fraction of a millimeter thick. The seams occur right along the sutured boundaries. **Clayey** and **carbonaceous seams** are composed of the insoluble residue that was not flushed out of the host rock in solution (Heald, 1955). The amount of insoluble residue that is lodged along each stylolitic joint varies according to the bulk composition of the host rock and the degree of dissolution.

There is no question that rock volume is lost along stylolitic joints. Fossils in contact with stylolitic joints commonly are incomplete, their flanks and edges removed by dissolution. If the incompleteness of fossils along stylolitic joints were due simply to truncation and offset along joints or microfaults, the missing parts would be evident in adjacent wallrock. But they are not to be found. Where fossils record the extent of dissolution, it sometimes can be shown that the height of stylolite teeth and/or cones is approximately equal to the thickness of rock removed by pressure solution (Fletcher and Pollard, 1981).

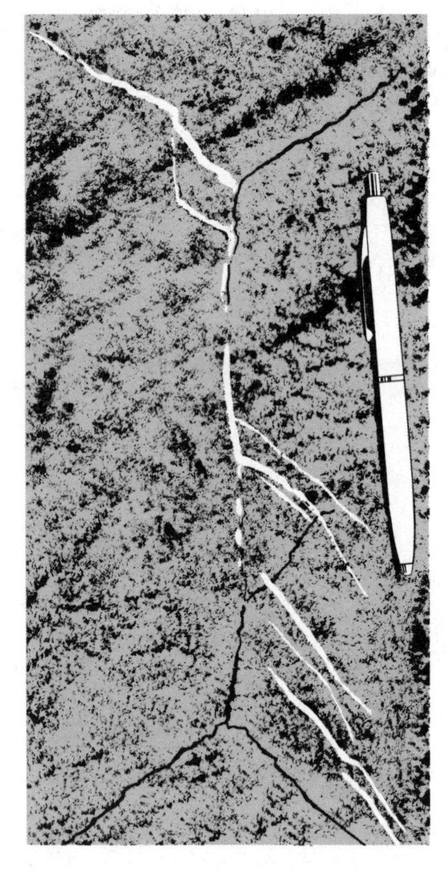

Figure 10.21 Remarkable display of the perfect coordination that can exist among joints and joint-related structures. Microfault connects an antisymmetric coupling of stylolitic joints and veins. This structure was discovered by René Rispoli (see Rispoli, 1981). [Tracing of a photograph by Groshong (1981). Published with permission of Geological Society of America and the author.]

Shortening by dissolution along stylolitic joints can produce faultlike offsets of preexisting features. Dissolution and removal of vein and host rock material adjacent to each joint leads to displacements that can easily be misinterpreted as products of ordinary faulting.

Stylolitic suture patterns may be completely removed during prolonged pressure solution, leaving only smooth, jointlike discontinuities lined with residue. Such solution surfaces may still be called stylolitic, provided it can be shown with reasonable assurance that wall rock has been lost to dissolution (Groshong, 1975b). In some strongly folded rocks, the solution surfaces are so closely spaced and parallel that they actually comprise a **cleavage**. (see Chapter 11).

In places it can be shown, even at the outcrop scale, that volume loss due to pressure solution is compensated by volume gain due to mineral precipitation in veins. Host rock that is dissolved along stylolitic joints is ultimately redeposited in **sinks** characterized by available open space and a chemical environment favorable to precipitation. Some of the removed material is clearly redeposited in fractures as veins. For example, Rispoli (1981) described a **volume-constant deformation** achieved by the formation of mutually perpendicular, mutually bisecting veins and stylolitic joints. And Fletcher and Pollard (1981) have called attention to **fracture couplings** that feature beautifully coordinated, **antisymmetric patterns** of stylolitic joints, veins, and faults. Groshong (1981) photographed an especially fine example in Rene Rispoli's map area (Figure 10.21). It features a left-slip fault that serves as a kind of transform fault connecting zones of volume loss (the stylolitic joints) and volume gain (the veins). The kinematic displacements balance one another: net slip on the fault, the maximum aperture of each vein, and the maximum interpenetrations along each stylolitic joint are all the same. Fracture couplings underscore the fundamental role of jointing— to avoid unnecessary and unsightly overlaps and gaps.

REGIONAL ASSOCIATIONS

Joints and joint-related structures are developed to different degrees in different regional geologic environments. Flat-lying sedimentary rocks in cratonic and foreland terranes are marked by through-going systematic sets of joints that display conspicuous preferred trends over large areas. Plumose joints are very common, but as in any terrane most joints are featureless. Striated surfaces, stylolitic joints, and veins tend to be scarce in comparison to their number in regions that have suffered greater deformation.

Jointing is likely to be more highly developed in miogeoclinal and foreland basin terranes where thick sequences of sedimentary rocks have been deformed by folding and thrusting. Through-going systematic joints, including plumose joints, are abundant and conspicuous, but in addition there is greater nonsystematic jointing. Striated surfaces, veins, and stylolitic joints are often abundantly developed in and around fold structures and fault zones.

Plutons and volcanic flows, so abundant in arc terranes, are commonly very strongly jointed. Certain plutons, in fact, become shattered in their late stages of emplacement, especially where abundant volatiles (notably water) enhance the ease of fracturing. Veins and alteration selvages are especially common along joints in igneous rocks. The expression of jointing in plutons and volcanic flows ranges from simple, systematic patterns to complicated patterns of interfering systematic and nonsystematic jointing. Some joints in crystalline rocks, like columnar joints in basalt and cross joints in granitic plutons, can be recognized as primary structures. But in addition to primary fracturing there is almost always a secondary jointing related to younger, regional deformation.

Joint systems in metamorphic terranes are generally dominated by fractures formed in the late stages of metamorphism. These are commonly distinguished by abundant veining. Both the veining and the close-spaced nature of joints in metamorphic rocks are favored by the availability of fluids driven off by the heat and pressure of metamorphism. Rocks in metamorphic terranes seldom escape pounding by superposed younger deformations. As a result, joint systems in metamorphic basement terranes tend to be very complicated.

METHODS OF STUDY

ESTABLISHING STRUCTURAL DOMAINS

Joints do not exist in isolation. Rather they are members of enormously large families of fractures with literally millions of members. Every regional rock assemblage, like a granitic batholith or a miogeoclinal prism of sedimentary rock, is pervaded by jointing. It is impossible to try to explain the origin of every joint in an outcrop, let alone every joint within a regional rock assemblage. Instead, we try to explain the origin of dominant **sets** of joints that can be identified through statistical analysis of the orientations and physical properties of joints within a given system.

To begin to discover order among millions upon millions of joints, it is essential to subdivide regional rock assemblages into **structural domains**, each of which may be thought of as containing its own **joint system**. Strictly speaking, two or more **joint sets** comprise a joint system. But I find it preferable to view the term "joint system" as the entire family of joints within a given structural domain, regardless of whether all the individual fracture surfaces can be neatly packaged into sets of like orientation.

Structural domains are designated on the basis of geographic boundaries, lithologic contacts, structural subdivisions, ages of rock formations, and combinations of these and other factors. The criteria vary according to the scope of the investigation. Nickelsen and Hough (1967), analyzing jointing in the Appalachian Plateau region of Pennsylvania and New York, subdivided domains on the basis of lithology: one domain was restricted to all joints measured in coal; another to all joints measured in sandstone. Rehrig and Heidrick (1972) analyzed joints in Laramide plutons in southern Arizona, establishing domains of mineralized versus unmineralized plutons. They wanted to assess the degree to which copper mineralization was associated with preferred fracture orientations. Later Rehrig and Heidrick compared joint patterns in Laramide versus mid-Tertiary plutons, attempting to recognize differences in fracture directions that might disclose differences in the regional tectonic stress patterns that existed in Laramide versus mid-Tertiary time (Rehrig and Heidrick, 1976). To evaluate the influence of joint orientation(s) on the trends of tunnels and shapes of rooms in the Carlsbad Caverns, Jagnow (1979) subdivided the caverns into domains on the basis of cave anatomy.

Once structural domains have been assigned, the work begins. The substance of detailed joint anaysis lies in evaluating the geologic characteristics of the joint system that occupies each structural domain. There is no single conventional procedure for doing this, even though efforts have been made to standardize nomenclature and methods (Commission on Standardization of Laboratory and Field Tests on Rock, 1978; McEwen, 1980). Most studies are based on structural analysis of joints at selected **stations** within each structural domain. Less commonly, and usually for very specific applied purposes, **fracture-pattern maps** are generated.

MAPPING OF JOINTS

In regions where joints are distinctly expressed in the weathered landscape, the mapping of joints is best achieved photogeologically. Aerial images often display remarkable portrayals of jointing (Figure 10.22). With patience and care, the intricacies of joint patterns can be accurately reproduced in inked tracings (Figure 10.23). A masterful example of regional photogeologic mapping of jointing is found in Kelley's (1955) fracture-pattern map of the Colorado Plateau. Conventional large-scale aerial photography (either black and white or color) serves nicely as a base for most studies. But for regional analysis of joint systems, it is advantageous to use some of the extraordinary space-satellite imagery that is available. Advances in remote-sensing techniques that have accompanied the proliferation of small-scale imagery make photogeologic analysis of jointing even more attractive.

There are shortcomings associated with using the photogeologic approach, exclusively, in the mapping of joints. Low-dipping joints are automatically screened from view, thus biasing the data toward moderate to steep-dipping fractures. Furthermore, it is impossible to measure from photographs the dip and dip direction of joints and to gather information regarding the types of joint and joint-related structure that exist in each structural domain.

Where jointing is not well expressed in the landscape, or is too fine to be resolved on aerial images, the "mapping" of joints is reduced to an exercise in systematically measuring the orientations of representative joints in the field. Under these circumstances, aerial photos do not afford any special leverage in analysis, and thus either a topographic map or an aerial photograph can be used as a base map. No attempt is made to show the actual physical traces of joints, for they are generally far too numerous and much too short to portray at reasonable map scales. Instead, the orientations of the dominant systematic sets of joints are portrayed through standard joint sym-

Figure 10.22 Aerial photograph of jointing at Arches National Park, Utah. The conspicuous, through-going joints cut the Moab Member of the Entrada Sandstone (Jurassic). Spacing of joints is 25 to 40 m. (Photograph by R. Dyer.)

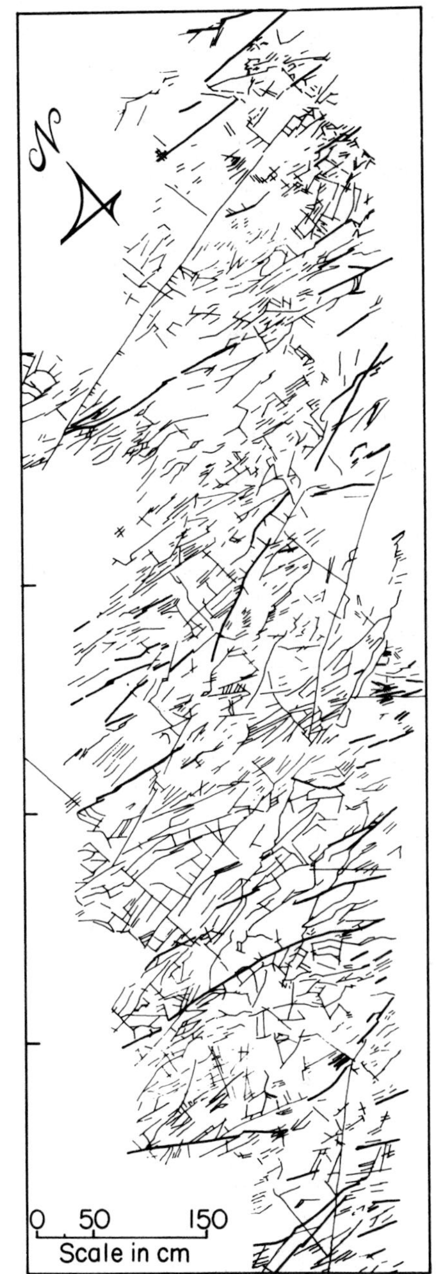

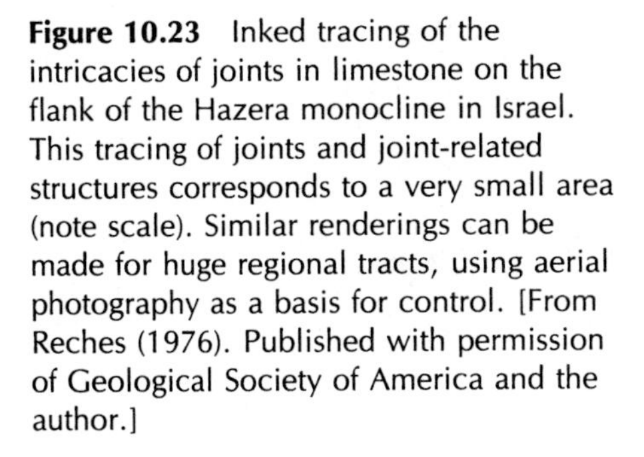

Figure 10.23 Inked tracing of the intricacies of joints in limestone on the flank of the Hazera monocline in Israel. This tracing of joints and joint-related structures corresponds to a very small area (note scale). Similar renderings can be made for huge regional tracts, using aerial photography as a basis for control. [From Reches (1976). Published with permission of Geological Society of America and the author.]

bology, plotted as close as possible to the locations where the joints are measured. The symbols are drafted in dense overlapping clusters that literally fill the map to overflowing. Under ideal circumstances such fracture-pattern maps show at a glance the chief types of joint and their respective orientations. Under less favorable circumstances, the maps are a collage of confusion, doing little to clarify the basic elements of the fracture pattern.

It is common practice to measure eye-catching joints and joint-related structures as a part of the normal geologic mapping process. Representative structures are plotted according to type and orientation. The few and scattered joint data that are typically posted on geologic maps do not constitute a basis for structural analysis. Instead, they simply provide a preliminary forecast of the dominant joint sets that might exist in the area of investigation.

MAP SYMBOLOGY

The map symbology that I find useful in representing joint data on fracture-pattern maps and normal geologic maps is presented in Table 10.1. The joints and joint-related structures that are distinguished by the symbols include ordinary featureless joints, plumose joints, veins, crystal fiber veins, en echelon veins and fractures, striated surfaces, and stylolitic joints. The symbol for plumose joints includes portrayal of the direction of convergence of the feather markings. The parallel-line strike symbol for ordinary veins can be color coded to distinguish veins according to mineralogy and/or alteration assemblages. The symbol for crystal fiber veins allows the trend of the fibers and their curvature, if any, to be shown. The symbol for en echelon veins and fractures includes a portrayal of the trend of the line of bearing. Symbols for striated surfaces distinguish whether the lineation is a crystal fiber linea-

Table 10.1
Map Symbology for Joints and Joint-Related Structures

76	Ordinary Featureless Joint
85	Plumose Joint w/Direction of Convergence of Plumes
(3) 75	Vein, w/Mineralogy Color-Coded & Aperture Noted, in cm.
80 60	Crystal Fiber Vein w/Orientation of Fibers
76	En Echelon Joints, Line of Bearing Dashed
56	Striated Surface
72	Striated Surface w/Crystal Fiber Lineation
84 60	Stylolitic Joint, w/Trend & Plunge of Teeth

tion or a scratching produced by frictional sliding. And finally the symbol for stylolitic joints allows the trend and plunge of the stylolite teeth to be shown. The symbols, taken as a whole, convey significant geometric and kinematic information.

CHOOSING STATIONS FOR STRUCTURAL ANALYSES

It is impossible to examine all joints and joint-related structures that are contained in a given structural domain. Thus, standard practice is to evaluate jointing through detailed structural analysis at selected **stations**. The strategy is to learn the nature of the overall joint system through systematic examination of representative **subareas** within the domain.

A ***sampling station*** established for joint analysis is very small compared to the size of the structural domain within which it lies. It is a site of well-exposed jointed bedrock where joints and joint-related structures are classified and measured. In most studies the stations are simply outcrop areas, of varying size and shape. In a restricted sense, stations can be designated as circular or square **inventory areas** of specified dimension, or as relatively short **sample lines** of specified traverse length and direction.

MEASURING JOINT ORIENTATIONS

In practice, there are two basic approaches that are used in collecting orientation data at sample stations. One, which we might call the **selection method**, involves selecting only certain joints for measurement and study. The second approach, which we might call the **inventory method**, requires measuring and classifying every single joint at a station site.

The basis of the selection method is to restrict analysis to joints and joint-related structures that are continuous and through-going, and are conspicuously associated with other fractures of similar appearance and orientation. It is neither very easy nor very objective to decide which structures should be measured at each station, and which should be left alone. But in spite of this difficulty, many workers have found the selection method to be practical and useful. Parker (1942) restricted his joint-orientation measurements in the Appalachian Plateau region of New York to those fractures that appeared straight and continuous. Hodgson (1961), in analyzing the fracture pattern of part of the Colorado Plateau, measured only smooth planar continuous joints, especiallly those that displayed plumose markings. Rehrig and Heidrick (1972, 1976) measured joints and veins that occurred in sets of three or more, ignoring all others.

In contrast to the selection method, the inventory method requires measuring *all* the joints and joint-related structures that occupy a sampling station. Measurements are taken within an inventory area or along a sample line. In sampling along a line it is common practice to stretch out a tape measure to some prescribed length (normally less than 20 m), and to classify and measure the orientation of every joint that is intercepted by the tape.

The inventory approach that I favor is the **circle-inventory method**. First a circle of some known and predetermined diameter (normally less than 3 m) is traced out on a bedrock surface of perfectly exposed jointed rock (Figure 10.24*A*). The circle is drawn with a piece of carpenter's chalk attached to a string of suitable length. Then the orientation and trace length of each fracture within the circle are measured. Data are recorded in the manner shown in Table 10.2. To avoid measuring the same joint twice, it is helpful to trace out the full length of each joint with chalk after it is measured (Figure 10.24*B*). Outcrops are generally available in which the three-dimensional expression of each joint surface is clear, allowing joint orientations to be

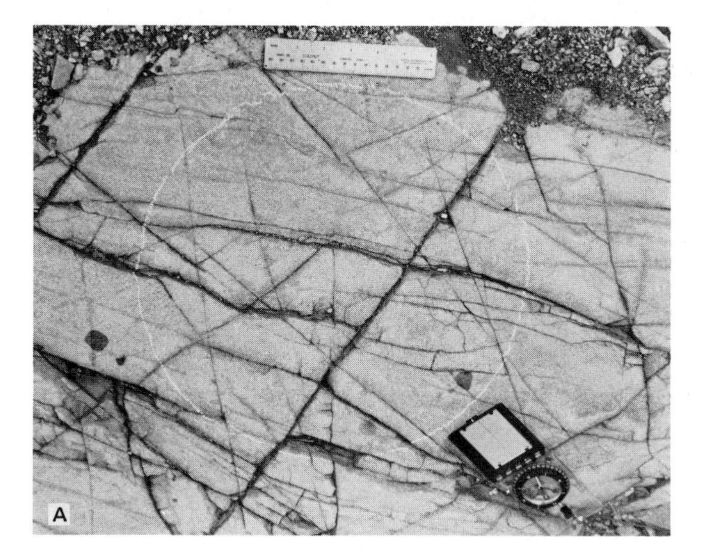

Figure 10.24 Circle-inventory method for sampling the orientations of joints and joint-related structures. (*A*) The circle, drawn on bedrock with chalk. (*B*) With joints measured and traced, Matt Davis and "Katie" relax on the outcrop. (Photographs by G. H. Davis.)

measured in terms of strike and dip. But where outcrops are so smooth and flat that only the straight-line traces of the joints are evident in the bedrock surface, trends alone can be measured.

The time and tedium of the circle-inventory operation depends largely on the size of the sampling circle and the abundance of the joints that must be measured. Some initial trial-and-error planning may permit selection of an optimum circle diameter. When analyzing joints in sedimentary and volcanic rocks, it useful to select a circle radius whose magnitude is some function of layer thickness. In fact, setting the radius equal to layer thickness works out reasonably well in thin–to medium-bedded rocks. However, if the host rock for the joints is very thick bedded or massive, for example 50 to 500 ft (15 to 150 m) or more, it becomes necessary to adapt the circle-inventory method to a photogeologic approach . . . unless there is plenty of chalk, string, and time available. The photogeologic adaptation requires drawing an inventory circle of appropriate diameter onto an aerial photograph, or a transparent overlay of the photograph, then measuring the trends and trace lengths of joints directly from the photograph. North arrow and scale provide the control. Subsequent field investigations can focus on measuring dip magnitudes of fracture sets and classifying the joints according to their physical and kinematic characteristics.

MEASURING JOINT DENSITY

The abundance of jointing at a given station is described through the evaluation of **joint density**. Joint density can be measured and described in a number of ways: average spacing of joints; number of joints in a given area; total cumulative length of joints in a specified area; surface area of all joints within a given volume of rock. The measure of joint density used in conjunction with the circle-inventory method is the summed length of all joints within an inventory circle, divided by the area of the circle:

$$\rho_j = \frac{L}{\pi r^2}$$

Where

$$\rho_j = \text{joint density},$$
$$L = \text{cumulative length of all joints, and}$$
$$r = \text{radius of inventory circle.}$$

Table 10.2
Circle-Inventory Method for Evaluating Fracture Density

Station #34, Tucson Mountains
Lower Cretaceous Sandstone
Radius of Inventory Circle = 19 cm; Area = 1134 cm^2

Trend of Fracture	*Length of Fracture (cm)*
N79E	35
N46W	36
N04W	30
N32E	25
NS	22
EW	20
N16E	20
N42W	39
N41W	13
N44W	17
N43W	11
N30W	28
N41E	18
N82W	13
N64W	13
NS	6
N12W	20
N23E	14
N11E	7
	387cm = Cumulative Fracture Length

$$\text{Fracture Density} = \frac{387\text{ cm}}{1134\text{ cm}^2} = \boxed{.34\text{ cm}^{-1}}$$

An example of the computation is shown in Table 10.2. The joint density is expressed in units of length/area, (e.g., ft/ft^2, cm/cm^2, m/m^2, km/km^2). In practice, the values of joint density are converted to the reciprocal form (ft^{-1}, cm^{-1}, m^{-1}, and km^{-1}). Thus a joint density of 0.57 ft/ft^2 would be expressed as 0.57 ft^{-1}.Comparative analysis of fracture density, however expressed, is especially useful in applied studies. Petroleum geologists Harris, Taylor, and Walper (1960) compared joint density to degree of curvature of the flanks of two oil-producing domes in Wyoming. They were interested in evaluating the relationship between fracture-induced permeability and degree of folding. As part of their study they found it necessary to "normalize" the natural variations in joint density that are due to differences in rock type and bedding thickness.

Haynes and Titley (1980) compared quantitative differences in fracture density in veined, mineralized rocks of the Sierrita porphyry copper deposit south of Tucson. Their purpose was to explore for centers of intrusion and/or mineralization, using joint-density variations as a guide. Wheeler and Dickson (1980) sought to evaluate whether systematic changes in joint density can disclose the locations of known or hidden faults. They prepared contour maps showing variations in joint density in a part of the Central Appalachians, and compared these maps with the known fault distribution as revealed on geologic maps of the same region.

RECORDING THE DATA

It is important to keep a systematic record of the descriptive characteristics of joints that are examined at each sampling station. Notebook entries should include site location, rock type, orientation of bedding or foliation (if present), bed or layer thickness (if applicable), type of joint or joint-related structure, orientation of joint or joint-related structure, and measurements pertinent to computing joint density. Block diagrams that schematically portray the array of joints and joint-related structures are also a good idea (Figure 10.25).

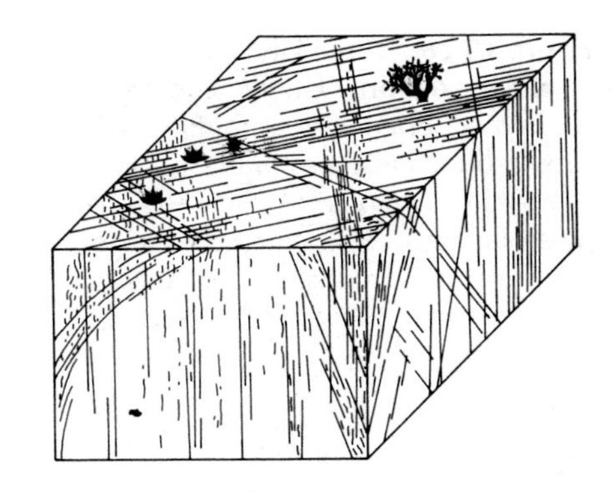

Figure 10.25 Block diagram portrayal of the nature and orientation of dikes, joints, and joint-related structures in Laramide granitic plutons in southern Arizona. (By permission, T. L. Heidrick and S. R. Titley, "Fracture and Dike Patterns in Laramide Plutons and Their Structural and Tectonic Implications," fig. 4.1, in *Advances in Geology of the Porphyry Copper Deposits: Southwestern North America*, S. R. Titley, editor. University of Arizona Press, Tucson, copyright ©1982.)

Systematically recorded joint data provide a basis for answering the typical questions that are raised in the structural analysis of joints. Are certain joints and joint-related structures associated with specific rock types? How does joint density vary according to lithology and layer thickness? Does joint density and/or orientation change according to structural location? Are rocks of different ages characterized by different joint patterns? Can favored directions of mineralization be recognized? The data that are collected and recorded, station by station, provide a playground for comparative analysis of a statistical nature.

The recording of data is time-consuming, especially when working alone. Working in pairs aids the flow of data from outcrop to notebook immeasurably. Rehrig and Heidrick (1972, 1976) accelerated their research even further by recording data directly on IBM mark-sense cards, which they carried with them in the field in plywood clipboards equipped with special holding compartments. Using #2 pencils, they entered the requisite physical and geometric information directly onto the cards, column by column, according to a prearranged coding system. The data, in this form, were transformed *directly* by computer into statistical plots, thus eliminating intermediate steps in data reduction.

PREPARING JOINT-ORIENTATION DIAGRAMS

Orientation data collected during the course of joint analysis may be summarized in **pole diagrams**, **pole-density diagrams**, **rose diagrams**, and **strike histograms**. Pole diagrams (and contoured pole-density diagrams) are three-dimensional stereographic displays of strike-and-dip data (Figure 10.26*A*). Where three-dimensional control on the attitude of joints is not attainable, either due to the nature of the bedrock surfaces or because the joint orientations are gathered from aerial photographs, it is appropriate to present the orientation data on rose diagrams or strike histograms, that is, on two-dimensional plots (Figure 10.26*B*, *C*).

In preparing rose diagrams and strike histograms, the trend and/or strike data are first organized into **class intervals** of 5° or 10°, encompassing the orientation range from west through north to east. The number and percentage of readings that fall within each class interval are then tallied. Data thus arranged are plotted in one of two ways. For rose diagrams, class intervals are distinguished by rays subtending arcs of 5° or 10° that extend outward from a common point (see Figure 10.26*B*). A family of concentric circles provides scaled control for the number (or percentage) of joint readings that occupy each class interval. For strike histograms, class intervals are plotted along the y axis of an x–y plot; numbers (or percentages) of readings are plotted along the x axis (see Figure 10.26*C*).

Joint-orientation diagrams allow sets of joints to be recognized. Preferred orientations of joints emerge from pole diagrams as dense clusterings of poles to joints (see Figure 10.26*A*). The orientation of the center of distribution represents the average orientation of the set. Joint sets of distinctive preferred orientation are more clearly recognized and documented using contoured pole-density diagrams. Trends of dominant joint sets are even more readily apparent in rose diagrams and strike histograms. Preferred orientations coincide with high-frequency peaks. Rehrig and Heidrick

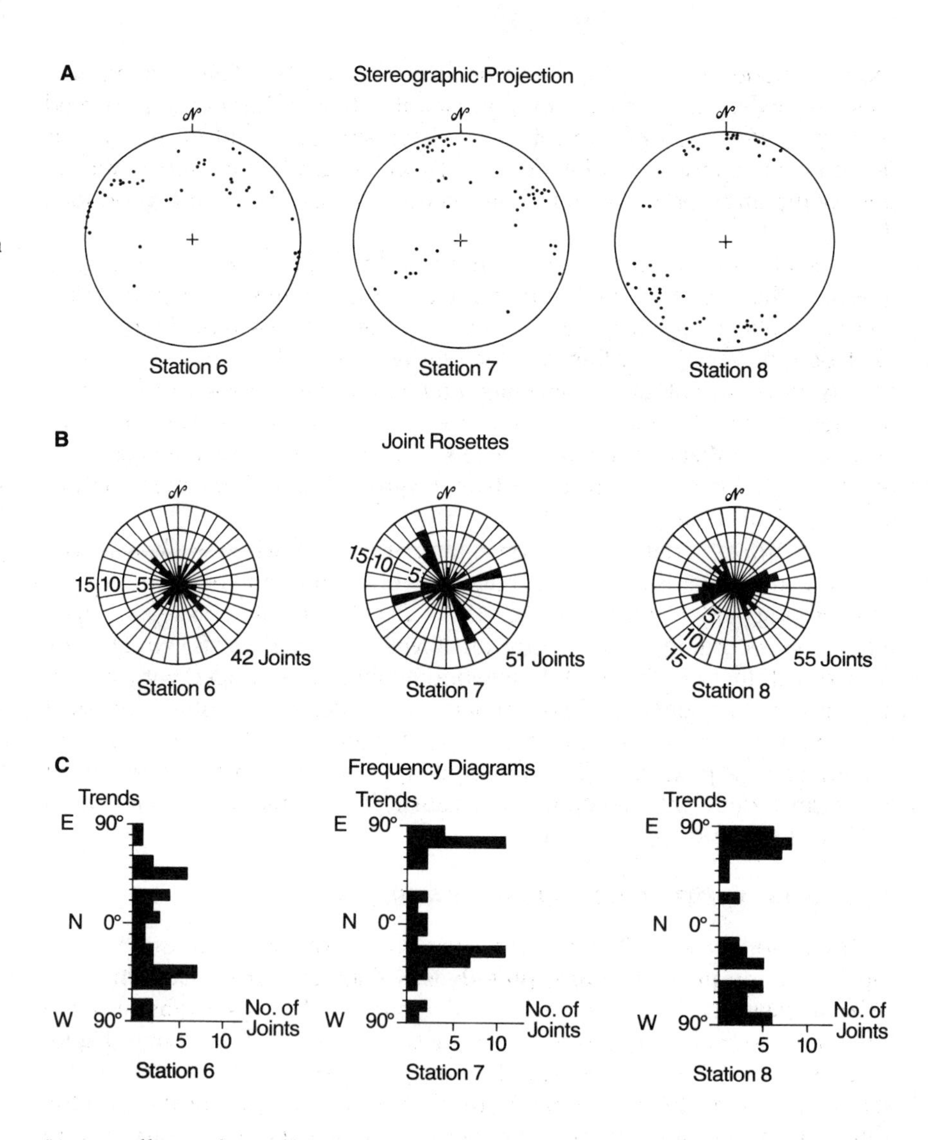

Figure 10.26 Three kinds of plots for displaying the orientations of joints: the data reflect orientations of joints in Cretaceous sandstones in the Tucson Mountains, Arizona. (*A*) Stereographic pole diagrams. (*B*) Joint rosettes (rose diagrams). (*C*) Frequency diagrams. (Data collected and plotted by R. Chavez.)

(1972) effectively used rose diagrams in presenting the orientations of steeply dipping Laramide dikes, veins, and elongate plutons and veins in the southern Basin and Range of Arizona and New Mexico. Where structures are not uniformly steeply inclined, rose diagram and strike histogram presentations can be ambiguous: a given peak can hide two sets of joints of distinctly different inclination and/or dip direction.

Preferred orientations of joint sets are commonly estimated by eye from joint-orientation diagrams. This is easy to do where preferred orientations are obvious (Figure 10.27), but is difficult and subjective where orientations are diffuse. Fortunately, joint data and joint-density diagrams lend themselves to rigorous statistical analysis. Joint enthusiasts are not without computer packages that integrate orientation analysis and statistical analysis of the data. An excellent example of the integration of orientation analysis and statistical analysis of structure data is revealed in an illustration published by Heidrick and Titley (1982) (Figure 10.28). Pole-density diagrams display the orientations of 12,412 mineralized joints, veins, dikes, and faults in Laramide plutons in the American Southwest. The data were subdivided according to three domains: productive plutons, ore-related plutons, and nonproductive plutons. The data in each diagram were contoured according to **statistical probability of random distribution**.

N
W E
125 100 75 50 25 0 25 50 75 100 125
1 Reading Per 1 cm Trace Length
• 387 cm Cumulative Fracture Length

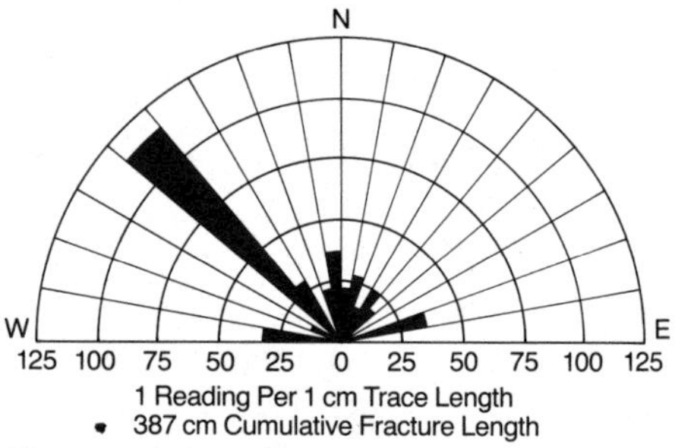

Figure 10.27 Rose diagram of joint orientation data presented in Table 10.2.

It is beneficial to distinguish the orientations of different types of joint and joint-related structure. Separate joint-orientation diagrams can be made for each class of structure. Alternatively, the orientations of the physically distinctive types of joint can be shown on a common diagram through the use of coded symbols. Depending on the purpose of study, individual joint-orientation diagrams can be prepared for each station, for a given array of stations, for an entire structural domain, and/or for an entire region.

Although joint-orientation diagrams can be presented as discrete illustrations in a technical report, it is more effective to post the diagrams on a structural geologic map of the region within which the joint analysis was carried out. The diagrams are positioned as close as possible to the locations of sampling stations. To provide greater visual impression of the dominant joint trends, **joint trajectories** can be constructed whose orientations are everywhere parallel to the average trends of high-angle joint sets recognized in each of the joint-orientation diagrams (Figure 10.29). The lines intersect at the locations of the control points, the sample stations. Interpretations of the trends of fracture trajectories *between* sites can be augmented by photogeologic analysis (see Hodgson, 1961).

EVALUATING STRAIN SIGNIFICANCE

ROLE OF JOINTING

Joints and joint-related structures exist because the rocks in which they are found were forced to undergo dilation, distortion, or both. Imagine some enormous regional rock assemblage, like a miogeoclinal prism of sedimentary strata, and picture its countless joints as tiny pervasive discontinuities that have permitted the mass to flow as a granular or blocky aggregate (Price, 1967). Within such a mass, individual joint-bounded blocks are free to move imperceptibly in relation to one another when forced to do so.

The formation of individual joints and joint-related structures achieves a specific kinematic function. In some cases the kinematic role of a given joint is obvious: veins accommodate positive dilation; stylolitic joints accommo-

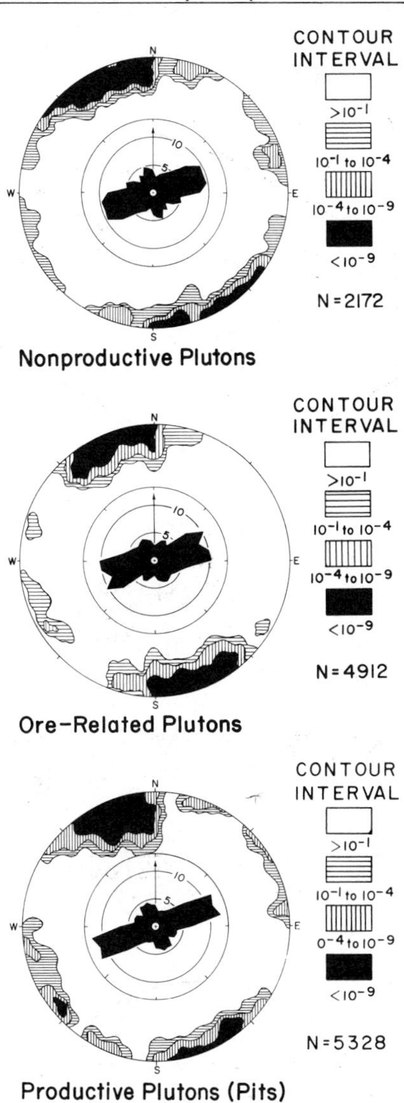

Figure 10.28 Pole-density diagrams showing the orientations of mineralized joints, veins, dikes, and faults in Laramide plutons in the Southwest. (By permission, T. L. Heidrick and S. R. Titley, "Fracture and Dike Patterns in Laramide Plutons and Their Structural Tectonic Implications," fig. 4.6 in *Advances in Geology of the Porphyry Copper Deposits: Southwestern North America,* S. R. Titley, editor. University of Arizona Press, Tucson, copyright ©1982.)

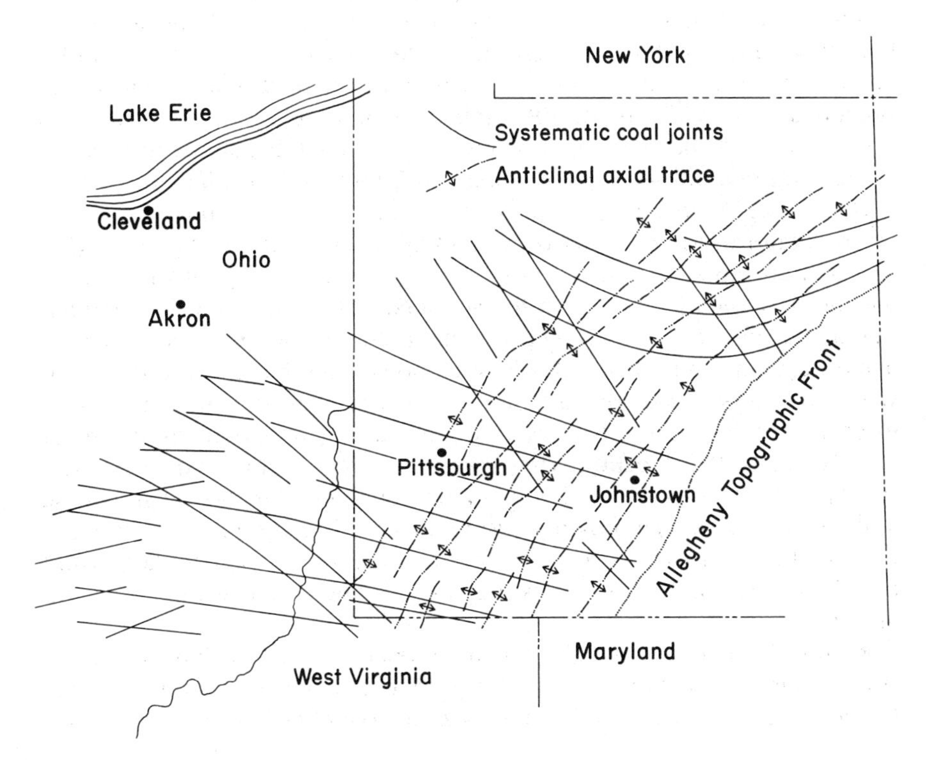

Figure 10.29 Use of joint trajectories in picturing the dominant joint orientations in coal in the Appalachian Plateau. [From Nickelsen and Hough (1967). Published with permission of Geological Society of America and the authors.]

date negative dilation; and striated surfaces accommodate translational slip. In most cases, however, the kinematic function of an individual joint is not at all clear. **Ordinary featureless joints** that are devoid of vein fillings, striations, and stylolitic solution surfaces hide their kinematic functions very effectively. And plumose structures, however distinctive they may be, do not have an agreed-on origin. Some geologists believe that plumose joints form as a response to shear; others believe that they reflect tensional separation. Only a small fraction of joints and joint-related structures in any study area can be classified convincingly according to the kinematic role(s) they have played.

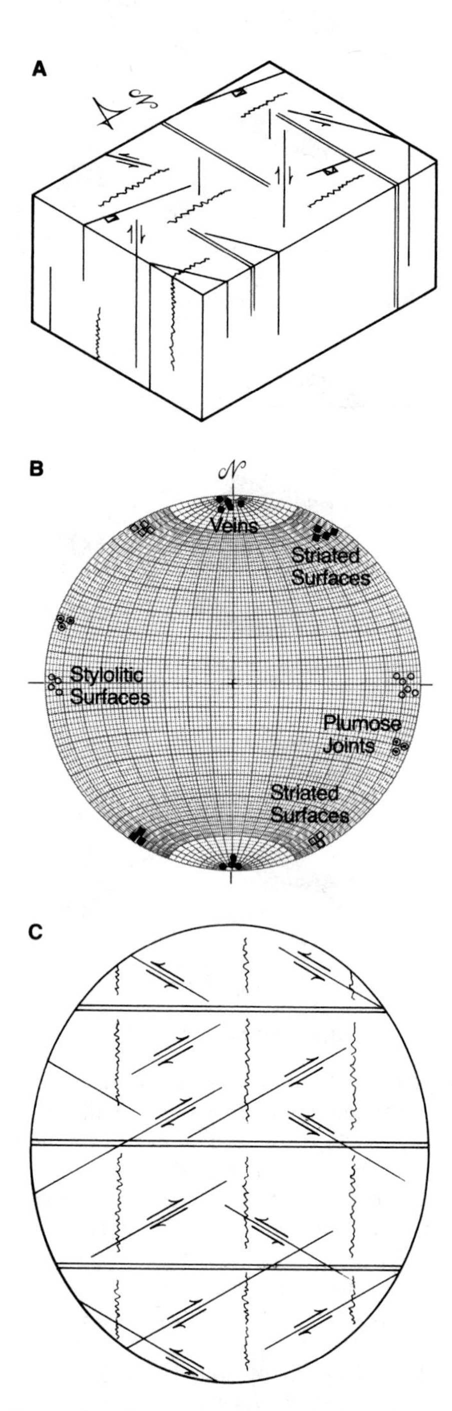

Figure 10.30 Steps in the strain analysis of joints and joint-related structures. (*A*) Block diagram showing the nature of jointing at an inventory station. (*B*) Stereographic representation of five sets of joints and joint-related structures. (*C*) Schematic rendering of the kinematic function(s) of four of the sets of joints and joint-related structures.

EXAMPLES OF STRAIN ANALYSIS OF JOINTS

Strain analysis of joints involves combining joints and joint-related structures into sets of *like* orientation and kinematic function. An attempt is made to identify sets of structures that display compatible geometric and kinematic relationships. The conceptual basis for this brand of fracture analysis has been discussed by Price (1967).

A simplified example of the recommended steps in strain analysis of joints and joint-related structures is pictured in Figure 10.30. Figure 10.30*A* shows a block diagram of fractures observed at an inventory station. The orientations of the structures measured at the inventory station are represented stereographically in Figure 10.30*B*. Five sets of steeply dipping joints can be distinguished on the basis of common orientation and physical properties: set 1, east–west striking veins; set 2, north–south-striking stylolitic surfaces; set 3, N60°W-striking striated surfaces with subhorizontal striations; set 4, N60°E-striking striated surfaces with subhorizontal striations; and set 5, N20°E-striking plumose joints. Some of the northwest-striking striated surfaces reveal tiny right-slip offsets; and some of the northeast-striking striated surfaces reveal tiny left-slip offsets.

A reasonable kinematic interpretation of the fracture sets, thus arranged, would proceed as follows: set 1 accommodated north–south stretching; set 2 accommodated east–west shortening; sets 3 and 4, together, accommodated east–west shortening and north–south extension. The kinematic function of set 5 is unknown. Given these interpretations, sets 1 to 4 can be combined into a common system based on the **compatibility of kinematic functions** of each (Figure 10.30C). Joints and joint-related structures of sets 1 to 4 appear to have accommodated an east–west crustal shortening accompanied by north–south stretching. Set 5 remains unexplained.

Under exceptional circumstances it is possible to interpret the actual magnitude of distortion caused by jointing. Lockwood and Moore (1979) summed up separations along conjugate strike–slip microfaults in a study area north of Mount Whitney in the Sierra Nevada, and then calculated the size and shape of the strain ellipse that would best describe the distortion achieved by the microfaulting. They demonstrated that a reference circle would be transformed into a strain ellipse characterized by a maximum extension of 2.3% in a N61°W direction, and a minimum extension of −2.3% in a N29°E direction. They went on to show that the axis of greatest principal strain (X) is oriented parallel to the inferred direction of stretching in the Basin and Range province just to the east. The presence of abundant microfaulted veins, schlieren, and inclusions made it possible for Lockwood and Moore to push their work beyond routine, qualitative, geometric analysis.

Reches (1976) quantitatively evaluated separation offsets along joints and striated surfaces in limestone and dolomite at a locality known as "the Gorge" on the southeast flank of the Hazera monocline in Israel. The station site was selected for its unusual bedding-plane exposures of anastomosing trails left by bottom-dwelling creatures back in Paleozoic time. The trails are displaced along joints that cut the bedrock. Like Lockwood and Moore, Reches concluded on the basis of his analysis that the strain accomplished by the jointing was mighty small: only −1.5% shortening.

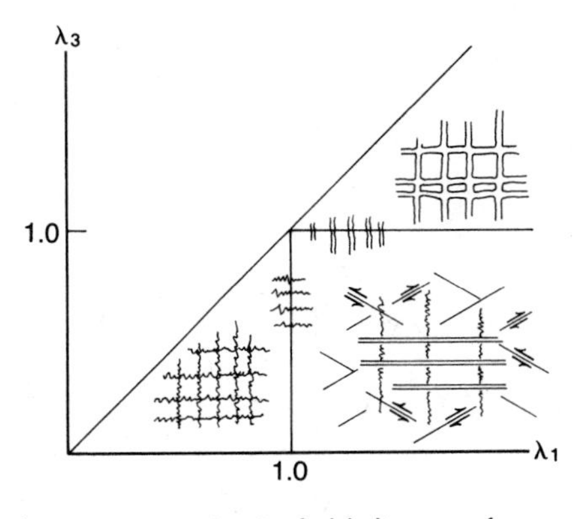

Figure 10.31 Strain-field diagram for joints and joint-related structures.

STRAIN-FIELD DIAGRAMS FOR JOINTING

Strain-field diagrams can be prepared to describe the strain functions of joints and joint-related structures. The ways in which the structures are combined in sets and systems determines how the rock bodies are distorted and/or dilated. The strain-field diagram in Figure 10.31 shows just some of the countless ways in which joints and joint-related structures can be combined. The field of expansion is illustrated by two sets of veins that intersect at high angles to one another. In contrast, the field of contraction is portrayed by two sets of mutually perpendicular stylolitic surfaces. The field of linear stretching is illustrated by a single set of subparallel veins; the field of linear shortening shows a single set of subparallel stylolitic surfaces. The field of compensation features four kinematically coordinated sets of joint structures that are able to accommodate simultaneous shortening and stretching: shortening along stylolitic surfaces is perpendicular to extension due to veining; and movements on the conjugate microfaults cause mutually perpendicular shortening and stretching.

The strain-field diagram presented in Figure 10.31 is a highly selective one, drawing together structures of clear-cut kinematic function and excluding ordinary featureless joints whose kinematic function is unknown. Unfortunately, ordinary featureless joints are the rule, not the exception.

ORIGIN OF JOINTS AND JOINT-RELATED STRUCTURES

PRIMARY JOINTING

Primary joints are related to the primary formation of the rocks in which they are found. They form in not-yet-consolidated sediments and in not-yet-congealed magma or lava. Many primary joints have kinematic functions that are clearly related to dynamic processes involved in the very formation of the rocks in which they are contained. For example, geometry and symmetry of columnar jointing in a volcanic flow reveal the intimate coordination that can exist between the properties of primary joints and the internal stresses that are generated during the formation of a rock body. Columnar joints are like the mud cracks created during the buildup of tensile stresses in a contracting layer of drying mud. The lava contracts as a response to loss of heat; the mud contracts as a response to loss of water.

DIASTROPHIC JOINTING

The most readily interpretable secondary joints are those that display a clear-cut *spatial*, *geometric*, and *kinematic* relation to local or regional structures, especially folds and faults. These are here called **diastrophic joints**. The term **diastrophism**, from which "diastrophic" is extracted, refers to conspicuous distortional deformation accompanying mountain building or regional crustal warping. Interpreting the dynamics of formation of diastrophic joints is often a byproduct of interpreting the structures with which the joints are associated. Examples of diastrophic joints and joint-related structures include en echelon veins in a shear zone, conjugate strike–slip

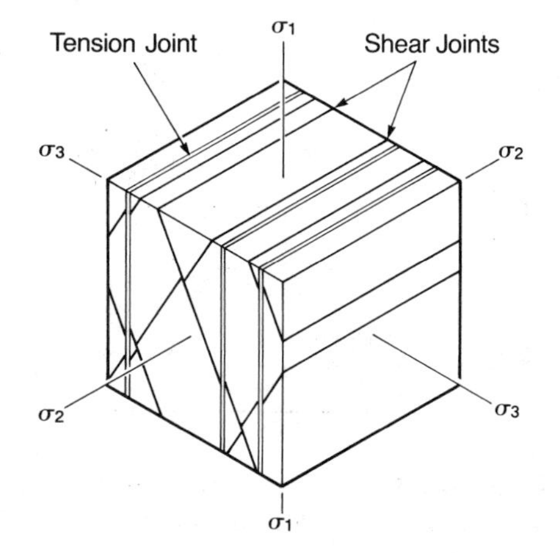

Figure 10.32 Relation of tension fractures and conjugate joints to the principal stress directions, assuming plane strain.

microfaults and striated surfaces in a fold/thrust belt, joints and veins oriented perpendicular to the trend of folds in a fold system, and stylolitic solution surfaces aligned perpendicular to the direction of flattening in folds.

The key to interpreting diastrophic joints lies in understanding the dynamics of faulting and folding. Just like faults, conjugate joints and conjugate striated surfaces can form under plane strain conditions of brittle failure in an environment of differential stress (Figure 10.32). Classical interpretation states that conjugate joints intersect in the axis of intermediate stress (σ_2), with the axis of greatest principal stress (σ_1) splitting the acute angle between the conjugate fractures. Striations, if present, form at the intersection of each joint surface and the σ_1/σ_3 plane. Joints interpreted in this manner, as if they were microfaults, are called **shear joints**. **Tension joints** may also form during brittle failure, breaking at right angles to the direction of least principal stress (σ_3).

In environments of three-dimensional strain, rocks tend to fracture in orthorhombic patterns composed of three or more joint sets (Reches, 1978a, 1983; Reches and Dieterich, 1983). As discussed in dynamic analysis of faulting (Chapter 9), the configuration of fractures formed in this manner reflects both the angle of internal friction (ϕ) of the rock body and the strain ratio of deformation. Multiple sets of fractures develop to accommodate deformation in a way that minimizes the dissipation of deformation and strain energy (Reches, 1978a).

Simple shear created by faulting along a discrete surface of shear, or within a zone of shear, can create local environments of simple shear rotational strain within which diastrophic joints can form. En echelon joints, conjugate joints, and/or tension joints and veins may develop in response to the simple shear (Figure 10.33*A*). Early-formed joints and joint-related structures rotate out of their original orientation(s) during progressive simple shear deformation (Figure 10.33*B*). At the instant of formation of joints in a shear zone, the principal axes of strain that describe the distortion due to fracturing are **coaxial** with the principal axes of stress responsible for the fracturing. But as the joints and joint-related structures rotate out of their original attitude during progressive simple shear, the principal axes of strain depart from parallelism with the principal stress axes (Figure 10.33*B*).

It will become clear in Chapter 11 ("Folds") that folding produces systems of joints and joint-related structures that are intimately coordinated with the geometry and kinematics of the folding process. The coordination is so conspicuous that it is generally straightforward to identify which fractures of a given system are products of fold-forming processes and which are not. Orientations and kinematic functions of joints in folds reflect the sense and magnitude of the simple shear and pure shear movements that were forced to operate in the various parts of the fold's anatomy. The movements, in turn, result from the dynamics of buckling, bending, flexing, and flattening.

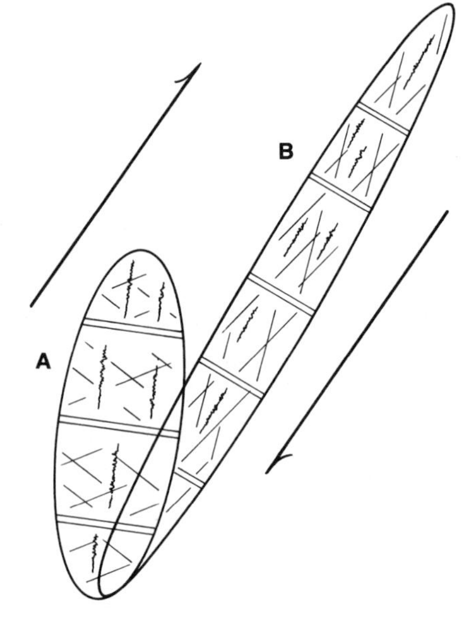

Figure 10.33 Formation of joints and joint-related structures through simple shear kinematics. (*A*) Initial orientations of the structures at the onset of simple shear. (*B*) Rotation of the early-formed structures as a result of progressive simple shear.

DISTINGUISHING DIASTROPHIC AND NONDIASTROPHIC JOINTS

Joints whose kinematic functions are not understood are usually very difficult to explain. And unfortunately these comprise most of the fractures in rocks. To interpret such joints, it is useful on the basis of geometry and physical properties to assign joints of uncertain origin to the established sets of clear-cut primary and/or diastrophic origin. In doing so it must be demonstrated that the joints have orientations that are statistically identical to those of the sets to which they are assigned. And it must be shown that the physical properties of these joints are compatible with those of the sets to which they are assigned.

NONDIASTROPHIC JOINTS

The countless joints that cannot be assigned with certainty to sets of diastrophic joints are grouped together as **nondiastrophic joints**, whose origin(s) must still be explained. The pattern displayed by nondiastrophic joints in a given region is typically composed of several steeply dipping sets, each characterized by a certain preferred orientation, or by a systematically changing (curving) preferred orientation (Hodgson, 1961; Nickelsen and Hough, 1967). When the joint sets are represented on maps as joint trajectories, the trajectories are seen to overlap one another in simple to complex arrays of intersecting straight and curved lines (see Figure 10.29). In some parts of a given region the sets of nondiastrophic joints may intersect in conjugate patterns, but such conjugate joints do not necessarily share the same dynamic significance of true conjugate faults.

Systematic nondiastrophic joints seem to form very early in the life history of the strata in which they are contained. It has been demonstrated by many workers that sets of nondiastrophic joints commonly form before regional folding and faulting, and that they are subsequently locally rotated out of their original attitudes by diastrophic movements. Kelley (1955) and Hodgson (1961) showed this to be the case in the Colorado Plateau province, and Parker (1942) and Nickelsen and Hough (1967) documented the early formation of nondiastrophic joints in the Appalachian Plateau region.

Most structural geologists are convinced that a significant amount of nondiastrophic jointing may actually be a primary jointing due to burial and compaction of sediments. As a given layer of sediment is buried and compacted, it acquires a stiffness that allows it to fracture in response to tensional stresses. Joints are propagated upward through piles of accumulating sediments as each successive layer attains the requisite stiffness to fracture (Hodgson, 1961). Compacted sediments and rocks have such low values of tensile strength that very little tensile stress is required to cause the jointing.

The gravitational tug of the Moon on the Earth is thought to achieve fracturing of semiconsolidated sediments. **Earth tides** pass twice a day across any given location on the face of the globe. The tides have the effect of "lifting" the Earth's outer surface, including its sediment veneer. The lift is maximum in equatorial regions, amounting to about 20 cm (Kendall and Briggs, 1933). Although 20 cm of displacement is minuscule compared to the size of the Earth, its continuous operation over hundreds of thousands of years may lead to fatigue and fracturing of stiffened layers. Kendall and Briggs (1933) and Hogdson (1961) have argued that torsional waves of stress generated by Earth tides are responsible for the formation of primary joints.

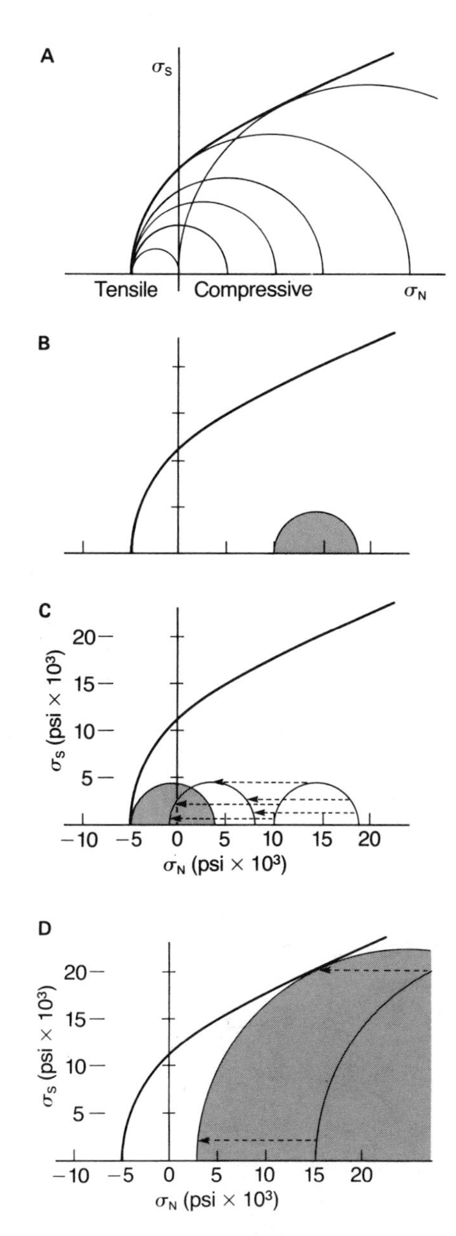

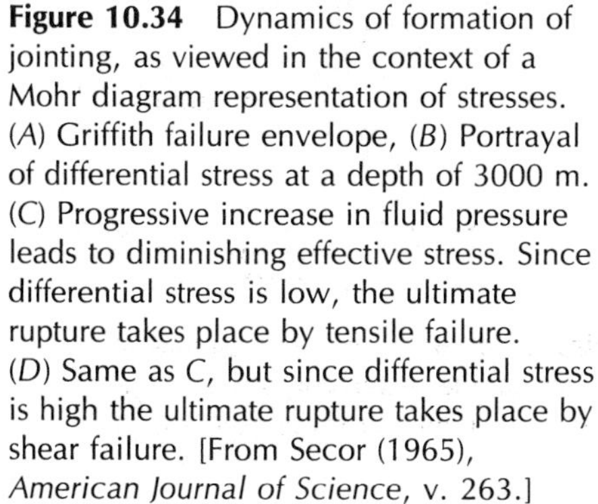

Figure 10.34 Dynamics of formation of jointing, as viewed in the context of a Mohr diagram representation of stresses. (*A*) Griffith failure envelope, (*B*) Portrayal of differential stress at a depth of 3000 m. (*C*) Progressive increase in fluid pressure leads to diminishing effective stress. Since differential stress is low, the ultimate rupture takes place by tensile failure. (*D*) Same as *C*, but since differential stress is high the ultimate rupture takes place by shear failure. [From Secor (1965), *American Journal of Science*, v. 263.]

ROLE OF FLUID PRESSURE IN JOINTING

Fluid pressure contributes significantly to the development of joints, both in sedimentary and igneous rocks. Secor (1965) has shown that elevated fluid pressures can permit tensile stresses to exist at surprising depths, as great as 3000 m and more; and that the tensile stresses can overcome the inherently weak tensile strength of rocks to produce joints. Secor's dynamic model is based on the Griffith fracture criterion (see Chapter 5). The **Griffith failure envelope** is essentially straight lined in the compressive field and strongly parabolic in the tensional field (Figure 10.34*A*). The **tensile strength** of a given rock is represented by the intersection of the envelope with the normal stress axis. For many rocks, tensile strength is less than 5000 psi (340 bars). Thus for joints to form under tension, the level of tensile stress generally must exceed 5000 psi. How can such a dynamic state exist at depths of 5000 psi (351 kg/cm^2), 10,000 psi (702 kg/cm^2), or more?

Shown in Figure 10.34*B* is a stress circle that displays the values of greatest and least principal stress that might exist at a depth of 3000 m in a region of weak horizontal compressive stress. Both σ_1 and σ_3 are compressive. The differential stress value ($\sigma_1 - \sigma_3$) is not great enough to cause failure by fracture, for the stress circle does not intercept the Griffith failure envelope (Figure 10.34*B*). But if the rocks at this depth possessed high fluid pressure, the state of stress would be radically modified. The respective values of σ_1 and σ_3 would each be reduced by the exact value of the fluid pressure: the effective value of σ_1 would equal σ_1 minus fluid pressure; and the effective value of σ_3 would equal σ_3 minus fluid pressure. When the ***effective stress*** levels of σ_1 and σ_3 are then plotted, the stress circle shifts leftward toward the field of tensional stress (Figure 10.34*C*). If the magnitude of fluid pressure is great enough, and the differential stress is small, the stress circle can be driven into collision with the Griffith failure envelope in the tensional field. As soon as σ_3 achieves the value of tensile strength of the rock, tension joints form perpendicular to σ_3. In dynamic situations where differential stress is high, the gradual elevation of fluid pressure will drive the stress circle into collision with the Griffith failure envelope in the compressional field (Figure 10.34*D*). In this situation, shear joints form.

Rehrig and Heidrick (1976) have applied Secor's model in explaining the formation of diastrophic joints in Laramide plutons in southern Arizona. They concluded that Laramide plutons were intruded into the upper levels of a crust that was transmitting horizontal compressive stresses. Differential stress was relatively weak. Fluid pressure buildup in the cooling plutons had the effect of decreasing the magnitude of greatest (σ_1) and least (σ_3) principal stress. This produced tensional fracturing by jointing. Fracturing served to dissipate the elevation of fluid pressure for a time, and then pressure built again, culminating in yet another phase of jointing.

JOINTING—A MEANS OF STRESS RELEASE

Yet another factor contributes to the formation of nondiastrophic joints. When strata are deeply buried and horizontally compressed, elastic strain energy can be stored up in the rocks. The energy may remain pent up in the rocks for long periods of time, despite some degree of jointing. When strata with stored strain energy are raised from very deep levels to very high structural levels by block uplift and/or thrusting, the strata are permitted to expand in response to progressively decreasing confining pressure. Lateral expansion allows some of the strain energy to be dissipated along closely spaced joints (Price, 1959). Observed density of jointing may be directly related to the magnitude of stored energy. A simple geometric/strain image of the lateral expansion that must accompany the uplift of a large regional sequence of sediments is portrayed in Figure 10.35.

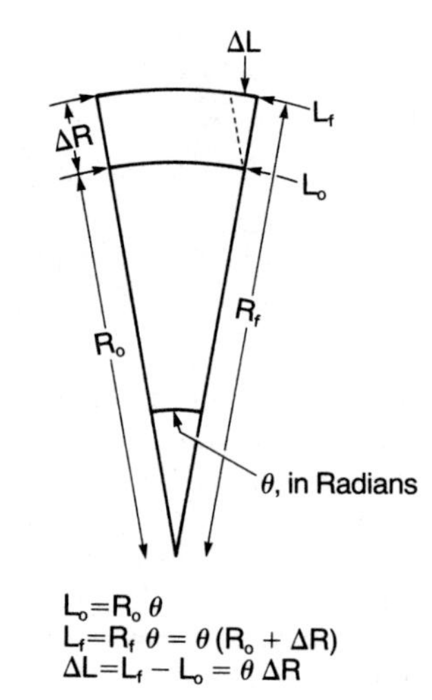

Figure 10.35 Geometric picture of layer-parallel extension accompanying regional uplift. [From Price (1959).]

Residual stress in Precambrian basement rocks can be relieved through the release of stored strain energy. Wise (1964) measured 6500 microjoints in mountains of Precambrian basement in the Wyoming province. The Precambrian rocks are parts of enormous fault blocks that were uplifted by thousands of meters in the Laramide orogeny in early Tertiary time. Wise interpreted the microcracks to be tension joints that helped to relieve ancient locked-in stresses. The microjoints, although visible, are hairline and spaced less than 3 mm apart. Wise believes that these nearly invisible, subtle structures may constitute the **rift** and **grain** of granites in New England, the weaknesses that quarry workers have long used to split out commercial building blocks (Dale, 1923).

OPPORTUNITIES IN JOINT ANALYSIS

Joints and joint-related structures readily lend themselves to many aspects of detailed structural analysis. There is never a problem finding jointed rocks to examine. Field-and-lab projects involving fracture analysis can foster firsthand experience in orientation analysis and the sorting out of structures. Furthermore, fracture analysis can provide valuable experience in photogeologic interpretation, experimental deformation, kinematic analysis, and mathematical inquiry into dynamics. Since the origin of jointing is so commonly elusive, there is plenty of room for independent, creative inquiry.

FOLDS *chapter* 11

INCENTIVES FOR STUDY

VISUAL IMPACT

Folds are visually the most spectacular of Earth's structures. They are extraordinary displays of strain, conspicuous natural images of how the original shapes of rock bodies can be changed during deformation. The physical forms and orientations of folds seem limitless. Some are upright (Figure 11.1*A*); others lie on their sides (Figure 11.1*B*). Some show neatly

Figure 11.1 Some of the many geometric renditions of folds. *(A)* Upright anticline in Silurian sandstones and shales, 3 mi west of Hancock, Maryland. (Photograph by I. C. Russell. Courtesy of United States Geological Survey.) *(B)* Folds in Oligocene–Miocene turbidites in Kii Peninsula, southwest Japan. (Photograph by W. R. Dickinson.)

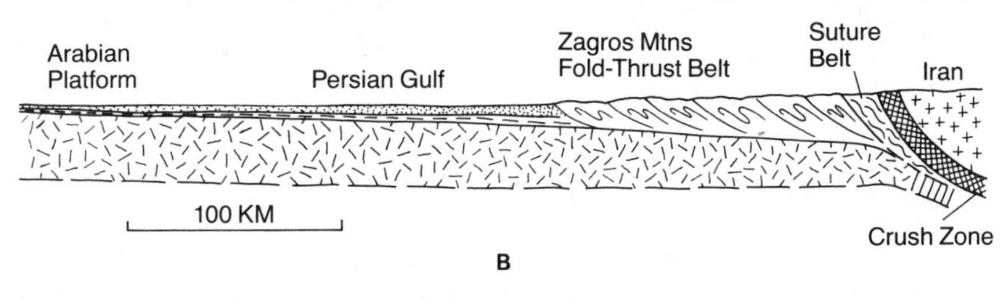

Figure 11.2 *(A)* Apollo 7 view of anticlines in the Zagros Mountains, Iran. The dark circular forms are the surface expressions of salt domes. Persian Gulf in foreground. (Courtesy of National Aeronautics and Space Administration.) *(B)* Structure section shows the plate tectonic configuration responsible for the formation of the Zagros Mountains fold-thrust belt. The belt lies at the convergent boundary of the Arabian and Iran plates. [(From W. R. Dickinson (unpublished).]

arranged, uniformly thick layers; others are sloppy. Fold size varies too, from anticlines that fit into the palm of a hand to regional folds best seen through the eyes of a satellite (Figure 11.2). Mapping the forms of folds is pure pleasure, unless of course the folds turn into a geometric nightmare.

MECHANICAL CONTRADICTION OF FOLDING

It is almost impossible to view waves of solid rock without wondering how materials that we regard as strong can be folded in a manner that makes them seem so weak. Bailey Willis (1894), in trying to model the structures of the Appalachian fold belt, found it necessary to represent sedimentary strata with layers of *very* soft materials like clay, putty, cheese, and wax (see Figure 2.16). Regional rock bodies yield effortlessly to folding by penetrative movements along preexisting weaknesses, like bedding, and/or by the generation of secondary penetrative weaknesses, like cleavage.

Where strata are very well bedded and relatively stiff, folding is achieved by layer-parallel slippage (Donath and Parker, 1964). Like the pages of a slick magazine, beds can easily slip past one another along bedding surfaces when folded. Where the mechanical influence of bedding or other layering is not as dominant, and the layers tend to be relatively soft, folding is achieved by flow or slip or even the pressure-solution loss of material along cleavage surfaces created during the deformation. The cleavages themselves testify to the weakness of rock in the face of tectonic stresses: cleavages cut across the lithological layering of folded beds as if the beds did not exist (Figure 11.3), as if the mechanical influence of layering could hardly be felt, as if the rocks had no strength to resist!

Figure 11.3 Folded beds cut by strongly penetrative, closely spaced cleavage surfaces. These slates are located at "old quarry #2" at Slatington, Lehigh County, Pennsylvania. (Photograph by E. B. Hardin. Courtesy of United States Geological Survey.)

GEOMETRIC PLEASURES

Analysis of the geometry of folds invites us to probe more deeply into stereographic analysis of structural data. This, in turn, leads to a more expansive awareness of geometrical concepts and methods. Simple stereographic procedures permit us to calculate and describe the orientations of folds, to measure fold tightness, to unfold folds and reconstruct original orientations of primary structures, and to describe the total integrated geometry of fold systems. The full use of stereographic analysis in modern geometric analysis of folds and fold-related structures was inspired by Turner and Weiss (1963) in their classic text, *Structural Analysis of Metamorphic Tectonites*.

INFORMATIVE MINOR STRUCTURES

Folding results in the formation of a delightful and curious array of cognate structures, typically referred to as **minor structures**. The adjective "minor" belies the usefulness of minor structures in evaluating the kinematics of folding. Among the family of minor structures are small faults of all kinds; a variety of cleavages; innumerable lineations, joints, and joint-related structures; and boudins. Different kinds of minor structure form in different parts of folds because different parts of folds are marked by different local strain environments. Some parts of folded beds are stretched; others are shortened; still other parts suffer no strain whatsoever.

Most students of geology enjoy the challenge of trying to interpret the orientations and structural locations of minor structures in light of the overall geometry and kinematics of folding. Learning to do so is fundamental to mapping and unraveling regional assemblages of folded rocks. For example, a change from Z-shaped to S-shaped minor folds within uniformly dipping strata (Figure 11.4*A*) may signal the presence of an otherwise-hidden fold structure (Figure 11.4*B*).

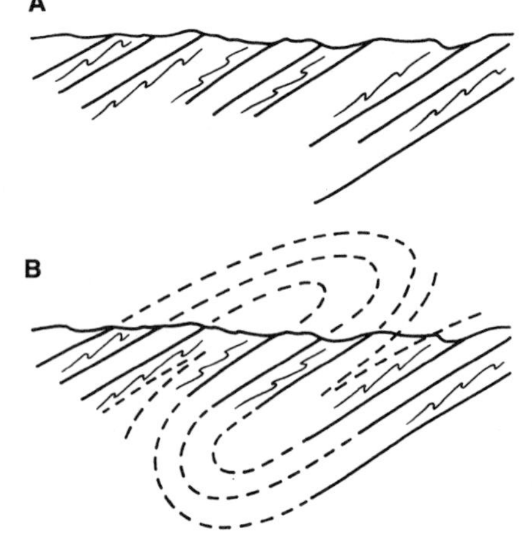

Figure 11.4 The asymmetry of folds can be used to disclose hidden structures. (*A*) Z-Shaped and S-shaped minor folds in homoclinally dipping strata denote the limbs of (*B*) a hidden anticline.

OPPORTUNITIES IN KINEMATIC AND DYNAMIC ANALYSIS

Interpreting folds, including the formation of minor structures, is a fruitful exercise in applying principles and methods in kinematic and dynamic analysis. Kinematic analysis of folding requires integration of *all* kinematic movements: translation, rotation, dilation, and distortion. Dynamic analysis requires identifying the physical and mechanical variables that influence the nature of folding. Dynamic analysis is pursued both theoretically and experimentally.

Fold analysis provides an unusual opportunity for integrating and synthesizing principles of structural geology. This is the main reason we have not begun our serious study of folds until this eleventh hour. Friends of folds have John Ramsay (1967) to thank for *Folding and Fracturing of Rocks*, which presents the foundation for modern quantitative fold analysis.

TECTONIC CONSIDERATIONS

To understand the origin of folding we must move beyond geometric and kinematic analysis, beyond experimental and theoretical studies, to search for dynamic circumstances at the tectonic scale that are likely to generate folding. In this regard, it is especially useful to view the locations of the world's fold systems in the context of reconstructions of plate configurations through time. In doing so we find that folds and fold systems tend to be found in settings that are or were convergent plate margins.

Many different tectonic circumstances can give rise to folds. Consequently, it is generally difficult to interpret the ultimate cause of folding within any given tectonic assemblage in any given region. Folds found in subduction complexes can form as products of shearing due to underthrusting and underplating accompanying subduction; but they also can form as a response to gravity-induced flow of oversteepened, not-yet-consolidated sediments. Strata in forearc, interarc, and backarc basins can fold during overall crustal shortening and/or strike–slip shearing created by plate convergence. However, the same sediments and volcanics, deposited in and around magmatic arcs, may contain folds that owe their existence to the forceful shouldering aside of strata during emplacement of granitic plutons, and/or to gravity-induced gliding of strata off the tops of rising diapirs. The origin of folds and fold systems in foreland settings, inboard from the edges of active plate margins, has been interpreted by some as the direct expression of significant crustal shortening, a shortening due to stresses generated at plate margins and transmitted through continental lithosphere for great distances horizontally. But such folds and fold systems have been interpreted by others as products of the draping of strata over the edge(s) of vertically uplifted fault blocks, in a manner not related to crustal shortening.

Thus, where strain is intense, deformations superimposed, and exposures limited, it may not always be simple to interpret the dynamic conditions responsible for folding. The geologic literature, bulging with contradictory interpretations regarding the origin of folds and fold systems in specific regions, bears this out.

GUIDES TO EXPLORATION AND MINING

There are practical incentives to come to know folds and their properties. Foremost, there is a legacy of discovery of oil and gas in **structural traps** created by folding (Figure 11.5). Folds commonly serve as collection sites for oil and gas that migrate up the dip of strata from hydrocarbon **source beds** below. The migration of oil and gas takes place within a permeable **reservoir rock**, like a porous sandstone or a vuggy limestone. If the reservoir bed is

Figure 11.5 Folds commonly trap petroleum accumulations. This fold occurs in thin-bedded Monterey formation exposed along Soto Street, north of Alhambra Avenue, Los Angeles. (Photograph by M. N. Bramlette. Courtesy of United States Geological Survey.)

overlain depositionally by a tight, impermeable layer, like shale, the top of the reservoir is effectively sealed to prevent escape of the oil and gas during up-dip migration. Where a sealed reservoir is folded into a dome or anticline, the **closure** of the fold prevents further upward migration of the oil and gas. The fluids may collect there to form a rich "pool."

Where ore deposits are known to be localized within folded rocks, there are clear exploration and mining incentives to understand the three-dimensional geometry of folds. Of special importance is learning methods for projecting the forms and trends of folds to depth as an aid in estimating reserves and in planning underground mining. **Saddle reef deposits** represent perhaps the most intimate of the associations between folding and mineralization. Saddle reefs are **lodes** of quartz and precious metals that occupy the cores of folds, in openings where bedding and/or foliation has been separated by fold-forming movements (Figure 11.6). Although saddle shaped in cross-sectional view, they are pipelike in three dimensions, trending parallel to the axis of folding. The most renowned saddle reef deposits are in the Bendigo goldfields of Victoria, Australia (Park and MacDiarmid, 1964; McKinstry, 1961) and in the Salmon River gold district of Nova Scotia (Malcolm, 1912).

Strata-bound ore deposits are commonly associated with folded, metamorphosed rocks. The lead–zinc–silver deposits of Broken Hill, Australia, constitute a fine example (Gustafson, Burrell, and Garretty, 1950) (Figure 11.7). Shaped like ordinary beds of volcanic or sedimentary rocks, these deposits are extraordinary mineralogically, abundantly rich in sulfides and/or precious metals. Although traditionally viewed as replacements of chemically favorable host rocks, strata-bound sulfide deposits are now considered by most workers to represent **submarine chemical sediments** and/or **volcanic–exhalative precipitates**. In the course of the normal geologic history of orogens, these unusually valuable rock layers become folded, faulted, and metamorphosed, just like their ordinary sedimentary and volcanic counterparts.

Folded ore bodies, and ore bodies in folds, are challenging to exploration and mining geologists. The step-by-step underground geologic mapping of ore bodies in folded rocks, carried out during the normal course of mining, has yielded illuminating documentaries on the three-dimensional anatomy of folds.

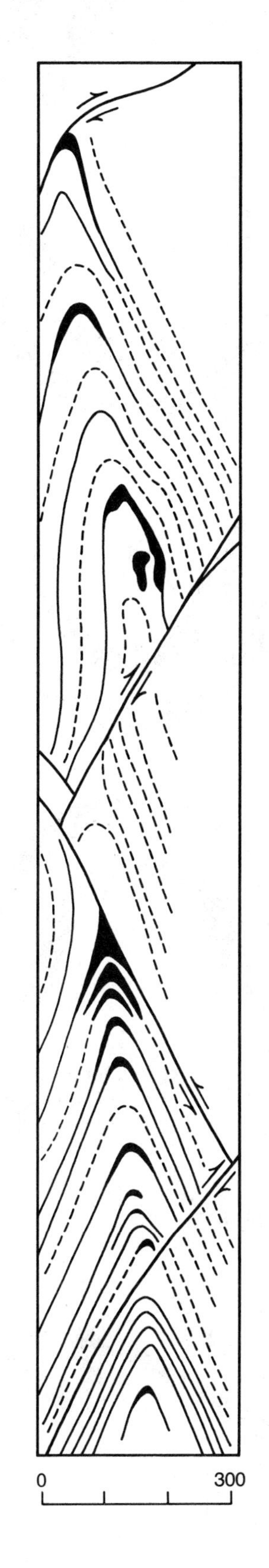

Figure 11.6 Saddle reef deposit in the Bendigo goldfield, Bendigo, Victoria, Australia. (From *Mining Geology* by H. E. McKinstry. Published with permission of Prentice-Hall, Inc., Englewood Cliffs, New Jersey, copyright ©1961.)

ANTICLINES AND SYNCLINES

BASIC DEFINITIONS

"Anticline" and "syncline" are beloved terms to all who have been introduced to geology. An **anticline** is a fold that is convex in the direction of the youngest beds in the folded sequence. Ordinarily we think of anticlines as upright, convex-upward folds, like that shown in Figure 11.8*A*. But actually the term can apply to folds that are convex in the direction of youngest beds, *regardless* of the absolute configuration of the fold. Even the upside-down anticline shown in Figure 11.8*B* is called an anticline, albeit a special kind of anticline known as a **synformal anticline**. The adjective "synformal" means that the fold is concave upward.

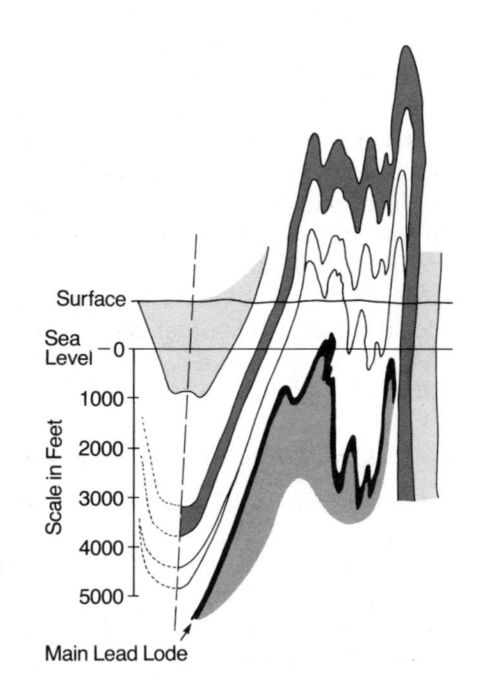

Figure 11.7 Structure section through part of the folded, strata-bound Broken Hill ore deposit. [From Gustafson, Burrell, and Garretty (1950), Geological Society of America.]

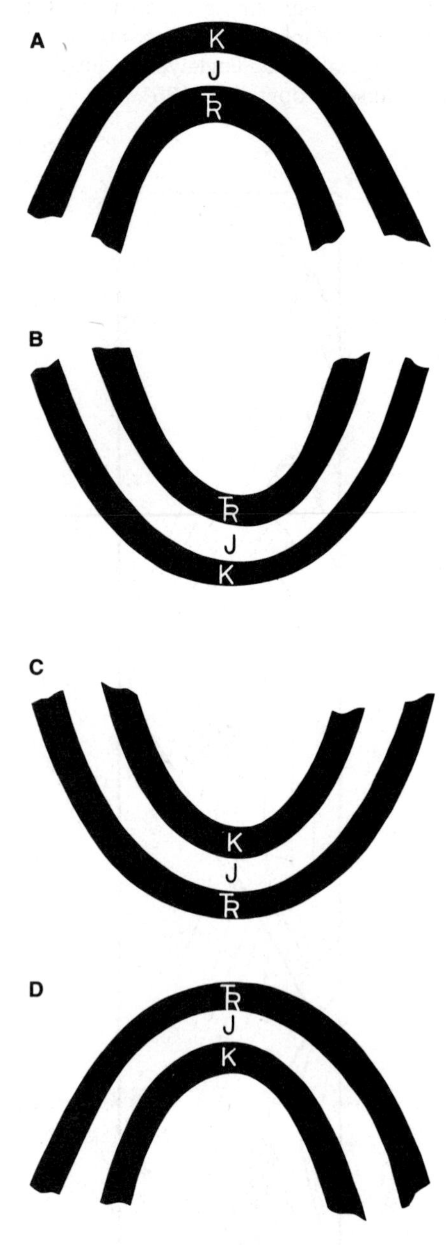

Figure 11.8 "Anticlines" and "synclines," the cornerstones of fold terminology. (*A*) Anticline. (*B*) Synformal anticline. (*C*) Syncline. (*D*) Antiformal syncline.

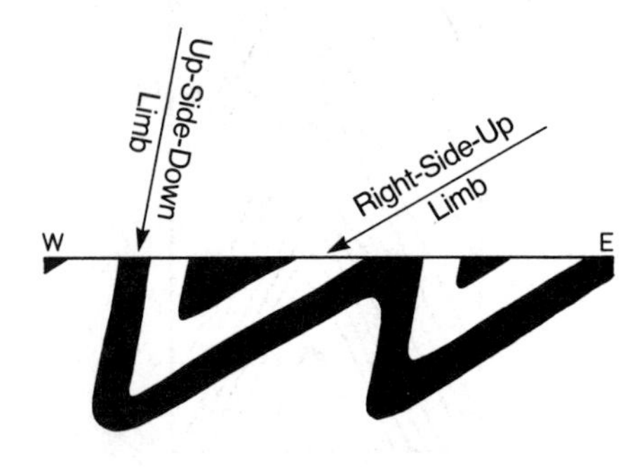

Figure 11.9 Cross-sectional view of overturned anticline and syncline.

A **syncline** is a fold that is convex in the direction of the oldest beds in the folded sequence (Figure 11.8*C*). These too can be oriented in any way, from perfectly upright synclinal folds to convex-upward **antiformal synclines** (Figure 11.8*D*).

OVERTURNED FOLDS

It is very rare for anticlines and synclines to be completely upside down, but it is not at all uncommon for them to be **overturned**. The distinction between upside down and overturned is this: upside-down folds are totally inverted, like the antiformal syncline shown in Figure 11.8*D*. In contrast, a fold is considered to be **overturned** if at least one of its **limbs** (i.e., flanks) is overturned. Saying that the limb of a fold is overturned does not mean that the fold is completely inverted (upside down). It simply means that the limb has been rotated beyond vertical such that the facing direction of the limb points downward at some angle. A system of overturned anticlines and synclines is shown in Figure 11.9. Each fold is marked by a rightside-up limb dipping at about 30°W and an overturned limb dipping at about 80°W.

ANTIFORMS AND SYNFORMS

Use of the term anticline (or syncline) implies that stratigraphic succession within the folded sequence has been worked out on the basis of the established geological column and/or the evaluation of facing. Where facing and stratigraphic order cannot be established, these terms must be scrapped, at least temporarily, in favor of **antiform** and **synform**. An antiform is, very simply, a fold that is convex upward. A synform is a fold that is concave upward.

The terms antiform and synform are normally used in reference to folds in sedimentary and/or volcanic sequences within which facing and/or stratigraphic order are either unknown or uncertain. But "antiform" and "synform" are also appropriate in describing folds in plutonic rocks and metamorphic rocks. If layering and/or foliation in a rock body is not related to normal depositional process, nor to conventional stratigraphic succession, it is meaningless to attempt to describe facing and stratigraphic order.

The terms anticline and syncline may have only limited application in describing **superposed folds** in a sedimentary and/or volcanic sequence. By way of example, the early-formed fold shown in structure profile view in Figure 11.10*A* may be described as a **recumbent, isoclinal** anticline. "Recumbent" means that the fold lies on its side. "Isoclinal" means that the limbs of the fold are equally inclined. Figure 11.10*B* shows the early-formed isoclinal fold after it has been modified by a second folding. The anticline is no longer perfectly recumbent. Rather, it is deflected into convex-upward and convex-downward superposed folds. These late folds cannot be described as anticlines and synclines because there is no overall internal coherence of stratigraphic order and facing within the layers that define the folds. The second-stage folds are appropriately described as antiforms and synforms.

ANTICLINORIA AND SYNCLINORIA

Regional fold belts contain very large anticlines and synclines, kilometers across, that are themselves marked by the presence of reasonably systematically spaced smaller anticlines and synclines (Figure 11.11). When describing these regional structures it is sometimes useful to refer to them as **anticlinoria** and **synclinoria**. The flank of an **anticlinorium** (or **synclinorium**) is typically marked by a set of approximately equal-sized **second-order**

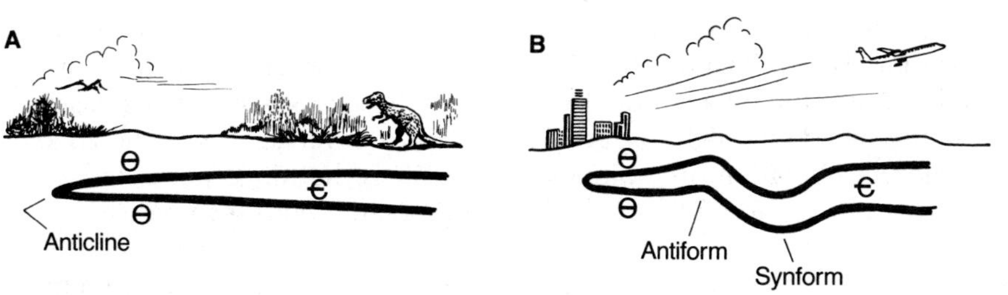

Figure 11.10 An example of superposed folding. (*A*) Recumbent, isoclinal anticline, of pre-Mesozoic formation. (*B*) Antiformal and synformal folding of the recumbent, isoclinal anticline.

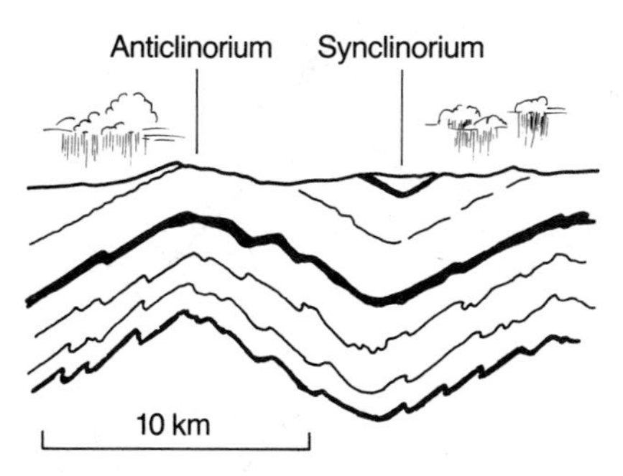

Figure 11.11 Schematic rendering of an anticlinorium and a synclinorium.

anticlines and synclines. These in turn may contain sets of **third-order** folds, and so it goes.

GEOMETRIC ANALYSIS OF FOLDS

GEOMETRIC PROPERTIES OF FOLDED SURFACES

LIMBS, HINGES, AND INFLECTIONS. Folded surfaces vary tremendously in size and form. Furthermore, they come in every conceivable orientation and configuration. They may occur as single isolated structures, or as part(s) of a system of repeated wave forms. The infinite variety of folds, and their three-dimensional complexity, compels us to learn a special vocabulary for descriptive analysis and geologic mapping.

We can begin to dissect the anatomy of a fold by extracting a single **folded surface** from the deformed sequence of which it is a part (Figure 11.12*A*, *B*). By stripping away the rock layers that serve as boundaries to the folded surface, top and bottom, we can focus on the geometric properties of the folded surface itself. We start the process two-dimensionally, by describing the folded surface as seen in **normal profile view**, at a right angle to the axis of rotation of the folded surface (Figure 11.12*C*).

Considered two-dimensionally in normal profile view, folded surfaces can be subdivided into **limbs** and **hinges** (Figure 11.13). Limbs are the flanks of folds, and these are joined at the hinge. The hinge of a folded surface is sometimes a single point, called the **hinge point**. More commonly the hinge is a zone, the **hinge zone**, distinguished by the maximum curvature achieved along the folded surface (Ramsay, 1967). Figure 11.13*A* shows an angular fold marked by planar limbs and an easily identifiable hinge point that separates the limbs. Figure 11.13*B*, on the other hand, pictures a fold marked by planar limbs connected by a hinge zone within which the folded surface displays uniformly high curvature. Strictly speaking, the hinge zone of a folded surface does not possess a unique hinge point. But for descriptive purposes a hinge point can be arbitrarily posted at the midpoint of the hinge zone.

Limbs of folds are commonly curved. Curved limb segments of opposing convexity join at locations known as **inflection points** (Figure 11.13*C*) (Ramsay, 1967). A special case of continuously curved folded surfaces is shown in Figure 11.13*D*, a surface that has been folded into a series of perfectly circular arcs. Folded surfaces of this type lack fold limbs, per se. Instead, each discrete circular arc may be thought of as a hinge zone marked by uniform curvature. The hinge point of each fold is taken as the midpoint of each circular arc. The inflection points of the folded surface separate circular arcs of opposing convexity.

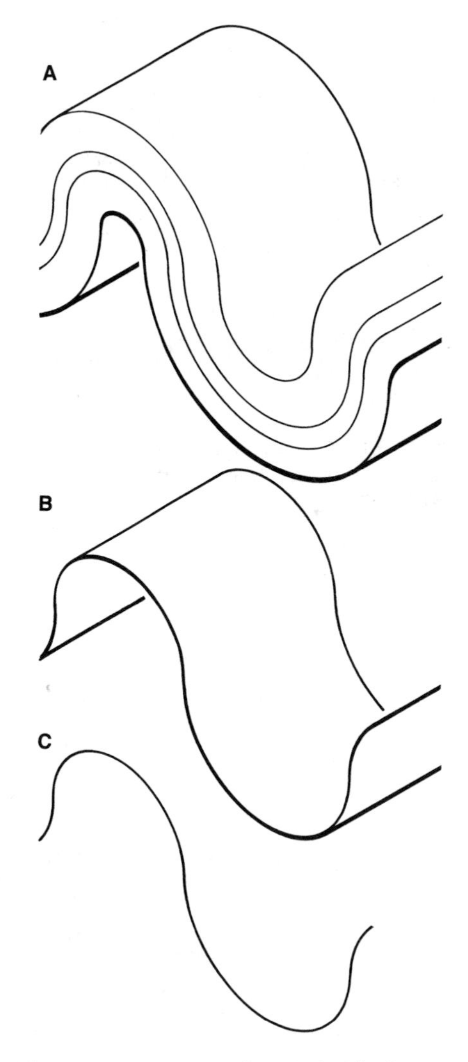

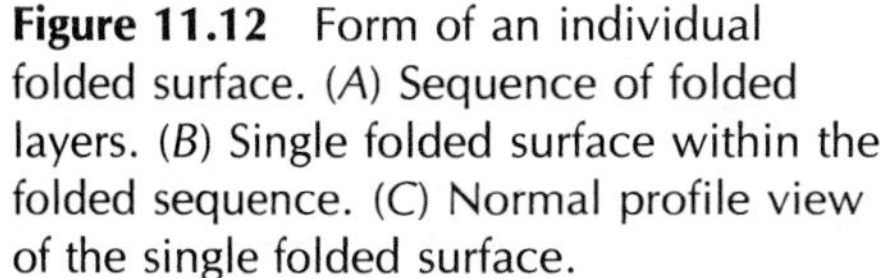

Figure 11.12 Form of an individual folded surface. (*A*) Sequence of folded layers. (*B*) Single folded surface within the folded sequence. (*C*) Normal profile view of the single folded surface.

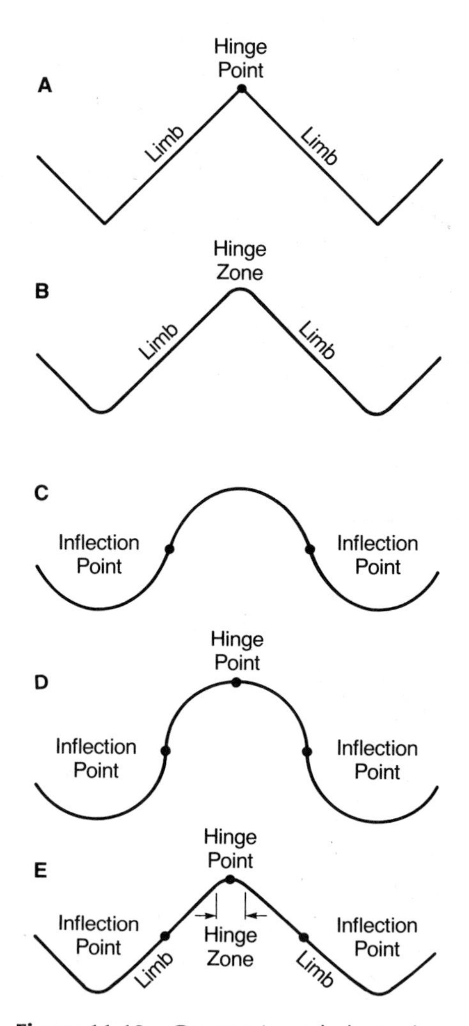

Figure 11.13 Geometric and physical elements of single folded surfaces. See text for discussion.

The vocabulary used in describing folded surfaces in normal profile view is summarized in Figure 11.13*E*, a fold marked by slightly curved to planar limbs connected by a narrow hinge zone within which the surface displays a pronounced and high degree of curvature. The hinge zone is the zone of maximum curvature, the midpoint of which is the hinge point. Inflection points are taken to be the midpoints of the planar limb segments.

HINGELINES, AXIAL SURFACES, AND AXIAL TRACES. Three-dimensionally, folded surfaces display geometric characteristics that invite a yet fuller nomenclature (Figure 11.14). The hinge points along a single folded surface, taken together, define a **hingeline**. In some cases the hingeline of a folded surface is perfectly straight (Figure 11.14*A*); more commonly hingelines are systematically curved (Figure 11.14*B*) or, more rarely, terribly irregular (Figure 11.14*C*) (Turner and Weiss, 1963).

The orientation of a folded surface can in part be specified by the orientation of its hingeline. Like all geologic/geometric lines, the orientation of a hingeline is described conventionally in terms of trend and plunge. A single measurement of trend and plunge is adequate for hingelines that are perfectly straight. But where folded surfaces are marked by hingelines that are not straight, it is necessary to document the variations in hingeline orientation through a number of measurements of representative segments of the hinge. In practice this is achieved by subdividing the folded surface into domains within which the hingeline approaches a straight line (Figure 11.14*D*).

Knowing the orientation of the hingeline of a fold does not uniquely establish the attitude of the fold. Folds having the same hingeline orientation can have strikingly different configurations (Figure 11.15). To describe unambiguously the orientation of a fold, it is necessary to measure yet another structural element, a geometric element known as an **axial surface**. The axial surface of a fold passes through successive hingelines in a stacking of folded

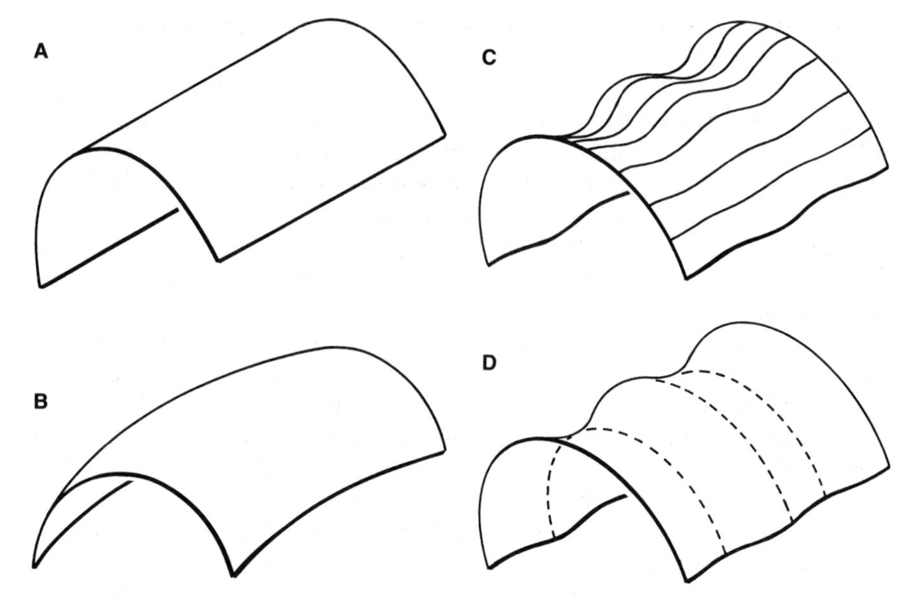

Figure 11.14 (*A*) Fold with straight hingeline. (*B*) Fold with systematically curved hingeline. (*C*) Fold with irregularly curved hingeline. (*D*) Subdivision of fold with irregularly curved hingeline into domains marked by nearly straight hingelines.

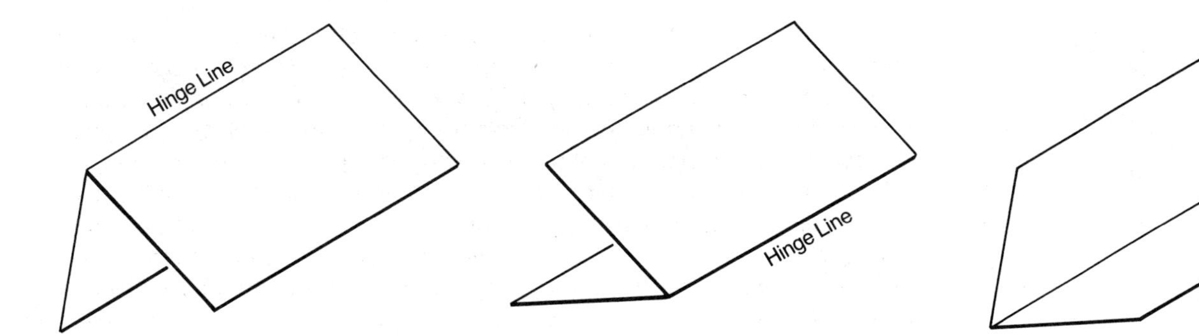

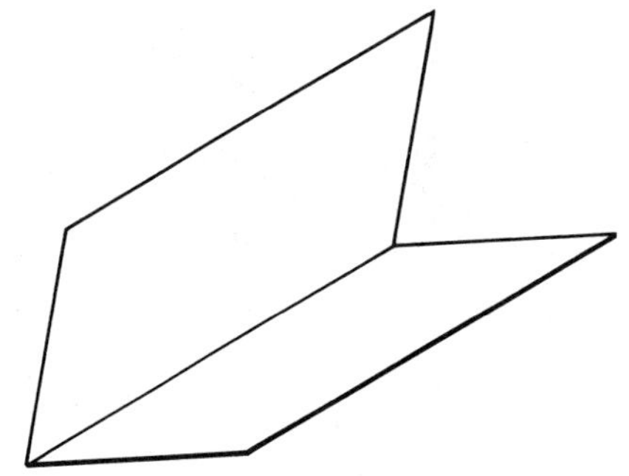

Figure 11.15 The trend and plunge of the hingeline of a fold does not uniquely define the overall orientation of the fold. See for yourself.

surfaces (Figure 11.16) (Ramsay, 1967; Dennis, 1972). The axial surface of a fold may be planar (Figure 11.16*A*), in which case it is called an **axial plane**. More commonly the axial surface of a fold is either systematically curved (Figure 11.16*B*), or nonsystematically irregular (Figure 11.16*C*), in which case "axial surface" is the appropriate term (Turner and Weiss, 1963).

In normal profile view, the trace of the axial surface of a fold can be seen to pass through successive hinge points in the stacking of folded surfaces (Figure 11.16*D*). This line is called the **axial trace** of the fold in profile view. In a more general sense, "axial trace" refers to the line of intersection of the axial surface with *any* other surface, whether it be the ground surface, the cut of an open-pit mine, the steep flank of a mountain, or the faces of a block diagram of folded layers.

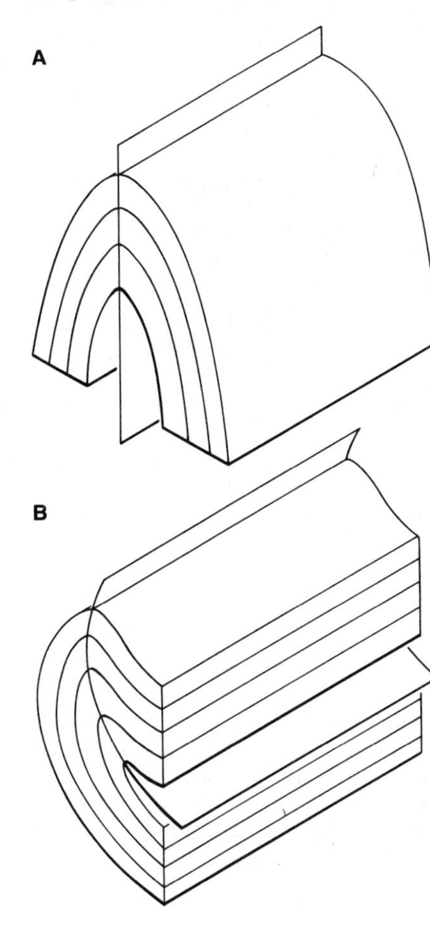

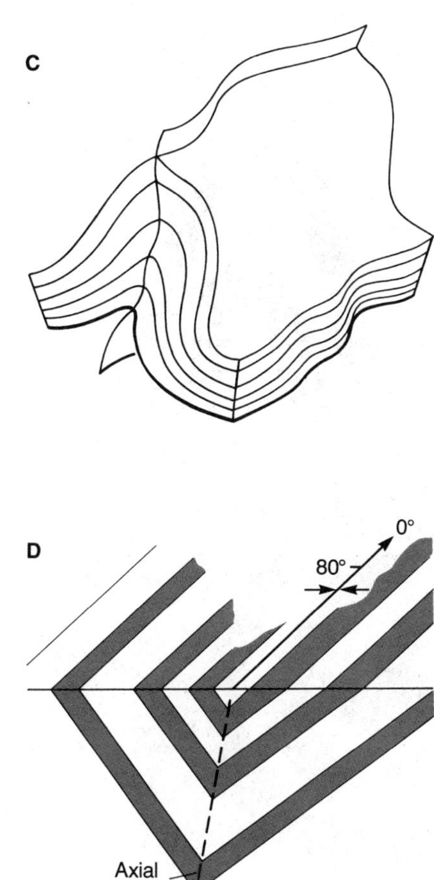

Figure 11.16 (*A*) Fold with a planar axial surface. (*B*) Fold with a systematically curviplanar axial surface. (*C*) Fold with an irregularly curviplanar axial surface. (*D*) Axial trace of a fold, as seen in cross section and in map view.

The orientation of the axial surface of a fold is described in terms of strike and dip. A single strike-and-dip measurement is all that is necessary to describe the orientation of the **axial plane** of a fold. For a nonplanar axial surface, however, a number of strike-and-dip measurements are required to document the full spectrum of orientations of the axial surface. Knowing the orientation of the axial surface does not fix uniquely the attitude of the fold. Folds having a common axial surface orientation can have radically different configurations (Figure 11.17). Only when the orientations of both the hingeline *and* the axial surface of a fold are known can the configuration of the fold be firmly established.

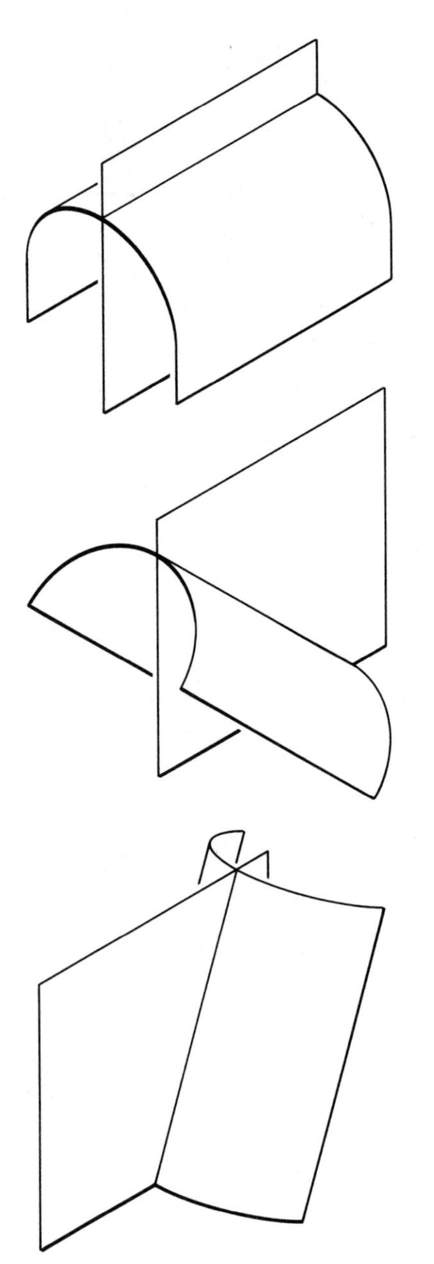

Figure 11.17 The strike and dip of the axial surface of a fold does not uniquely define the overall orientation of the fold.

GEOMETRIC COORDINATION OF HINGELINES AND AXIAL SURFACES. Because of the manner in which "hingeline" and "axial surface" are defined, the hingeline of a fold must lie within the fold's axial surface. This constraint notwithstanding, hingelines and axial surfaces can be combined in many more ways than we might at first imagine. It is relatively easy to picture a hingeline that is parallel to the strike of the axial surface (Figure 11.18*A, B*). Although this arrangement is common, it is a very special case: for in general *the trend of the hingeline of a fold may be parallel, perpendicular, or oblique to the strike of the axial surface* and still remain within the axial surface (Figure 11.18*C*). Said another way, the rake of the hingeline in an axial surface can range from 0° to 90°.

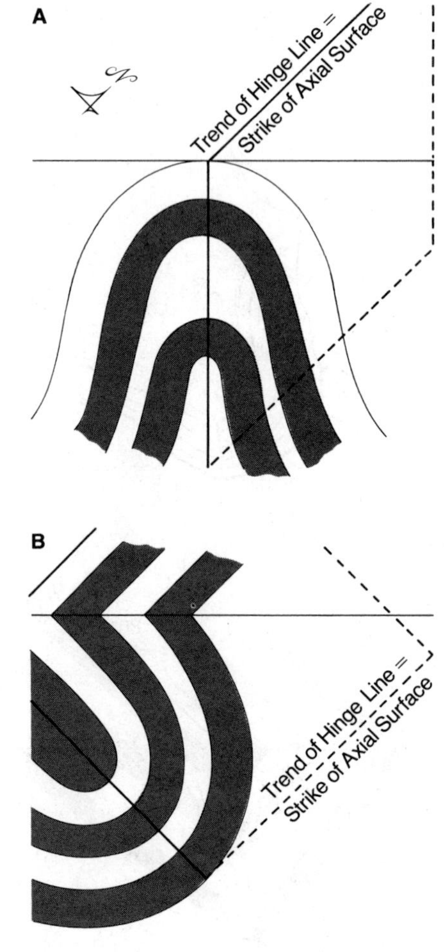

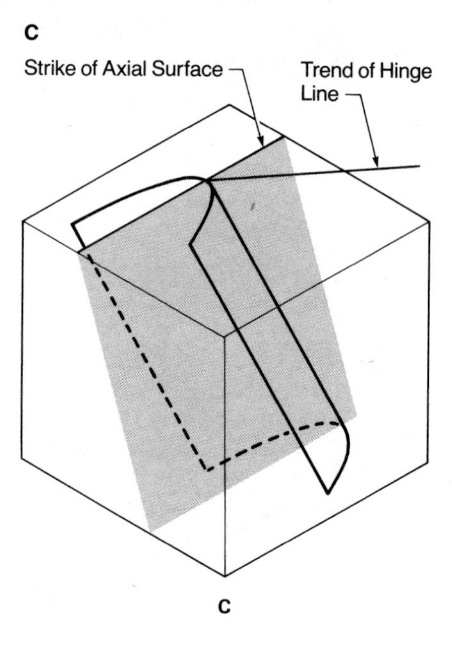

Figure 11.18 (*A*) Fold marked by hingeline whose trend is parallel to the strike of the axial surface. (*B*) Another example. (*C*) Fold marked by hingeline whose trend is discordant to the strike of the axial surface.

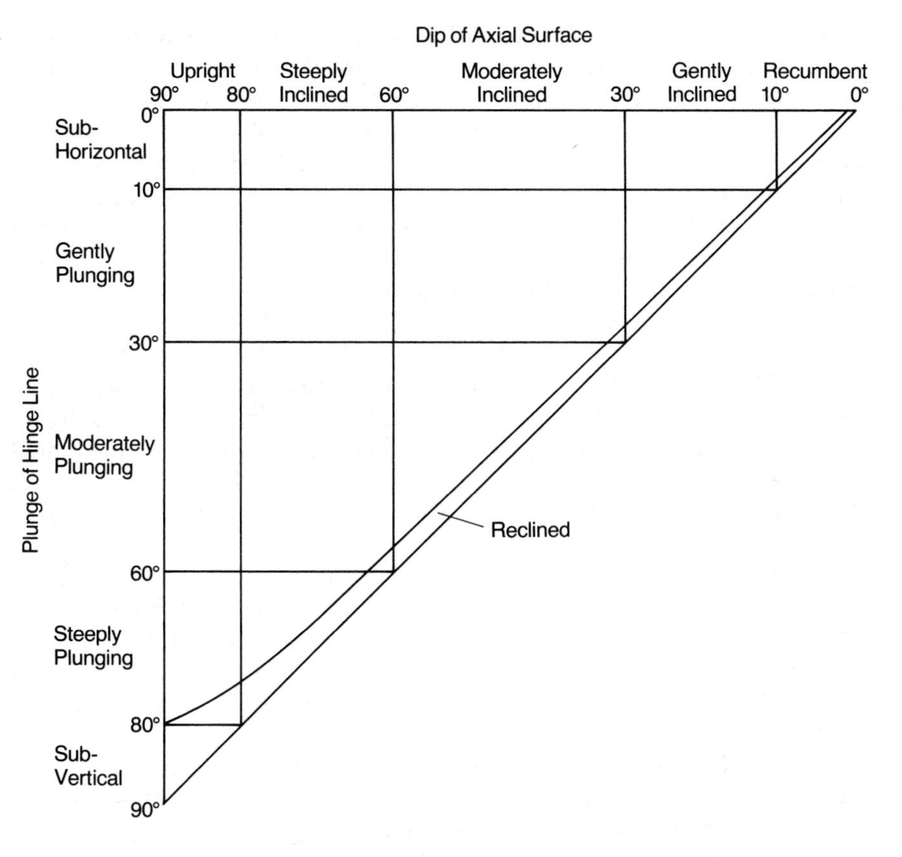

Figure 11.19 Fleuty diagram used for describing folds on the basis of the geometric interrelationship of axial surface and hingeline. [After Fleuty (1964). From *Folding and Fracturing of Rocks* by J. G. Ramsay. Published with permission of both the Geologists Association and McGraw-Hill Book Company, New York, copyright ©1967.]

The breadth of the geometrically permissive *relative* orientations of hingelines and axial surfaces, combined with the even greater range of *absolute* orientations that hingelines and axial surfaces can assume, present us with limitless possible fold configurations. To deal with such broad-ranging geometries, Fleuty (1964) created a useful classification scheme, one that is based on both the relative orientations and the absolute inclinations of hingelines and axial surfaces. The classification scheme, in the form of a diagram (Figure 11.19), permits folds to be named according to fold configuration. Along the *x* axis of Fleuty's diagram is plotted the **dip** of the axial surface; along the *y* axis is plotted the **plunge** of the hingeline. On the basis of cutoffs of 0°, 10°, 30°, 60°, 80°, and 90°, eighteen categories of folds are identified. A fold whose hingeline plunges 5° N38°E and whose axial surface orientation is N40°E, 82°NW can be described as a **subhorizontal upright fold** (Figure 11.19). A fold whose hingeline plunges 50° S72°W and whose axial surface orientation is N80°W, 70°SW is classified as a **moderately plunging, steeply inclined fold**. A fold with hingeline and axial surface orientations of 20° N65°E and N30°W, 20°NE, respectively, is called a **gently plunging, gently inclined, reclined fold**. A **reclined** fold is one whose hingeline plunges directy down the dip of the fold's axial surface.

THE DIFFERENCE BETWEEN A HINGELINE AND A FOLD AXIS. It is easy to fall into the trap of using the terms **hingeline** and **fold axis** interchangeably. But beware. Strictly speaking, a fold axis is a geometric (thus imaginary) linear structural element that does not possess a fixed location. It is the closest approximation to a straight line that when moved parallel to itself, generates the form of the fold (Figure 11.20) (Donath and Parker, 1964; Ramsay, 1967).

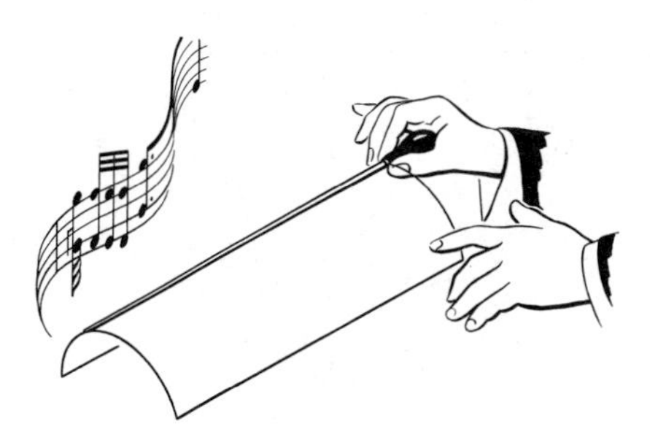

Figure 11.20 Fold axis, an imaginary straight line, which when moved parallel to itself, orchestrates the form of the fold.

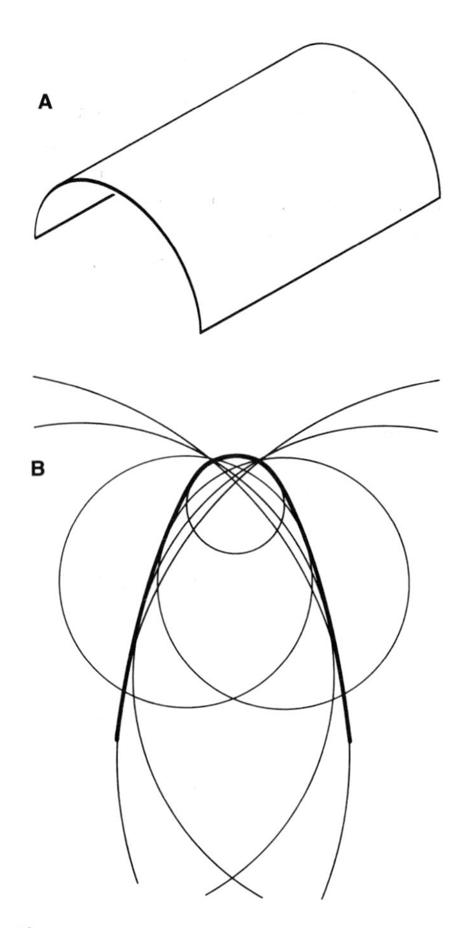

Figure 11.21 Two perfectly acceptable cylindrically folded surfaces. (*A*) One conforms to the outline of a single, perfect cylinder. (*B*) The other is composed of a coaxial arrangement of parts of cylindrical surfaces of different curvatures.

Folds that possess axes are **cylindrical folds**. Cylindrical folds should bring to mind parts of tin cans and pipes. A cylindrical fold can have a form best described as part of a single, perfect cylinder (Figure 11.21*A*). But more commonly a cylindrical fold has the form of a **coaxial** arrangement of parts of cylinders of different diameters (Figure 11.21*B*). The distinctive geometric characteristic of cylindrical folds is that *every part of the folded surface is oriented such that it contains a line whose orientation is identical to that of the hingeline*. The orientation of this single, unique line, common to all parts of the folded surface, is that of the fold axis.

A close geometric coordination exists between the form of a perfectly cylindrical fold and the orientation of its fold-generating axis. The coordination is best pictured stereographically. Poles to great circles representing the strike-and-dip orientations of cylindrically folded bedding lie exactly on a common great circle, perpendicular to the trend and plunge of the hingeline (Figure 11.22*A*). This means that "every part of the folded surface is oriented such that it contains a line whose orientation is identical to that of the hingeline" of the fold.

Truly cylindrical folds are rare in nature, but many folds so closely approximate purely cylindrical forms that they are considered to be cylindrical, and thus they are considered to possess axes. The term cylindrical, then, can be broadened to include **near-cylindrical folds**. Poles to great circles representing the strike-and-dip orientations of near-cylindrically folded bedding do not lie exactly on a common great circle (Figure 11.22*B*).

Noncylindrical folds do not possess fold axes. The limbs and hinge zones are so irregular in orientation, and the hingelines so crooked and/or curved, that there does not exist a single straight line that when moved parallel to itself, generates the form of the fold. When the geometrical attributes of a noncylindrical fold are plotted, the results are messy (Figure 11.22*C*). Poles to great circles representing the strike-and-dip orientations of different parts of noncylindrically folded bedding display a bewildering array of points that cannot be fit to a common great circle. To penetrate the chaos of noncylindrical folding in the course of detailed structural analysis, it is necessary to subdivide noncylindrical folds into domains of cylindrical or near-cylindrical folds, each of which is marked by a relatively short, relatively straight hingeline. When this is done, and the strike-and-dip orientations of the folded surfaces are plotted domain by domain, the stereographic patterns clean up immeasurably.

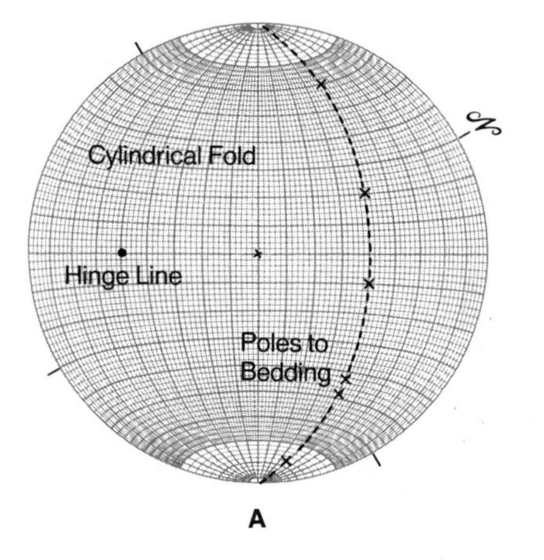

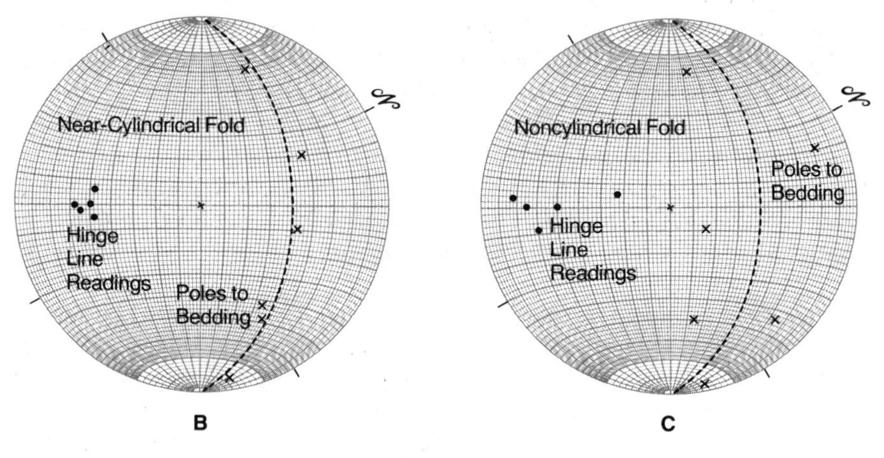

Figure 11.22 Stereographic geometry of (*A*) a perfectly cylindrical fold, (*B*) a near-cylindrical fold, and (*C*) a noncylindrical fold.

FIELD AND STEREOGRAPHIC DETERMINATIONS OF FOLD ORIENTATIONS

DIRECT MEASUREMENT OF AXES AND AXIAL SURFACES. It might be useful to review from start to finish the steps involved in measuring the orientations of axes and axial surfaces of folds. Where folds are so small that they are completely exposed in single outcrops, the orientations of fold axes and axial surfaces are usually easy to visualize and can be measured directly with a compass. Conditions of measurement are ideal when the hinge of one or more of the folded layers has weathered out as a pencillike or rodlike linear form. The hinge is taken to be a line whose orientation is parallel to the axis of the fold. Its orientation is measured and described in terms of trend and plunge and is entered in the field notebook under the "fold axis" orientation. To help visualize the hingeline when it is not weathered out in full three-dimensional relief, it is useful to align a pencil parallel to the orientation of the hingeline. The pencil provides a tangible, physical guide in shooting the trend and plunge.

Measuring the orientation of the axial surface of a fold in outcrop requires a degree of physical dexterity. Since axial surfaces are geometric elements, there is no real physical surface on which to directly measure the axial surface orientation, unless by chance or by Nature's design, there are cleavage surfaces or joint surfaces that are exactly parallel to the axial surface. In the absence of a natural physical surface on which to take a direct measurement, the strike-and-dip orientation is measured on a clipboard or field notebook held in one hand, parallel to the axial surface. To be sure that the clipboard or notebook is aligned properly, the trace of the axial surface must be visible on at least two surfaces of the outcrop. Using as reference two (or more) axial traces common to a single axial surface, the clipboard (or notebook) can be aligned in the unique attitude that satisfies the trends of each. Axial surface orientation is measured and then posted in the field notebook in terms of strike and dip.

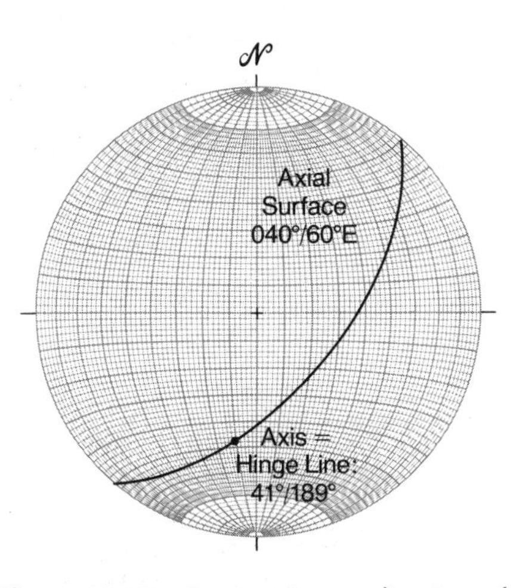

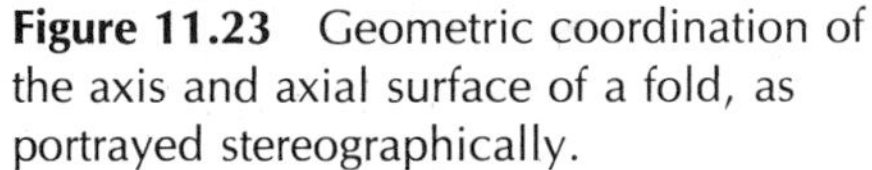
Figure 11.23 Geometric coordination of the axis and axial surface of a fold, as portrayed stereographically.

STEREOGRAPHIC DISPLAY OF AXIS AND AXIAL SURFACE. A stereographic plot of the axis and axial surface orientations measured for a single fold at a single outcrop displays the geometric coordination of these elements. An axial surface, by definition, passes through the hingelines of successive folded surfaces within a given fold. Expressed stereographically, the point representing the trend and plunge of the hingeline (or fold axis) lies on the great circle that describes the orientation of the axial surface (Figure 11.23).

STEREOGRAPHIC DETERMINATION OF FOLD AXIS ORIENTATION. It is usually impossible to measure directly in an accurate way the axis and axial surface orientations of a large fold that spills beyond the expanse of a single outcrop. Instead, the trend and plunge of the fold axis and the strike and dip of the axial surface must be calculated stereographically on the basis of representative strike-and-dip measurements of the folded surface.

In the simplest case, the trend and plunge of the axis of a fold can be computed from merely two strike-and-dip measurements, one for each limb of the folded surface. The orientations of the two limbs are plotted stereographically as great circles (Figure 11.24). The intersection of the great circles, labeled $\boldsymbol{\beta}$, represents a close approximation to the trend and plunge of the hingeline of the folded surface. The hingeline orientation, in turn, is taken to be a close approximation to the fold axis orientation. This specific stereographic construction is called a **beta ($\boldsymbol{\beta}$) diagram**, where β refers to the

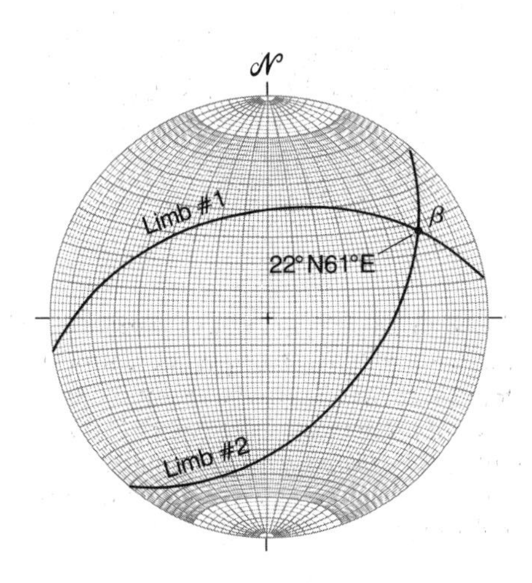

Figure 11.24 Simple β diagram. The orientation of the axis of the fold (β) is the line of intersection of the fold limbs.

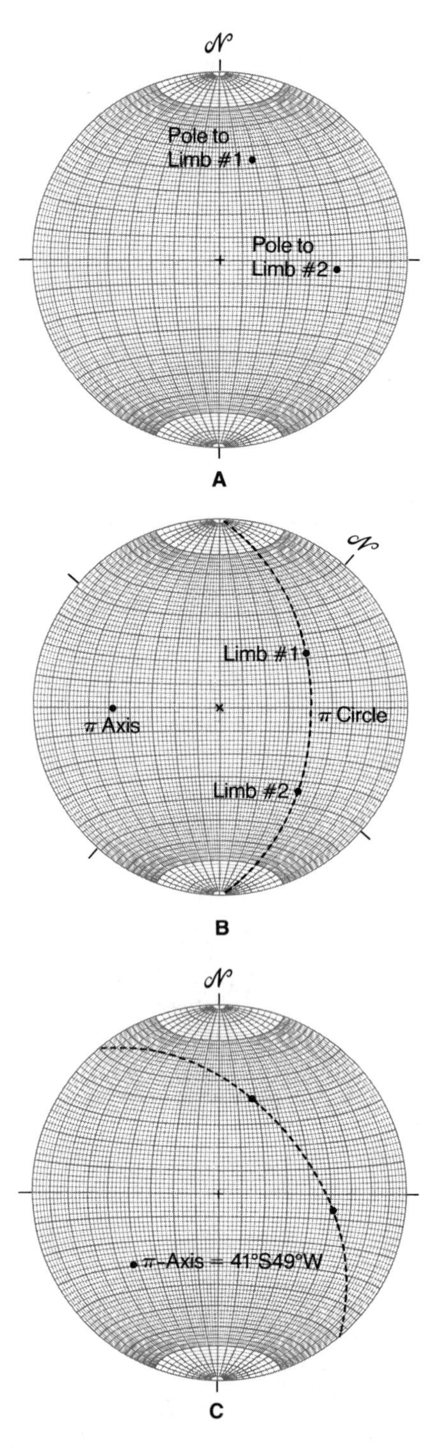

Figure 11.25 Steps in the construction of a π diagram. (*A*) Stereographic portrayal of poles to limbs. (*B*) The fitting of the poles to a common great circle (the π circle). Identification of the pole to the π circle (the π axis). (*C*) Trend and plunge of the π axis.

trend and plunge of a fold axis deduced stereographically in the manner described.

Another way to calculate stereographically the orientation of a fold is through the construction of a **pi (π) diagram**. This requires plotting the limbs of the folded surface as *poles*, not as great circles (Figure 11.25*A*). Once plotted, the poles to each limb are fitted to a common great circle, known as a **π circle** (Figure 11.25*B*). The special geometric property of a π circle is that it represents the strike and dip of a plane that is perfectly perpendicular to the hingeline of the fold. The pole to this great circle, known as the **π axis** (Figure 11.25*B*, *C*), expresses stereographically the orientation of the fold axis; π axis refers to the trend and plunge of a fold axis as deduced stereographically in this manner.

The key to comfort in understanding the π-diagram construction is visualizing that poles to a cylindrically folded surface indeed lie geometrically in a plane oriented at a right angle to the hingeline of the fold. Figure 11.26, an extraordinary view along the mined-out hinge of an anticline, can help us picture this relationship. Each of the timbers that supports the mine roof is oriented as a pole to the bedding it supports. The orientations of the timbers, taken together, define a plane at right angles to the hingeline. The miner with hands on hips in the deep recesses of the tunnel is smiling because he recognizes how closely his timber support system captures the inherent stereographic geometry of π diagrams.

STEREOGRAPHIC DETERMINATION OF AXIAL SURFACE ORIENTATION. Although the axial surfaces of small folds can normally be measured directly in outcrop, stereographic procedures are required to evaluate the axial surface orientations of large folds. The simplifying premise in the stereographic calculation is that the **bisecting surface** of a fold is a close approximation to the axial surface. For a given folded surface, the bisecting surface passes through the hingeline and splits the angle (the **interlimb angle**) between the limbs (Figure 11.27*A*). When the bisecting surface of a single folded surface is compared to the axial surface of the fold as a whole (Figure 11.27*B*), minor differences in orientation may sometimes be evident. But generally the differences are so slight that the strike and dip of the bisecting surface can be taken as the strike and dip of the axial surface.

Figure 11.26 Visual image of the geometry of a π diagram. (Photograph by C. D. Walcott. Courtesy of United States Geological Survey.)

The stereographic procedure in computing the orientation of the bisecting surface of a folded surface is reasonably straightforward. Although shortcuts can be taken, the full flavor of the method emerges by combining a β diagram and a π diagram on a common projection. First a simple β diagram is constructed by plotting the attitudes of the fold limbs as great circles, then identifying the intersection of the great circles as β (Figure 11.28*A*). Next a π diagram is added, by plotting the poles to the fold limbs and fitting these poles to a common great circle (Figure 11.28*B*). The π axis of this great circle is coincident with β. The next critical step is a visual one: recognizing that the π circle corresponds geometrically to a perfect normal profile view of a fold, like the view that was pictured in Figure 11.27*A*. Traces of bedding in the profile view are represented stereographically as the points of intersection of each limb of the fold and the π circle. The bisecting surface of the fold is the great circle that passes through the hingeline (through β) and perfectly bisects the angle between the traces of bedding as measured in the profile-view great circle (Figure 11.28*C*). In the example we are considering, the strike and dip of the bisecting surface proves to be N60°W, 66°SW (Figure 11.28*D*).

There is always some ambiguity in stereographically computing the orientation of a bisecting surface. There are actually two different points on the π circle that serve as bisectors to the limbs (see Figure 11.28*C*): one bisects the acute angle between the bedding traces; the other bisects the obtuse angle. One of these bisectors, but not both, must be used along with β as a reference point for constructing the great circle representing the orientation of the bisecting surface. To select the appropriate bisector, it is necessary to keep in mind the fold form whose orientation is being sought. If the fold is

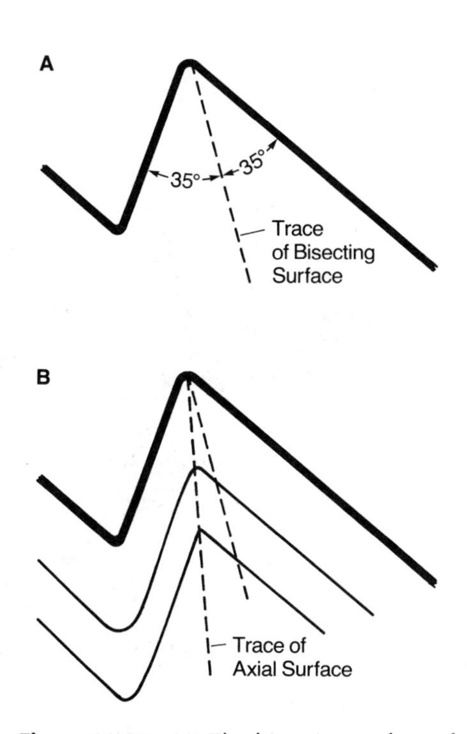

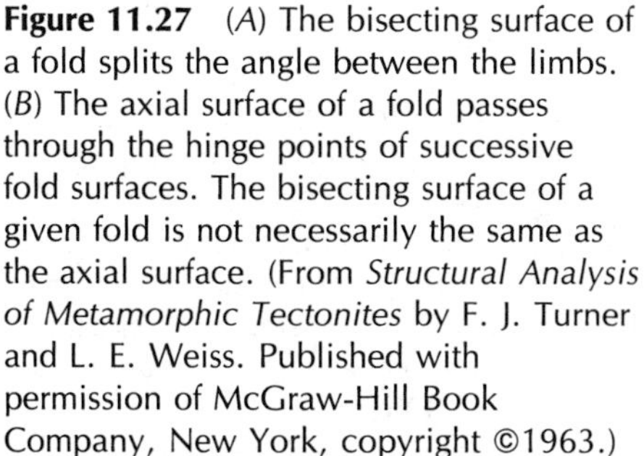
Figure 11.27 (*A*) The bisecting surface of a fold splits the angle between the limbs. (*B*) The axial surface of a fold passes through the hinge points of successive fold surfaces. The bisecting surface of a given fold is not necessarily the same as the axial surface. (From *Structural Analysis of Metamorphic Tectonites* by F. J. Turner and L. E. Weiss. Published with permission of McGraw-Hill Book Company, New York, copyright ©1963.)

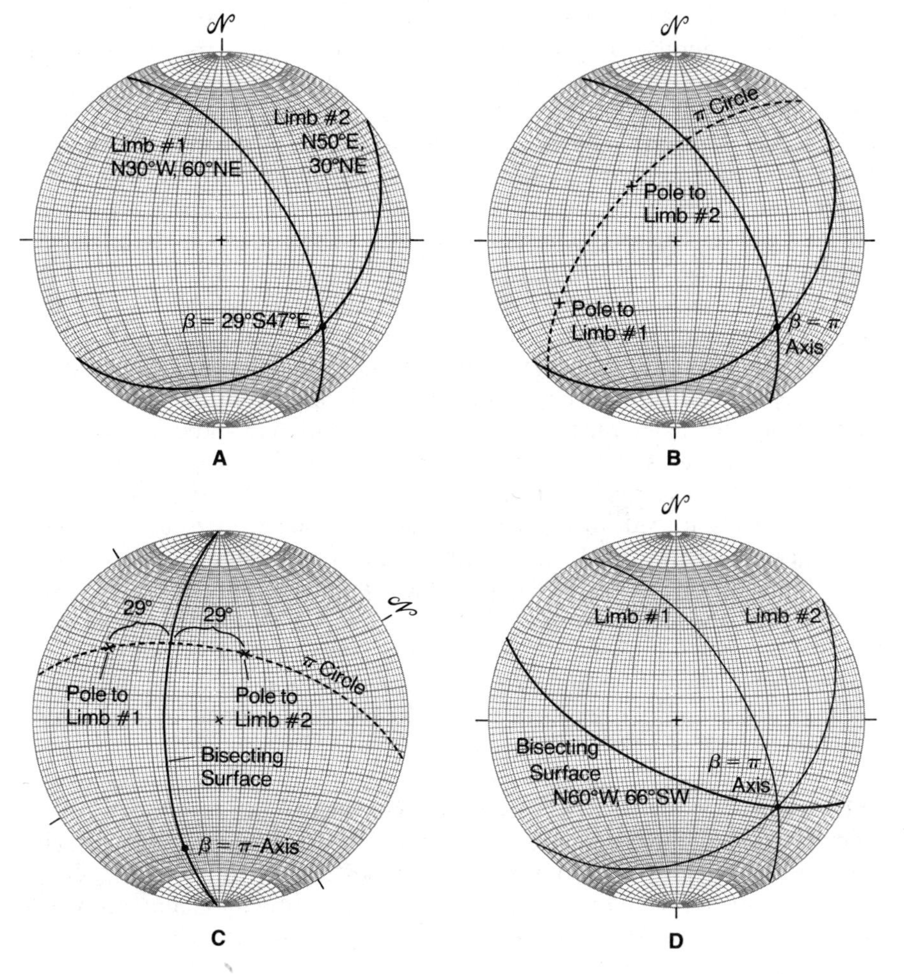

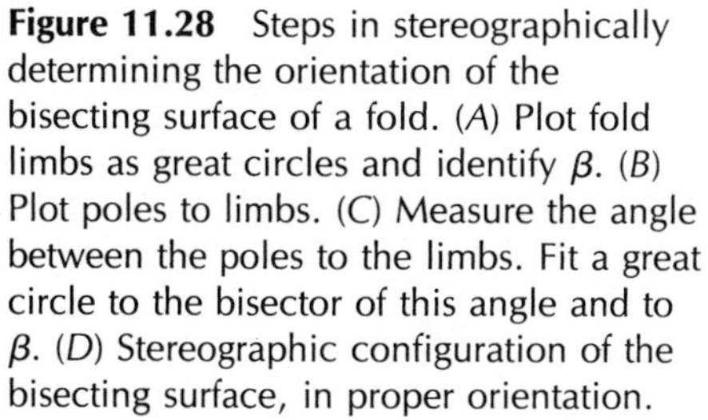
Figure 11.28 Steps in stereographically determining the orientation of the bisecting surface of a fold. (*A*) Plot fold limbs as great circles and identify β. (*B*) Plot poles to limbs. (*C*) Measure the angle between the poles to the limbs. Fit a great circle to the bisector of this angle and to β. (*D*) Stereographic configuration of the bisecting surface, in proper orientation.

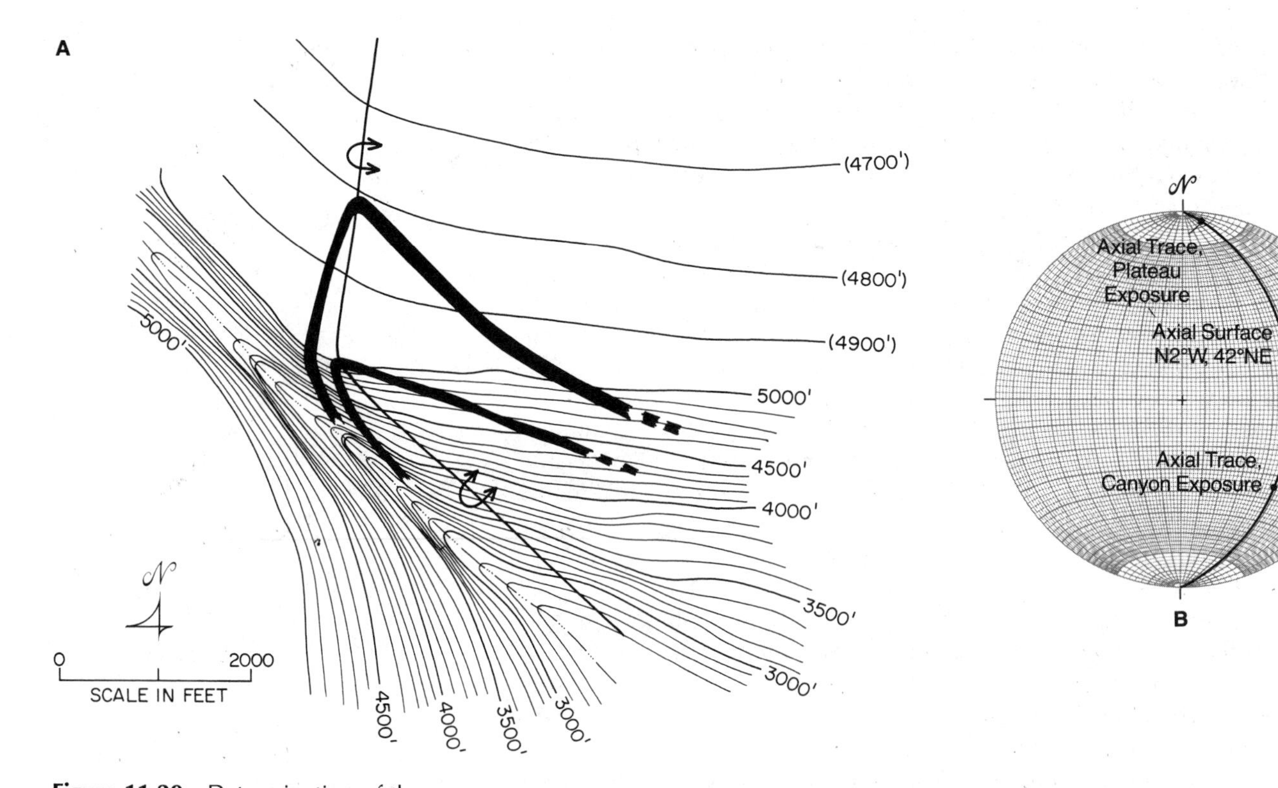

Figure 11.29 Determination of the orientation of the axial surface of a fold on the basis of axial trace orientations. (*A*) Geologic map expression of folded beds. Two nearly straight axial traces can be identified, one crossing the high, flat tableland, the other exposed along the canyon wall. Orientations of these axial traces can be measured from the map relationships. (*B*) The axial surface orientation is determined stereographically by fitting the two axial trace orientations to a common great circle.

upright, the proper bisector is one that yields a relatively steeply inclined bisecting surface. If the fold is overturned or recumbent, the proper bisector is one that yields a relatively low-dipping bisecting surface.

There is yet another way to calculate, stereographically, the orientation of the axial surface of a large fold. And this method does not depend on measuring the orientation of the bisecting surface as an approximation to the axial surface. The procedure is to plot stereographically the trend and plunge of two or more axial traces of the fold as points, and to fit these points (representing the orientations of lines) to a common great circle. Two lines define a plane. The great circle, constructed in this manner, describes the orientation of the axial plane.

An example of the use of this method is shown in Figure 11.29. The data base is a simplified geologic map of an overturned anticline. As revealed by the topographic contour lines, the fold is exposed in a plateaulike terrain cut by a steep-walled canyon (Figure 11.29*A*). The topography affords excellent exposures of the axial trace of the anticline, both in map and cross-sectional views. On the plateau surface the axial trace of the fold is seen to trend N5°E along a reasonably straight line that passes, as it should, through hinge points of successively folded layers. The plunge of the axial trace, as calculated trigonometrically (or graphically) from the elevation control available, is 5° NE. The axial trace of the fold exposed to view in the northeast wall of the canyon trends S47°E. Its plunge is 32° SE (Figure 11.29*A*). The two axial trace orientations are lines, one plunging 5° N5°E, and the other plunging 32° S47°E. When these two lines are plotted stereographically and fitted to a common great circle, the overall attitude of the axial surface (axial plane) is found to be N2°W, 42°NE (Figure 11.29*B*).

π DIAGRAMS AND β DIAGRAMS, IN PRACTICE. Because folds are seldom perfectly cylindrical, the construction of π and β diagrams usually requires the plotting of more than two strike-and-dip orientations. More readings, typically five or ten, are required to describe the degree to which the geometry of the folded surface departs from an ideally cylindrical form.

Based on the **scatter** of readings, an estimate of the average/approximate fold axis orientation can be made.

Rarely will five or ten great circles that are plotted on a β diagram intersect in a tightly constrained bulls-eye, let alone in a single point. The greater the dispersion of intersection points, the more difficult it is to estimate the average trend and plunge of the hingeline of the fold. The estimation may even require statistical methods to locate the exact center of distribution of the intersection points (Ramsay, 1967, p. 17–18). If the intersection points are tightly grouped in a dense cluster, the fold is considered to be cylindrical or near-cylindrical. If the intersection points spray all over the diagram, the fold is considered to be noncylindrical and the fold axis orientation is indeterminate.

π Diagrams do not eliminate the problem of scatter. When tens of poles to a folded surface are plotted on a common π diagram, the poles seldom (rarely) lie right on a discrete π circle. Instead they are dispersed to one degree or another and must be fitted by eye and/or by statistical methods to a **best-fit** great circle (Ramsay, 1967, p. 18–20). If the poles to the folded surface tend to lie within 20° of the best-fit great circle, the fold is considered to be **cylindrical** or **near-cylindrical**. Where they scatter well beyond the 20° limit, the fold is considered to be **noncylindrical** and determinations of π-circle and π-axis orientations are not very meaningful. π Diagrams displaying more than 50 poles can be contoured as an aid in defining the best possible fit to a common great circle (Figure 11.30).

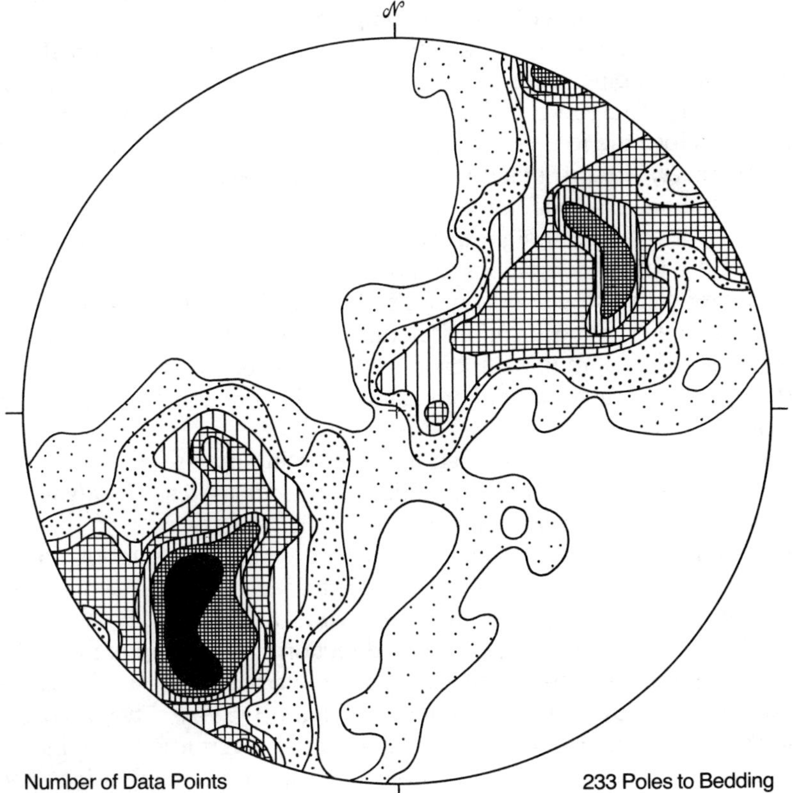

Figure 11.30 Use of contoured pole-density diagram as basis for determining best-fit great circle to scores of poles to folded bedding in Cretaceous strata in the Santa Rita Mountains, southern Arizona.

It is generally easier to fit an array of poles to a common great circle than to estimate the center of distribution of a broad scattering of intersection points. For this reason alone, π diagrams are generally favored over β diagrams in the plotting of more than several strike-and-dip orientations. β diagrams escalate in complexity as great circles are added, one by one. Whitten (1966, p. 50) has pointed out how rapidly the intersection points multiply. Here is the relationship.

$$i = \frac{n(n-1)}{2} \tag{11.1}$$

where i = number of intersections and n = number of great circles.

Thus when 34 different strike-and-dip attitudes of a folded surface are plotted as 34 different great circles, the intersection points created among these great circles number 561!

PLOTTING FOLD-ORIENTATION DATA ON GEOLOGIC MAPS. Fold-orientation data, whether measured directly in the field or computed stereographically from representative strike-and-dip orientations, should become part of the geologic map record. How many of the orientation data are plotted depends largely on the size and scale of the final, rendered map. Orientation data that do not become part of the final geologic map are not lost. Rather, they constitute an integral part of stereographic displays of the preferred orientations of bedding, axes, and axial surfaces, displays that support the map relationships in a complementary way.

Formation contacts and marker beds that are plotted in the normal course of geologic mapping serve to outline the general form of map-scale folds. Strike-and-dip data representing the orientations of bedding and/or foliation help to disclose the fold configurations. The main fold-orientation information, however, is represented by symbology for axial trace, axis, and axial surface attitude. The axial trace of each fold is drawn not as a ruled straight line but as a smooth, straight to curved line that passes through the hinge points of the folded formations and marker beds (Figure 11.31). The trend and plunge of the fold axis and the strike and dip of the axial surface are plotted at one or more locations along each axial trace. Plotting axis and axial surface data at multiple locations along a given axial trace is necessary to describe the changes in fold orientation and configuration across the area of investigation.

The conventional symbols used in representing map-scale folds on geologic maps are listed in Table 11.1. The symbols are used to identify whether a given fold is an anticline or syncline, or an antiform or synform. Overturned folds receive special attention. Like the standard symbols for anticlines and synclines, the symbols for overturned anticlines and synclines represent the dip directions of the limbs with small arrows. For overturned folds, these arrows point in a common direction, like the dip of the limbs of the overturned folds. The arrowheads point toward the U-shaped closure on the symbol of an overturned syncline (see Figure 11.29*A*). They point away from the U-shaped closure on the symbol for an overturned syncline.

Representative minor folds are plotted on geologic maps using special symbols (see Table 11.1). Each symbol portrays the strike and dip of the axial surface, the trend and plunge of the fold axis, and the generalized form of the folded surface as it is viewed **down plunge**. Symmetrical folds are shown as M-shaped forms. Asymmetrical folds are shown as Z-shaped and S-shaped forms. It is visually useful to orient the M's, S's, and Z's on the map in such a way that the axial trace of the minor fold bisects the form of the fold. If the

Table 11.1
Geologic Map Symbology for Folds and Fold Elements

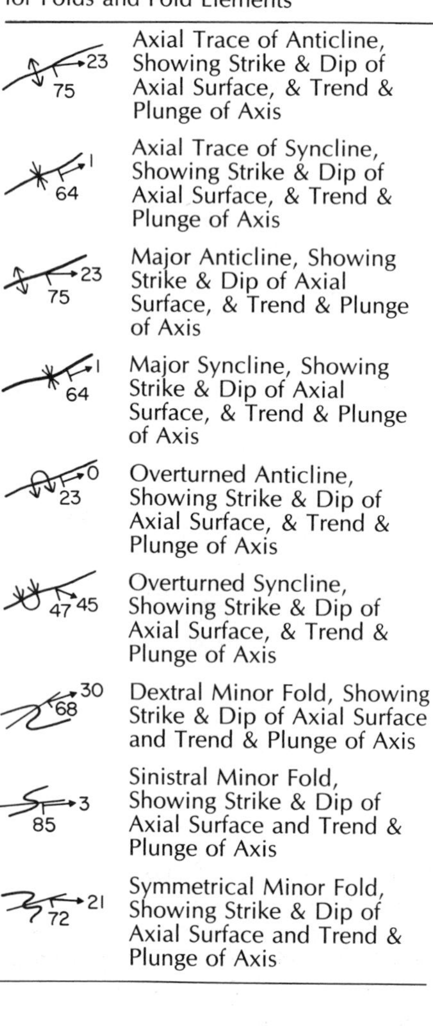

23, 75	Axial Trace of Anticline, Showing Strike & Dip of Axial Surface, & Trend & Plunge of Axis
1, 64	Axial Trace of Syncline, Showing Strike & Dip of Axial Surface, & Trend & Plunge of Axis
23, 75	Major Anticline, Showing Strike & Dip of Axial Surface, & Trend & Plunge of Axis
1, 64	Major Syncline, Showing Strike & Dip of Axial Surface, & Trend & Plunge of Axis
0, 23	Overturned Anticline, Showing Strike & Dip of Axial Surface, & Trend & Plunge of Axis
47, 45	Overturned Syncline, Showing Strike & Dip of Axial Surface, & Trend & Plunge of Axis
30, 68	Dextral Minor Fold, Showing Strike & Dip of Axial Surface and Trend & Plunge of Axis
3, 85	Sinistral Minor Fold, Showing Strike & Dip of Axial Surface and Trend & Plunge of Axis
21, 72	Symmetrical Minor Fold, Showing Strike & Dip of Axial Surface and Trend & Plunge of Axis

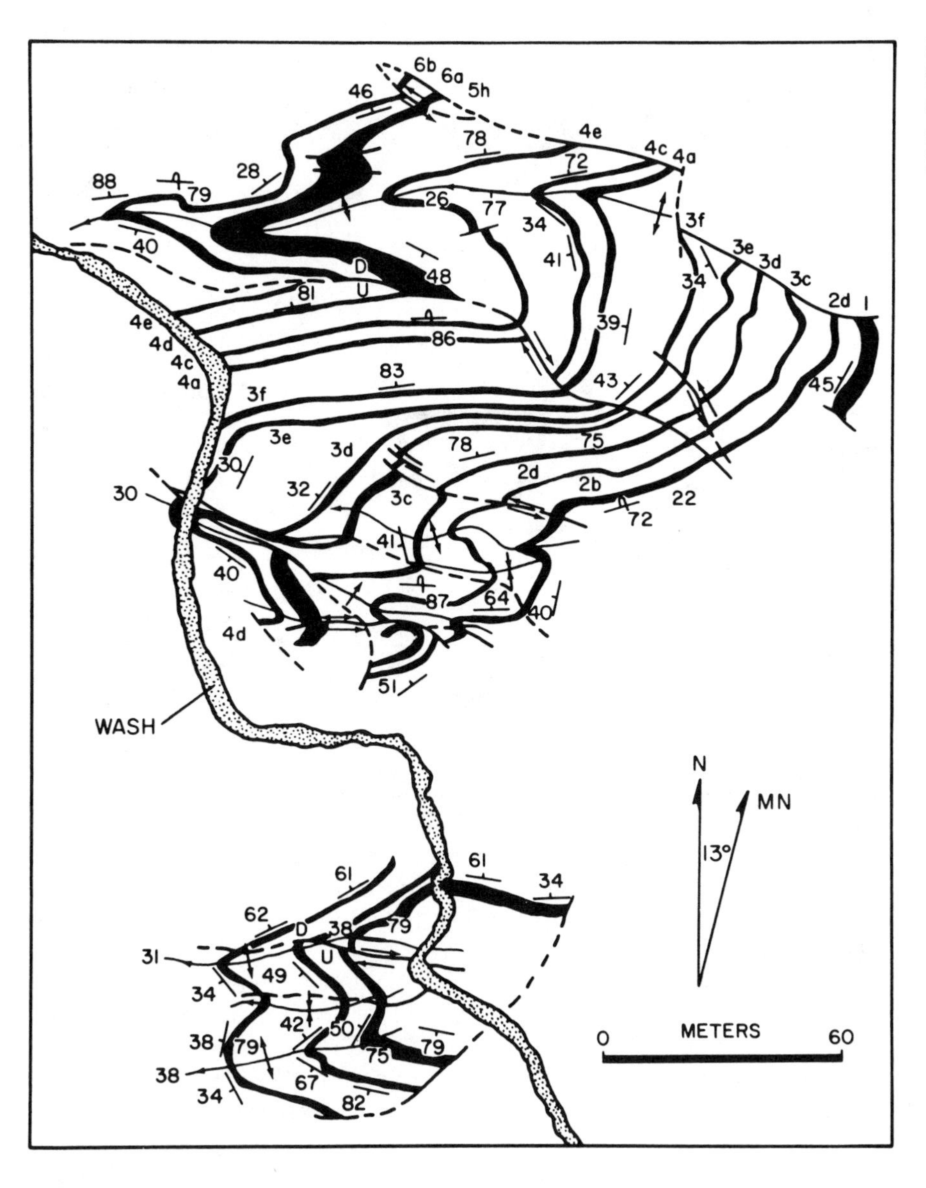

Figure 11.31 Structural geologic map of folded strata at the Agua Verde locality near Tucson, Arizona. Alphanumeric symbols like "3f" serve to identify marker beds. (Map based on work by D. Gossett.)

scale of mapping permits, strike-and-dip orientations of the limbs of the minor folds should be posted as well.

DESCRIBING THE SHAPE AND SIZE OF A FOLDED SURFACE

COMMON FOLD SHAPE. As part of the overall description of a folded surface it is useful to convey a sense of the shape of the fold, including its tightness. Fold shape is described in normal profile view. Normal profile views of folded surfaces are afforded by appropriately oriented outcrop exposures, photographs, geologic cross sections, and rock slabs or thin sections.

All the conventional terms for describing the profile shape of a folded surface attempt to convey a picture of the form and the configuration of limbs and hinge (Figure 11.32). A **chevron fold**, for example, is marked by planar limbs that meet at a discrete hinge point or at a very restricted subangular hinge zone (Figure 11.32*A*). A **cuspate fold** exhibits curved limbs that are opposite in sense of curvature to those of most ordinary folds (Figure 11.32*B*). An upright, cuspate anticline displays limbs that are concave up-

Figure 11.32 Some common fold shapes.

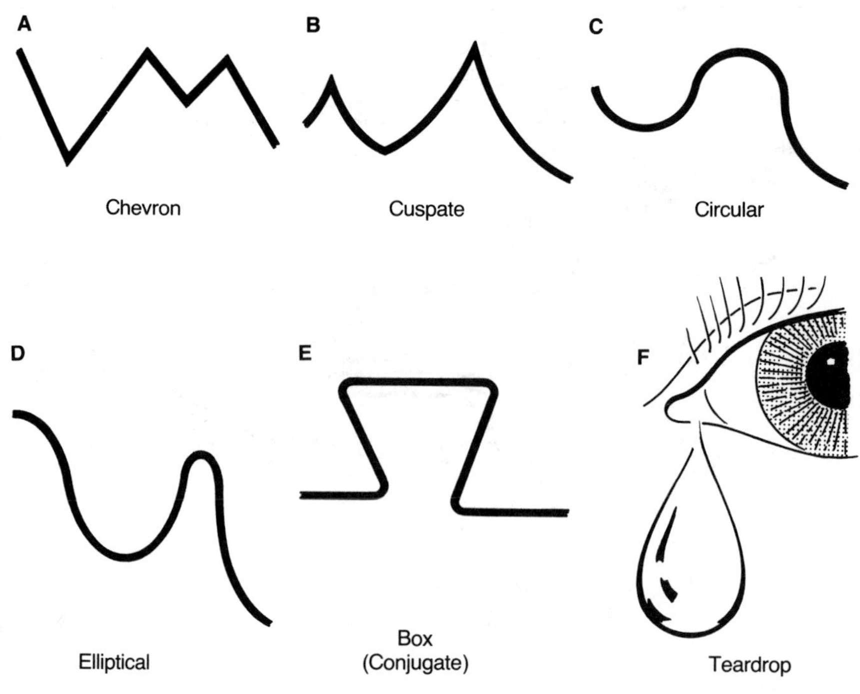

ward; an upright, cuspate syncline has limbs that are concave downward. Oddly enough, there is no conventionally used term to describe a folded surface whose profile form is wholly part of a circular arc (Figure 11.32*C*), nor is there a term to describe a folded surface whose profile form is part of an ellipse (Figure 11.32*D*). For the first, I would propose the term **circular fold**; for the second, **elliptical fold**.

Some folded surfaces have two hinges. **Box folds** (or **conjugate folds**) are composed of three planar limbs connected by hinge points or narrow, restricted subangular hinge zones (Figure 11.32*E*). **Teardrop folds** are continuously curved folded surfaces shaped, of course, like teardrops or mushrooms. They are involuted and curve back on themselves (Figure 11.32*F*).

FOLD TIGHTNESS. Fold tightness is described in terms of **interlimb angle** (Ramsay, 1967), the internal angle between the limbs of the folded surface. Although the interlimb angle of a folded surface can be measured with a protractor on the surface of a profile exposure of a small fold, or from a profile-view photograph of a large fold, profile views of folds are the exception, not the rule. Consequently it is usually necessary to calculate interlimb angles stereographically. This is achieved by taking the strike and dip of the folded surface at each inflection point, plotting the orientations stereographically as poles, fitting the poles to a common great circle, and measuring the angle between the poles along the common great circle (Figure 11.33). To know whether the acute or obtuse angle between the poles is the appropriate interlimb angle, it is necessary to keep clearly in mind the general form of the fold. The interlimb angle of a very tight fold is acute. The interlimb angle of a very open fold is obtuse.

The measured value of interlimb angle provides a basis for choosing an adjective that describes fold tightness. Figure 11.34 shows a classification scheme adapted but slightly modified from the nomenclature proposed by Fleuty (1964). **Gentle folds** are marked by interlimb angles ranging from 170° to 180°. **Open folds** have interlimb angles ranging from 90° to 170°. Folds are considered to be **tight** if they display interlimb angles in the range

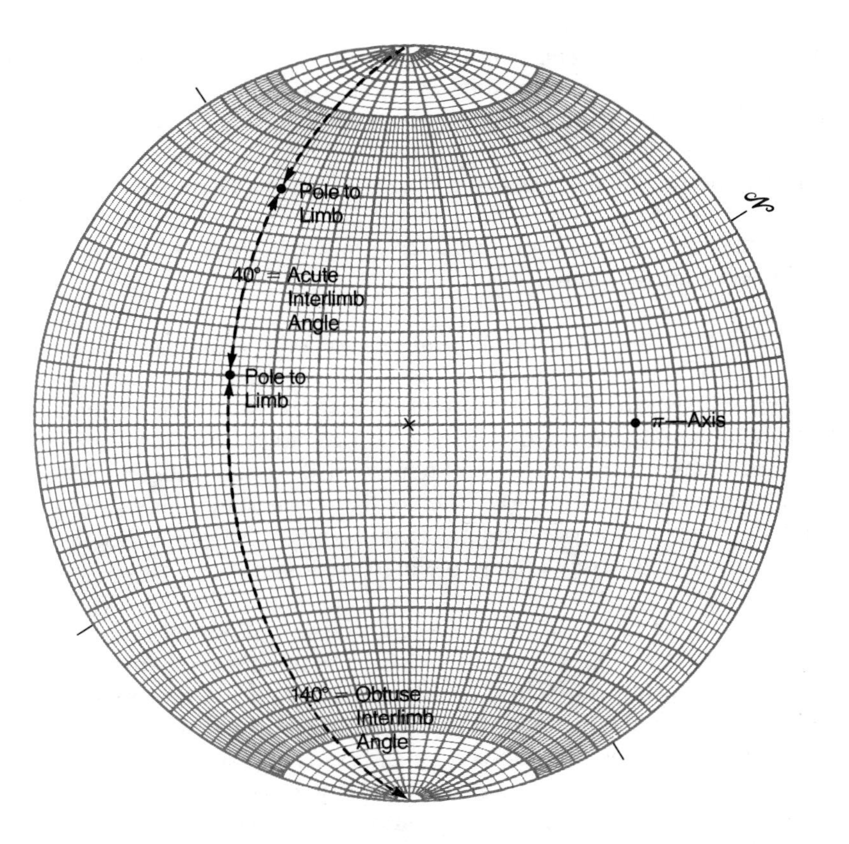

Figure 11.33 Stereographic determination of the interlimb angle of a fold.

of 10° to 90°. And **isoclinal folds** are marked by interlimb angles in the range of 0° to 10°. The cutoffs for isoclinal, tight, open, and gentle are easy to remember: 10°, 90°, and 170°.

FOLD SIZE. Fold size is surprisingly difficult to describe. The standard measures of **wavelength** and **amplitude** can seldom be employed because so many folds occur as solitary, isolated structures and not as obvious parts of continuous, repeated, **sinusoidal** wave forms. Many folds encountered in the field are not linked structurally to other folds: they are **rootless**, cut off on either side by faults and/or shear zones. Some folds that appear to be rootless may in fact be continuous with other folds, but the connection cannot be demonstrated because of the fortunes of erosion and/or the quality of exposure(s). Even where a fold can be shown to occur within an interconnected system of folds, the shapes and sizes of individual folds within a wave train may vary tremendously, not at all like an ideal wave. Given these problems and limitations, I find it practical to describe the size of a folded surface in terms of two measures: **fold height** and **fold width**, as measured in profile view.

To describe exactly what is meant by fold height and fold width it is necessary to introduce the term **median surface**. A median surface of a fold is an imaginary, geometric surface that passes through all the inflection points of a given folded surface (Figure 11.35*A*) (Ramsay, 1967). Both fold height and fold width are measured with respect to the **median trace** of the folded surface, that is, the trace of the median surface as seen in a profile view of the fold (Figure 11.35*B*). I find it practical to describe **fold height** as the distance between the median trace and the hinge point of the folded surface as measured along the axial trace of the fold, and **fold width** as the distance between inflection points on a folded surface as measured along the median trace (Figure 11.35*C*, *D*).

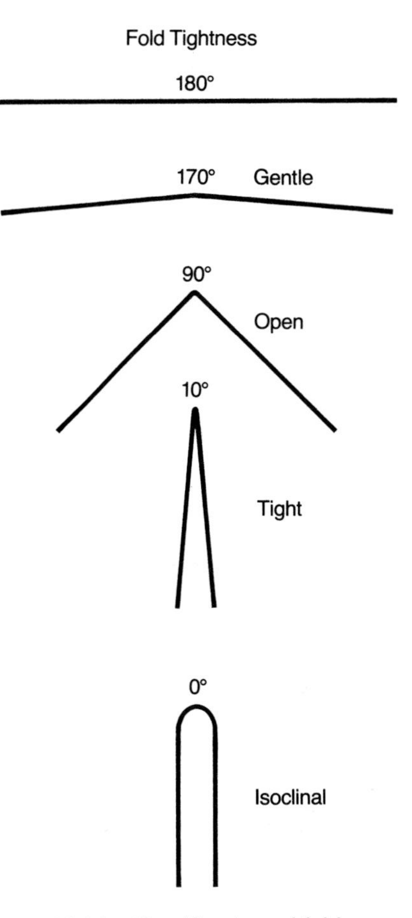

Figure 11.34 Classification of folds according to tightness, based on the size of the interlimb angle. [Modified from Fleuty (1964). Published with permission of the Geologists Association.]

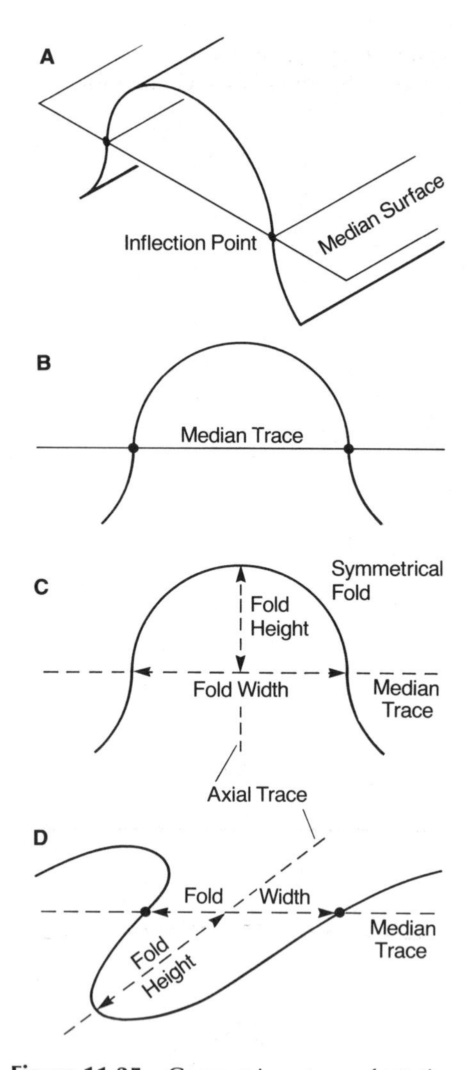

Figure 11.35 Geometric nature of (*A*) the median surface and (*B*) the median trace of a fold. Convention for measuring fold height and fold width of (*C*) symmetrical and (*D*) asymmetrical folds.

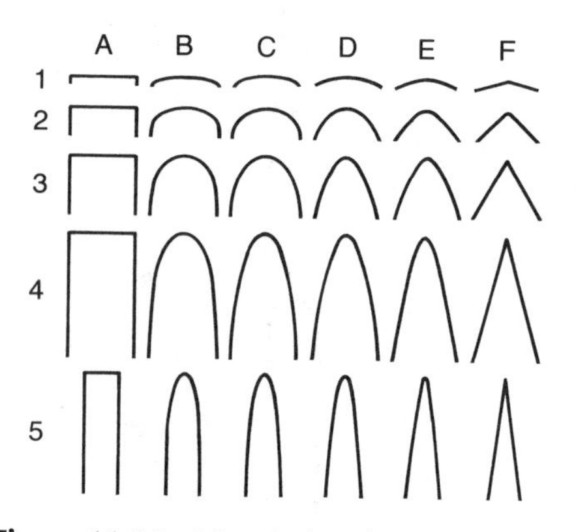

Figure 11.36 Visual classification of the shape(s) of individual folded surfaces. [From Hudleston (1973), *Tectonophysics*, v. 16. Published with permission of Elsevier Scientific Publishing Company, Amsterdam.]

FOLD SYMMETRY. The overall **symmetry** of a fold can be described in terms of the angular relationship of its median trace and axial trace. **Symmetrical folds** are characterized by a median trace and an axial trace that are mutually perpendicular; thus fold height and fold width are measured along mutually perpendicular lines (see Figure 11.35*C*). **Asymmetrical folds**, on the other hand, are marked by limbs of different lengths; thus the median trace and the axial trace of an asymmetric fold intersect at some oblique angle (see Figure 11.35*D*).

OVERALL FORM. The overall form of a folded surface owes its character to a combination of factors, including shape, tightness, symmetry, and ratio of height to width. Thus, strings of adjectives are normally required to describe adequately the profile form of a given folded surface: for example, "the fold is best described as a tight symmetrical cuspate anticline." If we add adjectives bearing on the geometric configuration of the folded surface, the string of adjectives becomes unmanageably long: "the fold is best described as a gently plunging moderately inclined tight attenuated symmetrical cuspate anticline."

With the goal of trying to convey more detail about fold form in fewer words, Hudleston (1973) devised a **visual classification scheme** that aids in categorizing the forms of folded surfaces. Using the Hudleston classification, the shape of a folded surface, from hinge point to inflection point, is compared to 30 idealized fold forms arranged systematically by number (1 to 5) and letter (*A* to *F*) (Figure 11.36). In using Hudleston's scheme it is simplest to reproduce the 30 basic fold forms on a plastic or Mylar template, then compare the forms to the folded surface in question by peering through the template toward the outcrop, photograph, or geologic cross section portraying the fold. The payoff in using this technique comes in discovering that certain rock types and/or structural domains are characterized by certain specific fold shapes.

SUGGESTIONS FOR DESCRIBING FOLDS IN OUTCROP. Analysis of the orientations of shapes of folded surfaces demands care in systematically collecting and recording the pertinent data. Suppose we encounter a well-exposed fold in profile view in outcrop, of the type shown in Figure 11.37*A*. What should we measure and describe before leaving the outcrop? Here are my suggestions. First, make a sketch of the fold in profile view, showing as clearly as possible the shape of one or two or more of the individual surfaces (like bedding surfaces) that separate the folded layers (Figure 11.37*B*). Label each end of the profile sketch with a direction indicator, like NW (northwest), and place a simple bar scale below the sketch for approximate size reference. If fold analysis is a serious part of the ongoing study, photograph the fold, in normal profile view if possible, making sure that there is some reference scale in the picture. The photograph will capture details of fold shape, fold size, and fold tightness, which can be studied later in the laboratory.

Next, measure representative strike-and-dip attitudes of the folded layers. For a house-size fold, measure two or three readings on each limb, and three or four more within the hinge zone. For desk-size folds, one reading on each limb and two readings in the hinge zone, if possible, will suffice. Record on the profile sketch of the fold the locations of each reading (Figure 11.37*B*). Finally measure the orientations of the axial surface and the hingeline of the fold, recording these in the field notebook along with the rest of the data.

With this information in hand, the fundamental field operations bearing on shape, size, and orientation analysis are completed. The only remaining step

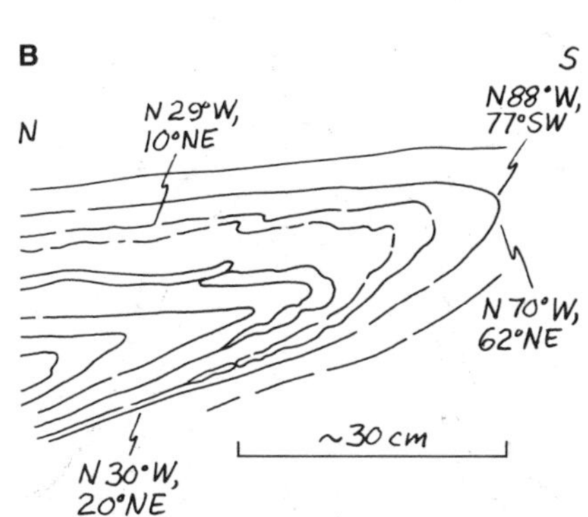

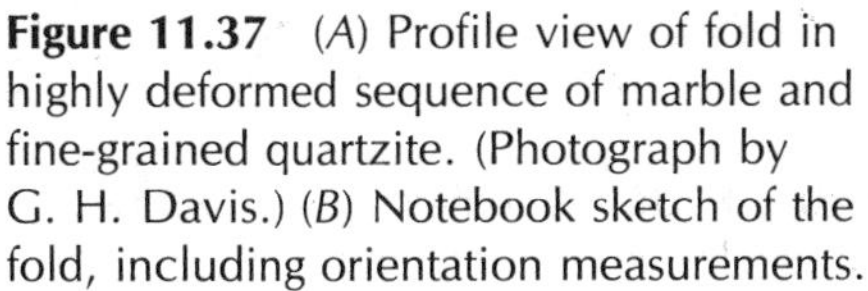
Figure 11.37 (*A*) Profile view of fold in highly deformed sequence of marble and fine-grained quartzite. (Photograph by G. H. Davis.) (*B*) Notebook sketch of the fold, including orientation measurements.

before heading to the next outcrop is posting the orientation data on the geologic map.

CLASSIFYING FOLDS BY CHANGES IN LAYER THICKNESS

THICKNESS CHANGES, A REFLECTION OF DISTORTION. Some **folded layers** maintain uniform thickness across the full profile view of the fold. Other layers show striking, systematic variations in layer thickness. Whether the thickness of a rock layer is modified during folding depends on the internal stresses that it is forced to bear and the rock's strength to resist. Remarkably, the degree of distortion from one layer to the next is always somehow perfectly regulated to assure a perfectly compatible fit among layers within the folded sequence. Seldom are there gaps or overlaps! Such **strain compatibility** from folded layer to folded layer is yet more of the magic of strain.

CONCENTRIC FOLDS. Individual folded layers that are marked by uniform thickness are known as **concentric folds** (Van Hise, 1896). The profile forms commonly are circular or elliptical. Surfaces that separate individual folded layers in an ideal concentric fold are perfectly parallel to one another, like the rails of a curved train track at a bend in the line. Because of this distinctive geometric characteristic, concentric folds are also known as **parallel folds**.

An unexpected geometric peculiarity arises from parallel folding: the profile form of folded layers must continuously change upward and downward within the folded sequence, until the folds gradually disappear altogether. For example, an upright anticline becomes progressively tighter downward within a concentrically folded sequence, ultimately transforming into a narrow, pinched, cuspate anticline before completely dying out (Figure 11.38). Upward, the concentric anticline progressively flattens into a very gentle arc before vanishing. Synclines behave in the opposite manner. Upright synclines pinch out upward in very tight cuspate folds. Downward they gradually become gentle dishlike folds before subtly merging with deeper, unfolded layers.

Figure 11.38 Geometric properties of an ideally concentric anticline.

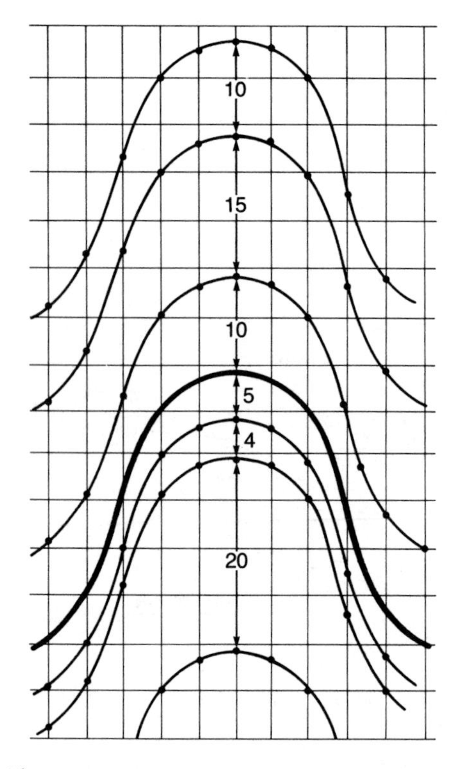

Figure 11.39 Geometric construction of an ideally similar anticline.

The geometric idiosyncrasies of concentric folding can be more fully appreciated by graphically constructing a structure profile view of an upright, circular concentric anticline. Concentric circular arcs representing folded surfaces are drawn with a drafting compass (as in Figure 11.38). The circular arcs serve to define the boundaries of individual folded layers, which maintain uniform thickness *as measured perpendicular to layering*. As the arcs are drawn, one by one, it becomes more and more difficult to propagate the form of the anticline to depth. A room problem develops and it becomes impossible to fit a decent circular arc into the available space. The room problem is satisfied by replacing the folded layers above with unfolded flat-lying layers below. In essence, the folded layers are **decoupled** from their underlying foundation. Nature achieves decoupling through formation of a surface of "unsticking," a decollement zone of layer-parallel slippage and rock flowage. The last remaining vestige of the concentric anticline that can be constructed is a tiny cuspate fold. Between the cuspate anticline and the flat-lying strata below, a small amount of open space is created. In natural systems this open space is immediately filled by soft incompetent rock, capable of distortional flow during folding.

SIMILAR FOLDS. Individual folded layers that display significant thickening in the hinge and thinning on the limbs are known as **similar folds**. Perfect similar folds are marked by layers whose upper and lower surfaces are virtually identical in shape (Van Hise, 1896). Because this is so, the form of an ideally similar fold can be propagated upward and downward for *any* distance without change. The secret to the geometry of an ideal similar fold is that *layer thickness measured parallel to the axial trace of the fold remains constant.*

The geometric intrigue of similar folding can be appreciated through another graphical construction (Figure 11.39). The first step is to draw, arbitrarily, the profile form of a single folded surface in the middle of a long sheet of paper. The next step is to construct folded layers above and below this folded surface, carefully building each folded layer by measuring and maintaining a constant thickness *parallel to the axial trace*. When this construction is carried out with care and precision, there is *never* a departure of individual folded surfaces from the starting profile form. The ancient road builders in northern Italy knew and used this geometric principle (Figure 11.40). Specific fold forms made of stone are propagated along ribbons of highway without change. The stones in individual layers were chosen and/or cut in such a way that lengths measured parallel to the roadways are of constant value.

FULL RANGE OF SHAPES OF FOLDED LAYERS. Concentric folds and similar folds are simply two special cases within a broad range of possible shapes of folded layers. Ramsay (1967, p. 359–372) was able to demonstrate that fundamental classes of folded layers can be distinguished on the basis of **relative thickness** of the folded layer in the hinge versus the limbs. He showed that the **relative curvature** of the upper and lower bounding surfaces of an individual folded layer is also a sensitive index to systematic variations in layer thickness.

Three classes of folds were distinguished by Ramsay on the basis of the relative curvature of the upper and lower bounding surfaces—that is, upper and lower "**arcs**"—of a folded layer (Figure 11.41). **Class 1 folds** are marked by a curvature of the inner arc that is greater than that of the outer arc. **Class 2 folds** are ideal similar folds, distinguished by identical curvatures of the

Figure 11.40 Expression of the fundamental geometry of similar folding in an ancient roadway of northern Italy. (Photograph by G. H. Davis.)

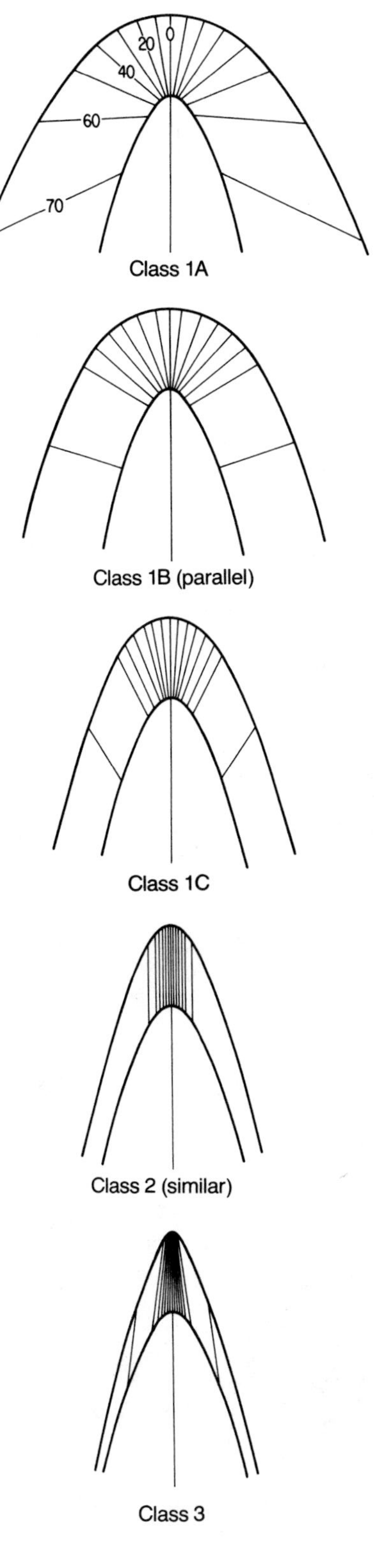

Figure 11.41 The fundamental classes of shapes of folded layers. (From *Folding and Fracturing of Rocks* by J. G. Ramsay. Published with permission of McGraw-Hill Book Company, New York, copyright ©1967.)

inner and outer arcs. **Class 3 folds** are marked by curvature of the outer arc that is greater than that of the inner arc.

Ramsay further subdivided Class 1 folds into three types on the basis of thickness variations (Figure 11.41). **Class 1A folds** are marked by a layer thickness in the hinge that is less than layer thickness on the limbs. **Class 1B folds** are ideal concentric folds, distinguished by uniform layer thickness across the whole fold profile. **Class 1C** are intermediate between ideal concentric folds (Class 1B) and ideal similar folds (Class 2). They show a modest thickening in the hinge, and a modest thinning on the limbs.

DISTINGUISHING FUNDAMENTAL FOLD CLASSES. The assignment of a given folded layer to one of the five fundamental fold classes can be carried out in a qualitative way on the basis of an eyeball estimate of relative curvature and relative thickness. But the power of Ramsay's approach is best appreciated through the actual measurement of relative curvature and relative thickness. Normal profile views of the folded layers are used as the data base for carrying out the necessary constructions and measurements.

By convention, the fold profile under study is rotated into the orientation of a perfectly upright antiform (Figure 11.41). Then **dip isogons** connecting points of equal inclination on the outer and inner bounding surfaces of the folded layer are constructed graphically. Once constructed, the dip isogon pattern sensitively reveals differences in outer arc and inner arc curvature, thus providing a basis for assigning the folded layer to Class 1, 2, or 3 (Figure 11.41).

Class 1 folds are distinguished by dip isogons that converge downward, signifying that the curvature of the outer arc is less than that of the inner arc. Dip isogons drawn for Class 2 folds are strictly parallel to one another, revealing that curvatures of the outer arc and the inner arc of the fold are identical. Class 3 folds are marked by dip isogons that diverge downward, because outer arc curvature exceeds inner arc curvature (Figure 11.41).

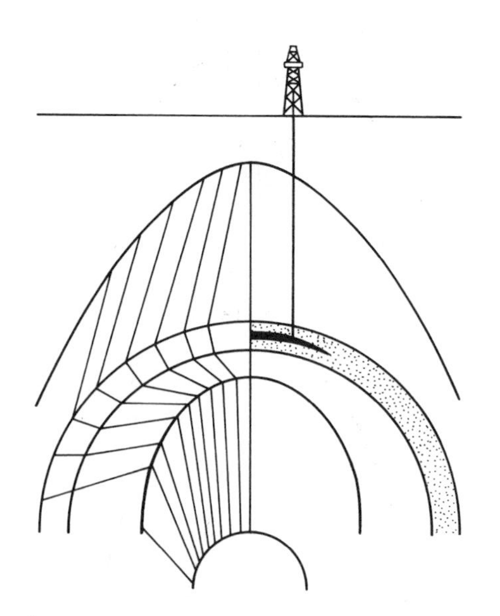

Figure 11.42 Schematic diagram showing how changes in inclination of dip isogons reflect changes in shape(s) of folded layers.

Dip isogon patterns are especially revealing when they are drawn for a series of folded layers of different shapes (Figure 11.42). The divergence, convergence, and parallelism of dip isogons as they cut through a folded sequence of layers draws attention to the variety of classes of folded layers that can be represented in a single structure. Dip isogon "maps" of folds call attention to layer shape distortion as a function of rock type.

Ramsay (1967) presented an even more sensitive approach to sorting out the shapes of folded layers according to classes. The basis for analysis is a geologic cross section or a profile-view photograph (Figure 11.43*A*). The method requires comparing layer thickness measured at a number of locations on the limbs of a fold with layer thickness measured in the hinge. The thickness that is measured is **true thickness. Limb thickness (t_α)** is identified according to **limb inclination (α)** as measured when the fold is positioned as a perfectly upright antiform (Figure 11.34*B*). Limb thickness is taken to be the spacing between tangents drawn through control points of equal inclination value on the upper and lower surfaces of the folded layer (Figure 11.43*B*). The thicknesses thus measured are recorded and become the basis for constructing a graph allowing fold class to be precisely assigned.

The graph which reveals layer shape is a plot of **relative thickness (t')** versus **inclination angle (α)** (Figure 11.43C). Relative thickness (t') is the ratio of thickness (t_α) measured at a location on the fold limb to thickness (t_o) measured in the hinge of the fold:

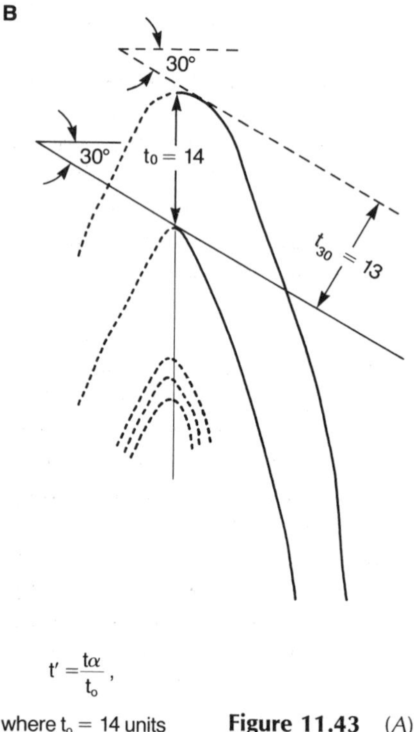

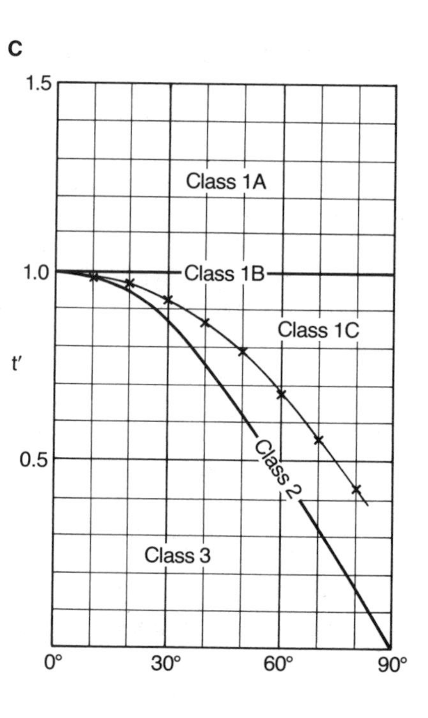

$$t' = \frac{t\alpha}{t_o},$$

where t_o = 14 units

α	$t\alpha$	t'
10°	13.8	0.98
20°	13.5	0.96
30°	13.0	0.93
40°	12.0	0.86
50°	11.0	0.79
60°	9.5	0.68
70°	7.8	0.56
80°	6.0	0.43

Figure 11.43 (*A*) Normal profile view of a tightly folded sequence of layers. (*B*) Construction steps used in determining layer thickness at layer inclinations of 0° and 30°. (C) The standard graphical expression of the variations in layer thickness that characterize each of the major classes of folds. Note that the fold shown in *A* plots as a Class 1C fold. (From *Folding and Fracturing of Rocks* by J. G. Ramsay. Published with permission of McGraw-Hill Book Company, New York, copyright ©1967.)

$$t' = \frac{t_\alpha}{t_o} \tag{11.2}$$

where t' = relative thickness; t_α = limb thickness, where inclination = α; t_o = layer thickness in hinge, where $\alpha = 0°$. Using the measured values of t_α and t_o, t' is calculated and recorded in the manner shown in Figure 11.43*B*. Then each set of (t', α) values is plotted to form the graphical construction shown in Figure 11.43*C*. The plotted points for a given fold layer are connected by a smooth curve. The locus of the curve within the graph can then be compared with the subareas and lines occupied by the fundamental classes of folds. Graphs of this type show that the full range of layer shapes of folds is very expansive and that concentric folds (Class 1B) and similar folds (Class 2) are indeed very special cases.

CONSTRUCTING GEOLOGIC CROSS SECTIONS IN NORMAL PROFILE VIEW

NEED FOR NORMAL PROFILE VIEWS. Geometric analysis of the shapes of individual folded surfaces and folded layers requires views of folds in normal profile. Although certain canyon walls, road cuts, outcrops, and mine tunnels may fortuitously display perfect normal profile views of folds, geologic maps of folded regions are seldom that cooperative. Usually the interference of fold configuration and topography creates map patterns that are **oblique profile views**. Transforming geologic map patterns into normal profile views requires the preparation of **geologic cross sections**. When constructed for purposes of structural geologic or tectonic analysis, geologic cross sections are often called **structure sections**.

PREPARATION OF VERTICAL STRUCTURE SECTIONS. In areas or regions where folds are nonplunging, normal profile views are obtained by constructing **vertical** structure sections. The base of control is a geologic map (Figure 11.44). The **line of section** along which the structure section is to be fashioned is laid out perpendicular to the average trend of fold axes, an average determined stereographically if necessary. End points of the line of section are marked on the map (*A, A'*, Figure 11.44). The line of section is positioned where strike-and-dip data for the folded layering are reasonably abundant and where the fold patterns are especially interesting and informative.

Once the line of section has been chosen and positioned on the geologic map, a topographic profile is constructed using elevation control afforded by topographic contour lines (Figure 11.44). The topographic profile, like the structure section itself, is drawn without vertical exaggeration, to avoid distortion of the true form of the folded layers. To the topographic profile are added the exact locations where contacts between formations and marker beds cross the line of section.

Strike-and-dip data posted on the geologic map nearest the line of section provide a means to gauge the direction and angle of inclination of each geologic contact that is to be portrayed in the structure section. Where the strike of layering is perpendicular to the line of section, **true dip** is plotted (by protractor) in the structure section. Where the strike of layering is oblique to the line of section, **apparent dip** must be plotted. Apparent dip can be computed stereographically.

The inclinations of the upper and lower surfaces of each folded layer, as measured in outcrop, are represented on the structure section by short **guide lines** that are plotted with a protractor at appropriate locations (Figure 11.44). No matter how the pattern of folded layers is portrayed at depth,

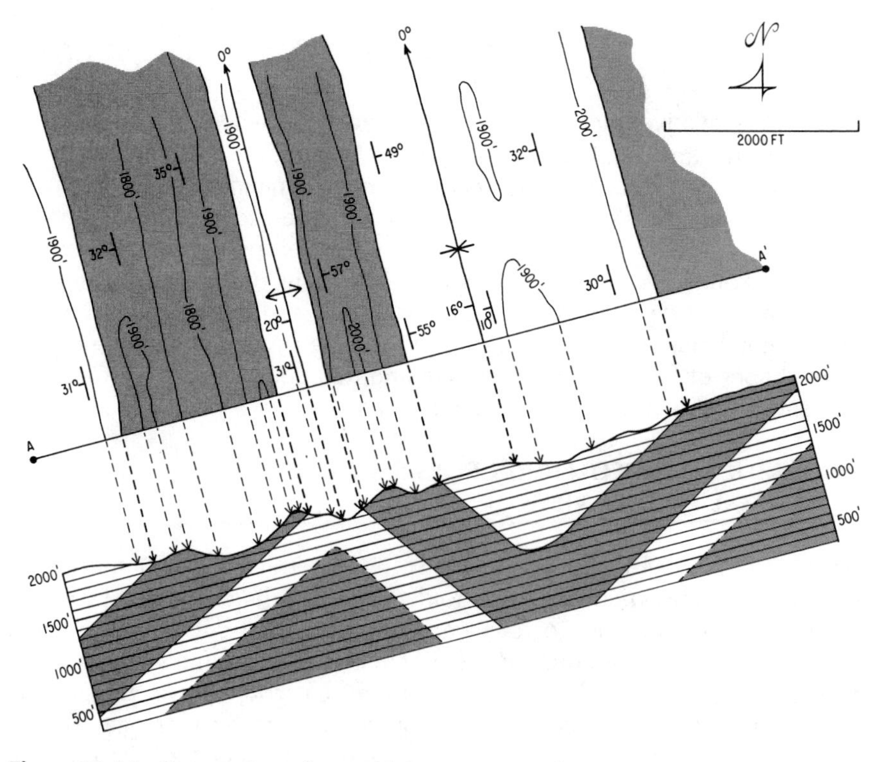

Figure 11.44 Preparation of a vertical structure section on the basis of geologic map relationships.

each contact of each folded layer must emerge at the surface of the section along one of the guide lines.

The manner in which folded layers are portrayed in the subsurface is guided by field observations regarding the response of each folded layer to the distortional influence of folding. Layers that deform by Class 1B concentric folding are drawn in a way that serves to preserve a uniform thickness for each. Layers that tend to deform by Class 2 similar folding are drawn such that thickness measured parallel to the axial trace of each fold remains constant within each folded layer. As a final check on the internal consistency of the subsurface interpretation, the trace lengths of the **midlines** of each of the folded layers are measured to determine whether the section is **balanced**.

PREPARATION OF INCLINED STRUCTURE SECTIONS. Normal profile views of folds are more difficult to construct in regions where the folds are plunging. Vertical structure sections drawn through terranes of plunging folds yield **oblique** views of the shapes of folded layers and folded surfaces. To obtain normal profile views of plunging folds it is necessary to construct **inclined structure sections**. The data base for constructing inclined sections is provided by geologic map relationships, such as those diagrammatically presented in Figure 11.45*A*.

The first step in preparing an inclined structure section is to determine the trend and plunge of the axis of folding. This can be achieved by preparing a π diagram of poles to the folded layering (Figure 11.45*B*). In this example, the orientation of the fold axis is found to be 40° N26°E. The orientation of the inclined structure section that would serve as a normal profile view of a fold of this orientation is N64°W, 50°SW.

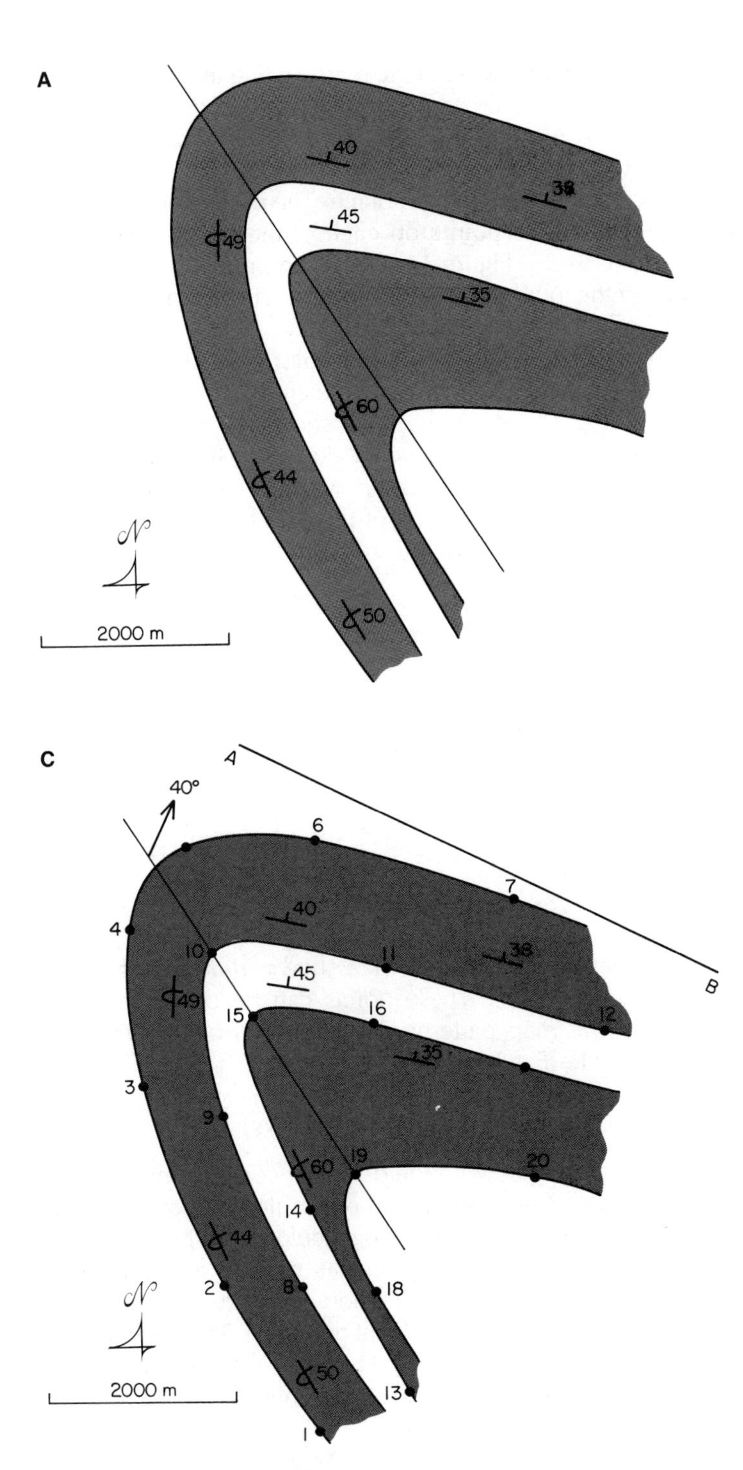

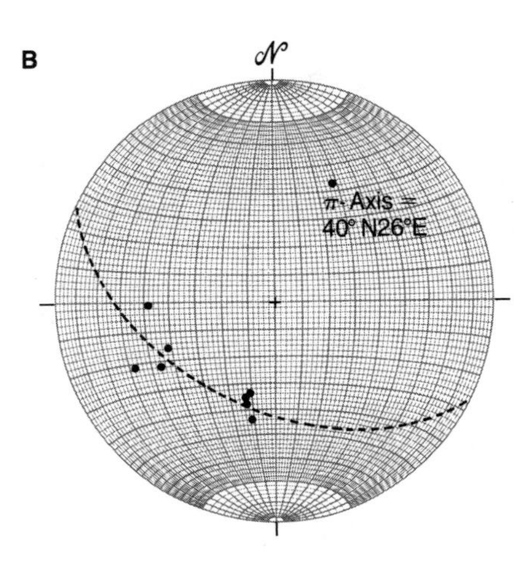

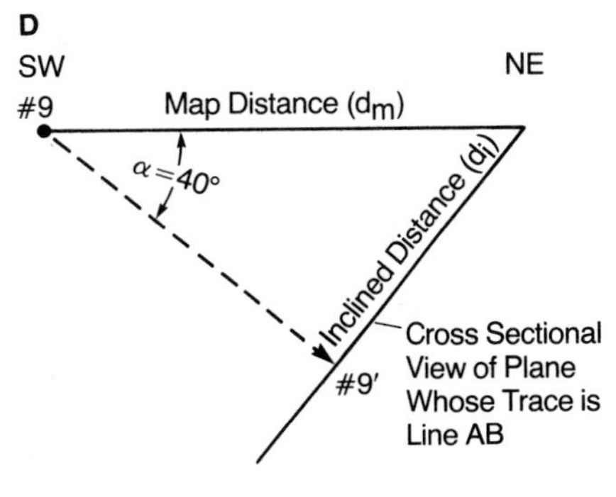

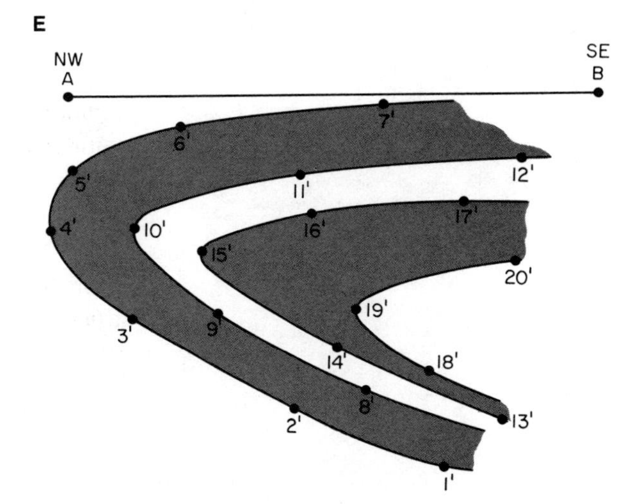

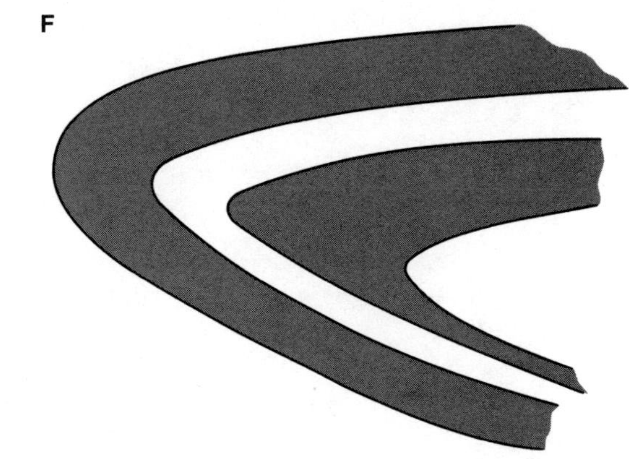

Figure 11.45 Preparation of an inclined structure section. (*A*) Geologic map relationships. (*B*) Stereographic determination of the orientation of the fold axis. (*C*) Identification of line of section (line *AB*) and reference points (1–20). (*D*) The geometry of subsurface projection. (*E*) The projected distribution of reference points, as viewed in the subsurface. (*F*) The final structure section.

The next step is to position on the geologic map the **line of section** where the inclined structure section is to be constructed. Trending N64°W, the line of section is shown as line *AB* in Figure 11.45*C*. Line *AB* is the trace of a plane that dips 50°SW. In preparing the structure section, the entire map pattern must be projected into this inclined plane. The projection is accomplished by identifying **reference points** on each of the contacts between folded layers (e.g., points 1–20, Figure 11.45*C*), then projecting each control point to the plane of the inclined structure section. The line of projection is parallel to the axis of the fold, namely 40° N26°E.

The geometry of projection of each reference point is pictured in Figure 11.45*D*, a vertical cross section passing through reference point 9 along the trend of the axis of folding. The vertical section shows the 50°SW-dipping trace of the inclined structure section. Reference point 9 on the map projects to reference point 9′ in the inclined section. A simple trigonometric relationship relates the **inclined distance (d_i)** of point 9′ beneath the line of section to the **map distance (d_m)** of point 9 from the line of section:

$$\sin \alpha = \frac{d_i}{d_m} \tag{11.3}$$

where

$$d_i = \text{inclined distance}$$
$$d_m = \text{map distance}$$
$$\alpha = \text{plunge of fold axis}$$

Thus,

$$d_i = d_m (\sin \alpha) \tag{11.4}$$

Point 9′, along with other reference points, are shown *in the plane of the inclined section* in Figure 11.45*E*. These points can be connected in a manner consistent with the map pattern. Once this has been done, the normal profile springs to life (Figure 11.45*F*).

DOWN-STRUCTURE METHOD OF VIEWING FOLDS

One of the incentives to struggle with the details of constructing inclined structure sections is to gain a full appreciation of the elegance of J. Hoover Mackin's "down-structure method of viewing geologic maps" (Mackin, 1950). Normal profile views of plunging folds can be seen at a glance by viewing the geologic map patterns in the direction of plunge, at an angle of inclination of view corresponding to the amount of plunge. Try it on the map pattern shown in Figure 11.46*A*, comparing what is seen to the graphically constructed geologic cross section (Figure 11.46*B*). Then try out the method on the hieroglyphics shown in Figure 11.46*C*.

KINEMATICS OF FOLDING

ROLE OF LAYERING

Folding of rocks cannot be accomplished without an infinite number of outcrop-scale to microscopic kinematic adjustments. The kinematic adjustments are recorded in the full range of geometric and physical properties of folds. A paper by Donath and Parker (1964), entitled *Folds and Folding*, provides a very useful basis for understanding fold properties in the context of kinematics.

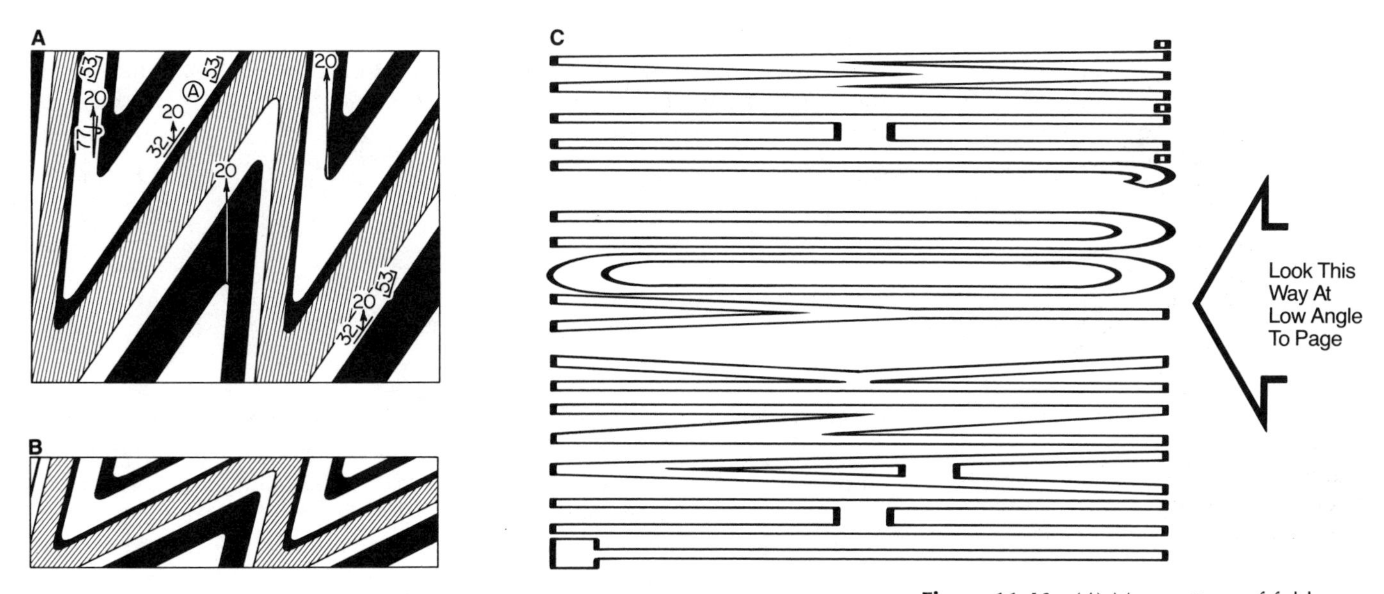

Figure 11.46 (*A*) Map pattern of folds and (*B*) down-structure view of same. [From *Structural Geology of Folded Rocks* by E. T. H. Whitten, after Mackin (1950). Originally published in 1966 by Rand-McNally and Company, Skokie, Illinois. Published with permission of John Wiley & Sons, Inc., New York] (C) Glance at this pattern in down-structure view.

Donath and Parker recognized two fundamental mechanisms of folding: **flexural folding** and **passive folding**. Flexural folding takes place when the mechanical influence of layering in a rock is very strong. The layers *actively* participate in the folding by bending and flexing. Passive folding is the favored mechanism when the mechanical influence of layering in a sequence of rocks is very weak. Passive folding can be thought of as a fake folding: layers take on a folded form without really having been flexed or bent. The layering is *passive*. It is not active, but is acted on. It submissively endures distortion, apparently without much resistance.

Many passive folds are marked by the presence of penetrative **cleavage**, that is, the presence of an array of closely spaced aligned secondary discontinuities that cut the folded layers in a direction parallel or subparallel to the axial surfaces of folds. Full discussion of passive folding is deferred to Chapter 12, where it is presented in the context of the physical properties and origins of cleavages and foliations.

FLEXURAL FOLDING

FLEXURAL-SLIP FOLDING. Flexural folding can take place in two ways, by flexural slip and/or by flexural flow (Donath and Parker, 1964). Depending on the mechanical properties of the layered sequence, one or both of these mechanisms are initiated when layer-parallel resistance to shortening is overcome, and the layers of rock begin to actively buckle. **Flexural-slip folding** accommodates the buckling by **layer-parallel slip** along contacts between layers (Figure 11.47). If the layers are sedimentary, the flexural slip between layers is called **bedding-plane slip**.

Whenever I use a telephone book, I cannot resist flexing it and thinking about the *sense* of layer-parallel slip. If I flex the book by flexural-slip folding into the form of an upright antiform (or synform), each page moves up-dip with respect to the page(s) beneath (Figure 11.48). *And the direction of relative slip of the pages is perfectly perpendicular to the hinge of the fold.*

Flexural-slip displacements between layers (or pages) are tiny when viewed individually, but the *sum* of the displacements is always enough to accommodate a true bending of a rock body (or book). The actual amount of slippage along the top of any layer is easy to calculate (Ramsay, 1967, p. 392–393). As in the analysis of layer shape, the fold form to be analyzed is

Figure 11.47 The kinematic character of flexural-slip folding.

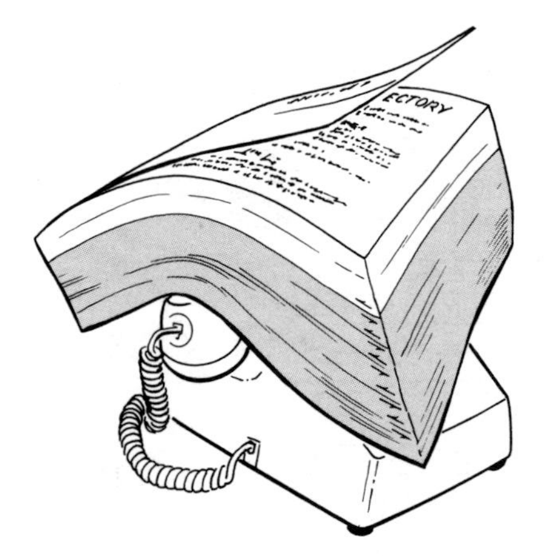

Figure 11.48 Flexural slip of pages in a book. The direction of interbed slip is perpendicular to the axis of folding.

rotated into the orientation of a perfectly upright antiform. Then the locations where slip is to be calculated are specified by the inclination values (α) of the top of the folded layer at the chosen sites (Figure 11.49). Slip is then determined using the following formula:

$$s = t\alpha \tag{11.5}$$

where

s = slip

t = thickness of the folded layer

α = inclination, in radians ($1° = 0.0175$ radian)

For the fold shown in Figure 11.49, we can use Equation 11.5 to calculate slip at 10 sites on the top of layer A (thickness = 9 cm) and 10 more sites along the top of layer B (thickness = 3 cm). The sites for each layer correspond to 10° inclination values, from 0° to 90°. The calculations demonstrate that *the amount of layer-parallel slip increases both with layer thickness and with distance from the hinge*. In fact, the calculations show that no interlayer slip whatsoever takes place at the actual hinge point of a folded layer.

The shear strain (γ) due to interbed slip can be calculated too (Ramsay, 1967, p. 393),

$$\gamma = \alpha \tag{11.6}$$

where

γ = shear strain

α = inclination (radians)

Figure 11.49 shows calculations of shear strain for layer A. The distribution of values of shear strain reveal that *shear strain due to flexural-slip folding is greatest at the inflection of a fold but is negligible at the hinge.*

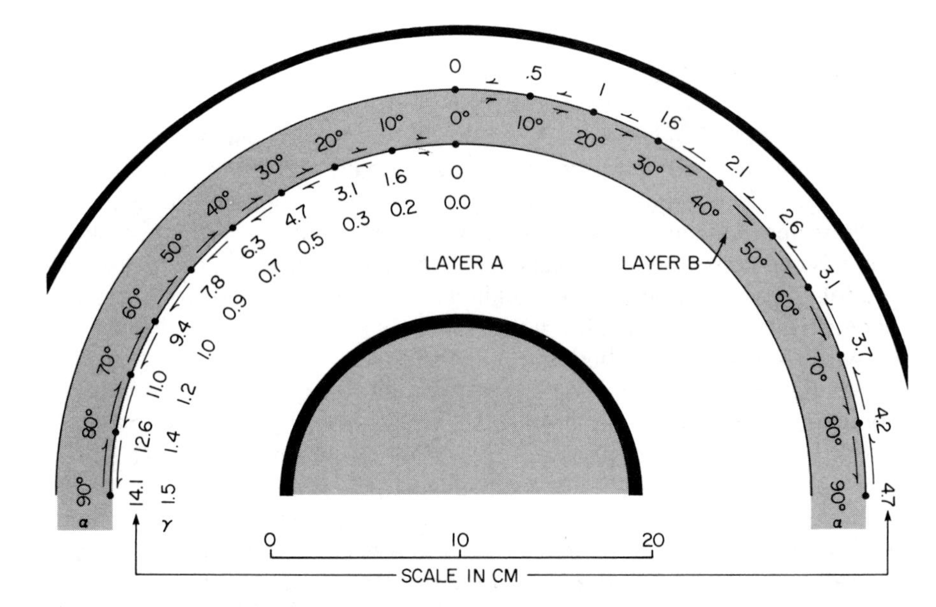

Figure 11.49 The amount of slip between layers of a flexural-slip fold depends on layer thickness and limb inclination. Shear strain depends on limb inclination alone.

Figure 11.50 Flexural-slip folding of ribbon cherts in the Cook Inlet region, Alaska. (Photograph by M. W. Higgins. Courtesy of United States Geological Survey.)

Figure 11.51 Layer-parallel stretching and layer-parallel shortening associated with folding.

Donath and Parker (1964) have emphasized that layered sequences that readily fold by flexural slip are marked by strong, stiff layers, the contacts of which are marked by **low cohesive strength**. Thin- to medium-bedded sandstone, siltstone, and limestone sequences are especially vulnerable to flexural slip (Figure 11.50). Individual layers that are folded by the flexural-slip mechanism tend to retain their primary, original thicknesses, in the same way that the pages of a telephone directory neither thicken nor thin out when the directory is flexed. Thus layer shape of flexural-slip folds tends to be Class 1B, that is, concentric.

Even though individual layers tend to retain their original thickness during flexural-slip folding, they nonetheless generally endure some internal distortion. The distortion takes place mainly in the hinge zone of the folded layer, where curvature is greatest. When an individual layer is actively buckled, rock on the outer arc of the hinge undergoes **layer-parallel stretching**, and rock on the inner arc of the hinge experiences **layer-parallel shortening** (Kuenen and DeSitter, 1938) (Figure 11.51). Layer-parallel strain decreases toward the middle of each folded layer, toward the **neutral surface** of no strain. The neutral surface separates an outer arc domain of layer-parallel stretching from an inner arc domain of layer-parallel shortening. When thinning of outer arc rocks by layer-parallel stretching is perfectly compensated by thickening of inner arc rocks by layer-parallel shortening, the folded layer retains a Class 1B form.

MINOR STRUCTURES CREATED DURING FLEXURAL-SLIP FOLDING. Flexural-slip folding creates an informative array of minor structures. The minor structures reflect a combination of four complementary mechanisms of deformation: overall layer-parallel shortening, layer-parallel slip on the fold limbs, layer-parallel stretching on the outer arc of the hinge zone, and layer-parallel shortening on the inner arc of the hinge.

Overall layer-parallel shortening before the onset of significant buckling can create minor folds and thrust-slip faults in thin-but-stiff members within a layered sequence (Figure 11.52*A*). The minor folds initially are symmetrical, with axial surfaces perpendicular to the direction of layer-parallel shortening.

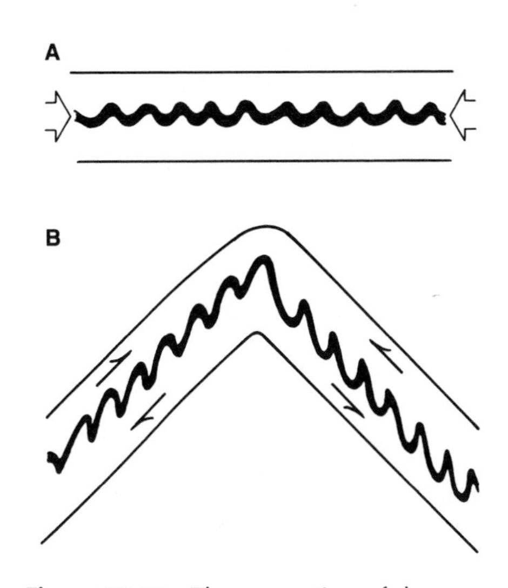

Figure 11.52 The generation of drag folds. (*A*) Layer-parallel shortening before buckling creates an array of upright, symmetrical anticlines and synclines. (*B*) Buckling and the onset of flexural-slip folding transforms the symmetrical folds into asymmetrical folds.

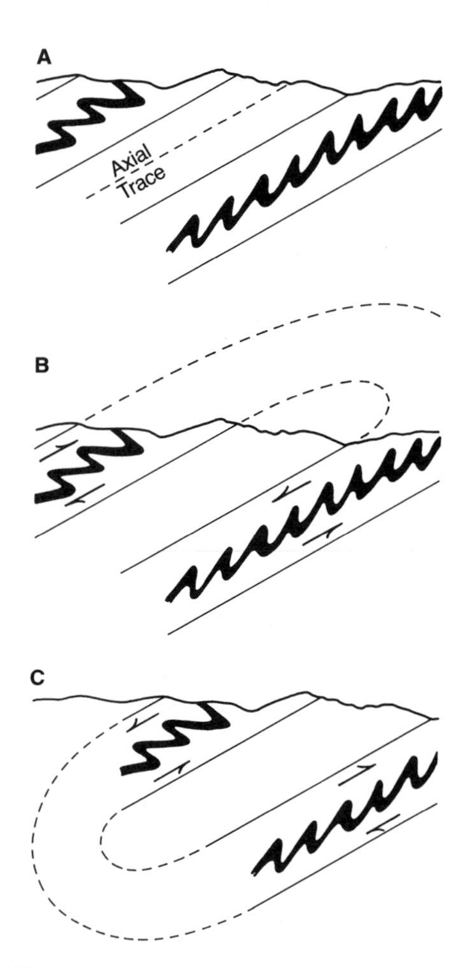

Figure 11.53 (*A*) Is the fold an overturned synform or an overturned antiform? (*B*) The fold is not an antiform, for the asymmetry of drag folds contradicts the sense of bedding-plane movements that would characterize the limbs of an antiform. (*C*) The drag fold pattern conforms perfectly to flexural slip on the limbs of an overturned synform.

When buckling ensues, and with it **layer-parallel slip between layers**, minor folds that formed during overall layer-parallel shortening may be transformed into asymmetrical folds on the limbs of the major structure(s) (Figure 11.52*B*). The opposite limbs of a common fold are marked by minor folds of contrasting asymmetry because the sense of layer-parallel slip is different on opposing limbs.

Called **drag folds** (or **parasitic folds**), asymmetric minor folds formed in this way are valuable for (at least) three reasons. First, the axis of a drag fold is subparallel to the axis of the larger fold with which it is associated. Thus, even when the hinge of the major fold is not exposed, an approximation of its axis orientation can be deduced in the field, before going to the trouble of constructing a π or β diagram. Second, in terranes where folds are isoclinal and poorly exposed, the location of the axial trace of a major fold can be identified on the basis of the shift in minor-fold asymmetry from Z to S, or S to Z (compare Figures 11.4 and 11.52*B*). Third, upon discovery of a "hidden" isoclinal fold, its antiformal versus synformal nature can be interpreted on the basis of the **sense** of layer-parallel slip reflected by the drag folds on each limb.

The use of drag folds to interpret fold patterns is pictured in Figure 11.53. Figure 11.53*A* shows the predicament. The axial trace of a major isoclinal fold is discovered and mapped on the basis of asymmetry of drag folds. What remains uncertain is whether the fold is an antiform or a synform. When an antiformal fold form is fitted to the configuration of the axial trace and the fold limbs (Figure 11.53*B*), the expected sense of layer-parallel slip on each limb of the antiform is contradicted by the sense of asymmetry of the drag folds. On the other hand, when a synformal fold form is fitted to the axial trace/fold limb configuration, the observed drag fold pattern is wholly consistent with the expected sense of layer-parallel slip (Figure 11.53*C*).

Minor structures that form in response to flexural-slip folding conform in orientation, location, and strength of development with the state of strain from inflection to hinge (Figure 11.54). En echelon tension fractures commonly develop as a response to stretching parallel to the local direction of greatest extension (*X*). Early-formed tension fractures may become distorted into sigmoidal gash fractures and veins as the fold becomes tighter and tighter. Such distortion is a response to **progressive simple shear**. The axial surfaces of drag folds remain perpendicular to the direction of greatest shortening (*Z*) during progressive deformation. Cleavage may develop in weak layers such that its orientation is subparallel to the axial surfaces of the minor folds. The preferred cleavage orientation is perpendicular to the direc-

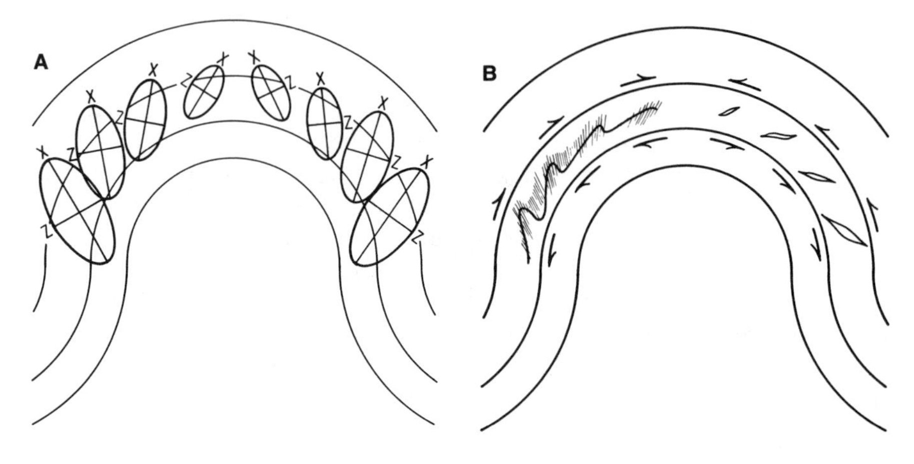

Figure 11.54 (*A*) Schematic portrayal of the state of strain that is generated within layers subjected to flexural-slip folding. (*B*) Minor structures naturally emerge as expression of the state of strain produced by flexural-slip folding. Schematically shown on the left limb are asymmetric folds cut by axial plane cleavage. Sigmoidal gash veins and tension fractures are pictured on the right limb.

tion of shortening; the formation of cleavage accommodates shortening (Chapter 12).

Layer-parallel stretching on the outer arc of a folded layer can be accommodated in a number of ways, depending on the strength of the layer (Figure 11.55). Stiff layers respond to the stretching by the formation of tension fractures and normal-slip faults. Tension fractures, including veins, form perpendicular to the direction of layer-parallel stretching. Conjugate normal-slip faults form in such a way that their line of intersection is parallel to the axis of folding. **Keystone grabens** are classic expressions of stretching on the outer arc of a folded layer.

If the layering in the outer arc of a fold is a composite of soft and stiff layers, stretching is commonly achieved by boudinage and pinch-and-swell. **Boudins** form in sequences of alternating soft and stiff layers that have been subjected to flattening and extension. Stiffer layers tend to break or neck, and the softer layers tend to flow and fill in, wherever required. The forms of boudins are endlessly variable, depending on the ductility contrast between the layers that are flattened and stretched. Some boudins are symmetrical, **pinch-and-swell structures**. Some, however, are like bricks that have been pulled apart, the soft layers filling in like mortar.

Layer-parallel shortening on the inner arc of a folded layer gives rise to symmetrical folds, thrust-slip faults, and/or cleavage (Figure 11.55). These structures work together to accommodate the room problem created when the inner arc of a layer closes in on itself. The minor symmetrical folds are coaxial with the axis of the major fold. Conjugate thrust-slip faults intersect in a line parallel to the axis of folding. And the cleavage that forms as a response to layer-parallel shortening is typically aligned parallel to the axial surface of the major fold.

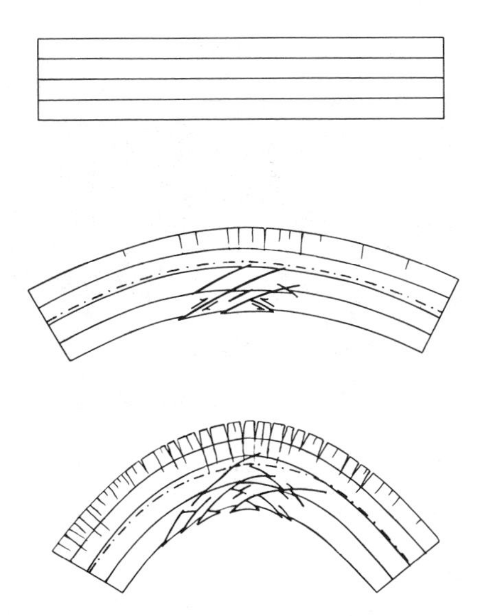

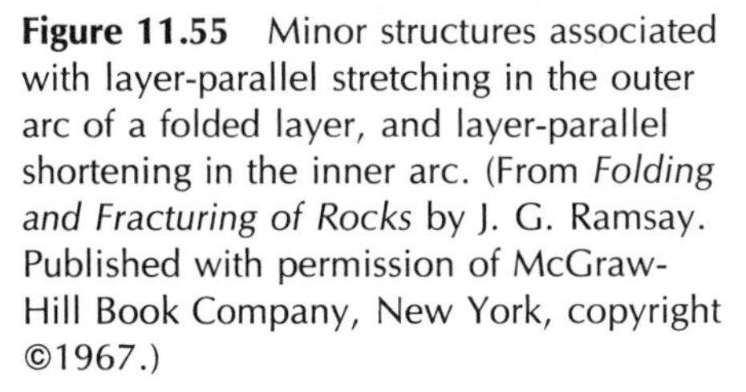

Figure 11.55 Minor structures associated with layer-parallel stretching in the outer arc of a folded layer, and layer-parallel shortening in the inner arc. (From *Folding and Fracturing of Rocks* by J. G. Ramsay. Published with permission of McGraw-Hill Book Company, New York, copyright ©1967.)

JOINTING AND FLEXURAL-SLIP FOLDING. Joints that develop during the flexural-slip folding of layers are often closely coordinated, geometrically, with the orientation properties of the folds with which they are associated. Three classes of joints can be distinguished: **cross joints, longitudinal joints**, and **oblique joints** (Figure 11.56). Cross joints ideally are aligned perpendicular to the axis of folding. They reflect stretching of brittle rock during **hinge-parallel elongation** of folded layers. Hinge-parallel elongation partly compensates for the room problems that can develop in the inner arc of a folded layer as it becomes more and more tightly appressed. Cross joints are unusually planar and are unusually regularly spaced (Figure 11.57). Oftentimes they are vein filled, clearly expressing stretching.

Longitudinal joints are subparallel to the axial surfaces of folds (see Figure 11.56). They tend to be through-going, planar, continuous structures. The kinematic reason for the development of longitudinal joints is not always clear. Billings (1972) interpreted longitudinal joints as **release joints** that open up in folded layers when fold-forming shortening stresses are relieved. However, some longitudinal joints may reflect shortening perpendicular to their average attitude, a shortening taken up by pressure-induced dissolution along stylolitic surfaces.

Oblique joints ideally comprise two conjugate sets that are symmetrically disposed to the hinge and axial surface of a given fold. The oblique joint sets are arranged such that the the axial surface of the fold bisects the obtuse angle of intersection of the joint sets (see Figure 11.56). Oblique joints are classically interpreted as **conjugate shear joints** that form in folded layers as a response to shortening perpendicular to the axial surface of a fold. The acute angle of intersection of the joints is thus bisected by a line that describes the direction of shortening.

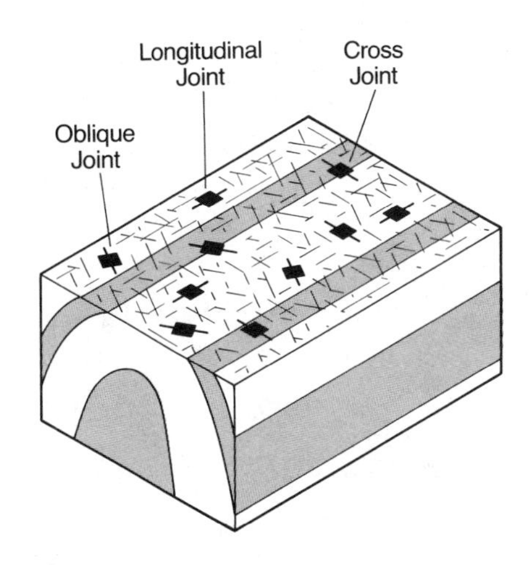

Figure 11.56 Idealized picture of the relation of joints to folds. Cross joints, longitudinal, and oblique joints are distinguished.

Figure 11.57 (*A*) Cross joints in folded metamorphic rocks along the Bay of Fundy, New Brunswick, Canada. (*B*) Close-up view of one of the joint faces shows the nature of the folded layering. (Photographs by G. H. Davis.)

FLEXURAL-FLOW FOLDING. **Flexural-flow folding** achieves a true bending or flexing of layers, but without slippage between the layers. The bending and flexing is accommodated by distributed simple shear of material within layers of low strength (Donath and Parker, 1964). To visualize the flexural-flow process, we might replace the telephone directory with an extravagantly large ice cream sandwich, then proceed to try to bend it. If the ice cream is frozen solid, the sandwich will break but not bend. But in the desert sun, bending of the sandwich can be achieved through layer-parallel flow of the softened ice cream. If we could view the "flow" of the ice cream at microscopic scale, we would see that it is achieved by tiny discrete simple shear translations on microscopic slip surfaces. Again we are reminded of the scale dependency of what is called slip versus what is called flow.

Sequences of rocks that lend themselves to flexural-flow folding must possess a special combination of mechanical properties. They need to contain layers of very high ductility, like salt, gypsum, and clay, which have the inherent weakness and capacity to flow. At the same time, there must be stiff, strutlike members like limestone, sandstone, and dolomite that serve to channel the flow in layer-parallel fashion. In the language of Donath and Parker (1964), flexural-flow folding requires a starting sequence marked by high **ductility contrast** among the layers.

Folded sequences that have deformed by flexural-flow folding are characterized by distinctive layer shapes (Figure 11.58). The stiff layers tend to retain original thickness as Class 1B layers. In contrast, the soft layers undergo differential thickening and display Class 1C, Class 2, and sometimes even Class 3 layer forms.

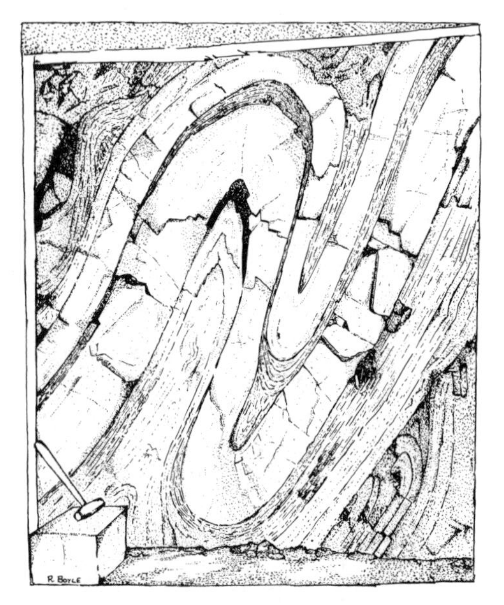

Figure 11.58 Flexural-flow fold in thick- and thin-bedded shale, Black Rock mine, northwestern Queensland, Australia. [From Donath and Parker (1964). Published with permission of Geological Society of America and the authors.]

DYNAMICS OF FOLDING

MECHANICAL ANALYSIS

Dynamic analysis of folding is an inquiry into the mechanics of folding. The goal is to interpret the details of folding in terms of the sum total of specific physical properties of a given sequence of layered rocks. The language of mechanical analysis is very precise.

Mechanical analysis of folding traditionally has focused on analysis of **buckling**. The very best analyses of folds and folding have combined theory and experiment (Biot, 1957; Biot, Ode, and Roever, 1961; Ramberg, 1967; Johnson, 1977). The heart of dynamic analysis is mathematical analysis pursued from an engineering–physics perspective. Experimental analysis provides a testing ground for the theory. Analysis of fold mechanics is clearly

Figure 11.59 Buckled, intestinelike "ptygmatic" fold. The fold developed in a stiff pegmatite dike (white) that was free to shorten within a viscous granitic magma. Santa Catalina Mountains near Tucson, Arizona. [Tracing by D. O'Day of photograph by G. H. Davis. From Davis (1980). Courtesy of Geological Society of America and the author.]

one of the most advanced topics in structural geology, one that we barely scratch in the context of this presentation.

INSTABILITY AND THE DOMINANT WAVELENGTH

It can be shown both theoretically and experimentally that **instability** develops when layers of different mechanical properties are subjected to layer-parallel stresses (Biot, 1957). The instability gives rise to a buckling of the stiffest layer(s) in the sequence of rocks, like a stiff pegmatite dike within a plastically deforming granite (Figure 11.59). The fold that emerges is of some particular **dominant wavelength**, the fold wave that can be created with the least amount of layer-parallel stress. Buckling instability is not confined to rocks. An interesting buckle emerged in the trolley tracks of the San Francisco streets during the great earthquake of 1906 (Figure 11.60).

Knowledge gained from mechanical analysis makes it possible to predict the dominant wavelength that will emerge when a **single folded layer**, or a **multilayer sequence**, is shortened. Predictions are based on hard-earned mathematical descriptions that relate dominant wavelength to the strength and thickness properties of the layers to be deformed.

Figure 11.60 Buckling of rails by compression on Howard Street (South Van Ness Avenue) near 17th Street, San Francisco. The buckling was caused by movements related to the earthquake of 1906. (Photograph by T. L. Youd. Courtesy of The United States Geological Survey.)

FOLDING OF A SINGLE LAYER, IN THEORY

The dependence of wavelength on thickness and strength characteristics is most simply expressed in equations that describe the buckling of a stiff layer embedded in a softer medium. Dominant wavelength depends not only on the thickness and strength of the stiff layer, but also on the strength of the weak, confining medium. Thickness is easy to deal with both mathematically and experimentally. But how is "stiffness" of a layer modeled quantitatively? As it turns out, the mathematical description of layer strength depends on whether the mechanical properties of the single stiff layer and its confining medium are viewed as elastic or viscous. If an elastic model of deformation is applied, the strengths of layers are described in terms of the fundamental elastic moduli, **Young's modulus (E)** and **Poisson's ratio (υ)** (Figure 11.61*A*). However, if a viscous model of deformation is used, the strengths of the layer and its confining medium are expressed in terms of

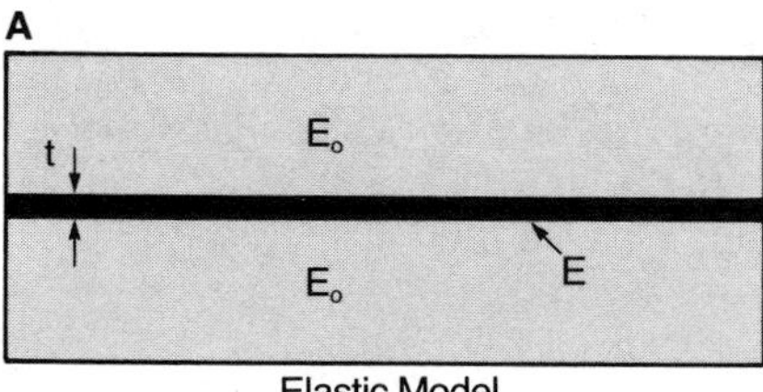

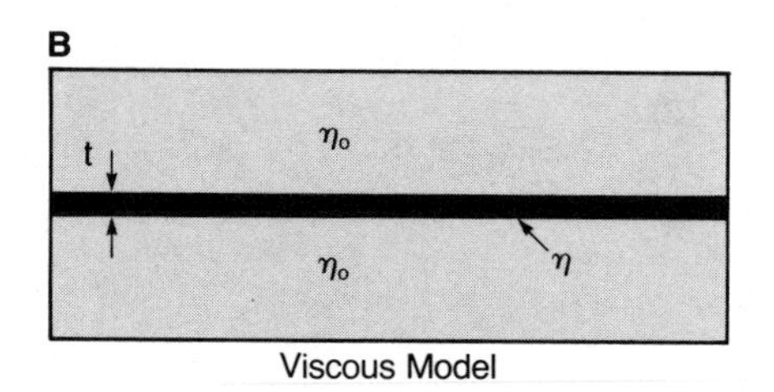

Figure 11.61 The modeling of the mechanical properties of layers about to be deformed by layer-parallel shortening and buckling. (*A*) Elastic model. (*B*) Viscous model.

viscosity coefficients (Figure 11.61*B*). In hybrid models where a stiff elastic layer is considered to be embedded in a soft, nonelastic confining medium, the strength of the stiff layer is described in terms of elastic moduli, and the strength of the confining medium is specified by a coefficient of viscosity.

Bijlaard (1946) modeled the mechanics of folding of a single layer in terms of a stiff elastic plate in a soft elastic medium. What he discovered was a surprisingly straightforward relationship:

$$L = 2\pi t \sqrt[3]{\frac{B}{6B_0}} \tag{11.7}$$

where

L = dominant wavelength
t = thickness of stiff layer
B = elastic modulus of stiff layer
B_0 = elastic modulus of confining medium

Elastic moduli B and B_0 in Equation 11.7 are not mystery variables pulled from the sky. Rather they express strength in terms of Young's modulus and Poisson's ratio:

$$B = \frac{E}{1 - v^2} \tag{11.8}$$

where

E = Young's modulus
v = Poisson's ratio

Currie, Patnode, and Trump (1962) reexamined the mechanics of folding of a stiff elastic layer in a soft, elastic confining medium. In doing so they chose to eliminate Poisson's ratio as a variable, arguing that the influence of Poisson's ratio on folding is small, especially considering the uncertainty in trying to describe precisely its value for real rock layers at the time of folding. The resulting equation has the form of Equation 11.7.

$$L = 2\pi t \sqrt[3]{\frac{E}{6E_0}} \tag{11.9}$$

where

E_0 is Young's modulus of the stiff layer
E is Young's modulus of the confining medium

Biot (1959) and Ramberg (1959) treated the folding of a single folded layer in the perspective of viscous deformation. Their independently derived mathematical analyses uncovered the same kind of relationship reported for the case of elastic deformation:

$$L = 2\pi t \sqrt[3]{\frac{\eta}{6\eta_0}} \tag{11.10}$$

where

L = dominant wavelength
t = thickness of stiff layer
η = coefficient of viscosity of stiff layer
η_0 = coefficient of viscosity of confining medium

Coefficients of viscosity are expressed in **poises**, the standard measure of **resistance to flow** of a viscous material.

Just because Equations 11.8, 11.9, and 11.10 have the same form does not mean that they are the last word on the folding of single layers. Folds and folding continue to keep us humble. Sherwin and Chapple (1968), for example, have shown that the dominant fold wavelength that arises during layer-parallel shortening is responsive to the amount of layer-parallel strain that the layer absorbs *before* buckling. Thus in addition to strength and thickness, **layer-parallel strain** emerges as an important variable that must be taken into account.

Layer-parallel strain is specified in terms of a parameter that is familiar to us: quadratic elongation (λ). Sherwin and Chapple found that it is necessary to describe quadratic elongation in two directions within the plane of layering, both parallel *and* perpendicular to the direction of layer-parallel shortening. As pointed out by Hobbs, Means, and Williams (1976), Hudleston (1973) rewrote the Sherwin-Chapple equation in a form that can be directly compared with the Biot (1959) and Ramberg (1962) equations,

$$L = 2\pi t \sqrt[3]{\frac{\eta\,(s-1)}{6\eta_0\,(2s^2)}} \qquad (11.11)$$

where

$$s = \sqrt{\lambda_1/\lambda_3}$$

λ_1 = quadratic elongation perpendicular to shortening

λ_3 = quadratic elongation parallel to shortening

As in all matters of science, closer and closer scrutiny of the Earth at work always seems to lead to a greater appreciation of the delicacy of dynamic process. We learn that Earth processes are influenced by a much broader range of variables than originally perceived.

FOLDING OF A SINGLE LAYER, IN PRACTICE

Part of the fun of dynamic analysis is testing equations to see if they really work. Biot did not wait for others to test the equation he derived for the folding of a single layer, viewed viscously (Equation 11.8). Instead, he teamed up with two colleagues, Ode and Roever, to check it himself (Biot, Ode, and Roever, 1961). Together the investigators set up a series of experiments that included the layer-parallel shortening of single layers of stiff pitch (i.e., tar), which they deformed in a confining medium of corn syrup (Figure 11.62). Layers of pitch of different thicknesses were fabricated in molds of different depths. Viscosities of both the pitch and the syrup were carefully measured before the start of the experiments.

On the basis of strength and thickness data (Table 11.2), Biot, Ode, and Roever calculated the dominant wavelengths predicted by Biot's (1959) equation. Then they subjected each of the three pitch layers to layer-parallel shortening and measured the range of wavelengths of folds that emerged in

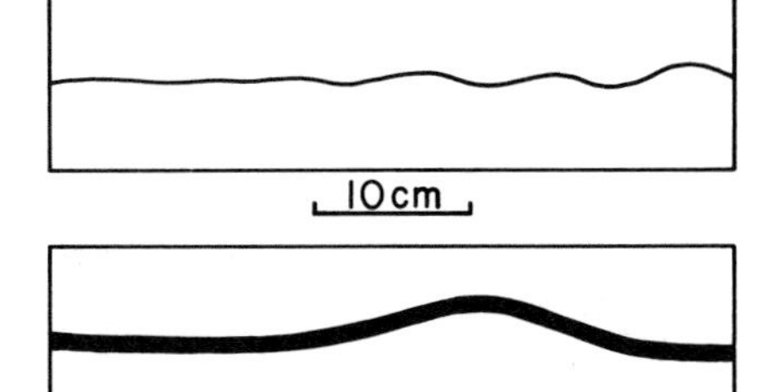

Figure 11.62 Layer-parallel shortening of pitch layers of different thicknesses in a medium of syrup. [From Biot, Ode, and Roever (1961), Geological Society of America.]

Table 11.2
The Testing of Biot's Equation

$$L = 2\pi t \sqrt[3]{\mu/6\mu_0}$$

Thickness (t) of Pitch Layer	*Viscosity (μ) of Pitch Layer*	*Viscosity (μ_0) of Corn Syrup*	*Predicted Fold Wavelength (L_p)*	*Observed Fold Wavelength (L_0)*
0.35 cm	3×10^7 poise	1.35×10^4 poise	15.78 cm	12.4–18.0 cm
0.87 cm	3×10^7 poise	1.35×10^4 poise	39.24 cm	34.0–41.0 cm
1.08 cm	3×10^7 poise	1.35×10^4 poise	48.71 cm	38.0–52.0 cm

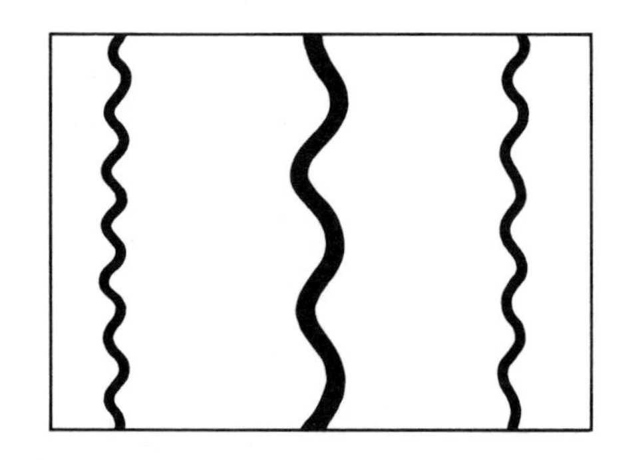

Figure 11.63 Layer-parallel shortening of gum-rubber strips in a medium of gelatin. The outside strips are 1/64 in. thick. The middle strip is 1/32 in. thick. [From Currie, Patnode, and Trump (1962). Published with permission of Geological Society of America and the authors.]

each buckled layer. Experimental results were found to be quite consistent with the predictions of theory!

As part of their research, Currie, Patnode, and Trump (1962) experimentally tested the equation they generated to describe folding of a single elastic plate in an elastic medium (Equation 11.9). They found it practical to deform thin gum-rubber strips of known thickness within a medium of gelatin (Figure 11.63). The gum rubber used for their experiments yielded, upon testing, a Young's modulus (E) of 100 psi. The gelatin in which the gum-rubber layers were embedded was mixed from scratch, in such a way that the Young's modulus (E_0) for each gelatin specimen could be predetermined, within a range of 1 to 10 psi.

The work of Biot, Ode, and Roever (1961) and Currie, Patnode, and Trump (1962) lends itself to two challenging and informative exercises in structural geology. The "forward" problem is predicting dominant wavelength(s) on the basis of starting information regarding strength and thickness. The "backward" problem, which has significant geological ramifications, is interpreting ductility contrast on the basis of dominant wavelength and thickness of single folded layers observed in nature.

FOLDING OF MULTILAYERS

Complicated mathematical expressions are required to describe the behavior of **multilayer sequences** containing layers of widely different strength and thickness. The mathematical expressions must include variables above and beyond those already mentioned, notably the spacing of stiff layers within the sequence and the degree of bonding between layers within the sequence. Degree of bonding is expressed in terms of cohesive strength.

The ratio of dominant wavelength to thickness of a folded stiff layer is greatly reduced when the layer belongs to and is analyzed as part of a multilayer sequence (Bijlaard, 1946; Johnson, 1977). Gum-rubber and gelatin experiments carried out by Currie, Patnode, and Trump (1962) reveal this quite clearly (Figure 11.64). Widely separated gum-rubber strips display short-wavelength fold waves, but the dominant wavelength steadily increases as the gum-rubber strips are brought into closer and closer contact.

One of the mechanical idiosyncrasies of layer-parallel shortening of multilayers is that thinner layers in the sequence may buckle into short-wavelength folds before the folding of the entire sequence (Ramberg, 1963). Such "minor" folding is an expression of the layer-parallel strain that constitutes the preliminary step in the formation of most drag folds. As discussed earlier, when buckling of the entire multilayer sequence occurs, and flexural-slip folding is initiated, these originally symmetric, short-wavelength minor folds are transformed into asymmetric drag folds by layer-parallel simple shear (see Figure 11.52).

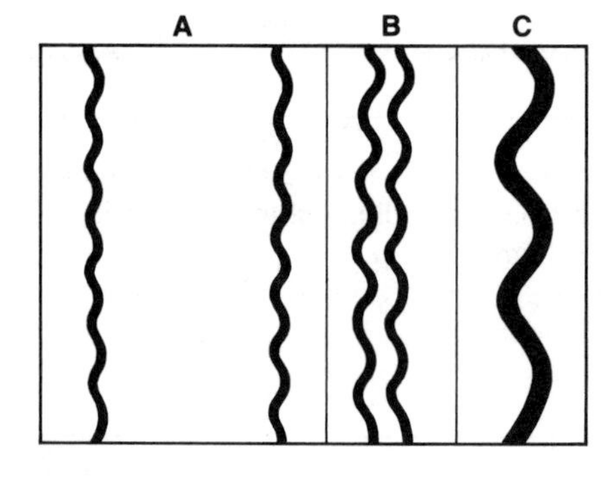

Figure 11.64 Experiments by Currie, Patnode, and Trump demonstrated that the spacing of stiff layers within a multilayer sequence has a significant influence on dominant wavelength. [From Currie, Patnode, and Trump (1962). Published with permission of Geological Society of America and the authors.]

Dynamic analysis of folds has led to some very unexpected discoveries, and none illustrate this quite as well as Ghosh's (1968) analysis of layer-parallel shortening of multilayer sequences. Ghosh discovered through experimentation that the degree of cohesion between layers profoundly influences the profile forms of folds. By simply spreading different amounts of grease between layers of modeling clay, he was able to create a striking array of fold forms, without even changing the strength or the thickness or the spacing of layers within the sequence. When layers are liberally greased, layer-parallel shortening creates smooth, rounded, sinusoidal folds (Figure 11.65*A*). But when layers are placed in frictional contact with one another, without grease in between, layer-parallel shortening creates **kink bands** and **kink folds** (Figure 11.65*B*), identical to those so abundantly found in multiply deformed well-foliated rocks like schists (Figure 11.66*A*, *B*). Paterson and

Weiss (1966) eliminated most of the mystery of kink folding by successfully producing the phenomenon in highly foliated, real rock specimens, which they subjected to layer-parallel shortening under confining pressure. Furthermore, the work of Paterson and Weiss revealed the close relation between the configuration of kink bands, kink folds, and chevron folds to the angular relation between the orientation of foliation and the direction of loading. The structural geologic community learned from this experimentation the degree to which the formation of kink folds and kink bands depends on the presence of highly laminated materials and layer-parallel stress. But it took Ghosh's contribution to drive home the realization that kinking also depends on cohesive bonding between layers.

Tampering with the bonding between layers can be achieved mathematically as well as experimentally. Experts handle these mathematics as easily as we might squeeze grease out of a tube.

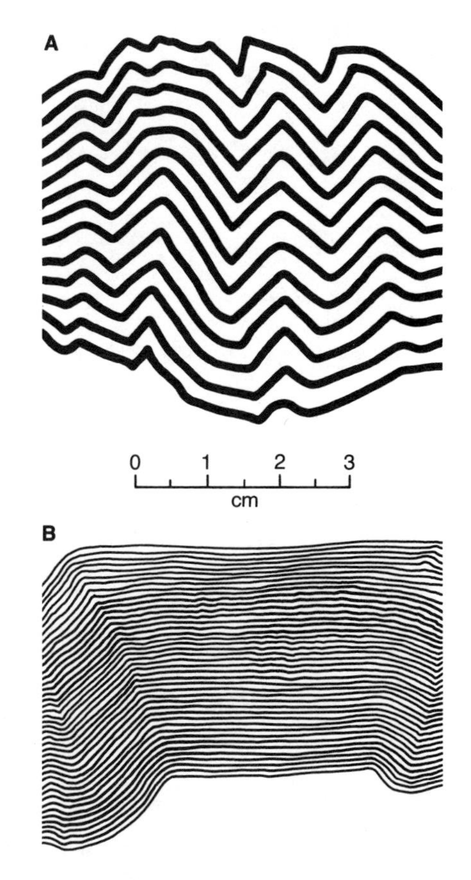

Figure 11.65 The profile forms of folds in multilayer sequences are strongly influenced by the degree of cohesion between layers. (*A*) Where little cohesion between layers exists, folds tend to be smooth, rounded, and sinusoidal. (*B*) Where a high level of cohesion characterizes the contacts between layers, kink folds and kink bands develop. [From Ghosh (1968), *Tectonophysics*, v. 6. Published with permission of Elsevier Scientific Publishing Company, Amsterdam.]

APPLICATION OF DYNAMIC PRINCIPLES TO REGIONAL FOLDS

FAULT-RELATED FOLDS AND FOLD SYSTEMS. Regional fold structures in multilayer sequences in the real world invite the application of principles of fold mechanics. Regional folds and fold systems are generally intimately coupled with faults and fault systems. So close and so common is the association that *the presence and the role of faults* must be added to the ever-growing list of variables that influence folds and folding.

Three major regional occurrences of **fault-related folds** have received the bulk of scrutiny, insofar as fold dynamics are concerned. These include **monoclines**, especially those of the Colorado Plateau; great symmetric **basement-cored anticlines**, like those in the Wyoming province; and folds and fold belts associated with thick, thrusted sedimentary prisms, like those of the Canadian Rockies and the central Appalachians. Monoclines and the great basement-cored anticlines differ from the **folds of thrust belts** in that they do not occur as parts of through-going, continuous fold belts within which the folds (and faults) are aligned. Instead they stand as discrete, relatively isolated folds within a regional network that lacks a conspicuous internal order.

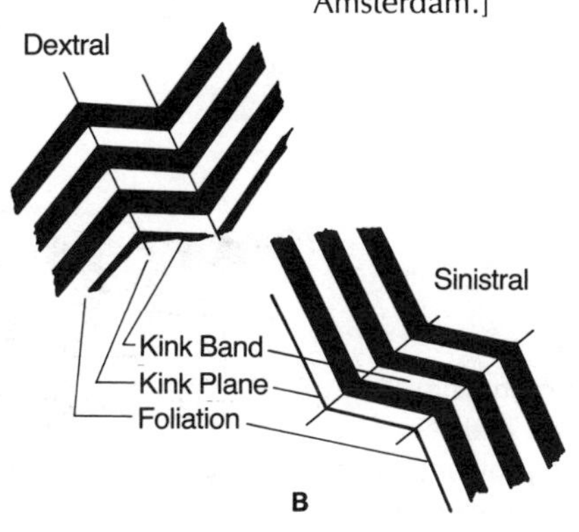

Figure 11.66 (*A*) Kink bands in Devonian phyllite exposed near Morthoe, Devonshire, England. From *The Minor Structures of Deformed Rocks: A Photographic Atlas* by L. E. Weiss. Published with permission of Springer-Verlag, New York, copyright ©1972.) (*B*) Elements of kink folds.

COLORADO PLATEAU MONOCLINES

Monoclines are regional, steplike folds in which otherwise horizontal or *very* shallowly dipping strata abruptly bend to a steeper inclination within a very narrow zone (Figure 11.67*A*). Asymmetric in profile form, monoclines are marked by two hinges (one anticlinal, one synclinal) connected by a **middle limb** (Figure 11.67*B*). The middle-limb strata are generally smoothly curved and continuous, but sometimes they are broken by faults. The most

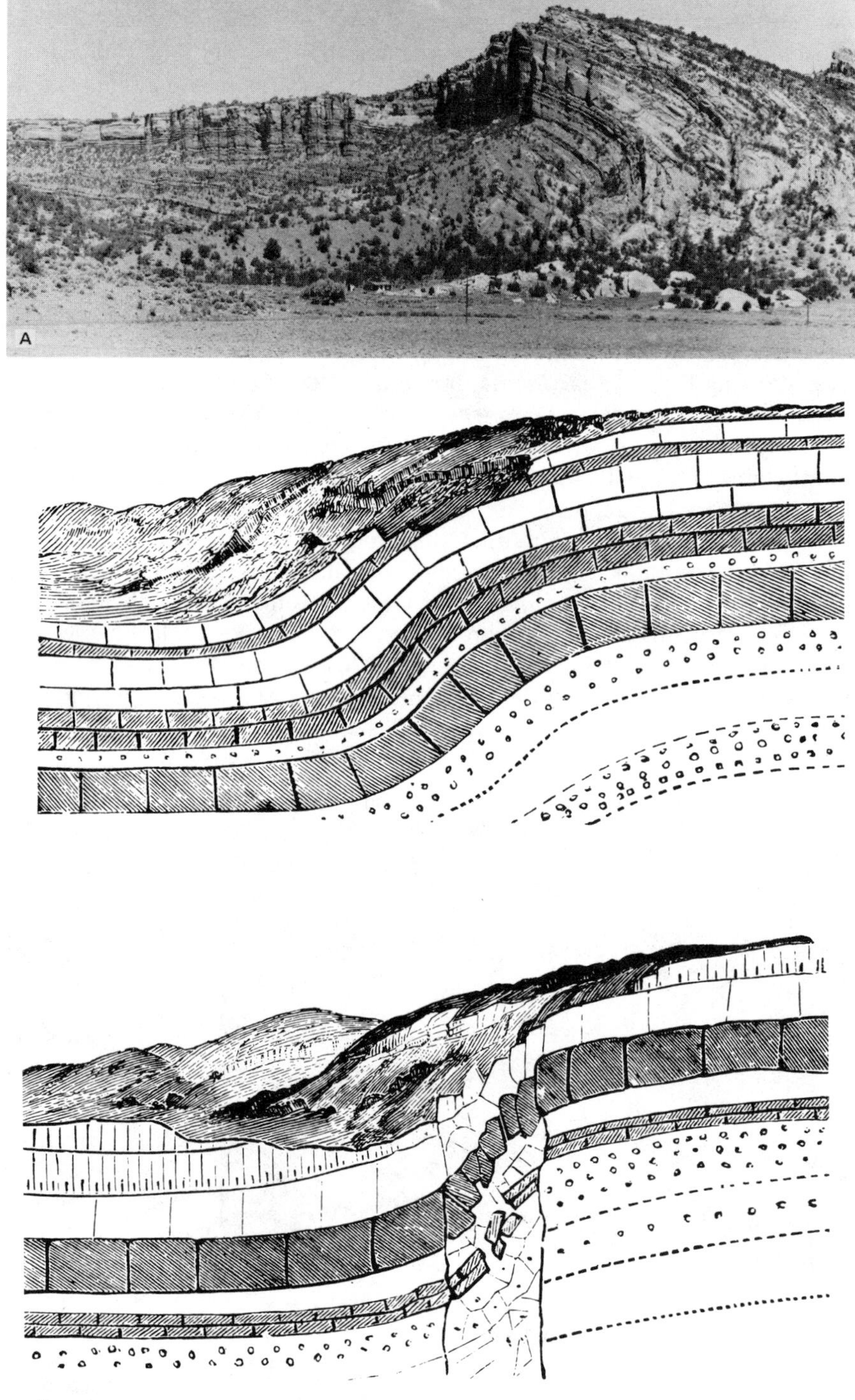

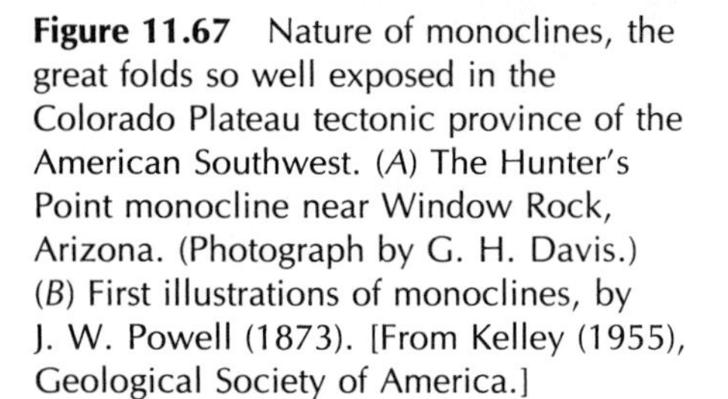

Figure 11.67 Nature of monoclines, the great folds so well exposed in the Colorado Plateau tectonic province of the American Southwest. (*A*) The Hunter's Point monocline near Window Rock, Arizona. (Photograph by G. H. Davis.) (*B*) First illustrations of monoclines, by J. W. Powell (1873). [From Kelley (1955), Geological Society of America.]

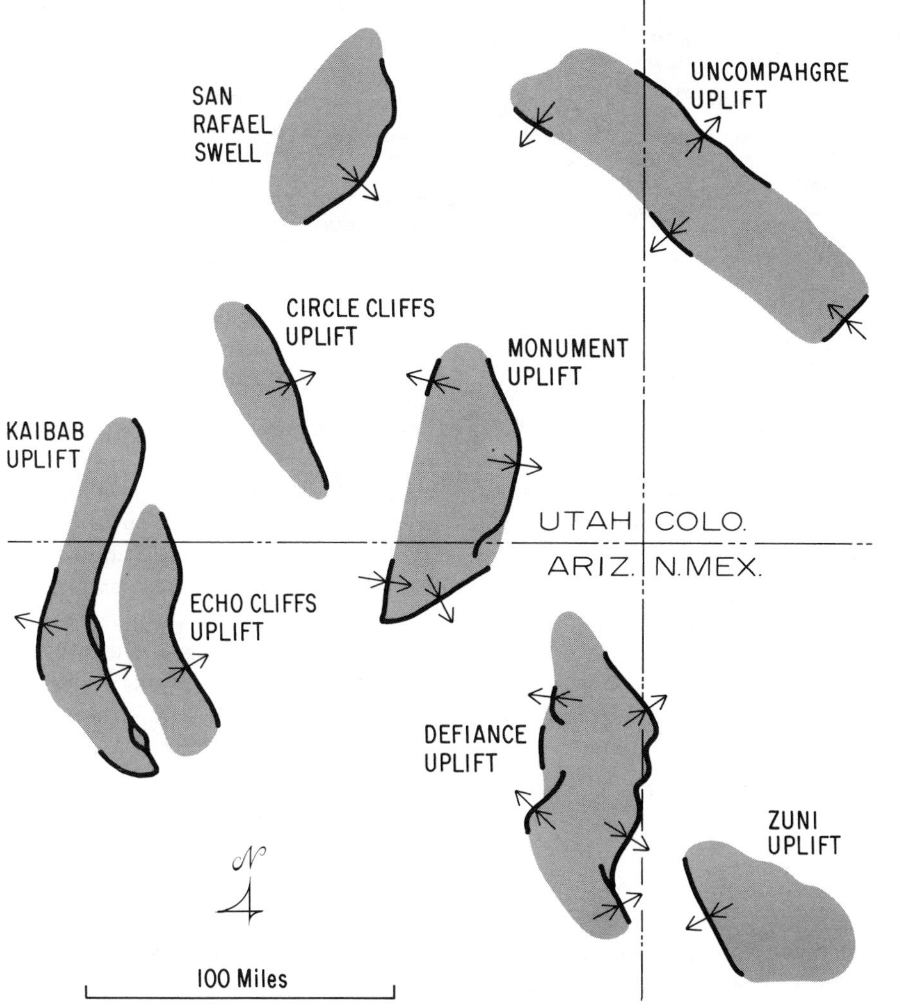

Figure 11.68 Pattern of monoclinal folding in the Colorado Plateau. Monoclines mark the boundary between the great uplifts and basins of the plateau.

spectacular monoclines show off middle-limb dips of 90°! Structural relief on the Colorado Plateau monoclines commonly exceeds 1 km. The very largest monoclines reflect displacements that approach 3 km.

The pattern of monoclines in the Colorado Plateau is marked by sinuous multidirectional, branching folds (Figure 11.68) (Kelley, 1955; Davis, 1978). Individual folds are up to hundreds of kilometers long. Monoclines in the western half of the Colorado Plateau generally **face** eastward (i.e., middle limbs dip easterly); those in the eastern half generally face westward (Kelley, 1955; Kelley and Clinton, 1960).

The major monoclines serve to mark the boundaries between the great uplifts and basins of the Colorado Plateau (Figure 11.68). For example, the Waterpocket monocline in Utah marks the eastern edge of the Circle Cliffs uplift and the western edge of the Henry basin. And the Comb Ridge monocline in Arizona and Utah separates the Monument uplift and the Blanding basin. Locally monoclines interfere with one another and constructively compound their respective displacements (Barnes, 1974; Barnes and Marshall, 1974).

Monoclines appear to be associated with ancient, reactivated, steeply dipping fault zones. This association is especially clearly revealed in the Grand Canyon, where deep erosion has exposed the "roots" of the East and West Kaibab monoclines. Separation relationships disclose the presence of ancient, reactivated Precambrian faults. Offsets among markers of different

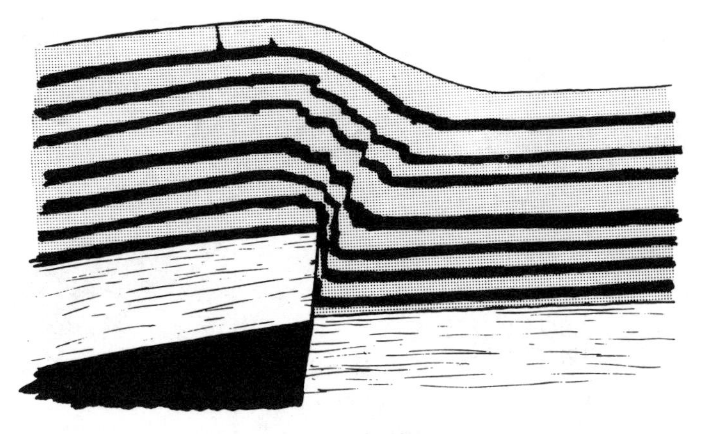

Figure 11.69 Experimental model picturing the relation between monoclinally folded layers and an underlying fault in basement. [From Davis (1978). Published with permission of Geological Society of America and the author.]

ages are *inconsistent* in magnitude of separation and sometimes in sense of separation as well.

The typical profile form of monoclines, including the spatial relationship of monoclines to preexisting basement faults, is portrayed in Figure 11.69. The upper reaches of monoclines are marked by gentle flexure. Middle-limb inclinations steepen downward to 90°, whereupon folding gives way to faulting. Fault zones that cut the basal strata of monoclinally folded multi-layer sequences coincide with preexisting but reactivated basement faults. In the model shown in Figure 11.69, the multilayers are represented by alternating layers of modeling clay and dry, powdered kaolinite. The basement is represented by a pine board. The reactivated ancient fault zone is a sawcut through the pine board.

BASEMENT-CORED ANTICLINES OF THE WYOMING PROVINCE

The basement-cored anticlines of the Wyoming province are larger and geometrically more complicated than monoclines, although both technically might be called basement-cored anticlines. The basement-cored folds of the Wyoming province are enormous asymmetrical anticlines (Figure 11.70), each of which is marked by a gently dipping flank and a steep to overturned flank (Berg, 1981). The cores of these folds are virtual mountains of Precambrian basement. The steepened, overturned limbs of the folds give way to faults that penetrate basement. At surface levels of exposure, the fold-related faults range from steep to shallow in dip.

The configuration of basement-cored anticlines in the Wyoming province is peculiar, both with respect to the distribution of the folds and their orientations (Figure 11.71). Many of the basement-cored uplifts with which the folds are associated trend north–northwest. But the full range of orientation is very broad, including east–west, north–south, and northeast–southwest. The distribution of the uplifts may reflect the influence of compartmental faulting. There *must* be order in this apparent disorder. Finding it will require a careful summing of strains in all directions.

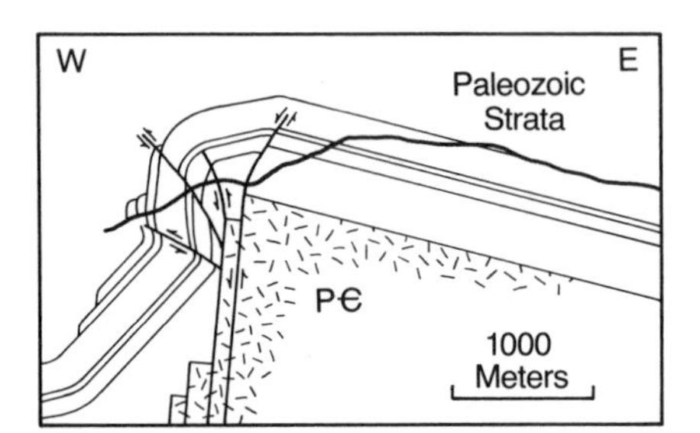

Figure 11.70 Basement-cored uplift in Wyoming. [From Stearns (1978). Published with permission of Geological Society of America and the author.]

FOLDS AND THRUST BELTS

Folds and fold belts associated with thick, thrusted sedimentary prisms have enviable continuity and alignment. Although classically pictured as rounded to circular sinusoidal fold waves, the folds are now most commonly regarded as angular, almost kinklike; for example, see Figure 9.35. Details of the three-dimensional profile forms of the folds have emerged from the combination of detailed structural mapping, exploration drilling, and seismic-reflection profiling. As emphasized earlier (Chapter 9), the forms of the folds may relate in part to the shapes of the thrusts along which the strata were being translated at the time of folding.

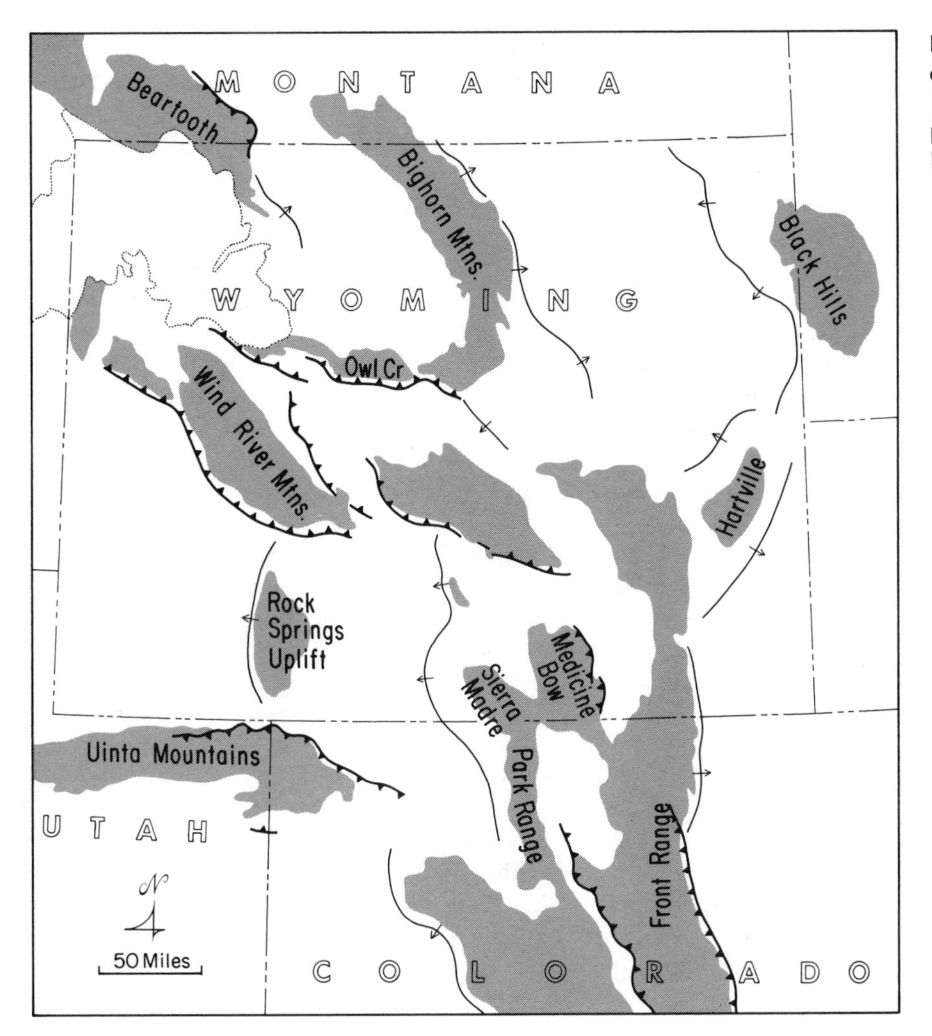

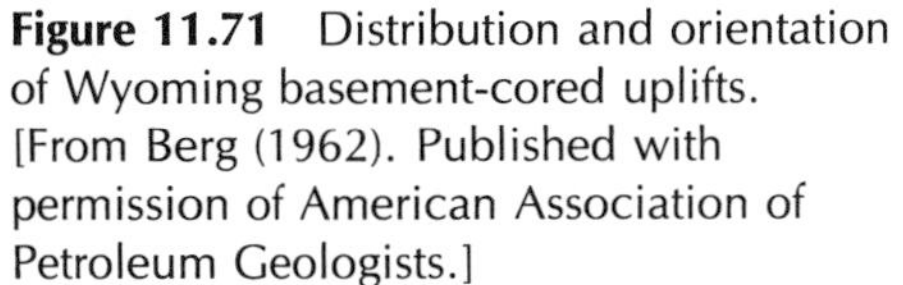
Figure 11.71 Distribution and orientation of Wyoming basement-cored uplifts. [From Berg (1962). Published with permission of American Association of Petroleum Geologists.]

FORCED FOLDING VERSUS FREE FOLDING

Monoclines and basement-cored anticlines can be explained, mechanically, in two strikingly different ways. On the one hand, they have been interpreted as products of **forced folding**, in which the form and the orientation of the monocline or anticline are "forced upon" the sedimentary layers from below by an uplifted basement block (Stearns, 1968). When this occurs the overall size, shape, and trend of the fold in the layered sequence reflects the size, shape, and trend of the basement block (Figure 11.72). The specific geometry of forced folding depends on factors like the ductility contrast between basement and cover, degree of bonding between basement and cover, and the absolute ductilities of basement and cover. In its simplest

Figure 11.72 Experiment in forced folding. Precut basement block is forced to accommodate differential vertical uplift. Overlying layered materials passively fold. [From Friedman and others (1976). Published with permission of Geological Society of America and the authors.]

form, forced folding is a **drape folding** in which fault-dominated movements below give way to fold-dominated movements above. Friedman and others (1976) have modeled forced folding, both experimentally and theoretically, with specific emphasis on the role of **differential vertical uplift** along vertical faults and/or steeply dipping reverse faults. Details of their models match details of fault-related folding in the Wyoming province to a remarkable degree.

At the other extreme of dynamic interpretation, monoclines and basement-cored anticlines can be viewed as products of **free folding**. Free folding is achieved by layer-parallel shortening of multilayers. As reviewed earlier, the folds that result from layer-parallel shortening display profile forms that depend on the physical/mechanical properties of the multilayer sequence.

Reches and Johnson (1978) have analyzed monoclinal folding experimentally and theoretically, in the contexts of both forced folding and free folding. They concluded that drape folding and buckling are both viable mechanisms for generating monoclines. In fact, they concluded that **kink folding** can also generate monoclines. Kink folding is neither layer parallel nor "layer perpendicular," but represents yet a third mechanical option in which the compression responsible for folding is *inclined* to layering.

Each mechanism of monoclinal folding—whether it be draping, buckling, or kinking—provides fold forms that closely match those seen in Nature. Consequently, the interpretation of a specific fold in terms of a specific mechanism of folding requires a thorough knowledge of fold mechanics and a complete inventory of the geometric and physical properties of the fold under scrutiny.

Perhaps the most important theme that emerges from the work of Reches and Johnson (1968) is that monoclines initiate at a site of a preexisting flaw or mechanical disturbance. Without some flaw in a multilayer sequence, there is no theoretical basis to initiate a monocline by free folding. Dynamical consideration then give us yet additional insight regarding why monoclines and ancient faults might be associated with each other.

THE LAST STRAW

Given our present state of knowledge, all folds, not just monoclines and basement-cored anticlines, can be interpreted in a number of different and contradictory ways. Folds in thrust belts bear this out. Their angular kinklike geometries may be the "simple" products of **buckling** by layer-parallel shortening of cohesively bonded multilayers, of **kinking** by layer-inclined shortening of cohesively bonded multilayers, of **sinusoidal folding** by layer-parallel shortening of weakly bonded multilayers, or of **forced folding** in which the forms of folds reflect the forms of thrust faults over which they ride.

Furthermore the sites of specific folds may not relate so much to predictable dominant wavelengths as to unpredictable sites of flaws in the multilayer sequence. Willis (1894) recognized through experimental modeling that 1° to 2° changes in the initial dip of sedimentary layers can predetermine the sites where fold hinges will emerge (see Johnson, 1970). Such observations underscore one of the great contradictions that emerges from the mechanical analysis of folds: *buckling cannot occur in perfectly planar multilayers that are shortened by stresses that are perfectly layer parallel* (Biot, 1959). Fortunately for fold enthusiasts, the smallest imperfections in the primary geometry of layering can trigger the fold-forming process(es).

chapter 12 CLEAVAGE, FOLIATION, AND LINEATION

NATURE OF CLEAVAGE

GENERAL OUTCROP APPEARANCE

Folded sedimentary and metamorphic rocks often display a fundamental internal grain known as **cleavage**. The presence of cleavage in a rock permits the rock to be split into thin plates and slabs. The term "cleavage" is difficult to define: it broadly refers to closely spaced, aligned, planar to curviplanar **discontinuities** that tend to be associated with folds and oriented parallel to subparallel to the axial surfaces of folds (Figure 12.1). As will become apparent, the discontinuities can take many physical forms. Cleavage is always penetrative at the outcrop scale, and commonly it is penetrative at the microscopic scale as well (Figure 12.2). Cleavage typically cuts bedding discordantly, without much regard to the orientation of bedding.

When a rock possessing cleavage is smacked with a hammer, the rock will typically break along the cleavage. Similarly, when rocks possessing cleavage are subjected to scores of centuries of persistent weathering, the worn-down rock that survives in outcrop is commonly marked by sharp-edged, finlike projections that express the presence and general orientation of its internal grain (Figure 12.3). The slabby, platy nature of cleaved outcrops

Figure 12.1 Well-developed cleavage exposed in folded rocks from the South Stack formation in North Wales. (Reprinted with permission from *Journal of Structural Geology*, v. 2, "The Tectonic Implications of Some Small Structures in the Mona Complex of Holy Isle, North Wales", J. W. Cosgrove. Pergamon Press, Ltd., Oxford, copyright ©1980.)

Figure 12.2 Penetrative cleavage as seen in photomicrograph of quartz–sericite schist from the Caribou mine area, New Brunswick, Canada. Folded black layer is composed of fine-grained pyrite. (Photograph by G. H. Davis.)

Figure 12.3 Weathered expression of cleaved metaconglomerates in Precambrian basement of northwestern Arizona. Stretched and flattened pebbles in the cleaved rocks are unusually interesting. From top to bottom of the photo are Mike Davis, Sue Beard, Ji Xiong, and "Mom," the dog. (Photograph by G. H. Davis.)

sometimes misleads us into thinking that cleavage is akin to fracturing. In truth, cleavage forms *without apparent loss of cohesion*, and in this respect alone cleavage discontinuities are far different from fracture discontinuities.

GEOMETRIC RELATIONSHIP OF CLEAVAGE TO FOLDING

Regardless of the specific petrographic and structural characteristics that give rise to cleavage in a rock, the presence of cleavage is expressed in outcrop as planar **cleavage surfaces**. Detailed mapping has shown that a close geometric coordination exists between the orientation(s) of cleavage surfaces and the configuration of folded bedding. Ordinarily cleavage surfaces are either perfectly parallel to the axial surface of folding, or they are disposed symmetrically about the axial surface in a **fan** of orientations (Figure 12.4). In either case, the cleavage surfaces comprise an **axial plane cleavage**. Folded bedding in an upright fold is cut by cleavage surfaces that everywhere are steeper than the inclination of bedding. In overturned folds, cleavage can dip less steeply than bedding. *An axial plane cleavage that dips in the same direction as bedding, but less steeply than bedding, is a warning signal that bedding may be overturned.*

The relationship of axial plane cleavage to bedding in folded rocks can often be used to evaluate the likely facing of a bed, and to construct the

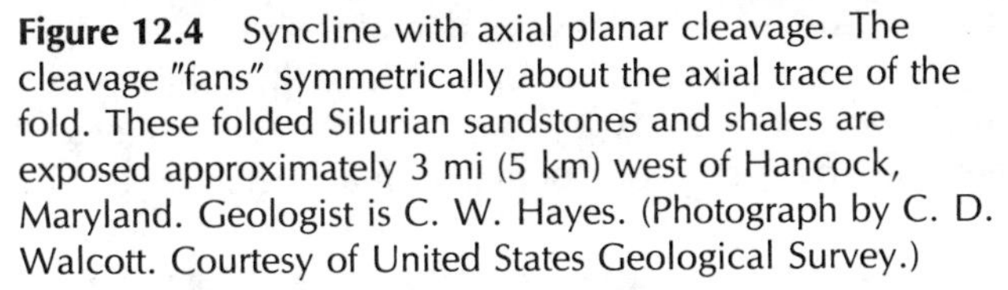

Figure 12.4 Syncline with axial planar cleavage. The cleavage "fans" symmetrically about the axial trace of the fold. These folded Silurian sandstones and shales are exposed approximately 3 mi (5 km) west of Hancock, Maryland. Geologist is C. W. Hayes. (Photograph by C. D. Walcott. Courtesy of United States Geological Survey.)

configuration of folds in profile view. A useful application of the fundamental geometric relationship between bedding and axial plane cleavage is illustrated in Figure 12.5. In the outcrop shown in Figure 12.5*A*, bedding is cut by cleavage surfaces that are known to be axial planar. Bedding dips 80°E, and the cleavage surfaces dip 45°E. Knowing that the cleavage is an axial plane cleavage, it is possible to determine the fold configuration of which the bed is a part. *This is achieved by drawing a folded surface that maintains an axial planar relationship to the cleavage surfaces.* If a fold profile is drawn in such a way that the east-dipping bedding represents the west limb of an upright syncline (Figure 12.5*B*), the form of the syncline cannot be fit in an axial planar manner to the cleavage surfaces. If, on the other hand, the east-dipping bedding is considered to be part of the overturned west limb of an overturned anticline (Figure 12.5*C*), the form of the fold is perfectly compatible with a 45°E-dipping axial plane cleavage.

Bedding and cleavage surfaces are carefully distinguished on geologic maps. The common map symbol for cleavage is shown in Figure 12.6, a simplified geologic map of a plunging anticline/syncline pair. Cleavage symbols in combination with bedding symbols serve to highlight the interrelationships of bedding and cleavage across folds. Where cleavage surfaces cut through the hinge of a fold, there is a maximum discordance between bedding and cleavage. At the hinge point proper, the discordance is fully 90°. At each point on the limb of a fold, cleavage surfaces generally cut bedding at some small, acute angle. The actual magnitude of the angle of intersection steadily *decreases* from the hinge to the inflections of a fold. Isoclinal folds present the special case in which cleavage surfaces and bedding on the fold limbs are perfectly parallel to each other.

The orientation of cleavage surfaces in the hinge of a fold is a close

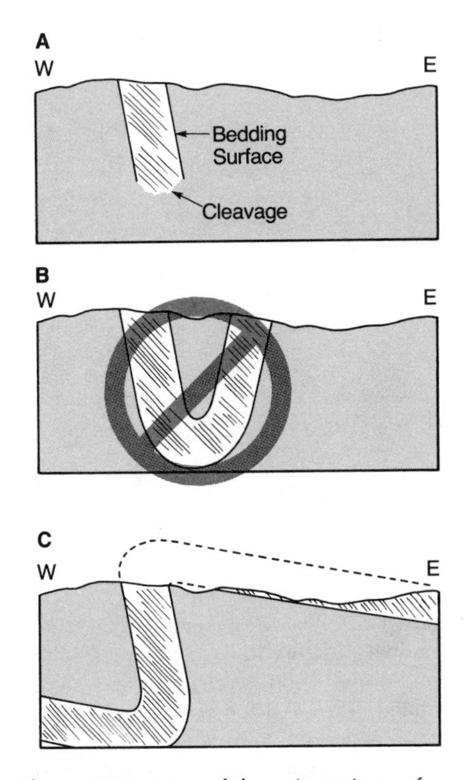

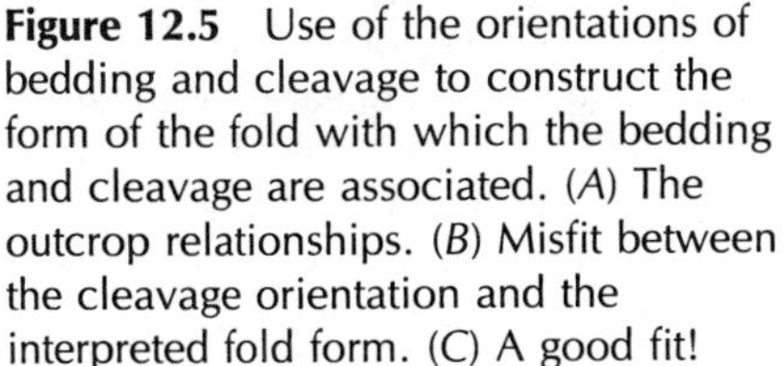

Figure 12.5 Use of the orientations of bedding and cleavage to construct the form of the fold with which the bedding and cleavage are associated. (*A*) The outcrop relationships. (*B*) Misfit between the cleavage orientation and the interpreted fold form. (*C*) A good fit!

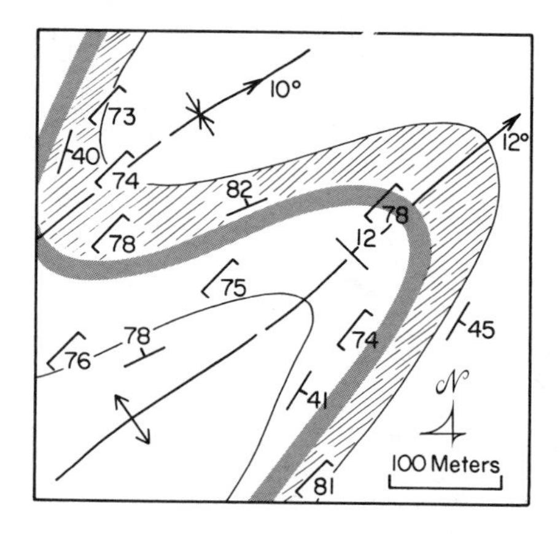

Figure 12.6 Geologic map expression of the relationship(s) between cleavage and folded bedding.

approximation to the axial surface of the fold. However, the orientation of cleavage surfaces at any one point on the limb of a fold usually does not reflect the orientation of the axial surface of folding, simply because cleavage surfaces generally display a fanning spectrum of orientations across the folded surfaces.

DOMAINAL CHARACTER OF CLEAVED ROCKS

When any cleaved rock is examined closely, the property called cleavage is found to be an expression of systematic variations in mineralogy and **fabric**. "Fabric" refers to the total sum of grain shape, grain size, and grain configuration in a rock. The systematic variations in mineralogy and fabric that give rise to cleavage are not primary features related to the formation of the rock; rather, they are expressions of the changes in mineralogy and fabric that were required to accommodate distortion of the rock body within which the cleavage is found.

The systematic variation in mineralogy and fabric in cleaved rock gives expression to the presence of what may be called **domainal structure**, that is, a kind of structural lamination composed of alternating **cleavage domains** and **microlithon domains** (Figure 12.7). Cleavage domains are thin, anastomosing to subparallel, mica-rich laminae within which the fabric of the original host rock has been strongly rearranged and/or partially removed. Minerals and mineral aggregates within cleavage domains show a strongly preferred dimensional and/or crystallographic orientation. Microlithon domains, or simply **microlithons**, are narrow lensoidal to trapezoidal slices of rock within which the mineralogy and fabric of the original host rock remain

Figure 12.7 Excellent example of domainal structure in a quartz–mica schist exposed near Loch Leven, Inverness-shire, Scotland. The cleavage domains are the dark, fine-grained micaceous zones. The microlithon domains are the light-colored, coarser grained zones of crenulated laminae of quartz and mica. (From *The Minor Structures of Deformed Rocks: A Photographic Atlas* by L. E. Weiss. Published with permission of Springer-Verlag, New York, copyright ©1972.)

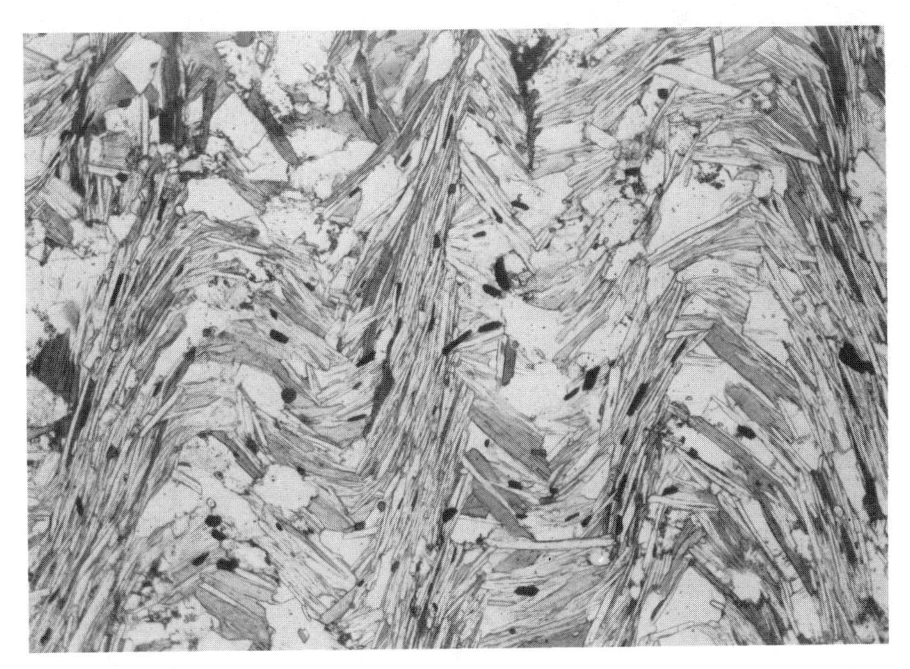

Figure 12.8 Photomicrograph showing domainal structure in mica schist. Oriented micas comprise cleavage domains. The cleavage domains separate microlithon domains of quartz, feldspar, and mica. (Photograph by D. M. Sheridan. Courtesy of United States Geological Survey.)

essentially preserved. Unless the microlithons are composed of rock that contains a preexisting cleavage, minerals and mineral aggregates in microlithons tend to be equigranular, lacking a conspicuous preferred orientation. Microlithons are sharply or gradationally bounded on either side by cleavage domains.

The domainal structure of some cleaved rocks is apparent in outcrop and/or thin section. However, in many cases the domainal structure is visible only when the cleaved rock is scrutinized microscopically at high levels of magnification (Figure 12.8).

TYPES OF CLEAVAGE

MAIN SUBDIVISIONS. There are many ways to name and classify cleavage, and here I present yet another, one that is anchored in the insightful work of Dennis (1972) and Powell (1979). Both Dennis and Powell recognized that it is practical to subdivide cleavage into two classes on the basis of the **scale** at which the domainal character of cleavage can be recognized. Where the distinction between cleavage domains and microlithons can be made with the unaided eye, the cleavage can be described as **discontinuous cleavage**. Where the domainal character of a cleaved rock is too fine to be resolved without the aid of a petrographic or an electron microscope, the cleavage is described as **continuous cleavage**.

CONTINUOUS CLEAVAGE. The main types of continuous cleavage are **slaty cleavage, phyllitic structure**, and **schistosity**. All three are associated with strongly folded and distorted metasedimentary and metavolcanic rocks. Although slaty cleavage, phyllitic structure, and schistosity are distinctively different from one another in outcrop expression, the only real difference among them is grain size and the scale of development of domainal structure.

Slaty cleavage is typically associated with very fine-grained (< 0.5 mm) pelitic rocks metamorphosed to low grade. Where well-developed, slaty cleavage imparts to rocks an exquisite splitting property. Indeed, the presence of slaty cleavage allows a rock to be cleaved into perfectly tabular, thin

Figure 12.9 The phenomenal splitting capacity of slates, as displayed in the "middle quarry" of the Penrhyn Slate Company, Washington County, New York. (Photograph by C. D. Walcott. Courtesy of United States Geological Survey).

plates or sheets (Figure 12.9). Roofing slates and slate blackboards owe their existence and usefulness to slaty cleavage.

The splitting capacity of schist is not nearly as elegant as that of slates, but it is nonetheless very pronounced. Schistosity is best developed in pelitic sedimentary rocks and certain volcanic rocks metamorphosed to medium or high grade. Grain size is typically medium (1–10 mm), expressing the combined influence of initial grain size and recrystallization accompanying metamorphism. The most obvious outcrop characteristic of schistosity is the parallel, planar alignment of micas, including muscovite, biotite, chlorite, and sericite. Schists seldom split cleanly and evenly when struck with a hammer. Instead, they break off in the form of discoidal to crudely tabular handspecimens or slabs.

Phyllitic structure is intermediate in grain size and overall character between slaty cleavage and schistosity. In outcrop phyllites display a soft, pearly, satiny luster. They exhibit the capacity to split neatly but not perfectly.

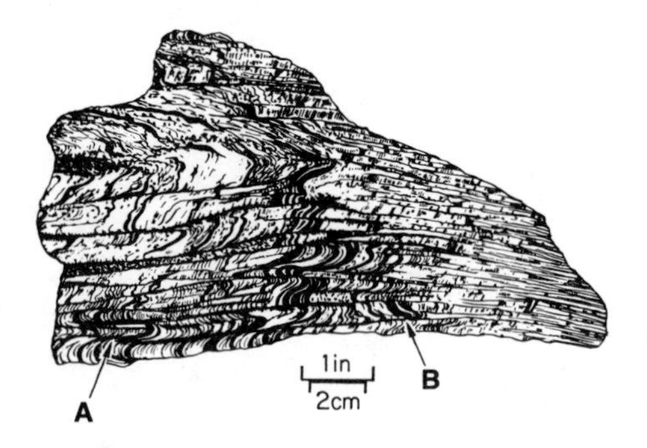

Figure 12.10 Specimen showing examples of both (*A*) discrete and (*B*) zonal crenulation cleavage. [From *Structural Geology of Folded Rocks* by E. T. H. Whitten, after Balk (1936). Originally published by Rand-McNally and Company, Skokie, Illinois, copyright ©1966. Published with permission of John Wiley & Sons, Inc., New York.]

DISCONTINUOUS CLEAVAGES. There are two main types of discontinuous cleavage, **crenulation cleavage** and **spaced cleavage**. Crenulation cleavage is very distinctive in that it cuts a host rock that possesses a preexisting continuous cleavage, especially phyllitic structure or schistosity. The preexisting continuous cleavage is typically crenulated into microfolds, thus the name "crenulation cleavage." Two kinds of crenulation cleavage are recognized: discrete and zonal (Gray, 1977a). **Discrete crenulation cleavage** is a discontinuous cleavage in which very narrow cleavage domains sharply truncate the continuous cleavage of the microlithons, almost like faults (Figure 12.10*A*). **Zonal crenulation cleavage**, on the other hand, is marked by wider cleavage domains that coincide with tight, appressed limbs of microfolds in the preexisting continuous cleavage preserved within microlithons (Figure 12.10*B*). Whether discrete or zonal, cleavage domains in rocks possessing crenulation cleavage are closely spaced, generally between 0.1 mm and 1 cm. Discrete crenulation cleavage tends to form in slate. Zonal crenulation cleavage tends to form in schist.

Spaced cleavage is a second type of discontinuous cleavage, one that has classically been described as **fracture cleavage**. Few workers presently honor the term "fracture cleavage" because it conveys, in a misleading way,

Figure 12.11 Spaced cleavage in a strongly cleaved impure limestone. Arrows point out thrust-fault imbrication of insoluble black chert layer that was incapable of shortening by pressure solution. [From Alvarez, Engelder, and Lowrie (1976). Published with permission of Geological Society of America and the authors.]

the image that this kind of discontinuous cleavage forms through brittle fracture. More and more geologists, myself included, are simply using the expression **spaced cleavage** to name and describe what has historically been called fracture cleavage. Spaced cleavage consists of an array of parallel to anastomosing, stylolitic to smooth, fracturelike partings that are often occupied by clayey and carbonaceous matter (Nickelsen, 1972). Spaced cleavage is typically found in folded but unmetamorphosed sedimentary rocks, especially impure limestone and marl (Figure 12.11). Spacing of the partings (i.e., the cleavage domains) typically ranges from 1 to 10 cm, and thus the microlithons are quite thick compared with all other cleavages. Thickness of the partings often is on the order of 0.02 to 1 mm, although they may be as thick as 1 cm or more.

A fundamental characteristic of spaced cleavage is the **offset** (i.e., separation) of bedding markers along the cleavage, except where cleavage and bedding are perfectly perpendicular to each other. Offset of bedding along spaced cleavage is commonly seen in outcrop. Although the offsets associated with spaced cleavages are faultlike, the cleavage surfaces are certainly not faults. Cleavage domains associated with bedding offsets are never marked by striae or polish. And fossils truncated at the boundaries of microlithons are never found in the adjacent cleavage domain, nor across the cleavage domain in the next adjacent microlithon (Groshong, 1975a).

MICROSCOPIC PROPERTIES OF CLEAVAGE

SLATY CLEAVAGE

Microscopic, high-magnification examination of rocks possessing slaty cleavage reveals a fabric marked by discoidal to lenticular aggregates of quartz, feldspar, and minor mica enveloped by anastomosing, discontinuous mica-rich laminae (Figure 12.12). The micaceous laminae (known as **M-domains**, that is, mica-rich domains) constitute cleavage domains. The discoidal, lenticular quartz–feldspar aggregates (known as **QF-domains**, that is, domains rich in quartz–feldspar) comprise microlithons. The scale of development of domainal structure in slaty cleavage is mighty small. Thickness of the QF microlithons typically ranges from 1 mm to less than 10 μ. The M-domains are typically only 5 μ thick (Roy, 1978).

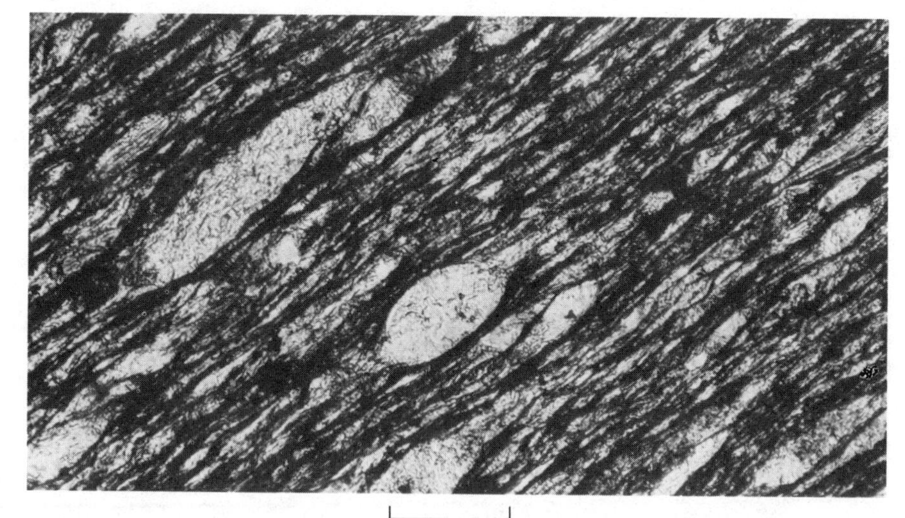

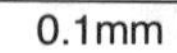

Figure 12.12 Domainal microfabric in slaty cleavage from the Ribagorzana Valley area, Spanish Pyrenees. Mica-rich domains (M-domains) are the black laminae that anastomose around large quartz grains and aggregates of the QF-domains. (Photograph by W. C. Laurijssen. From *An Outline of Structural Geology* by B. E. Hobbs, W. D. Means, and P. F. Williams. Published with permission of John Wiley & Sons, Inc., New York, copyright ©1976).

The QF-domains in rocks possessing slaty cleavage provide a glimpse of the nature of the original host rock. Except for micas, individual grains and mineral aggregates tend to be equigranular, lacking a conspicuous preferred orientation. In sharp contrast to the QF-domains, the M-domains are zones within which the original fabric of the rock is almost completely reconstituted, transformed into strongly oriented intergrowths of aligned mica, quartz, and feldspar. The micas show the most conspicuous alignment, but hidden among the micas are flat to lensoidal quartz and feldspar grains, aligned parallel to the overall orientation of micas and the M-domains. The "flattened" nature of individual grains is further accentuated by **overgrowths** of chlorite and quartz. The overgrowths are like **beards**, growing from the "chins" of relatively large grains of quartz, feldspar, and pyrite (Roy, 1978). Crystal fiber beards grow in the "plane" of cleavage, as a response to the influence of directed stress.

One of the surprising revelations of high-magnification examination of slaty clavage is that the M-domains are so curviplanar and anastomosing. Such conspicuous microscopic irregularity seems to be inconsistent with the capacity of slates to split along "perfectly" planar and parallel surfaces. This apparent inconsistency reminds us once again of the influence of scale on our geologic observations.

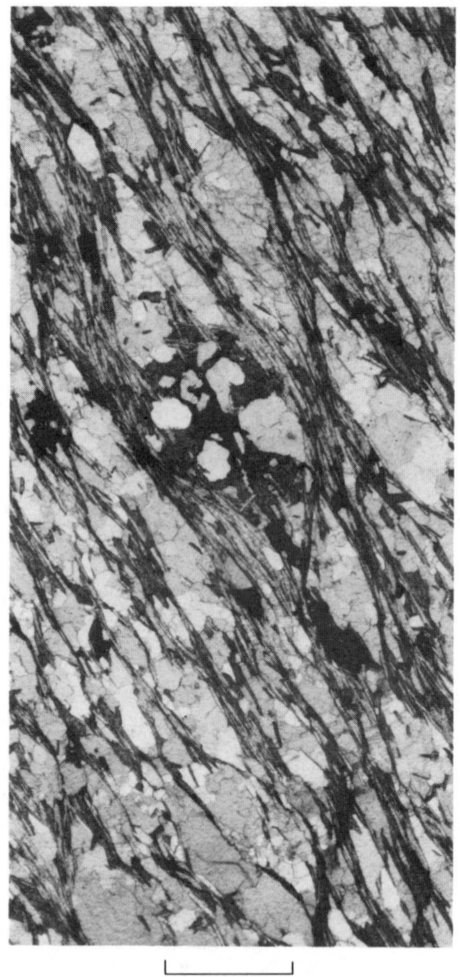

0.1mm

Figure 12.13 Domainal microfabric in schist from Ducktown, Tennessee. Micas form films that envelope aggregates composed principally of quartz. (Photograph by W. C. Laurijssen. From *An Outline of Structural Geology* by B. E. Hobbs, W. D. Means, and P. F. Williams. Published with permission of John Wiley & Sons, Inc., New York, copyright ©1976.)

SCHISTOSITY AND PHYLLITIC STRUCTURE

Like the fabric of slaty cleavage, the microscopic fabric that gives expression to schistosity is composed of anastomosing M-domains and lenticular QF-domains, the cleavage domains and microlithon domains, respectively. The parallelism of micas within the M-domains imparts to schists and phyllites their fundamental splitting capacities (Figure 12.13). These mica-rich cleavage domains contain lenslike and disclike quartz and feldspar grains, which are commonly overgrown by chlorite and quartz at their tips. Quartz and feldspar grains within the QF-domains contain a reasonably preserved record of the fabric of the original host rock, albeit one that may be slightly reconstituted by the effects of recrystallization.

The fundamental distinction between phyllitic structure and schistosity is simply one of grain size. Phyllites tend to be fine grained, with average grain diameter less than 1 mm. Schists tend to be medium grained, with average diameter ranging from 1 to 10 mm. Wispy anastomosing M-domains in schist and phyllite are typically 0.05 mm or less.

CRENULATION CLEAVAGE

The microscopic fabric of crenulation cleavage is quite distinctive. The cleavage domains are M-domains packed with aligned, interlocking micas surrounding lensoidal quartz and feldspar grains as well as opaque minerals and clots of carbonaceous material (Gray, 1979) (see Figure 12.8). The microlithons are QF-domains composed of a preexisting continuous cleavage, like slaty cleavage, phyllitic structure, or schistosity. The continuous cleavage that makes up the microlithons of crenulation cleavage is typically "crenulated" into unbroken wave forms of tiny folds. Axial surfaces of the folds are subparallel to cleavage.

The physical and geometic relation of M-domains to QF-domains depends on whether the crenulation cleavage is discrete or zonal. M-Domains associated with discrete crenulation cleavage are relatively narrow micaceous laminae along which the continuous cleavage of adjacent microlithons is abruptly and sharply truncated. They are faultlike. In contrast, zonal crenulation cleavage is marked by M-domains that are relatively wide and serve as unbroken fold limbs connecting microfolds in the continuous cleavage of adjacent microlithons (Marlow and Etheridge, 1977; Gray, 1979) (Figure 12.14). Whether associated with discrete or zonal crenulation cleavage, micas within M-domains are oriented within 5° of the orientation of the cleavage domains as a whole.

Microlithons in rocks marked by crenulation cleavage are very rich in quartz and feldspar but very poor in micas (Marlow and Etheridge, 1977; Gray, 1979). Cleavage domains in rocks marked by crenulation cleavage are very rich in micas but very poor in quartz and feldspar. To be sure, the segregation in crenulation cleavage of mica-rich domains and domains rich in quartz–feldspar is strikingly conspicuous, reflecting a strain-induced differentiation that we consider shortly.

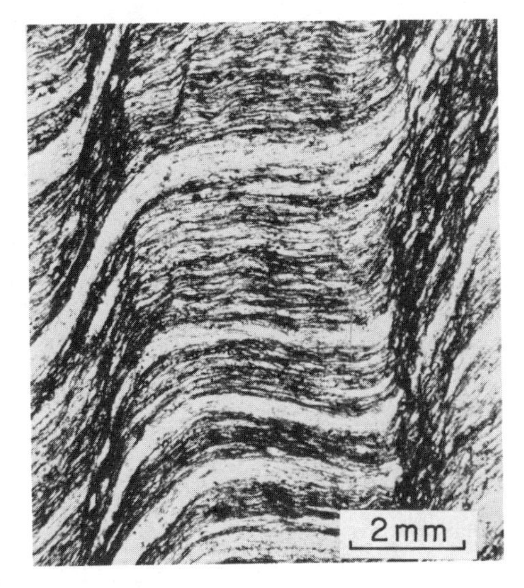

Figure 12.14 Photomicrograph of zonal crenulation cleavage (vertical) coincident with the steep limbs of asymmetric folds in schistosity. The zonal cleavage domains are carbonaceous and micaceous. They have a distinctively lower proportion of quartz than that of the initial fabric. [From Gray (1979), *American Journal of Science*, v. 279.]

SPACED CLEAVAGE

Microscopic examination of cleavage domains in rocks cut by spaced cleavage reveals that they are fracturelike discontinuities lined with seams or films of clayey and/or carbonaceous material (Figure 12.15). Sometimes the cleavage domain discontinuities are stylolitic, but more often they are smooth, planar to curviplanar, parallel to anastomosing surfaces.

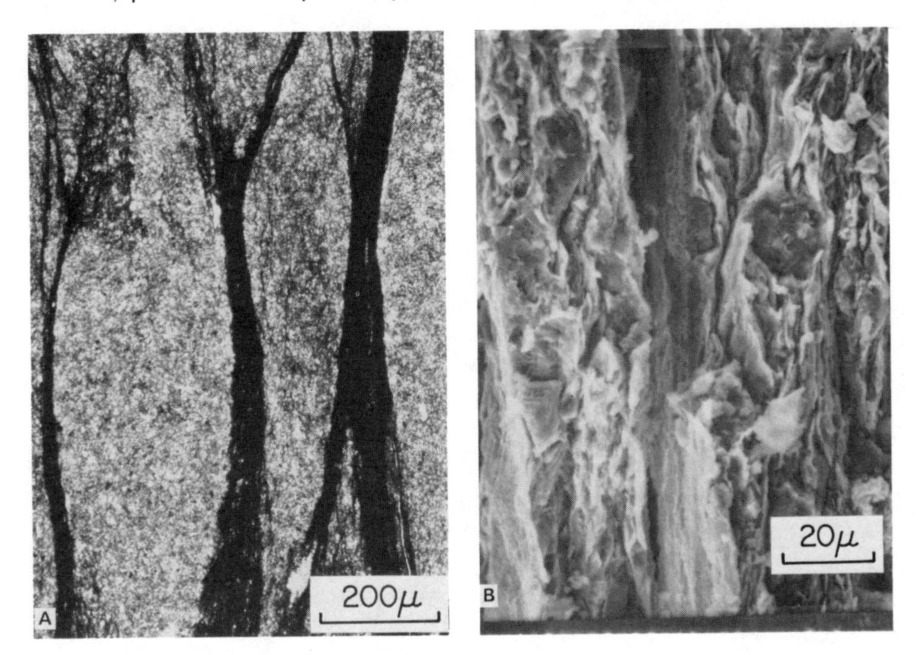

Figure 12.15 (*A*) Photomicrograph of anastomosing, dark, undulating spaced cleavage seams. (*B*) Scanning electron micrograph of the spaced cleavage seams. Composed of densely packed clays, these seams are markedly straight in their trace expression (vertical). Clays in intervening microlithons are more loosely packed, not as preferentially oriented. [From Gray (1981), *Tectonophysics*, v. 78. Published with permission of Elsevier Scientific Publishing Company, Amsterdam.]

Microlithons between cleavage domains in rocks cut by spaced cleavage typically lack a preexisting continuous cleavage. Rather, the host rock for most spaced cleavage is simply unmetamorphosed sedimentary rock, especially limestone and marl.

STRAIN SIGNIFICANCE OF CLEAVAGE

THE ISSUES

The exact role of cleavage has been debated vigorously for more than a century. So many of the descriptive characteristics of cleavage invite explanation! These include the geometric coordination of folds and cleavage, the mechanical role of cleavage in the folding process, the kinematic development of preferentially oriented mineral grains and cleavage domains, the segregation of micas and quartz–feldspar aggregates into domainal structure, the presence of clay-filled partings in rocks possessing spaced cleavages, the development of oriented beardlike overgrowths of chlorite and quartz, the whereabouts of the missing parts of truncated fossils, and the overall strain significance of cleavage.

Traditionally, geologists have attempted to explain the presence and nature of cleavage in terms of **constant volume mechanisms** of deformation, especially rigid body rotation of mineral constituents within the original host rock, mineral recrystallization within a directed stress field, and simple shear translation along close-spaced fracturelike discontinuities. Although each of these mechanisms may contribute to cleavage development, it now seems clear that the formation of cleavage requires significant pressure-solution removal of original host rock. Wholesale removal of rock by pressure solution is perhaps the supreme strain response to directed stress.

SLATY CLEAVAGE

EXPRESSION OF SHORTENING. There is unmistakable evidence that slaty cleavage forms as a distortional response to extreme crustal shortening. Slaty cleavage is always associated with strongly folded rock. And the intimate coordination of the geometry of folding to the orientation(s) of cleavage leaves little doubt that folding and slaty cleavage development are in part synchronous processes. Thus, if folds are considered to be products of shortening, slaty cleavage must be considered a product of shortening as well.

Where fossils, reduction spots, and other primary structures are found preserved in slates, they are typically flattened in the "plane" of cleavage. Flattened fossils and reduction spots thus provide a dramatic statement that slaty cleavage is indeed an expression of severe shortening (Figure 12.16).

The deformation of fossils in slates was recognized and appreciated more than a century ago in the slate quarries of Wales. Phillips (1844) was the first to point out the close relation between fossil distortion and slaty cleavage. Sharpe (1847) went a step further than Phillips, calling attention to the fact that the most highly distorted fossils are associated with the most highly cleaved slates. Noting that the fossils are flattened in the cleavage surfaces and that the cleavage surfaces tend to be parallel and/or symmetrically disposed about the axial surfaces of associated folds, Sharpe concluded that *slaty cleavage forms perpendicular to the direction of greatest shortening* of the rocks in which the cleavage is formed.

To estimate the amount of shortening accommodated by the formation of slaty cleavage, Sorby (1853, 1856) cleverly used reduction spots as a guide to distortion. He concluded that the presence of slaty cleavage can signal

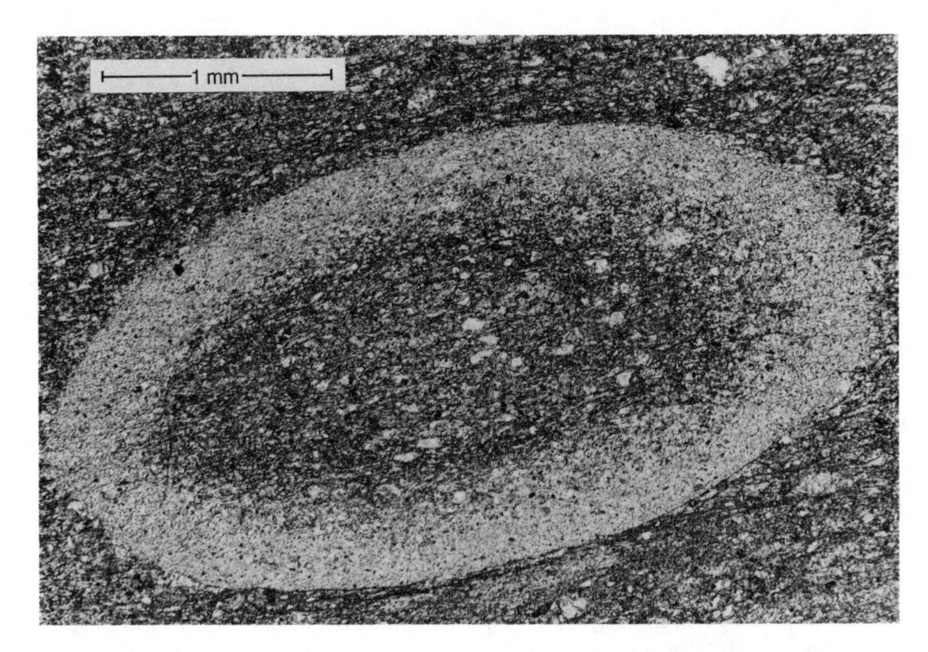

Figure 12.16 Distorted lapillus in tuff from the Lake district of England. [From Oertel (1970). Published with permission of Geological Society of America and the author.]

levels of distortion as great as 75%. Always ahead of his time, Sorby was able to demonstrate through strain analysis that the plane of cleavage in slates is statistically perpendicular to the direction of greatest shortening.

Since the time of the classic work by Phillips, Sharpe, and Sorby, many other geologists have addressed the strain significance of cleavage. The results are the same, time and time again. Oertel (1970), for example, analyzed slaty cleavage in a volcanic tuff unit in the Lake district of England, using ellipsoidal lapilli as guides to strain. Oertel assumed that the lapilli were initially spherical, but were transformed into ellipsoids during folding and the development of slaty cleavage. He proceeded to show that cleavage surfaces in the tuff developed perpendicular to the direction of greatest shortening (Figure 12.16). Tullis and Wood (1975), like Sorby long before, used reduction spots to evaluate the state of strain in Cambrian slates in north Wales. They concluded that the direction of greatest shortening responsible for the formation of the slaty cleavage was oriented precisely perpendicular to the cleavage. Shortening averaged about 65% in the rocks they examined.

EARLY ATTEMPTS TO EXPLAIN THE ALIGNMENT OF MICAS. Measuring the nature and degree of distortion in rocks possessing slaty cleavage is one thing. Determining exactly how the development of slaty cleavage allows rocks to shorten is yet another. Flattened fossils in slate apparently reflect some form of nonrigid body deformation. But how is it achieved?

Early studies on the origin of slaty cleavage focused on explaining the alignment of platy minerals, notably mica and clay, for the preferred orientation of platy minerals was considered to be the chief contributing factor to the physical expression of slaty cleavage. Explaining how platy minerals in slate are brought into preferred alignment was perceived as the means to understanding how the development of slaty cleavage achieves distortion.

Some of Sorby's contemporaries believed that mica and clay minerals in slate are brought into preferred, parallel alignment by (re)crystallization. Advocates of the role of recrystallization in the formation of slaty cleavage emphasized that micas and clay minerals grow preferentially in a common direction as a response to directed stress during metamorphism. Specifically, the micas and clays were pictured as growing perpendicular to the direction of greatest stress.

Although Sorby recognized that recrystallization was a contributing factor to the development of slaty cleavage, he believed that the alignment of platy minerals was mainly due to rigid body rotation of these minerals during distortion of the rock in which these minerals were contained. To demonstrate the mechanism that he envisioned, Sorby (1856) experimentally deformed wax blocks containing evenly distributed and randomly oriented flakes of metals. Sorby verified that the rigid metal flakes progressively rotated into subparallel alignment as the shortening was accomplished. The greater the shortening of the wax block, the better the alignment. By measuring the initial and final orientations of the metal flakes with respect to the direction of shortening, and by comparing the initial and final orientations of the flakes in light of the magnitude(s) of shortening, Sorby anticipated one of the fundamental strain equations (Equation 4.9). Sorby clearly understood that each rigid metal flake, and perhaps each mica or clay in an evolving slate, rotates by an amount that is related to the initial orientation of the metal flake, the direction of shortening, and the percentage of shortening.

The role of grain rotation in the formation of slaty cleavage has been emphasized by modern workers as well, notably Maxwell (1962) and Tullis and Wood (1975). Rigid body rotation of platy minerals into an alignment perpendicular to shortening provides a way to explain alignment of micas in M-domains. However, mechanical rotation alone fails to explain why micas are so densely concentrated in the M-domains, and it fails to explain the presence of flattened, discoidal to lensoidal quartz in the M-domains.

ROLE OF RECRYSTALLIZATION. In seeking additional contributing mechanisms to explain the characteristics of slaty cleavage, recrystallization returns as a mechanism for consideration. The role of **directional crystallization** has its place in understanding the formation of slaty cleavage. It contributes to slaty cleavage by altering grain shape and enhancing the flattened, elongated appearance of minerals and mineral aggregates. In essence, new mineral growth takes place in the plane of cleavage, in **pressure shadows** next to relatively large rigid mineral grains that can provide shelter from the harsh directed stresses that would otherwise inhibit the growth of new minerals. Pressure shadows of chlorite and fiber quartz are very common in M-domains in slates (Figure 12.17*A*). The shadows grow as microscopic beards from the tips of pyrite, feldspar, and quartz grains. The direction of crystal growth is the direction of greatest elongation, regardless of the attitude of bedding cut by the cleavage (Figure 12.17*B*).

Like Sorby's model of grain rotation, the mechanism of recrystallization is by itself inadequate to explain many of the distinctive characteristics of slaty cleavage. It fails to explain the domainal character of slaty cleavage, and it fails to explain the extreme thinness of quartz and feldspar grains within the M-domains of slaty cleavage.

PRESSURE-SOLUTION ORIGIN OF SLATY CLEAVAGE. Significant insights regarding the formation of slaty cleavage have been derived from fuller appreciation of the role of pressure solution in crustal deformation. Based on an explosion of research since the early 1970s, it is now recognized that pressure solution can accommodate significant distortion in sedimentary and metamorphic rocks. *Many of the attributes of slaty cleavage seem to be mineralogical and textural by-products of a shortening achieved through pressure solution.*

We should not be disheartened to learn that Sorby's emphasis on the dominant role of rigid body rotation in the formation of slaty cleavage is now dwarfed by models emphasizing the pressure-solution removal of material.

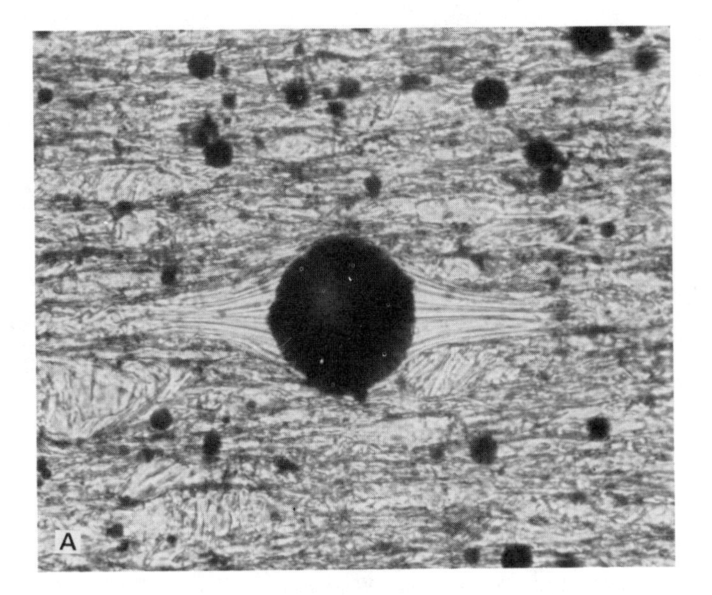

Figure 12.17 (*A*) Photomicrograph of pressure shadows containing fibrous quartz and chlorite. The pressure shadows are "attached" to a spherical pyrite aggregate. Diameter of pyrite is 36 μm. From fold in Martinsburg Slate, Delaware Water Gap, New Jersey. (Photograph by E. C. Beutner.) (*B*) Photomicrograph of feathery pressure shadows (crystal fiber beards) of quartz at the ends of pyrite crystals and calcareous slate. Note the faint horizontal trace of bedding in the matrix of this rock. The pyrite occurs mostly along the bedding, but the pressure shadows have formed parallel to cleavage. (Photograph by L. Pavlides. Courtesy of United States Geological Survey.)

In truth, Sorby was the first to point out the role of pressure solution in geological process, and he was the first to suggest that pressure solution contributes to the formation of slaty cleavage. These insights were not derived simply from examination of handspecimens or outcrops of slate. Rather they were based on microscopic observation, the primary source for fundamental appreciation of the structural significance of slaty cleavage. The saga of scientific achievements by Sorby is humbling: Sorby prepared the very first thin section of a rock, and thus was the first to study the petrography of rocks microscopically using transmitted light. He attacked the microscopic characteristics of slate not by accident, but by design!

Awareness that minerals or parts of minerals can be removed from rock in solution undermines the commonly held premise of **constant volume deformation**. We soon see that volume losses of 50% and more are not unusual in highly cleaved rocks. Clear insight into the role of pressure solution is afforded by the deformational characteristics of quartzite–pebble conglomerates. It has been known for some time that when adjacent quartzite pebbles in a conglomerate are forced into contact during strong, penetrative deformation, the quartz in one or both pebbles is capable of dissolving (or diffusing) at the site of contact. The quartzite pebbles thus interpenetrate one another as a means of accommodating the requisite shortening. The pebbles, when extracted from the bedrock, display indentations or "dimples," which testify to their unseemly behavior. Rims of the dimples may be marked by stylolitic halos. It is the presence of a **stress-induced chemical potential gradient** that drives quartz away from sites of high stress concentration. The quartz reprecipitates into shelters of low stress concentration, where the quartz can recrystallize as pressure shadows, veins, and/or beardlike crystal fiber overgrowths.

Microscopic characteristics of slaty cleavage show all the signs of pressure solution (Figure 12.17). Lenslike and trapezoidal grains of quartz and feldspar are corroded relics of what were once larger, more equant grains. The cleavage-parallel flanks of the lensoidal grains are facets marking the extent of advancement of pressure solution. So are the cleavage-parallel flanks of the lensoidal mineral aggregates that comprise QF-domains. Some of the material missing from individual grains and grain aggregates may be accounted for in the presence of overgrowths and pressure shadows. But much of the dissolved rock must have passed completely out of the system, per-

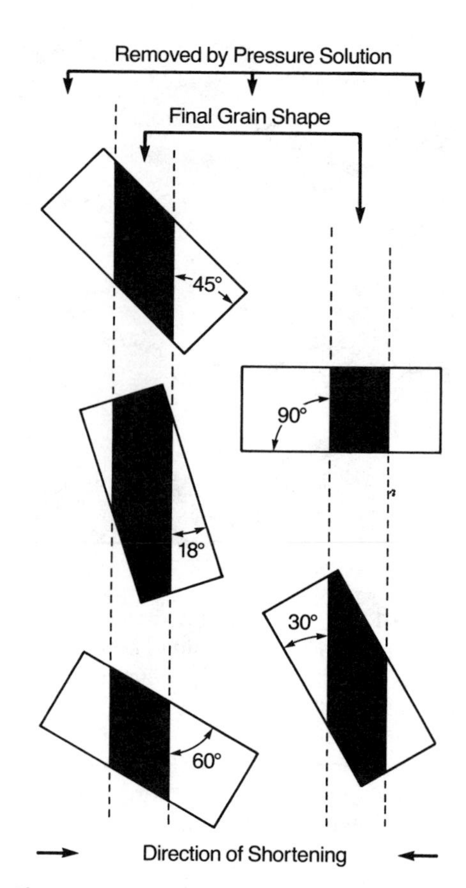

Figure 12.18 The formation of a preferred dimensional alignment of chlorite through progressive pressure solution. [From Beutner (1978), *American Journal of Science*, v. 278.]

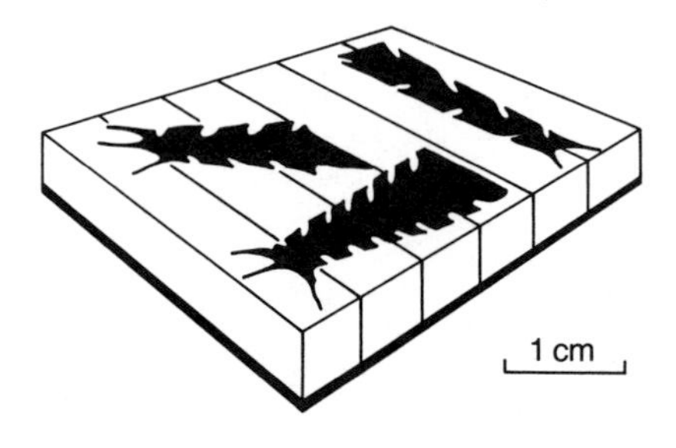

Figure 12.19 Telltale signs of pressure solution of graptolite-bearing slate. The observed size and shape of each distorted graptolite are systematically related to the orientation of each graptolite with respect to the direction of shortening. Graptolites in the plane of bedding that are oriented parallel to the trace of cleavage are long and narrow. Graptolites in the plane of bedding that are perpendicular to the trace of cleavage are short and fat. [From Wright and Platt (1982), *American Journal of Science*, v. 282.]

haps along the M-domains and along fractures. The densely packed concentrations of micas and carbonaceous matter in **cleavage seams** represent the accumulations of less soluble to insoluble residue of the original host rock. The strongly preferred orientation of the micas in the M-domains may, in some cases, be interpreted as the natural result of strain-induced progressive rotation of micas in response to the stress-induced removal of the surrounding rock matrix.

DEVELOPMENT OF PREFERRED MICA ORIENTATIONS WITHOUT GRAIN ROTATION. One of my childhood next-door neighbors, Ed Beutner (1978), demonstrated that the preferred orientation of micas in slaty cleavage can take place *without* grain rotation. Beutner analyzed the structural geology and structural petrology of the Martinsburg Slate, a favorite target of structural geologists working in the eastern part of the central Appalachians.

Beutner's work focused on the chlorite grains, which he interpreted to be part of the mineral assemblage of the original pelite from which the slate was derived. Orientation analysis of the chlorite grains led him to conclude that the grain alignment of the chlorite was a by-product of the progressive systematic, selective corrosion of each original chlorite grain. The general mechanism is pictured in Figure 12.18. Beutner originally suggested that randomly oriented chlorite grains can become dimensionally aligned by virtue of preferential grain-size reduction at right angles to the direction of greatest shortening. Depending on the original orientation of an individual chlorite grain with respect to the direction of greatest shortening, final grain shape due to corrosion can be a parallelogram, a rectangle, or a diamond. The resulting fabric is marked by preferred dimensional orientation of grains, but *not* by a preferred crystallographic orientation of grains.

ESTIMATES OF VOLUME LOSS. The amount of shortening and volume loss accommodated by the pressure-solution removal of host rock can be evaluated quantitatively in rocks containing abundant primary objects of known original shape and size. Wright and Platt (1982) analyzed the Martinsburg formation shales in this regard, using the geometric and dimensional properties of fossil graptolites in the slate as a guide to distortion. The deformed graptolites proved to be magnificent strain indicators, for Wright and Platt had developed a bank of data containing the size and shape attributes of the same graptolites in the undeformed state. Graptolites oriented parallel to the trace of cleavage in the plane of bedding were found to be narrower than normal (Figure 12.19). Graptolites oriented perpendicular to cleavage in the plane of bedding were found to be shorter than normal. (We would be too if acted on by the stresses that created the slates of the eastern Appalachians.)

Examining deformed graptolites oriented at all angles to the direction of cleavage, Wright and Platt were able to show that the Martinsburg formation shales were shortened by an average of 50%, perpendicular to the orientation of the slaty cleavage. Because shortening was accommodated by volume loss, not by constant volume deformation, the rock was not required to stretch out in any direction as a compensation for the shortening. Consequently, the shortening by an average of 50% reflects an average volume loss of 50%!

Continued studies of slaty cleavage will undoubtedly take structural geologists and structural petrologists into the twilight zone of submicroscopic investigations and thermodynamical considerations. Surely there are some surprises yet to surface. In the meantime we can rest assured that the presence of slaty cleavage testifies to significant levels of crustal shortening and, in some cases, significant amounts of volume loss. Upon coming to know slaty cleavage, we can no longer limit our vision to processes of constant volume deformation. Slaty cleavage teaches us to recognize the interplay of **dilational** and **nondilational strain**.

CRENULATION CLEAVAGE

The evaluation of the strain significance of crenulation cleavage serves to further underscore the importance of pressure solution in the formation of cleavage. Pressure solution nicely explains the conspicuous domainal fabric of crenulation cleavage, both discrete and zonal. Furthermore, it provides a means of understanding why micas are concentrated in thin bands and laminae (the M-domains), while the quartz–feldspar mineralogy is for the most part physically separated from the micas in separate domains (the QF-domains).

One of the most enlightening papers about the strain significance and mechanism(s) of formation of crenulation cleavage is that by Gray and Durney (1979). They emphasized that the development of crenulation cleavage involves a physical/chemical redistribution of minerals as a function of relative solubilities and chemical mobilities. They picture the cleavage domains as locations where pressure solution removes substantial amounts of host rock, leaving behind insoluble residues of clayey and carbonaceous material. According to Gray and Durney, the pressure solution takes place on grain and/or layer boundary discontinuities oriented perpendicular to the direction of greatest shortening. Movement of dissolved material follows paths controlled by chemical potential gradients that relate in magnitude and direction to the local stress environment. Cleavage domains emerge along the limbs, or the former positions of limbs, of microfolds in the continuous cleavage of the host rock. Spacing of the cleavage is related to the dominant wavelength of the microfolds and to the amount of solution-induced shortening across the limbs of the microfolds (Gray, 1977b, 1979).

Shortening accommodated by the formation of crenulation cleavage is achieved through a kind of progressive deformation through time, as illustrated in Figure 12.20. Continuous microfold wave forms are buckled into existence by layer-parallel and/or layer-inclined shortening of the preexisting continuous cleavage (Marlow and Etheridge, 1977; Gray, 1979). The fold forms that emerge reflect the influence of strength and thickness of continuous cleavage laminae, degree of cohesion between the multilayers, and magnitude of shortening. Where the shortening required by the strain envi-

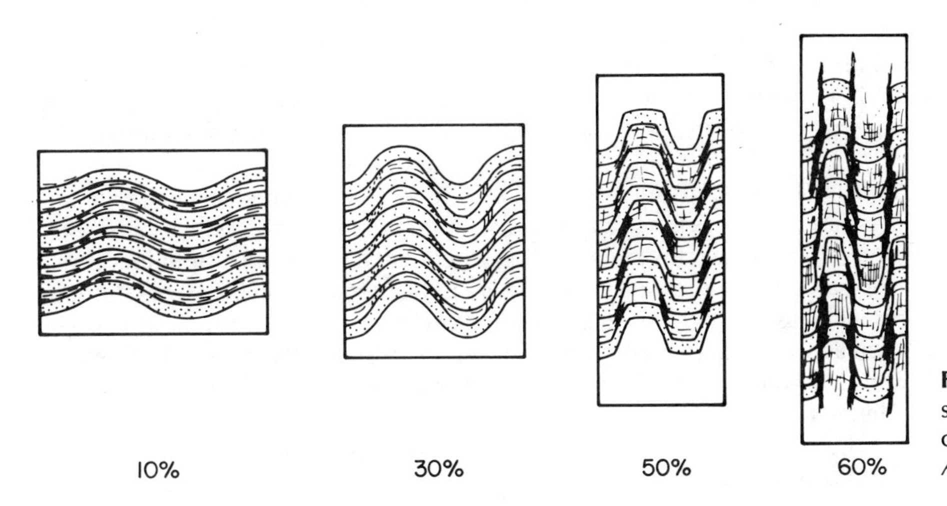

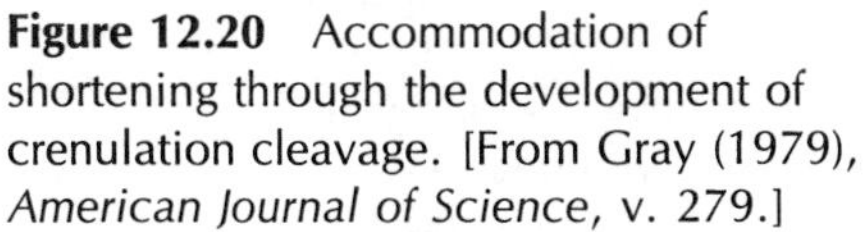
Figure 12.20 Accommodation of shortening through the development of crenulation cleavage. [From Gray (1979), *American Journal of Science*, v. 279.]

Figure 12.21 Photomicrograph of incomplete fusulinid fossil truncated along dark stylolitic seam of insoluble residue. (Photograph by A. Bykerk-Kauffman.)

ronment surpasses what can be achieved through folding alone, the rock begins to shorten by pressure-solution loss of material. Dissolution takes place along the loci of fold limbs, and soon cleavage domains emerge within which quartz and feldspar become relatively depleted compared to that found in adjacent microlithon (QF-) domains. When soluble mineral phases are dissolved along the fold limbs, they are transported in solution along chemical potential paths to fold hinges, where new minerals are deposited in the form of overgrowths and/or thin laminae. Some of the dissolved material leaves the system altogether.

Cleavage domains (M-domains) may initially form at a variety of angles with respect to the direction of overall shortening, but progressive strain eventually brings the cleavage domains into subparallel alignment, perpendicular to the direction of greatest shortening. Substantial pressure solution can lead to the complete removal of fold limbs, leaving faultlike truncations of continuous cleavage within the microlithon (QF-) domains.

SPACED CLEAVAGE

SIGNIFICANCE OF STYLOLITIC SURFACES. Spaced cleavage appears to be yet another product of pressure solution. Indeed, a revolution of thought has emerged from the discovery of the significant role of pressure solution in the formation of spaced cleavage in *unmetamorphosed* strata, especially impure limestones and marls. It is not surprising to find spaced cleavages associated with tightly folded strata. It *is* surprising, however, to discover that spaced cleavage can develop in essentially flat-lying strata, lacking conspicuous folds.

Although not always present, one of the clearest indications of pressure solution is the presence of **stylolitic surfaces**, telltale signs of dissolution and volume loss. Stylolitic cleavage surfaces tend to be occupied by clayey and carbonaceous matter (Nickelsen, 1972), the insoluble residue of pressure solution.

Stylolitic surfaces typically are axial planar with respect to associated folds, a geometric relationship that is quite compatible with the notion that stylolitic surfaces accommodate shortening (Nickelsen, 1972). Teeth and cones of stylolitic surfaces tend to be oriented perpendicular to axial surfaces of folding (Alvarez, Engelder, and Lowrie, 1976), an observation that supports the premise that dissolution proceeds in the direction of greatest principal stress.

Not all spaced cleavage surfaces are stylolitic. In fact it would seem that spaced cleavage surfaces become smoother and smoother as dissolution proceeds. Even in the absence of stylolitic surfaces, the role of pressure solution can be recognized. For example, a clear signature of pressure solution is the abrupt truncation of fossils along cleavage surfaces (Figure 12.21). The incompleteness of **truncated fossils** is due to a stealing away of material by pressure solution, not to faulting or extensional fracturing.

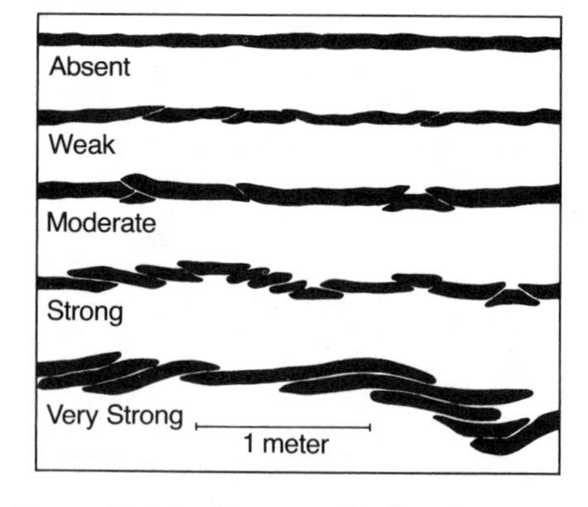

Figure 12.22 Degree of fault imbrication of insoluble chert layers (black) corresponds to the intensity of development of cleavage. See Figure 12.11 for actual field example. [From Alvarez, Engelder, and Geiser (1978). Published with permission of Geological Society of America and the authors.]

STRAIN RESPONSE OF INSOLUBLE BEDS. Another clear signature of pressure solution is the strain response of insoluble layers, like chert, that are forced to undergo shortening within a sequence of **pressure-soluble layers**, like marl and impure limestone. Whereas the pressure-soluble layers shorten through loss of rock volume and the synchronous development of spaced cleavage, insoluble chert layers are obliged to shorten by old-fashioned, constant volume deformational mechanisms, namely folding and thrusting (Figure 12.22).

Where faulted chert layers are driven into adjacent pressure-soluble rock, the pressure-soluble rock responds by dissolving away, leaving a **wad** of preferentially oriented insoluble residue as a record of its former existence. The pressure-soluble layer responds to the stress at the leading edge of an advancing chert layer in the same way that a glacier, feeling the "stress" of rising temperatures, retreats by melting and dumping its unmeltable residue.

KINEMATIC SIGNIFICANCE OF OFFSET BEDDING. Yet another signature of pressure solution in rocks with spaced cleavage is the **offset of bedding** or other laminae along the spaced cleavage surfaces (Figure 12.23). The offset is due to **dilational closing**, the converse of dilational opening along a vein or dike. The magnitude and sense of offset depend on a number of factors, including the orientation of the cleavage surface relative to the orientation of the marker bed, the orientation of the direction of dilational closing with respect to the orientation of the marker bed, and the magnitude of dilational closing. There are only two conditions wherein pressure solution along a cleavage surface will not cause offset of the marker bed that is cut by the cleavage: namely, when cleavage and bedding are mutually perpendicular or strictly parallel, and the direction of dilational closing is perpendicular to the cleavage surface. Otherwise, anything goes!

An example of bedding offset due to pressure solution is portrayed in Figure 12.24. Figure 12.24*A* is a geologic map of a nonplunging, overturned anticline cut by an axial planar spaced cleavage. At station 36 on the rightside-up western limb of the fold, there is a beautiful exposure of the interrelationship of bedding and cleavage (Figure 12.24*B*). Both bedding and cleavage strike 350°, but they dip westerly by different degrees. The bedding dips 20°W and the cleavage dips 50°W. Bedding is repeatedly offset along the cleavage by very small amounts. Separation is in all cases normal, averaging 0.5 cm.

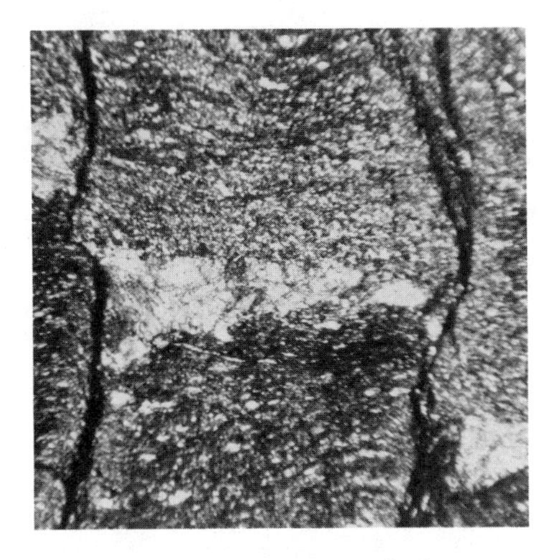

Figure 12.23 Photomicrograph showing offset of once-continuous calcite lamina (white) by pressure solution along stylolitic cleavage seam (black). (Photograph by A. Bykerk-Kauffman.)

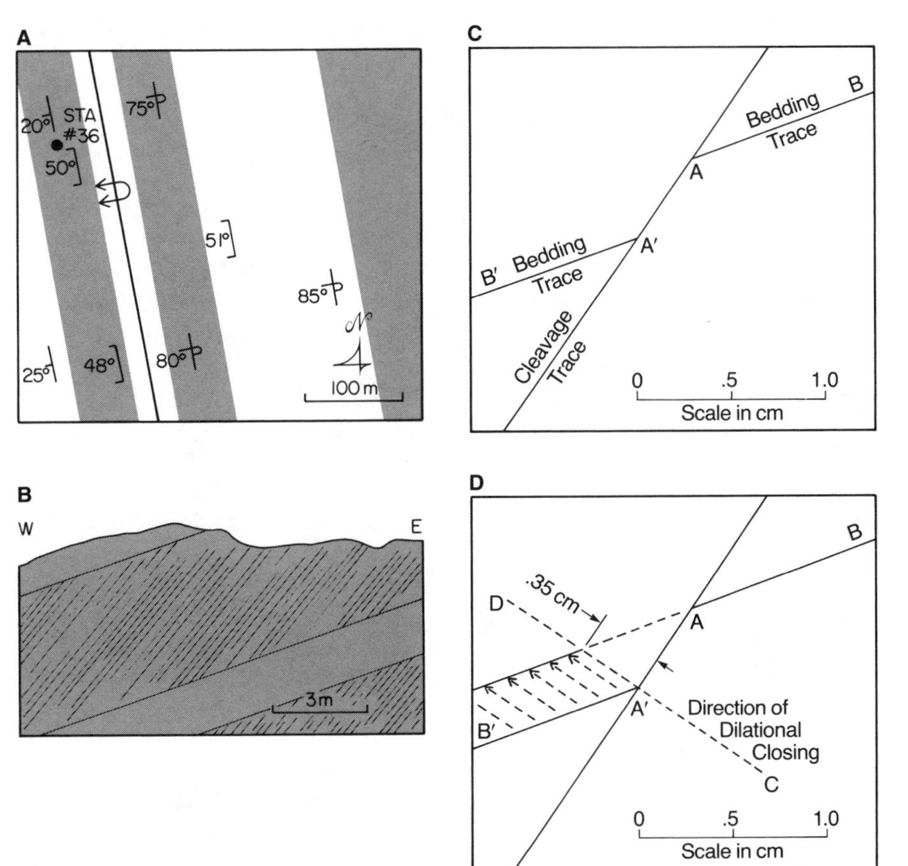

Figure 12.24 Kinematic analysis of bedding offset due to shortening by pressure solution. (*A*) Map showing overturned fold cut by axial plane cleavage. (*B*) Bedding/cleavage relationship exposed at station 36. (*C, D*) Graphical determination of the magnitude of "dilational closing" required to explain the offset of bedding along the cleavage surface.

Assuming that the direction of dilational closing was perpendicular to the orientation of cleavage, it is possible to graphically determine the amount of dissolution required to account for the offset along any given cleavage surface. Figure 12.24C shows the relationships that need to be explained. Lines *AB* and *A'B'* are bedding traces that formerly were in alignment. Distance *AA'* is the magnitude of normal separation, namely 0.5 cm.

To determine the magnitude of dilational closing, simply back off bedding trace *A'B'* along the direction of dilation closing (path *DC*, Figure 12.24*D*), maintaining the 20° dip of *A'B'*. When aligned with the projection of bedding trace *AB*, line *A'B'* is situated in its restored position, *relative to AB*, before deformation. The magnitude of dilational closing is measured along the direction of dilational closing (*DC*) between the restored location of *A'* and the trace of cleavage. It measures 0.35 cm.

Calculations of this type, when averaged over an array of cleavage surfaces in a fold, provide the basis for estimating total volume loss due to dissolution.

CLASSIFICATION OF SPACED CLEAVAGE. Alvarez, Engelder, and Geiser (1978) designed a classification of spaced cleavage, one that provides the means to estimate shortening within a rock layer on the basis of the nature and spacing of cleavage surfaces. In effect, Alvarez, Engelder, and Geiser systematically correlated the properties of spaced cleavage with quantitative estimates of shortening deduced from bedding offsets, truncated fossils, and the degree of folding and thrusting of insoluble chert layers.

The categories of spaced cleavage that Alvarez, Engelder, and Geiser recognized are described as **weak, moderate, strong**, and **very strong**. These correspond with shortening percentages of 0 to 4%, 4 to 25%, 25 to 35%, and greater than 35%, respectively. Cross-sectional and bedding-plane views of two of the intensities of cleavage (moderate and strong) are shown in Figure 12.25.

The strongest cleavages are the most closely spaced. The weakest cleavages are the most stylolitic. Moderate and strong spaced cleavages tend to display clear-cut intersecting sets that are symmetrically disposed about the direction of greatest shortening. Very strongly developed cleavage is marked by sigmoidal cleavage and abundant calcite veining. Veining is yet another

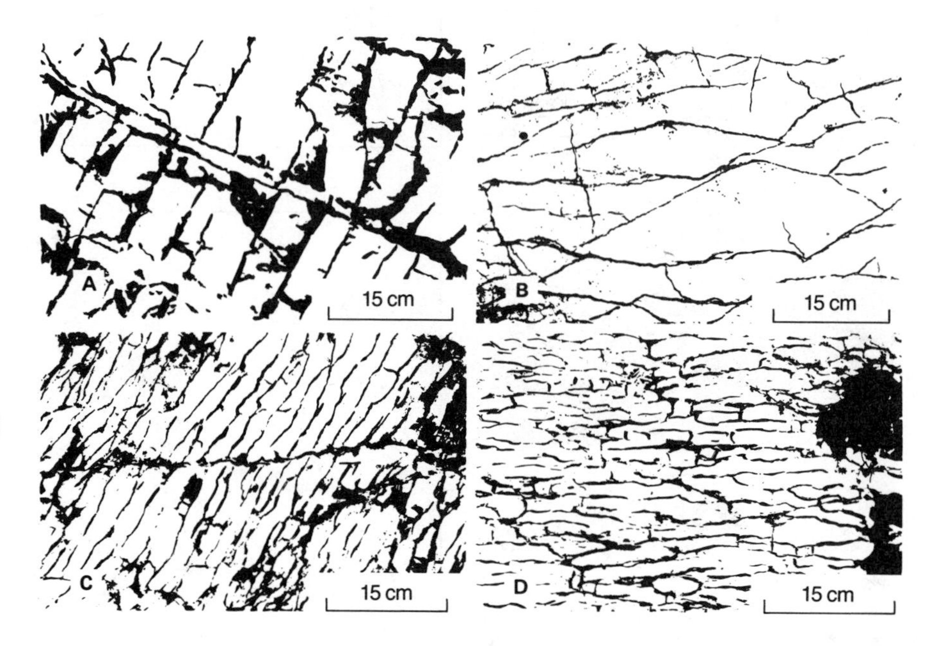

Figure 12.25 Geometry and spacing of moderate and strongly developed spaced cleavage. (*A*) Cross-sectional view of "moderate" cleavage; (*B*) bedding-parallel view of same. (*C*) Cross-sectional view of strong "cleavage"; (*D*) bedding-parallel view of same. [From Alvarez, Engelder, and Geiser (1978). Published with permission of Geological Society of America and the authors.]

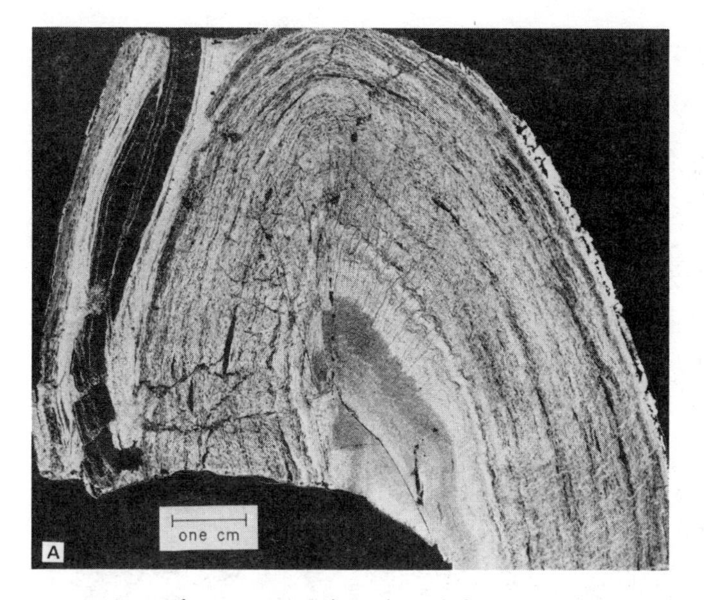

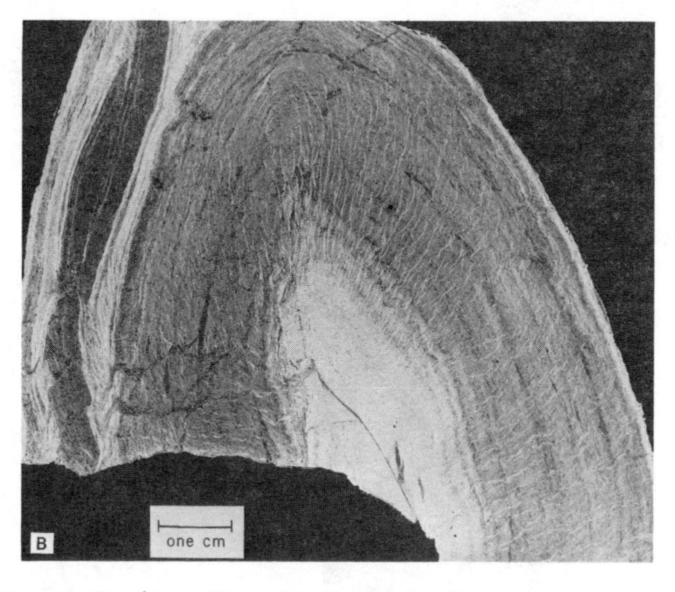

Figure 12.26 Small fold in impure limestone. (*A*) Photograph taken in normal lighting emphasizes nature of bedding. (*B*) Photograph taken in polarized light emphasizes nature of cleavage. [From Groshong (1975a). Published with permission of Geological Society of America and the author.]

expression of pressure-induced mobilization of rock constituents, in this case, calcite.

PALINSPASTIC RECONSTRUCTION. Palinspastic reconstructions of folds and fold belts marked by spaced cleavage requires not only the stretching back of folded layers, it also requires an ***accounting of losses*** due to pressure solution. The folds must be stretched out like an accordion, to account for pressure-solution losses, before daring to rotate bedding up to the horizontal, thereby restoring deformation due to flexural-slip folding and/or buckling.

A miniature example of palinspastic reconstruction of folded layers cut by spaced cleavage was presented by Groshong (1975a). His analysis focused on a single buckle fold in impure limestone. Spacing of cleavage in the fold was found to be 0.2 to 0.5 cm, on average (Figure 12.26*A, B*). Bedding in the fold is repeatedly offset along the cleavage surfaces, especially in the inner arc, core region of the fold. Small fossils in the limestone were seen to be truncated, corroded along the cleavage surfaces. After evaluating the average displacements and pressure-solution losses along the cleavage surfaces, Groshong palinspastically restored the fold to a more open configuration by separating the microlithons. Minimum volume loss was estimated to be 18%.

Volume losses well above 18% are recorded in highly deformed folded terranes. Magnitudes of 40 to 50% dissolution are not at all uncommon. Significant volume losses are also recorded in certain foreland terranes, like the Appalachian foreland. Bedding in the Appalachian foreland region of Pennsylvania and New York is essentially flat lying, folded about very gentle folds with limb dips less than 3° to 4°. The innocent, flat-lying nature of these foreland strata is sharply contradicted by the measured strain state of these rocks. Engelder and Engelder (1977) have demonstrated a layer-parallel shortening of 10 to 15%, most of which has been accommodated by the formation of spaced cleavage.

PASSIVE FOLDING

CHARACTERISTICS OF PASSIVE FOLDS

Passive folds are folds whose development was accomplished by the development of penetrative axial plane cleavage (Figure 12.27). Passive folds characteristically display profile forms that are Class 1C, 2, or 3,

Figure 12.27 (*A*) Passive fold in polished slab of pyritic ore from the Caribou strata-bound sulfide deposit in the Bathurst mining district of New Brunswick, Canada. Cleaved black layers represent original bedding. Cleavage is axial planar to the folded layering. (Photograph by G. Kew.) (*B*) Outcrop-scale passive-flow folds in hornblende–plagioclase gneiss, Medicine Bow Mountains. [From Donath and Parker (1964). Published with permission of Geological Society of America and the authors.] (*C*) Passive-flow fold in Barton River Slate, north-central Vermont. [Photograph by C. G. Doll, United States Geological Survey. From Donath and Parker (1964). Published with permission of Geological Society of America and the authors.]

typified by some degree of hinge thickening and limb attenuation. The axial plane cleavages associated with passive folding run the full spectrum of slaty cleavage, phyllitic structure, schistosity, and spaced cleavage. The axis of a passive fold is simply the trend and plunge of the line of intersection of cleavage and bedding.

There are two classes of passive folding: **passive slip** and **passive flow** (Donath and Parker, 1964), and the distinction between them is scale dependent. If the axial plane cleavage can be seen in outcrop or hand-specimen, the fold is described as a passive-slip fold. However, if the presence of axial plane cleavage can only be discerned microscopically, the fold is considered to be a passive-flow fold.

CONDITIONS FAVORING PASSIVE FOLDING

Sequences that are especially susceptible to passive folding are distinguished by uniformly soft, weak layers. Donath and Parker (1964) describe this mechanical condition as one of **high mean ductility** and **low ductility contrast**. Rocks in this state are capable of profound internal distortion without the loss of cohesion. Passively folded sequences of rocks lack stiff, strong, competent members that, if present, would favor the development of flexural-flow folding.

There are several common geologic circumstances that can lead to the development of layered rock bodies marked by high mean ductility and low ductility contrast. Some not-yet-consolidated hydroplastic sediments are uniformly weak, completely devoid of the mechanical influence of layering. Given half a chance, such weak sediments deform readily by passive-slip and/or passive-flow folding. Certain lithified rocks, like salt or gypsum, are so inherently weak that they behave as rheids and deform by passive folding (Carey, 1962). Some rocks that are perfectly stiff and strong under "normal" temperature and pressure conditions may lose most of their strength under metamorphic conditions of elevated temperature and confining pressure. The heat and pressure blot out the mechanical influence layering might have had under ordinary, nonmetamorphic conditions of deformation.

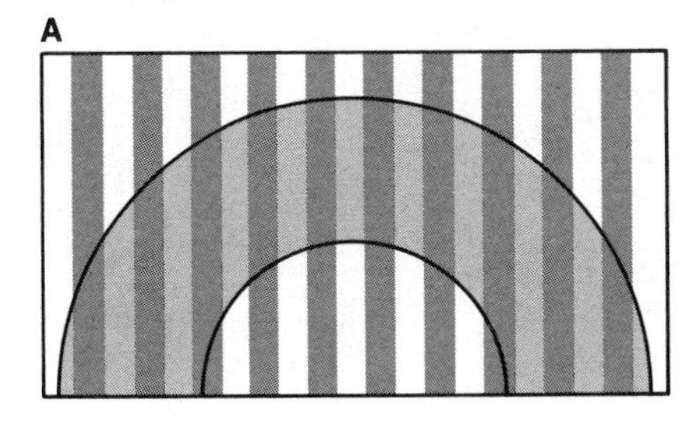

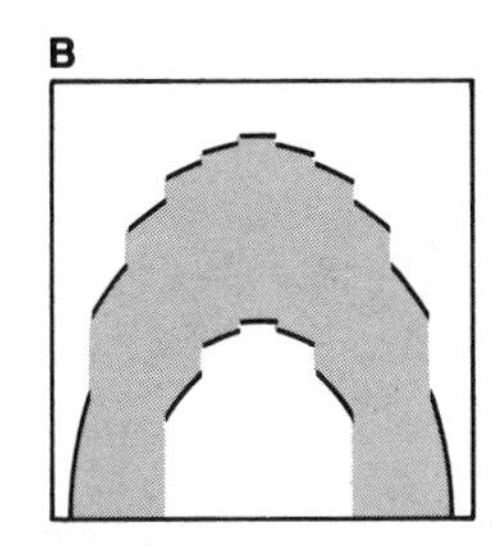

Figure 12.28 Schematic rendering of the transformation of a Class 1B fold to a Class 1C fold by pressure solution. Such a transformation can be simulated easily with a deck of computer cards, not by displacing the cards in simple shear fashion, but by removing domains of material at spaced intervals within the deck.

ORIGIN OF PASSIVE FOLDS

It is inviting, but wrong, to imagine that passive-slip and passive-flow folds form by microfaulting, that is, by simple shear along the close-spaced cleavage surfaces. To be sure, shearlike offsets of bedding along cleavage are seen at many scales of observation. And passive-fold geometries can be created instantly through the shearing of a deck of computer cards (see Figure 2.12) (Carey, 1962; Ragan, 1973). Although simple shearing is one way to create Class 1C, 2, and 3 profile forms, it is impossible to explain in mechanically realistic terms why shear displacements should be systematic enough to produce regular antiforms and synforms (Johnson, 1977). In lieu of a viable mechanism to explain the deformation of wavelike passive folds by simple shear, the computer card analogy is very misleading. And in light of the abundant evidence that the formation of cleavage involves a major component of pressure solution, the concept that cleavage domains are surfaces of simple shear becomes untenable.

The onset of cleavage development and passive folding probably takes place in most cases after a significant amount of buckling and flexural folding has already been achieved. The onset of pressure solution permits a rock to shorten to a degree not possible by further rotation of the limbs of a fold. In effect, a point is reached in the folding process where material needs to be forced out of the inner arc region of the fold. **Flattening** is required to eliminate the room problem.

Flattening can be accomplished by pressure-induced removal of material. Class 1B folds can be transformed into Class 1C, 2, or 3 folds by removal of narrow, parallelogram-shaped sections of rock along directions parallel to the axial surface of the folds with which the cleavages are associated (Figure 12.28). As pressure solution takes place and material is removed, shortening keeps pace so that no open space is created. Adjacent microlithons bordering on a common cleavage domain move toward one another and interpenetrate one another in a direction that is parallel to the direction of greatest shortening. The effect of this, as we have already learned, is to produce shearlike separations of bedding along cleavage surfaces.

The effect of pressure solution during passive folding results in the natural steepening of fold limbs. Although the limbs steepen, bedding within each microlithon *does not* necessarily steepen (see Figure 12.29).

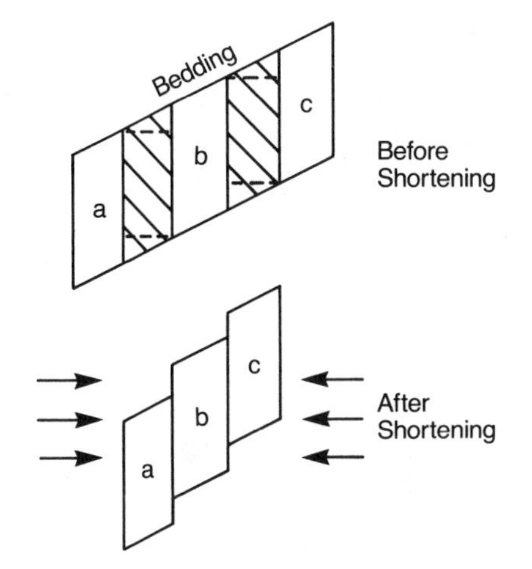

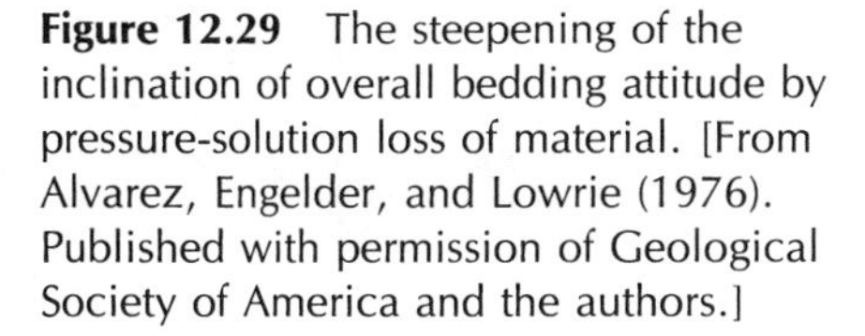
Figure 12.29 The steepening of the inclination of overall bedding attitude by pressure-solution loss of material. [From Alvarez, Engelder, and Lowrie (1976). Published with permission of Geological Society of America and the authors.]

FOLIATION

DEFINITION

The term **foliation** refers to the presence of closely spaced, planar, parallel discontinuities in a rock, discontinuities that are **penetrative** at the scale of outcrop and/or handspecimen (Turner and Weiss, 1963). Foliations have a wide variety of physical expressions, including cleavage (all types), flattened-pebble conglomerate, flow banding in volcanic rocks, eutaxitic structure in ash flow tuffs, flow foliations in intrusive igneous rocks, and fluxion structure in mylonites and other fault rocks.

METAMORPHIC FOLIATIONS

Foliation is a fundamental characteristic of regionally metamorphosed rocks. The chief types of metamorphic foliations are gneissic structure, schistosity, phyllitic structure, slaty cleavage, crenulation cleavage, and flattened-pebble conglomerate. The foliation known as **gneissic structure** is found in middle- to high-grade metamorphic rocks (Figure 12.30). It is defined by compositional bands of different mineralogy, color, and/or texture. The compositional banding arises through the combination of recrystallization, mechanical shearing, and dissolution.

The planar, parallel alignment of platy micas, and the presence of thin compositional laminae is expressed by the foliation in **schistosity** and **phyllitic structure**. At the outcrop scale, rocks possessing **slaty cleavage** display a foliation defined by very thin plates and slabs of rocks that are bounded by planar, parallel discontinuities. As we have learned, the actual physical expression must be explored microscopically.

The foliation called **crenulation cleavage** is defined by subparallel cleavage domains within which pressure solution has forced the concentration of abundant micas. In handspecimen view, crenulation cleavage takes the form of faultlike discontinuities and/or the systematically aligned loci of fold hinges. The foliation of **flattened-pebble conglomerate** is defined by the alignment of distorted, flattened pebbles (see Figure 1.12). Flattened-pebble conglomerates are among the most spectacular foliated rocks. They underscore visually the fact that the mechanical function of foliation is to accommodate distortion.

FOLIATIONS IN SEDIMENTARY AND IGNEOUS ROCKS

Although we commonly associate foliations with metamorphic rocks and metamorphic environments, foliations are found in sedimentary and igneous rocks as well. Weak incompetent sedimentary layers like salt, gypsum, and mudstone typically become foliated as a response to folding. The foliations may take the form of either a bedding-parallel cleavage or an axial plane cleavage, depending on whether the folding mechanism was flexural flow or flexural slip. **Spaced cleavage** is another foliation that can form in sedimentary rocks. So common in highly deformed impure limestones and marls, it is defined by aligned fracturelike partings, sometimes stylolitic, lined with carbonaceous and clayey residue.

Igneous rocks often display penetrative foliation(s). Viscous lava flows of rhyolitic composition characteristically possess **flow banding**, which is itself a kind of foliation. Flow banding is physically expressed in layers and laminae of distinctive mineralogy, color, and/or texture. Ash flow tuffs are distinguished by **eutaxitic structure**, a foliation created through compaction-induced flattening of pumice fragments and the rinds of gas bubbles (Figure 12.31).

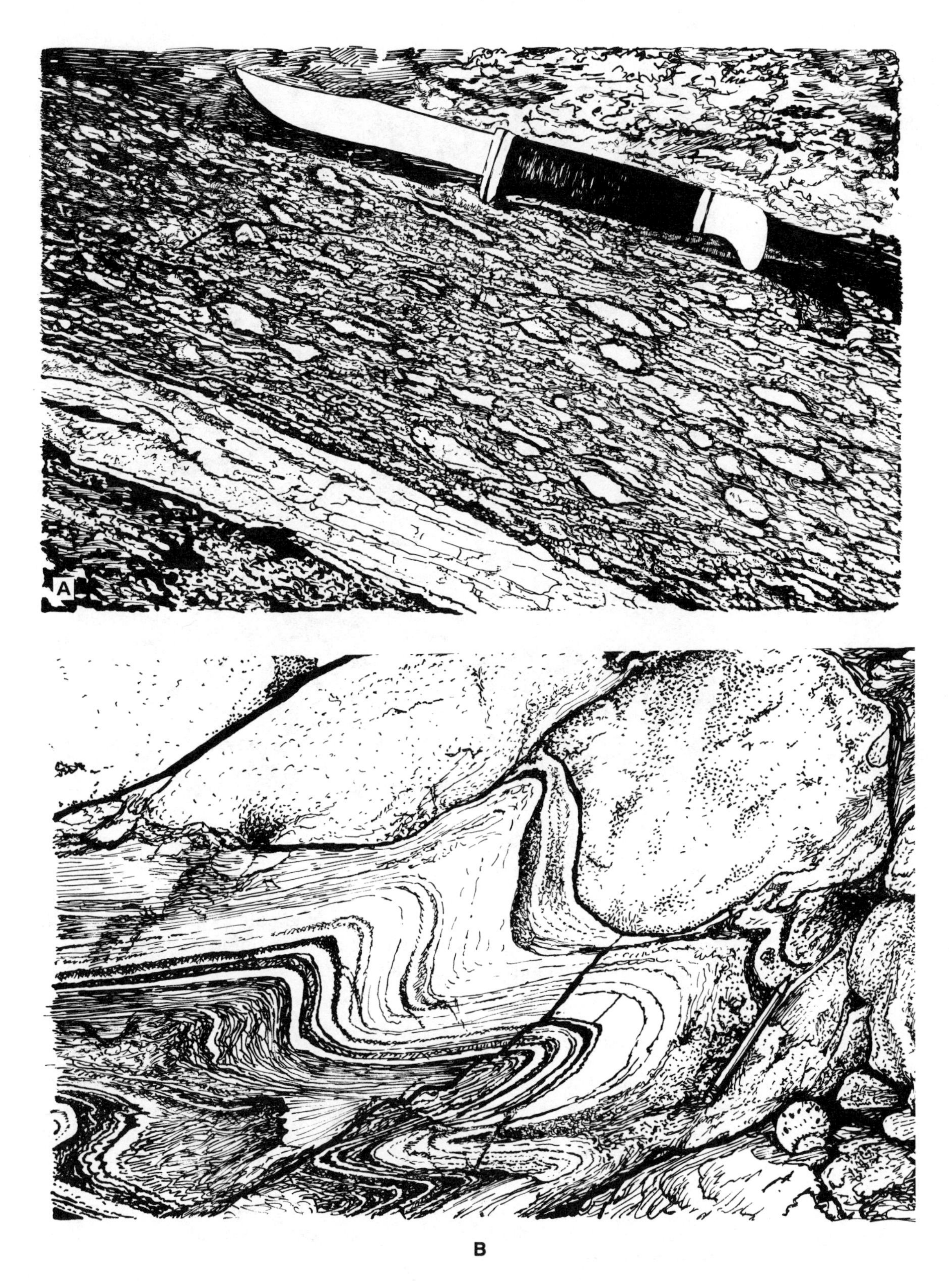

Figure 12.30 Two examples of gneissic structure. (*A*) Augen gneiss in the Tanque Verde Mountain, Tucson, Arizona. The aligned augen ("eyes") are composed of feldspar (white). The matrix is composed of quartz, feldspar, and biotite mica. The white layer that contributes to the gneissic structure is composed of quartz. (*B*) Banded gneiss composed of layers and laminae of quartz–feldspar (white) and biotite mica (black), Little Rincon Mountains, Arizona. [Tracings by D. O'Day of photographs by G. H. Davis. From Davis (1980). Published with permission of Geological Society of America and the author.]

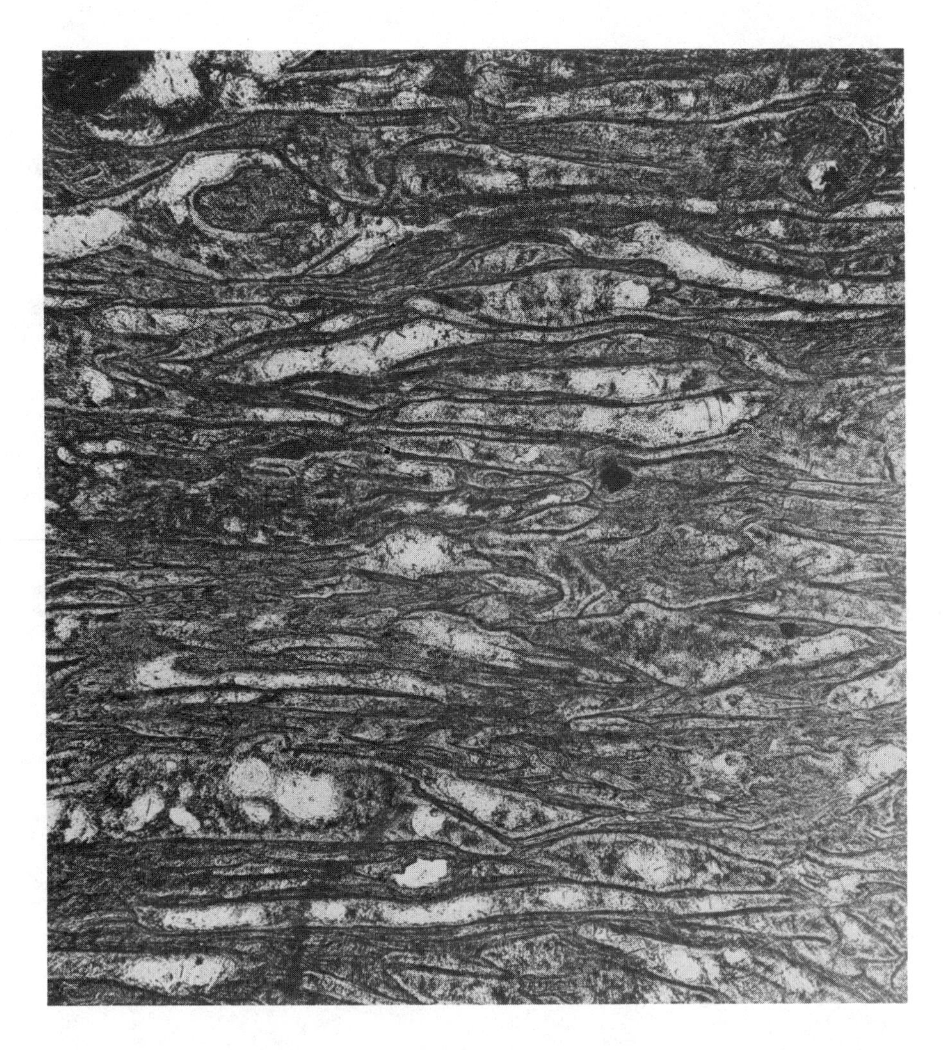

Figure 12.31 Photomicrograph of primary igneous foliation in rhyolite from the Creede caldera, San Juan Mountains, Colorado. (Photograph by J. C. Ratte. Courtesy of United States Geological Survey.)

Intrusive igneous rocks, both large and small, commonly possess **flow foliation**. For example, the margins of dikes, sills, and plugs may take on a strongly foliated appearance as a result of viscous flow of magma past wall rock. And as we learned in the discussion of granite tectonics (Chapter 8), granitic rocks sometimes retain a record of internal flow movements in the form of subtle foliations defined by aligned crystals, phenocrysts, inclusions, and xenoliths.

FOLIATIONS IN FAULT ZONES

Where fault zones or shear zones cut through rocks of any type, whether metamorphic, sedimentary, or igneous, the rocks may be rendered into gouge or some other **fault rock**. Distributed simple shear can impart to fault rocks two kinds of foliation: close-spaced shear surfaces parallel to the walls of the fault or shear zone, and cleavage oriented in the plane of shear-induced flattening. Mylonites and ultramylonites are the ultimate in fault rocks, and each is marked by a strongly developed foliation known as **fluxion structure**.

Careful examination of shear zones by Berthé, Choukroune, and Jegouzo (1979) revealed that mylonitic rocks commonly possess *two* foliations that intersect at 45° or less (Figure 12.32). Named **C surfaces** and **S surfaces** by Berthé, Choukroune, and Jegouzo, these foliations reflect two distinctly different kinematic functions. The C surfaces are planes of relative movement aligned parallel to the shear zone walls. The S surfaces are planes of

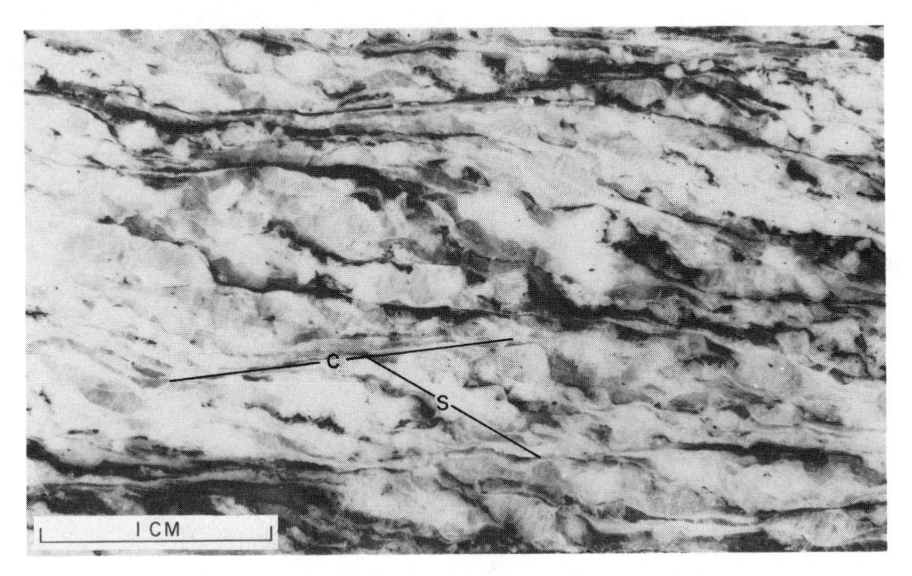

Figure 12.32 Polished slab of mylonitic granodiorite from the South Mountains, Arizona, showing excellent development of C and S surfaces. The configuration of the C and S surfaces reveals a left-handed sense of shear, relative to the orientation of the slab. (Photograph by S. J. Reynolds and K. Matesich.)

flattening. Initially oriented at 45° to the C surfaces, the S surfaces rotate toward parallelism with the C surfaces as progressive simple shear transpires. The relative configuration of the C and S surfaces permits the sense of simple shear to be determined (Figure 12.32).

LINEATION

DEFINITION AND EXPRESSION

Lineation is the subparallel to parallel alignment of elongate, linear elements in a rock body, elements that are penetrative at the outcrop and/or handspecimen scale of observation. Just like foliation, lineation has many physical expressions and can form in metamorphic, igneous, and sedimentary rocks (Cloos, 1946; Turner and Weiss, 1963; Weiss, 1972). The dominant classes of lineation are intersection lineation, crenulation lineation, and mineral lineation. These lineations are typically so penetrative that the lineated rock looks fibrous, like wood. Other lineations are so coarse, expressed in the form of such large elements, that it might be best to describe them as **linear structure**. Examples of linear structure include stretched-pebble conglomerate, rodding, mullion, pencil structure, and boudins.

TYPES OF LINEATION

INTERSECTION LINEATION. Perhaps the most common lineation in metamorphic rocks is **intersection lineation**, a lineation composed of geometric lines created by the intersection of two (or more) foliations, or the intersection of foliation with the outcrop surface. As we soon learn, multiple foliations are a characteristic of strongly deformed metamorphic rocks. The intersection of two closely spaced, penetrative, parallel foliations can result in well-developed lineation, a linear grain if you will, marked in outcrop by a myriad of closely spaced subparallel to parallel lines. The orientation of intersection lineation quite naturally is the trend and plunge of the intersection of the "planes" of foliation; this can be confirmed stereographically.

CRENULATION LINEATION. **Crenulation lineation** is a lineation expressed in the form of bundles of closely spaced tiny fold hinges. Crenulation lineation is especially well developed in phyllites and schists that have been repeatedly deformed by folding. Viewed in the plane of cleavage (Figure

Figure 12.33 Crenulation lineation, defined by crests and troughs of minor folds in quartz–sericite schist. (Photograph by G. H. Davis.)

12.33), crenulation lineation is an array of straight to slightly curved, discontinuous **crests** and **troughs** of folds, more folds than anyone could ever hope to measure.

Crenulation lineation is especially pronounced in mica schists, where the conspicuousness of the phenomenon is enhanced by contrasts in light reflected from the variously oriented micaceous surfaces. It is not surprising that phyllites and schists so characteristically display crenulation lineation. Phyllitic structure and schistosity impart to rocks an **anisotropy** that makes the rock highly vulnerable to kink folding in the presence of stresses that are approximately layer parallel. Mica-rich phyllites and schists readily crinkle, almost like paper.

MINERAL LINEATION. **Mineral lineation** forms in the plane of foliation of many rocks, especially slates, phyllites, schists, gneisses, and felsic volcanic flows. Mineral lineation is typically marked by a **streaky**, fiberlike lineation (Figure 12.34), the actual physical expression of which is hard to discern in outcrop or handspecimen. When viewed microscopically, however, mineral lineation is typically found to be composed of a combination of things: aligned crystal fibers made up of aggregates of fine-grained minerals, especially quartz and mica (Figure 12.35); aligned inequant mineral grains, like feldspar and hornblende; beardlike pressure shadows growing in a preferred orientation from the tips of a sheltering mineral grain; or a subtle crenulation lineation produced by the wrapping of thin mineral laminae over relatively large, aligned, inequant grains or mineral aggregates.

Figure 12.34 Mineral lineation in strongly deformed quartzite from the Coyote Mountains, southern Arizona. (Photograph by G. Kew.)

Some mineral lineations resemble fault striations. But unlike fault striae, mineral lineation is not restricted to a single surface or a thin zone of faulting or shearing. Instead it pervades a substantial part of the body of rock with which it is associated (Figure 12.36). Much of the expression of mineral lineation is due to preferred directional crystallization of minerals. However, the linear alignment of minerals and mineral aggregates can in part be derived by mechanical breakdown, that is, **comminution**, of once-larger elements. The relative roles of recrystallization and comminution are a function of the mineralogy of the rock and the conditions under which deformation was achieved.

STRETCHED-PEBBLE LINEATION. **Stretched-pebble conglomerate** is composed of closely packed elongate clasts, a linear structure fashioned through the distortion of cobbles and pebbles, and in some cases boulders. Many different stretched-pebble shapes are possible: cigar-shaped pebbles (Figure 12.37), and pebbles shaped like tongue depressors (ahhh, what pebbles!) are especially fun to collect. The expression of stretched-pebble linear structure is often enhanced by the presence of mineral lineation in the rock matrix and by the development or crystal fiber beards emanating from the tips of the stretched clasts. The longest stretched pebble that I ever extracted from an outcrop measured 20 in. (50 cm). Longer ones are common, but they are difficult to remove from outcrop in one piece.

Figure 12.35 Mineral lineation defined by aligned biotite "fibers" in schist from Gooseberry Gulch, Colorado. The mineral lineation trends from upper right to lower left. The dark band is an intersection lineation, the trace of bedding (?) in the plane of schistosity. (Photograph by G. Kew.)

RODDING. **Rodding** is a linear structure that is defined by a penetrative array of straight, parallel, highly elongate, rodlike bodies of minerals (Wilson, 1961). Most are made of milky or icy quartz. Rods typically vary in size from pencils to walking sticks. Viewed end on, they are circular, oblate, or lensoidal.

Quartz rods and stretched quartz–pebble conglomerate clasts can resemble each other to a remarkable degree. In fact, "rodding" is a perfectly suitable descriptive term for a rodlike linear structure that is suspected to be, but not proved to be, a stretched quartz–pebble conglomerate.

Figure 12.36 Streaky mineral lineation in the plane of foliation in augen gneiss in Tanque Verde Mountain, Tucson, Arizona. (Photograph by G. H. Davis.)

Quartz rods can be formed in a number of ways. Some may be the boudined, necked expressions of once-continuous layers of quartz. Some may be thought of as the linear equivalent of veins, products of open-space filling and/or replacement, not along fractures but rather along the hinge zones of penetrative folds. Quartz in such rods may in part represent the reprecipitation of quartz made available by local pressure solution of the same host rock in which the rods were found.

Figure 12.37 Cigar-shaped stretched pebble, enjoyed by Stan Keith in the Tortolita Mountains, southern Arizona. (Photograph by G. H. Davis.)

MULLION. The term "**mullion**" reminds us again of the kinship of structural geology and architecture. In the architectural sense, mullions are the long, vertical stone members that separate adjacent window openings of Gothic churches (Figure 12.38*A*) (Holmes, 1928; Hobbs, Means, and Williams, 1976). The mullion face that projects into open air is convex outward. If the mullions were laid out and aligned on the ground in the manner shown in Figure 12.38*B* before being attached to a church, the array of stone would look exactly like mullion structure in rock.

Viewed in cross section, mullion structure displays regular, repeated, foldlike forms, ranging in wavelength from centimeters to meters. The foldlike forms are very distinctive, consisting of linked circular or elliptical arcs. They bear a likeness to oscillation ripple marks but are much more linear and systematic, and they are usually larger. Mullions are never composed of

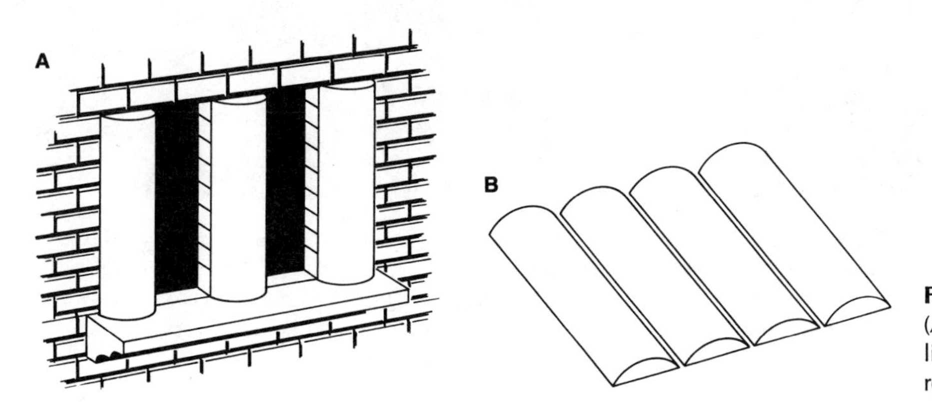

Figure 12.38 Architectural mullions (*A*) adorning a Gothic church and (*B*) lined up on the ground in a way resembling geologic mullions.

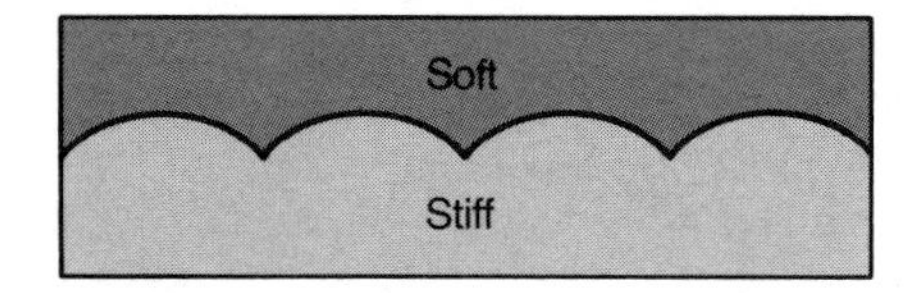

Figure 12.39 Schematic, cross-sectional view of mullion structure.

newly introduced minerals like quartz, but rather are always fashioned from the host rock itself.

Perhaps the most important descriptive relationship bearing on mullion structure is its occurrence along the **interface** between a mechanically soft and a mechanically stiff layer—argillite and quartzite, respectively, for example. The foldlike mullion forms are *convex* in the direction of the mechanically soft layer, with the pinched, cuspate foldlike forms pointing toward the mechanically stiff layer (Figure 12.39). Mullion structures arise from **buckling instability** produced by layer-parallel shortening of a contact separating two rock layers of contrasting mechanical strength. Verification of this can be achieved in the laboratory by subjecting two-layer models to strong layer-parallel shortening (Ramsay, 1967). Experiments of this type demonstrate that **dominant wavelength** of mullion structure is determined by the viscosity ratio of the stiff and soft layers.

Mullion structures relate both to rocks and regions. Ramsay (1963, 1967) suggested that layer-parallel shortening of the basement/cover interface between stiff granitic rocks (below) and layered sedimentary rocks (above) can result in buckling instability that gives rise to broad, domelike granite arches separated by pinched, cuspate synforms of infolded sedimentary rocks.

PENCIL STRUCTURE. Pencil structure is a very distinctive linear structure associated with folded and cleaved mudstones and siltstones. Outcrops pervaded by pencil structure are strewn with unnatural-looking "pencils" of rock (Figure 12.40*A*). Tiny ones are more like brads. Larger ones are like magic wands (Figure 12.40*B*). Pencils of rock are irresistible to collectors of Nature's oddities.

Pencil structure is almost always found in folded strata. In fact, the orientation of pencil structure is dependably subparallel with the axes of associated folds. The actual physical expression of pencil structure is formed by the intersection of **bedding fissility** and cleavage (Reks and Gray, l982). Bedding-parallel fissility is a common characteristic of incompetent layers within folded strata. The fissility is an expression of layer-parallel shear. Where layers containing bedding fissility are cut by penetrative axial plane cleavage, the intersection of the two foliations serves to isolate millions of parcels of pencils. The shapes and sizes of the pencils in cross-sectional view

Figure 12.40 Pencil structure. (*A*) Outcrop of pencil structure in fine-grained calcareous siltstone at Agua Verde near Tucson, Arizona. (*B*) A record-breaking pencil structure, proudly displayed by Ralph Rogers. The pencil was extracted from a wonderful display in Devonian phyllite in central Peru. (Photographs by G. H. Davis.)

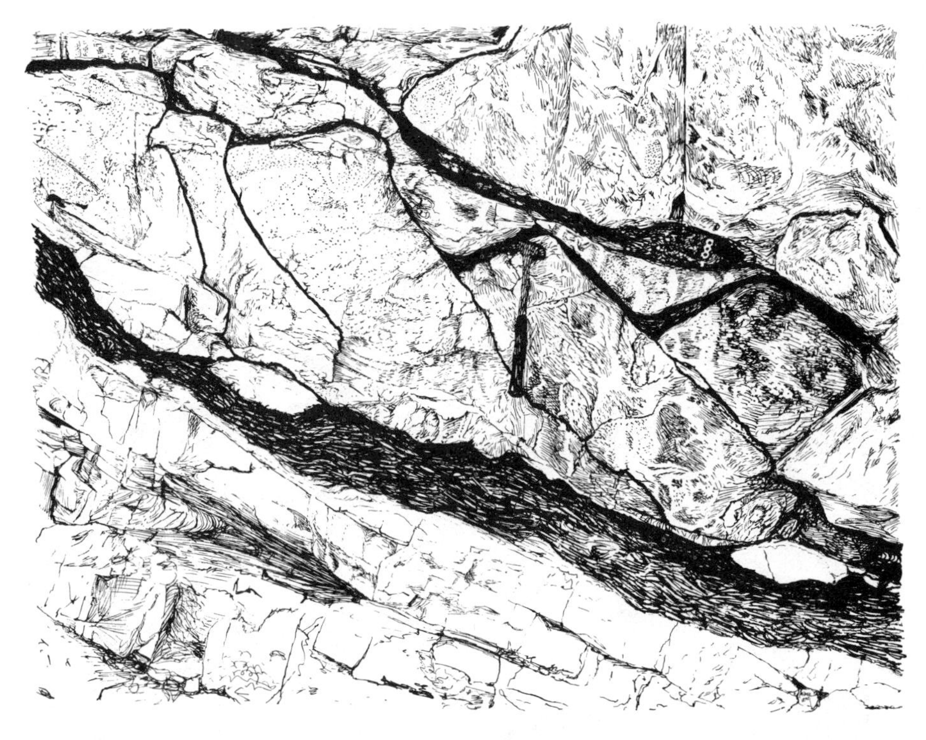

Figure 12.41 Outcrop-scale example of boudins in quartzofeldspathic gneiss. Stiff layer of boudined gneiss is broken and offset by normal faults. The overlying mechanically soft layer accommodated the stretching through plastic deformation. [From Davis (1980). Published with permission of Geological Society of America and the author.]

reflect the spacing and geometric characteristics of the fissility and cleavage surfaces. The lengths of the pencils are determined by spacing of cross fractures. Individual pencils are not hexagonal like "USA Readibond Wallace CONQUEST pencils"; rather, they are irregularly faceted along smoothly curved interfering surfaces.

Reks and Gray (1982) have pointed out that pencil structure is a potentially useful strain marker. Their work indicates that pencils form parallel to the axis of intermediate strain (Y) in rocks that have been shortened by 9 to 26%. It is within this strain range that bedding fissility and axial plane cleavage are equally well developed.

BOUDINS. A final linear structure of significant importance is one we have already encountered—the **boudin**. As discussed earlier, boudins form as a response to layer-parallel extension (and/or layer-normal flattening) of stiff layers enveloped top and bottom by mechanically soft layers (Figure 12.41). The way in which the stiff layer stretches depends mainly on the strength differences of the participating layers and the values of quadratic elongation describing the level of strain parallel and perpendicular to layering (Ramsay, 1967). Where the strength difference (i.e., viscosity contrast) is very high, the stiff layer is likely to break into boudins whose cross sections are rectangular. Each rectangular boudin is bounded by tension fractures, which allow the layer to separate. Potential openings are filled by plastic flow of the enveloping soft layers. Where the strength difference is moderate, the stiff layer will tend to deform into boudins whose cross-sectional forms are lensoidal or sausagelike. Where the strength contrast is very small, the relatively stiff layer will deform by simple pinch and swell (see Figure 8.18).

Irrespective of the profile forms of boudins, the process of **boudinage** creates in rock a linear structure in the plane of layering (Figure 12.42). The linear structure produced by **pinch and swell** is an array of parallel to subparallel furrows. The linear structure produced by true boudins in which stiff layers are repeatedly disrupted consists of a subparallel array of stripes that reflect the presence and structural configurations of the stiff and soft layers.

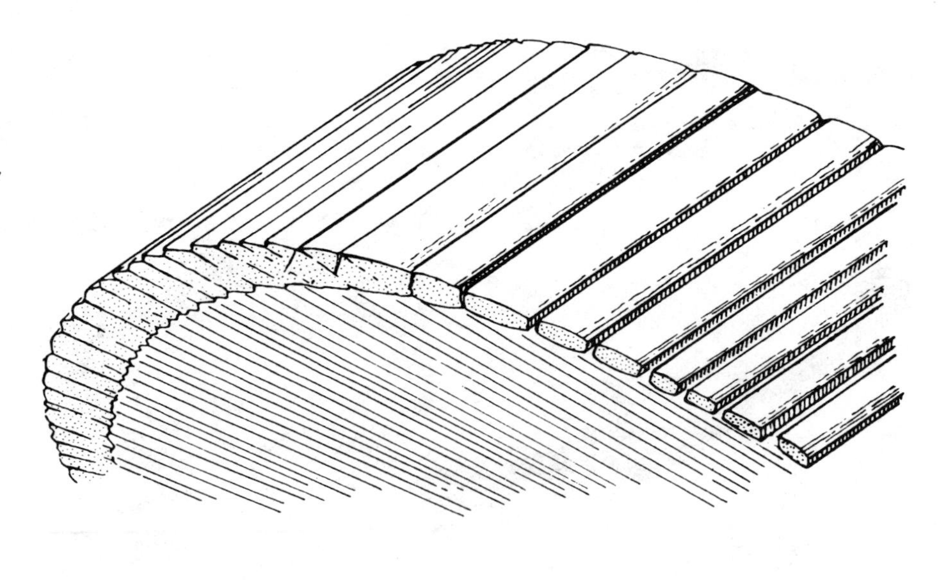

Figure 12.42 Boudins on the flank of a fold. (From *Introduction to Small-Scale Geological Structures* by G. Wilson. Published with permission of George Allen & Unwin (Publishers) Ltd., London, copyright © 1982.)

TECTONITES

THE CONCEPT

Rocks that are pervaded by foliation and/or lineation are known as **tectonites**. Tectonites are rocks that have flowed in the solid state in such a way that no part of the rock body escaped the distortional influence of flow, at least when observed at the scale of a single handspecimen and/or outcrop. Tectonites are like the penny that has been flattened by a train.

Foliation and lineation are the kinds of structure that can accommodate the distortion a tectonite is forced to endure. The extraordinary alignment of foliation and/or lineation in a tectonite is an expression of the geometric requirements of the **state of strain**. Although tectonites, by definition, are rocks that have been able to flow in the solid state (Turner and Weiss, 1963), we now are fully aware that what we perceive as flow is scale dependent. The flow of tectonites is seen microscopically to be a combination of slip and/or crystallization and/or dissolution along exceedingly closely spaced discontinuities.

Most tectonites, and thus most foliations and lineations, form in environments of elevated temperature and confining pressure. Metamorphic and igneous environments are ideal. However, tectonites also form in sedimentary environments, during soft-sediment distortion before lithification, through distortion of excessively weak lithologies like salt or gypsum, or through distortion of lithologies that are especially vulnerable to stress-induced dissolution.

TYPES OF TECTONITE

Several classes of tectonites can be distinguished on the basis of whether the tectonite contains foliation, lineation, or both. These three types are known as **S-tectonites, L-tectonites**, and **LS-tectonites** (Figure 12.43), where "L" and "S" refer to lineation and foliation, respectively. L-Tectonites are tectonites marked by lineation but not foliation. S-Tectonites are tectonites marked by foliation but not lineation. The use of the letter **S** is based on the long-established convention of employing "**S-surface**" in reference to the penetrative, planar, parallel elements that constitute foliation (Turner and Weiss, 1963). The unprepossessing "S-surface" is handy because it places no limits on the physical and/or geometric element that may contribute to the expression of a foliation in a rock. S-Surfaces in a schist are the aligned micas and cleavage domains. S-Surfaces in a gneiss are the planar, parallel, compositional bands. S-Surfaces in a flattened-pebble conglomerate are the aligned discoidal pebbles. And so it goes.

Of all the tectonites, LS-tectonites are the most common. These are the tectonites marked both by foliation and lineation. Lineation in LS-tectonites lies in the plane of foliation.

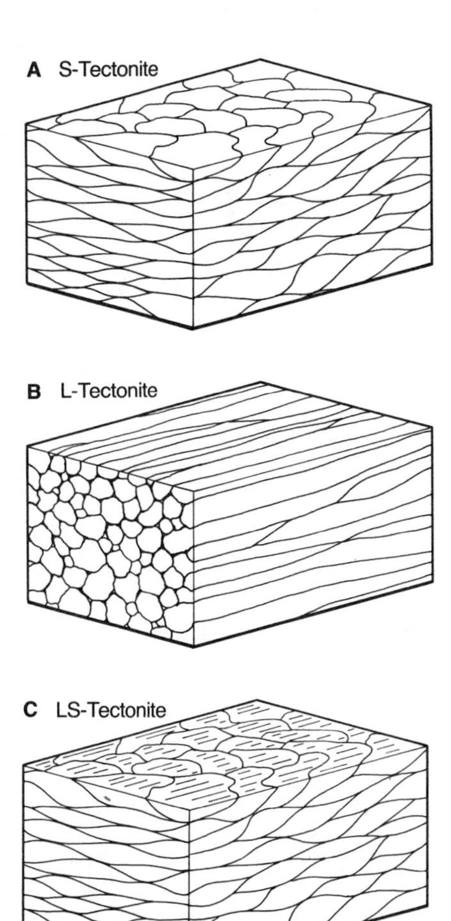

Figure 12.43 Schematic portrayal of S-, L-, and LS-tectonites. (*A*) S-Tectonites are marked by a single, penetrative foliation. (*B*) L-Tectonites are marked by pervasive lineation, but no foliation. (*C*) LS-Tectonites are marked both by foliation and lineation. The lineation in LS-tectonite lies in the plane of foliation.

STRAIN SIGNIFICANCE OF TECTONITES

OVERALL OBJECTIVE. The presence of foliation and/or lineation in a rock is a signal that the rock in which these structural elements are contained has undergone significant distortion, with or without dilation. One of the goals of detailed structural analysis is to interpret the strain significance of tectonites, that is, to try to interpret the magnitudes and directions of distortion and/or dilation that were accommodated by the development of foliation and/or lineation. This, of course, is not a simple task. Success in interpreting the strain significance of tectonites depends to a large extent on discovering and analyzing distorted primary objects, of known original shape and/or size, as guides to the extent of dilation and/or distortion. Finding such treasures is more the exception than the rule.

FLATTENING, CONSTRICTION, AND PLANE STRAIN. The evaluation of distorted primary objects in tectonites has revealed that **S-tectonites** tend to be products of **flattening**, a state of strain in which $\lambda_1 = \lambda_2 > \lambda_3$. Flattening is the kind of distortion that transforms an original sphere into an **oblate strain ellipsoid** (Figure 12.44*A*). **L-Tectonites** tend to be products of unidirectional stretching or **constriction**, a state of strain in which $\lambda_1 > \lambda_2 = \lambda_3$. Constriction transforms an original sphere into a **prolate strain ellipsoid** (Figure 12.44*B*). **LS-Tectonites** ideally form through **plane strain simple shear** such that stretching in one direction is compensated by flattening at right angles to the direction of stretching. Plane strain simple shear transforms an original sphere into a **triaxial ellipsoid** in which the state of strain is marked by $\lambda_2 = 1$, and $\lambda_1 > \lambda_2 > \lambda_3$ (Figure 12.44*C*). Flattening, constriction, and plane strain are distortional strains that may or may not be accompanied by gains or losses in volume of the rock that is converted to tectonite.

It is not difficult to imagine structural environments in which tectonites can be formed through flattening, constriction, and plane strain. Some slaty cleavage is undoubtedly a product of distortion by flattening. An axial planar slaty cleavage could form through significant stress-induced shortening perpendicular to the "plane" of cleavage, unaccompanied by appreciable stretching or shortening within the plane of cleavage. Examples of the development of an L-tectonite through constriction could include the emplacement of a salt diapir, and the magmatic intrusion of a rhyolitic plug. Plane strain is a state of strain that typifies many shear zones. Progressive simple shear can create LS-tectonites whose physical and geometric characteristics express the nature of the simple shear process itself. Foliation would occupy the direction of flattening within the shear zone. Lineation would form in the plane of flattening, oriented in the direction of greatest elongation.

FLINN DIAGRAMS

The full range of three-dimensional strains that may be reflected in the physical and geometric properties of tectonite goes well beyond the special cases of pure flattening, pure constriction, and plane strain. As we know,

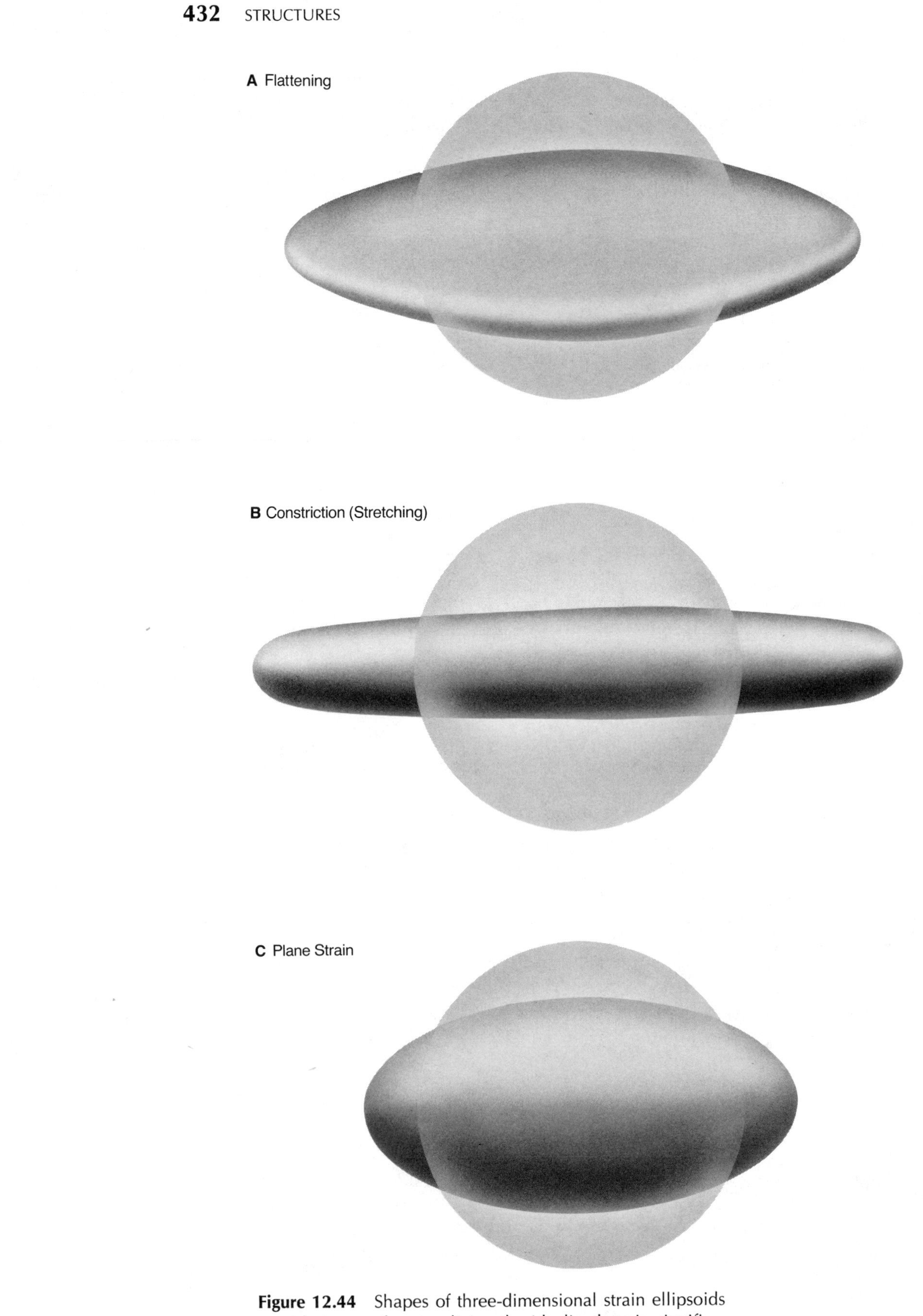

Figure 12.44 Shapes of three-dimensional strain ellipsoids provide images for visualizing the idealized strain significance of (*A*) S-tectonites (an accommodation to flattening), (*B*) L-tectonites (an accommodation to unidirectional stretching), and (*C*) LS-tectonites (an expression of plane strain).

each of these strain states can be achieved with or without volume changes. Furthermore, stretching and flattening may team up in a spectrum of combinations that are limitless.

A convenient way to visualize the possibilities of three-dimensional strain, with or without volume change, is through a device known as a **Flinn diagram**. Introduced by Zingg (1935) and expounded by Flinn (1962), the Flinn diagram is a simple x–y graph that pictures ellipsoids that result from distortion and/or dilation of an original reference sphere. Along the y axis of a Flinn diagram is plotted the strain ratio X/Y, a value equal to $\sqrt{\lambda_1}/\sqrt{\lambda_2}$ (Figure 12.45). Along the x axis of a Flinn diagram is plotted the strain ratio Y/Z, a value equal to $\sqrt{\lambda_2}/\sqrt{\lambda_3}$. Values of X, Y, and Z are derived from strain analysis of distorted primary objects in the tectonite under study, objects like pebbles, fossils, and reduction spots.

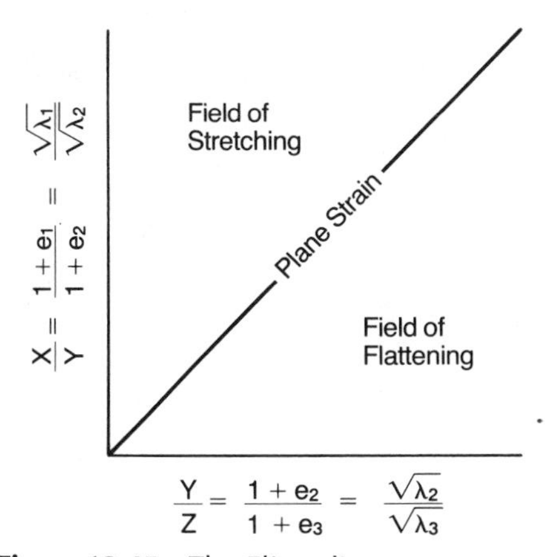

Figure 12.45 The Flinn diagram, a device for portraying the state of strain of deformed rock.

If a tectonite forms through pure flattening, the coordinates X/Y, Y/Z that describe the state of strain of primary objects in the tectonite should plot along the x axis of a Flinn diagram. Conversely, if a tectonite forms through pure constriction, the coordinates X/Y, Y/Z should plot along the y axis. If a tectonite forms by plane strain, the coordinates X/Y, Y/Z should plot along a 45°-sloping line that intersects the origin of the plot. The lines of pure flattening, pure constriction, and constant volume plane strain are simply three in number. There is plenty of room in a Flinn diagram to plot the limitless combinations of flattening and constriction that one might encounter in nature.

LOGARITHMIC FLINN DIAGRAM

An even more useful way to present the state of strain of tectonite is through the use of a logarithmic Flinn diagram. Introduced by Ramsay (1967), it is a modification of the Flinn diagram and can be used to keep track of changes in volume that might accompany distortion.

Logarithmic strain (ϵ), otherwise known as **natural strain** or **true strain**, is equal to the logarithm of the quantity of (1 + e) (Ramsay, 1967):

$$\epsilon = \log_e (1 + e) = \log_e (\sqrt{\lambda}) \tag{12.1}$$

where

ϵ = logarithmic strain

e = extension

λ = quadratic elongation

Logarithmic strain (ϵ) might at first look more intimidating than extension (e) and quadratic elongation (λ), but ϵ can be readily calculated if values of e and λ are known. To represent on a logarithmic Flinn diagram the state of strain of a given tectonite, values of e_1, e_2, and e_3 must first be transformed into values of ϵ_1, ϵ_2, and ϵ_3. Once these conversions have been made, $\epsilon_1 - \epsilon_2$ is plotted along the y axis of the logarithmic strain Flinn diagram, and $\epsilon_1 - \epsilon_3$ is plotted along the x axis (Figure 12.46). As in the ordinary Flinn diagram, pure flattening is described by values of ($\epsilon_1 - \epsilon_2$, $\epsilon_2 - \epsilon_3$) that plot along the x axis, and pure constriction is represented by values of $\epsilon_1 - \epsilon_2$ and $\epsilon_2 - \epsilon_3$ that plot along the y axis of the diagram. Points representing the condition of constant volume plane strain fall along a 45°-sloping line that intersects the origin of the plot (Figure 12.46). If a tectonite forms through plane strain accompanied by dilation, the coordinates $\epsilon_1 - \epsilon_2$ and $\epsilon_2 - \epsilon_3$ describing the state of strain of primary objects in the tectonite will plot along 45°-sloping lines that do not intersect the origin of the plot. Instead they intersect the x or y axis, depending on whether the dilation was accompanied by a volume decrease or a volume increase.

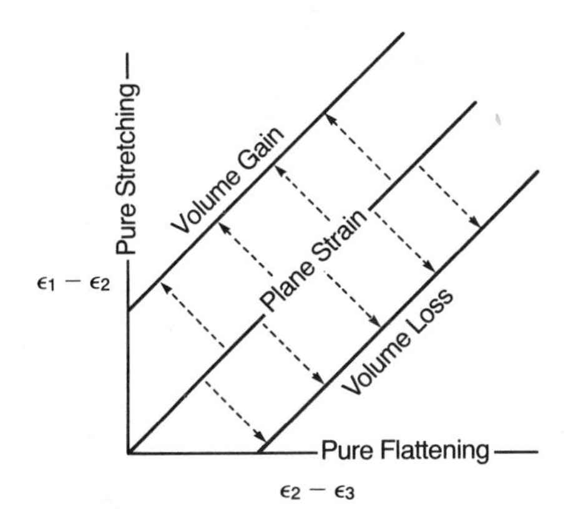

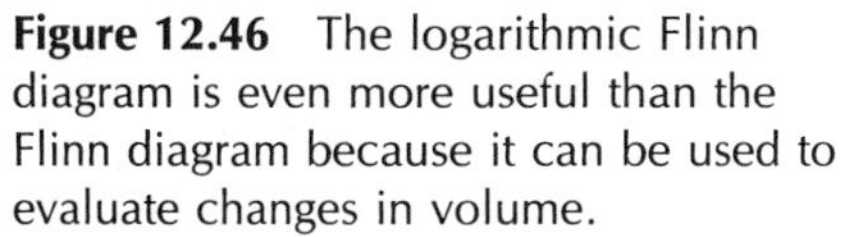

Figure 12.46 The logarithmic Flinn diagram is even more useful than the Flinn diagram because it can be used to evaluate changes in volume.

DEFORMATION PATHS

Flinn diagrams and logarithmic Flinn diagrams take on value and meaning when we put them to use. Let us consider a simple example of plane strain that we can model with a computer deck. Figure 12.47 shows progressive simple shear of a deck embossed with a reference circle. Five stages of incremental, progressive deformation are shown—*A* through *E*. At each stage of the deformation, the value of e_2, measured perpendicular to our view of the simple shear, is zero. In other words, there is neither shortening nor stretching in the third dimension. Knowing the size of the original reference circle, the values of e_1 and e_3 at each stage of the deformation can be calculated. These values in turn can be converted to logarithmic strain (ϵ) (Figure 12.47).

With these strain data in hand, it becomes a simple matter to plot on Flinn and logarithmic Flinn diagrams the state of strain representing deformational stages *A* through *E* (Figure 12.48). Portrayal of these states of strain on both diagrams pictures what we know to be true: that the deformation was achieved by constant volume plane strain. Points *A* through *E* fall along a 45°-sloping line that intersects the origin of the graph. These points, taken together, represent the **deformation path** that was followed during the progressive deformation.

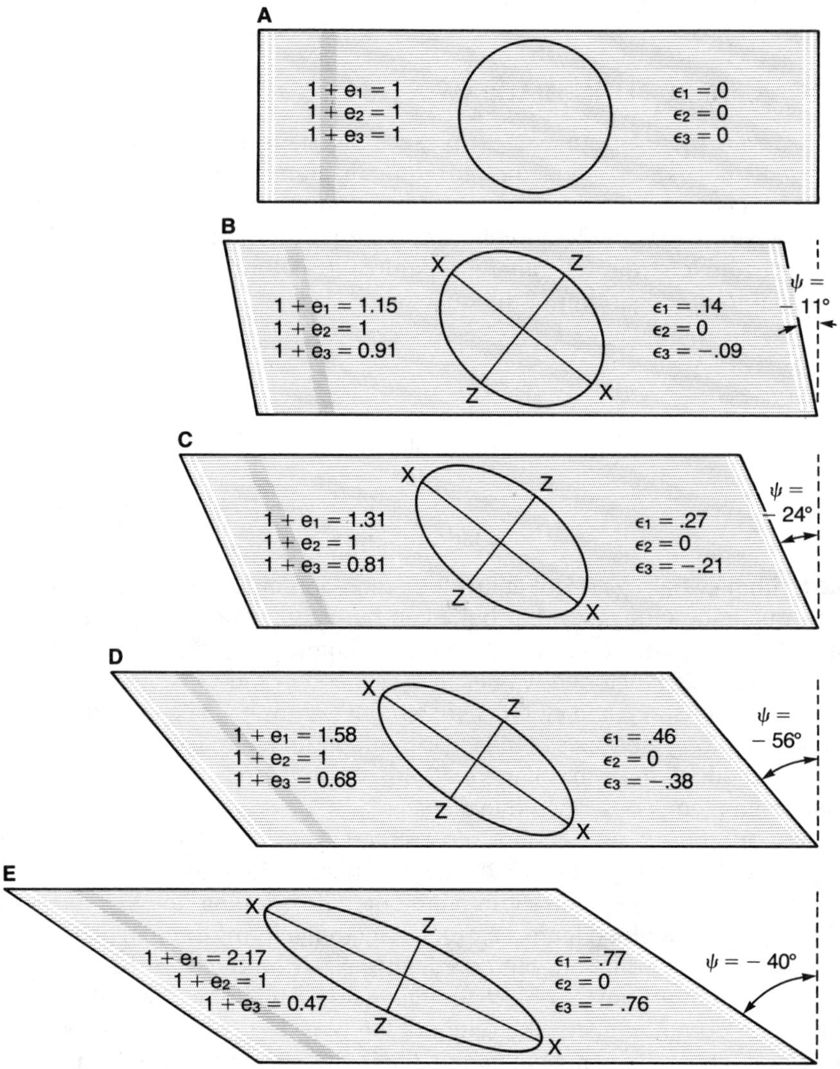

Figure 12.47 Computer deck modeling of simple shear provides a data base for preparing Flinn diagrams and logarithmic Flinn diagrams, to see how they work. Extension (e) values and logarithmic strain (ϵ) are computed for five stages of progressive simple shear—*A*–*E*.

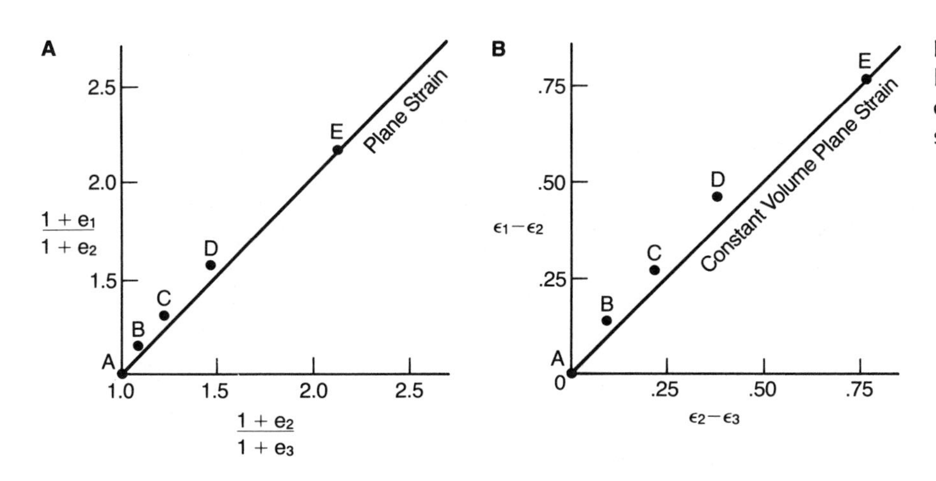

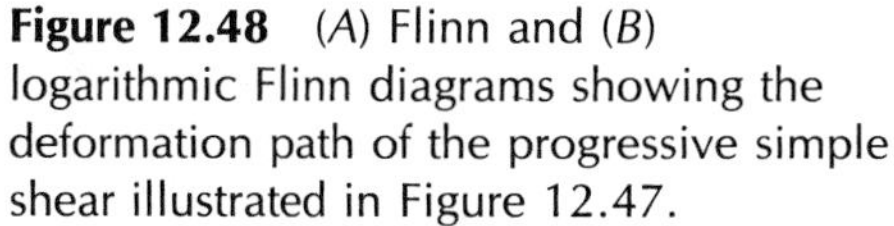
Figure 12.48 (*A*) Flinn and (*B*) logarithmic Flinn diagrams showing the deformation path of the progressive simple shear illustrated in Figure 12.47.

Progressive simple shear accompanied by volume changes would follow a different deformational path than that represented by path *A–E* in Figure 12.48*A, B*. By way of example, the logarithmic Flinn diagram pictured in Figure 12.49 shows two deformation paths (1 and 2), each representing a different combination of distortion and dilation. Path 1 describes a progressively increasing distortion accompanied by steady volume loss. Path 2 portrays a progressively increasing distortion accompanied by steady volume gain.

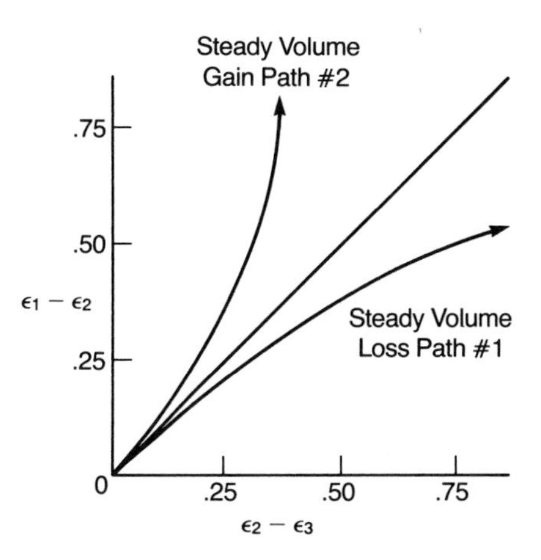

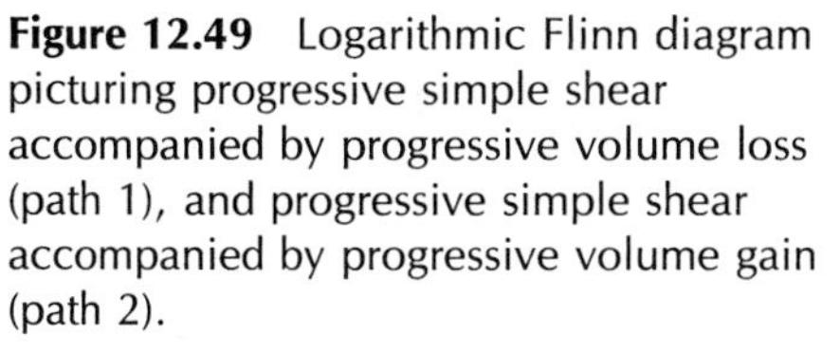
Figure 12.49 Logarithmic Flinn diagram picturing progressive simple shear accompanied by progressive volume loss (path 1), and progressive simple shear accompanied by progressive volume gain (path 2).

Deformation paths of the type represented in Figure 12.49 may be simulated with computer card decks. It requires sleight of hand. For example, to simulate a 20% volume loss followed by simple shear, first remove every fifth card from the computer card deck. This automatically transforms the original reference circle on the flank of the deck to an ellipse. Then deform the deck by progressive simple shear. To simulate an increase in volume, add cards at an appropriate even spacing through the deck. To portray simple shear accompanied by steady volume loss, prepare a computer deck by drawing a reference circle on the flank of the deck when the deck is "leaning" in the configuration shown in Figure 12.50. Then begin to deform the deck by progressive simple shear, while at the same time steadily rotating the cards to an upright position. The effect of rotating the deck to an upright position is to decrease the surface area of the deck, and thus to decrease the area of the original circular object. With the help of a friend it is possible to measure the lengths of the axes of the strain ellipse *at each stage* of the progressive simple shear. These data can be converted to values of logarithmic strain (ϵ) describing each stage of the progressive deformation. When plotted in the proper manner on a logarithmic diagram, the array of points defines a deformation path reflecting a deformation marked by steadily increasing distortion and steadily decreasing volume.

IDENTIFYING DEFORMATION PATHS OF NATURAL TECTONITES

Within any tectonite found in Nature, the state of strain is never uniform and homogeneous. Rather, different parts of a body of tectonite are distorted and/or dilated by different amounts. The specific magnitude of distortion and/or dilation at each point in the body will never be fully known, but representative values can be gleaned from preserved but distorted primary objects. When the states of strain thus derived are plotted on a common logarithmic Flinn diagram, it becomes possible to evaluate the strain significance of the tectonite, whether it be an S-tectonite, an L-tectonite, or an LS-tectonite. Furthermore, the array of plotted points makes it possible to

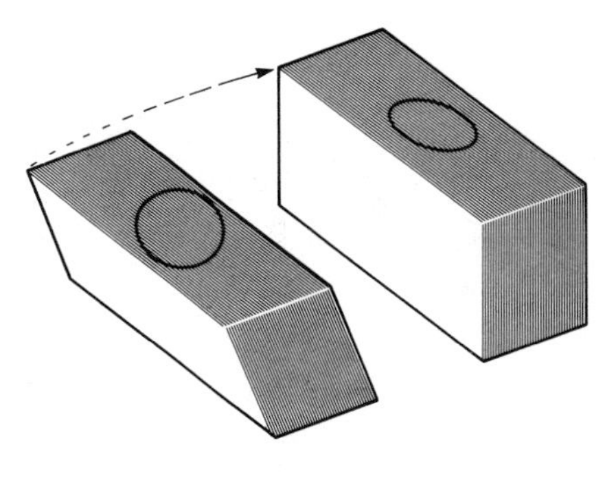
Figure 12.50 Computer card tricks to simulate volume loss.

interpret the path of deformation that was following during the distortion and/or dilation.

REGIONAL TERRANES OF TECTONITES

FORMATION OF TECTONITES IN OROGENS. Orogens are tectonite factories. There are several reasons for this statement. First of all, tectonites are by definition, expressions of penetrative distortion. What better place does the Earth offer for full-body distortion to be accomplished? Second, the deep reaches of orogens are the dominant sites of melting of crustal and/or upper mantle materials, and thus orogens receive huge invasions of magma in the form of igneous intrusions and volcanic extrusions. Viscous magmas and lavas are commonly converted to tectonites as part of the emplacement process, for they cannot resist accomodating penetrative internal movements as they pass through tight places and over (or around) immovable obstacles. Finally sedimentary and igneous rocks that occupy sites of orogens are subjected to the deforming influence of stress under conditions of elevated temperature and pressure, a dynamic state that favors penetrative internal distortion and the formation of tectonites.

The interpretation of the strain significance of tectonites in any given orogen is no easy matter, nor is the interpretation of the specific tectonic conditions responsible for the structural evolution of the given terrane of tectonite under study. The first step in interpreting the regional strain significance of tectonite is the one that is commonly omitted: describing the state of strain from point to point as carefully as possible. Data bearing on the state of strain are then integrated within the larger body of general geologic information, to try to unravel the kinematic, dynamic, and tectonic evolution of the orogen within which the tectonite is found. What complicates matters is that a given state of strain can be produced in different ways, and the state of strain will vary, quite naturally, within different parts of any structural system. Moreover, it is often difficult to pin down the timing of formation of a given terrane of tectonite because the thermal events associated with and following the deformation(s) confound most radiometric clocks. For these reasons, unique solutions to the interpretation of the strain significance of tectonites are often elusive.

PLATE TECTONIC ENVIRONMENTS OF TECTONITE FORMATION. The three fundamental classes of plate tectonic movements are all quite capable of producing regional terranes of tectonites. Ridge spreading is achieved to some extent by stretching of upwelling mafic intrusions that issue from asthenosphere and move into the dynamically active lower part of oceanic lithosphere at spreading centers. Undoubtedly S- and LS-tectonites with low-dipping foliation are fashioned in these zones. Where spreading centers unabashedly cut across continents, the deep reaches of continental crust may be rendered into tectonite as a response to penetrative layer-parallel stretching. If the truth were known, deep continental crust along the Red Sea rift might be found to be exquisite LS-tectonites. Although not located at a spreading center, the LS-tectonites that occupy the **metamorphic core complexes** of the Western Cordillera of North America may be rare glimpses of continental crustal rocks that have been rendered into tectonite by regional stretching (Figure 12.51) (Crittenden, Coney, and Davis, 1980; Davis, 1983; Davis and Hardy, 1981).

Strike–slip movements along transform boundaries between plates can create regional terranes of tectonite through simple shear. Tectonites thus fashioned tend to be LS-tectonites characterized by steep foliation and sub-

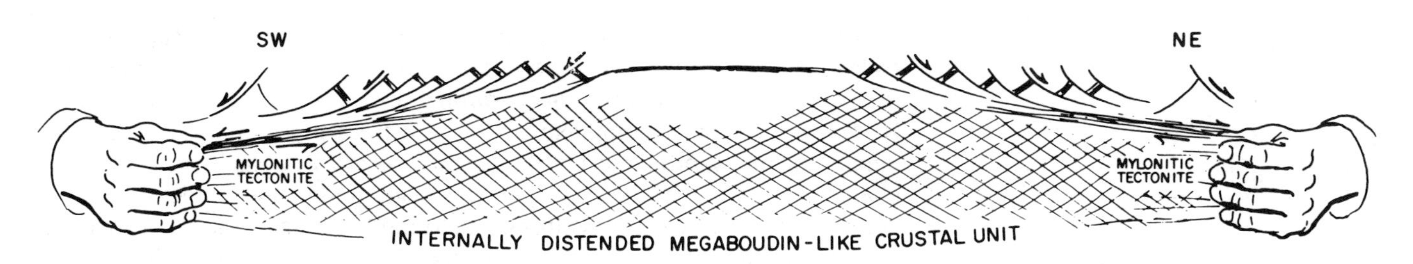

Figure 12.51 Schematic portrayal of the evolution of LS-tectonites and contact relationships in metamorphic core complexes. Normal ductile shear at deep levels creates LS-mylonites. [From Davis and Hardy (1981). Published with permission of Geological Society of America and the authors.]

horizontal lineation. The formation of tectonites along transform faults and strike–slip faults is especially favored at depth, where simple shear affects rocks that are situated in an environment of elevated temperature and pressure. Shear zone kinematics apply beautifully to transform boundaries. If the deformational conditions are right, the simultaneous flattening and stretching so characteristic of simple shear will be recorded in the formation of thick zones of mylonitic LS-tectonites.

The most abundant tectonites on the face of the Earth seem to have been created along convergent plate margins. The tectonic movements that operate at convergent margins all favor the development of tectonite. These include the formation of magmatic arcs through voluminous intrusion of granitic batholiths, the simple shear subduction of oceanic slabs beneath continental margins, the accretion of plates or parts of plates against continental edges, and the accommodation of major crustal shortening in response to continental collision and/or subduction-related processes. Whether a given part of a continental margin responds by simple shear or pure shear deformation, by constant volume or variable volume deformation, or by mild or substantial distortion, is ultimately a function of many, many factors, including structural position within the tectonite-forming system and the scale at which the products of strain are being observed and described.

Some parts of fold belts that are created in convergent settings are marked by S-tectonites. The S-tectonites may include axial planar cleavages in slate and schist, and axial planar spaced cleavages in sedimentary rocks. Tectonites of this type might be regarded as products of simple flattening, a response to horizontal crustal shortening generated by plate collision. In contrast to flattening fabrics in fold belts, some parts of fold belts are distinguished by LS-tectonites in the form of expansive terranes of low-dipping schistosity and low-plunging, unidirectional lineation. The tectonites are intimately associated with imbricate thrust nappes and fold nappes. Structural systems like these reflect thrust-related simple shear on a crustal scale (Mattauer, 1975), a structural accommodation to profound crustal shortening. LS-Tectonite terranes of this type mark young and old orogens alike, such as the tectonically active Himalayan orogen (Mattauer, 1975) and the ancient Moine thrust complex in Scotland (Elliot and Johnson, 1980; McClay and Coward, 1981).

The formation of tectonite terranes in some convergent settings is influenced significantly by the role of igneous intrusion, including diapirism. The Precambrian shield provinces that lie within the world's continents contain, among other things, **greenstone belts** distinguished by deep exposures of enormous batholithic complexes of granitic gneiss that separate strongly foliated and deeply infolded regionally metamorphosed metavolcanic and metasedimentary rocks (Figure 12.52). Batholithic rocks in **magmatic arc terranes** of continental margins contain voluminous granitic rocks, some of which are tectonites by virtue of penetrative internal flowage accompanying emplacement. And within the batholiths are extensive roof

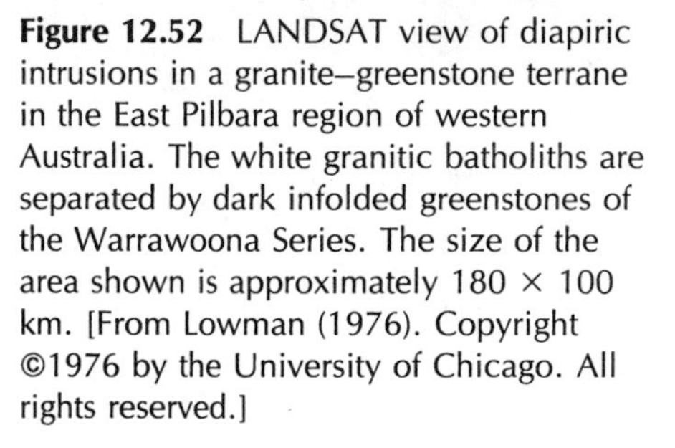

Figure 12.52 LANDSAT view of diapiric intrusions in a granite–greenstone terrane in the East Pilbara region of western Australia. The white granitic batholiths are separated by dark infolded greenstones of the Warrawoona Series. The size of the area shown is approximately 180 × 100 km. [From Lowman (1976). Copyright ©1976 by the University of Chicago. All rights reserved.]

pendants of folded and foliated metasedimentary and metavolcanic rocks, tectonites fashioned by the combination of prebatholith deformation and the rigors of intrusion. **Mantle gneiss domes**, classically described by Eskola (1949), are tectonite complexes that occur from place to place along the lengths of many orogenic belts, like the eastern Appalachians. They are characteristized by domes and arches of high-grade gneissic tectonites that are concordantly overlain by strongly deformed, high-grade layered gneisses and schists. In some ways mantle gneiss domes appear to be the tops of enormous diapiric intrusions of magma and remobilized basement. However, workers who have addressed the geology of gneiss domes uniformly conclude that the histories of some are long and complex, and may involve crustal shortening, not just diapirism.

To sum up, regional terranes of tectonites force us to apply principles of strain on a vast scale. They require us to think of flattening, constriction, and plane-strain simple shear at scales that are almost unimaginable.

DESCRIPTIVE AND GEOMETRIC ANALYSIS OF TECTONITES

THE PROBLEM

Tectonites present some special challenges in geologic mapping and detailed structural analysis. A wide variety of physical and geometric forms can give rise to foliation(s) and lineation(s), and we must be prepared to identify and describe them and to measure their orientations. Furthermore, some tectonites possess multiple, **superposed** foliations and lineations. Unscrambling the interrelationships of these is sometimes geometrically very difficult. Some tectonites display a half-dozen foliation and lineation elements in a single exposure!

THE CODING SYSTEM

FOLIATIONS. There are established methods for sorting and classifying foliations and lineations in the course of mapping and analyzing tectonites (Turner and Weiss, 1963). Foliations are coded with the letter "**S**," meaning S-surface. Each S-surface is subscripted according to the **apparent relative order** of formation within a body of tectonite. For example, a given outcrop might contain two foliations, a schistosity and a crenulation cleavage. Suppose that close examination of the outcrop reveals that the relative order of formation of the foliations is schistosity (first) followed by crenulation cleavage (second). Schistosity is entered into the field notebook as S_1. Crenulation cleavage is entered as S_2. The physical properties and orientations of each foliation are described and posted before moving on to the next outcrop. If, at the next outcrop, the crenulation cleavage (S_2) is seen to be cut by yet another foliation, perhaps a spaced cleavage, the crenulation cleavage retains the status S_2 and the spaced cleavage is awarded the symbol S_3. If, at yet another outcrop, schistosity (S_1) is seen to cut across the original bedding of the rock from which the tectonite was derived, the bedding is symbolized as S_o. The subscript "o" is reserved for original bedding.

From time to time in the course of a field investigation, new foliations are discovered that must be inserted between already "established" subscripted S-surfaces. Just when we feel confident that there are three, and only three, foliations within the body of tectonite under examination, we roll back the moss on an outcrop and discover another crenulation cleavage, one that postdates S_2 but predates S_3. When this happens, the subscripts in the entire notebook must be edited according to the change. Alternatively, the newly discovered crenulation cleavage is temporarily assigned the notation S_{2A}, with the understanding that eventually all the foliations must be reordered.

LINEATIONS. Lineations are coded with the letter "**L**," meaning lineation or linear structure. As in the case of foliations, each lineation is subscripted according to the apparent relative order of development within the body of tectonite under study. The symbol L_o is reserved for primary lineations or primary linear structure within the rock from which the tectonite was derived.

Some lineations in tectonite are simply intersection lineations, produced as the passive geometric product of the intersection of two (or more) foliations. It is important to recognize intersection lineations as such, and not to confuse them with bona fide physical lineations. The presence of a conspicuous intersection lineation commonly forces us to recognize the presence of the foliations whose intersection they reflect, foliations that in some cases might otherwise be subtle and hard to spot. Intersection lineations are seldom entered as part of the hierarchy of subscripted L's; L-symbols are reserved for real, physical lineations.

FOLDS. Folds in tectonite are coded "**F**." Since most bodies of tectonite are marked by superposed folds, it is necesary to assign subscripts to identify the relative order of the development of folds. As with S and L, F_o is reserved for folds of primary, penecontemporaneous origin that under unusual circumstances might be found preserved within the tectonite. Folds formed during and after the development of the tectonite in which they are found are coded F_1, F_2, F_3, . . . , in the order in which they formed.

Within fold belts in regionally metamorphosed terranes, it is the rule, not the exception, to find two clearly defined fold sets in the tectonite. The earliest fold set (F_1) is typically composed of tight to isoclinal passive folds that are axial planar to the earliest formed foliation (S_1) (Figure 12.53). These F_1 folds, along with S_1, are typically refolded by a set of younger folds (F_2) that is composed of tight to open flexural-slip folds, especially kink folds. Key outcrops show the interrelationship and relative timing between two (or more) fold sets in multiply deformed areas.

Fold structures entered into the field notebook as F_1, F_2, and F_3 are described carefully according to physical and geometric properties. Furthermore, the orientations of representative folds are posted. Axial surface orientations are presented in terms of strike and dip; axis orientations are recorded in terms of trend and plunge.

THE GROUPING OF STRUCTURAL ELEMENTS. Foliations, lineations, and folds that are believed to have formed at the same time are given the same subscript rank. If a field notebook bears entries that discuss and describe S_1, L_1, and F_1, it is understood that all these structural elements formed contemporaneously in the tectonite (Figure 12.54). A typical scenario is this: S_1, a schistosity, formed as an axial plane cleavage to F_1, a passive fold; and L_1, a crenulation lineation, formed parallel to the axis of F_1 folding.

Lineations (or foliations or folds) that are quite dissimilar in physical expression may be assigned the same subscript rank on the basis of compatibility of physical expression as well as compatibility of geometric orientation(s). A given lineation that emerges within a body of tectonite as a result of simple shear distortion may have a physical expression that varies from lithology to lithology because of the texture and mineralogy of the host rock(s) from which the tectonite was derived. The clue that the physically

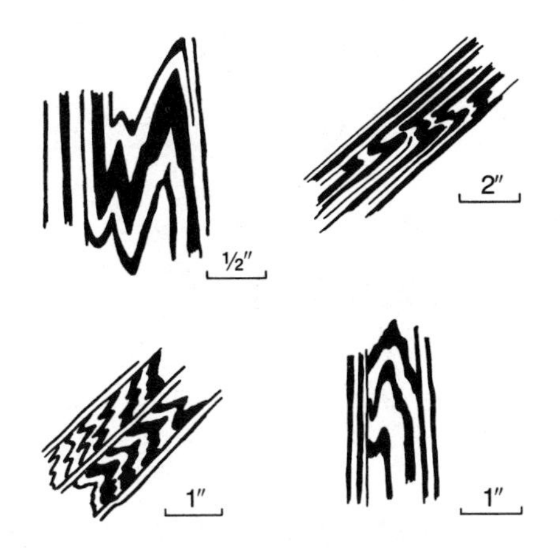

Figure 12.53 Some sketches of intrafolial folds in low-grade schists. [From Davis (1972). Published with permission of Society of Economic Geologists.]

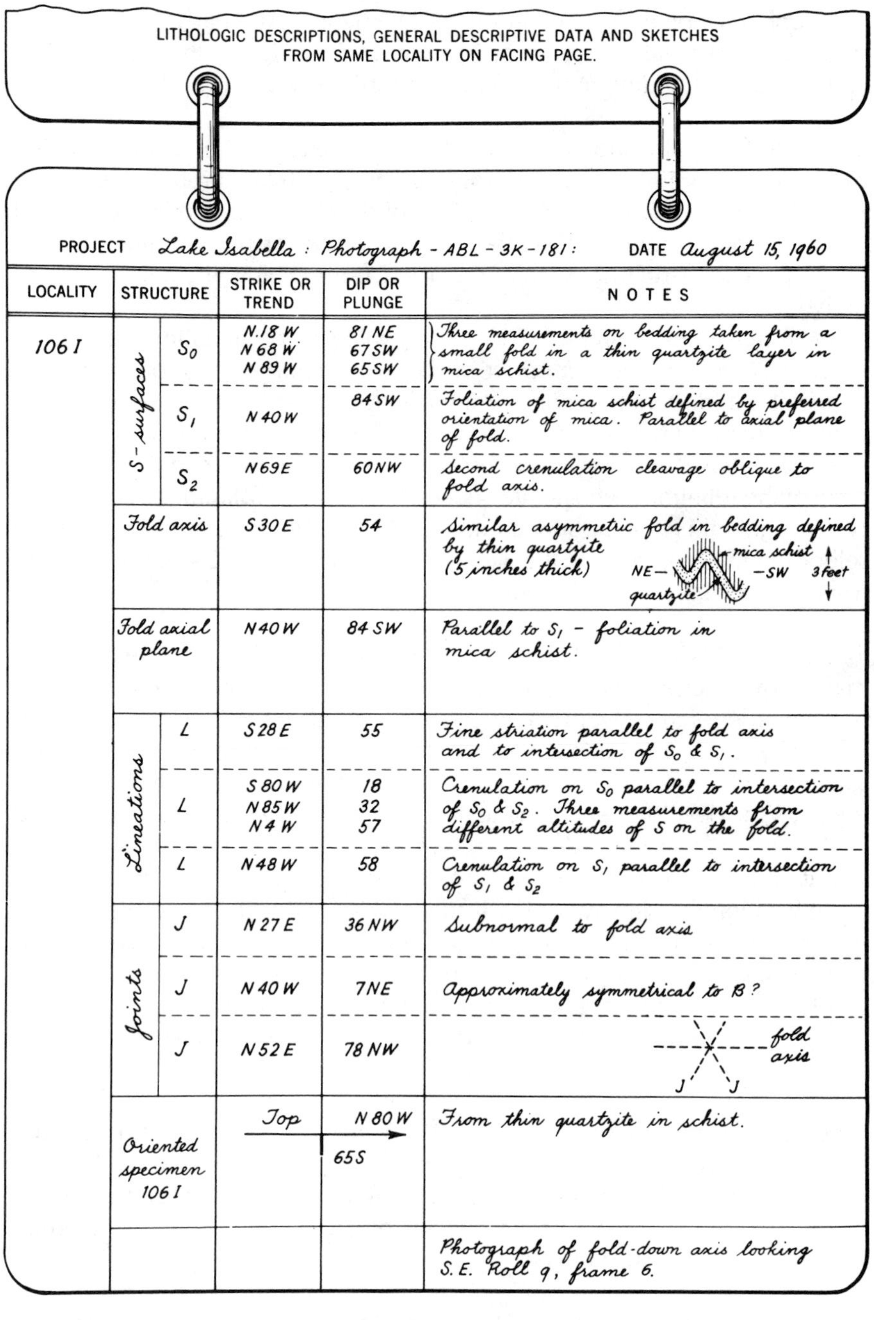

LITHOLOGIC DESCRIPTIONS, GENERAL DESCRIPTIVE DATA AND SKETCHES FROM SAME LOCALITY ON FACING PAGE.

PROJECT *Lake Isabella : Photograph - ABL - 3K - 181 :* DATE *August 15, 1960*

LOCALITY	STRUCTURE		STRIKE OR TREND	DIP OR PLUNGE	NOTES
106 I	*S - surfaces*	S_0	N.18 W N 68 W N 89 W	81 NE 67 SW 65 SW	*Three measurements on bedding taken from a small fold in a thin quartzite layer in mica schist.*
		S_1	N 40 W	84 SW	*Foliation of mica schist defined by preferred orientation of mica. Parallel to axial plane of fold.*
		S_2	N 69 E	60 NW	*Second crenulation cleavage oblique to fold axis.*
	Fold axis		S 30 E	54	*Similar asymmetric fold in bedding defined by thin quartzite (5 inches thick)* NE – *mica schist* – SW 3 *feet* *quartzite*
	Fold axial plane		N 40 W	84 SW	*Parallel to S_1 – foliation in mica schist.*
	Lineations	L	S 28 E	55	*Fine striation parallel to fold axis and to intersection of S_0 & S_1.*
		L	S 80 W N 85 W N 4 W	18 32 57	*Crenulation on S_0 parallel to intersection of S_0 & S_2. Three measurements from different altitudes of S on the fold.*
		L	N 48 W	58	*Crenulation on S_1 parallel to intersection of S_1 & S_2*
	Joints	J	N 27 E	36 NW	*Subnormal to fold axis*
		J	N 40 W	7 NE	*Approximately symmetrical to B?*
		J	N 52 E	78 NW	*fold axis* J J
	Oriented specimen 106 I		*Top*	N 80 W 65 S	*From thin quartzite in schist.*
					Photograph of fold-down axis looking S.E. Roll 9, frame 6.

Figure 12.54 Field notebook entries showing the record-keeping of foliation, lineation, and folding. (Modified from *Structural Analysis of Metamorphic Tectonites* by F. J. Turner and L. E. Weiss. Published with permission of McGraw-Hill Book Company, New York, copyright ©1963.)

dissimilar lineations formed as a set is expressed in the compatibility of geometric orientation(s). The geometric compatibility may be disclosed as a constancy of preferred orientation, or as a systematically changing array of orientations. The final check on whether the lineations should be grouped together is based on interpreting whether all the lineations could have formed in the same temperature–pressure environment by a common structural process. Even though fault striae and mineral lineation in high-grade gneiss may have a common orientation, they are not compatible physical elements. Fault striae are products of brittle faulting; mineral lineation in high-grade gneiss is a product of deformation and/or recrystallization under metamorphic conditions of elevated temperature and pressure.

Yet another sorting process is required to decide which foliations, lineations, and folds may have formed together as a system in response to a common strain event. Decisions for this brand of sorting are based on the geometric interrelationships among the elements and, once again, the compatibility of physical forms. Correctly interpreting the structural compatibility of foliations, lineations, and folds is a critical step in beginning to evaluate the strain significance of particular tectonite fabrics. It is this step that determines whether a given tectonite is described as an S-tectonite, an L-tectonite, or an LS-tectonite.

CORRELATION OF TECTONITE ELEMENTS IN TIME AND SPACE

The subscripted rankings of foliation, lineation, and folds give the impression that structures of different rankings formed during different structural events, perhaps even different tectonic events. This is not necessarily the case. A given set of penetrative folds (F_1) might be systematically refolded by a second set of folds (F_2) during a continuum of **progressive deformation**. In similar fashion, a given cleavage (S_1) that develops in the early stages of progressive simple shear may be systematically rotated to near-parallelism with the plane of shear and be cut by a newly formed cleavage (S_2) that is genetically related to the same simple shear continuum of deformation. Determining whether structural elements in a tectonite are products of a continuum of deformation or are products of discrete events widely separated in time is based, once again, on physical and geometric compatibility. In some cases radiometric age dating can help resolve the problem.

Subscripted rankings may be misleading in yet another way. It is tempting to assume that an S_2 structure defined by a geologist working in one part of a deformed belt may correlate with an S_2 structure mapped and defined by some other geologist working in the same deformed belt, but in a different area. Keep in mind that one person's S_2 may be another person's S_3! Correlation of structural elements from place to place within a region requires careful appraisal of descriptive and geometric properties of the elements.

A final interpretive pitfall is linked to the assumption that a foliation–lineation–fold suite in one area formed synchronously with an *identical* foliation–lineation–fold suite in another area within the same orogen. This is not necessarily the case. Distortion can act in a **diachronous** fashion. Distortion can move as a wave across a region, leaving an orderly sequence of identical tectonite elements in its wake.

THE TRICK OF TRANSPOSITION

While on the subject of misleading relationships in regional terranes of tectonite, it is important to identify the most misleading of all. Strongly foliated sequences of metasedimentary and metavolcanic rocks in fold belts of orogens characteristically display a parallelism of foliation and lithologic layering. The lithologic layers, distinguished on the basis of color, texture, mineralogy, and general outcrop appearance, tend to pinch out along strike, giving way to other combinations of lithologic layers (Figure 12.55). The mapped relationships give the impression that foliation developed parallel to **bedding** in the original host rock, a host rock that was marked by abundant abrupt facies changes.

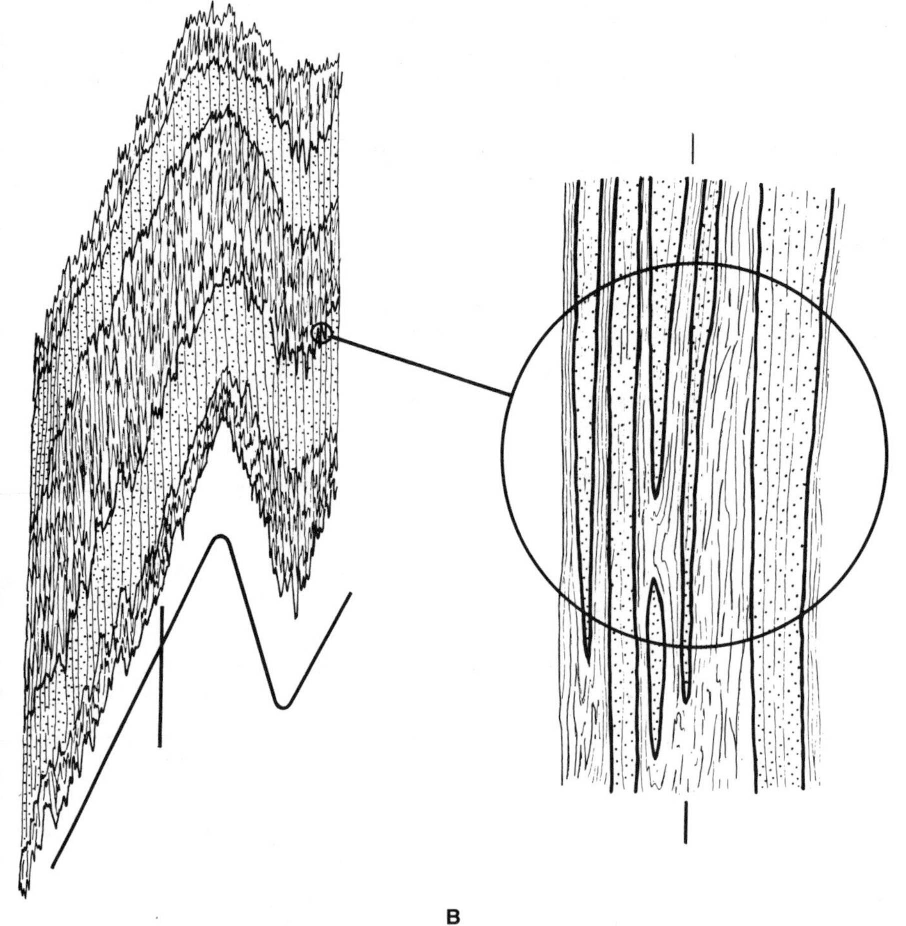

Figure 12.55 (*A*) Pseudostratigraphy in tectonite in metasedimentary rocks in the Happy Valley region of the Rincon Mountains, near Tucson, Arizona. (Photograph by S. H. Lingrey.) (*B*) Pseudostratigraphy within folded tectonite. (*From Structural Analysis of Metamorphic Tectonites* by F. J. Turner and L. E. Weiss. Published with permission of McGraw-Hill Book Company, New York, copyright © 1963.)

Structural relationships are not always what they seem. Careful inspection of terranes containing "bedding-plane foliation" usually uncovers rare exposures of **tight to isoclinal intrafolial folds** that are cut by the dominant foliation in axial planar fashion (Figure 12.56). The folds are passive, Class 1C to 3 in form, and they are sandwiched by through-going foliation and lithologic layering. According to Turner and Weiss (1963) and Whitten (1966), intrafolial folds in metamorphic tectonites commonly reflect **bedding transposition** in which tight folding of the original beds is accompanied by slip parallel to the axial planes of developing flexures (Figure 12.57*A, B*). Individual fold limbs are attenuated by progressive slip and pressure solution. Ultimately they are separated from their hinge zones (Figure 12.57*C, D*). The planes of slip and dissolution are in fact the foliation (S_1).

Transposition creates a **pseudostratigraphy** containing disrupted and rotated segments of once-continuous beds. The entire sequence is pervaded by structural discontinuities, one expression of which is apparent "bedding-plane cleavage." Transposed sequences show no internal consistency of facing: rightside-up and upside-down beds are stacked in a common sequence (Whitten, l966). The main clue that all this has happened is the presence of intrafolial folds. The concept of bedding transposition teaches us

Figure 12.56 Some nearly hidden, preserved isoclinal folds in transposed strata cut by axial plane foliation. (*A*) Folds in melange, San Juan Islands, Washington. (Photograph by G. H. Davis.) (*B*) Small recumbent isoclinal fold in tectonite derived from Cretaceous mudstone, Rincon Mountains, near Tucson, Arizona. Breadth of exposure is about 20 cm. (Photograph by G. H. Davis.) (*C*) Isoclinal folds in tectonite derived from Paleozoic limestone, Rincon Mountains near Tucson, Arizona. [From Davis (1980). Published with permission of Geological Society of America and the author.]

Figure 12.57 Transposition of bedding. (*A*) Flexural folding of bedded sequence of stiff (black) and soft (white) layers. (*B*) Tight folding and onset of cleavage development. (*C*) Attenuation and rupture of fold limbs. (*D*) Flattening of sequence and creation of pseudostratigraphy. (Modified from *Structural Analysis of Metamorphic Tectonites* by F. J. Turner and L. E. Weiss. Published with permission of McGraw-Hill Book Company, New York, copyright ©1963.)

to be very cautious in awarding the subscript rank S_o. Lithologic layering in metamorphic tectonites is always described as S_1, unless it can be firmly demonstrated that the tectonite contains an internal, undisrupted, coherent stratigraphy that fundamentally is identical to that of the original host before it was transformed into tectonite.

THE DELIGHT OF TECTONITES: SUPERPOSED FOLDING

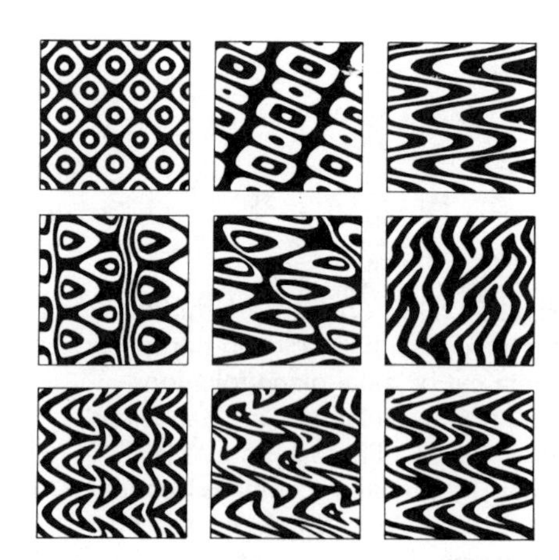

Figure 12.58 Some of the patterns of refolding that can emerge in multiply deformed tectonites. (From *Folding and Fracturing of Rocks* by J. G. Ramsay. Published with permission of McGraw-Hill Book Company, New York, copyright ©1967.)

One of the joys in carrying out mapping and detailed structural analysis in regional terranes of metamorphic tectonites is trying to unravel the geometrical artwork of superimposed folding (Figure 12.58). The superposition of passive folding is a natural product of progressive deformation in metamorphic and igneous environments. The products are almost unbelievable: egg carton patterns, boomarang folds, isoclinally folded isoclines . . . each a special product of the geometric relationship of the marker layers to the overall kinematic picture.

Less spectacular, but more abundant, are flexural-slip folds that distort preexisting passive folds. The formation of passive folds is a natural by-product of the transformation of sedimentary and/or volcanic rock to tectonite. The passive folds distort original bedding, and they are cut by an axial plane foliation that is the dominant foliation in the tectonite. The presence of penetrative foliation in any rock makes the rock especially vulnerable to flexural-slip folding. Situated in orogenic belts at or near a plate margin, the penetratively foliated tectonites may rest undisturbed for eons, only to be subjected to layer-parallel or layer-inclined stresses generated by renewed plate interference. The S-tectonites, including the early-formed passive folds, are refolded by kinklike flexural folding as an accommodation to crustal shortening.

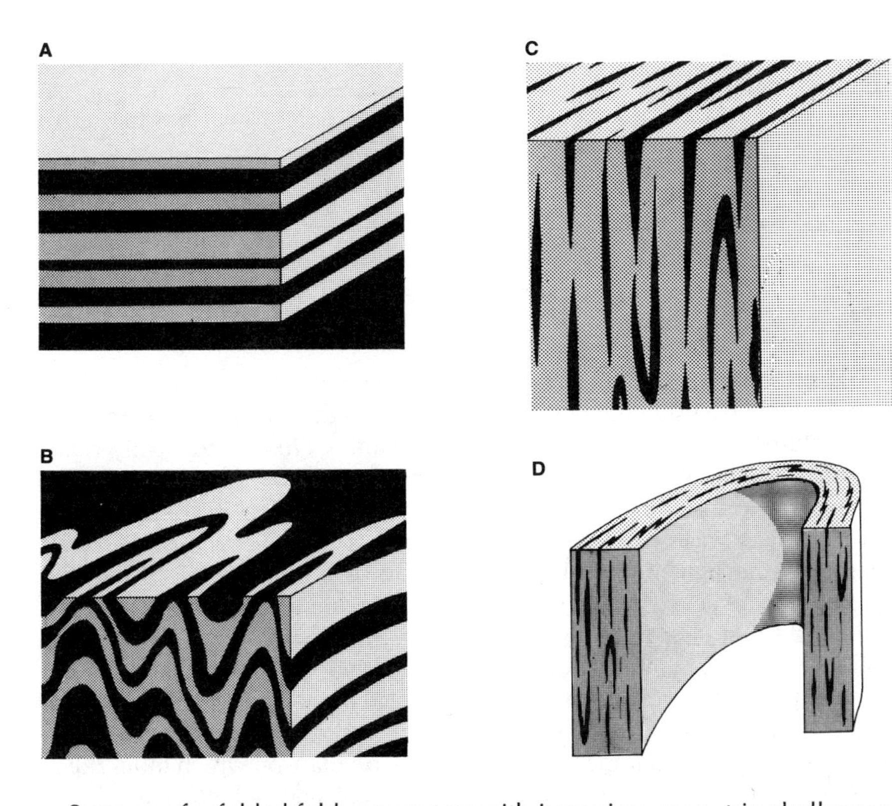

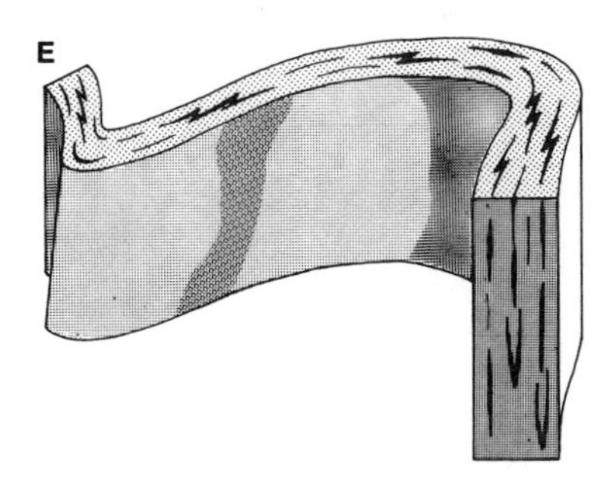

Figure 12.59 Schematic rendering of superposed folding. (*A*) Horizontal bedding, before folding. (*B*) Flexural-flow folding. (*C*) Transposition of bedding during tight passive folding (F_1) and development of penetrative axial plane foliation (S_1). (*D*) Folding (F_2) of foliation into steeply plunging anticlinorium. (*E*) Refolding (F_3) of limb of anticlinorium. [From Davis (1972). Published with permission of Society of Economic Geologists.]

Systems of refolded folds present us with imposing geometric challenges. We become forced to monitor the changes in orientation of early-formed foliations, lineations, and folds and to track the changes in structural geometry that are brought about by superposed passive and/or flexural folding (Turner and Weiss, 1963; Ramsay, 1967). Refolded folds are a symbol to end on (Figure 12.59). More than all other structures, their beautifully complicated paradoxical forms invite us to yet higher levels of detailed structural analysis. They compel us to try to read more closely what the rocks are trying to say (Figure 12.60).

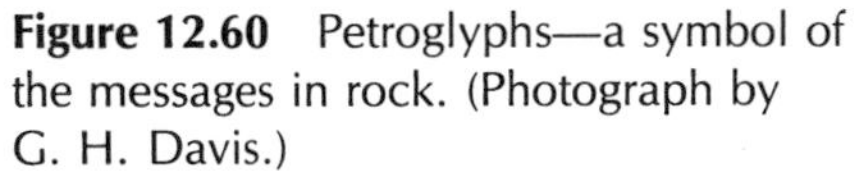

Figure 12.60 Petroglyphs—a symbol of the messages in rock. (Photograph by G. H. Davis.)

CONCLUDING THOUGHTS

The presence of structures in rocks and regions of the Earth's crust is a manifestation of sustained dynamic activity through time. The very presence of structures underscores the fact that the Earth's crust is not really rigid. Rather, it is capable of extravagant distortion.

Were we to instantly place ourselves on a distant planet, 40 quintillion miles away, and from there view the Earth as a whole with a powerful telescope, the light from Earth would bring us a delayed telecast of the deformation achieved during the past 70 million years . . . provided we were willing to sit and watch the whole scenario. A time-lapse telecast would in a shorter time reveal an Earth distorted by continuous nonrigid body movements, including distortion of the western edge of the South American plate as the Pacific plate flowed beneath it, and the crumpling of the Eurasian block as a response to impact of the lithospheric plate on which India rides. Western North America would be a terrific episode, which, like all the others, would be "continued next week." Although impressed by the flow of the Earth's crust through time, we would remind ourselves that what we see as flow from the distant view is achieved by nonrigid body translation and rotation, as seen up close.

The immediate cause of much of the 70 million years of distortion would be evident in our distant view: the continual interference of plates through time. However, the ultimate origin of the movements of plates would not be particularly obvious from the distant view, nor was it from the closer view. To be sure, the movement of plates is a response to achieve a mechanical/thermal equilibrium. After all, the distortion of the crust, present and past, is a reflection of the striving after equilibrium—a state I hope never comes.

Far from Earth, we might speculate that two complementary factors help to stimulate and sustain plate movements—the cooling of the Earth's interior and the work of gravity. Cooling, a thermodynamic process that transforms thermal energy into mechanical energy, is capable of setting up convection circulation in rheidlike mantle material. It has long been suggested that such movement could move lithospheric plates in conveyor belt fashion. The force of gravity is ultimately responsible for the movement of plates, even if its role merely is to power convective circulation. Its larger function may be the constant monitoring and adjusting of the distribution of lithospheric mass. If a dab of grease or spit on a baseball can perturb the flight of such a sphere, the uneven distribution of dabs of continental lithosphere on the sphere called Earth can surely, in the presence of gravity and an asthenosphere capable of continuous flow, create and sustain plate movements.

The longer we view the drama of deformation from our far-distant perch, the more we may appreciate the vast amount of time that is available to create deformational systems. We have "time" to think about the role of strain rate and its impact, if any, on the laws of deformation, laws which were created and tested during 1/22,500,000 of Earth time (i.e., about 200 years).

If the distant view influences our thinking about structures at all, it leads us to conclude that structures like faults, joints, folds, cleavages, foliations, and lineations do not operate in isolation. Instead, structures work together as a system to achieve the translation, rotation, distortion, and dilation that are required by dynamic and thermodynamic conditions.

As interesting as the distant view may be, I need to return to the Earth, to mountain- and outcrop-scale features, where I can examine again the primary sources from which we learn so much. Perhaps the greatest stimulus in structural analysis is the opportunity to travel to and from close-up views to distant views and return, to gain new insights for creative work.

Some of my goals for the journey into the structure of rocks and regions were these: to describe the properties of structures in sufficient detail that recognition of structure becomes reasonably straightforward; to equip us with a variety of skills for use in descriptive analysis and measurement; to provide a perspective for identifying relationships that are critical to interpretation; to build confidence in using basic mathematic skills to more fully describe, understand, and appreciate the elegance of strain in natural structural systems; and to present a basis for probing the mechanical significance of structures and structural systems. The best part is yet to come: the opportunity, on your own, to explore the structure of rocks and regions. Possibilities for discovery are limitless.

REFERENCES

Allis, R. G., 1981, Continental underthrusting beneath the southern Alps of New Zealand: Geology, v. 9, p. 303–307.

Alvarez, W., Engelder, T., and Lowrie, W., 1976, Formation of spaced cleavage and folds in brittle limestone by dissolution: Geology, v. 4, p. 698–701.

Alvarez, W., Engelder, T., and Geiser, P. A., 1978, Classification of solution cleavage in pelagic limestones: Geology, v. 6, p. 263–266.

Anderson, E. M., 1951, The dynamics of faulting and dyke formation with applications to Britain: Oliver & Boyd, Edinburgh, 206 p.

Anderson, R. E., 1971, Thin-skin distension in Tertiary rocks of southeastern Nevada: Geological Society of America Bulletin, v. 82, p. 43–58.

Anderson, R. E., 1973, Large magnitude Late Tertiary strike slip faults, north of Lake Mead, Nevada: United States Geological Survey Professional Paper 794, 18 p.

Anderson, R. N., and Noltimier, H. C., l973, A model for the horst and graben structure of midocean ridge crests based upon spreading velocity and basalt delivery to the oceanic crust: Geophysical Journal of the Royal Astronomical Society, v. 34, p. 137–147.

Anderson, T. H., Burkart, B., Clemons, R. E., Bohnenberger, O. H., and Blount, D. N., 1973, Geology of the western Altos Cuchumatanes, northwestern Guatemala: Geological Society of America Bulletin, v. 84, p. 805–826.

Antoine, J. W., Martin, R. G., Jr., Pyle, T. G., and Bryant, W. R., 1974, Continental margins of the Gulf of Mexico, *in* Burke, C. A., and Drake, C. L. (eds.), The geology of continental margins: Springer-Verlag, New York, p. 683–694.

Armstrong, R. L., 1964, Geochronology and geology of the eastern Great Basin in Nevada and Utah: Ph.D. dissertation, Yale University, New Haven, Connecticut, 202 p.

Armstrong, R. L., 1968a, Sevier orogenic belt in Nevada and Utah: Geological Society of America Bulletin, v. 79, p. 429–458.

Armstrong, R. L., 1968b, The Cordilleran miogeosyncline in Nevada and Utah: Utah Geological and Mineralogical Survey Bulletin 78, 58 p.

Armstrong, R. L., l972, Low-angle (denudation) faults, hinterland of the Sevier orogenic belt, eastern Nevada and western Utah: Geological Society of America Bulletin, v. 83, p. 1729–1754.

Arthaud, F., and Mattauer, M., 1969, Exemples de stylolites d'origine tectonique dans le Languedoc, leurs relations avec la tectonique cassante: Société Géologique de France, Bulletin, v. 11, p. 738–744.

Atwater, T. M., 1970, Implications of plate tectonics for the Cenozoic tectonic evolution of western North America: Geological Society of America Bulletin, v. 81, p. 3513–3536.

Balk, R., 1936, Structural and petrologic studies in Dutchess County, New York: Geological Society of America Bulletin, v. 47, p. 685–774.

Balk, R., 1937, Structural behaviour of igneous rocks: Geological Society of America Memoir 5, 177 p.

Balk, R., 1949, Structure of Grand Saline salt dome, Van Zandt County, Texas: American Association of Petroleum Geologists Bulletin, v. 33, p. 1791–1829.

Bally, A. W., Gordy, P. L., and Stewart, G. A., 1966, Structure, seismic data, and orogenic evolution of southern Canadian Rockies: Canadian Association of Petroleum Geologists Bulletin, v. 14, p. 337–381.

Bally, A. W., and Snelson, S., 1980, Realms of subsidence, *in* Miall, A. D. (ed.), Facts and principles of world petroleum occurrence: Canadian Society of Petroleum Geologists Memoir 6, p. 9–94.

Barnes, C. W., 1974, Interference and gravity tectonics in the Gray Mountain area, Arizona, *in* Karlstrom, T. N. V., Swann, G. A., and Eastwood, R. L. (eds.), Geology of northern Arizona, Part II—area studies and field guide: Northern Arizona University, Flagstaff, Arizona, p. 442–453.

Barnes, C. W., and Marshall, D. R., 1974, A dynamic model for monoclinal uplift and gravity gliding: Geological Society of America Abstracts with Programs, v. 6, p. 424.

Batiza, R., 1978, Geology, petrology, and geochemistry of Isla Tortuga, a recently formed tholeiitic island in the Gulf of California: Geological Society of America Bulletin, v. 89, p. 1309–1324.

Becker, C. L., 1949, Progress and power: Alfred A. Knopf, Inc., New York, 116 p.

Benioff, H., 1954, Orogenesis and deep crustal structure: additional evidence from seismology: Geological Society of America Bulletin, v. 65, p. 385–400.

Berg, R. R., 1962, Mountain flank thrusting in Rocky Mountain foreland, Wyoming and Colorado: American Association of Petroleum Geologists Bulletin, v. 46, p. 2019–2032.

Berg, R. R., 1981, Review of thrusting in the Wyoming foreland, *in* Boyd, D. W., and Lillegraven, J. A. (eds.), Rocky Mountain foreland basement tectonics: Contributions to geology, v. 19: University of Wyoming, Laramie, Wyoming, p. 93–104.

Berthé, D., Choukroune, P., and Jegouza, P., 1979, Orthogneiss, mylonite and noncoaxial deformation of granites: the example of the South Armorican shear zone: Journal of Structural Geology, v. 1, p. 31–42.

Beutner, E. C., 1978, Slaty cleavage and related strain in Martinsburg Slate, Delaware Water Gap, New Jersey: American Journal of Science, v. 278, p. 1–23.

Bijlaard, P. P., 1946, On the elastic stability of thin plates supported by a continuous medium: Royal Dutch Academy of Science Proceedings, v. 49, p. 1189–1199.

Billings, M. P., 1972, Structural geology (3rd edition): Prentice-Hall, Englewood Cliffs, New Jersey, 606 p.

Biot, M. A., 1957, Folding instability of a layered viscoelastic medium under compression: Royal Society of London Proceedings, Series A, v. 242, p. 211–228.

Biot, M. A., 1959, On the instability of folding deformation of a layered viscoelastic medium under compression: Journal of Applied Mechanics, v. 26, p. 393–400.

Biot, M. A., Ode, H., and Roever, W. L., 1961, Experimental verification of the folding of stratified viscoelastic media: Geological Society of America Bulletin, v. 72, p. 1621–1630.

Bishop, R. S., 1977, Shale diapir emplacement in south Texas: Laward and Sherriff examples: Transactions of the Gulf Coast Association of Geological Societies, v. 27, p. 20–31.

Bishop, R. S., 1978, Mechanism for emplacement of piercement diapirs: American Association of Petroleum Geologists Bulletin, v. 62, p. 1561–1583.

Blake, D. R., and Roy, C. J., 1949, Unusual stylolites: American Journal of Science, v. 247, p. 779–790.

Bronowski, J., 1973, The ascent of man: Little, Brown, & Company, Boston, 448 p.

Brown, W. G., 1975, Casper Mountain area (Wyoming)—structural model of Laramide deformation (abs): American Association of Petroleum Geologists Bulletin, v. 59, p. 906.

Bruce, C. H., 1972, Pressured shale and related sediment deformation: Gulf Coast Association of Geological Societies Transactions, v. 22, p. 23–31.

Bucher, W. H., 1944, The stereographic projection, a handy tool for the practical geologist: Journal of Geology, v. 52, p. 191–212.

Burchfiel, B.C., and Davis, G. A., 1975, Nature and controls of Cordillera orogenesis, western United States: extensions of an earlier synthesis: American Journal of Science, v. 275-A, p. 363–395.

Carey, S. W., 1953, The rheid concept in geotectonics: Geological Society of Australia Journal, v. 1, p. 67–117.

Carey, S. W., 1962, Folding: Journal of Alberta Association of Petroleum Geologists, v. 10, p. 95–144.

Carver, R. E., 1968, Differential compaction as a cause of regional contemporaneous faults: American Association of Petroleum Geologists Bulletin, v. 52, p. 414–419.

Chapman, D. S., and Pollack, H. N., 1977, Regional geotherms and lithospheric thickness: Geology, v. 5, p. 265–268.

Chilingarian, G. V., and Wolf, K. H. (eds.), 1975, Compaction of coarse-grained sediments. I: Developments in Sedimentology, v. 18A, 552 p.

Choukroune, P., 1969, Un exemple d'analyse microtectonique d'une série calcaire affectée de plis isopaques ("concentriques"): Tectonophysics, v. 7, p. 57–70.

Cloos, E., 1946, Lineation: A critical review and annotated bibliography: Geological Society of America Memoir 18, 122 p.

Cloos, E., 1947, Öolite deformation in the South Mountain Fold, Maryland: Geological Society of America Bulletin, v. 58, p. 843–918.

Cloos, E., 1955, Experimental analysis of fracture patterns: Geological Society of America Bulletin, v. 66, p. 241–256.

Cloos, E., 1968, Experimental analysis of Gulf Coast fracture patterns: American Association of Petroleum Geologists Bulletin, v. 52, p. 420–444.

Cloos, H., 1922, Über Ausbau und Anwendung der granittektonischen Methode: Preussischen Geologischen Landesanstalt, v. 89, p. 1–18.

Cloos, H., 1925, Einführung in die tektonische Behandlung magmatischer Erscheinungen (Granitektonik): Borntraeger, Berlin, 194 p.

Cloos, H., 1936, Einführung in die Geologie, ein Lehrbuch der inneren Dynamik: Borntraeger, Berlin, 503 p.

Cloos, E., and Cloos, H., 1927, Die Quellkuppe des Drachenfels am Rhein: Zeitschrift für Vulkanologie, bd. II, p. 33–40.

Commission on Standardization of Laboratory and Field Tests on Rock, 1978, Suggested methods for the quantitative description of discontinuities in rock masses: International Society for Rock Mechanics, v. 15, p. 319–368.

Compton, R. R., 1962, Manual of field geology: John Wiley & Sons, New York, 378 p.

Coney, P. J., 1978, Mesozoic–Cenozoic Cordilleran plate tectonics, *in* Smith, R. B., and Eaton, G. P. (eds.), Cenozoic tectonics and regional geophysics of the western Cordillera: Geological Society of America Memoir 152, p. 33–50.

Coney, P. J., 1981, Accretionary tectonics in western North America, *in* Dickinson, W. R., and Payne, W. D. (eds.), Relations of tectonics to ore deposits in the southern Cordillera: Arizona Geological Society Digest, v. 15, p. 23–37.

Coney, P. J., Siberling, N. J., Jones, D. L., and Richter, D. H., 1981, Structural relations along the leading edge of Wrangellia terrane in the Clearwater Mountains, Alaska: United States Geological Survey Circular 823-B, p. 56–58.

Conybeare, C. E. B., and Crook, K. A. W., 1968, Manual of sedimentary structures: Commonwealth of Australia, Department of National Development, Bureau of Mineral Resources, Geology and Geophysics Bulletin, v. 102, 327 p.

Cook, F. A., Brown, L. D., and Oliver, J. E., 1980, The southern Appalachians and the growth of continents: Scientific American, v. 243, p. 156–168.

Cooper, J. R., 1960, Some geologic features of the Pima mining district, Pima County, Arizona: United States Geological Survey Bulletin 1112-C, p. 63–103.

Cooper, J. R., and Silver, L. T., 1964, Geology and ore deposits of the Dragoon quadrangle, Cochise County, Arizona: United States Geological Survey Professional Paper 416, 196 p.

Cosgrove, J. W., 1980, The tectonic implications of some small scale structures in the Mona Complex of Holy Isle, North Wales: Journal of Structural Geology, v. 2, p. 383–396.

Coulomb, C. A., 1773, Sur une application des regles de maximus et minimis à quelques problèmes de statique relatifs à l'architecture: Académie Royale des Sciences, Mémoires de Mathématique et de Physique par divers Savants, v. 7, p. 343–382.

Cox, A. V., Dalrymple, G. B., and Doell, R. R., 1967, Reversals of the Earth's magnetic field: Scientific American, v. 216, p. 44–45.

Crittenden, M. D., Jr., Coney, P. J., and Davis, G. H. (eds.), 1980, Cordilleran metamorphic core complexes: Geological Society of America Memoir 153, 490 p.

Crowell, J. C., 1959, Problems of fault nomenclature: American Association of Petroleum Geologists Bulletin, v. 43, p. 2653–2674.

Crowell, J. C., 1974, Origin of late Cenozoic basins in southern California, *in* Dickinson, W. R. (ed.), Tectonics and sedimentation: Society of Economic Paleontologists and Mineralogists Special Publication 22, p. 190–204.

Crowell, J. C., and Ramirez, V. R., 1979, Late Cenozoic faults in southeastern California, *in* Crowell, J. C., and Sylvester, A. G. (eds.), Tectonics of the juncture between the San Andreas fault system and the Salton trough, southeastern California—a guidebook for Geological Society of America meeting, San Diego, 1979, p. 27–39.

Currie, J. B., Patnode, A. W., and Trump, R. P., 1962, Development of folds in sedimentary strata: Geological Society of America Bulletin, v. 73, p. 655–674.

Dahlstrom, D. C. A., 1969, Balanced cross sections: Canadian Journal of Earth Science, v. 6, p. 743–757.

Dale, T. N., 1923, The commercial granites of New England: United States Geological Survey Bulletin 738, 488 p.

Dana, S. W., 1980, Analysis of gravity anomaly over coral-reef oil fields: Wilfred pool, Sullivan County, Indiana: American Association of Petroleum Geologists Bulletin, v. 64, p. 400–413.

Davis, G. A., Anderson, J. L., Frost, E. G., and Shackelford, T. J., 1980, Mylonitization and detachment faulting in the Whipple–Buckskin–Rawhide Mountains terrane, southeastern California and western Arizona, *in* Crittenden, M. D., Jr., Coney, P. J., and Davis, G. H. (eds.), Cordilleran metamorphic core complexes: Geological Society of America Memoir 153, p. 79–129.

Davis, G. H., 1972, Deformational history of the Caribou strata-bound sulfide deposit, Bathurst, New Brunswick, Canada: Economic Geology, v. 67, p. 634–655.

Davis, G. H., 1978, The monocline fold pattern of the Colorado Plateau, *in* Matthews, V. (ed.), Laramide folding associated with basement block faulting in the western United States: Geological Society of America Memoir 151, p. 215–233.

Davis, G. H., 1979, Laramide folding and faulting in southeastern Arizona: American Journal of Science, v. 279, p. 543–569.

Davis, G. H., 1980, Structural characteristics of metamorphic core complexes, *in* Crittenden, M. D., Jr., Coney, P. J., and Davis, G. H. (eds.), Cordilleran metamorphic core complexes: Geological Society of America Memoir 153, p. 35–77.

Davis, G. H., 1981, Regional strain analysis of the superposed deformations in southeastern Arizona and the eastern Great Basin, *in* Dickinson, W. R., and Payne, W.D. (eds.), Relation of tectonics to ore deposits in the southern Cordillera: Arizona Geological Society Digest, v. 14, p. 155–172.

Davis, G. H., 1983, Shear-zone model for the origin of metamorphic core complexes: Geology, v. 11, p. 348–351.

Davis, G. H., Eliopulos, G. J., Frost, E. G., Goodmundson, R. C., Knapp, R. B., Liming, R. B., Swan, M. M., and Wynn, J. C., 1974, Recumbent folds—focus of an investigative workshop in tectonics: Journal of Geological Education, v. 22, p. 204–208.

Davis, G. H., Phillips, M. P., Reynolds, S. J., Varga, R. J., 1979, Origin and provenance of some exotic blocks in lower Mesozoic red-bed basin deposits, southern Arizona: Geological Society of America Bulletin, Part I, v. 90, p. 376–384.

Davis, G. H., and Hardy, J. J., Jr., 1981, The Eagle Pass detachment, southeastern Arizona: Product of mid-Miocene listric(?) normal faulting in the southern Basin and Range: Geological Society of America Bulletin, Part I, v. 92, p. 749–762.

Davis, G. H., Showalter, S. R., Benson, G. S., McCalmont, L. S., and Cropp, F. W., 1981, Guide to the geology of the Salt River Canyon region, Arizona: Arizona Geological Society Digest, v. 13, p. 48–97.

Dennis, J. G., 1967, International tectonic dictionary, English terminology: American Association of Petroleum Geologists Memoir 7, 196 p.

Dennis, J. G., 1972, Structural geology: Ronald Press, New York, 532 p.

DeSitter, L. U., 1964, Structural geology (2nd edition): McGraw-Hill Book Company, New York, 551 p.

Dewey, J. F., 1972, Plate tectonics: Scientific American, v. 226, p. 56–72.

Dewey, J. F., and Bird, J. M., 1970, Mountain belts and the new global tectonics: Journal of Geophysical Ressearch, v. 75, p. 2625–2647.

Dewey, J. R., and Bird, J. M., 1971, Origin and emplacement of the ophiolite suite: Appalachian ophiolites in Newfoundland: Journal of Geophysical Research, v. 76, p. 3179–3206.

Dickinson, W. R., 1973, Reconstruction of past arc–trench systems from petrotectonic assemblages in the island arcs of the western Pacific, *in* Coleman, P. J. (ed.), The western Pacific: island arcs, marginal seas, geochemistry: University of Western Australia Press, Nedlands, West Australia, p. 569–601.

Dickinson, W. R., 1974, Plate tectonics and sedimentation, *in* Dickinson, W. R. (ed.), Tectonics and sedimentation: Society of Economic Paleontologists and Mineralogists Special Publication 22, p. 1–27.

Dickinson, W. R., 1977, Tectono-stratigraphic evolution of subduction-controlled sedimentary assemblages, *in* Talwani, M., and Pitman, W. C. (eds.), Island arcs, deep-sea trenches, and back-arc basins: American Geophysical Union Maurice Ewing Series 1, p. 33–40.

Dickinson, W. R., 1980, Plate tectonics and key petrologic associations, *in* Strangway, D. W. (ed.), The continental crust and its mineral deposits: Geological Association of Canada Special Paper 20, p. 341–360.

Dickinson, W. R., and Seely, D. S., 1979, Structure and stratigraphy of forearc regions: American Association of Geologists Bulletin, v. 63, p. 2–31.

Dickinson, W. R., and Payne, W. D. (eds.), 1981, Cover illustration, Relation of tectonics to ore deposits in the southern Cordillera: Arizona Geological Society Digest, v. 14, 288 p.

Dietz, R. S., 1961, Continent and ocean basin evolution by spreading of sea floor: Nature, v. 190, p. 854–857.

Dietz, R. S., 1963, Collapsing continental rises: an actualistic concept of geosynclines and mountain building: Journal of Geology, v. 71, p. 314–333.

Dietz, R. S., and Holden, J. C., 1966, Miogeoclines in space and time: Journal of Geology, v. 74, p. 566–583.

Donath, F. A., 1961, Experimental study of shear failure in anisotropic rocks: Geological Society of America Bulletin, v. 72, p. 985–989.

Donath, F. A., 1962, Analysis of basin–range structure, south-central Oregon: Geological Society of America Bulletin, v. 73, p. 1–16.

Donath, F. A., 1970a, Rock deformation apparatus and experiments for dynamic structural geology: Journal of Geological Education, v. 18, p. 1–12.

Donath, F. A., 1970b, Some information squeezed out of rock: American Scientist, v. 58, p. 54–72.

Donath, F. A., and Parker, R. B., 1964, Folds and folding: Geological Society of America Bulletin, v. 75, p. 45–62.

Droxler, A., and Schaer, J. P., 1979, Deformation cataclastique plastique lors du plissement, sous faible couverture, de states calcaires: Eclogae Geologicae Helvetiae, v. 72/2, p. 551–570.

Dubey, A. K., 1980, Model experiments showing simultaneous development of folds and transcurrent faults: Tectonophysics, v. 65, p. 69–84.

Durney, D. W., and Ramsay, J. G., 1973, Incremental strains measured by syntectonic crystal growths, *in* De Jong, K. A., and Scholten, R. (eds.), Gravity and tectonics: John Wiley & Sons, New York, p. 67–96.

Dutton, S. P., 1982, Pennsylvanian fan–delta and carbonate deposition, Mobeetie field, Texas Panhandle: American Association of Petroleum Geologists Bulletin, v. 66, p. 389–407.

Eaton, G. P., 1980, Geophysical and geological characteristics of the crust of the Basin and Range province, *in* Continental tectonics: National Academy of Sciences, Washington, D.C., p. 96–113.

Edwards, M. B., 1981, Upper Wilcox Rosita delta system of south Texas: American Association of Petroleum Geologists Bulletin, v. 65, p. 54–73.

Elliot, D., 1965, The quantitative mapping of directional minor structures: Journal of Geology, v. 73, p. 865–880.

Elliot, D., and Johnson, M. R. W., 1980, Structural evolution in the northern part of the Moine thrust belt, northwest Scotland: Royal Society of Edinburgh Transactions, v. 71, p. 69–96.

Engelder, T., and Engelder, R., 1977, Fossil distortion and decollement tectonics of the Appalachian plateau: Geology, v. 5, p. 457–460.

England, P., and Wortel, R., 1980, Some consequences of the subduction of young slabs: Earth and Planetary Science Letters, v. 47, p. 403–415.

Ernst, W. G., 1975, Metamorphism and plate tectonic regimes: benchmark papers in geology: Halsted Press, New York, 440 p.

Eskola, P. E., 1949, The problem of mantled gneiss domes: Geological Society of London Quarterly Journal, v. 104, p. 461–476.

Euler, R. C., Gummerman, G. J., Karlstrom, T. N. V., Dean, J. S., and Hevly, R. H., 1979, The Colorado Plateau: cultural dynamics and paleoenvironment: Science, v. 205, p. 1089–1101.

Feininger, T., 1978, The extraordinary striated outcrop at Saqsaywaman, Peru: Geological Society of America Bulletin, v. 89, p. 494–503.

Fink, J., 1980, Surface folding and viscosity of rhyolite flows: Geology, v. 8, p. 250–254.

Fletcher, R. C., and Pollard, D. D., 1981, Anticrack model for pressure solution surfaces: Geology, v. 9, p. 419–425.

Fleuty, M. J., 1964, The description of folds: Geological Association Proceedings, v. 75, p. 461–492.

Flinn, D., 1962, On folding during three-dimensional progressive deformation: Geological Society of London Quarterly Journal, v. 118, p. 385–433.

Folk, R. L., 1965, Henry Clifton Sorby (1826–1908), the founder of petrography: Journal of Geological Education, v. 13, p. 43–47.

Friedman, M., Handin, J., Logan, J. M., Min, K. D., and Stearns, D. W., 1976, Experimental folding of rocks under confining pressure. Part III. Faulted drape folds in multilithologic layered specimens: Geological Society of America Bulletin, v. 87, p. 1049–1066.

Geiser, P. A., and Sansone, S., 1981, Joints, microfractures, and the formation of solution cleavage in limestone: Geology, v. 9, p. 280–285.

Ghosh, S. K., 1968, Experiments of buckling of multilayers which permit interlayer gliding: Tectonophysics, v. 6, p. 207–249.

Goodwin, A. M., 1976, Giant impacting and the development of continental crust, *in* Windley, B. F. (ed.), The early history of the Earth: John Wiley & Sons Ltd., Chichester, England, p. 77–95.

Gray, D. R., 1977a, Morphologic classification of crenulation cleavages: Journal of Geology, v. 85, p. 229–235.

Gray, D. R., 1977b, Some parameters which affect the morphology of crenulation cleavages: Journal of Geology, v. 85, p. 763–780.

Gray, D. R., 1979, Microstructure of crenulation cleavages: An indication of cleavage origin: American Journal of Science, v. 279, p. 97–128.

Gray, D. R., 1981, Compound tectonic fabrics in singly folded rocks from southwest Virginia, U.S.A.: Tectonophysics, v. 78, p. 229–248.

Gray, D. R., and Durney, D. W., 1979, Investigations on the mechanical significance of crenulation cleavage: Tectonophysics, v. 58, p. 35–79.

Griffith, A. A., 1924, Theory of rupture: Proceedings of the First International Congress on Applied Mechanics, Delft, the Netherlands, p. 55–63.

Griggs, D. T., Turner, F. J., and Heard, H. C., 1960, Deformation of rocks at 500° to 800°C, *in* Griggs, D. T., and Handin, J. (eds.), Rock deformation: Geological Society of America Memoir 79, p. 39–104.

Groshong, R. H., Jr., 1975a, "Slip" cleavage caused by pressure solution in a buckle fold: Geology, v. 3, p. 411–413.

Groshong, R. H., Jr., 1975b, Strain, fractures, and pressure solution in natural single-layer folds: Geological Society of America Bulletin, v. 86, p. 1363–1376.

Groshong, R. H., Jr., 1981, Cover photograph: Geology, v. 9, no. 9.

Gustafson, J. K., Burrell, H. C., and Garretty, M. D., 1950, Geology of the Broken Hill ore deposit, New South Wales, Australia: Geological Society of America Bulletin, v. 61, p. 1369–1437.

Hafner, W., 1951, Stress distribution and faulting: Geological Society of America Bulletin, v. 62, p. 373–398.

Halbouty, M. T., 1969, Hidden and subtle traps in Gulf Coast: American Association of Petroleum Geologists Bulletin, v. 53, p. 3–29.

Hall, A. L., 1932, The Bushveld igneous complex of the Central Transvaal: Geological Society of South Africa Memoir 28.

Hamblin, W. K., 1965, Origin of "reverse drag" on the downthrown side of normal faults: Geological Society of America Bulletin, v. 76, p. 1145–1164.

Hamblin, W. K., Damon, P. E., and Bull, W. B., 1981, Estimates of vertical crustal strain rates along the western margin of the Colorado Plateau: Geology, v. 9, p. 293–298.

Hamilton, W., 1978, Mesozoic tectonics of the western United States, *in* Howell, D. G., and McDougall, K. A. (eds.), Mesozoic paleography and geography of the western United States: Society of Economic Paleontologists and Mineralogists, Pacific Coast Paleogeography Symposium, p. 33–70.

Hamilton, W., 1979, Tectonics of the Indonesian region: United States Geological Survey Professional Paper 1078, 345 p.

Hamilton, W., and Myers, W. B., 1967, The nature of batholiths: United States Geological Survey Professional Paper 554-C, 30 p.

Handin, J., 1969, On the Coulomb–Mohr failure criterion: Journal of Geophysical Research, v. 74, p. 5343–5348.

Handin, J., and Hager, R. V., 1957, Experimental deformation of sedimentary rocks under confining pressure: tests at room temperature on dry samples: American Association of Petroleum Geologists Bulletin, v. 41, p. 1–50.

Hardin, F. R., and Hardin, G. C., 1961, Contemporaneous normal faults of Gulf Coast and their relations to flexures: American Association of Petroleum Geologists Bulletin, v. 45, p. 238–248.

Harding, T. P., 1973, Newport–Inglewood trend, California—an example of wrenching style of deformation: American Association of Petroleum Geologists Bulletin, v. 57, p. 97–116.

Harris, J. F., Taylor, G. L., and Walper J. L., 1960, Relation of deformation features in sedimentary rocks to regional and local structure: American Association of Petroleum Geologists Bulletin, v. 44, p. 1853–1873.

Harris, L. D., 1979, Similarities between the thick-skinned Blue Ridge anticlinorium and thin-skinned Powell Valley anticline: Geological Society of American Bulletin, Part I, v. 90, p. 525–539.

Hawley, J. E., 1962, The Sudbury ores: their mineralogy and origin: Canadian Mineralogist, v. 7, p. 1–145.

Haynes, F. M., and Titley, S. R., 1980, The evolution of fracture-related permeability within the Ruby Star granodiorite, Sierrita porphyry copper deposit, Pima County, Arizona: Economic Geology, v. 75, p. 673–683.

Heald, M. T., 1955, Stylolites in sandstones: Journal of Geology, v. 63, p. 101–114.

Heard, H. C., 1963, Effect of large changes in strain rate in the experimental deformation of Yule marble: Journal of Geology, v. 71, p. 162–195.

Heezen, B. C., 1960, The rift in the ocean floor: Scientific American, v. 203, p. 98–110.

Heidrick, T. L., and Titley, S. R., 1982, Fracture and dike patterns in Laramide plutons and their structural and tectonic implications: American Southwest, *in* Titley, S. R. (ed.), Advances in geology of the porphyry copper deposits: Southwest North America: University of Arizona Press, Tucson, Arizona, p. 73–91.

Hedry, H. E., and Stauffer, M. R., 1977, Penecontemporaneous folds in cross-bedding: inversion of facing criteria and mimicry of tectonic folds: Geological Society of America Bulletin, v. 88, p. 809–812.

Hess, H. H., 1962, History of ocean basins, *in* Engel, A. E. J., James, H. L., and Leonard, B. F. (eds.), Petrologic studies: a volume in honor of A. F. Buddington: Geological Society of America, p. 599–620.

Higgins, M. W., 1971, Cataclastic rocks: United States Geological Survey Professional Paper 687, 97 p.

Hill, J. G. (ed.), 1976, Geology of the Cordilleran hingeline: Rocky Mountain Association of Geologists, 432 p.

Hill, M. L., 1959, Dual classification of faults: Geological Society of America Bulletin, v. 43, p. 217–221.

Hill, M. L., and Dibblee, T. W., Jr., 1953, San Andreas, Garlock, and Big Pine faults, California: Geological Society of America Bulletin, v. 64, p. 443–458.

Hills, E. S., 1972, Elements of structural geology (2nd edition): John Wiley & Sons, New York, 502 p.

Hobbs, B. E., Means, W. D., and Williams, P. F., 1976, An outline of structural geology: John Wiley & Sons, New York, 571 p.

Hodgson, R. A., 1961, Classification of structures on joint surfaces: American Journal of Science, v. 259, p. 493–502.

Holmes, A., 1928, The nomenclature of petrology (2nd edition): Thomas Murby & Co., London, 284 p.

Hoover, J. R., Malone, R., Eddy, G., and Donaldson, A., 1969, Regional position, trend, and geometry of coals and sandstones of the Monongahela Group and Waynesburg formation in the Central Appalachians, *in* Donaldson, A. C. (ed.), Some Appalachian coals and carbonates: models of ancient shallow-water deposition: West Virginia Geological and Economic Survey, Morgantown, West Virginia, p. 157–192.

Hubbert, M. K, 1937, Theory of scale models as applied to the study of geologic structures: Geological Society of America Bulletin, v. 48, p. 1459–1519.

Hubbert, M. K., 1951, Mechanical basis for certain familiar geologic structures: Geological Society of America Bulletin, v. 62, p. 355–372.

Hubbert, M. K., and Rubey, W. W., 1959, Role of fluid pressure in mechanics of overthrust faulting. Part I: Geological Society of America Bulletin, v. 70, p. 115–166.

Hudleston, P. J., 1973, Fold morphology and some geometrical implications of theories of fold development: Tectonophysics, v. 16, p. 1–46.

Humphris, C. C., Jr., 1979, Salt movement on continental slope, northern Gulf of Mexico: American Association of Petroleum Geologists Bulletin, v. 63, p. 782–798.

Hunt, C. B., Averitt, P., and Miller, R. L., 1953, Geology and geography of the Henry Mountain region, Utah: United States Geological Survey Professional Paper 228, 234 p.

Huntoon, P. W., and Richter, H. R., 1979, Breccia pipes in the vicinity of Lockhart Basin, Canyonlands area, Utah, *in* Baars, D. L. (ed.), Permianland: Four Corners Geological Society Guidebook, 9th Field Conference, p. 47–53.

Hutchinson, D. R., Grow, J. A., Klitgord, C. D., and Swift, B. A., 1983, Deep structure and evolution of the Carolina Trough, *in* Watkins, J. S., and Drake, C. L. (eds.), Studies in continental margin geology: American Association of Petroleum Geologists Memoir 34, p. 129–152.

International Nickel Company Staff, 1946, The operations and plants of International Nickel Company of Canada Limited: Canadian Mining Journal, v. 67, p. 322–331.

Isacks, B., Oliver, J., and Sykes, L. R., 1968, Seismology and the new global plate tectonics: Journal of Geophysical Research, v. 73, p. 5855–5899.

Jaeger, J. C., and Cook, N. G. W., 1976, Fundamentals of rock mechanics: Halsted Press, New York, 585 p.

Jaeger, E., and Hunziker, J. C. (eds.), 1979, Lectures in isotope geology: Springer-Verlag, Berlin, 312 p.

Jagnow, D. H., 1979, Cavern development in the Guadalupe Mountains: Adobe Press, Albuquerque, New Mexico, 55 p.

Johnson, A. M., 1970, Physical processes in geology: Freeman, Cooper, and Company, San Francisco, 577 p.

Johnson, A. M., 1977, Styles of folding: mechanics and mechanisms of folding of natural elastic materials: Elsevier Scientific Publishing Company, Amsterdam, 406 p.

Jordan, W. M., 1965, Regional environmental study of the Early Mesozoic Nugget and Navajo Sandstones: Ph.D. dissertation, University of Wisconsin, Madison, Wisconsin, 206 p.

Kamb, W. B., 1959, Theory of preferred orientation developed by crystallization under stress: Journal of Geology, v. 67, p. 153–170.

Karig, D. E., 1971, Origin and development of marginal basins in the western Pacific: Journal of Geophysical Research, v. 76, p. 2542–2561.

Kelley, V. C., 1955, Monoclines of the Colorado Plateau: Geological Society of America Bulletin, v. 66, p. 789–804.

Kelley, V. C., and Clinton, N. J., 1960, Fracture systems and tectonic elements of the Colorado Plateau: University of New Mexico Publication in Geology 6, 104 p.

Kendall, P. F., and Briggs, H., 1933, The formation of rock joints and the cleat of coal: Royal Society of Edinburgh Proceedings, v. 53, p. 167–187.

King, P. B., 1959, The evolution of North America: Princeton University Press, Princeton, New Jersey, 189 p.

Knopf, E. B., and Ingerson, E., 1938, Structural petrology: Geological Society of America Memoir 6, 270 p.

Kuenen, P. H., and DeSitter, L. U., 1938, Experimental investigation into the mechanism of folding: Leidse Geological Mededlingen, v. 9, p. 217–239.

Le Pichon, X., 1968, Sea floor spreading and continental drift: Journal of Geophysical Research, v. 73, p. 3661–3697.

Le Pichon, X., and Sibuet, J., 1981, Passive margins: a model of formation: Journal of Geophysical Research, v. 86, p. 3708–3720.

Lockwood, J. P., and Moore, J. G., 1979, Regional deformation of the Sierra Nevada, California, on conjugate microfault sets: Journal of Geophysical Research, v. 84, p. 6041–6049.

Lowell, J. D., 1968, Geology of the Kalamazoo ore body, San Manuel district, Arizona: Economic Geology, v. 63, p. 645–654.

Lowell, J. D., and Genik, G. J., 1972, Sea-floor spreading and structural evolution of southern Red Sea: American Association of Petroleum Geologists Bulletin, v. 56, p. 247–259.

Lowell, J. D., Genik, G. J., Nelson, T. H., and Tucker, P. M., 1975, Petroleum and plate tectonics of the southern Red Sea, *in* Fisher, A. G., and Judson, S. (eds.), Petroleum and global tectonics: Princeton University Press, Princeton, New Jersey, p. 129–153.

Lowman, P. D., Jr., 1976, A satellite view of diapiric Archean granites in western Australia: Journal of Geology, v. 84, p. 237–238.

Lowman, P. D., Jr., 1981, A global tectonic activity map with orbital photographic supplement: NASA Technical Memorandum 82073, 117 p.

Lowry, H. F., 1969, *in* College Talks, Blackwood, J. R. (ed.): Oxford University Press, New York, 177 p.

Luyendyk, B. P., 1976, Dips of downgoing lithospheric plates beneath island arcs: Geological Society of America Bulletin, v. 81, p. 3411–3416.

Macdonald, G. A., 1967, Forms and structures of extrusive basaltic rocks *in* Hess, H. H., and Poldervaart, A. (eds.), Basalts—the Poldervaart treatise on rocks of basalt composition: Wiley-Interscience Publishers, New York, v. 1, p. 1–61.

Macdonald, G. A., 1972, Volcanoes: Prentice-Hall, Englewood Cliffs, New Jersey, 510 p.

Mackin, J. H., 1950, The down-structure method of viewing geologic maps: Journal of Geology, v. 58, p. 55–72.

McClay, K. R., and Coward, M. P., 1981, The Moine thrust zone: an overview, *in* McClay, K. R., and Price, N. J. (eds.), Geological Society of London Special Publication 9, p. 241–260.

McEwen, T. J., 1980, Fracture analysis of crystalline rocks: field measurements and field geomechanical techniques: Institute of Geological Sciences Report No. ENPU 80-11, 69 p.

McGill, G. E., and Stromquist, A. W., 1979, The grabens of Canyonlands National Park, Utah: geometry, mechanics, and kinematics: Journal of Geophysical Research, v. 4, p. 4547–4563.

McKenzie, D. P., and Morgan, W. J., 1969, The evolution of triple junctions: Nature, v. 224, p. 125–133.

McKinstry, H. E., 1961, Mining geology: Prentice-Hall, Englewood Cliffs, New Jersey, 680 p.

Malcolm, W., 1912, Gold fields of Nova Scotia: Canadian Geological Survey Memoir 20-E, 331 p.

Marlow, P. C., and Etheridge, M. A., 1977, Development of a layered crenulation cleavage in mica schists of the Kanmantoo Group near Macclesfield, South Australia: Geological Society of America Bulletin, v. 88, p. 873–882.

Martin, R. G., 1978, Northern and eastern Gulf of Mexico continental margin, stratigraphic and structural framework, *in* Bouma, A. H., Moore, G. T., and Coleman, J. M. (eds.), Framework, facies, and oil-trapping characteristics of the upper continental margin: American Association of Petroleum Geologists Studies in Geology 7, p. 21–42.

Marvin, R. F., Stern, T. W., Creasey, S. C., and Mehnert, H. H., 1973, Radiometric ages of igneous rocks from Pima, Santa Cruz, and Cochise counties, southeastern Arizona: United States Geological Survey Bulletin 1379, 27 p.

Mattauer, M., 1975, Sur le mechanisme de formation de la schistosité dans l'Himalaya: Earth and Planetary Science Letters, v. 8, p. 144–154.

Maxwell, J. C., 1962, Origin of slaty and fracture cleavage in the Delaware Water Gap area, New Jersey and Pennsylvania, *in* Petrologic Studies: a volume in honor of A. F. Buddington: Geological Society of America, p. 281–311.

Mayo, E. B., 1941, Deformation in the interval Mt. Lyell–Mt. Whitney, California: Geological Society of America Bulletin, v. 52, p. 1001–1084.

Means, W. D., 1976, Stress and strain: Springer-Verlag, New York, 339 p.

Melosh, H. J., and Raefsky, A., 1980, The dynamical origin of subduction zone topography: Geophysical Journal of the Royal Astronomical Society, v. 60, p. 333–354.

Menges, C. M., 1981, The Sonoita Creek basin: implications for late Cenozoic evolution of basins and ranges in southeastern Arizona: M.S. thesis, University of Arizona, Tucson, Arizona, 239 p.

Michener, J. A., 1959, Hawaii: Random House, New York, 937 p.

Milnes, A. G., 1979, Albert Heim's general theory of natural deformation (1878): Geology, v. 7, p. 99–103.

Miyashiro, A., 1973, Metamorphism and metamorphic belts: William Clowes & Sons, Ltd., London, 492 p.

Miyashiro, A., 1974, Volcanic rock series in island arcs and active continental margins: American Journal of Science, v. 274, p. 321–355.

Mohr, O. C., 1882, Über die Darstellung des Spannungszustandes und des Deformationes-Zustandes eines Korperelementes und über die Anwendung derselben in der Festigkeitslehre: Civilingenieur, v. 28, p. 113–156.

Mohr, O. C., 1900, Welche Umstande bedingen die Elastizitätsgrenze und den Bruch eines Materials?: Zeitschrift der Vereines Deutscher Ingenieure, v. 44, p. 1524–1530 and p. 1572–1577.

Molnar, P., and Tapponier, P., 1975, Cenozoic tectonics of Asia: effect of a continental collision, Science, v. 189, p. 419–425.

Moore, J. C., 1978, Orientation of underthrusting during latest Cretaceous and earliest Tertiary time, Kodiak Islands, Alaska: Geology, v. 6, p. 209–213.

Morgan, W. J., 1968, Rises, trenches, great faults, and crustal blocks: Journal of Geophysical Research, v. 73, p. 1959–1982.

Morgan, W. J., 1971, Convection plumes in the lower mantle: Nature, v. 230, p. 42–43.

Morgan, W. J., 1972, Convection plumes and plate motions: American Association of Petroleum Geologists Bulletin, v. 56, p. 203–213.

Mosher, S., 1981, Pressure solution deformation of the Purgatory Conglomerate from Rhode Island: Journal of Geology, v. 89, p. 35–55.

Muehlberger, W. R., and Clabaugh, P. S., 1968, Internal structure and petrofabrics of Gulf Coast salt domes, *in* Braunstein, J., and O'Brien, G. H. (eds.), Diapirism and diapirs: American Association of Petroleum Geologists Memoir 8, p. 90–98.

Murray, G. E., 1961, Geology of the Atlantic and Gulf Coastal province of North America: Harper and Brothers, New York, 692 p.

Nettleton, L. L., 1934, Fluid mechanics of salt domes: American Association of Petroleum Geologists Bulletin, v. 18, p. 1175–1204.

Nevin, C. M., and Sherrill, R. E., 1929, Studies in differential compaction: American Association of Petroleum Geologists Bulletin, v. 13, p. 1–22.

Nickelsen, R. P., 1972, Attributes of rock cleavage in some mudstones and limestones of the Valley and Ridge province, Pennsylvania: Pennsylvania Academy of Science Proceedings, v. 46, p. 107–112.

Nickelsen, R. P., and Hough, V. D., 1967, Jointing in the Appalachian Plateau of Pennsylvania: Geological Society of America Bulletin, v. 78, p. 609–630.

Nicolas, A., and Le Pichon, X., 1980, Thrusting of young lithosphere in subduction zones with special reference to structures in ophiolitic peridotites: Earth and Planetary Science Letters, v. 46, p. 397–406.

Ocamb, R. D., 1961, Growth faults of South Louisiana: Gulf Coast Association of Geological Societies Transactions, v. 11, p. 139–175.

O'Driscoll, E. S., 1962, Experimental patterns in superposed similar folding: Journal of Alberta Society of Petroleum Geologists, v. 10, p. 145–167.

O'Driscoll, E. S., 1964a, Cross fold deformation by simple shear: Economic Geology, v. 59, p. 1061–1093.

O'Driscoll, E. S., 1964b, Interference patterns from inclined shear fold systems: Canadian Petroleum Geologists Bulletin, v. 12, p. 279–310.

Oertel, G., 1965, The mechanism of faulting in clay experiments: Tectonophysics, v. 2, p. 343–393.

Oertel, G., 1970, Deformation of a slaty, lapillar tuff in the Lake District, England: Geological Society of America Bulletin, v. 81, p. 1173–1188.

Park, C. F., and MacDiarmid, R. A., 1964, Ore deposits: W. H. Freeman, San Francisco, 475 p.

Parker, J. M., 1942, Regional systematic jointing in slightly deformed sedimentary rocks: Geological Society of America Bulletin, v. 53, p. 381–408.

Parker, T. J., and McDowell, A. N., 1955, Model studies of salt-dome tectonics: American Association of Petroleum Geologists Bulletin, v. 39, p. 2384–2470.

Paterson, M. S., and Weiss, L. E., 1966, Experimental deformation and folding of phyllite: Geological Society of America Bulletin, v. 77, p. 343–374.

Phillips, F. C., 1971, The use of stereographic projection in structural geology: Edward Arnold, London, 90 p.

Phillips, J., 1844, Orientation movements in the parts of stratified rocks: British Association for the Advancement of Science Report, 1843, p. 60–61.

Plafker, G., 1965, Tectonic deformation associated with the 1964 Alaska earthquake: Science, v. 148, p. 1675–1687.

Plafker, G., 1976, Tectonic aspects of the Guatemala earthquake of 4 February 1976: Science, v. 193, p. 1201–1208.

Powell, C. McA., 1979, A morphological classification of rock cleavage, *in* Bell, T. H., and Vernon, R. H. (eds.), Microstructural processes during deformation and metamorphism: Tectonophysics, v. 58, p. 21–34.

Powell, J. W., 1873, Geological structure of a district of country lying to the north of the Grand Canyon of the Colorado: American Journal of Science, v. 5, p. 456–465.

Powell, J. W., 1875, Exploration of the Colorado River of the west and its tributaries: United States Government Printing Office, Washington, D.C., 291 p.

Price, N. J., 1959, Mechanics of jointing in rocks: Geological Magazine, v. 96, p. 149–167.

Price, R. A., 1967, The tectonic significance of microscopic subfabrics in the southern Rocky Mountains of Alberta and British Columbia: Canadian Journal of Earth Science, v. 4, p. 39–70.

Price, R. A., and Mountjoy, E. W., 1970, Geologic structure of the Canadian Rocky Mountains between Bow and Athabasca rivers—progress report, *in*

Wheeler, J. O. (ed.), Geological Association of Canada Special Paper 6, p. 7–25.

Price, R. A., Mountjoy, E. W., and Cook, G. G., 1978, Geologic map of Mount Goodsir (west half), British Columbia: Geological Survey of Canada, map 1477A, 1:50,000.

Proffett, J. M., Jr., 1977, Cenozoic geology of the Yerington district, Nevada, and its implications for the nature and origin of Basin and Range faulting: Geological Society of America Bulletin, v. 88, p. 247–266.

Ragan, D. M., 1969, Introduction to concepts of two-dimensional strain and their application with the use of card-deck models: Journal of Geological Education, v. 17, p. 135–141.

Ragan, D. M., 1973, Structural geology, an introduction to geometrical techniques (2nd edition): John Wiley & Sons, New York, 208 p.

Ragan, D. M., and Sheridan, M. J., 1972, Compaction of the Bishop Tuff, California: Geological Society of America Bulletin, v. 83, p. 95–106.

Ramberg, H., 1955, Natural and experimental boudinage and pinch-and-swell structures: Journal of Geology, v. 63, p. 512–526.

Ramberg, H., 1959, Evolution of ptygmatic folding: Norsk Geologisk Tidsskrift, v. 39, p. 99–151.

Ramberg, H., 1962, Contact strain and folding instability of a multilayered body under compression: Geologische Rundschau, v. 51, p. 405–439.

Ramberg, H., 1963, Evolution of drag folds: Geological Magazine, v. 100, p. 97–106.

Ramberg, H., 1967, Gravity, deformation and the Earth's crust as studied by centrifuged models: Academic Press, New York, 214 p.

Ramberg, H., 1973, Model studies in gravity-controlled tectonics by the centrifuge technique, *in* De Jong, K. A., and Scholten, R. (eds.), Gravity and tectonics: John Wiley & Sons, New York, p. 49–66.

Ramsay, J. G., 1963, Stratigraphy, structure, and metamorphism in the western Alps: Geologists Association Proceedings, v. 74, p. 357–391.

Ramsay, J. G., 1967, Folding and fracturing of rocks: McGraw-Hill Book Company, New York, 560 p.

Ramsay, J. G., 1969, The measurement of strain and displacement in orogenic belts, *in* Kent, P. E., Satterthwaite, G. E., and Spencer, A. M. (eds.), Time and place in orogeny: Geological Society of London Special Publication 3, p. 43–79.

Ramsay, J. G., 1980, Shear zone geometry: a review: Journal of Structural Geology, v. 2, p. 83–99.

Reches, Z., 1976, Analysis of joints in two monoclines in Israel: Geological Society of America Bulletin, v. 87, p. 1654–1662.

Reches, Z., 1978a, Analysis of faulting in three-dimensional strain field: Tectonophysics, v. 47, p. 109–129.

Reches, Z., 1978b, Development of monoclines. Part I. Structure of the Palisades Creek branch of the East Kaibab monocline, Grand Canyon, Arizona, *in* Matthews, V. (ed.), Laramide folding associated with basement block faulting in the western United States: Geological Society of America Memoir 151, p. 235–272.

Reches, Z., 1983, Faulting of rocks in three-dimensional strain fields: II. Theoretical analysis: Tectonophysics, v. 95, p. 133–156.

Reches, Z., and Dieterich, J. H., 1983, Faulting of rocks in three-dimensional strain fields: I. Failure of rocks in polyaxial, servo-control experiments: Tectonophysics, v. 95, p. 111–132.

Reches, Z., and Johnson, A. M., 1978, Development of monoclines. Part II. Theoretical analysis of monoclines, *in* Matthews, V. (ed.), Laramide folding associated with basement block faulting in the western United States: Geological Society of America Memoir 151, p. 273–311.

Rehrig, W. A., and Heidrick, T. L., 1972, Regional fracturing in Laramide stocks of Arizona and its relationship to porphyry copper mineralization: Economic Geology, v. 67, p. 198–213.

Rehrig, W. A. and Heidrick, T. L., 1976, Regional tectonic stress during the Laramide and late Tertiary intrusive periods, Basin and Range province, Arizona: Arizona Geological Digest, v. 10, p. 205–228.

Reks, I. J., and Gray, D. R., 1982, Pencil structure and strain in weakly deformed mudstone and siltstone: Journal of Structural Geology, v. 4, p. 161–176.

Rich, J. L., 1934, Mechanics of low-angle overthrust faulting as illustrated by Cumberland thrust block, Virginia, Kentucky, and Tennessee: American Association of Petroleum Geologists Bulletin, v. 18, p. 1584–1596.

Ridgeway, J., 1920, Preparation of illustrations for the reports of the United States Geological Survey: with brief descriptions of processes of reproduction: United States Geological Survey, Washington, D.C., 101 p.

Riedel, W., 1929, Zur Mechanik geologischer Brucherscheinungen. Ein Beitrag zum Problem der "Fiederspalten": Centralblatt für Mineralogie, Geologie, und Paleontologie, Part B, p. 354–368.

Rispoli, R., 1981, The stress fields about strike–slip faults inferred from stylolites and tension gashes: Tectonophysics, v. 75, p. 29–36.

Roberts, J. C., 1961, Feather fractures and the mechanics of rock jointing: American Journal of Science, v. 259, p. 481–492.

Ross, C. S., and Smith, R. L., 1960, Ash-flow tuffs: their origin, geologic relations, and identification: United States Geological Survey Professional Paper 366, 81 p.

Roy, A. B., 1978, Evolution of slaty cleavage in relation to diagenesis and metamorphism: a study from the Hunsruckschiefer: Geological Society of America Bulletin, v. 89, p. 1775–1785.

Royse, F., Jr., Warner, M. A., and Reese, D. L., 1975, Thrust belt structural geometry and related stratigraphic problems, Wyoming–Idaho–northern Utah, *in* Bolyard, D. W. (ed.), Deep drilling frontiers of the central Rocky Mountains Symposium: Rocky Mountain Association of Geologists, p. 41–54.

Ryan, M. P., and Sammis, C. G., 1978, Cyclic fracture mechanisms in cooling basalt: Geological Society of America Bulletin, v. 89, p. 1295–1308.

Sander, B., 1930, Gefügekunde der Gesteine: Springer-Verlag, Vienna, 352 p.

Sanford, A. R., 1959, Analytical and experimental study of simple geologic structures: Geological Society of America Bulletin, v. 70, p. 19–52.

Schmincke, H., 1967, Flow directions in Columbia River basalt flows and paleocurrents of interbedded sedimentary rocks, south-central Washington: Geologische Rundschau, v. 56, p. 992–1020.

Schreiber, J. F., Jr., 1974, Field descriptions of sedimentary rocks, *in* Davis, G. H. (ed.), Geology field camp manual: University of Arizona, Tucson, Arizona, p. 97–110.

Secor, D. T., Jr., 1965, Role of fluid pressure in jointing: American Journal of Science, v. 263, p. 633–646.

Shafiqullah, M., Damon, P. E., Lynch, D. J., Kuck, P. H., and Rehrig, W. A., 1978, *in* Callendar, J. F., Wilt, J. C., and Clemons, R. E. (eds.), Land of Cochise: New Mexico Geological Society, 29th Field Conference, Socorro, New Mexico, p. 231–242.

Sharp, R. P., and Carey, D. L., 1976, Sliding stones, Racetrack Playa, California: Geological Society of America Bulletin, v. 87, p. 1704–1717.

Sharpe, D., 1847, On slaty cleavage: Geological Society of London Quarterly Journal, v. 3, p. 74–105.

Sherwin, J. A., and Chappel, W. M., 1968, Wavelengths of single layer folds: a comparison between theory and observation: American Journal of Science, v. 266, p. 167–179.

Sibson, R. H., 1980, Transient discontinuities in ductile shear zones: Journal of Structural Geology, v. 1, p. 165–171.

Silver, L. T., 1960, Age determinations on Precambrian diabase differentiates in the Sierra Ancha, Gila County, Arizona (abs): Geological Society of America Bulletin, v. 71, p. 1973–1974.

Silver, L. T., 1963, The use of cogenetic uranium–lead isotope systems in geochronology: Radioactive Dating, International Atomic Energy Agency, Athens, November 1962, p. 279–285.

Silver, L. T., 1978, Precambrian formations and Precambrian history in Cochise County, southeastern Arizona, *in* Callendar, J. F., Wilt, J. C., and Clemons, R. E. (eds.), Land of Cochise: New Mexico Geological Society, 29th Field Conference, Socorro, New Mexico, p. 157–163.

Simpson, C., and Schmid, S. M., 1983, An evaluation of criteria to deduce the sense of movement in sheared rocks: Geological Society of America Bulletin, v. 94, p. 1281–1288.

Sloss, L. L., 1963, Sequences in the cratonic interior of North America: Geological Society of America Bulletin, v. 74, p. 93–114.

Smiley, T. L., 1964, On understanding geochronological time: Arizona Geological Society Digest, v. 7, p. 1–12.

Smith, R. L., 1960a, Ash flows: Geological Society of America Bulletin, v. 71, p. 795–842.

Smith, R. L., 1960b, Zones and zonal variations in welded ash flows: United States Geological Survey Professional Paper 354-F, p. 149–159.

Smithson, S. B., Brewer, J., Kaufman, S., Oliver, J., and Hurich, C., 1978, Nature of Wind River thrust, Wyoming, from COCORP deep reflection data and from gravity data: Geology, v. 6, p. 648–652.

Sorby, H. C., 1853, On the origin of slaty cleavage: Edinburgh New Philosophical Journal, v. 55, p. 137–148.

Sorby, H. C., 1856, On slaty cleavage as exhibited in the Devonian limestones of Devonshire: Philosophical Magazine, v. 11, p. 20–37.

Sorby, H. C., 1859, On the structure produced by the currents present during the deposition of stratified rocks: The Geologist, v. 2, p. 137–149.

Spry, A. H., 1969, Metamorphic textures: Pergamon Press, Oxford, 350 p.

Stearns, D. W., 1968, Certain aspects of fractures in naturally deformed rocks, *in* Riecker, R. E. (ed.), National Science Foundation Advanced Science Seminar in Rock Mechanics for College Teachers of Structural Geology: Terrestrial Sciences Laboratory, Air Force Cambridge Research Laboratories, Bedford, Massachusetts, p. 97–118.

Stearns, D. W., 1978, Faulting and forced folding in the Rocky Mountain foreland, *in* Matthews, V. (ed.), Laramide folding associated with basement block faulting in the western United States: Geological Society of America Memoir 151, p. 1–37.

Stewart, J. H., and Suczek, C. A., 1977, Cambrian and latest Precambrian paleogeography and tectonics in the western United States, *in* Stewart, J. H., Stevens, C. H., and Fritsche, A. E. (eds.), Paleozoic paleogeography of the western United States: Society of Economic Paleontologists and Mineralogists, Pacific Coast Paleogeography Symposium 1, p. 1–17.

Stockdale, P. B., 1922, Stylolites: their nature and origin: Indiana University Studies, v. 9, 97 p.

Stockdale, P. B., 1926, The stratigraphic significance of solution in rocks: Journal of Geology, v. 34, p. 399–414.

Stokes, W. L., Judson, S., and Piccard, M. D., 1978, Introduction to geology: physical and historical (2nd edition): Prentice-Hall, Englewood Cliffs, New Jersey, 656 p.

Suppe, J., 1980a, A retrodeformable cross section of northern Taiwan: Geological Society of China Proceedings, no. 23, p. 46–55.

Suppe, J., 1980b, Imbricated structure of western foothills belt, south-central Taiwan: Petroleum Geology of Taiwan, no. 17, p. 1–16.

Swan, M. M., 1976, The Stockton Pass fault: an element of the Texas lineament: M.S. thesis, University of Arizona, Tucson, Arizona, 119 p.

Sykes, L. R., 1967, Mechanism of earthquakes and nature of faulting on the mid-oceanic ridges: Journal of Geophysical Research, v. 72, p. 2131–2153.

Syme Gash, P. J., 1971, A study of surface features relating to brittle and semi-brittle fractures: Tectonophysics, v. 12, p. 349–391.

Tobisch, O. T., Fiske, R. S., Sacks, S., and Taniguchi, D., 1977, Strain in metamorphosed volcaniclastic rocks and its bearing on the evolution of orogenic belts: Geological Society of America Bulletin, v. 88, p. 23–40.

Tullis, J., and Schmid, S., 1982, Notes: General remarks on flow and deformation mechanisms: short course on ductile deformation mechanisms and microstructures: Structural Geology Division, Geological Society of America, 28 p.

Tullis, T. E., and Wood, D. S., 1975, Correlation of finite strain from both reduction bodies and preferred orientation of mica in slate from Wales: Geological Society of America Bulletin, v. 86, p. 632–638.

Turner, F. J., and Weiss, L. E., 1963, Structural analysis of metamorphic tectonites: McGraw-Hill Book Company, New York, 560 p.

Uyeda, S., 1978, The new view of the Earth: W. H. Freeman & Company, San Francisco, 217 p.

Van der Voo, R., Mauk, F. J., and French, R. B., 1976, Permian–Triassic continental configurations and the origin of the Gulf of Mexico: Geology, v. 4, p. 177–180.

Van Hise, R., 1896, Principles of North American pre-Cambrian geology: United States Geological Survey 16th Annual Report, Part 1, p. 581–844.

Vine, F. J., and Matthews, D. H., 1963, Magnetic anomalies over oceanic ridges: Nature, v. 199, p. 947–949.

Waters, A. C., 1960, Determining direction of flow in basalts: American Journal of Science, v. 258a, p. 350–366.

Webster's new collegiate dictionary, 1973: G. C. Merriam Co., Springfield, Massachusetts, 1526 p.

Weiss, L. E., 1972, The minor structures of deformed rocks: a photographic atlas: Springer-Verlag, New York, 431 p.

Wernicke, B., and Burchfiel, B. C., 1982, Modes of extension tectonics: Journal of Structural Geology, v. 4, p. 105–115.

Wheeler, R. L., and Dickson, J. M., 1980, Intensity of systematic joints, methods, and application: Geology, v. 8, p. 230–233.

Whitten, E. T. H., 1966, Structural geology of folded rocks: Rand-McNally, Skokie, Illinois, 663 p.

Wilcox, R. E., Harding, T. P., and Seely, D. R., 1973, Basic wrench tectonics: American Association of Petroleum Geologists Bulletin, v. 57, p. 74–96.

Willemse, J., 1969, The geology of the Bushveld igneous complex: the largest repository of magnetic ore depositions in the world: Economic Geology Monograph 4, p. 1–22.

Williams, H., 1942, The geology of Crater Lake National Park, Oregon, Carnegie Institute Publication 540, 162 p.

Willis, B., 1894, The mechanics of Appalachian structure: United States Geological Survey 13th Annual Report, Part 2, p. 213–281.

Wilson, G., 1961, The tectonic significance of small-scale structures and their importance to the geologist in the field: Annales de la Société Géologique de Belgique, v. 84, p. 424–548.

Wilson, G., 1982, Introduction to small-scale geologic structures: George Allen & Unwin (Publishers) Ltd., London, 128 p.

Wilson, J. T., 1965, A new class of faults and their bearing on continental drift: Nature, v. 207, p. 343–347.

Wise, D. U., 1964, Microjointing in basement, Middle Rocky Mountains of Montana and Wyoming: Geological Society of America Bulletin, v. 75, p. 287–306.

Woodbury, H. O., Murray, I. B., Jr., Pickford, P. J., and Akers, W. H., 1973,

Pliocene and Pleistocene depocenters, outer continental shelf, Louisiana and Texas: American Association of Petroleum Geologists Bulletin, v. 57, p. 2428–2439.

Woodworth, J. B., 1896, On the fracture system of joints, with remarks on certain great fractures: Boston Society of Natural History Proceedings, v. 27, p. 163–184.

Worrall, D. M., 1977, Structural development of Round Mountain area, Uinta County, Wyoming, *in* Heisey, E. L. (ed.), Rocky Mountain thrust belt, geology and resources: Wyoming Geological Association Guidebook 29, p. 537–541.

Wright, L. A., and Troxel, B., 1973, Shallow-fault interpretation of Basin and Range structure, southwestern Great Basin, *in* De Jong, K. A., and Scholten, R. (eds.), Gravity and tectonics: John Wiley & Sons, New York, p. 397–407.

Wright, T. O., and Platt, L. B., 1982, Pressure dissolution and cleavage in the Martinsburg Shale: American Journal of Science, v. 282, p. 122–135.

Zingg, T., 1935, Beitrag zur Schotteranalyze: Schweizer Mineralogische und Petrographische Mitteilung, 15, p. 39–140.

AUTHOR INDEX

Numbers in **boldface** denote figure and/or table.

Akers, W. H., 217
Albee, A. L., **9**
Allis, R. G., 303
Alvarez, W., **89, 407,** 416, **418,** 418, **429**
Anderson, E. M., 314, 318
Anderson, J. L., 100, 293, **294**
Anderson, R. E., **94,** 94, **100,** 100, **222,** 293
Anderson, R. N., 170
Anderson, T. H., 303
Antoine, J. W., 220
Armstrong, R. L., 60, **61,** 293
Arthaud, F., 338
Atwater, T. M., 176, 303–304, **304, 305**
Averitt, P., 216

Balk, R., 217–219, **218,** 220, 253, **254,** 256, 258, **406**
Bally, A. W., 167, 188, **189,** 283
Barnes, C. W., 397
Batiza, R., **246**
Beard, S., **402**
Becker, C. L., 31
Benioff, H., 177
Benson, G. S., **34, 107**
Berg, R. R., 321, **322, 399**
Berthé, D., 424
Beutner, E. C., **413, 414,** 414
Bijlaard, P. P., 392, 394
Billings, M. P., 4, 94, 389
Biot, M. A., 391, **393,** 393–394, 400
Bird, J. M., 173, 189, **190**
Bishop, R. S., **216,** 217–220, **219, 220**
Blake, D. R., 338
Blew, R. M., 157
Blount, D. N., 303
Bohnenberger, O. H., 303
Boyd, A., **224**
Bramlette, M. N., **358**
Brewer, J., **321**
Briggs, H., 351
Brogan, G., **223**
Bronowski, J., 4
Brown, L. D., **265,** 266–267
Brown, R. D., Jr., **92**
Brown, W. G., 285
Bruce, C. H., **243,** 243–244, **244**
Bryant, W. R., 220
Bucher, W. H., 68
Bull, W. B., **34, 164, 234, 293**
Burchfiel, B. C., 279, 296
Burkart, B., 303
Burrell, H. C., **359,** 359
Bykerk-Kauffman, A., **416, 417**

Carey, D. L., **90,** 90
Carey, S. W., 158, 421
Carhuft, N. W., **206**
Carver, R. E., 244
Cashman, K., **117**
Chapman, D. S., 165
Chappel, W. M., 393
Chilingarian, G. V., 240
Choukroune, P., 337, 424
Claybaugh, P. S., 217
Clemons, R. E., 303
Clinton, N. J.,397
Cloos, E., 124–126, **125,** 244, 253–254, **288,** 287–289, **280,** 297–298, 425
Cloos, H.,253–254, **256, 258,** 258, 287–289
Cluff, L., **223**
Compton, R. R., 45
Coney, P. J., 183, **184,** 200, 436
Conybeare, C. E. B., 237–238
Cook, F. A., **265,** 266–267
Cook, G. G., **285**
Cook, N. G. W., 107, 132, 139, 145, 160
Cooper, J. R., **44,** 297
Cosgrove, J. W., **401**
Coulomb, C. A., 309
Cowan, D., 182
Coward, M. P., 437
Cox, A. V., **171,** 183
Creasey, S. C., **34**
Crespi, J. M., **331**
Crittenden, M. D., Jr., 436
Crook, K. A. W., 237–238
Cropp, F. W., **34, 107**
Crowell, J. C., 267–268, **300,** 300, **301,** 302, 304
Currie, J. B., 392, **394,** 394
Currier, D., **51**

Dahlstrom, D. C. A., 280–281, 283–284, **284, 285**
Dale, T. N., **337,** 352
Dalrymple, G. B., **171,** 183
Damon, P. E., **34**
Dana, S. W., **58**
Davis, G. A., 100, 279, 293, **294**
Davis, G. H., **9, 24, 34, 48,** 100, **107, 115, 131, 226, 228, 242, 261,** 291, **294, 391,** 397, **398, 423, 429,** 436, **437, 439, 443, 445**
Dean, J. S., **34**
Dennis, J. G., 216, 363, 405
DeSitter, L. U., 281, 387
Dewey, J. F., 173, **188,** 189, **190**

Dibblee, T. W., Jr., 303
Dickinson, W. R., 171–172, **172, 177, 185, 186, 187,** 185–186, **354, 355**
Dickson, J. M., 344
Dieterich, J. H., **320,** 320, 350
Dietz, R. S., 171, 173–174
Doell, R. R., **171,** 183
Doll, C. G., **420**
Donaldson, A., **60**
Donath, F. A., **134, 135,** 135, **136, 151, 153, 155, 319,** 355, 365, 384–385, 387, **390,** 390, 420
Droxler, A., 337
Dubey, A. K., **28**
Durney, D. W., 334, **336, 337,** 336–337
Dutton, C., 291
Dutton, S. P., **58**
Dyer, R., **326, 330, 341**

Eaton, G. P., 291, **292**
Eddy, G., **60**
Edwards, M. B., 242–244
Eliopulus, G. J., **49**
Elliot, D., 238, 437
Emery, K. O., **235**
Engel, A. E. J., **335**
Engelder, R., **23,** 419
Engelder, T., **23, 89,** 407, **416,**416, **418,** 418–419, **429,** 429
England, P., 183
Ernst, W. G., 183
Eskola, P. E., 438
Etheridge, M. A., 409, 415
Euler, R. C., **34**

Feininger, T., **11**
Fink, J., **247**
Fiske, R. S., **23, 262**
Fletcher, R. C., 337–338
Fleuty, M. J., **365,** 365, 374, **375**
Flinn, D., 433
Folk, R. L., 233
French, R. B., 100–101
Friedman, M., **27,** 321, **399,** 400
Fries, C., 250
Frost, E. G., **49, 89,** 100, 293, **294**

Garretty, M. D., **359,** 359
Geiser, P. A., 338, **416, 418,** 418
Genik, G. J., 289, **290, 291,** 289–291
Ghosh, S. K., 394, **395**
Gilbert, G. K., **93, 325**
Gillette, R., **54**
Glass, C., **223**
Goodmundson, R. C., **49**
Goodwin, A. M., 170, **174**
Gordy, P. L., 283
Gossett, D., **373**
Gray, D. R., 406, **409,** 409, **415,** 415, 428–429
Greeley, R., **248**
Griffith, A. A., 318
Griggs, D. T., **155**
Groshong, R. H., Jr., **338,** 338, **419,** 419
Grow, J. A., **209**
Gummerman, G. J., **34**
Gustafson, J. K., **359,** 359

Hafner, W., **321,** 321
Hager, R. V., 155
Halbouty, M. T., 216, 243
Hall, A. L., 216
Hamblin, W. K., **37, 271,** 271
Hamilton, W., **181,** 215, **233,** 296
Handin, J., **27,** 155, 318, 321, **399,** 400
Hardin, E. B., **356**
Hardin, F. R., 243
Hardin, G. C., 243
Harding, T. P., **299,** 299–300, 304
Hardy, J. J., Jr., 100, 436, **437**
Harris, J. F., 331, 344
Harris, L. D., 282, **283**
Hawley, J. E., 216
Haynes, F. M., 344
Heald, M. T., 338
Heard, H. C., **155,** 156, **157**
Heezen, B. C., 168–169
Heidrick, T. L., 340, 342, **345,** 345–346, **347,** 352
Heim, A., 110
Hendry, H. E., **242**
Hess, H. H., 171, 174
Hevly, R. H., **34**
Higgins, M. W., 226, **227, 387**
Hill, J. G., 60
Hill, M. L., 267, 275, 303
Hillers, J. K., **28**
Hills, E. S., 234
Hobbs, B. E., 160–162, 393, **408,** 427
Hodgson, R. A., **333,** 333, 342, 351
Holden, J. C., 173
Holmes, A., 427
Holzer, T. L., **241**
Hoover, J. R., **60**
Hough, V. D., 339–340, **347,** 351
Hubbert, M. K., 155, 257, 289, 311–314, 322–324
Huber, N. K., **6**
Hudleston, P. J., 373, **376,** 393
Humphries, C. C., Jr., 220
Hunt, C. B., 216
Huntoon, P. W., **58**
Hunziker, J. C., 33
Hurich, C., **321**
Hutchinson, D. R., 209

Ingerson, E., 16
Isacks, B., **167,** 171–172, 175

Jaeger, E., 33
Jaeger, J. C., 107, 132, 139, 145, 160
Jagnow, D. H., 327, 340
Jegouza, P., 424
Johnson, A. M., 390, 394, 400, 421
Johnson, M. R. W., 437
Jones, D. L., 183, **184**
Jordon, W. M., **60**
Judson, S., 173

Kamb, W. B., 86
Karig, D. E., 187
Karlstrom, T. N. V., **34**
Kaufman, S., **321**
Keith, A., **8, 10**
Kelley, V. C., 340, 351, **396,** 397
Kendall, P. F., 351
King, P. B., **58, 59,** 29
Klitgord, C. D., **209**
Knapp, R. B., **49**
Knopf, E. B., 16
Krantz, R. W., **27, 212, 311**
Krauskopf, K., **44**
Kuck, P. H., **34**
Kuenan, P. H., **387**

Laurijssen, W. C., **408**
Lee, W. T., **159**
Le Pichon, X., 171–172, 187, 196
Lepry, L. A., **11, 204**
Liming, R. B., **49**
Lingrey, S. H., **225, 442**
Lipman, P. W., **252**
Lockwood, J. P., **6,** 348
Logan, J. M., **27,** 321, **399,** 400
Lovering, T. S., **159**
Lowell, J. D., 289–291, **289, 290, 291**
Lowell, J. David, 297
Lowman, P. D., Jr., **46, 168, 169, 181, 437**
Lowrie, W., **89, 407, 416,** 416, **421**
Lowry, H. F., vii, 12
Luyendyk, B. P., 183
Lynch, D. J., **74, 205, 206, 215, 245, 257**

McBride, E. F., **238**
McCalmont, L. S., **34,** 107
McClay, K. R., **437**
MacDiarmid, R. A., 358
Macdonald, G. A., 244–246, **247, 248**
McDowell, A. N., 217
McEwen, T. J., 340
McGill, G. F., **7, 327**
McKenzie, D. P., 198, 304
Mackin, J. H., 384, **385**
McKinstry, H. E., 358, **359**
Malcolm, W., 358
Malone, R., **60**
Marlow, P. C., 409, 415
Marshall, D. R., 397
Martin, R. G., Jr., **217,** 220, **243,** 243
Marvin, R. F., **34**
Matesich, K., **425**
Mattauer, M., 338, 437
Matthews, D. H., 171
Mauk, F. J., 100–101
Maxwell, J. C., 412
Mayo, E. B., **4, 211, 213, 214,** 253, **255,** 258
Means, W. D., 107, 127, 141–142, 145, 160–162, 393, **408,** 427
Mehnert, H. H., **34**
Melosh, H. J., **108**
Menges, C. M., **54**
Michener, J. A., 31
Miller, R. L., 216
Milnes, A. G., **110,** 110
Min, K. D., **27,** 321, 398, 400
Miyashiro, A., 185
Mohr, O. C., 121
Molnar, P., **91**
Moore, J. C., 182
Moore, J. G., 348
Morgan, W. J., 191, 196, 198, 304
Mosher, S., **9**
Mountjoy, E. W., 18–19, **19,** 25, 279, **280, 281, 285,** 286
Muehlberger, W. R., 217
Murray, G. E., 243
Murray, I. B., Jr., 217
Myers, W. B., 215

Nelson, T. H., 289
Nettleton, L. L., 217, 220
Nevin, C. M., 240
Nicholas, A., 171
Nickelsen, R. P., 339–340, **347,** 351, 407, 416
Noltimier, H. C., **170**

Ocamb, R. D., 243
Ode, H., 390, **393,** 393–394
O'Driscoll, E. S., 25, **26**
Oertel, G., 319–320, 411
Oliver, J. E., **167,** 171–172, 175, **265,** 266–267, **321**

Park, C. F., 358
Parker, J. M., 342, 351
Parker, R. B., 355, 365, 385, 387, **390,** 390, **420,** 420
Parker, T. J., 217
Paterson, M. S., 394–395
Patnode, A. W., 392, **394,** 394
Pavlides, L., **413**
Payne, W. D., **185**
Phillips, F. C., 68, 72, 81
Phillips, J., 411
Phillips, M. P., **226, 242**
Piccard, M. D., 173
Pickford, P. J., 217
Plafker, G., **221, 301, 302,** 302, **303**
Platt, L. B., **414,** 414
Pollack, H. N., 165
Pollard, D. D., 337–338
Powell, C. McA., 405
Powell, J. W., **206, 396**
Price, N. J., **352,** 352
Price, R. A., 18–19, **19,** 279, **280, 281, 285,** 286, 347
Proffett, J. M., Jr., 100, 295–296, **296**
Pyle, T. G., 220

Raefsky, A., **180**
Ragan, D. M., 112, 113, 127, 128, 190–191, 252, 421
Ramberg, H., **217,** 217, **257,** 257, 390, 392–394
Ramirez, V. R., 304

Ramsay, J. G., 16, 90, 107, 109, 111, 113, 116, 118–119, 127–129, **128, 130,** 145, 148, 155, 228, 229, **229, 241,** 334, **336,** 336–337, 357, 361, 363, **365,** 365, 371, 375, 378–380, **379, 380,** 386, **389,** 429, 433, **444,** 445
Ratte, J. C., **424**
Reches, Z., 319–320, **320,** 348–350, **341,** 400
Reese, D. L., 279, 283–284, 286
Rehrig, W. A., **34,** 340, 342, 345–346, 352
Reks, I. J., 428–429
Reynolds, S. J., **226, 242, 425**
Rich, J. L., **282,** 282
Richter, D. H., 183, **184**
Richter, H. R., **58**
Ridgeway, J., **47,** 47
Riedel, W., 255, 256
Riggs, N., **54**
Rispoli, R., **338,** 338
Roberts, J. C., 332–333
Roever, W. L., 390, **393**, 394
Ross, C. S., **251**
Roy, A. B., **408**
Roy, C. G., 338
Royse, F., Jr., 279, 283–284, 286
Rubey, W. W., 155, 322–324
Rusmore, M., **354**
Russell, I. C., **221**
Ryan, M. P., **249**

Sacks, S., **23**
Sammis, C. G., **249**
Sander, B., 16
Sanford, A. R., **321,** 321
Sansone, S., 338
Schaer, J. P., 337
Schmid, S. M., 160, 161–162, 226, **230,** 230
Schmidt, R. G., **88, 97**
Schmincke, H., 249
Schmitt, H., 203
Schreiber, J. F., Jr., 37
Secor, D. T., Jr., **351,** 351–352
Seely, D. R., **299,** 299–300, 304
Seely, D. S., 186
Shackelford, T. J., 100, 293, **294**
Shafiqullah, M., **34**
Sharp, R. P., **90,** 90
Sharpe, D., 410–411
Sheridan, D. M., **405**
Sheridan, M. J., 252
Sherrill, R. E., 240
Sherwin, J. A., 393
Showalter, S. R., **34, 107**
Siberling, N. J., 183, **184**
Sibson, R. H., **227,** 227
Sibuet, J., 172
Siepert, A. F., **97**
Silver, L. T., **34, 44**
Simpson, C., **230,** 230
Sloss, L. L., 235
Smiley, T. L., 32
Smith, R. L., 186, 250, **251**
Smithson, S. B., **321**
Snelson, S., 167, 188, **189**
Sorby, H. C., 233, 410–413
Spry, A. H., **161**
Stacy, J. R., **3, 100, 103, 222**
Stauffer, M. R., **242**
Stearns, D. W., **27,** 321, **398, 399,** 400
Stern, T. W., **34**
Stewart, G. A., 283
Stewart, J. H., 173
Stockdale, P. B., 337
Stokes, W. L., 173
Stopper, R., **44**
Stose, G. W., **48, 211, 239**
Stromquist, A. W., **327**
Suczek, C. A., 173
Suppe, J., 282, **286,** 286
Swan, M. M., **49,** 270, **271**
Swanson, D. A., 117
Swift, B. A., **209**
Sykes, L. R., **167,** 171–172, 175
Syme Gash, P. J., 333

Taniguchi, D., **23**
Tapponnier, P., **91**
Taylor, G. L., 331, 334
Titley, S. R., 344–345, **345,** 346, 347
Tobisch, O. T., **23,** 232
Trent, D., **257**
Troxel, B., 295
Trump, R. P., 392, **394,** 394
Tucker, P. M., 289
Tullis, J., 160, 162, 226
Tullis, T. E., 411, 412
Turner, F. J., 16, 23, 36, **37,** 155, **162,** 162, 356, 362–363, **369,** 422, 425, 430, 438, **440, 442,** 442, **444,** 445

Uyeda, S., 196

Van de Voo, R., 101
Van Hise, R., 377–378
Varga, R. J., **226, 242**
Vine, F. J., 171

Walcott, C. D., **8, 368, 403, 406**
Wallace, R. E.,**7, 93**
Walper, J. L., 331, 334
Warner, M. A., 279, 283–284, 286
Waters, A. C., 248, 249
Weiss, L. E., 16, 23, 36, **37, 88, 112, 162,** 162, **330,** 356, **359,** 362–363, 394–395, **395, 404,** 422, 425, 430, 438, **440, 442,** 442, 444, **445**
Wernicke, B., 296
Wheeler, R. L., 344
Whitten, E. H. T., **72, 82, 125,** 372, **385,** 406, 442
Wilcox, R. E., **299,** 299–300, 304
Willemse, J., 216
Williams, H., 186
Williams, P. F., 160–162, 393, **418,** 427
Willis, B., **28,** 355, 400
Wilson, G., 426, **430**
Wilson, J. T., 174–175, **176,** 194, **195,** 302
Windley, B. F., **174**
Wise, D. U., 352

Wolf, K. H., 240
Wood, D. S., 411–412
Woodbury, H. O., 217
Woodworth, J. B., 331
Worrall, D. M.,**50**
Wortel, R., 183
Wright, L. A., 295
Wright, T. O., **414,** 414
Wust, S., **45**
Wynn, J. C., **49**

Xiong, J., 402

Yeats, K., **45**
Youd, T. L., **391**
Young, J., **17, 18, 22, 24, 239**

Zingg, T., 433

SUBJECT INDEX

Numbers in **boldface** denote figure and/or table.

Absolute motion(s), 190–191
Abyssal plain sediments, 182
Accreted terrane(s), 183, **184**
Accretion, 176, 183
Accretionary wedge, **181,** 181
Active margin(s), 181
African plate, 187, **188,** 190, 193, 198–199
Aleutian arc, 176
Aleutian Trench, 192
Allochthon, 279, **280**
Allochthonous rocks, 279, **280**
Alpine fault, New Zealand, 301, 303
American–Antarctica Ridge, **169,** 169
Amygdules, 248
Andean batholith, 216
Angle of internal friction, 309–310, 350
Angular shear, 105–107, **111, 112**
Angular unconformity, 204–207, **205, 206, 207**
Angular velocity, 197
Anisotropy, 154
  influence on faulting, 318–319
Antarctic plate, 187, **188,** 198–199
Anticline, 359–360, **360**
Anticlinorium, 360–361, **361**
Antiform, 360
Apophyses, **211,** 211
Appalachian Plateau, 339–340, 342, 351
Apparent dip, 64–66, 96, 381
Arabian plate, 187, **188**
Arc, 176, 185–187
Arches National Monument, **326,** 326–327, **330,** 341
Ash flow(s), 186, **251, 252**
  compaction strain, 250–252
Asthenosphere, 89, **166,** 166, **167,** 170, 172, 185
Atlantic seaboard, 209
Attitude, 63
Aureole, 213
Autochthon, 279
Autochthonous rocks, 279
Axial plane, 362–364, **363, 364**
  field measurement, 367
  orientation, 362–364
  stereographic determination, 368, 370
Axial stress, 136–137
Axial surface, **21,** 362–364, **363, 364, 367**
  field measurement, 367
  orientation, 362–365
  stereographic determination, 368–370, **370**
Axial trace, **21,** 362–363, **370,** 370, 372
Azimuth, 52, **53**

Backarc basin(s), **186,** 186–187, 357
Backarc spreading, 187
Back-limb thrust, 313
Balanced cross section, 280–281, 283–284, **284**
Basal conglomerate, 207
Base maps, 40–47
  aerial photographs, 45–47
  grid-line, 42, **43**
  pace-and-compass, **42,** 42
  picket-line, 42–43
  tape-and-compass, 42
  topographic, 45
Basement, **173–174,** 205, **437**
Basin(s), **59,** 173
Basin and Range province, 291–297, **292, 293**
Batholith, **215,** 215, **437,** 437
Beard(s), 408
Bearing, 52
Bedding, **39, 234,** 234–235
  map symbols, **51**
  transposed, 441–444
Bedding fissility, 428
Bendigo goldfields, 358
Benioff zone(s), 177, 183, 185
Beta diagram, **367,** 367–368, 370–372
Bissecting surface, 368–371
Bismarck plate, **188,** 188
Blanding basin, 397
Blueschist, 183
Body force, 136
Boudin(s), **241,** 241, 389, **429,** 429, **430**
Boudinage, **241,** 241, 389, 429
Box fold, 374
Breakup unconformity, **172, 209,** 209
Breccia, 224–225, **226**
Brittle failure, 150
Broken Hill deposit, 359
Bryce National Park, Utah, 326–327
Buckling, 385, 390–395, **391,** 400, 428
  multilayer sequence, 391, **394,** 394–395, **395**
  single layer, 391–394, **393, 394**
Bushveld complex, 216
Buttress unconformity, **209,** 209

Cactolith, 216
Caldera, 186, **246**
Canadian Rockies, 18–19, **19,** 279–282, **280, 281,** 286

Canyon de Chelly, Arizona, **5**
Canyonlands, Utah, **7, 59, 327**
Carbonate platform, 173
Caribbean plate, **188,** 301
Carlsbad Caverns, New Mexico, 327, 340
Carlsberg Ridge, **169,** 169, 190, 198, 289
Carrizo dome, Arizona, 216
Cataclastic flow, 160, 226
Channeling, 208
Chatter marks, **264,** 264
Chile Ridge, **169**
Chill zone, 213
Chiricahua National Monument, Arizona, 326–327, **327**
Chocolate tablet structure, 241
Circle Cliffs uplift, Utah, 397
Clastic dike(s), 211, **212**
Clastic sill(s), 211
Cleavage(s), 9–10, 385, **401, 402,** 401–419, 422
  axial plane cleavage, **10, 356, 401,** 402–404, **402, 403**
  classifications, 405, 418–419
  cleavage domains, **404,** 404, **405**
  cleavage seams, 414
  continuous cleavage, 405–406
  crenulation cleavage, **404, 405, 406,** 406, **409,** 409, 415–416, 422
  discontinuous cleavage, 405–407
  domainal character, 404–405
  domainal structure, 404, **404, 405**
  fan configuration, 402, **403**
  fracture cleavage, **331,** 331, 406–407
  insoluble residue, 331, 414
  map symbols, **51**
  M-domains, 407–408
  microlithon domains, **404,** 404–405, , **405**
  outcrop expression, 401–402, **402, 403**
  origin, 411–418
  phyllitic structure, 405–406, 408, 422
  pressure shadows, 412, **413**
  QF-domains, 407–408
  relation to bedding, 401–404, **403**
  relation to folding, **388,** 401–404, **403,** 410–411, 415–416
  scale of observation, 405
  schistosity, 405–406, **408,** 408, 422
  sigmoidal, 229
  slaty cleavage, **10,** 405–406, **406,** 407–408, **408,** 410–415, 422
  spaced cleavage, **331,** 331, 406–407, **407, 409,** 409–410, 416–419, **418.** 422
  strain significance, 410–419
  volume loss, 414–415
Cleavage domains, **404,** 404, **405**
Cleavage seams, 414
Cleavage surface(s), 402
Closure, 358
Coast Range batholith, 216
Cockscomb structure, **226,** 226, 334
COCORP, 266–267, 322
Cocos plate, 187, **188**
Coefficient of internal friction, 309–311
Coefficient of sliding friction, 309–310, 319
Collision, 167, 176, 183, 187
Collisional tectonics, 180
Colonnade, 249
Colorado Plateau, 216, 291, 351, 396–398
Colorado Plateau uplifts, **397,** 397
Columbia River Plateau, 249
Columnar jointing, **249,** 249–250, **250**
Comb Ridge monocline, Arizona, Utah, 397
Comminution, 226, 426
Compaction fold, 240
Compass, 52–57
  Brunton, **52,** 52–57
  clinometer, 53
  Silva, **52,** 52–57
  strike and dip, 54–57, **55, 56, 57**
  trend-and-plunge, 52–54, **54, 55**
Compatibility, 348, 377
Competency, 154
Compositional banding, 422
Concentric folds, **377,** 377–378
  decollement, 378
  geometry, 377–378
Confining pressure, 136, 151–153
Contact(s), 14, **203,** 203–230
  depositional, 203–204, **205**
  ductile shear zone(s), 203, **228,** 228–230, **229, 230**
  fault, 203, 221–227, **221, 222, 223, 224**
  intrusive, 203, 210–221, **211, 212, 213, 214, 215, 216, 217, 218, 219, 220**
  map symbols, **50, 203,** 203
  unconformities, 204–210, **205, 206, 207**
Contact metamorphism, 213
Continental drift, 165, **189,** 189
Continental interior, 173
Continental margin, 173, 186
  sedimentary prism, **172,** 173
Continental shelf, 173
Continental slope, 173
Convergence, 166, 177, 183, 187
Cooling unit(s), 249
Country rock, 210
Crater Lake National Park, Oregon, 186
Craton, 173, 207
Creep, **158,** 158
  dislocation, 160, 162
  primary, 158
  secondary, 158
  tertiary, 158
Crenulation cleavage, **404, 405, 406,** 406, **409,** 409, 415–416, 422
  discrete, **404,** 406
  microscopic character, 409
  outcrop character, 406
  progressive deformation, **415,** 415–416
  role of pressure solution, 415–416
  strain significance, 415–416
  zonal, **405,** 406, **409**
Crenulation lineation, **11,** 425–426, **426**

Critical shear stress, 309–311
Cross-bedding, **232, 235,** 235, **236**
Cross fractures, 358–359
Cross-lamination, 235
Cross stratification, 235–236
Crust, 164–166
  continental, 164–166, **166,** 173, 185, **172**
  oceanic, 164–166, **166, 172,** 173
  transitional, **172,** 173
Crustification, **334,** 334
Crystal defects, 160–161
Crystal fiber growths, **334,** 334–337, **335, 337**
Crystal plasticity, 162
C-surfaces, 424–425
Current direction, 235–237
  cross bedding, 235
  parting lineation, 236
  pebble imbrication, 237
  ripple marks, 236
  sole marks, 237
Current ripple marks, 236

Datum, 59
Decollement, 280–282, **377**
Defiance uplift, Arizona, **46**
Deformation, 2
  constant volume, 410
  crystal-plastic, 162, 226
  diachronous, 441
  ductile, 153
  elastic, 148, 151–152, 156, 391–392
  elasticoviscous, 156
  failure, 135
  heterogenous, 107
  homogenous, 107
  intraplate, 164
  nonrigid body, 87–89, **88,** 107, 161
  plastic, 150–153, 156, 171
  progressive, 130, **131,** 388–389, 424–425, 441
  rigid body, 87, **88,** 161
  soft-sediment, 182, 238–244, **238, 239**
  superposed, 130
  thin-skinned, 280–282
  viscous, 156, 158, 391–392
  volume constant, 338
Deformational mechanisms, 159–162
  cataclastic flow, 160
  dislocation creep, 160–162
  dislocation glide, 160–162, **161**
  mechanical twinning, 160–162, **162**
  microcracking, 160
  pressure solution, 160, 162, 331, 337–338, 410
Deformation path, 434–435
Deformed fossils:
  belemnite, **110,** 110
  crinoid, **11, 23**
  trilobite, **88, 112**
Dendrochronology, 32
Density:
  continental crust, 165
  oceanic crust, 165
  salt, 216–217
  uncompacted mud, 216–217
  unconsolidated sand, 216–217
Density inversion, 216–217
Depositional contact(s), 203–204
  interformational, 204
  intraformational, 204
Descriptive analysis, 14, 16–19, 36–61
Detachment faulting, 293–297
  Death Valley, California, 295
  Rincon Mountains, Arizona, 293, **294**
  Whipple Mountains, California, 293, **294**
Detailed structural analysis, 14, 16
Diachroneity, 31, 441
Diapir(s), 210–211, **211,** 216, **217**
  nonpiercement, **216,** 216
  piercement, **216,** 216
Diastrophism, 349–350
Differential compaction, **240,** 240–241, 243–244, 250, 252
Differential stress, 148, 162
Differential vertical uplift, 321, 399–400
Dike(s), 96–98, **97,** 170, **211,** 211, **212, 213, 215**
  aplite, **55**
  computing dilation of, 96–98, **97, 98, 99**
Dikelet(s), 211, **212**
Dike swarm(s), 211, **213, 215**
Dilation, **17, 22,** 22, 87–88, 107
  change in area, **127,** 127–128
  change in volume, 125, **127,** 127–128
Dilational separation, 330
Dip, 54
Dip isogons, **379,** 379–380, **380**
Direction of transport, 90–92, 233
Disconformity, **205,** 205, 207
Discontinuities, 21, **89,** 89, 401
Dislocation, 161
Dislocation creep, 160, 162
Dislocation glide, 160–162, **161**
Displacement vector, **90,** 90–92, **91,** 175
Distance of transport, 90–92
Distortion, **17, 22,** 22–23, 87–88, 107
Divergence, 166–172, 196–197
Domainal structure, **404,** 404, **405**
Dome(s), **58, 59,** 173
Donath apparatus, **135,** 135–136
Down-structure method, 384–385, **385**
Drachenfels, West Germany, 254–255, **256**
Drape folding, 399–400
Ductile, 150
Ductile shear zones, **228,** 228–230, **229, 230,** 242
Ductility, 155
Ductility contrast, **241,** 241, 390
Dynamic analysis, 14, 16, 26–28
  faults, 306, 314–318, 319–324
  folds, 357, 390–395
  joints, 351–353
Dynamic models, 26–27

Earth cracks, 240–241, **241**
Earthquakes:
  Alaskan, **221**
  deep focus, 166–167, 183
  first motion, 175
  Hebgen Lake, Montana, **3,** 3, **222**
  Motagua, Guatemala, 301–303, **302, 303**
  San Francisco, California, **93, 391,** 391
  seismic noise, **168,** 175
  shallow-focus, 167, 170, 174
Earth tides, 351
East African rift zone, 289
East Kaibab monocline, Arizona, 397
East Pacific Rise, **169,** 169, 176, **190,** 190, **194,** 194
Effective stress, 323, 352
Elastic limit, **150,** 150, 152
Elastic modulus, 148, 391–392
  Poisson's ratio, 148–149, **149,** 391–392
  Young's modulus, 148, **149,** 391–392
Emperor Seamounts, **191,** 191–192
Entablature, 249
Envelope of failure, 307, 311–312, 318, 351–352
Epeirogenic movement, 235
Eurasian plate, 187–188, **188**
Eutaxitic structure, 250–252, **251, 252, 424**
Exotic block(s), 182
Experimental deformation:
  beer-can experiment, **323,** 323–324
  buckling of multilayers, **394,** 394–395, **395**
  buckling of single layer, **392, 393,** 393–394, **394**
  centrifuge models, **217, 257,** 257
  circle in clay cake, 117–118, **118, 120, 121**
  computer-card slip, **26, 112, 114, 434, 435**
  conjugate faulting, **306,** 306
  description of materials, 39, **393**
  diapirs, 216–217, **217,** 257
  faulting in three-dimensional strain field, **319,** 319–320, **320**
  folding, **28,** 355, 393–395, 400
  growth faults, 244
  intrusion, 254
  kink folding, 394–395, **395**
  lines in clay cake, 117–118, **181,** 119–121, **120, 121, 122**
  monoclinal folding, 398, **398**
  normal faulting, **287,** 287–289, 312–313
  plumose markings, 332–333
  reverse faulting, **27**
  rotational strain, 128–130
  rotation of platy minerals, 412
  sand box experiment, **311, 312, 313,** 311–313
  shearing of cards, 25–26, **26,** 111–114, **434, 435**
  shear strain, 112–113
  strain ellipse, 113–114
  strike-slip faulting, **297,** 297–300, **299**
  tear faulting, 28
  thrust faulting, 312–313
  triaxial deformation, **27,** 27, **134,** 134–138, **135,** 147–153, **151,** 306–308, **319**
Extension, 107–110, 115–116, 137

Fabric, 404
Facing, 231
  cross bedding, 236
  graded bedding, 327
  mud cracks, 239
  pillow basalt, **348,** 348
  ripple marks, 236
  sole marks, 237
  vesicles, 248
Failure, 135
Failure angle, 151–153
Failure envelope, 307, **308,** 311–312, 318, 351–352
Failure point(s), 307
Farallon plate, 282, 304
Fatigue, 158
Fault-line scarp, **222,** 222
Fault rock(s), 424
  breccia, 224–225, **226**
  cataclastic rocks, 226
  fluxion structure, 227, 424
  gouge, 226, **226**
  microbreccia, 226–227, 293
  mylonite, 226–227, **227, 425**
  pseudotachylite, **227,** 227
  ultramylonite, **227,** 227
Faults, 6–7, **53,** 261–324, **261**
  Alpine fault, New Zealand, **222,** 301, 303
  antithetic, **243,** 272, 288–289, 299
  back-limb thrust, 313
  basal shearing plane, 281
  braiding, 300
  branching, 300
  chatter marks, **264,** 264
  compartmental, **285,** 285
  conjugate, **298,** 298–299, 305
  consistency of displacement, 284
  curvature, 300–301
  decollement, **281–283**
  denudational faults, 293
  detachment faults, 293
  dip slip, 269
  double bend(s), 300
  drag folds, 270, 270–272, **271**
  duplexes, 286
  en echelon, 291, **300,** 300, 304
  extensional, 170, 171, 187
  external rotation, 298
  fault-line scarp, 222
  fault mullion, 295
  fault scarp, 222
  fault zone, 221–223
  gaps, 276
  gash fractures, 272

graben, **287, 288,** 288, 389
grooves, **54,** 224, 269
growth faults, 242–244, **243, 244**
Hamblin Bay fault, Nevada, **94,** 94
horsts, **287,** 288
Hurricane fault, Utah, 291
imbricate, 182, 284
influence of preexisting structure, 278–279
interchange, **284,** 284
internal rotation, 298
left-handed, **92,** 92, **266,** 267
left-lateral, 274
lineaments, 222
line of bearing, 300
listric, **100,** 100, **271,** 271
low-angle normal slip, **266,** 267
map symbology, **274,** 274–276, **275**
microfaults, 224, 261, 330
Motagua, Guatemala, **301,** 301–303, **302, 303**
nappes, 281
Newport-Ingelwood, California, 304
normal, **273,** 274
normal-slip, **266,** 267, 276–277
oblique-slip, **92, 266,** 268
offset, **6, 93, 94, 221,** 221
omission of strata, 262, **263**
overlap, 277
Paunsaugunt fault, Utah, **233,** 291
penecontemporaneous, **239,** 242
physical properties, 223–227, **224,** 263–264
Pine Mountain thrust, Tennessee, 282
ramp, 282
reactivation, 321
releasing bend(s), **300,** 300
repetition of strata, 262, **263**
replacement, **284,** 284
restraining bend(s), **300,** 300
reverse, 274, **321,** 321–322
reverse drag, 243, **271,** 321–322
reverse-slip, **266,** 268, 277
right-handed, **266,** 267
right-lateral, 271, **273**
rotational, 100–101, **266,** 268, 298
San Andreas fault, California, **7, 93,** 176, 190, 199, 301, 303–305
scissors, 268
separation, 267, 272–274, **273**
Sevier fault, Utah, 291, **293**
slickensided surfaces, 224
slip, 91–96, 267
sole fault, 281
splay faults, 284
step, 282
stereographic projection, 314–318
Stockton Pass fault, Arizona, 270, **271**
strain significance, 276–278, 286, 306
stratigraphic throw, **273,** 273
striations, **76,** 224, **225, 264,** 269
strike-slip, **266,** 267, 278, **299**
synthetic, 300
tear fault(s), **285,** 285
tectonic transport, 282
throw, 273
thrust, 187, 274
thrust-slip, **266,** 267, 277
turtleback structures, 295
tilted step blocks, 289
transfer zones, **285,** 285
transform, **174,** 174–176, **176,** 190
transcurrent, 174
triangle facet, **223,** 223
truncation and offset, 262, **263**
wedges, **301,** 301
Wind River fault, Wyoming, **321,** 321–322, **322**
Fault scarp(s), **222,** 222
Fault zone, 223–224, 261
Field notebook, **37,** 36–39, **440**
Fiji plate, **188,** 188
Fishtail structure, **240,** 240
Fissility, 428–429
Fissure(s), 261, **262,** 330
Fissure vein, 330
Flattened pebble conglomerate, **9,** 422
Flattening, 298, 421, 424–425, 431
Flinn diagram, 431–436, **433, 435**
Float, 42
Flow, 10, 25
Flow banding, 422
Flow direction:
columnar jointing, **249,** 249–250, **250**
intraformational folds, 249
lava tubes, 247
ropy lava, 246
spiracles, 248
Flow foliation, 253–254, 424
Flow structure:
plutonic rocks, 253–257, **255, 256**
volcanic rocks, 246–249
Fluid pressure, 155, 322–324, 328, 351–352
Flute casts, 237, **238**
Fluxion structure, 227, 424
Fold(s), 7–9, **8, 9,** 354–400
amplitude, 375
anticline, 359–360, **360**
anticlinorium, 360–361, **361**
antiform, 360
antiformal syncline, **360,** 360
arc(s), 378–379
asymmetrical, **376,** 376
axial plane, **363,** 363, **364**
axial surface, **21,** 362–365, **363, 364, 367, 370**
axial trace, **21,** 362–363, **369,** 370–372
axis, 365–366, **365, 367**
basement-cored, 395, 398–400, **398, 399**
basin(s), **59,** 173
beta diagram, **367,** 367–368, 370–372
bissecting surface, 368–370, **369**
boudins, 389
box, **374,** 374
chevron, **9, 21,** 21, 373, **374**

Fold(s) (*Continued*)
circular, **374,** 374
circular arcs, 361
closure, 358
coding, 439
Comb Ridge monocline, Arizona, Utah, 397
compaction fold, **240,** 240
concentric, **377,** 377–378
conjugate, 374
cuspate, 373, **374**
cylindrical, **366,** 366, 370–371
dip isogons, **379,** 379–380, **380**
disharmonic, **242,** 242
dome(s), **58, 59**
dominant wavelength, 391–394
down-structure viewing, 384–385, **385**
drag folds, **270,** 270–272, **271, 356,** 356, **387, 388,** 388
drape fold, **398, 399,** 399–400
dynamic analysis, 357, 390–395
East Kaibab monocline, Arizona, 397
elliptical, 374, **374**
en echelon, 304
fault-related, 395
field description, 376–377, **377**
field measurement, 367, 376–377
Fleuty diagram, **365,** 365
flexural, 385–387
flexural flow, 385–390, **390**
flexural-slip, 384–389, **385, 386, 387**
folded layer(s), 377
folded surface(s), **361,** 361
fold height, 375, **376**
fold width, 375, **376**
forced fold(s), **399,** 399–400
form, 376
gentle, 374, **375**
geologic mapping, 372–373
geometric classifications, 365–366, 374–375
Hazera monocline, Israel, 348–349
hinge, 361–362
hingeline, **362,** 362–366, **363**
hinge point, **21,** 361, **362**
hinge zone, 361, **362**
inclined, 365
inclined sections, 382–384
inflection point, 361, **362, 376**
interlimb angle, 368, 374–375, **375**
intrafolial, **439,** 442–444, **443**
intraformational, **238,** 241, **242,** 249
isoclinal, 360, **361,** 375, **375**
jointing, **389,** 389
keystone graben, 389
kinematic analysis, 357, 384–390
kink bands, 394–395
kink folds, **9,** 394–395, **395**
layer shape, 378–379, **379, 380**
layer shape classification, 377–381, **379, 380**
limb(s), 360–361, **362**
map symbology, **372,** 372–373
mechanics, 355, 390–395
median surface, 375
median trace, 375, **376**
minor faults, 387, 389
minor folds, 356, 372, 387–389
minor structures, 356, 387–389, **388, 389**
monocline, 99, 395–400, **396, 397, 398**
near-cylindrical, **366,** 366, 371
neutral surface, 387
noncylindrical, **366,** 366, 371
normal profile, **361,** 361, 373, 381–382, **382**
oil accumulations, 357–358, **358**
open, 374, **375**
ore deposits, 358–359, **359**
orientation, 362–365
overturned, **360,** 360
palinspastic reconstruction, 419
parallel, 377–378
parasitic, **356,** 356, 372, **387,** 387–388, **388**
passive, 385, 419–421, **420**
passive-flow, **420,** 420
passive-slip, **420,** 420
penecontemporaneous, **238,** 241–242, **242**
pi diagram, 367–368, **368,** 370–372, **371**
pinch and swell, 389
plunging, 365
profile, **361,** 361, 373–374, 381–384, **382**
Ramsay fold classes, 378–381
reclined, 365
recumbent, **354,** 360, **361**
regional systems, 395–400
regional tectonic occurrence, 357
right-handed, 300
rollover anticline, 243, **243, 271**
rootless, 375
S-folds, 372–373
saddle reef, 358
shape, 373–374, **374, 376**
shape classification, **376,** 376
similar, **378,** 378–379
size, 354–355, 360–361, 375
snakehead, 282
stereographic projection, 366–372
strata-bound ore deposits, 359
structural traps, 357–358
subhorizontal, 365
superposed, 360, **361,** 439, **444,** 444– – 445, **445**
symmetrical, **376,** 376
symmetry, 376
synclinorium, 360–361, **361**
syncline, 359–360, **360**
synform, 360, **360**
synformal anticline, 359
teardrop, **374,** 374
thrust belts, 279–286, **355,** 398
tight, 374–375, **375**
tightness, 374–375, **375**

upright, **354,** 365
visual impact, **354,** 354–355
Waterpocket monocline, Utah, 397
wavelength, 375
Z-folds, 372
Fold axis, **365,** 365–366, **367**
stereographic determination, **367,** 367–368, **368**
Folding:
bedding-plane slip, 385–387
boudinage, 389
buckling, 385, 390–395, **391, 393, 394, 395,** 400
cohesion between layers, 394–395
development of minor structures, 387–389, **388, 389**
dominant wavelength, 391, 392–394
drag folding, **387,** 387–388, **388,** 394
drape folding, **398, 399,** 400
ductility contrast, 390, 420–421
experimental deformation, 393–395, 398
fault-related, 395–400
flexural, 385–390
flexural-flow, 390
flexural-slip, **385,** 385–389, **386**
forced, **399,** 399–400
free, 399–400
instability, 391
joint development, **389,** 389
kink folding, 394–395, **395,** 400
layer-parallel shortening, **387,** 387, 389
layer-parallel slip, 385–389
layer-parallel strain, **388,** 393
layer-parallel stretching, **387,** 387, 389
mean ductility, 421
mechanics, 390–395
multilayer sequence, 391, 394–395
parasitic, 387–388
passive, 385, 419–421, **420**
passive-flow, **420,** 420
passive-slip, **420,** 420
pressure solution, 410, 415–416, **421,** 421
role of layering, 384–387
shear strain, 386
single layer, 391–394
superposed, 360, 439, **444,** 444–445, **445**
transposition, 441–444
Foliation, 9–10, 422–425, 438–439
coding, 438–439, **440**
compositional bands, 422
crenulation cleavage, **404, 405,** 406–407, **406, 409,** 415–416, 422
C-surfaces, 424–425, **425**
eutaxitic structure, 422, **424**
field analysis, 438–444, **440**
flattened pebble conglomerate, **9,** 422
flow banding, 422
flow foliation, **255,** 424
fluxion structure, 424
gneissic structure, 422, **423**
map symbols, **51**
metamorphic, 422
phyllitic structure, 405, 422
schistosity, 405, **408,** 422
sigmoidal, 229
slaty cleavage, **10,** 405–406, **406,** 407–408, **408,** 410–415, 422
spaced cleavage, 331, 406–407, **407, 409,** 409–410, **418,** 422
S-surfaces, 424–425, **425,** 438–439
strain significance, 431
superposed, 438–439
Footwall, **266,** 267
Force(s), 132, 136–137
balancing of, 141–142
body force, 136
load, 136
normal force, 311
relation to stress, 132
shear force, 311
units, 136
Forced folding, 399–400
Forearc basin(s), **185, 186,** 186, 357
Foreland, 187, 357
Foreland basin(s), **187,** 187
Fracture(s), 258–259
cross, **258,** 258, **389, 390**
gash, **24, 272,** 272, 298, 330, **388,** 388
longitudinal, **258,** 258, **389**
tension, 298, 300
Fracture cleavage, **331,** 331, 406–407
Fracture coupling(s), **338,** 338
Fracture surface, 223
Free folding, 399–400
Frictional sliding, **151,** 160, 174
Fundamental strength, 158

Gash fracture(s), **24, 272,** 272, 330, **388**
Geological column, **34,** 33–34
Geological cross section, 18–19, **19, 34**
line of section, 384
steps in construction, 381–384
Geological map, 47–51
borders, 50
colors, **48**
components, **47,** 47–51
contacts, **50,** 50, **203**
elements, **47**
explanation, 47–48
fold symbology, **372,** 372–373, **373, 404**
magnetic declination, 52
north arrow(s), 50
scale(s), 50
symbols, 47–50, **48, 50, 51, 203,** 203, 274–275, **275,** 341–342, **342, 372, 404**
title, 50
Geologic mapping, 39–57
aerial photo, 45–47, **46**
base maps, 41–47, **42, 43**
decisions, **40, 41**
equipment, 51
grid-line, 42–43, **43**
measuring orientations, 52–57

Geologic mapping (*Continued*)
- pace-and-compass, **42,** 42–43
- philosophy, 39–41
- picket-line, 42–43
- plane-table, **45,** 45
- stations, 51
- tape-and-compass, 42
- undergound, **44,** 44–45, **45**
- units, 51

Geologic time, 12, 31–35
- magnitude, 31–32
- measurement, 32–33

Geologic time scale, 34, **35**
Geometric analysis, 61
Geometric order, 16, 21
Geothermal gradient(s), 185
Glass shards, 186, 252
Gneiss dome(s), 438
Gneissic structure, 422, **423**
Gouge, **226,** 226
Graben, **287, 288,** 288, 389
Graded bedding, **237,** 237
Grain-size reduction, 160
Grand Canyon, Arizona, 205–206, **206,** 397
Grand Saline salt dome, Texas, 217–219, **218**
Granite tectonics, 253
Gravitational tectonics, 180, 182
Greenstone belt(s), **437,** 437
Griffith cracks, 318
Griffith failure envelope, 351–352
Groove(s), **11, 54,** 224, 269
Groove marks, 237, **238**
Growth faults, 242–244, **243, 244**
Gulf Coast basin, 216, 220, 243, **243, 244**
Gulf of California, 189
Gulf of Mexico, 220

Hamblin Bay Fault, Nevada, **94,** 94
Hamblin-Cleopatra volcano, Nevada, **94,** 94
Hanging wall, **266,** 267
Hawaiian-Emperor chain, **191,** 191–192
Hawaiian Islands, **191,** 191–192
Hazera monocline, Israel, **341,** 348–349
Hebgen Lake earthquake, Montana, **3,** 3
Henry basin, Utah, 397
Henry Mountains, Utah, 216
Hingeline, **362,** 362–367, **363**
- form, **362,** 362–363
- orientation, 362–363

Hinge point, **21,** 361, **362**
Homocline, 59, 240
Homogeneous deformation, 107, 113, 121
Hoodoo(s), 326–327
Hooke's Law, 148
Hot spot(s), **191,** 191
Hunter's Point monocline, Arizona, **99,** 99
Hurricane fault, Utah, 291, **293**
Hydroplastic behavior, 238
Hydrostatic stress, 136, **145,** 145–146
Hydrothermal circulation, 327
Hydrothermal deposition, 327
Hydrothermal fluids, 327
Hysteresis, **149,** 149–150

Idaho-Wyoming thrust belt, 279, 284–285
Igneous intrusion, 210
Ignimbrite, 186, **205,** 250–252, **327**
Inclination, 52, **53**
Inclusion(s):
- igneous, 211, **213,** 256
- tectonic, **182,** 182
- xenoliths, 211

Incompetency, 154
Indian-Australian plate, 187–188, **188,** 198–199, **199**
Interarc basin(s), **186,** 186, 357
Internal friction, 309
Intersection lineation, 425
Intraformational fold(s), 241
Intrusion(s), 210
- apophyses, **211,** 211
- batholith, **215,** 215
- cactolith, 216
- characteristics, 211
- clastic dikes, 211, **212**
- clastic sills, 211
- concordant, 215
- diapirs, **211,** 211
- dike(s), **211, 212,** 212, **213, 215**
- dikelets, 211, **212**
- dike swarms, 211, **213, 215**
- discordant, 215
- igneous, 210, **211**
- laccolith, **215,** 216
- lopolith, **215,** 216
- phacolith, **215,** 216
- plug, **215,** 215
- pluton, 215
- radial dikes, 211, **215**
- salt diapir, 211, **216, 217, 218, 219**
- sedimentary, **211,** 211, **216, 219, 220**
- shapes, 215–220
- sills, 211, **212**
- soft-sediment, 211
- stock, 215, **215**

Island arc(s), 176–177, **177,** 185
Isopach map, **60,** 60, **61**

Joint(s), **7,** 7, **325,** 325–353
- Appalachian Plateau, 339–340, 342, 351
- Arches National Monument, Utah, **326,** 326–327, **330, 341**
- border planes, 333
- Canyonlands, Utah, **7, 327**
- Carlsbad Caverns, New Mexico, 327
- Chiricahua National Monument, Arizona, 326–327, **327**
- Colorado Plateau, 340, 342, 351
- columnar, **249,** 249–250, **250**
- conjugate, 389
- conjugate shear joints, 389
- cross, **258, 389,** 389, **390**
- cross planes, 333

density, 343–346
diastrophic, 349–350
dynamic analysis, **351,** 351–352
exfoliation, **6**
fluid pressure, 328, **351,** 352–353
fringes, 333
grain, 352
groundwater circulation, 327–328
hydrothermal circulation, 327–328
interpretation difficulties, 325–326
joint face, 332–333, **333**
Laramide plutons, Arizona, 340, 345–346, 352
longitudinal, **258, 389,** 389
main joint face, 333
mapping, 340–342, **341**
map symbology, 341–342, **342**
methods of study, 339–347
microjoints, 331–352
nondiastrophic, 351
nonsystematic, 331
oblique, **389,** 389
open-space filling, 327, 334
origin, 349–353
photogeologic expression, **326,** 340
photogeologic mapping, 340
physical characteristics, 331–333
plumose markings, **332,** 332–333, **333**
practical value, 326–328
pressure solution, 337–338
primary, 349, 351
regional tectonic associations, 338–339
relation to folds, 389
release, 389
rift, 352
scenic value, 326–327
sets, **326,** 339
shape, 331–332
shattering, 331
shear, 350
simple-shear, 350
spacing, 331
strain analysis, 348–349
strain field diagram, **349,** 349
strain significance, 325, 347–349, **348, 349**
stress release, 352–353
stylolitic, 337–338
systematic, **330,** 330
systems, 339
tension, 350
tidal origin, 351
trajectories, 347, 351
Joint analysis:
circle-inventory method, 342–343, **343, 344**
density, 343–344
inventory areas, 342
inventory method, 342–345
joint-orientation diagrams, 345–347, **346,** 347
notebook data, 344–345
pole-density diagrams, 345–347, **347**
pole diagrams, 345–347
rose diagrams, 345, **346, 347**
sample lines, 342
sampling, 342–343
selection method, 342
stations, 340, 342
statistical probability, 346
strike histograms, 345, **346**
structural domains, 339–340
trajectories, **347,** 347
Joint face, 331–333, **332, 333**
Joint-related structures:
map symbology, 341–342, **342**
origin, 349–353
regional tectonic associations, 338–339
strain analysis, 348–349
striated surfaces, 336–337
stylolitic surfaces, **337,** 337–338, **338**
veins, **8, 97, 330, 333,** 333–336, **334, 335, 336**
Juan de Fuca plate, 187, **188,** 198–200, **199**
Juan de Fuca Ridge, **169,** 176, 190, 303
Jura Mountains, 281

Keystone graben, 389
Kinematic analysis, 14, 16, **20,** 22–26, 87–131
faults, 283–284, 290, 295–297, 299–301
folds, 357, 385–390
Kink folds, **9,** 394–395, **395,** 400
Klippen, **280,** 280

Laccolith, **215,** 216
Lahar(s), 186
Lake district, England, 411
Lake Mead region, 94, **100**
LANDSAT image, **46, 163**
Laramide orogeny, 200
Lava tubes, 247
Line(s):
changes in angle, 111–113, 116–121, **118, 120, 121**
changes in length, 107–110, 116–121, **120, 121**
geometric, 94
Lineaments, 222
Linear elements, 52–53
Linear structure, 425–429
boudins, 429
grooves, **11, 54,** 224, 269
mullion, **427,** 427–428, **428**
pencil structure, **428,** 428–429
pinch and swell, 429
quartz rods, 426–427
rodding, 426–427
stretched-pebble, 426, **427**
Linear velocity, 197
Lineation, 9–10, **11,** 425–426, 439
coding, 439
crenulation, 425–426, **426**
crystal alignment, 253
intersection, 425
map symbols, **51**

Lineation (*Continued*)
mineral, 253, **426,** 426, **427**
parting, 236–237
strain significance, 431
striations, **11, 76,** 224, 269, 426
superposed, 439
Line of bearing, 335
Listric faulting, **100,** 100, **101**
Lithology, rock strength, 154
Lithosphere, 89, 164, 165–166, **166, 167,** 170, 171–172, 174, 176, 180, 183, 187
Lithostatic stress, 136
Load, 136
Load-displacement curve, **148,** 148, **150**
Lode(s), 358
Logarithmic Flinn diagram, **433,** 433–435, **435**
Logarithmic strain, 433–435
Longitudinal fractures, 258–259
Lopolith, **215,** 216
Los Angeles basin, California, 304
Louann salt, 220, 243
Low-velocity zone, 166
LS-tectonites, 431–433
L-tectonites, 431

Magma, 210
viscosity, 244–246
Magmatic arc, **185,** 185–186, 357, 437
Magnetic declination, 52
Magnetic polarity, 33, **171**
Magnetic reversal(s), 33, 171, 198
Magnetic stripes, 171, 198
Mantle, 165–166, **166,** 176
Mantled gneiss dome(s), 438
Maps:
fracture pattern, 340
geologic, 47–51
isopach, 60
structure contour, **58,** 58–59, **59**
topographic, 45
Marginal fissures, 258–259
Marginal thrusts, 258–259
Marianas Trench, 183
Marker units, **48, 49,** 49, 51
Martinsburg slate, 414
Mass wasting, **328,** 328–329
M-domains, 407–409
Mechanical twinning, 160–162, **162**
Melange, 181–182, **182**
Melange wedge, **181,** 181
Mendocino fracture zone, **199,** 199
Mendocino triple junction, **199**
Metamorphic core complexes, 436
Metamorphism, 183
blueschist, 183
contact, 185, 213
high-pressure, 183, 185
low-pressure, 183, 185
low-temperature, 183
regional, 185, 213
Mexico Trench, 183
Michigan basin, **58, 59**
Microbreccia, 226–227
Microcracking, 160
Microfault(s), 224, 261, 330
Microjoint(s), 331, 352
Microlithon(s), **404,** 404–405, **405**
Microlithon domains, 404–405
Microstructure(s), 159–162, **230,** 230
Mid-Atlantic Ridge, **169,** 169, 190, **192,** 192–193, **193,** 193
Midway Islands, 191–192
Mineral lineation, 253–254, **426,** 426, **427**
Mining geology, 358–359
Miogeoclinal prism, 173, 207
Moho, 165
Mohorovicic discontinuity, 165, **166**
Mohr circle strain diagram, 121–124, **122**
Mohr circle stress diagram, 146–147, **147, 153,** 153, 306–308, **308,** 313–314
Mohr-Coulomb law of failure, 308–311, **310,** 322–324
Mohr envelope of failure, 306–307
Moine thrust, Scotland, 437
Monocline(s), 99, 395–400, **396, 397,** 398
Monument uplift, Arizona, Utah, 397
Monument Valley, Arizona, Utah, 327–328
Motagua fault, Guatemala, **301,** 301–303, **303**
Mountain(s), 163–164
Mountain building, 164
Mountain system(s), 163, **178, 179**
Alpine, 177, 281
Andean, 177
Apennine, 177
Carpathian, 177
continental margin, 177, 186
Himalayan, 177, 437
island arc, 176–177, **177,** 185
ocean ridge, 168–169
Mount Cook, New Zealand, **164**
Mount Mazama, 186
Mount St. Helens, Washington, **13,** 116–117, **117**
Mudcracks, **239,** 239
Mullion, **427,** 427–428, **428**
Mylonite, 226–227, **227, 425**

Nappes, 281, 437
Natural strain, 433–435
Navajo Mountain, Arizona, 216
Nazca plate, 187, **188,** 190, **194,** 194, 198
Newport-Inglewood fault, California, 304
Nonconformity, **205,** 205, **206, 208**
Nonrigid body deformation, 87–88, **88,** 107, **161,** 161
Normal faulting, 287–297
Basin and Range, 291–293, **292, 293**
Colorado Plateau, **293,** 293
Death Valley, **295,** 295
detachment faulting, 293–294, **294, 295**

kinematic analysis, 290, 295–297
Lake Mead region, 293
low-angle, 293–295, **294, 295**
Red Sea region, **289,** 289–291, **291**
San Manuel ore body, 296–297
tectonic setting, 289
Yerington district, 295, **296**
Normal force, 311
Normal stress, 138–139, 142, 160
compressive, 139
tensile, 139
North American plate, 187, **188,** 190, **199,** 199–200, 301, 304

Oil trap(s), 304, **358**
Olistostrome(s), 182
Open-space filling, 226, 327
Ophiolite, **170,** 170–171, 180, 183
Orbicules, 231, **232**
Ore deposits, 358–359, **359**
Orogen(s), 163–164, 171, 173, 176–177
Orogenic belt(s), 163–164
Orogenic movement, 235
Orthographic projection, 61–68
apparent dip, 64–66, **65, 66**
bedding offset by pressure solution, **417,** 417–418
bed thickness, 63
dilation along dike, 96–98
fault slip, **95,** 268–269, **269**
intersection of two planes, 66–68, **67**
reference plane(s), 63
slip on faults, 91–94
structural contour lines, 63–64, **64**
structural intercept(s), 63
structure profile, **62,** 61–63
vertical projection(s), 63
Oscillation ripple marks, 236
Outcrop breadth, 63
Overgrowth(s), 408, 413
Overlap, 207
Overthrusting, 322–323

Pacific-Antarctic Ridge, **169**
Pacific plate, 187, **188,** 190–192, **194,** 194, 198–200, **199,** 301, 304
Pahoehoe basalt, 246
Paleomagnetic dating, 33
Paleosols, 208
Palinspastic reconstruction, **280, 281,** 419
Pangaea, 188, 209
Partial melting, 185
Parting lineation, **236,** 236–237
Passive margin(s), **172,** 172–173, 209
Paunsaugunt fault, Utah, 291, **293**
Penecontemporaneous fold(s), **238,** 241
Pebble imbrication, **237,** 237
Pencil structure, **428,** 428–429
Penetrative deformation, 23–25, **24, 25,** 422
Peru-Chile Trench, **194,** 194
Phacolith, **215,** 216
Philippine plate, **188,** 188, 286
Phyllitic structure, 405–406, 422
microscopic character, 408
outcrop character, **395,** 405–406
Pi diagram, **368,** 368, 370–372, **371**
Pillow basalt, 171
Pillow lava, **248,** 248
Pinch-and-swell structure, 241, **241,** 389, 429
Pine Mountain thrust, Tennessee, **282,** 282
Pipe cylinders, 249
Pipe vesicles, 249
Piping, 241
Pitch, 76, **77**
Plane strain, 124, 128, 276, 434–435
Plastic flow, 152
Plastic materials, 156
Plate(s), 164–166, 187–188, **188**
African, 187, **188,** 190, **192,** 192–193, **193,** 198–199, **199**
Antarctic, 187, **188,** 198–199
Arabian, 187–188, **188**
Bismarck, **188,** 188
Caribbean, **188,** 301
Cocos, 187, **188**
Eurasian, 187, **188**
Farallon, 200, 304, **305**
Fiji, **188,** 188
Indian-Australian, 187, **188,** 198–199, **199**
Juan de Fuca, 187, **188,** 198–200, **199**
Nazca, 187, **188,** 190, 194, 198
North American, 187, **188,** 190, **199,** 199–200, 301, 304
Pacific, 187, **188,** 190–191, **194,** 198–200, **199,** 301, 304
Philippine, **188,** 188
Scotia, 187, **188**
Solomons, **188,** 188
South American, 187, **188,** 188, 190, **192,** 192–194, **193, 194,** 198
Plate boundary, 164–165, 167, **168,** 187
convergent, 176–187
divergent, 168–174
migration, **192,** 192–194
strike-slip, 174–176
Plate collision, 167
Plate kinematics, 187–200, **193, 194, 195, 197, 199**
Plate margin(s), 164, 166, 172–173
active, 180–181
convergent, 357, 437
passive, 172–173, 209
Plate motions, 166–167, 192–200, **193, 194, 195**
convergence, 166, 200
divergence, 167
strike-slip, 167
Plate tectonics, 14, 163–200, **290,** 446
Platform sediments, 173, 207
Plug, **215,** 215
Plume, 191
Plumose markings, **332,** 332–333, **333**
Plunge, 52–53, **77**
Pluton, 191
Poise(s), 156, 392

Poisson's ratio, 148–149, **149,** 392
Polarity, 180–181
Polarity of form, 231
Pole-density diagram, 85–86, 345–347
Pole diagram, 345–347
Pole of plate rotation, **196,** 196–198
Preferred orientation, **17, 22,** 81
  joints, 325
  unimodal distribution, 84
Pressure ridge(s), **247,** 247
Pressure shadow(s), 412, **413** ·
Pressure solution, **127,** 160, 162, 331, 337–338, 410
  crenulation cleavage, 415–416
  dimples in pebbles, 413
  insoluble chert layers, 416–417
  offset of bedding, 417–418, **421**
  slaty cleavage, 412–414
  spaced cleavage, 416–418
Pressurized shale masses, 243–244
Primary plutonic structure, 253–259
  cross fracture(s), **258,** 258
  crystal alignment, 253, **255**
  flow foliation, 253, **255, 256**
  fractures, **258,** 258–259
  longitudinal fractures, **258,** 258–259
  marginal fissures, 259
  marginal thrusts, 259
  mineral lineation, 253, **255**
  schlieren, 256, **257**
  stretching surface(s), **258,** 258–259
  xenolith alignment, 256
Primary sedimentary structure, 234–244
  bedding, **39,234**, 325–326
  cross-bedding, 235–236, **232, 235, 236**
  flute casts, 237, **238**
  graded bedding, **237,** 237
  groove marks, 237, **238**
  joints, 349, 351
  parting lineation, **236,** 236–237
  pebble imbrication, **237,** 237
  ripple marks, **236,** 236
  sole marks, 237, **238**
Primary strain, 231
Primary structure, 14, 87, 203, 231
Primary volcanic structure, 244–249
  ash-flow ellipsoids, **23**
  columnar jointing, **249,** 249–250, **250,** 349
  eutaxitic structure, 250–252, **251, 252**
  flow structure, 246–249
  glass shards, 252
  intraformational folds, **249,** 249
  joints, 349
  lava tubes, 247
  pillows, **248,** 248
  pressure ridge(s), **247,** 247
  ropy lava, **247,** 247
  spiracles, 248
  squeeze-ups, 247
  vesicles, 248
  welded tuff, 250
Principal strain directions, 115–116, **124,** 124
  relation to:
    cleavage, 388–389, 410–411, 415–416
    crenulation cleavage, 415–416
    crystal fibers, 334–335
    faults, **298,** 305, **307,** 319–320, **320**
    fault striae, 305
    folds, **272,** 388–389
    foliation, 431–433
    gash fractures, **272,** 305, 388
    gash veins, 335–336, 388
    joints, 350
    lineation, 431
    microfaults, 348–349
    slaty cleavage, 410–411
Principal stress directions, 143–144, 162
  relation to:
    faults, **307,** 307, **314,** 314, **315, 316, 317, 320,** 320
    joints, **350,** 350
    striations, 314–315, **316, 317**, 318
    stylolites, 337–338
    stylolitic surfaces, 416
    tension factures, 306, **350**
Progradation, 173
Progressive deformation, 130, **336,** 388, 434–436, 441
Pseudostratigraphy, **442,** 442
Pseudotachylite, **227,** 227
Pumice fragments, 186
Pure shear, **129,** 129–130

QF-domains, 407–408
Quadratic elongation, 109–110, 118–119, **127**
Quartz rods, 426–427

Racetrack Playa, California, **90**
Radial dikes, 211, **215**
Radiolarian chert(s), 182
Radiometric age dating, 32–33, **34**
  half-life, 32, **34**
  potassium-argon, 32
  problems in interpretation, 32–33
  radiocarbon, 32
  rubidium-strontium, 32
  uranium-lead, 32
Rake, 76, **77**
Recovery, 162
Recrystallization, 162, 226, 411–412
Red Sea, 170, 188–189
Red Sea rift zone, **169,** 170, 187–188, **289,** 289–291, **290, 291,** 436
Reduction spots, **232,** 411
Regional assemblages, 163
Relative motion(s), 190, 190–200
Replacement, 327, 334
Reservoir rock, 357–358
Residual stress, 352
Reverse drag, 243, **243, 271**

Rheid, 158–159, **159**
Ribbon chert(s), 182
Ridge(s), **169,** 169–172, 174–176
  America-Antarctica Ridge, **169,** 169
  Carlsberg Ridge, **169,** 169, 190, 198, 289
  Chile Ridge, **169**
  East Pacific Rise, **169,** 169, 176, 190, 194, 303
  Galapagos Ridge, **169,** 169
  Juan de Fuca Ridge, **169,** 176, 190, 303
  Mid-Atlantic Ridge, **169,** 169, 189–190, **192,** 192–193, **193**
  migration, 192–194
  Southeast Indian Ocean Ridge, **169,** 169, 198–199, **199**
  Southwest Indian Ocean Ridge, **169,** 169, 198–199, **199**
Rift(s), 170, 209
Rift-basin deposit(s), **170,** 170, **172**
Rifting, **172,** 172–173, 188, 209
Rift valley, 170
Rigid body deformation, 87, **88,** 161
Rincon Mountains, Arizona, **49,** 293, **294**
Ring of Fire, 185–186
Ripple marks, 102–106, **236,** 236
Rock descriptions, 37–38, **38**
Rock strength, influences:
  confining pressure, 151–153, **153, 155**
  fluid pressure, 155
  lithology, **154,** 154
  strain rate, 156–158, **157**
  temperature, **155,** 155–156
  time, 158–159
Rocky Mountains, 200
Rodding, 426–427
Rollover anticline, **243,** 243
Roof pendant(s), **214,** 214
Ropy lava, 246–247, **247**
Rose diagram, **255,** 345, 347
Rotation, **17, 22,** 22, 98–99, **99, 101**
  axis of rotation, 98–99
  magnitude of rotation, 98–99
  sense of rotation, 98–99
Round Mountain, Wyoming, **50**
Rupture, 152
Rupture strength, 152

Saddle reef deposit, 358
Salmon River gold district, 358
Salt diapir(s), 211, **216, 217, 218, 219**
Salt diapirism, 220
Salt dome(s), 159, 209, 211, **217, 218,***, **355**
  dynamics, 220
  Grand Saline, Texas, 217–219, **218**
  Gulf Coast, **217,** 217, **243**
  kinematics, 217–219
  shape, 217–220, **219**
Salt glaciers, 159
San Andreas fault, California, **7, 93,** 176, 190, **199,** 199, 301, 303–304, **304, 305**
Saqsaywaman, Peru, **11,** 203, **204**
Scale of observation, **18,** 18, 23–26, **25,** 31
Schistosity, 405–406, 422
  microscopic character, **408,** 408–409
  outcrop character, 405–406
Schlieren, 256, **257**
Scotia plate, **188,** 188
Screen(s), 214
Seafloor spreading, **171,** 171–172, 174–176, 188–189, 192–194
  rate(s), 172, 192–194
  stereographic portrayal, **197,** 197–198
Secondary strain, 231
Secondary structure, 14, 87, 203
Sedimentary basin(s), 186–187
  backarc, 186–187
  forearc, 186
  foreland, **187,** 187
  interarc, 186
Seismic discontinuity, 166
Seismic-reflection profiling, 265–267
Sense of transport, 90–91, 233
Separation on faults, 267, 272–275, **273, 274**
Septa, **214,** 214
Sequences, 235
Sets, 21–22
Sevier fault, Utah, 291
Shale diapir(s), 216–217, 219–220
  Laward diapir, **219**
  shape, 219
Shear, 111–112
Shear force, 311, **312**
Shear separation, 330
Shear strain, 112–113, 118–119
Shear stress, 138–139, 142, 160
  left-handed, 139
  right-handed, 139
Shear zone(s), **228,** 228–230, **229, 230,** *, 261, **272**
  brittle-ductile, 228
  contacts, 228–230
  properties, 228
  tectonite formation, 436–437
Shield province(s), 173, **174,** 437–438
Shortening, 137, 187
Sierra batholith, 216, 348
Sierra Nevada, California, **4, 6**
Sigsbee escarpment, 243
Sill(s), 211, **212**
Similar fold(s), **378,** 378
Simple shear, **129,** 129–130, 182, **434,** 434–437
Slaty cleavage, **10,** 405–408, 410–415, 422
  deformed graptolites, **414,** 414
  flattened fossils, 410–411
  M-domains, 407–408, **408**
  microscopic character, 407–408, **408,** 413–414
  outcrop character, 405–406, **406**

Slaty cleavage (*Continued*)
preferred orientation of micas, 411–412, 414
pressure shadows, 412, **413**
QF-domains, 407–408, **408**
relation to folding, 410–411
role of pressure solution, 412–415, **414**
role of recrystallization, 411–413
role of rotation of platy minerals, 412
strain significance, 410–411, **411**
volume loss, **414,** 414–415
Slickensides, 224
Sliding friction, **151,** 160, 174
Slip, 25, 91–92, **92,** 190
Slip on faults, 91–94, **92,** 267–272
dip slip, **92,** 92
net slip, **92,** 92
orthographic projection, 94–96, 268–269
strike-slip, **92,** 92, 113
Slip system, 160
Slope distance, **42**
Slumping, 182, 242, **328**
Soft-sediment deformation, 238–242
boudins, **241,** 241
compaction fold, **239,** 240
differential compaction, **240,** 240
ductile shear zones, 242
fishtail structure, **240,** 240
intraformational folds, **238,** 241
mudcracks, **239,** 239
penecontemporaneous faults, 242
penecontemporaneous folds, 241–242
pinch-and-swell, 241
Sole marks, 237, **238**
Solomons plate, **188,** 188
Source bed, 357
South American plate, 187, **188,** 189, **192,** 192–194, **193, 194,** 198
Southeast Indian Ocean Ridge, **169,** 169, 198–199, **199**
Southern California batholith, 216
South Mountain fold, Maryland, 124–125, **125**
Southwest Indian Ocean Ridge, **169,** 169, 198–199, **199**
Spaced cleavage, 331, 406–407, **407,** 409–410, 422
Appalachian Plateau, 419
classification, 418, **418**
insoluble chert layers, **407, 416,** 416–417
insoluble residue, 416–417, **417**
microscopic character, **409,** 409–410
offset markers, **407,** 407, **417,** 417–418
outcrop character, 406–407, **418**
palinspastic reconstruction, 419, **419**
relation to folding, 416, **417**
strain significance, 416–419
stylolitic surfaces, 416
truncated fossils, **416,** 416
Spanish Peak, Colorado, **211**
Spiracles, 248
Splitting properties, **39**
Spreading center, 173–174
Spreading rate(s), 192–195, 198–199
angular velocity, 197
average linear velocity, 197
half-spreading rate, 192, 194
total spreading rate, 192
Squeeze-ups, 247
S-surfaces, 424–425, 438–439
State of strain, 430
Station(s), 340, 342
S-tectonites, 430–431
Stereographic projection, 68–86
angle between two lines, 78, **79**
angle between two planes, **78–79**
axial surface, 368–370
Beta diagram, 367–368, 371–372
Biemesderfer counter, 81, **82**
bissecting surface, 368–370
center counter, 83
contouring, 83–86
cylindrical fold, 366
density distribution, 82
equatorial plane, 69
faults and principal stresses, 314–315, **315, 316, 317,** 318
fault slip, 268
fold axis, 367
folding, 102
geometry of projection, **67,** 68–70, **69**
great circle, 69–70, 102
intersection of two planes, 79–81, **80**
line in plane, **76,** 76–78
near-cylindrical fold, 366
noncylindrical fold, 366
north index, 72
peripheral counter, 84
pi axis, pi axis, 369
pi circle, 368
pi diagram, 368, 370–372
pitch, 76, **77**
plotting points, 81–82
pole-density diagram, **83,** 83–86, **84, 85, 86**
pole-density distribution, 83–86
preferred orientation, 81, 257
rake, 76, **77,** 78, **78**
rotation, 101–106
rotational faulting, **101,** 101
rotation of primary structure, 101–106, **102, 104, 106**
Schmidt net, **71,** 72, 81
small circle, 70–72
statistical tool, 81–86
stereogram, 69
stereonet, 70–72
strike and dip of plane, 74–76, **75, 76**
trend and plunge of line, 72–74, **73, 74, 77, 78**
Wulff net, **71,** 72, 81
zenith, 69
Stock, **215,** 215
Stockton Pass fault, Arizona, 270
Strain, 107–131
angular shear, 111–112, **112**

change in angle between lines, **111,** 111-113, 116–118, 119–121
change in area, 127–128
change in line length, 107–110, **108, 109,** 116–118, 119–121
compaction strain, 240
compensation, 128
constant volume, 609
constriction, 431–433
contraction, 128
dilation, **17, 22,** 107
distortion, **17, 22,** 107
expansion, 128
extension, 107–108, **109,** 110–111, 115, 137–138
external rotation, 129
finite, **130,** 130
flattening, 431–433
heterogeneous deformation, 107
homogeneous deformation, 107, 113, 121
incremental, 130, 335, 434–436
infinitesimal, **130,** 130, 334
internal rotation, 118, 129
lateral strain, 148
layer-parallel, 393
logarithmic strain, 433
longitudinal strain, 148
maximum extension, 115–116
minimum extension, 115–116
Mohr strain diagram, 121–124
natural strain, 433
nonrecoverable, 150
nonrigid body deformation, 107
nonrotational, 128–130, **129**
percentage, 148
permanent, 150–151
plane strain, 124, 128, 276, 431, 433–435
principal axes, 115–116, 121, 129
progressive deformation, 130, **131,** 434–436, 441
pure shear, **129,** 129–130
quadratic elongation, 107–110, 118–119
recoverable, 149
regional, 177
rotational, 128–130, **129**
shear strain, 112–113, 118–119
simple shear, **129,** 129–130, 388–389, 424–425, 434–436, 437
strain ellipse, **115,** 115–116
strain ellipsoid, 124
strain equations, 118–119
strain field diagram, **128,** 128, **278**
stretch, 107–110
stretching, 128
superposed, 130
three-dimensional, 319–320, 431–433
true strain, 433
volume change, 125, 127–128
Strain analysis, 22–23
ash flow tuff, 250–252
belemnite, **110,** 110
change in area, 127–128
circle with lines, **115, 116,** 116–118, **118,** 119–121, **120, 121**
differential compaction, 240
graptolites, 414–415
joints, 348–349
limitations, 126–127
normal faulting, **276,** 276–277, 290–291, 296
ooids, 124–126, **125**
pencil structure, 429
reduction spots, 411
reverse-slip faulting, 277
strike-slip faulting, 278
thrust-slip faulting, **277,** 277, 286
trilobite, 111, **112**
volcanic lapilli, 411
volume changes, 125–128
Strain compatibility, 348, 377
Strain ellipse, 113–118
lines of maximum shear strain, 121
lines of zero angular shear, 116
lines of zero length change, 121
principal axes, 115–116, 129
Strain ellipsoid, **124,** 124, **432**
oblate, 431, **432**
principal axes, 124
prolate, 431, **432**
triaxial, 431, **432**
Strain energy, 252
Strain equations, 118–119
Strain field diagram, 128, **278**
faulting, 278
jointing, 349
Strain hardening, **152,** 152
Strain rate, 156–158
Strata-bound ore deposits, 359
Stratigraphic throw, 273
Strehlen body, West Germany, 258
Strength:
angle of internal friction, 310–311
coefficient of internal friction, 309–311
coefficient of sliding friction, 309, 319
cohesive, 309
competency, 154
fundamental, 158
internal friction, 309
rupture, **150, 152,** 152
sliding friction, 309
tensile, 351–352
ultimate, **152,** 152, 155
yield, **150,** 150, 151–152, **152**
Stress, 132–133, 136–137
axial, 136–138
bars, 136
compressive, 138
computing stresses on plane, 139–143, **138, 140, 143, 145**
confining pressure, 136
critical shear, 309–311
differential, 148, 156, 162
effective, 155, 323, 352
hydrostatic, 136, 145, **145**
lithostatic, 136
Mohr stress diagram, 146–147, **147,** 153

Stress (*Continued*)
normal, 138–139, **139,** 160
principal stress directions, 143, 162
relation to area, 132, **133**
relation to force, 132, **133,** 136–137
residual, 352
shear, 138–139, **139,** 160
simple calculations, 132–133, **133,** 136–137
stress ellipse, **143,** 143–144, **144**
stress ellipsoid, 143–144, **144**
stress equations, 145–146
stress vector, **138,** 138–144, **142, 145**
tensile, 351
tensional, 138, 351–352
units, 132, 136
vector analysis, 138–143
Stress analysis, 133, 138–143
computing normal stress, 138–143, **139**
computing shear stress, 138–143, **139**
computing stress-vectors, 138–143, **138, 142, 145**
Stress ellipse, **143,** 143–144, **144**
Stress ellipsoid, 143–144, **144**
Stress fracture, 157
Stress release, 352
Stress-strain diagram, **148,** 148, 150
elastic behavior, 148, 152
elastic limit, 150, **150,** 152
failure, 150, 152
hysteresis loop, **149,** 149–150
nonrecoverable strain, 150
plastic deformation, 150–152
recovery, 149, 151–152
rupture, **150,** 152
strain hardening, **152,** 152
Stress trajectories, **321,** 321
Stress vector, 138–143, 160
Stretch, 107–110
Stretching, 172, 187
Stretching surface(s), 258–259
Striated surface(s), **11,** 336–337, 348
Striations, **11, 76, 224,** 224–225, **225, 264,** 269, 426
Strike, 56
Strike-and-dip, **55,** 56–57
measurement, **56,** 56–57, **57**
recording, 57
Strike histogram, 345–346
Strike-slip faulting, 297–305, 436–437
Guatemala, **301,** 301–303, **302, 303**
kinematic analysis, 299–301
New Zealand, 303
tectonic setting, 301
Strike-slip movements, 166–167, 174–175, 183
Structural analysis, 14, 16–35
Structural control, 41
Structural elements, 19–22, **21**
geometric, 19, **21,** 21–22
physical, 19, **21,** 21–22
Structural geology, 3–4
architecture, 4, 427
complexity, 12–14
nature, 3–15
purpose, 12
Structural profile, 61–63, **62**
steps in construction, 381–384, **382, 383**
Structural relief, 163, 183
Structural system, 22
Structural trap, 357–358
Structure contour lines, 58–59, 63
Structure contour map, **58,** 58–59, **59**
Structure profile, 61–63, **62**
steps in construction, 381–384, **382, 383**
Structure section, 18–19, 61–63, **62**
steps in construction, 381–384, **382, 383**
Stylolites, **337,** 337–338, **338**
Stylolite teeth, 337–338
Stylolitic surfaces, 337–338, 407, 409–410
clayey/carbonaceous seams, 338, 407, 409, 416
faultlike offset, 338, 417–418
strain significance, 347–348, 417–419
truncated fossils, 338, 416
Subarea(s), 342
Subduction, 167, 171, 176, 180–185, **181, 185,** 187
Subduction zone, 177, **177,** 180, 183, **185,** 195, 357
Submarine fan, 186
Subsidence, 173, 240
Sudbury complex, Ontario, 216
Suspect terrane(s), 183, **184**
Suture, 167
Suture zone, 167
Symmetry principle, 16, 217
Syncline, 359–360, **360**
Synclinorium, 360–361, **361**
Synform, 360
Syntectonic deposition, 279
System, 22

Taiwan fold-thrust belt, 286
Tectonic inclusion(s), 182
Tectonic transport, 282
Tectonite(s), 10, 430–445
analysis, 438–441
classes, 431
descriptive analysis, 438–445
folds, 439
foliation, 438–439
geologic occurrences, 430, 436–438
geometric analysis, 438–445
lineation, 439
LS-tectonites, 430–433, **431,** 436–437
L-tectonites, 430–433, **431**
S-tectonites, 430, 433, **431,** 437
strain significance, 431, **432**
tectonic significance, 436–438
Tethys Sea, 177
Thermal plume, 191
Three-point problem, 209–210, **210**

Throw, 273
Thrust faulting, 273, 279–286
  Appalachian Mountains, 282, **282, 283**
  Canadian Rockies, 18–19, **19,** 25, **280, 281,** 284, **285,** 286
  decollement, 281
  decollement faulting, 281
  dynamic analysis, 322–324
  Himalayan orogen, 437
  Idaho-Wyoming, 284, 286
  kinematic analysis, 283–286
  Moine thrust, 437
  ramp thrusting, **282,** 282–283, **283**
  regional characteristics, 279–280
  role of fluid pressure, 322–324
  strain significance, 277–278, 286
  Taiwan, **286,** 286
  tear faulting, **285,** 285
  tectonic environment, 279
  thin-skinnned overthrusting, 280–282
  transfer zones, **285,** 285
Topographic relief, 12, 163, 208
Transform fault(s), **174,** 174–176, **176,** 187–188, 190
  kinematics, 174–176, 194–195
  ridge-to-ridge, **174,** 174–176, **176,** 194–195, 197
  ridge-to-trench, 175, **176,** 195
  San Andreas, 176, 190
  trench-to-trench, 175, **176**
Transgression, 173
Translation, **17. 22.** 22. 88–98. 161
  nonrigid body, 161
  rigid-body, 161
Transposition, 441–444, **442, 444**
Tree-ring dating, **32,** 32
Trench(es), 175, **177,** 177, **180,** 180–183, **186,** 188
  Aleutian, **181,** 192
  inner wall, 180–182
  Middle America, **181**
  outer wall, 180–182
  Peru-Chile, **181,** 194
Trend, 52–53
Trend-and-plunge, 52–58
Triangular facet(s), 223
Triaxial deformation, **27, 135, 136**
  apparatus, **135, 136**
  axial stress, 136–137
  confining pressure, 136, 151–153
  copper jacket, 134–135
  Donath apparatus, **135,** 135–136
  failure angle, 151–153
  load cell, **135,** 136
  load-displacement curve, **148,** 148, **150**
  loading, 136
  pressure vessel, 135
  sample preparation, 134–135
  seating position, 137
  shortening, 137
  standard compression test, 147–153
  stress-strain diagram, **148,** 148, **150**
  X-Y recorder, 136–137, **137**
Triple junction(s), 169, **199**
  ridge-ridge-ridge, 169, 198
  transform-transform-trench, **198,** 198, **199**
  vector circuit diagrams, 198–200, **199**
True strain, 433
Turbidite(s), 173, 185, **354**
Turbidity current(s), 180
Turtleback structure, 295
Twinning, 160–162

Ultimate strength, 152
Ultramylonite, **227,** 227
Unconformities, 204–210, **205**
  angular, **205,** 205–207, **206, 207**
  breakup, **172, 209,** 209
  buttress, **209,** 209
  disconformity, **205,** 205, 207
  nonconformity, **205,** 205–207, **206,** 208
  outcrop characteristics, 207–208
  tectonic significance, 208–209, 235
Underthrusting, 176

Vectors:
  analysis, 138–139, **140,** 141–143
  displacement, 90
Vein(s), 7, **8, 97, 330,** 330, 333–337
  cockscomb texture, 334
  crustification, 334
  crystal-fiber growths, **334,** 334–337, **335, 337**
  crystal-fiber vein, 335–336
  en echelon, **335,** 336
  fissure vein, 330
  gash veins, 336
  kinematics of formation, 334–337, **335, 336, 337**
  replacement vein, 334
  sigmoidal, **336,** 336
  strain significance, 347
Vertical exaggeration, 180
Vertical projection, 53
Vesicles, 248
Viscosity, 39, 156, 244, 246
Viscosity coefficient(s), **156,** 391–392
Viscous materials, 155–156, **156**
Volcanic arc(s), 176–177, 185–186
Volcanic plateau(s), 180
Volcanic seamount(s), 180
Volcano(es), **245, 246**
  cinder cone(s), 186, **245**
  composite, 186
  volcanic dome(s), 186

Waterpocket monocline, Utah, 397
Welded tuff, 186, 250, **251,** 252
West Kaibab monocline, Arizona, 397
Whipple Mountains, California, 293, **294**
Windows, 279, **280**

Wind River fault, Wyoming, 321–322
Wind River uplift, Wyoming, **321,** 321–322, **322**
Work hardening, 152, 162
Wyoming province, 352, 398, **399**

Xenolith(s), 211–212, 256

Yerington district, Nevada, 295–296, **296**
Yield strength, 150
Young's modulus, 148, **149,** 391–392